RCN Conference on Pan American Biofuels and Bioenergy Sustainability 2014

Recife, Brazil
22-25 July 2014

ISBN: 978-1-5108-1574-2

Printed from e-media with permission by:

Curran Associates, Inc.
57 Morehouse Lane
Red Hook, NY 12571

Some format issues inherent in the e-media version may also appear in this print version.

Printed by Curran Associates, Inc. (2015)

AIChE
120 Wall Street, FL 23
New York, NY 10005-4020

Phone: (800) 242-4363
Fax: (203) 775-5177

www.aiche.org

Additional copies of this publication are available from:

Curran Associates, Inc.
57 Morehouse Lane
Red Hook, NY 12571 USA
Phone: 845-758-0400
Fax: 845-758-2634
Email: curran@proceedings.com
Web: www.proceedings.com

TABLE OF CONTENTS

ABSTRACTS

POSTERS

INTERROGATING SOCIAL SUSTAINABILITY IN THE BIOFUELS SECTOR IN LATIN AMERICA: GLOBAL STANDARDS AND LOCAL EXPERIENCES IN BRAZIL AND COLOMBIA

Dr. Theresa Selfa, Associate Professor
Department of Environmental Studies
SUNY- Environmental Science and Forestry
Syracuse, New York, USA

Presentation to RCN Conference,
Recife, Brazil, July 22-25, 2014

Based on Paper under Review:

Interrogating Social Sustainability in the Biofuels Sector in Latin America: Global Standards and Local Experiences in Mexico, Brazil and Colombia

Theresa Selfa, Department of Environmental Studies, SUNY ESF

Carmen Bain, Department of Sociology, Iowa State University

Renata Moreno, Department of Environmental Studies, SUNY ESF

Amarella Eastmond, Universidad Autonoma de Yucatan, Mexico

Sam Sweitz, Department of Social Sciences, Michigan Technological University

Conner Bailey, Department of Rural Sociology and Agricultural Economics, Auburn University

Gustavo Simas Pereira, Department of Environmental Sciences, Federal Rural University of Rio de Janeiro; Federal Institute of Education, Science and Technology of the State of Rio de Janeiro

Tatiana Martins, Environmental Management Lab, Department of Environmental Sciences, Federal Rural University of Rio de Janeiro

Rodrigo Medeiros, Department of Environmental Sciences, Federal Rural University of Rio de Janeiro / Conservation International, Americas Center for Sustainability

Presentation Outline

1. Background- Emergence of Biofuels
2. Environmental and Social Sustainability
3. Sugarcane and Palm Production in Colombia
4. Palm Oil Production in Brazil
5. Challenges for Sustainability/ Governance Tools
6. Conclusions

Background

- Dramatic growth in biofuel production around the world -2000s
  - Key role of governments (mandates for production and use)
  - EU 2009 Renewable Energy Directive (RED)
    - Interest in importing biofuels from Latin America
  - US 2008 Renewable Fuels Standard (RFS2)
- Expansion of Multi Stakeholder Initiatives (MSI) to certify environmental and social sustainability
  - MSI composed of non state, corporate and NGO actors; legitimacy through broad stakeholder participation
  - Examples: Bonsucro, Roundtable Sustainable Bioenergy (RSB), Roundtable Sustainable Palm (RSP), Roundtable Responsible Soy (RTRS)

Background

- MSI certification and standards are necessary for imports to the EU
- Standards emphasize environmental requirements:
  - reduction of GHG emissions
  - biofuels production does not occur on lands with high biodiversity or conservation areas
- Standards have fewer social requirements
  - Should ensure that labor and land tenure laws, human rights are respected
  - Should improve livelihoods for local communities

Research Objectives

- Multi-site comparative research interrogates:
  - How are global standards translated into local places?

(Silva-Castenada, 2012; Selfa, Bain and Moreno, 2014; Kohne, 2014; Klooster, 2011; German et al 2011; Bailis and Baka, 2011; Elgert, 2012)

  - How can we enact and measure social sustainability, i.e. livelihoods and resilience?

(Hunsberger, et al. 2013; Stattman and Mol, 2013; Wilkinson and Herrera, 2013)

Palm oil in the World

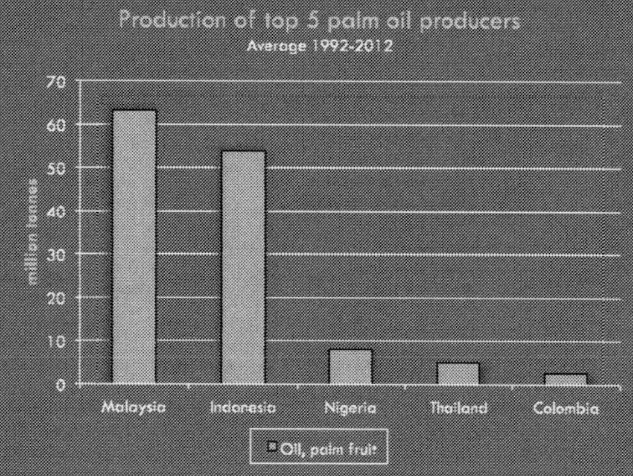

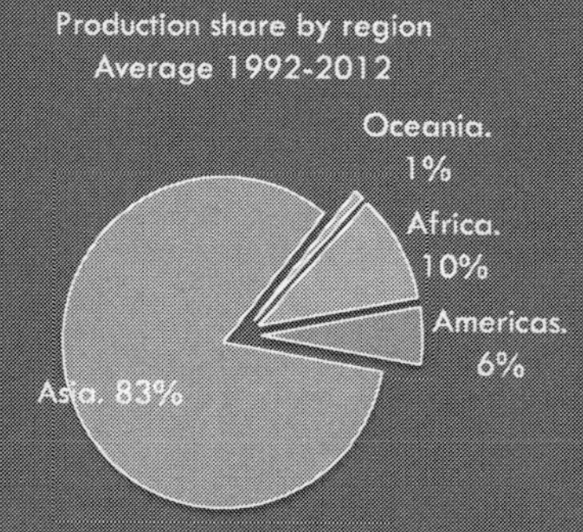

Source: FAO 2012 - http://faostat.fao.org

Colombia Case

- Colombia is one of larger biofuel producers in Latin America (both sugarcane ethanol and palm oil)
- President Uribe (2002-10) established biofuel production and use mandates; they not yet been met
  - Major strategy for rural development, context of armed conflict
  - Goal to become exporter of biofuels
- Long history of entrenched unequal land distribution patterns
- Expansion area into Los Llanos

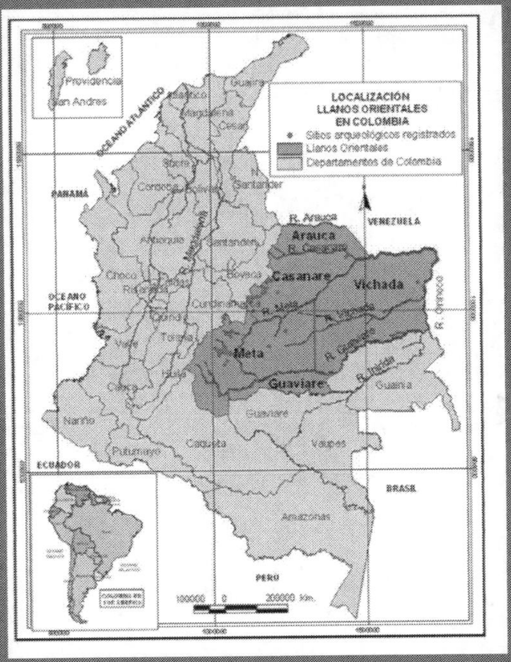

Colombia Los Llanos Region

Colombia Case: Los Llanos

- History of being a frontier, outside of government control
- Armed conflict and paramilitary groups
- Government and biofuel companies describe it as a region that lacks forests or reserve areas, with 'underutilized lands' that have been inefficiently used for cattle ranching
- Framed as 'Ideal' for biofuels because not there are no food crops

Colombia Case: Challenges

- Environmental groups contend the region contains a variety of ecosystems and cultural groups that have become invisible in push for biofuel expansion.
- Cultural diversity represented in an indigenous population estimated at 57,000
- Land grabbing by Colombian and foreign companies for palm oil production
- Easy to accomplish these grabs because of lack of state control, regulation and no rule of law

Colombia Case: Los LLanos

Territory includes tropical savannahs, wetlands, tropical forests and mountain ecosystems.

Colombia Case: Social Sustainability Challenges?

- Palm producers in Los Llanos are getting certification for Roundtable on Sustainable Palm (RSPO)
  - RSPO requires that land tenure, labor and human rights laws are respected and no conservation or high biodiversity areas are used for palm production
- In Colombia, palm companies are more concerned about gaining RSPO certification to access EU markets than they are about local sustainability
- Colombian state is ineffective at regulating companies to ensure they are adhering to sustainbility (local and global)

Brazil Case: Palm Oil and Social Inclusion

Brazil

- Long history of Sugarcane Ethanol production (1970s), second largest ethanol producer in the world
  - Brazil was criticized for labor and environmental conditions (i.e. burning) associated with sugarcane for ethanol
- In 2004, National Program of Biodiesel Use and Production (PNPB) was launched, with the goal to expand biodiesel production and increase its share in the Brazilian energy matrix.
- PNPB promotes social inclusion by targeting crops grown by small farmers for biodiesel (castor, palm oil)

Brazil

- 2010, Brazil launched the National Program of Sustainable Palm Oil Production (PSOP) to expand palm oil through more socially and environmentally sustainable production.
- Palm oil has high labor needs, and is targeted for production on 'degraded' pasture lands identified by zoning
- Para, Amazon has 86% of national palm production
- Big companies (Vale mining, Petrobras, ADM) buying or renting land for palm production

Palm Oil in Brazil

Area harvested (ha) of palm oil in Brazil and World

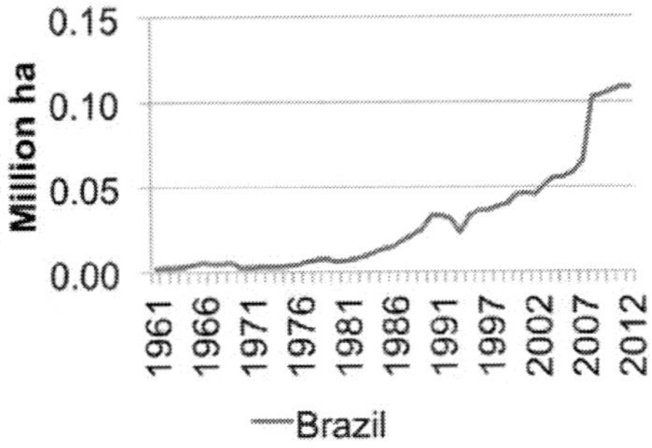

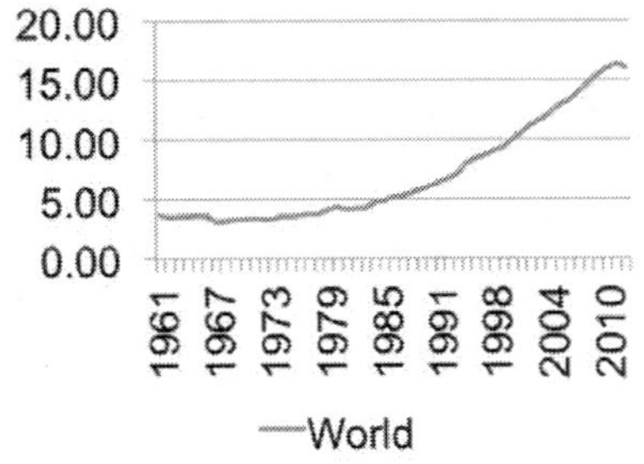

Source: FAO 2012 - http://faostat.fao.org

Agro-ecological zoning

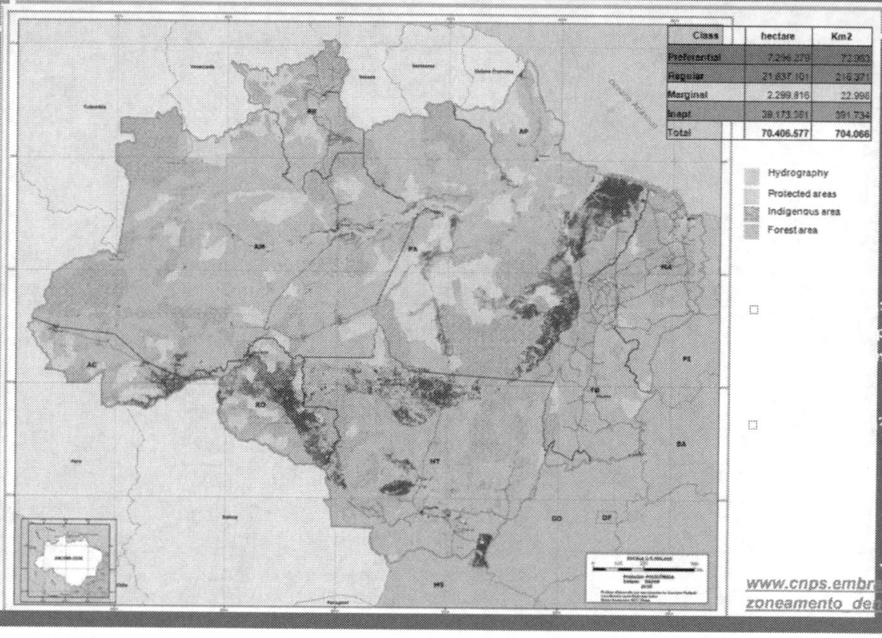

2
palm
n

2 in

– 2010

www.cnps.embr
zoneamento de

Pará State: Main production region

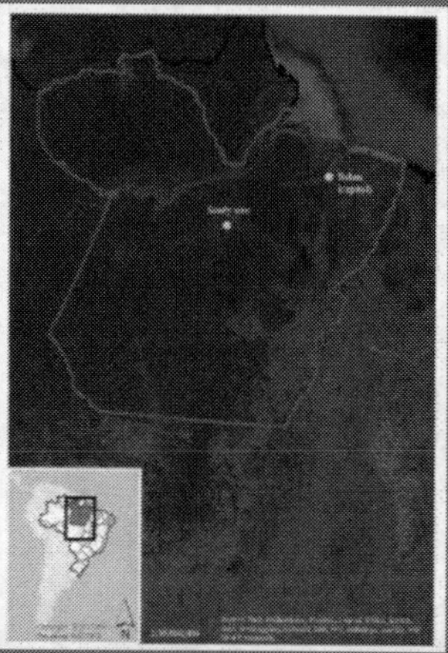

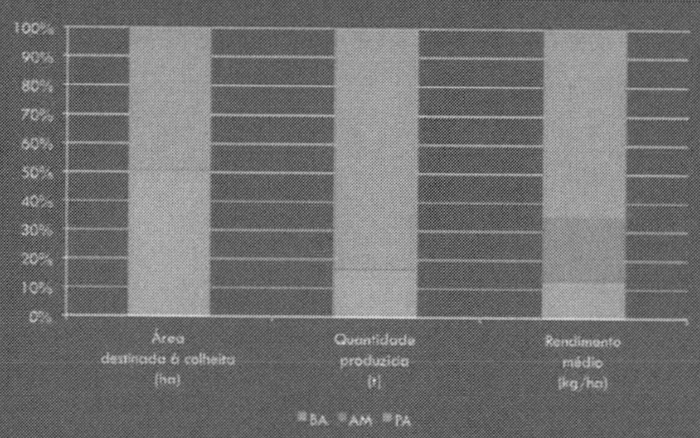

Source: PAM/IBGE 2011

Brazil Challenges

- Social Fuel Seal (SFS) given to companies who purchase diesel from small farmers (castor and palm oil)
- Challenges in past have been lack of production capacity by small farmers to produce enough oil to blend with diesel, so soybean oil is used instead
- Reinforces advantages of large soya producers

Brazil Challenges to Palm Expansion in Para

- Land use changes to palm monocultures
- Food security issues as small farmers lack subsistence food production and rely on income from palm to buy food
- Some small farmers abandon or sell land leading to greater land concentration
- Traditional communities (quilombolos and others) who do not want to participate in palm production face disruption of the livelihoods

Brazil

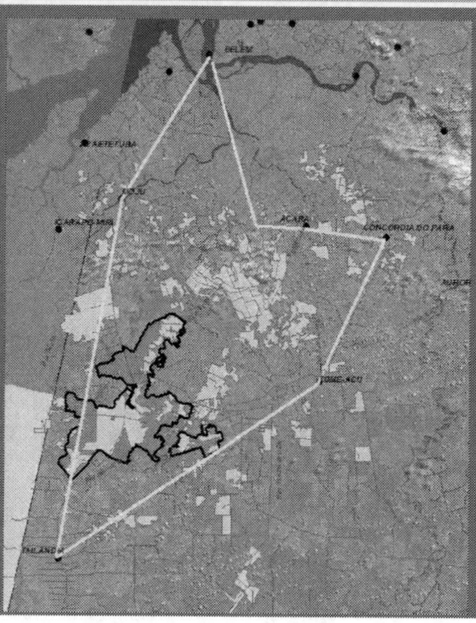

Field site locations in Para

Pilot Interviews

- 4 municipalities
  - Moju, Acará, Concórdia do Pará and Tomé-Açu
- 23 families interviewed
  - 15 with palm oil
  - 8 with other crops
- First contact with biodiesel companies
  - Belém Bioenergia (Galp and Petrobras)
  - Biopalma (Vale)
  - Dendetauá

Brazil Social Sustainability Challenges?

- Social Fuel Seal that was focused on 'social inclusion' of smaller farmers in the palm sector is not working as intended yet
- Resistance from traditional communities to palm
- Community concerns about impacts of pesticides on local water resources
- Jobs are created, unclear about future financial viability

Summary

- Governments throughout the Americas have incentivized biofuel production but have done less to manage social and environmental impacts
- 'Land grabbing' by national and foreign companies was justified by framing lands as 'marginal' 'abandoned', 'degraded', 'empty, 'underutilized', or 'inefficient
- Land use changes directly or indirectly impact the sustainable livelihoods of local peoples as access to land and natural resources was restricted
- Biofuels are not an overwhelming "win-win" for rural communities, though they may be for some (large) producers.

Conclusions:
Social and Environmental Sustainability of Biofuels?

- Governance tools like standards and certifications (RSPO, RSB, Bonsucro, and SFS) are imperfect
- Certifications are premised on compliance with global standard but often ignore local social and environmental conditions
- Need for developing robust policy and governance tools, such as MSIs, that can protect and enhance rural livelihoods
- Regardless of policy/governance instrument, need for incorporating local stakeholders into analysis, design and implementation of bioenergy projects

Acknowledgements

Research in Latin America funded by:

NSF OISE-PIRE: Sustainability, Ecosystem Services, and Bioenergy Development across the Americas

NSF Research Coordination Grant: Pan American Biofuels and Bioenergy Sustainability

I would like to thank my collaborators on the project and research participants in Brazil, Mexico and Colombia.

RCN CONFERENCE ON PANAMERICAN BIOFUEL AND BIOENERGY SUSTAINABILITY
Recife, July, 2014

GLOBAL AND LOCAL PERSPECTIVES ON THE JATROPHA PLANTATIONS FOR BIODIESEL IN SUCOPO, YUCATAN

MEXICAN CASE STUDY FOR PEER, PIRE and IAI PROJECTS

AMARELLA EASTMOND and SAMUEL SWEITZ

Unidad de Ciencias Sociales, CIR
Universidad Autónoma de Yucatán

Department of Social Sciences
Michigan Technological University

LOCALIZATION OF JATROPHA PLANTATIONS SUCOPO,
TIZIMIN, YUCATAN, MEXICO

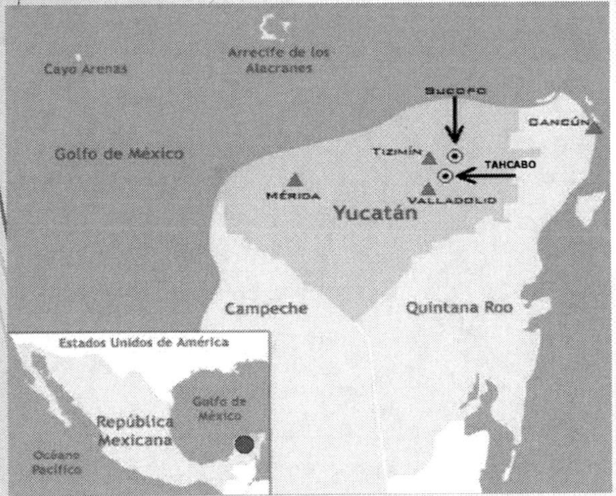

MAP OF SUCOPO`S EJIDO, AND THE JATROPHA PLANTATIONS

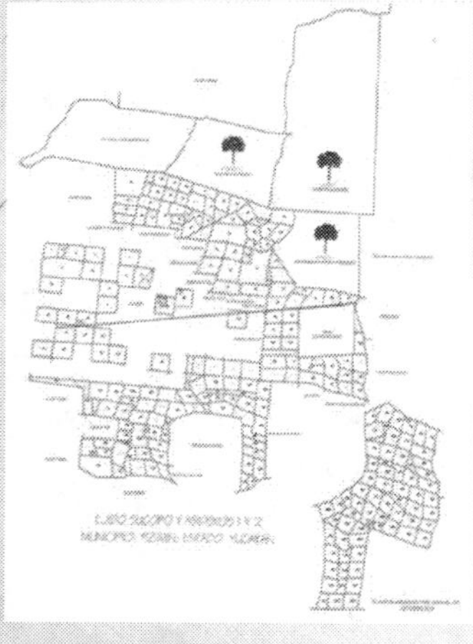

OBJECTIVE

ANALYZE LOCAL AND GLOBAL PERSPECTIVES ON:

1) Environment and ecosystem services

What do people get from the environment and what does it mean to them?

Land ownership, access to land, farming practices, crops, forests, wild animals, water, pests,

2) How sustainabilitly is understood

What is most important to people?

Keeping the forest, having a job?

3) What socioeconomic impacts did the jatropha plantations have?

METHODS

- Selection of control and treatment villages
- Qualitative in-depth interviews with around 160 households in the región, (100 in Sucopo)
- Life histories with key informants
- Interviews with local authorities
- Newspaper reports
- Public documents with information about the company (such as the Roundtable Sustainable Biofuels certificate)

ECOSYSTEM CONTEXT

- **Geology:** limestone plateau with no rivers in the north and very little surface water, sinkholes or *cenotes* are widespread
- **Soil:** very thin, stony and nutrient-deficient, mechanized agriculture very difficult
- **Climate:** Mostly warm, sub-humid
- Average annual temperature: **26^{0}C,**
- Average rainfall: **1,100 mm**, concentrated in the summer

ECOSYSTEM
Vegetation and land use

- **Natural vegetación:** Medium deciduous tropical forest

- **Present predominant land use:** Cattle ranching, small patches of traditional agriculture (maize, beans and squash), citrus and papaya plantations

LAND USE CHANGE
SLASH AND BURN (*MILPA)* AGRICULTURE AND EXTENSIVE CATTLE RANCHING

SOCIO-ECONOMIC CONTEXT

- **Maya peasant village**
- **Population in 2010: 1,517 (around 417 families)**
- **Languages: Maya and Spanish**
- **Education: 6 years**
- **Health services: Village health center**
- **Economic activities**: agriculture, very small businesses

  migration to Riviera Maya to work in construction industry. Subject to boom and bust cycles.
- **women** tend to stay at home, some work as farm laborers
- **Income:** Between $40 and 100 US dollars a week

SOCIO-ECONOMIC SUBTEAM ISSUES

- Land ownership, access to land and land use patterns
- Government policy on biofuels
- Economic activities
- Traditional agriculture and subsistence activities
- Values related to ecosystem services,
- Household incomes
- Employment opportunities and work conditions in and near the village
- Gender relations

VILLAGERS FROM SUCOPO

MEXICAN BIOFUEL POLICY

- **2008 Bioenergy Law to promote investment in biofuels**
1. Policy to reduce GHG emissions
2. Create jobs and boost rural development
3. Increase long term energy security

- Large number of posible feedstocks were identified for different regions. Maize was prohibited as a feedstock unless Mexico produced a surplus.

- jatropha selected for agriculturally marginal areas such as Yucatan

  Considered a poor man`s crop that will grow without irrigation, fertilizers and pesticides
- Although some varieties are native to Mexico and it was used in Sucopo for medicinal purposes, it is a **wild plant, that has never been domesticated**

GLOBAL CLEAN ENERGY HOLDINGS INC.

- In 2008 American company Global Clean Energy Holdings Inc.(GCE) created a Mexican subsidiary in association with Mexican **investors** and bought some 6,000 hectares of private land, near Sucopo´s ejido.
- Permission granted to plant jatropha on about half of the land by claiming that the land was "abandoned cattle ranches"
- Jatropha classified as a tree, so company got subsidies from Mexican Forestry Commission (CONAFOR) for the establishment stage and subsidies from the Min. of Agriculture (SAGARPA) for equipment and chemical inputs.
- It hired 500-600 people and promised the villagers that they would have long term, sustainable jobs (15 to 30 years) working on the plantations

JATROPHA PLANTATIONS

DEVELOPMENT OF JATROPHA PROJECT

- 500-600 people employed from Sucopo and around
- Pay was 195% of mínimum wage, benefits according to the law
- Program of free breakfasts for school children and support for local baseball team
- During 3-4 years things seemed to be going well but strict confidentiality about the companies´ activities
- Observations that jatropha leaves were turning yellow and falling off
- Aerial spraying and speculation about disease
- Poor yields, not even 1 ton per hectare
- End of 2012 drastic reduction in size of operation, but not total closure of company
- 500 people laid off and told the company was restructuring

RESULTS
LOCAL PERSPECTIVES

- Environment gives Sucopo people a strong sense of **identity**
- Villagers´ cosmovision: natural resources belong to the Gods, not to men, permission has to be sought to use them, through rituals and correct behavior
- Understanding exists that natural resources and ecosystem services are essential for production of crops, hunting wild animals and the production of flowers so their bees can produce honey.
- Recreation and interest in ecotourism for the future.
- Understanding that the environment is necessary for life but modernization, globalisation and more change have undermined the traditional self sufficient way of life so sustainability now means having a job and the ability to earn a decent wage.

RESULTS
LOCAL PERSPECTIVES

- Almost no one complained about GCE
- No changes in land ownership or quality and quantity of water
- Everyone appreciated having a local source of employment that paid better than any other company
- BUT huge disappointment when the company laid off almost all its workers. What happened to the sustainable jobs for 30 years?
- Only after long questioning, did the bee keepers complain that **old forest (30 to 50 year old trees)** had in fact been cut down for the plantations, leaving the wild animals and particularly the bees without their natural habitat and food sources.
- Animals became pests and many bees were killed by aerial spraying of pesticides.

LAND USE TRANSFORMATION FROM THE GLOBAL PERSPECTIVE

Written into the Roundtable Sustainable Biofuels certificate (Nov. 2012) and published in the newspapers

LAND USE TRANSFORMATION ON THE GROUND, REAL CONSEQUENCES

LACK OF TRANSPARENCY IN THE GLOBAL PERSPECTIVE

- **RHETORIC**
- Full compliance with the Roundtable Sustainable Biofuel criteria
- Helping the environment and the local people to be more sustainable

- **PRACTICE**
- No quantification of subsidies
- No transparency or quantification of agricultural practices (especially chemical inputs)
- No transparency about investors
- No transparency about environmental permits to change the land use
- No accountability to local people

PRELIMINARY CONCLUSIONS

- The jatropha plantations project appears to build on historical systems of resource extraction in the región such as henequén in which cheap local labor is used to produce an export, mostly benefitting large land owners and consumers in a foreign country and creating an asymmetric dependency between the local and global.
- Natural capital was reduced by the jatropha plantations because of deforestation of the secondary vegetation, negatively affecting subsistance farmers´ capacity to grow their own food and produce honey for sale.
- The jobs that were supposed to replace subsistance agricultura are no longer locally available.
- Sucopo`s peasants find themselves caught up in the unenviable situation of seeing both their sources of livelihood (formal and informal) becoming increasingly unstable and unsustainable

MUCHAS GRACIAS!

Agro-fuels and Rural Households income in Yucatan

Javier Becerril, Amarella Eastmond, Julio Sacramento & Sam Sweitz

NSF Research Coordination Network
Conference: 22-25 July 2014, Recife, Brazil

Outline

- Background
- Industry of Agro-Biofuels in Mexico
- Sources and Information
  - Research Area
  - Inhabitants Profile
- Impact Assessment (Propensity Score Matching)
- Findings

NSF Research Coordination Network
Conference: 22-25 July 2014, Recife, Brazil

"NSF-PIRE Collaboration: ***Sustainability Evaluation of Jatropha Oil Production in Yucatán*** **" USAID's PEER program**

Background

Social Science: Describe and evaluate the effect jatropha cultivation sites have had over surrounding communities
Dr. Amarella Eastmond, Dr. Sam Sweitz

Economics: Estimate the economic effect jatropha cultivation companies have had over surrounding communities
Dr. Javier Becerril

Process Engineering: Evaluate the overall sustainability of various design options of a jatropha-based biofuel industry (production pathways)
Dr. Julio Sacramento

Michigan Technological University

NSF Research Coordination Network
Conference: 22-25 July 2014, Recife, Brazil

"NSF-PIRE Collaboration: ***Sustainability Evaluation of Jatropha Oil Production in Yucatán*** **" USAID's PEER program**

Background

Social Science: Describe and evaluate the effect jatropha cultivation sites have had over surrounding communities
Dr. Amarella Eastmond, Dr. Sam Sweitz

Economics: Estimate the economic effect jatropha cultivation companies have had over surrounding communities
Dr. Javier Becerril

Process Engineering: Evaluate the overall sustainability of various design options of a jatropha-based biofuel industry (production pathways)
Dr. Julio Sacramento

Michigan Technological University

NSF Research Coordination Network
Conference: 22-25 July 2014, Recife, Brazil

Background

According to *Janda, Kristoufek and Zilberman* (2012).

- **First** generation biofuels: food-crops, rich in sugar, starch, or vegetable oil.
- **Second** generation: residual non-food parts of current crops, or novel non-food dedicated crops (jatropha!)
- **Third** generation: GMO mainly.
- **Fourth** generation?

NSF Research Coordination Network
Conference: 22-25 July 2014, Recife, Brazil

Energy Economics 34 (2012) 1922–1930

Contents lists available at SciVerse ScienceDirect

Energy Economics

journal homepage: www.elsevier.com/locate/eneco

Biofuels and economic development: A computable general equilibrium analysis for Tanzania

Channing Arndt [a], Karl Pauw [b], James Thurlow [c,*]

[a] Department of Economics, University of Copenhagen, Denmark
[b] International Food Policy Research Institute, Washington, DC, USA
[c] United Nations University, World Institute for Development Economics Research, Katajanokanlaituri 6 B, FI-00160 Helsinki, Finland

ARTICLE INFO

Article history:
Received 22 December 2009
Received in revised form 23 July 2012
Accepted 26 July 2012
Available online 11 August 2012

JEL classification:
D58
O13
Q42

Keywords:
Biofuels
Growth
Poverty
CGE model
Tanzania

ABSTRACT

Biofuels could offer new economic opportunities for low-income countries. We use a recursive dynamic computable general equilibrium model of Tanzania to evaluate different biofuels production options and estimate their impacts on growth and poverty. Our results indicate that maximizing the poverty-reducing effects of biofuels production in countries like Tanzania will require engaging and improving the productivity of smallholder farmers. Evidence shows that cassava-based ethanol production is more profitable than other feedstock options. Cassava also generates more "pro-poor" growth than sugarcane-based systems. However, if smallholder yields can be improved rather than expanding cultivated land, then both sugarcane and cassava out-grower schemes generate similar pro-poor outcomes. We conclude that, in so far as the public investments needed to establish a biofuels industry are consistent with other development needs, then producing biofuels will enhance economic development in countries like Tanzania.

According to *Arndt, Pauw and Thurlow* (2009).

Biofuels could offer new economic opportunities for low-income countries.

The findings suggest: economic growth and poverty alleviation in Tanzania.

Require engaging and improving the productivity of smallholder farmers.

NSF Research Coordination Network
Conference: 22-25 July 2014, Recife, Brazil

OMASS AND BIOENERGY 46 (2012) 409–418

Biofuels and employment effects: Implications for socio-economic development in Thailand

Thapat Silalertruksa [a,b], *Shabbir H. Gheewala* [a,b,*], *Katja Hünecke* [c], *Uwe R. Fritsche* [c]

[a] *Joint Graduate School of Energy and Environment, King Mongkut's University of Technology Thonburi, 126 Prachauthit Road, Bangkok, Thailand*
[b] *Center for Energy Technology and Environment, Ministry of Education, Thailand*
[c] *Öko-Institut, Energy & Climate Division, Darmstadt, Germany*

ARTICLE INFO

Article history:
Received 16 August 2011
Received in revised form
17 May 2012
Accepted 28 July 2012
Available online 30 August 2012

Keywords:
Biofuels
Socio-economic
Input–output analysis
Employment
Thailand

ABSTRACT

The study assesses the impacts of the biofuels production to the socio-economic development in Thailand. The four kinds of biofuels considered in the analyses are ethanol from cassava, molasses and sugarcane ethanol as well as palm biodiesel. The key elements of socio-economic development including employment generation, economic effects on GDP and trade balance are investigated based on a combination of process and input–output analysis. The results show that producing ethanol and biodiesel require about 17–20 and 10 times more workers than gasoline and diesel as per energy content, respectively. Direct employment in agriculture contributes to more than 90% of total employment. Nevertheless, there are the significant differences in the characteristics of employment in the agriculture and biofuel processing sectors. The overall impacts of ethanol production in Thailand in year 2022 are the employment generation of around 238,700–382,400 person-years, 150 M$ additional GDP, imported goods worth 1583 M$ but 2547 M$ of imports would be saved if compared to petroleum fuels. The other socio-economic aspects such as agricultural improvement and rural development due to biofuels policy in Thailand and some policy measures that need to be urgently promoted are also discussed in the study.

The findings from *Silalertruksa, Gheewala, Hunnecke and Fristche* (2012)
Key elements: jobs creation, economic effect GDP, trade balance, etc.
Findings: Producing Ethanol requires 17-20 times more workers than Gasoline and Diesel.
Direct employment in agriculture contributes to more than 90% of total employment.

Agro-Fuels Industry in Mexico

- The "Plan Nacional de Desarrollo 2007-2012", a proposal from the Federal Government established:
  - "...mediante la promoción y producción de bioetanol (se) dinamizarán las economías rurales y generará empleo al favorecer la puesta en marcha de un nuevo sector en el ámbito agrícola."
  - February 2008: Ley de Promoción y Desarrollo de Bioenergéticos (LPDB).

Agro-Fuels Industry in Mexico

- National Energy Strategy 2013 – 2027.
  - Policy 4: Energy transition
  - "México deberá lograr el correcto balance entre mantener al país económicamente competitivo, tecnológicamente innovador y diversificado, mejoramiento calidad ambiental, y cumplimiento de los compromisos ambientales globales, presente y futuros" p. 48.
  - The strategy for long term starting 2013:
    - Natural Gas resources available
    - Renewable energy sources: environmental benefits
    - Rethinking nuclear energy: security present and future

NSF Research Coordination Network
Conference: 22-25 July 2014, Recife, Brazil

Agro-Fuels Industry in Mexico

- Mexico has a portfolio of different sources:
  - Eolic, Solar, Geothermic, Biomass, and Hydraulic

- Specific Action:
  - To thrust the development of a competitive market, without affecting food security, as long as it be a viable alternative in environmental (as per a Life Cycle Assessment), economical, and social terms. (p. 53)

NSF Research Coordination Network
Conference: 22-25 July 2014, Recife, Brazil

Agro-Fuels Industry in Yucatan

- **KuoSol** group
- **LODEMO** group
- **The Global Clean Energy**
  - Jatropha (*curcas*)

NSF Research Coordination Network
Conference: 22-25 July 2014, Recife, Brazil

Sources & Information

- Data & Information from the research project: "NSF & PIRE ***Sustainability Evaluation of Jatropha Oil Production in Yucatán***" Financed by **USAID**
- **907** inhabitants (interviewed)
- **192** Households (surveyed)
- Three municipalities: **Tizimin$^{\varphi}$, Muna$^{\varphi}$, Santa Elena***
  $^{\varphi}$Treatment *Control

NSF Research Coordination Network
Conference: 22-25 July 2014, Recife, Brazil

Study area

Study area, characteristics

Municipality	Village	Maya-speaking population
Santa Elena	San Simón	93%
Muna	San José	40%
	Dzonot Carretero	77%
Tizimin	Tixcancal	82%
	Sucopo	67%

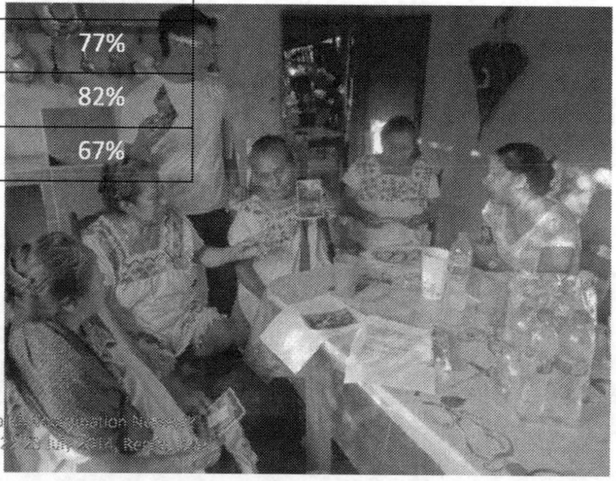

Inhabitants and Household Profiles

Description	Work for AF Industry	NO-Work for AF Industry	/ t /Value
Annual Income (Salary)	$37,950.0	$12,408.13	**6.00**
1 = Male	0.88	0.49	**3.78**
Age in years	34.58	40.23	**1.76**
Formal Education	6.45	5.80	0.76
n = 564 inhabitants over 18 years old.			

Mayan Household
and
Home-garden (Solar)

Inhabitants and Household Profiles

Description	Work for AF Industry	NO-Work for AF Industry	/ t / Value
Family Size (number of members)	4.33	4.62	0.58
Solar (Home garden) 1 = True	0.90	0.78	1.39
Milpa (farm with three crops: maize, Beans & Squash)	0.36	0.37	0.11
Parcela (farm with permanent trees)	0.27	0.16	1.24
Total crops (Solar + Milpa + Parcela)	5.27	4.68	1.21
Farm land endowment (in hectares)	12.02	24.92	**1.76**
Subsidy "Procampo" (cash transfer) 1 = True	0.41	0.25	1.62
"Seguro Popular" program (Public Social Security)	0.91	0.66	**2.41**
Income from Public Policy Programs	$5,834.55	$7,175.40	0.74
Owns motorcycle	0.55	0.29	**2.39**
Owns television	1	0.86	**1.85**
Owns sewing machine	0	0.05	**2.83**
n = 192 Households			

Impact Assessment

- How impact assessment the economic benefit to work at Agro-Fuels Industry?
- If we have:

  Y^1 = income if works in the Agro-Fuel Industry

  Y^0 = income if doesn't work in the Agro-Fuel Industry
- The economic benefit is the difference:

  $\Delta_i = Y^1 - Y^0$
- But, only know for each farmer: Y^1 or Y^0 that means, **Δ_i unknow**.
- To solve this prolem we applied the *Propensity Score Matching* quasi-experiment scenario.

NSF Research Coordination Network
Conference: 22-25 July 2014, Recife, Brazil

Impact Assessment

The main idea behind the *Propensity Score Matching*:

As similar as possible

Work for Agro-Fuel Industry
"Treatment Group"

No-Work for Agro-Fuel Industry
"Control Group"

NSF Research Coordination Network
Conference: 22-25 July 2014, Recife, Brazil

Impact Assessment: Economic Impact

"Matching" algorithm	Dependent Variable	ATT	Number of treatments	Number of controls
"Nearest Neighbor Matching"	Per-Capita Income Salary Annual	**$14,123.08 (1.75)**	13	231
n = 564 inhabitants over 18 years old.				

NSF Research Coordination Network
Conference: 22-25 July 2014, Recife, Brazil

Findings

- Positive Effects on annual per-capita income **$14,123.08** rate: US $1 dollar = MX $13.5 peso
- Considerable economic flow in salaries
- + Loss of Biodiversity (number of crops)
- + Attention to Own Farm Production (Solar, Milpa, and Parcela).
- - No evidence of Capital (productive) Investment evidence.
- - Investment on unproductive assets(Motorcycles and TVs)

NSF Research Coordination Network
Conference: 22-25 July 2014, Recife, Brazil

Thank you very much!

Questions and comments to:
javier.becerril@uady.mx

Presented by:
julio.sacramento@uady.mx

NSF Research Coordination Network
Conference: 22-25 July 2014, Recife, Brazil

Outline

1. Oil palm: current and potential use.

2. Study case: Chontalpa region.

3. Building scenarios

Oil palm: current use

Oil palm: a food source

Mexico produces only 14.1% of palm oil consumption and 11.2% of palm kernel oil

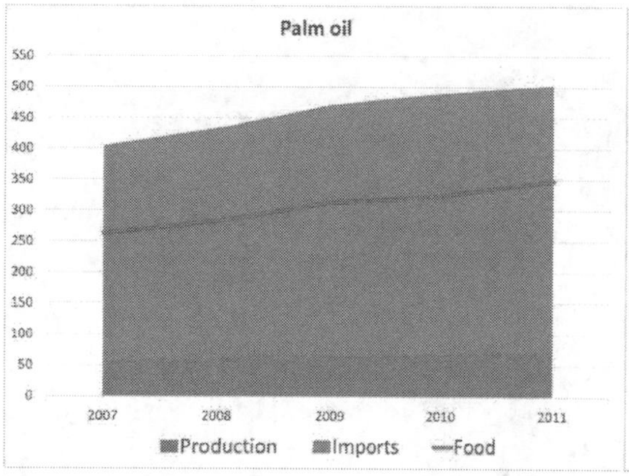

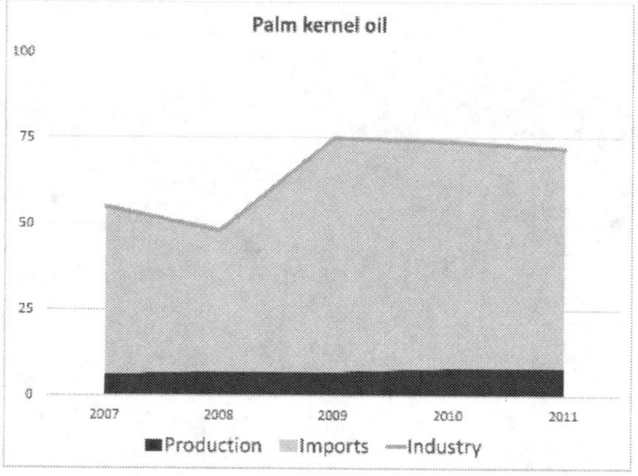

Oil palm: potential use

Oil palm: an energy source

Mexico produces 85% of its energy from oil and gas products. Almost half of the renovable energy comes from bioenergy (wood and sugar cane).

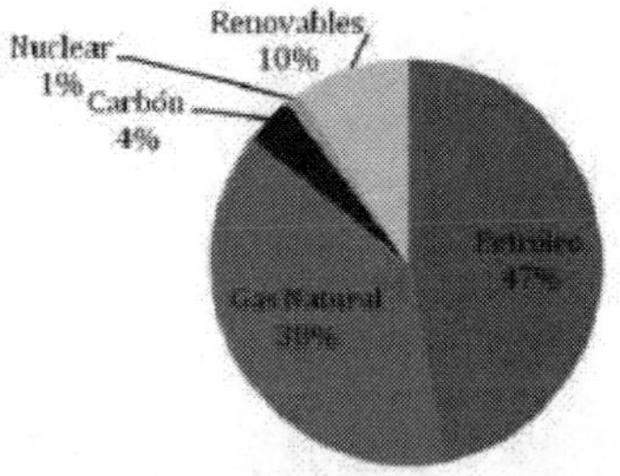

Sener 2009

Bioenergy potential
(petajouls per year)
3,569 PJ/a, 42% Total consumption

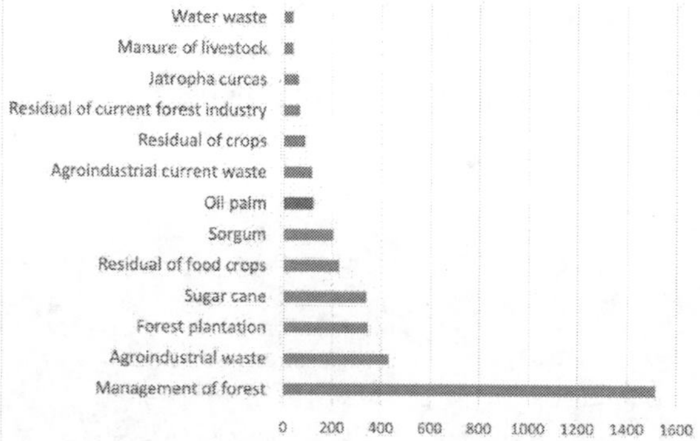

Johnson et al., 2009

Oil palm: potential use

Oil palm: an energy source

Most of the bioenergy crop potential can establish in low and flat land of Gulf Of Mexico, where precipitation is high and there are land without any agricultural use.

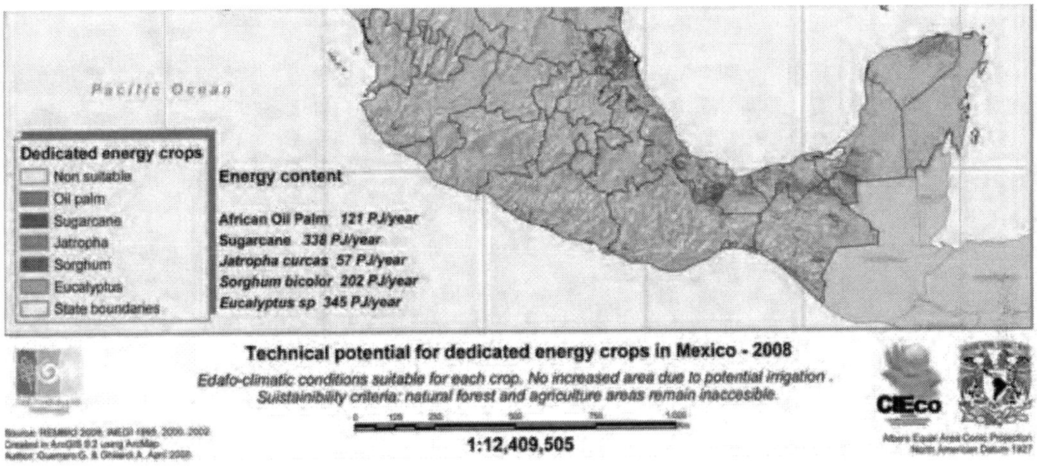

Region of the Chontalpa

Tabasco has a potential for food oil and bionergy production

Region of La Chontalpa

Tabasco has a potential for food oil and bionergy production

RLCh is located on the west part of Tabasco, including 4 municipalities: Cardenas, Huimanguillo, Paraiso and Cunducan. Total surface is 6,938 km^2 (25% of Tabasco's state).

RLCh comprises of 60% of agriculture land (cocoa, coconut, sugarcane, pineapple, cattle, maize, rubber, eucalyptus) and 40% of ecosystems (mangrove, hydrophytes, rainforest, swamps, acahual, water bodies)
Economy is based on oil industry and agriculture.
There are 500,000 inhabitants distributed across 140 localities.

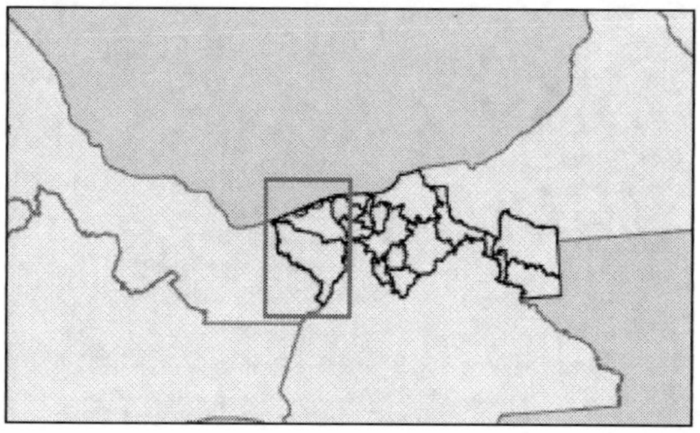

Region of the Chontalpa

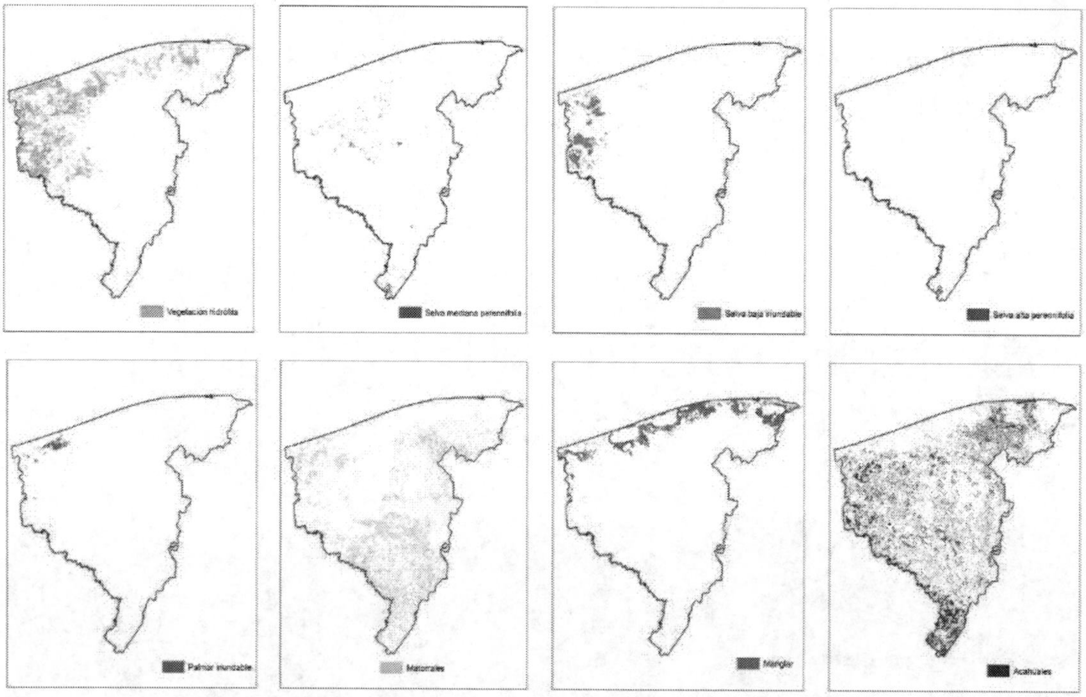

Region of the Chontalpa

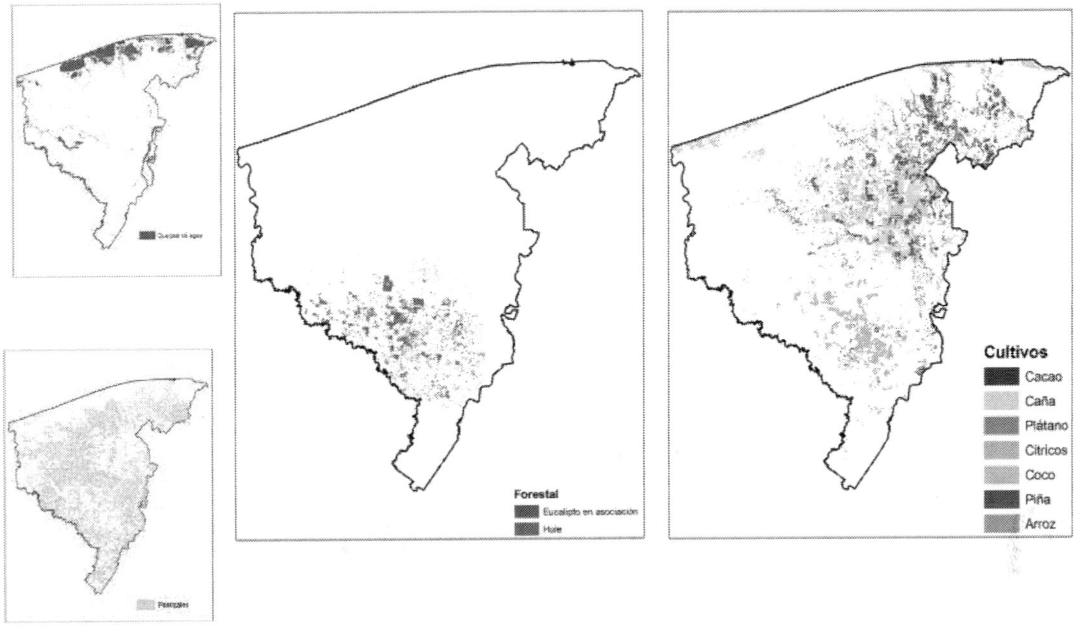

Building scenarios

Objective

To study changes of land use by including oil palm production in the Region of La Chontalpa.
To measure economic benefits/disbenefits for such change by building different scenarios

Assumptions

Public policy program uses the study of the oil palm potential in Tabasco state.
Ecosystem services values are considered for payment environmental program.
Simulation is based on production value rather than profit value.
Mangrove and water bodies are not considered in the simulation

Building scenarios

Methods

- Identification of potential área of oil palm
- Measurement of value of agricultural production
- Measurement of value of ecosystem services
- Building scenarios

Building scenarios

Identification of potential area of oil palm

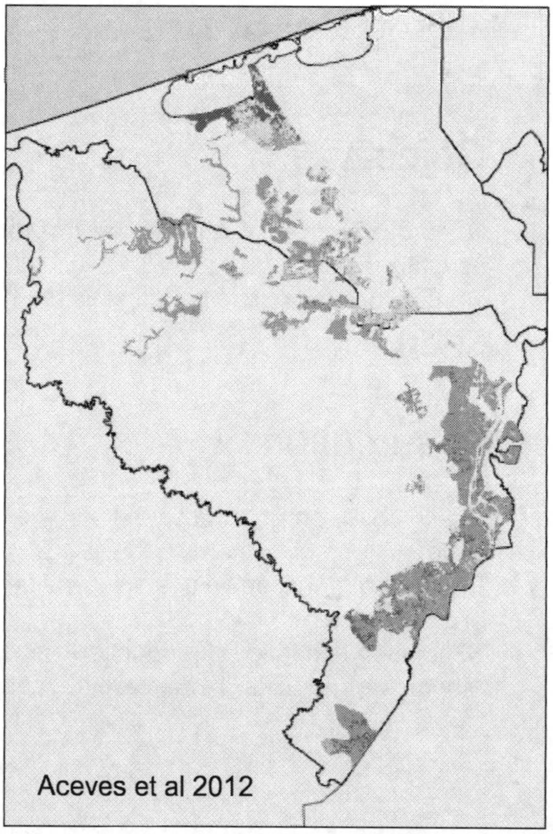

Building scenarios

Identification of potential area of oil palm

CATEGORÍAS	Uso de suelo	SUPERFICIE (Hectáreas)
Cultivos	Cultivo de arroz	509
	Cultivo de cacao	2894
	Cultivo de caña	4939
	Cultivo de cítricos	534
	Cultivo de coco	27
	Cultivo de hule	26
	Cultivo de piña	43
	Cultivo de plátano	319
	Pastizal	32077
	Plantaciones de eucalipto	860
	Superficie de Agroecosistemas	**42228** (65%)
Ecosistema	Acahual	6669
	Cuerpo de agua	732
	Manglar	3269
	Matorral	8053
	Palmar inundable	37
	Selva alta perennifolia	114
	Selva baja inundable	19
	Selva mediana perennifolia	398
	Vegetación hidrófila	2784
	Superficie de ecosistemas	**22075** (35%)
	Superficie total del área de estudio	**64303** (100%)

Building scenarios

Measurement of value of agricultural production

Production yield · Price payed to producer = value of production per hectare

TIPO	VALOR DE LA PRODUCCIÓN POR UNIDAD DE SUPERFICIE [USD hectárea^{-1}]	RENDIMIENTO [toneladas ha^{-1}]	PRECIO [USD tonelada^{-1}]
Arroz palay (milagro filipino)	707.40	2.846	248.56
Cacao seco	1127.51	0.429	2628.24
Caña de azúcar (industrial)	1973.59	52.996	37.24
Cítricos (limón persa)	2173.58	11.885	182.88
Copra	372.24	0.776	479.70
Hule hevea	1388.73	2.198	631.82
Piña (sin clasificar)	7365.31	32.314	227.93
Plátano (enano gigante)	12957.45	58.655	220.91
Palma de aceite (o africana)	1199.85	12.382	96.90
Bovino en pie (en pastizal)	429.39	0.329	1.30
Eucalipto (celulosa y corte a 10 años)	480.96	29.160	16.49

Building scenarios

Measurement of value of ecosystem services

Σ Quantification of ecosystem services · Price of ES = value of ES per hectare

$$VET_{i_{xyz}} = \sum_{x=1}^{x=n} ES_{i_{xy}} \cdot V_{i_z}$$

Servicio de provisión

Servicio de regulación

Servicio de sustento

Servicio cultural

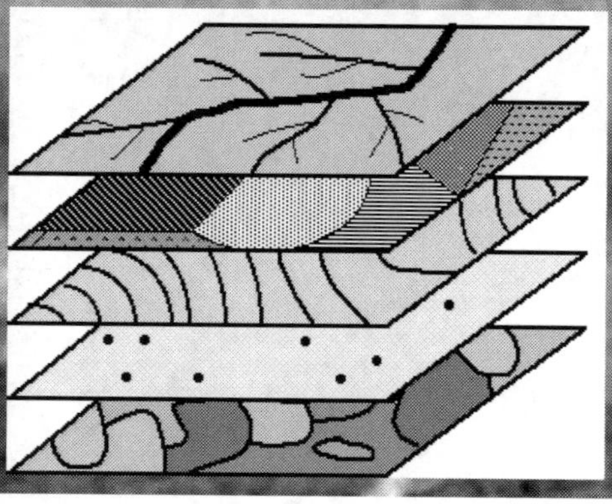

Building scenarios

Measurement of value of ecosystem services

Regulación de gases

Regulación del clima

Regulación de contingencias

Regulación del flujo de agua

Regulación de erosión

Ciclo de nutrientes

Purificación de agua y tratamiento de residuos

Suministro de agua

Building scenarios

Measurement of value of ecosystem services

Provisión de alimentos

Provisión de materia prima

Provisión de material genético

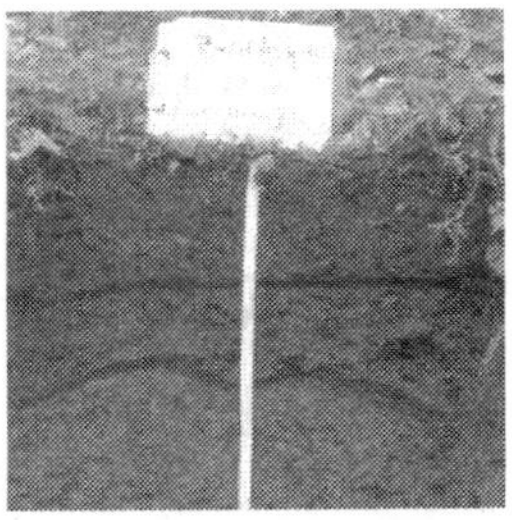
Formación de suelos

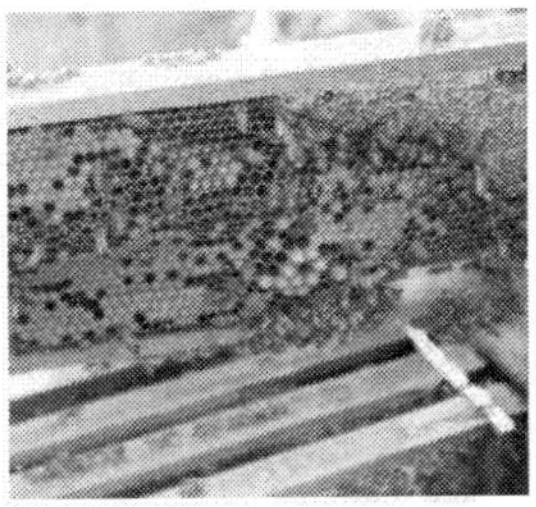
Servicio de polinización

Control biológico

Hábitat

Servicios recreativos, educativos y culturales

Building scenarios

Measurement of value of ecosystem services

TIPO	VALOR ECONÓMICO POR UNIDAD DE SUPERFICIE [USD $hectárea^{-1}$]
Acahual	3597.00
Cuerpo de agua	6149.00
Manglar	7832.00
Palmar inundable	7559.30
Matorral	2585.00
Selva alta perennifolia	7356.33
Selva baja inundable	3236.33
Selva mediana perennifolia	4365.33
Vegetación hidrófila	15118.60

Building scenarios

Oil palm scenarios

Current land use

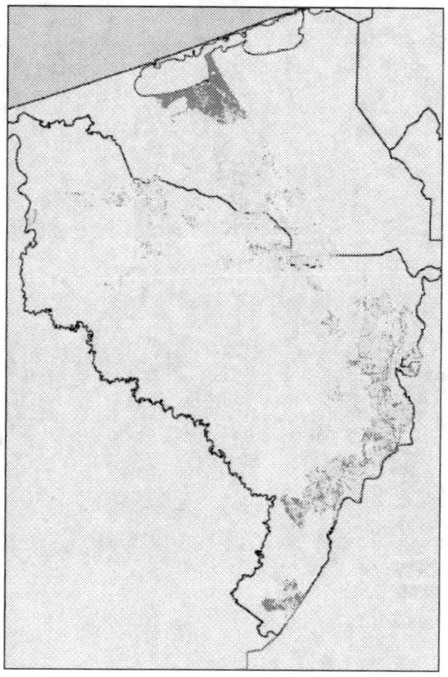

Oil palm production

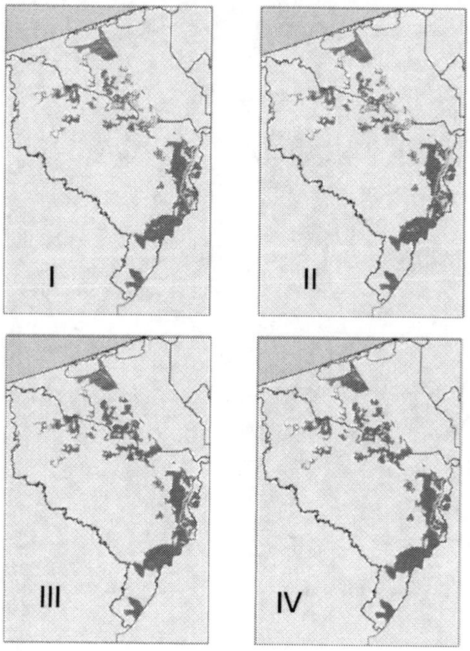

Building scenarios

Oil palm scenarios

		EFECTO	
ESCENARIO	**DESCRIPCIÓN**	**AGROECOSISTEMAS**	**ECOSISTEMAS**
0	v_{ae} v_{en}	*	*
I	$v_{pa} > v_{ae}$ $v_{pa} > v_{en}$	arroz palay, cacao seco, copra y bovino en pie	*
II	$1.5 \cdot v_{pa} > v_{ae}$ $1.5 \cdot v_{pa} > v_{en}$	arroz palay, cacao seco, copra, bovino en pie y hule hevea	*
III	$2 \cdot v_{pa} > v_{ae}$ $2 \cdot v_{pa} > v_{en}$	arroz palay, cacao seco, copra, bovino en pie, hule hevea, caña de azúcar y cítricos,	*
IV	v_{pa}	arroz palay, cacao seco, caña de azúcar cítricos, copra, hule hevea, piña, plátano, bovino en pie y eucalipto	acahual, palmar inundable, matorral, selva alta perennifolia, selva baja inundable, selva mediana perennifolia y vegetación hidrófila

Building scenarios

Oil palm scenarios

ESCENARIO	CAMBIO DE USO DE SUELO (hectáreas)		
	AGROECOSISTEMAS	ECOSISTEMAS	TOTAL
0	42 228	22 075	64 303
I	- 36 366 (86.1%)	0	- 36 366
II	- 36 392 (86.2%)	0	- 36 392
III	- 41 866 (99.1%)	0	- 41 866
IV	- 42 228 (100 %)	- 18 073 (81.8%)	- 60 301

Building scenarios

Oil palm scenarios

ESCENARIO	CAMBIO DE USO DE SUELO (millones de dólares)		
	AGROECOSISTEMAS	ECOSISTEMAS	TOTAL
0	33.210	119.910	153.119
I	+59.031 (178%)	119.910	+178.941
II	+80.859 (243%)	119.910	+200.769
III	+104.917 (316%)	119.910	+224.827
IV	+50.667 (153%)	-51.794 (43%)	-102.461

Building scenarios

Summary

Oil palm has an agronomic and economic potential in the Region of La Chontalpa, and this could change the land use.

Farmers probably improve their income by producing oil palm. However, establishment of plantation is an important issue.

Cattle production is one of the most vulnerable land use.

Converting ecosystems into oil palm has a negative economic effect.

The potential study of oil palm should consider other environmental factors, such as ecosystem services.

Ecosystem services approach can help decision makers to develop conservation and guardianship programs for particular ecosystems.

Muito Obrigado!

César J. Vázquez Navarrete
vcesar@colpos.mx

Effects of Use of Biofuels for Increasing Production and Productivity in Rural Communities in Northeast Brazil

José Ferreira Irmão, Prof. UFRPE (Ph. d. in economics)
Jacques Ribemboim, Prof. UFRPE (Ph. d. in economics)

Conference Program

RCN Conference on Pan American Biofuels & Bioenergy Sustainability
Golden Tulip Recife Palace,
Recife, Brazil July 22-25, 2014

PAN-AMERICAN BIOFUELS & BIOENERGY SUSTAINABILITY
AN NSF RESEARCH COORDINATION NETWORK

Wednesday, July 23, 2014: Socioeconomic Sustainability Dimensions of Biofuels in the Americas
08:30 AM - 11:30 AM (Boa Viagem Room)

Chair:	Sam Sweitz, Michigan Technological University, Michigan, USA
8:30 AM	Interrogating Social Sustainability in the Biofuels Sector in Latin America: Global Standards and Local Experiences in Mexico, Brazil and Colombia . **T. Selfa**, C. Bain, R. Moreno, A. Eastmond, S. Sweitz, C. Bailey, T. Martins, G. S. Pereira, and R. Medeiros
8:50 AM	Global and Local Perspectives on the Jatropha Plantations for Biodiesel in Sucopo, Yucatan, Mexico . **A. Eastmond** and S. Sweitz
9:10 AM	Agro-Fuels and Income of Rural Households in Yucatan, Mexico . **J. Becerril**, A. Eastmond, and S. Sweitz
9:30 AM	Bioenergy, Land Use Change and Ecosystem Services in Tabasco, Mexico . **C. J. Vazquez-Navarrete**
9:50 AM	Effects of Use of Biofuels for Increasing Production and Productivity in Rural Communities in Northeast Brazil . **J. Ferreira Irmao** and J. Ribemboim
10:10 AM	Energy Crops, Food Security, and Rural Development: The Role of Oilseed Production Around Brazil . **R. Bailis**
10:30 AM	Roundtable Discussion of Key Research Issues and Challenges

Recife location in the world map

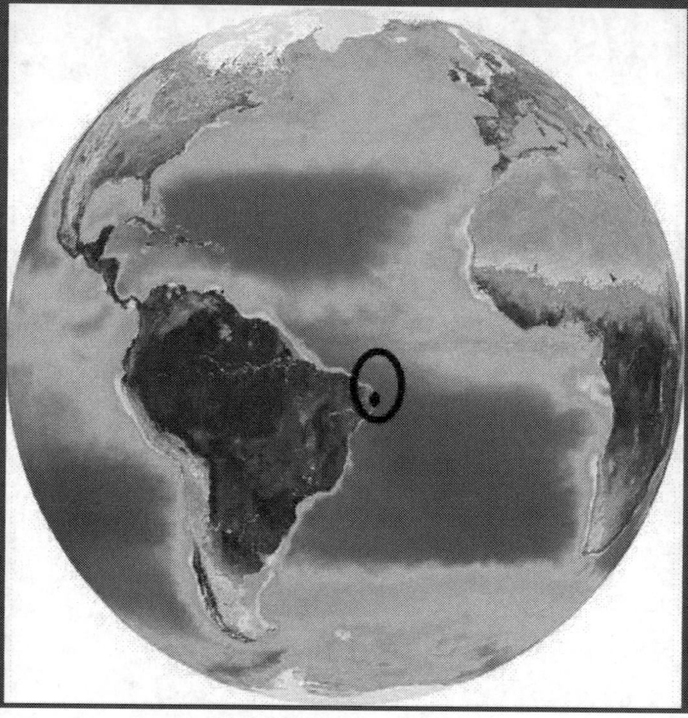

Context of the Paper

◆The paper contains results of a research undertaken for the ANEEL (National Agency for Electric Energy) on the sources of sustainable energy for production in small agriculture in Northeast Brazil. The aim was to identify forms of uses of electric energy in family farming agriculture for stimulating local development in rural communities. Methods of research included field work and application of questionnaires in rural communities and settlements of landless workers. Research covered seven of the nine Northeastern states in a sample of small landholdings. Results shown a very good potential for increasing production and productivity through the use of electric energy, in special, by means of irrigated agriculture when communities dispose of biofuel energy. Attention was also posed in the use of alternative sources of energy, those made available by sun and wind energy. Additional results also indicated that the use of energy in agro-industrial activities such as fruit processing and livestock production increase productivity, income and employment in rural communities mainly those outside main roads and means of transportation. But one further and less effective result was that the successful means for increasing production, productivity and income of small landholdings depend strongly on programs of education, promotion of rural non-agricultural activities and policies aiming reduction of urban-rural inbalances.

Context of the Research on the Theme

- 1) The role of pouwer in the development or rural communities in Northest Brazil;
- 2) Importance of alternative sources of energy for the local development – sun, wind and biofuel – for the rural development is very important in Northeast due to the fact that these resources are available;
- 3) Importance of these types of energy in the energetic profile of various countries;
- 4) Brazil has promoted the use of biofuels in the energetic matrix bringing an important contribution to the rural development namely in Northeast Brazil;
- 4) Relevance of important Government Programmes as Programa Nacional da Agroenergia, Luz para Todos e PRODEEM, in supporting rural development in poor rural communities in Norteast Brazil;
- 5) How the growing importance of these kinds of biofuels has been related to the growth and development of population in the Region.

Growth of Energy Consumption in Northeast

Increasing Pouwer Consumption in Northeast Brazil

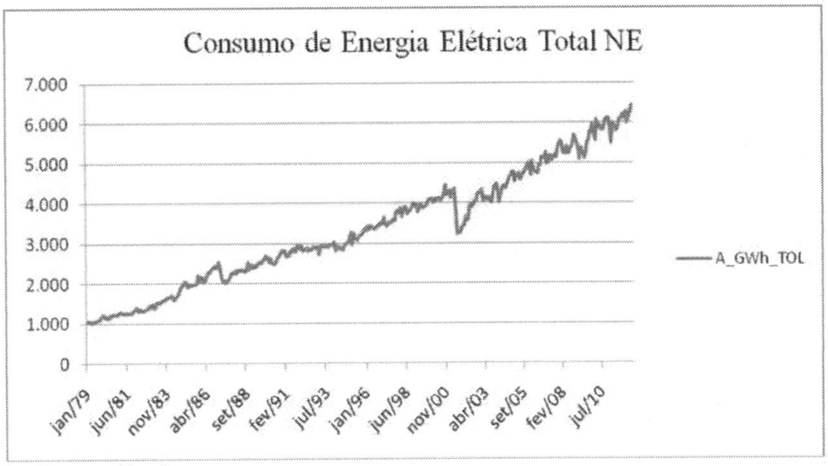

Source of Data : EPE

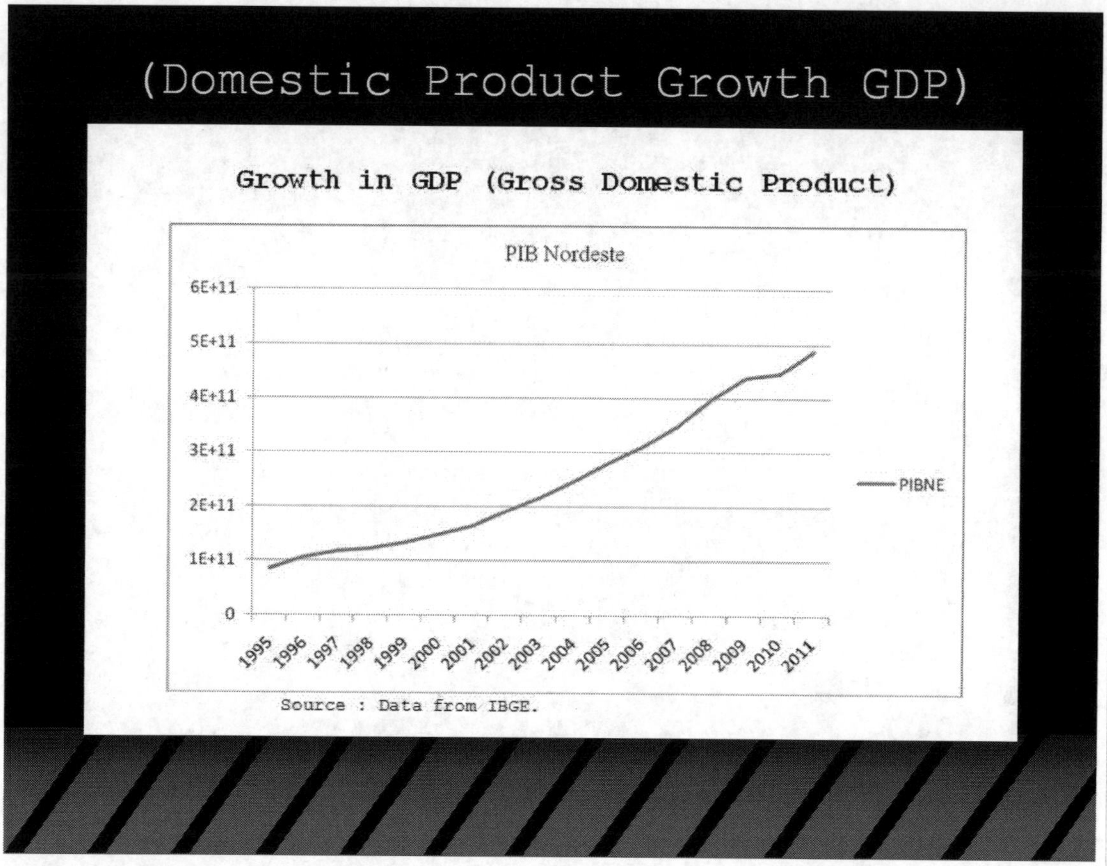
(Domestic Product Growth GDP)
Growth in GDP (Gross Domestic Product)
PIB Nordeste
6E+11
5E+11
4E+11
3E+11
2E+11
1E+11
0
1995 1996 1997 1998 1999 2000 2001 2002 2003 2004 2005 2006 2007 2008 2009 2010 2011
PIBNE
Source : Data from IBGE.

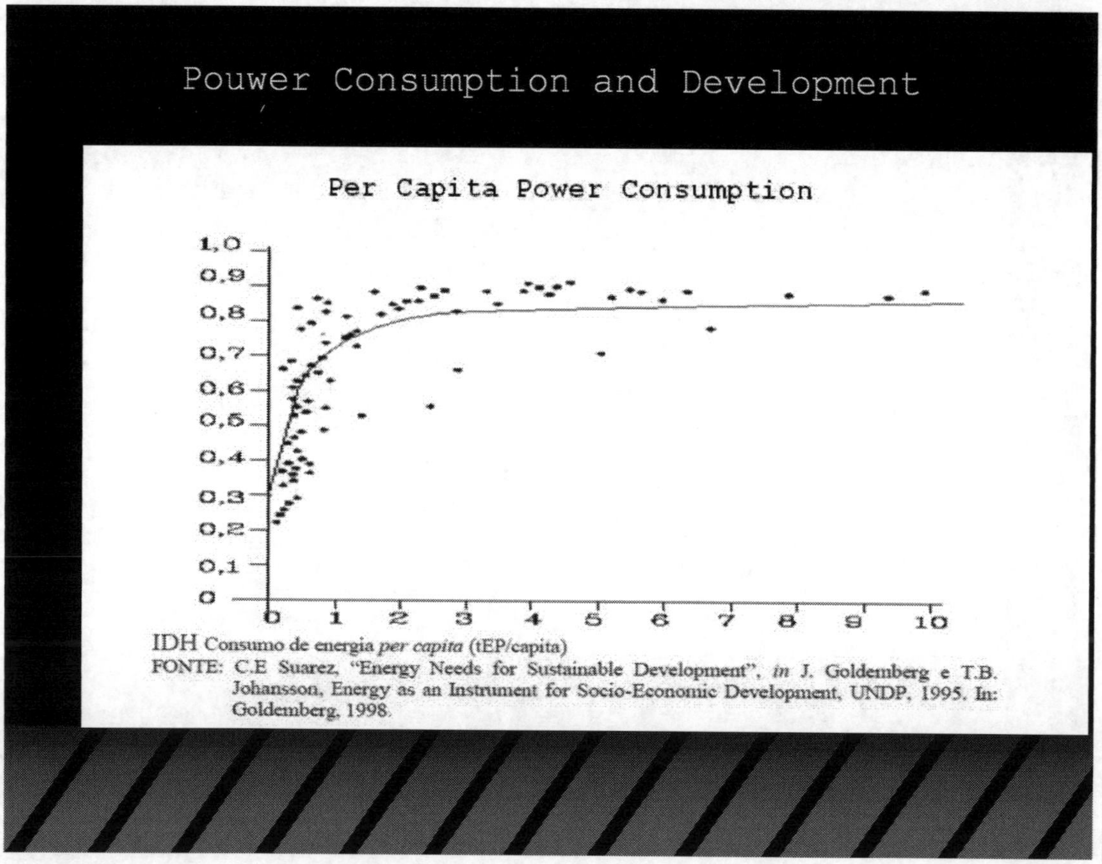
Pouwer Consumption and Development
Per Capita Power Consumption
1,0
0,9
0,8
0,7
0,6
0,5
0,4
0,3
0,2
0,1
0
0 1 2 3 4 5 6 7 8 9 10
IDH Consumo de energia per capita (tEP/capita)
FONTE: C.E Suarez, "Energy Needs for Sustainable Development", in J. Goldemberg e T.B. Johansson, Energy as an Instrument for Socio-Economic Development, UNDP, 1995. In: Goldemberg, 1998.

Research Context

- 1) Identify communities which benefit from Government energetic Programmes with the aim of pouwer consumpton and productivity investments in small scale farming agriculture;
- 2) Apply model to analyze productive chains in communities away from electric nets, communities producing raw materials and final products to the local, regional and national markets;
- 3) Text the theoretical relation between energetic consumption and development, leading to the increase in production, employment and income in small rural communities of Northeast Brazil.

Research Method (theoretical)

- 1) development and applciation of models aiming at identifying productive chains in small rural communities away from electric nets in Northeast Brazil;
- 2) Utilisaton of optimum socio-economic criteria to recommend state policies aiming to increase production, productivity and income in these small communitiesl;
- 3) Adoption of rational economic criteria in the development and application of models;
- 4) Policy recommendations incorporating these rational expectations in similar communites of the Region.

Research Method (empirical)

- 1) Identify communities distant from the electric nets with productive potencial;
- 2) Field visits for the application of questionnaires, and filing of problems and potentialities;
- 3) Computer processing of data and running of models developed;
- 4) Presentation of reports and papers in Congress and publication of papers to sharehold experiences.

(Model Development)

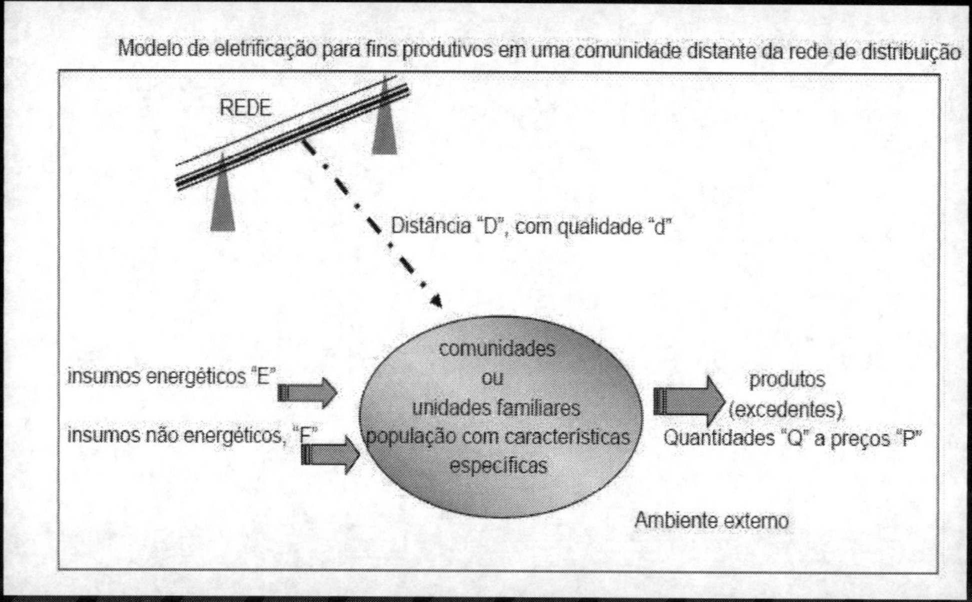

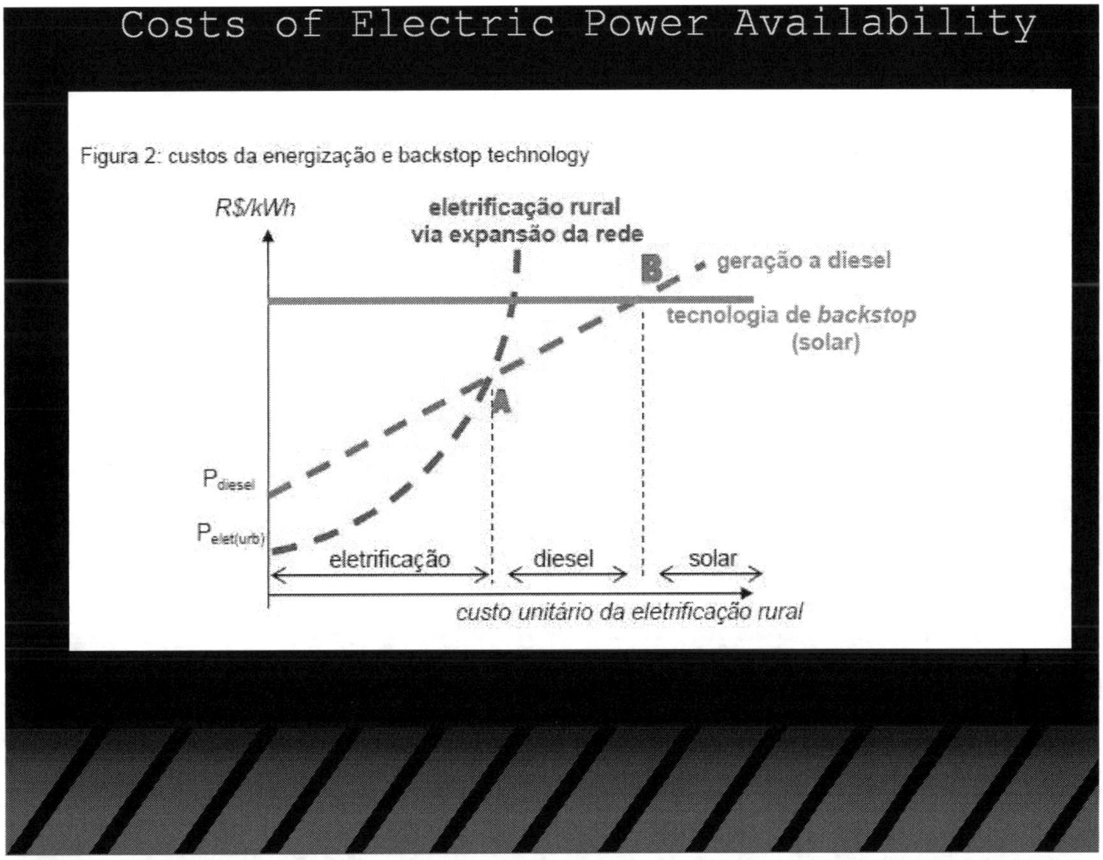
Costs of Electric Power Availability
Figura 2: custos da energização e backstop technology
R$/kWh
eletrificação rural
via expansão da rede
B
geração a diesel
tecnologia de backstop
(solar)
A
P_{diesel}
$P_{elet(urb)}$
eletrificação
diesel
solar
custo unitário da eletrificação rural

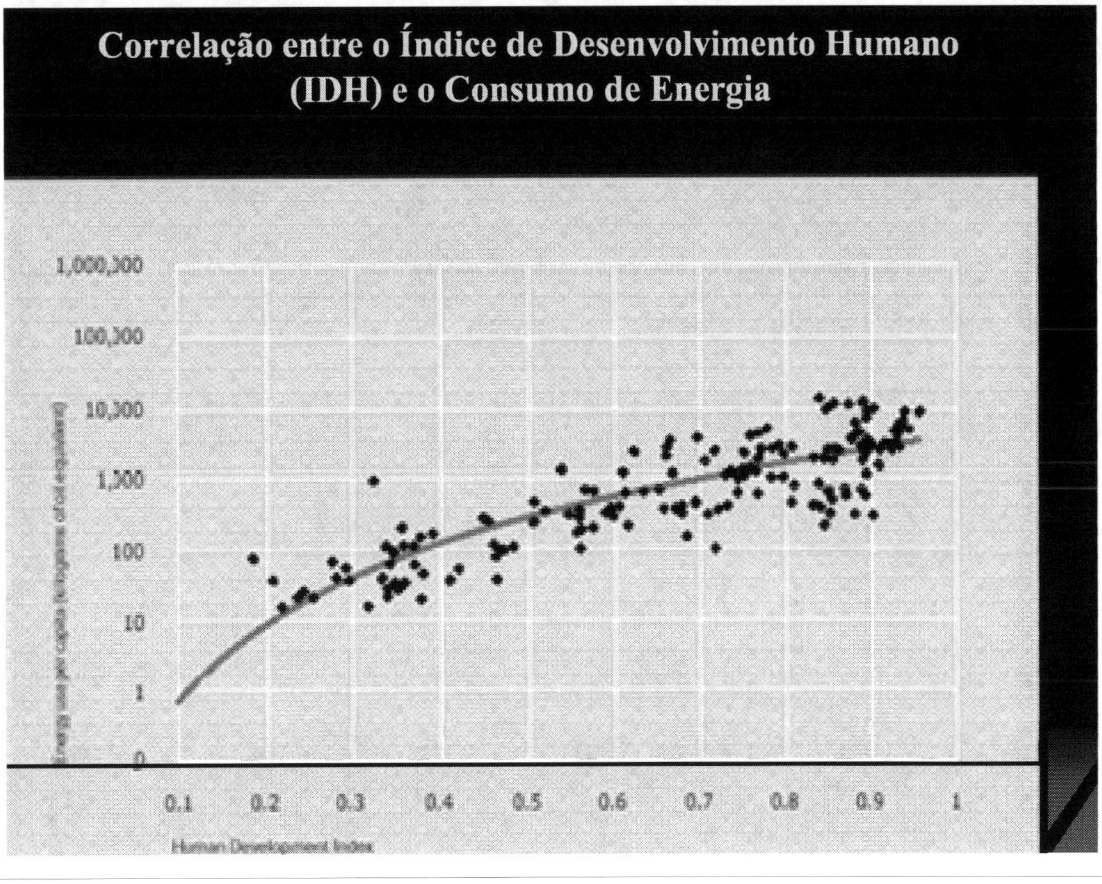
Correlação entre o Índice de Desenvolvimento Humano (IDH) e o Consumo de Energia
100
10
1
0
0.1
0.2
0.3
0.4
0.5
0.6
0.7
0.8
0.9
1
Human Development Index

Costs of Electrification

Tabela 3: Custos da eletrificação rural no Ceará (Programa Luz Para Todos)

Etapa do Programa	objetivo	Custo da ligação p/domicílio valores aproximados
FASE I	Atendimento de 70% das 73.000 unidades cadastradas	R$ 3.900,00
FASE II	Atendimento de 70% de 33.000 unidades cadastradas	R$ 4.500,00
FASE III	Atendimento de 70% de 21.000 unidades cadastradas	R$ 6.500,00

Fonte: Modelos de uso de fontes energéticas renováveis para produção rural em comunidades distantes da rede de distribuição elétrica. Chesf/Fadurpe/UFRPE, 2007

No. of Communities Visited (By Location)

Tabela 1.

Municípios e Comunidades Visitadas por Estado do Nordeste

Estados	Municípios	Comunidades	Famílias	Pessoas
Alagoas - AL	6	6	358	1660
Ceará - CE	3	6	214	1060
Paraíba - PB	8	9	266	1189
Pernambuco - PE	3	5	863	4532
Piaui - PI	4	6	1218	4882
Rio Grande do Norte - RN	4	5	508	2094
Sergipe - SE	5	5	300	1460
Total	33	42	3727	16877

Fonte: Modelos de uso de fontes energéticas renováveis para produção rural em comunidades distantes da rede de distribuição elétrica. Chesf/Fadurpe/UFRPE, 2007

Map of Communities Vsited

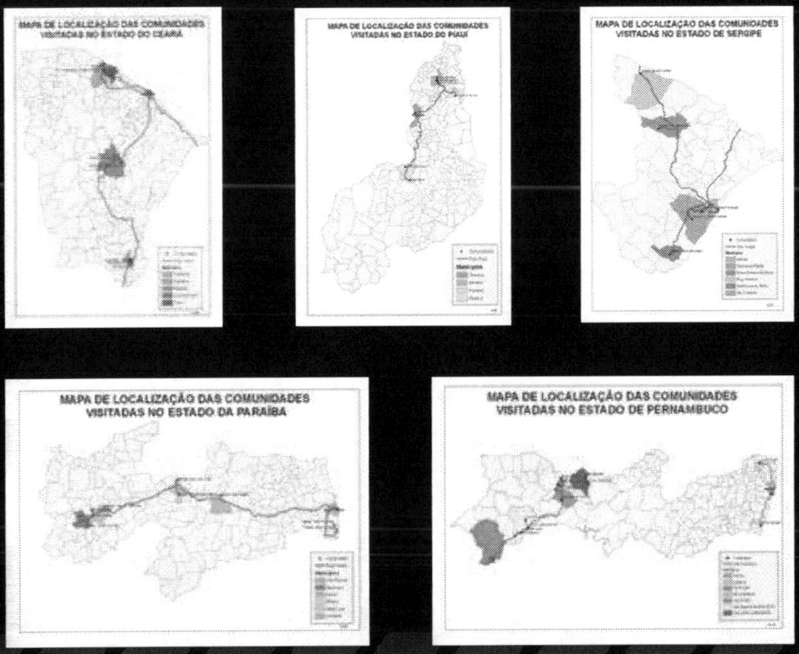

Community Features

- (i) **Low income generation (poverty);**
  (ii) **Subsistence agriculture;**
  (iii) **iIliteracy;**
  (iv) **No-organized structure;**
  (v) **No electric power;**
  (vi) **Exclusion of public services;**
  (vii) **Resource scarcity**
  (viii) **Low capital accumulation sources;**
  (ix) **Low health availability;**
  (x) **Low education availability.**

Community Features by Community

Tabela 2: características sócio-demográficas das comunidades estudadas

Comunidade	UF	Famílias	Pessoas	Tamanho médio da família	Número de pessoas que trabalham	Número de pessoas assalariadas
Saco dos Goitis	PB	12	60	5,00	15	5
Várzea dos Bois	PB	12	60	5,00	18	6
Timbó de Cima	PB	17	80	4,71	60	0
Timbó de Baixo	PB	10	75	7,50	55	0
Sítio Santo Estevão	PB	57	280	4,91	180	15
Sítio Maria Soares	PB	58	250	4,31	175	20
Serra Branca	PB	60	220	3,67	160	30
Murici	PB	10	55	5,50	30	5
Lajedo de Timbaúba	PB	30	109	3,63	80	26
Tabatinga	PI	22	132	6,00	65	15
Morcego	PI	80	400	5,00	160	120
Marinbondo	PI	55	275	5,00	250	12
Cachoeira	PI	35	175	5,00	150	0
Amolar	PI	1000	3800	3,80	1500	50
Papa Pombo	PI	26	100	3,85	65	0
Camará	CE	27	108	4,00	125	0
Bom Jesus	CE	53	250	4,72	85	15
Assentamento Alegre	CE	36	210	5,83	208	2
Comunidade Luciano	CE	17	72	4,24	72	0
Sítio Brejo Seco	CE	55	220	4,00	55	0
Catolé	CE	26	200	7,69	180	20
Frei Damião	RN	10	32	3,20	15	4
Santa Clara	RN	166	600	3,61	400	0
Palheiros	RN	100	500	5,00	250	25
José Milanez	RN	132	462	3,50	264	0
Guarapes	RN	100	500	5,00	225	20
Poço do Ico	PE	26	130	5,00	60	0
Pau Ferro	PE	26	104	4,00	78	0
Sítio do Mosquito	PE	16	68	4,25	68	0
Lagoa da Pedra	PE	45	230	5,11	50	0
Conceição das Creoulas	PE	750	4000	5,33	1750	250
Che Guevara	SE	20	120	6,00	100	0
D. Helder Câmara	SE	20	40	2,00	32	8
N. S. de Lourdes	SE	55	250	4,55	175	15
Vitória da União	SE	150	800	5,33	80	20
Florestan Fernandes	SE	55	250	4,55	250	0
Timbaúba	AL	22	150	6,82	60	5
Fazenda de Baixo	AL	100	600	6,00	300	0
Vitória da Conquista	AL	17	80	4,71	45	5
Luciana	AL	70	300	4,29	100	10
Tucuns	AL	29	180	6,21	30	20
Pedrão	AL	120	350	2,92	150	35
Total		3727	16877	4,52831	8190	758

Fonte: Modelos de uso de fontes energéticas renováveis para produção rural em comunidades distantes da rede de distribuição elétrica. Chesf/Fadurpe/UFRPE, 2007

Main Findings

- Os níveis de renda e demais indicadores de qualidade de vida têm melhorado sensivelmente no
- Nordeste rural durante os últimos dez anos. Para isto contribuíram, sem dúvidas, programas do
- tipo "renda mínima", tais como o Bolsa-Família, e outros programas de combate à pobreza rural.
- No setor de energia, esforços como o PRODEEM3 e o Programa Luz Para Todos também
- trouxeram melhorias significativas, o primeiro, fornecendo energia a escolas, enfermarias e
- centros comunitários e, o segundo, fornecendo eletrificação aos domicílios rurais, na maioria das
- vezes para consumo doméstico.

Main Findings

No caso de uso de fontes energéticas locais renováveis para fins produtivos, há um grande potencial de aumento da produção e da produtividade agropecuária nas comunidades visitadas, resultantes da abundância relativa (disponibilidade) de terra e mão-de-obra que, com a chegada da energia, poderiam ser melhor aproveitados, sobretudo com uso de irrigação. Outras atividades, contudo, foram identificadas com possibilidades de aproveitamento da energia para serviços de beneficiamento da produção, artesanato e turismo.

Main Findings

No caso de uso de fontes energéticas locais renováveis para fins produtivos, há um grande potencial de aumento da produção e da produtividade agropecuária nas comunidades visitadas, resultantes da abundância relativa (disponibilidade) de terra e mão-de-obra que, com a chegada da energia, poderiam ser melhor aproveitados, sobretudo com uso de irrigação. Outras atividades, contudo, foram identificadas com possibilidades de aproveitamento da energia para serviços de beneficiamento da produção, artesanato e turismo.

- **Many thanks for your audience**

Conclusões

- Contudo, é preciso que haja complementaridade entre as políticas de apoio a estas famílias de modo a explicar ao homem do campo como melhor aproveitar a nova energia disponibilizada.
Uma vez que a estrutura tarifária embute transferências reais de recursos das famílias urbanas
para as famílias rurais, é preciso cuidados especiais para se identificar em que momentos será
preferível o uso de sistemas autônomos de geração à eletrificação via expansão da rede, sobretudo na eletrificação de comunidades e propriedades rurais muito distantes.

Main Findings

Figura 3: Exemplo de potencialidade agrícola e aporte energético

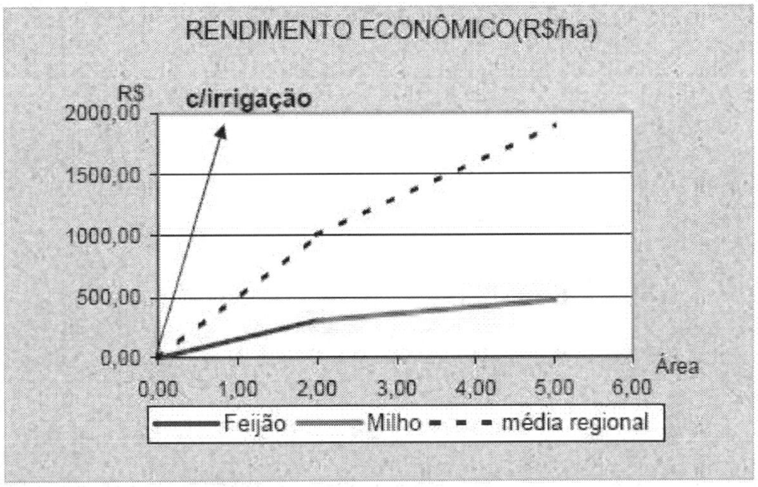

Energia e Inclusão Social: a posição do problema

- **Projeto CHESF/UFRPE/FADURPE: "Fontes Energéticas Locais Renováveis e Modelos de Uso Auto-sustentável para Cadeias Produtivas em Comunidades Distantes da Rede Elétrica"**

Projeto Chesf/ufrpe/fadurpe: conclusões preliminares

1) Desenvolvimento de modelo para identificação de comunidades distantes da rede, potencialmente beneficiárias de projetos e programas energéticos, segundo os ganhos econômicos e sociais que possam advir da energização. Este modelo servirá de orientação para a escolha das comunidades que serão objeto da pesquisa de campo da segunda etapa da pesquisa
2) desenvolvimento de modelo para estudo das cadeias produtivas das comunidades distantes da rede, onde serão abordados todos os aspectos mais importantes da cadeia produtiva, desde o suprimento de matérias-primas, até o fornecimento de itens intermediários ou finais ao mercado externo, incluindo os fluxos de distribuição;
3) adoção de formas de análise que priorize a relação entre energia e desenvolvimento local, dadas as características determinantes das comunidades estudadas, tais como nível de renda e emprego, perfil populacional e distributivo, potencial de incremento da produção etc

Projeto Chesf/ufrpe/fadurpe: andamento da pesquisa

- 1) O papel da eletrificação rural na promoção do desenvolvimento local sustentável tem variado, observando-se bons e maus resultados, sendo que as falhas em programas e projetos governamentais podem ser geralmente explicados pela falta de complementaridade em relação à oferta de outros bens públicos necessários ao bom aproveitamento social da chegada da energia elétrica;
- 2) A importância das fontes de energia alternativa – solar, eólica e a partir de biomassa – para o desenvolvimento das populações nordestinas é de fato muito grande, uma vez que, no Nordeste, estes recursos existem de forma relativamente abundante;
- 3) Há uma crescente participação relativa de geração eólica, térmica e de biomassa nas matrizes energéticas do Brasil e do resto do mundo;
- O uso de biomassa e, em particular, a produção de biodiesel pode se constituir em um excelente instrumento de desenvolvimento rural, sobretudo para elevação da renda das populações em algumas localidades do Nordeste;
- 4) A importância dos projetos e programas governamentais, notadamente aqueles previstos no Programa Nacional da Agroenergia, Luz para Todos e PRODEEM, para apoiar a energização das comunidades distantes da rede com potencial de desenvolvimento sustentável

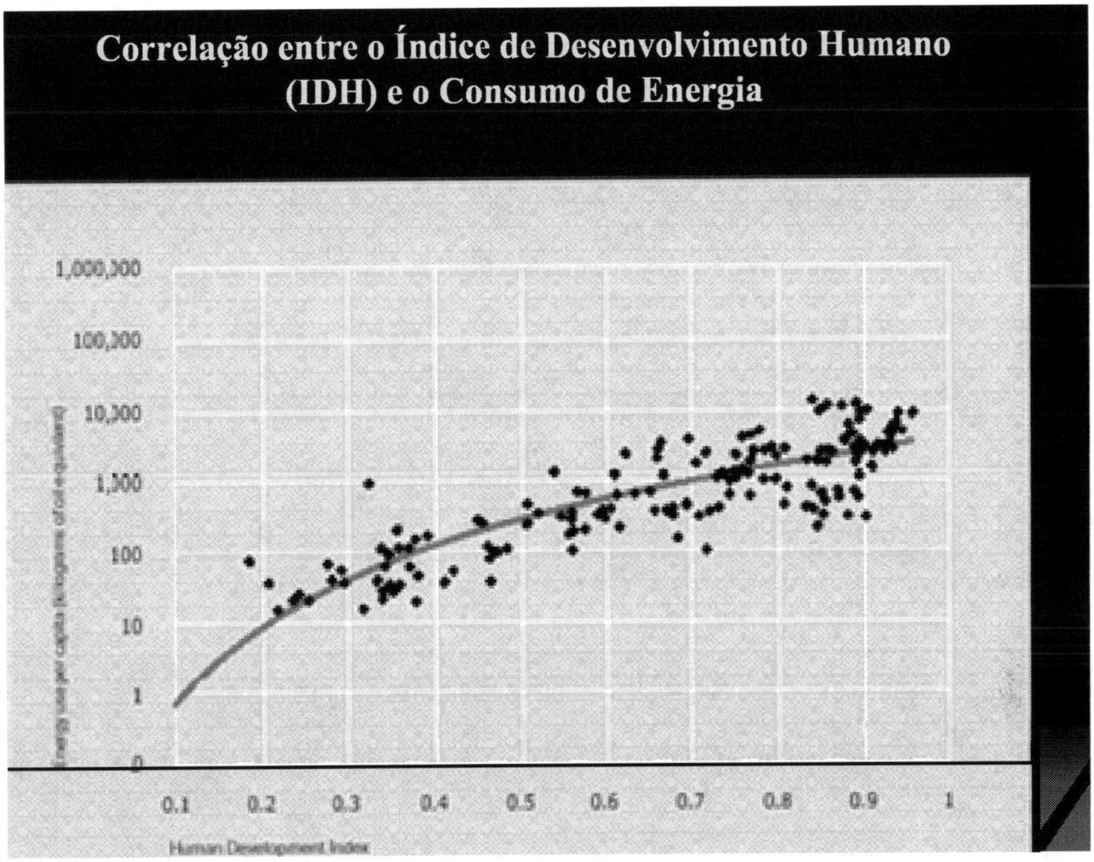
Correlação entre o Índice de Desenvolvimento Humano (IDH) e o Consumo de Energia
1,000,000
100,000
10,000
1,000
100
10
1
0
0.1
0.2
0.3
0.4
0.5
0.6
0.7
0.8
0.9
1
Human Development Index

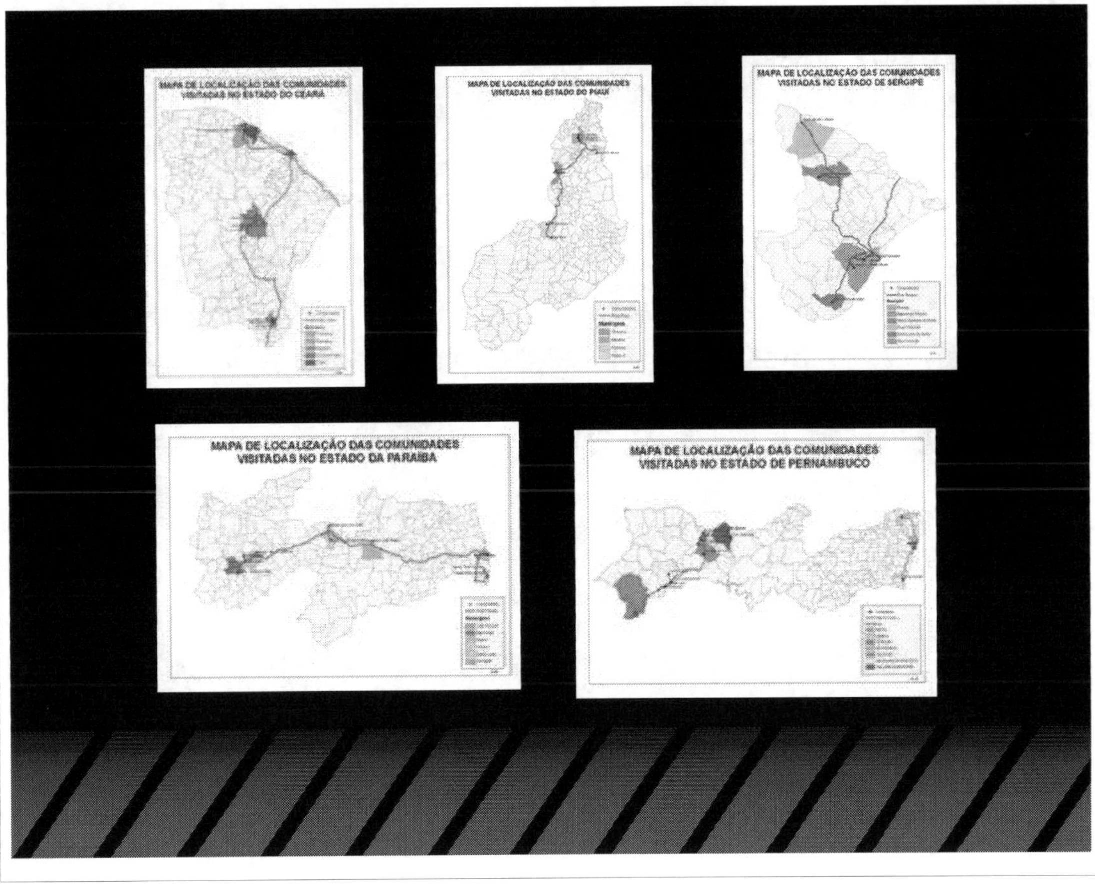
MAPA DE LOCALIZAÇÃO DAS COMUNIDADES VISITADAS NO ESTADO DO CEARÁ
MAPA DE LOCALIZAÇÃO DAS COMUNIDADES VISITADAS NO ESTADO DO PIAUÍ
MAPA DE LOCALIZAÇÃO DAS COMUNIDADES VISITADAS NO ESTADO DE SERGIPE
MAPA DE LOCALIZAÇÃO DAS COMUNIDADES VISITADAS NO ESTADO DA PARAÍBA
MAPA DE LOCALIZAÇÃO DAS COMUNIDADES VISITADAS NO ESTADO DE PERNAMBUCO

Características das Comunidades

I - Problemas

(i) Baixa renda, pobreza
(ii) Agricultura de subsistência
(iii) Analfabetismo
(iv) Desorganização
(v) Falta de energia para produção
(vi) Falta de políticas públicas
(vii) Pobreza de recursos
(viii) Pequena acumulação de capital
(ix) Saúde precária
(x) Educação de baixa qualidade

Características das comunidades

- **Potencialidades**

  (i) inserção no mercado
  (ii) presença do recurso água
  (iii) desenvolvimento da produção
  (iv)

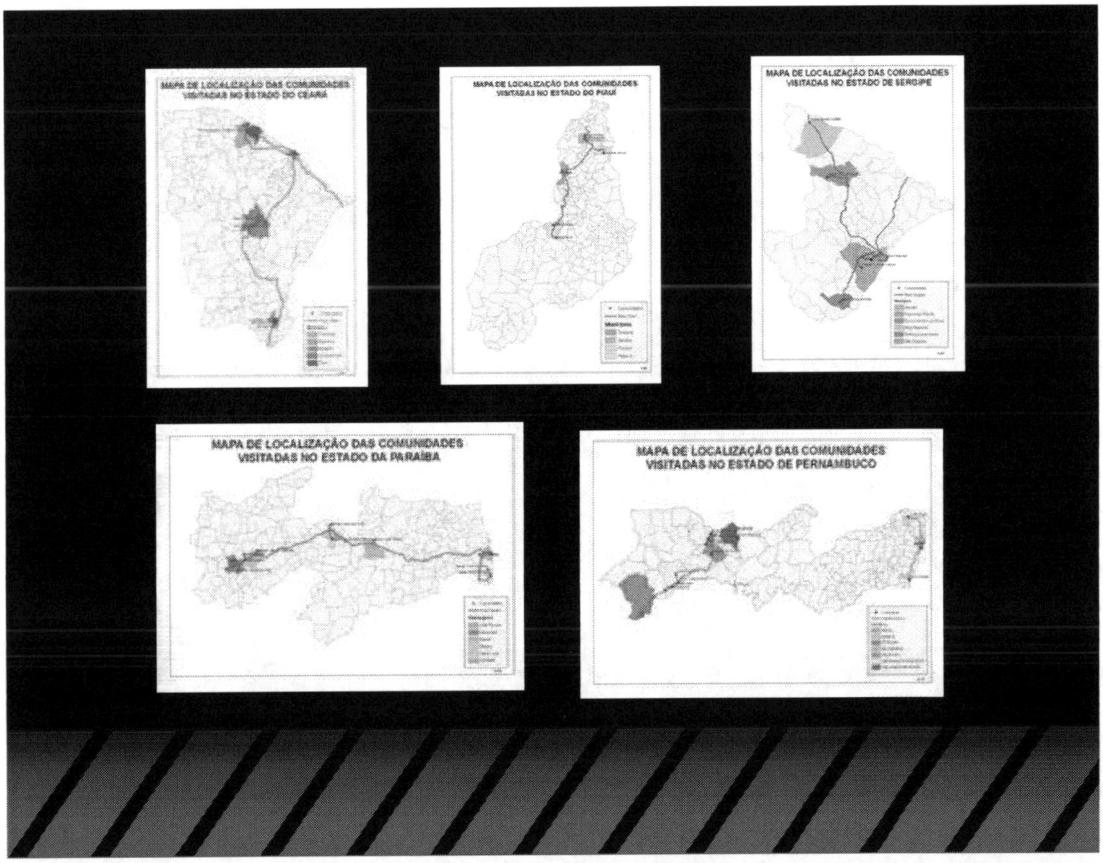

Equações de custo

- Qualquer que seja a tecnologia selecionada para eletrificação da comunidade, haverá custos de diversas naturezas. A necessidade do investidor de cobrir esses custos e auferir lucro é o que determinará o custo da energia para o consumidor. Considera-se que o custo da energia elétrica gerada pelo sistema é expresso por:
- CE = CCVa / Ec =, (eq. 15)
- onde CE é o custo da energia elétrica, CCVa é o custo do ciclo de vida anualizado do sistema e Ec é a energia elétrica consumida no período de um ano (Δt = 8.760 h). No denominador da eq. 15 é utilizado EC, e não EG, porque é do consumo de energia elétrica que virão as receitas para pagar o custo do sistema. Pode-se reescrever a eq. 15 em função de L como:
- L = CE CCVa / L x P t x × Δ

Equações de custo

- O custo do ciclo de vida anualizado é expresso por:

$$CCVa = [(CI+CRp)\times FRC(i,n)]+CO+CM, \text{ (eq. 17)}$$

onde CI é o custo inicial do sistema, CRp a soma dos valores presentes dos custos de reposição dos componentes do sistema, CO o custo de operação, CM o custo de manutenção e FRC(i, n) o fator de recuperação de capital, expresso por:

$$FRC\ (i,n) = i \times (1 + i)^n \ / \ (1 + i)^n$$

onde i é o custo do capital e n é o período de análise. Os custos que aparecem na eq. 17, compondo o custo total do sistema, são:

Equações de custo

- Os custos que compõem os custos totais expressos na fórmula

$$CCVa = [(CI+CRp)\times FRC(i,n)]+CO+CM$$

são dados por:

- Custos iniciais (CI)
- Custos de reposição trazidos a valor presente (CRp)
- Custos de operação (CO)
- Custos de manutenção (CM)

Equações de custo

- **Custos iniciais (CI).** Referem-se a projeto, aquisição, transporte e instalação de todo o aparato necessário à entrada em operação do sistema de energia elétrica;
- **Custos de reposição trazidos a valor presente (CRp).** Referem-se às substituições dos componentes principais do sistema, devido ao fim da vida útil dos mesmos.
- **Custos de operação (CO).** Referem-se aos custos necessários ao funcionamento do sistema.
- Custos de manutenção (CM). Referem-se a custos com manutenções preventivas e pequenas manutenções corretivas.

Sistemas de energização

- **Interligação à rede.** Trata-se de uma linha de transmissão trifásica em 13,8 kV, com transformador abaixador trifásico de 30 kVA. Supõe-se que o circuito mais próximo em 13,8 kV, pertencente ao sistema interligado, de onde se pode fazer a ligação, está uma distância d = 10 km da comunidade.
- **Sistema a diesel.** Trata-se de um gerador a diesel de 20 kVA, 220/127 V, 60 Hz, operando 24 h/dia.
- **Sistema solar fotovoltaico.** Sabe-se que um sistema fotovoltaico autônomo, baseado na tecnologia atualmente disponível, não alimenta uma carga de 15 kW de forma econômica.
- Sistema eólico.
- Sistema a biomassa.
- Sistema híbrido solar-eólico-diesel.

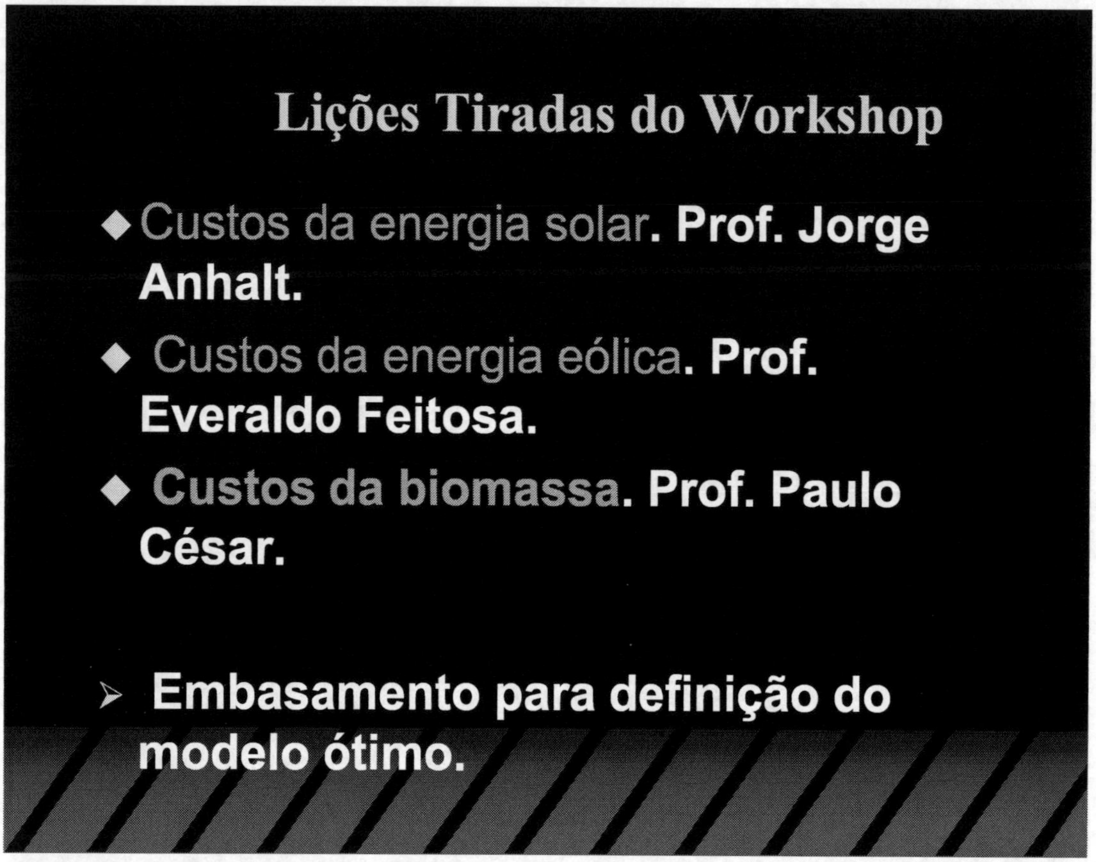
Lições Tiradas do Workshop
◆ Custos da energia solar. Prof. Jorge Anhalt.
◆ Custos da energia eólica. Prof. Everaldo Feitosa.
◆ Custos da biomassa. Prof. Paulo César.
➢ Embasamento para definição do modelo ótimo.

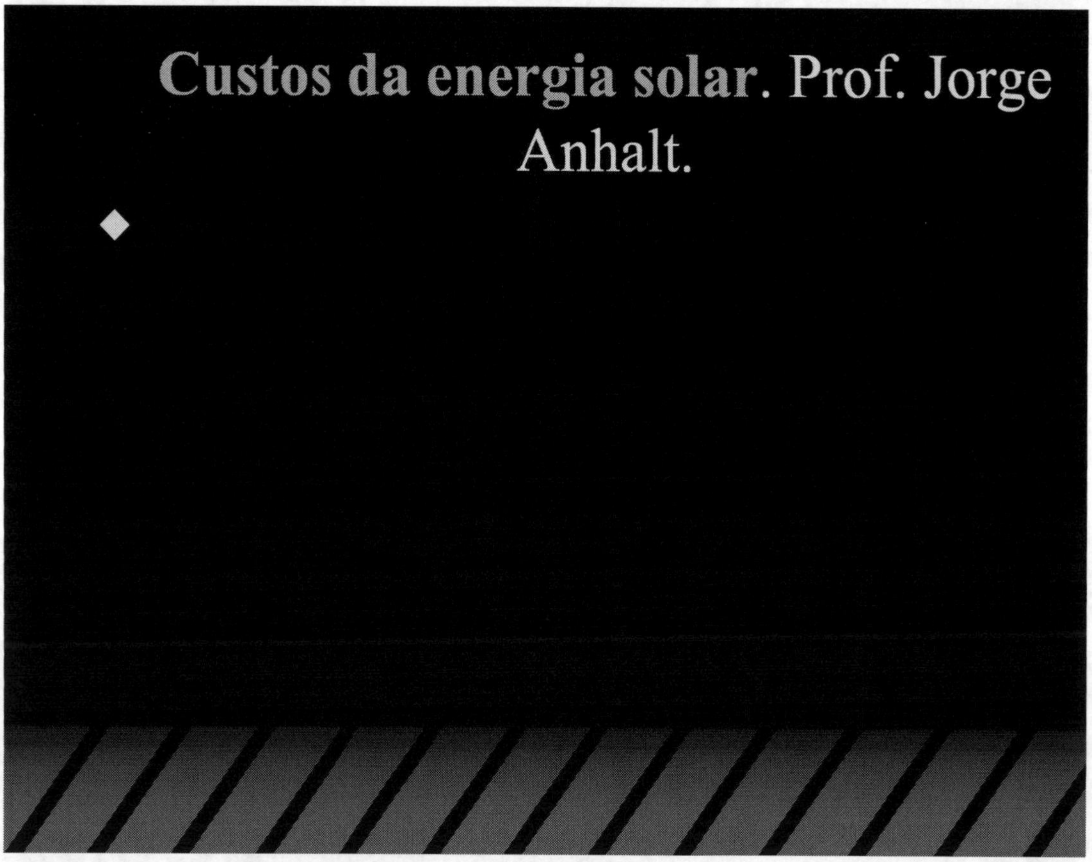
Custos da energia solar. Prof. Jorge Anhalt.
◆

Custos da energia eólica. Prof. Everaldo Feitosa

Custos da biomassa. Prof. Paulo César

Equações de custo

- Os custos que compõem os custos totais expressos na fórmula

  CCVa = [(CI+CRp)×FRC(i,n)]+CO+CM

  são dados por:

- Custos iniciais (CI)
- Custos de reposição trazidos a valor presente (CRp)
- Custos de operação (CO)
- Custos de manutenção (CM)

Efeito das Políticas sobre o Crescimento e o Desenvolvimento do Nordeste - II

- **Taxas de Crescimento do PIB *per capita* no Brasil, no Nordeste e nos Estados do Nordeste entre 1970 e 1999**
- **Brasil, NE, Estados 1970-80 1980-90 1990-97 1998-99**

Brasil, NE, Estados	1970-80	1980-90	1990-97	1998-99
Brasil	6,0	- 0,4	2,4	-0,3
Nordeste	6,5	1,3	3,3	2,0
Maranhão	6,2	6,2	5,5	2,8
Piaui	6,8	5,0	4,4	5,6
Ceará	8,8	2,9	5,2	2,9
RGN	8,1	5,0	5,6	0,8
Paraíba	4,9	4,1	3,8	4,8
Pernambuco	6,7	2,0	3,5	1,5
Alagoas	6,7	2,9	4,1	- 2,5
Sergipe	7,7	1,7	3,5	3,3
Bahia	6,3	1,5	3,4	2,7

Fatores Relevantes no Crescimento e Desenvolvimento do Nordeste

- Inserção na economia nacional e internacional;
- Mudanças na composição de fatores, de processos de produção tradicionais para processos de produção modernos;
- Mudanças na pauta de exportações, de produtos tradicionais para produtos com maior valor de mercado;
- Mudanças na qualificação do trabalho;
- Introdução de inovações tecnológicas;
- Intervenção do Estado na economia

Intervenção do Estado no Desenvolvimento Rural do Nordeste

- As primeiras intervenções se deram através da criação de órgãos públicos: IAA (1933), DNOCS (1945), CVSF (1948), CEPLAC (1957) , SUDENE (1959), BNB (1954);
- Depois vieram os Programas: PIN, PROTERRA, PDAN, POLONORDESTE, SERTANEJO, PRORURAL;
- Os Planos de Desenvolvimento Regional
- As Políticas de Integração Regional
- As políticas de aumento de gastos para a agricultura

A Política de Reforma Agrária

- A reforma agrária não é somente um processo social mas uma mudança econômica nas formas de produção;
- Início da reforma agrária com o Estatuto da Terra (1964);
- Até 1986, a reforma agrária quase não saiu do discurso governamental com poucos projetos realizados;
- Com a introdução da Nova República, em 1985, criou-se o I Plano Nacional de Reforma Agrária;
- Entre 1988 e 1994, a reforma agrária esteve quase parada;
- Em 1994, começa uma nova fase da reforma agrária com a implantação de assentamentos rurais;
- A ação dos movimentos sociais se torna mais forte;
- Até o ano de 1999 haviam se beneficiado com o processo de reforma agrária, no Brasil, 450 mil famílias

A Reforma Agrária em Pernambuco

- A reforma agrária é feita pelo INCRA e pelo FUNTEPE;
- Os assentamentos implantados no estado de Pernambuco se espalham pela Zona da Mata, do Agreste e do Sertão;
- A zona da Mata é a região que tem a terra mais concentrada no Estado e com maiores demandas de distribuição de terra pelos movimentos sociais;
- Os projetos de reforma agrária no estado compreendem atualmente um total de assentamentos com famílias beneficiadas
- Outro slide
- Outro slide

A Política de Irrigação

- Objetivos da Política de Irrigação
  - Aumento da renda dos produtores
  - Redução da migração rural-urbana
  - Expansão do emprego rural
  - Aumento da produtividade agrícola
  - Geração de excedente
  - Integração da produção ao mercado
- A irrigação aparece sempre ligada à intervenção do Estado;
- A política de irrigação só teve maior sucesso a partir da década de 1970

Resultados da Política de Irrigação no Pólo Petrolina-Juazeiro

- Área irrigada até o ano 2000 somente nos perímetros públicos: 38 mil hectares
- Empregos criados: 57 mil empregos
- Quantidade produzida de manga por ano: 500 mil toneladas
- Quantidade exportada por ano: Uva – 20 milhões de dólares, manga – 43 milhões de dólares
- A exportação de frutas é hoje uma atividade resultante da irrigação que está contribuindo efetivamente para o desenvolvimento do Nordeste
- As exportações de manga e uva do São Francisco representam a quase totalidade das exportações brasileiras

Os Pólos Agroindustrias

A Política da Previdência Rural

- **A Previdência Social chega ao meio rural somente a partir de 1990;**
- **A extensão da previdência ao campo significou o apoio do Estado à terceira idade;**
- **O impacto da previdência rural se fez sentir muito mais no Nordeste onde predomina a economia de natureza familiar;**
- **A grande parte dos beneficiários é do sexo feminino (63,0%) contra 36,0% do sexo masculino;**
- **Antes de receber o benefício, as pessoas se ocupavam principalmente na atividade rural. Depois do recebimento do benefício, existe uma grande diversificação das atividades e ocupações dos beneficiários**
- **A renda das famílias rurais depende hoje em dia fortemente do benefício social rural recebido da Previdência Social**

A Política da Previdência Rural (continua)

- A proporção das famílias rurais que depende em mais de 80,0% das transferências de renda da Previdência chega a 65,0%;
- As famílias que dependem do seguro previdenciário são as mais pobres. Em cerca de 50,0% dos domicílios que tem aposentadoria, o seguro previdenciário é a única fonte de renda;
- O seguro previdenciário reduziu as migrações do campo para a cidade;
- O seguro previdenciário contribuiu para a sustentabilidade das famílias que vivem na zona rural

A Política de Desertificação

- Ocorrência de áreas de desertificação em Pernambuco
- 

Política de Desertificação

- A política de desertificação no Estado se inicia em 2000 com a implantação do Projeto de Desertificação pela SECTMA;
- O Projeto já aplicou 3 milhões de reais para combater os efeitos da desertificação em comunidades rurais afetadas pelo processo;
- Até o presente, já foram beneficiadas 300 famílias com a assistência do projeto em oito municípios do Sertão;
- As estratégias do Projeto são: a) aumentar a renda das famílias rurais, b) desenvolver atividades produtivas adaptadas às áreas afetadas pela desertificação, c) aumentar o emprego, d) capacitar as famílias para manutenção da biodiversidade e do meio ambiente

Política de Combate aos Efeitos da Seca

- A seca e a desertificação, embora sejam processos distintos, estão relacionados;
- As secas são catástrofes naturais que assumem dimensões de calamidade pública em vista da pobreza das pessoas que são afetadas;
- As secas são desastres naturais que, em maior ou menor escala, atingem o semi-árido nordestino;
- Data de 1587 o relato de uma primeira seca no Nordeste;
- A seca afeta as atividades produtivas no semi-árido pois quase todas elas são dependentes de chuva;
- O crescimento do produto na agricultura é normalmente negativo nos anos de seca;
- As políticas públicas tem em vista atenuar os efeitos negativos da seca no semi-árido nordestino;

Política de Combate aos Efeitos da Seca (continua)

- Desde os anos 50, os governos vem aplicando programas e projetos para combate aos efeitos da seca na região;
- A SUDENE sempre esteve presente tradicionalmente na operação de programas de combate aos efeitos das secas;
- As políticas de irrigação e desertificação tem contribuído para aliviar os efeitos das secas na região;
- Os instrumentos foram tradicionalmente a construção de açudes e barragens, a implantação de infra-estrutura rural nos municípios e nas fazendas, além da assistência social através dos programas de emergên
- As tecnologias para armazenamento de água continuam sendo prioritárias;
- A implantação de eixos de produção agropecuária, através da transposição de bacias e exploração da caatinga, são instrumentos de políticas importantes para amenizar os efeitos das secas

Energy crops, food security, and rural development: the role of oilseed production around Brazil

Presented at RCN-SEES for Pan American Biofuels and Bioenergy Sustainability

"Socioeconomic Issues"

22 - 25 July 2014

Rob Bailis - Associate Professor; Yale School of Forestry & Environmental Studies

Outline

- Context – Brazil and biofuels
- Oilseeds in as "flex-crops"
- Experiences in Brazil
  - Soy
  - Jatropha, Babaçu, and Macaúba
- Pathways for future research...

Global biodiesel production (2000-2011)

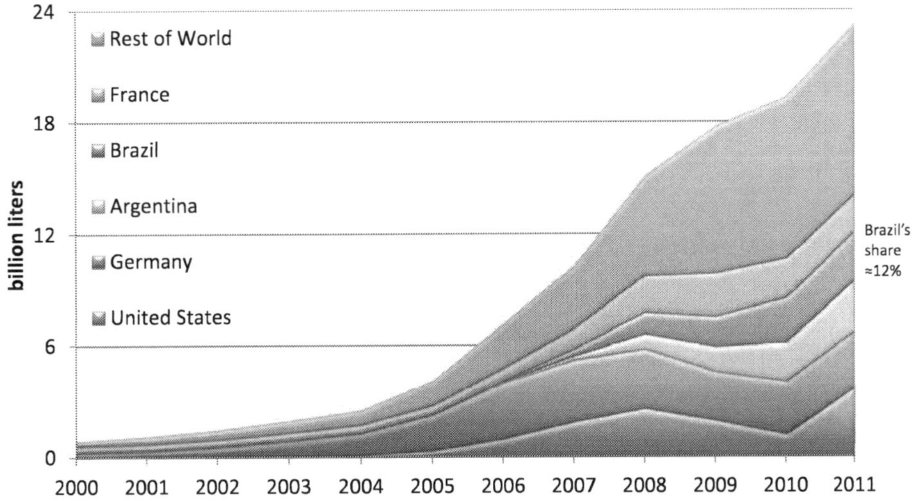

Source: http://tonto.eia.doe.gov/cfapps/ipdbproject/IEDIndex3.cfm?tid=79&pid=79&aid=1.

Biodiesel and oilseeds

- Biodiesel is 1/5 of global biofuel production
- Multiple pathways
- Nearly all commercial operations use flex crops from existing agro-industries

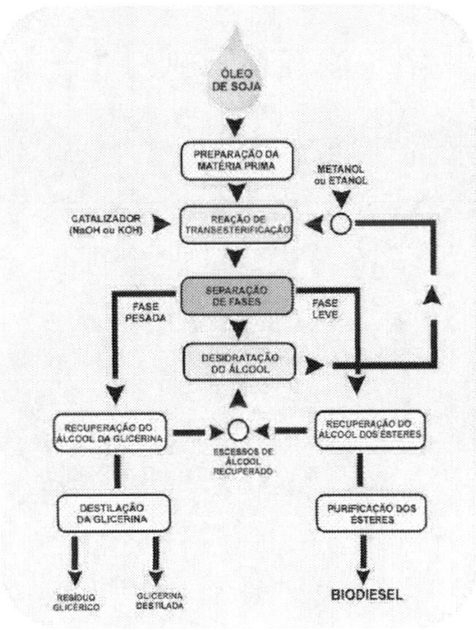

Brazil's soy complex (2010)

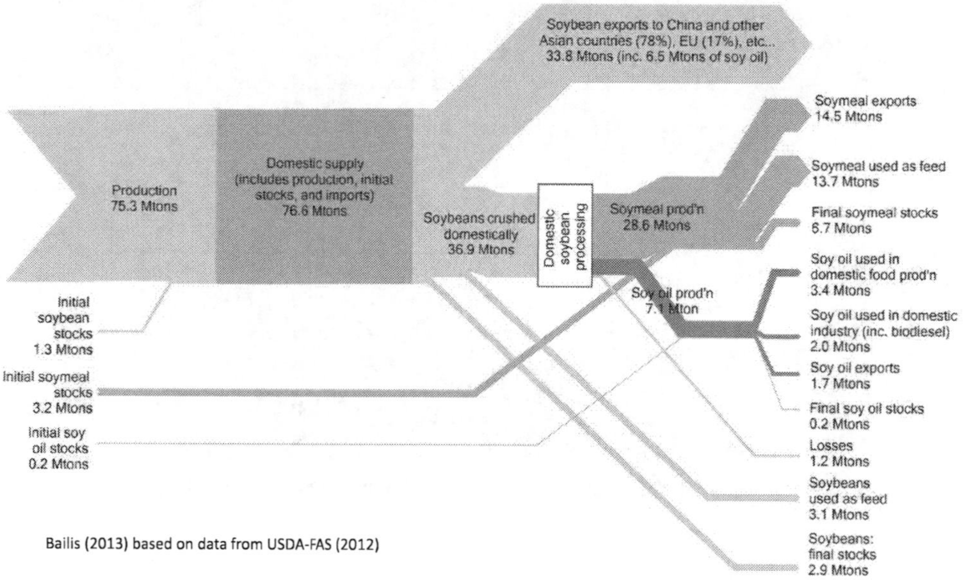

Bailis (2013) based on data from USDA-FAS (2012)

Who produces the world's edible oils?

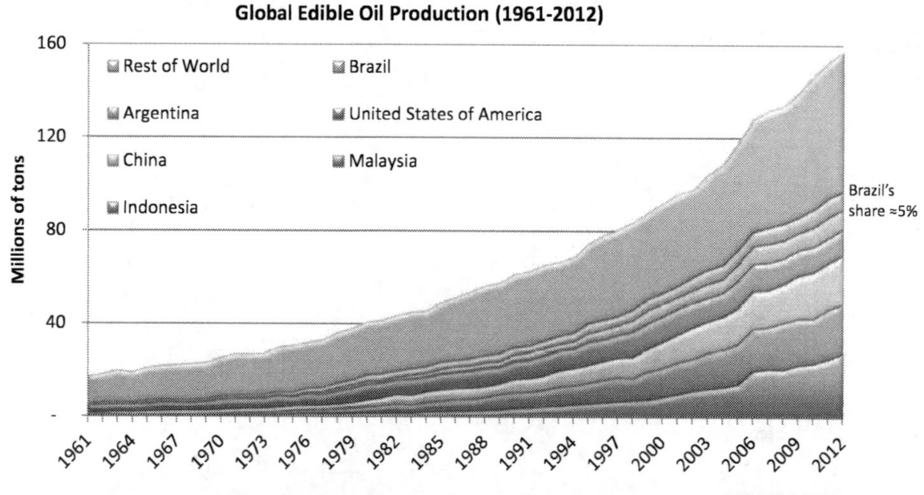

FAOSTAT, 2013

Oil as food

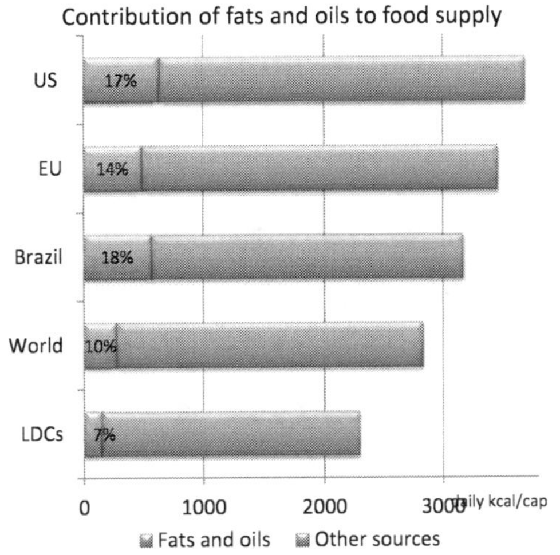

FAOSTAT, 2013

Fats and oils: increasingly important in Brazil's food supply

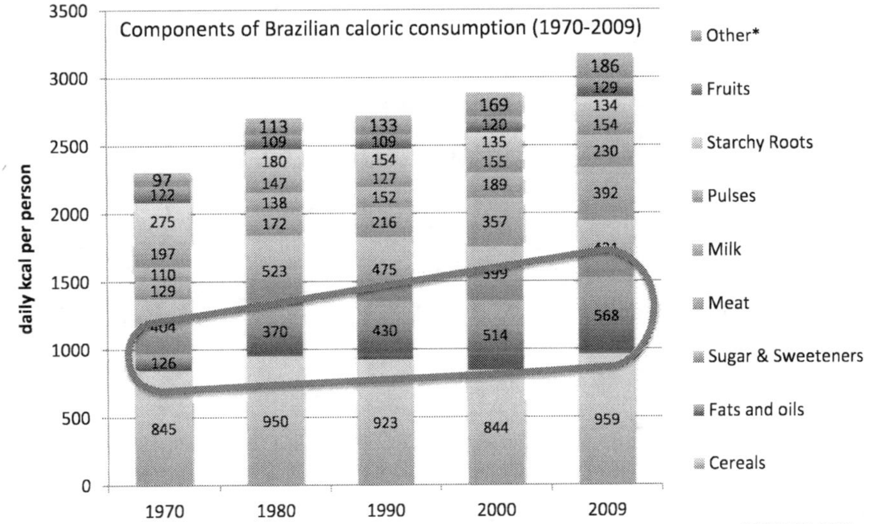

* Includes alcoholic beverages, vegetables, eggs, fish, stimulants, nuts, and spices.

FAOSTAT, 2013

Official biodiesel policy

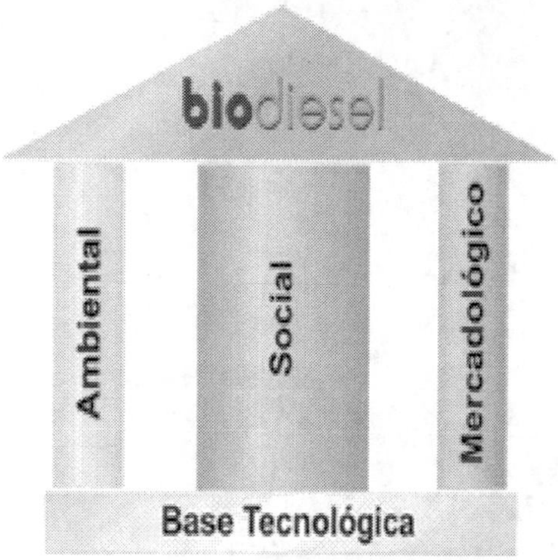

- Create a sustainable program promoting **social inclusion**
- Guarantee **competitive prices**, quality, and supply
- Produce biodiesel from **different sources** and from **diverse regions**

Are these objectives being met?

Social inclusion

Primary material	Region	Type of agriculture	Federal tax (R$/m^3)
Fossil diesel	Any region	NA	218
Any feedstock	Any region	Commercial farms	178
Any feedstock	North and semi-arid northeast	Commercial farms	151
Any feedstock	Any region	Family farms	70
Castor or oil palm	North and semi-arid northeast	Family farms	0

Tax structure incentivizes inclusion of family farmers and less developed regions

Sources of Brazil's Biodiesel

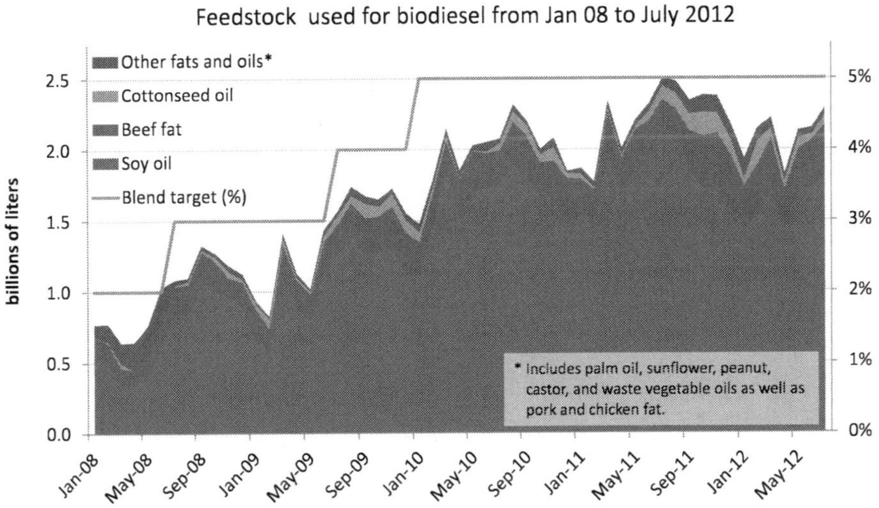

Bailis (2013) based on data from MME (2012)

Fat and oil production in Brazil

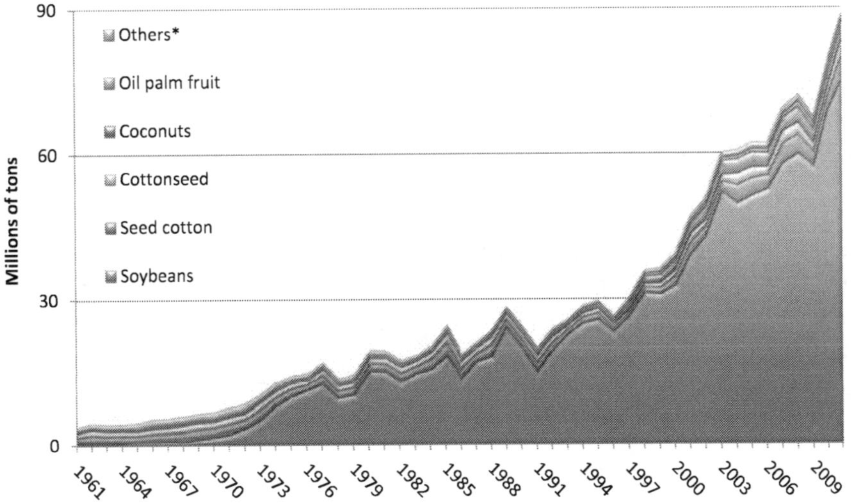

* Includes tallow, groundnuts, oil palm, castor, sunflower, rapeseed, linseed, sesame, tung and olives.

FAOSTAT, 2013

Material flows in Brazil's soy complex (2010)

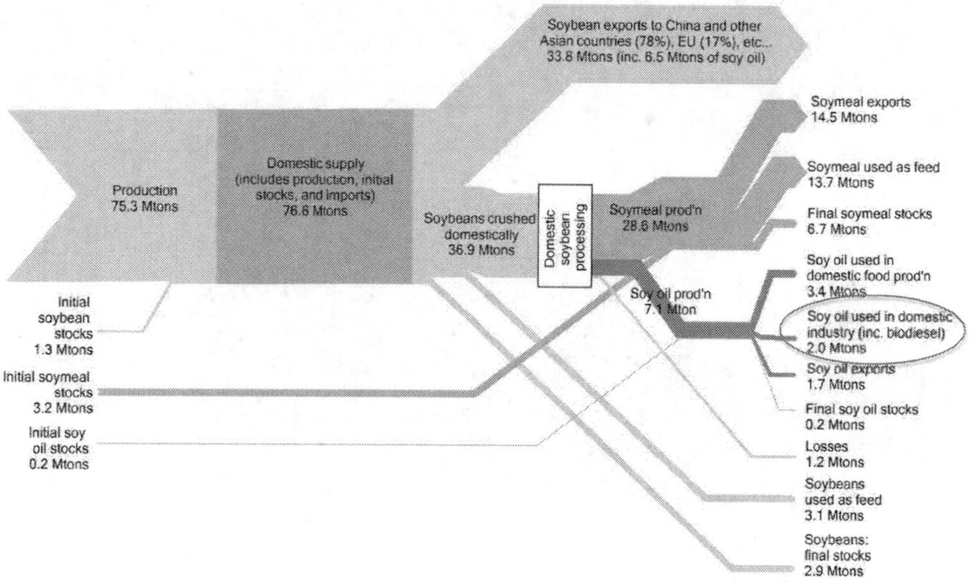

Is there a conflict of food vs. fuel?

Brazil's biodiesel program began in 2005

- Between 2005 and 2010:
  - Annual biodiesel production increased from 10 <u>million</u> to 2.4 <u>billion</u> liters (a 2400% increase)
  - Domestic supply of edible oils increased by just 37%
  - Prices were relatively unchanged (though volatile)

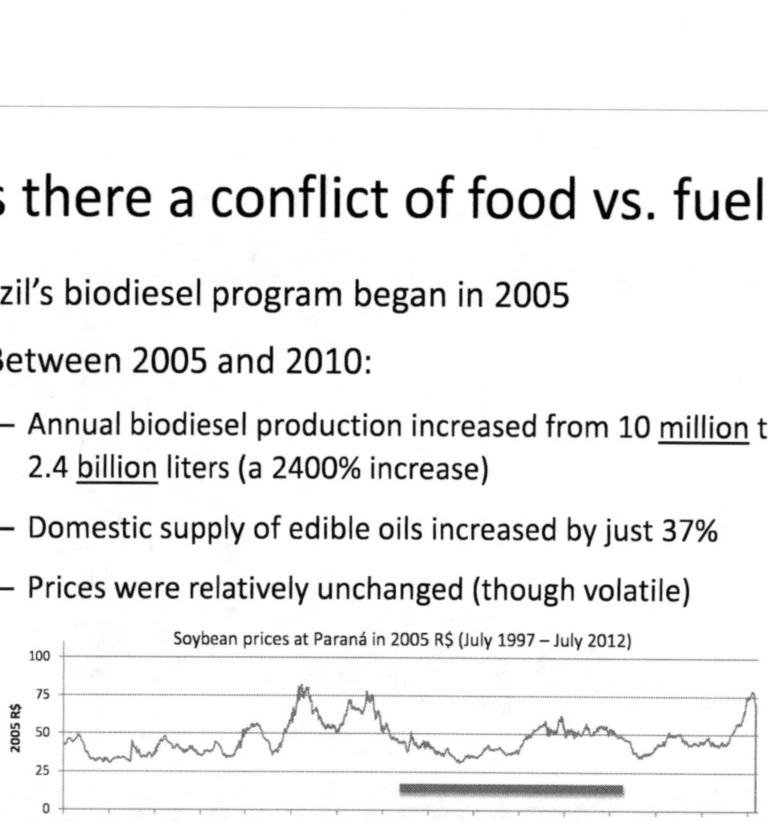

But there is a conflict…
Soy vs. Amazon

Tropical deforestation for cropland agriculture in Mato Grosso state (2001–2004) is concentrated along the existing agricultural frontier.

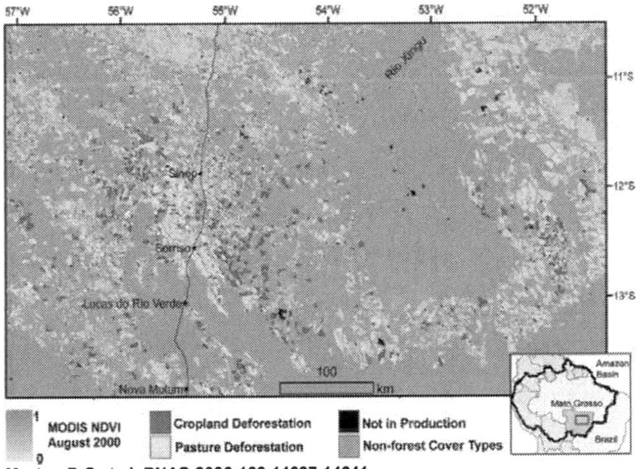

Morton D C et al. PNAS 2006;103:14637-14641

Soy in the Amazon region

Land use after deforestation in Mato Grosso from 2001 to 2010.

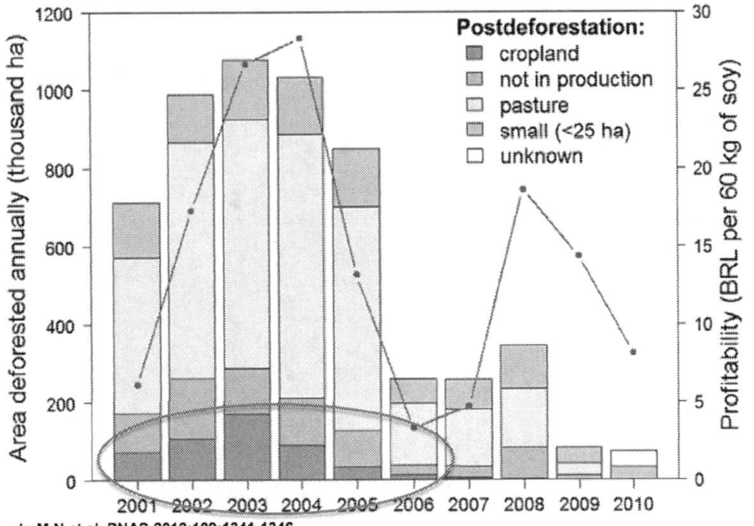

Macedo M N et al. PNAS 2012;109:1341-1346

PNAS

Soy and social inclusion

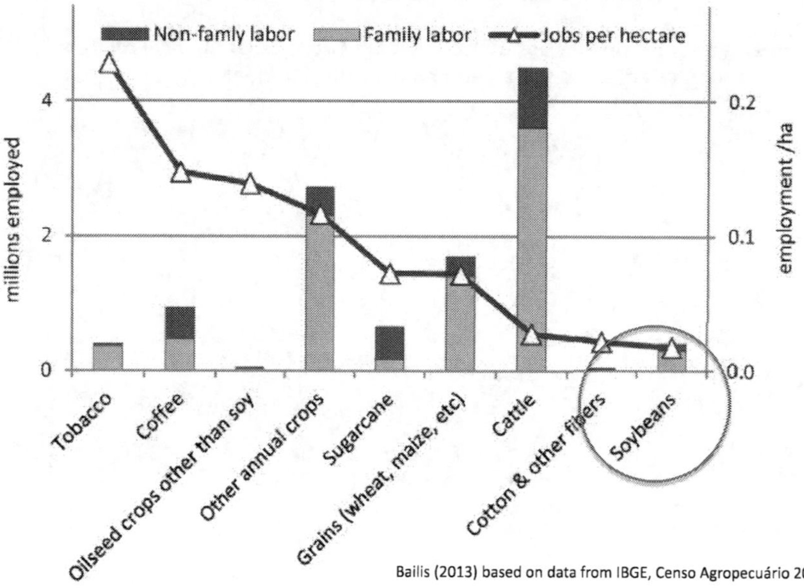

Bailis (2013) based on data from IBGE, Censo Agropecuário 2006

Soy and social inclusion

Tocantins soy case study:

- Family farmers producing soy are better capitalized than non-producers.
- Access to credit is limited to wealthy farmers, poor farmers are ineligible
- Some FFs are taking advantage of the "social inclusion" but...
  - Some absentee landowners pose as family farmers by paying poor families to use their DAP (Declaração de Aptidão do Produtor)

Based on Finco and Ribeiro (2013)

family farmer participation in feedstock sales to biodiesel producers

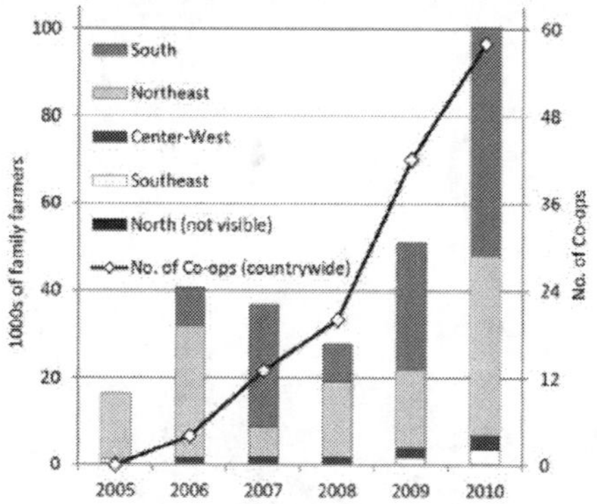

Bailis (2014)

Alternatives to soy?

...to reduce impacts on land use change?

...increase potential for social inclusion?

7/23/14 Bailis - RCN "Socioeconomic Issues" 19

Alternatives to soy?

...to reduce impacts on land use change?

...to improve for social inclusion?

7/23/14 Bailis - RCN "Socioeconomic Issues" 20

Jatropha

- A <u>toxic</u> perennial shrub native to C. America
- Hardy and drought tolerant
- 4-5 years gestation
- Planted in large areas 2007-2009
  - Included many family farmers
- Many projects have failed
  - Poor performance
  - Lack of agronomic knowledge
  - Poor economic prospects

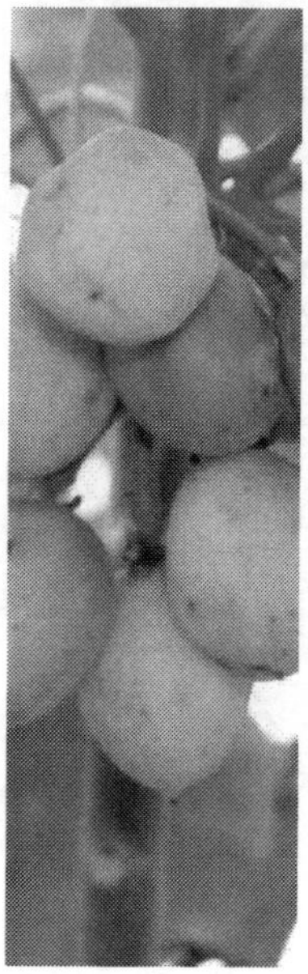

7/23/14 Bailis - RCN "Socioeconomic Issues" 21

Jatropha: social challenges

- Family farmers reduced food crop production and/or sold livestock to plant Jatropha
- Revenues from seed sales are too small to offset reduced food production
- Finco and Doppler showed family farmers adopting Jatropha in Tocantins were wealthier than non-adopters

From: Reporter Brazil (2009)

7/23/14 Bailis - RCN "Socioeconomic Issues" 22

"Jatropha 2.0"?

- Hybrid seeds
  - Higher/more consistent yield
- Intercropping
  - Silvipastoral system →
- Seedcake detoxification
  - Substitute with soy
- Cogeneration of electricity
  - Increase revenue and energy balance

Macaúba (Acrocomia aculeata)

- Long-lived native palm with sub-species in Brazil (MG, SP, MS) and Paraguay
- Tolerates dry periods and frost
- Precipitation needs > 800 mm
- Fruits and oil are edible
- Small market for cosmetics

Macaúba in natural stands

7/23/14 Bailis - RCN "Socioeconomic Issues" 25

Co-production with cattle

7/23/14 Bailis - RCN "Socioeconomic Issues" 26

Macaúba: multiple products

From Motoike (UFV)`

Small-scale processing at UBCM*

* Unidade de Beneficiamento do Coco Macaúba: Zona Rural de Montes Claros – MG

Used in bio-kerosene?

This month's GOL in-flight magazine…

"A Gol aircraft is fueled with bio-kerosene in Rio; in particular, macaúba fruit and cane sugar, used to produce biofuel."

7/23/14 Bailis - RCN "Socioeconomic Issues" 29

Babaçu (Orbignya phalerata)

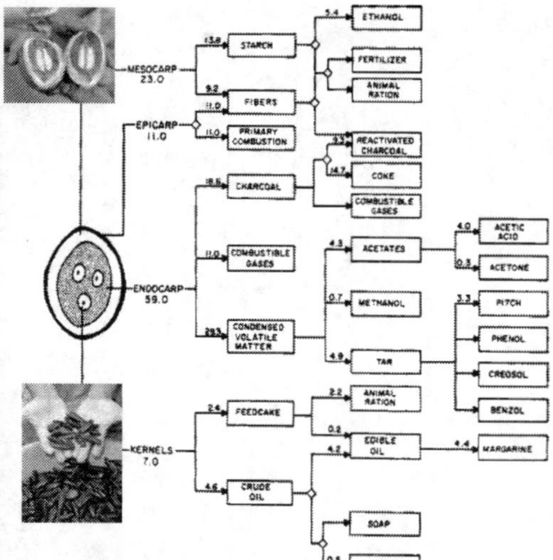

Figura 2: Componentes e subprodutos potenciais do coco babaçu, segundo estudos dos anos 70; em sistemas conjugados, alimentos, energia e múltiplos subprodutos (ANDERSON, et alii, 1991).

7/23/14 From Teixeira, An. 4. Enc. Energ. Meio Rural 2002 "Biomassa de Babaçu no Brasil" 30

A case study in Northern TO

- Quebradeiras depended on babaçu for generations but…
- Participation is declining
  - Access is more difficult
  - Bolsa Familia reduces the incentives to work as Quebradeiras
    - The work is very difficult, but also very social
  - Prices for are very low

Extractive Reserve in the north of Tocantins

A case study in Northern TO

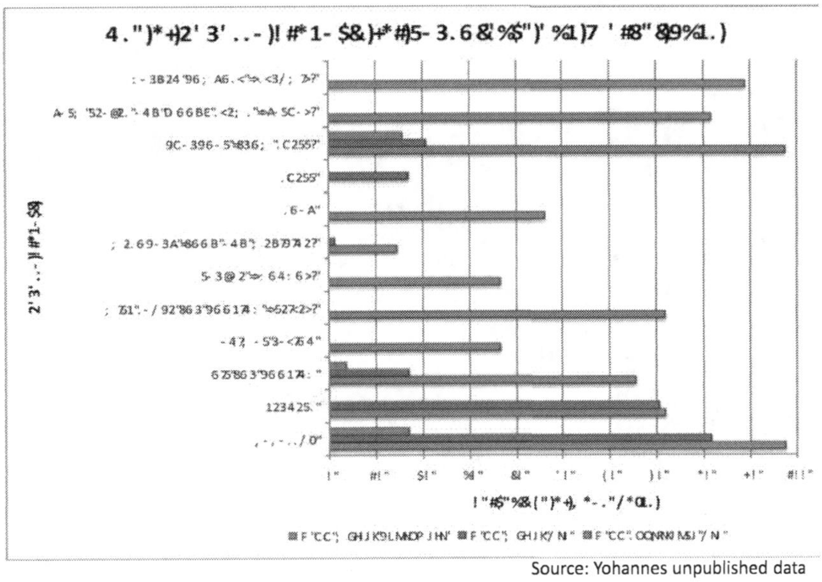

Source: Yohannes unpublished data

Some difficult questions…

Should we expect biofuel production to:

- contribute to the well-being of family farmers?

and

- improve environmental conditions?

or…

- at the very least <u>do no harm</u>?

Then soy is not a good choice…

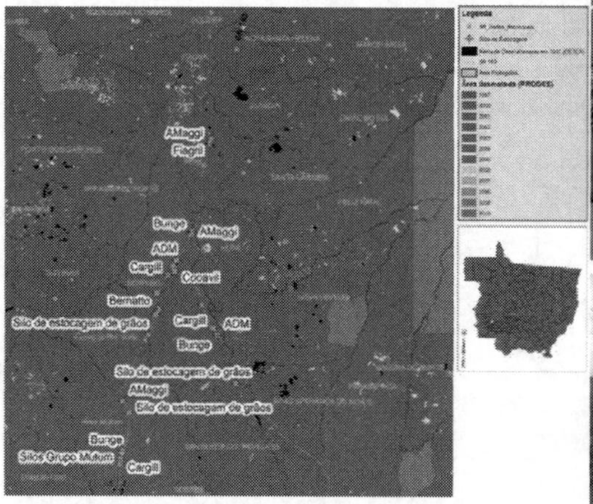

Trading companies' silos along the BR-163 federal road in Mato Grosso: the environmental degradation route (Reporter Brazil, 2011)

But the alternatives can't compete!

This raises additional questions…

1. Are biofuel crops an appropriate way to improve well-being among family farmers?
   – Are there better options?
2. Can we find ways to produce soy that are more socially equitable and less environmentally destructive?

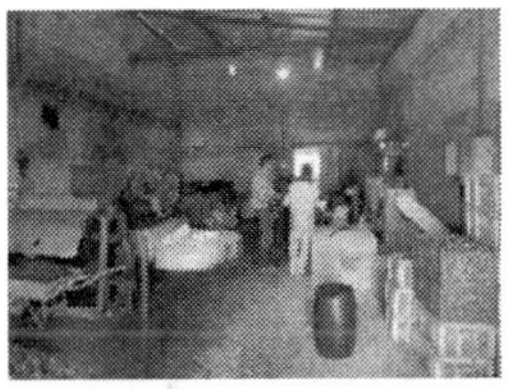

7/23/14 Bailis - RCN "Socioeconomic Issues" 35

Obrigado!

- and CNpQ for funding the research on babaçu

  In partnership Marcus Finco (UFT) and Lia Yohannes (Univ Maryland)

- for funding the research on macaúba

  In partnership with Dr Mirco Plath and Nepomuk Wahl (Luephana U)

- AIRBUS for funding the research on Jatropha

  In partnership with Goksin Kavlak (Yale), JetBio and Rio Pardo Bioenergia

7/23/14 Bailis - RCN "Socioeconomic Issues" 36

Extra slides

7/23/14 Bailis - RCN "Socioeconomic Issues" 37

Fats and oils in the Brazilian food supply

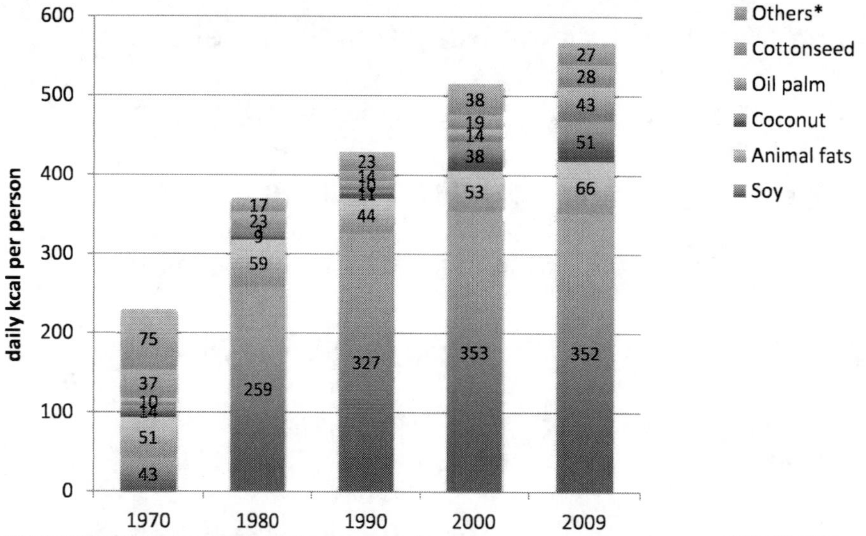

*Includes sunflower, olive, corn, groundnut, ricebran, and sesame oils

FAOSTAT, 2013

7/23/14 Bailis - RCN "Socioeconomic Issues" 38

Extractive Reserve in the north of Tocantins

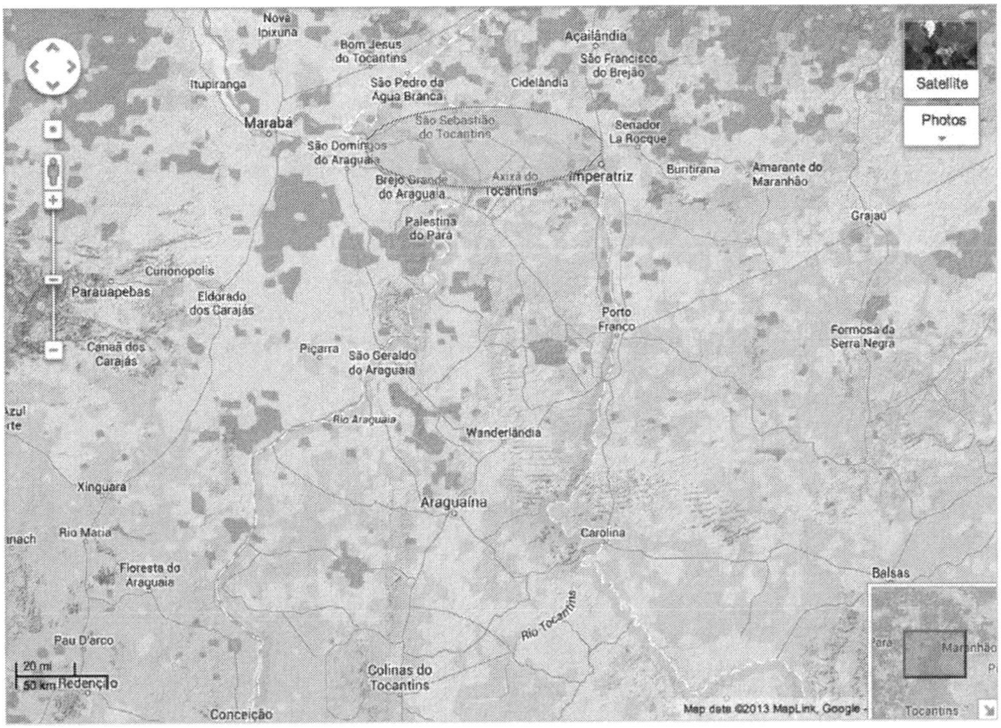

Use of fats & oils varies with income

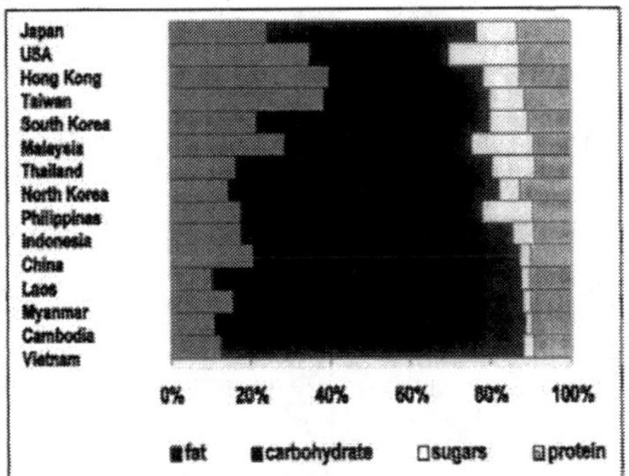

FIG 1. Incomes and diet structure for selected East Asian nations and the United States. The data are based on FAO food balance sheets (34) and 1994 Gross National Product (GNP) per capita as listed by the World Bank. The countries are arranged by their 1994 GNP.

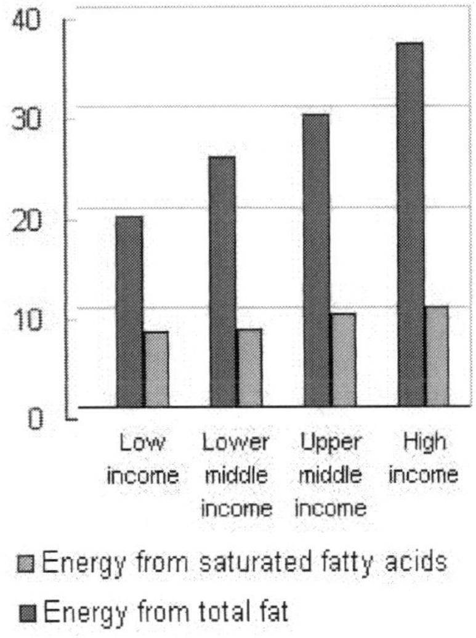

http://www.sciencedirect.com/science/article/pii/S0002822397007323

http://www.who.int/gho/ncd/risk_factors/unhealthy_diet_text/en/

Table 1. Oil yields for several oilseed crops

Plant	Approximate yields, kg ha^{-1} yr^{-1} Reported	Potential
Perennials		
African oil palm (*Elaeis guineensis*)	3000–5000[20]	6000–14,000[13]
Coconut palm (*Cocos nucifera*)	800[20]	3900[21]
Babassu palm (*Orbignya* spp.)	1100[22]	2150[22]
Chinese tallow tree (*Sapium sebiferum*)	300[23]	5000[23]
Physic nut (*Jatropha curcas*)	200–800[24]	4000[18]
Annuals		
Sunflower (*Helianthus annuus*)	600–750[25]	—
Groundnut (*Arachis hipogaea*)	1350–1700[26]	—
Soybean (*Glycine max*)	442[27]	—
Rapeseed (*Brassica napus*)	640–1388[28]	—
Cuphea species*	288–720[29]	—
Salicornia bigelovii	600[30]	—

Shay, E. Griffin. "Diesel fuel from vegetable oils: status and opportunities." *Biomass and Bioenergy* 4.4 (1993): 227-242.

"Jatropha 2.0"

- Shows improved environmental performance

But...

- Is it more suitable for social inclusion?
- Is it more likely to enhance food security?

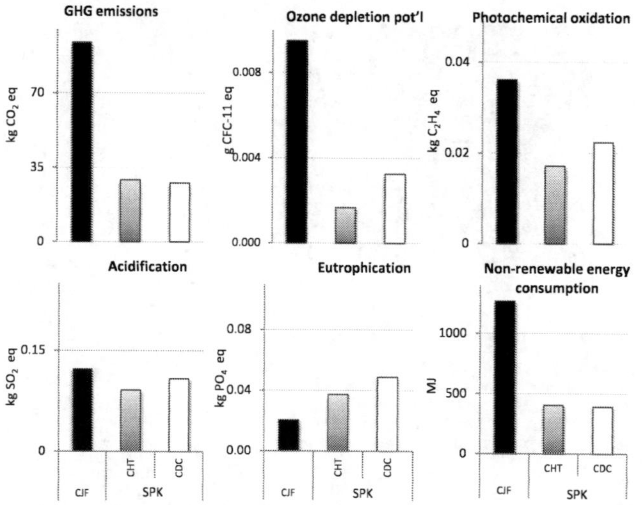

CJF – Conventional Jet Fuel
SPK – Synthetic paraffinic kerosene

From Bailis and Kavlak, 2013

Fluxograma das Inter-relações no sistema agroextrativista do Norte de Minas Gerais

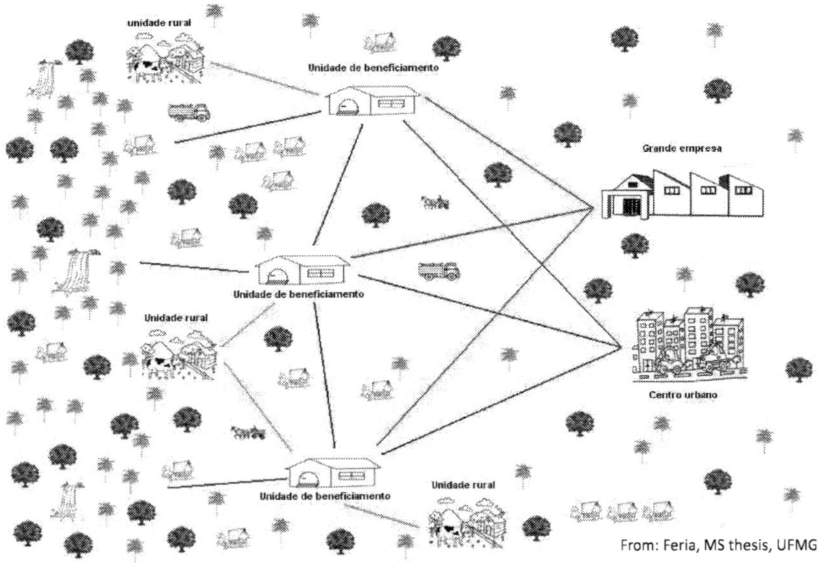

From: Feria, MS thesis, UFMG 2010

Bioenergy Development and Integrated Water-Energy Management

7 papers (2 hours)

Discussion of knowledge gaps and research plans (1 hr)

Co-Conveners:

David W. Watkins, Jr.[1] and Márcia M.G.A. Moraes[2]

[1]Dept. of Civil and Environmental Engineering, Michigan Technological University, Houghton, MI, USA

[2]Dept. of Economics, Universidade Federal de Pernambuco, Recife, BRA

Bioenergy Development and Integrated Water-Energy Management in Pan America

David W. Watkins, Jr.[1], Márcia M.G.A. Moraes[2], Heidi Asbjornsen[3], Alex S. Mayer[1], Julian Licata[4], Jose Gutierrez Lopez[3], Tom Pypker[5], Vivianna Gamez Molina[6], Guilherme Fernandes Marques[7], Ana Cristina Guimaraes Carneiro[2], Hector M. Nuñez[8], Hayri Önal[9], and Bruna da Nobrega Germano[2]

[1]Dept. of Civil and Environmental Engineering, Michigan Technological University, Houghton, MI, USA
[2]Dept. of Economics, Universidade Federal de Pernambuco, Recife, BRA
[3]Dept. of Natural Resources and Environment, University of New Hampshire, Durham, NH, USA
[4]Instituto Nacional de Tecnología Agropecuraria, Buenos Aires, ARG
[5]Dept. of Natural Resource Science, Thompson Rivers University, Kamloops, CAN
[6]Dept. of Civil and Environmental Engineering, University of Northern Arizona, Flagstaff, AZ, USA
[7]Dept. of Civil Engineering, Centro Federal de Educação Tecnológica de Minas Gerais, Belo Horizonte, BRA
[8]Dept. of Economics, Centro de Investigación y Docencia Económicas (CIDE), Aguascalientes, MEX
[9]Dept. of Agricultural and Consumer Economics, University of Illinois at Urbana-Champaign, IL, USA

Research Goals

- Review research on water resource impacts of bioenergy production.
- Improve understanding of fundamental ways bioenergy production affects the hydrologic cycle.
- Through case studies in Pan America, illustrate:
  - Relevant spatial and temporal scales for impact assessment,
  - Unique aspects of biofuel production compared to other agroforestry systems.
- Identify primary gaps in knowledge that hinder policy development for integrated management of water-bioenergy systems.

Integrated Approach

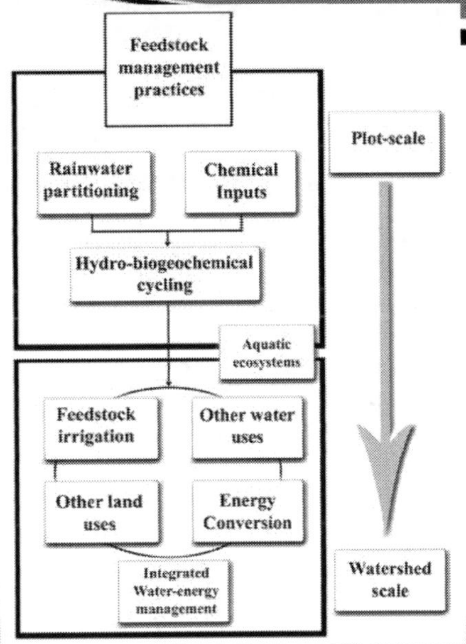

Feedstock Water Budget

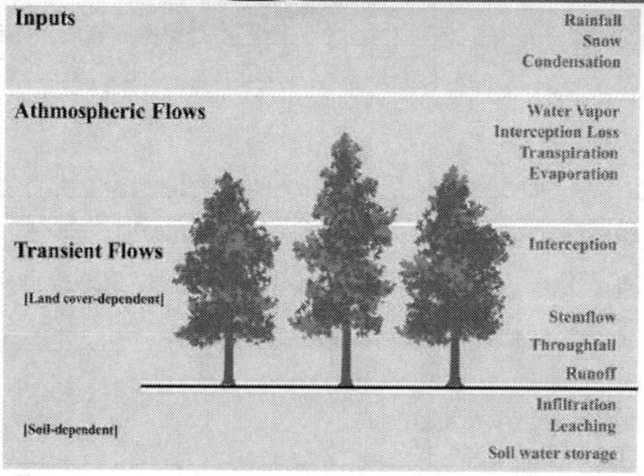

$$WY = P - T - I - E + \Delta S$$

where WY = water yield, P = precipitation, T = plant transpiration, I = rainfall interception by the canopy, E = soil evaporation, and ΔS = change in soil water (considered to be negligible over longer time periods)

Key Findings

- More research is needed to understand how different bioenergy systems vary in their ecohydrologic functions and the implications at the stand to the watershed scale.
- There is a knowledge gap of how bioenergy-oriented forestry will impact water use at the tree and stand levels.
  - Studies of monoculture and mixed-species of short-rotation bioenergy production systems are lacking.
- Research should be intensified on water efficiency in the energy conversion side of second-generation technologies.

Great Lakes Basin, USA

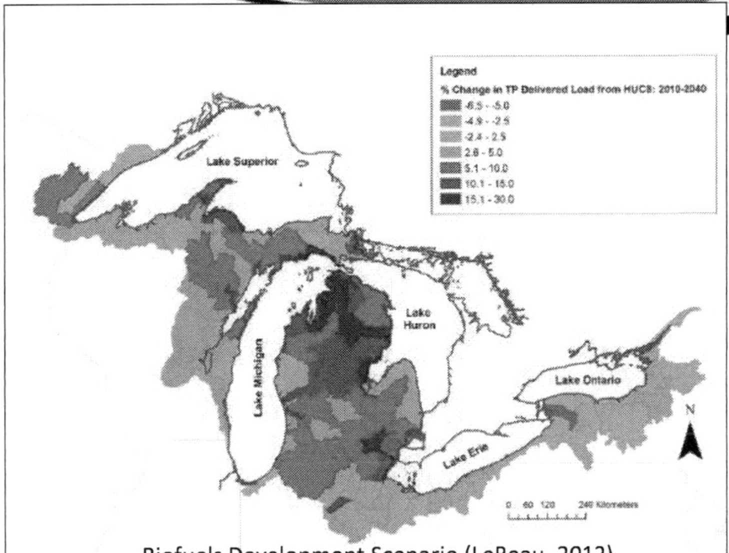

Biofuels Development Scenario (LeBeau, 2012)

Pirapama River Basin, Brazil

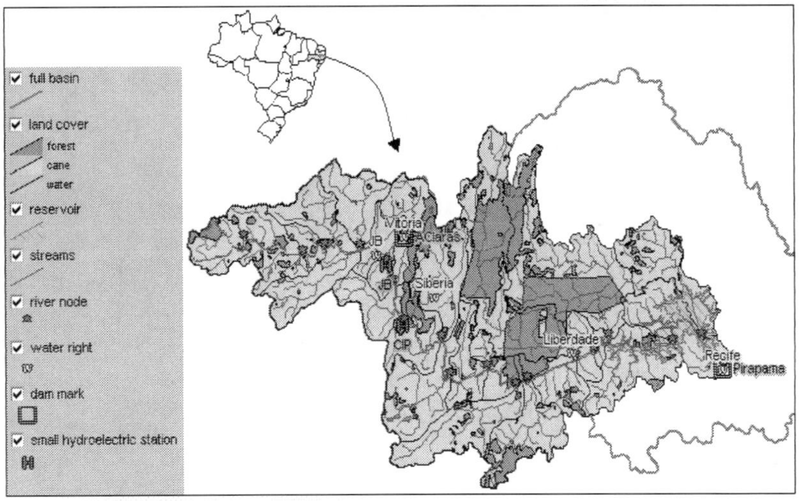

Source: Moraes et al. (2010).

Araguari River Basin, Brazil

Source: Marques et al. (2010)

Araguari River Water-Energy System

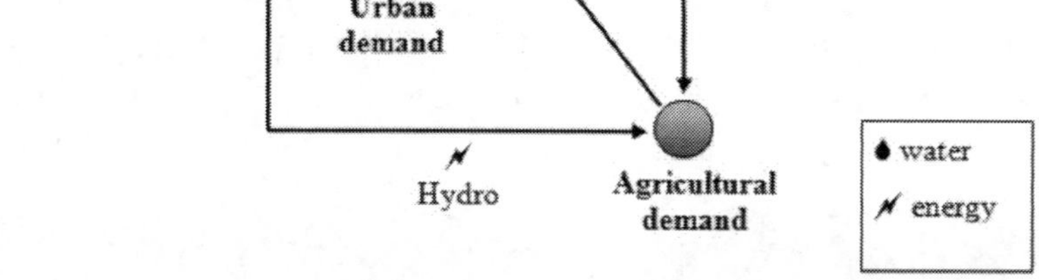

Source: Marques et al. (2010)

Land Use Change and Irrigated Biofeedstock Production in Brazil (Nuñez et al., 2013)

- Given vast pasture lands, Brazil may be able to help meet demands for bioethanol in domestic and international markets, but sugarcane expansion could be limited by water availability.
- An economic simulation model considers competition for limited resources between sugarcane and other major crops and determines the equilibrium in global commodity and fuel markets.
- Without water limitations, 21 million ha of additional land is allocated to crop production (50% increase) by 2030, 60% of which is used for sugarcane production.
- With water scarcity, lands allocated to corn and soybeans increase slightly, and the sugarcane area is reduced ~40%, mainly due to water scarcity in regions that are most suitable for sugarcane production.

South Florida, USA

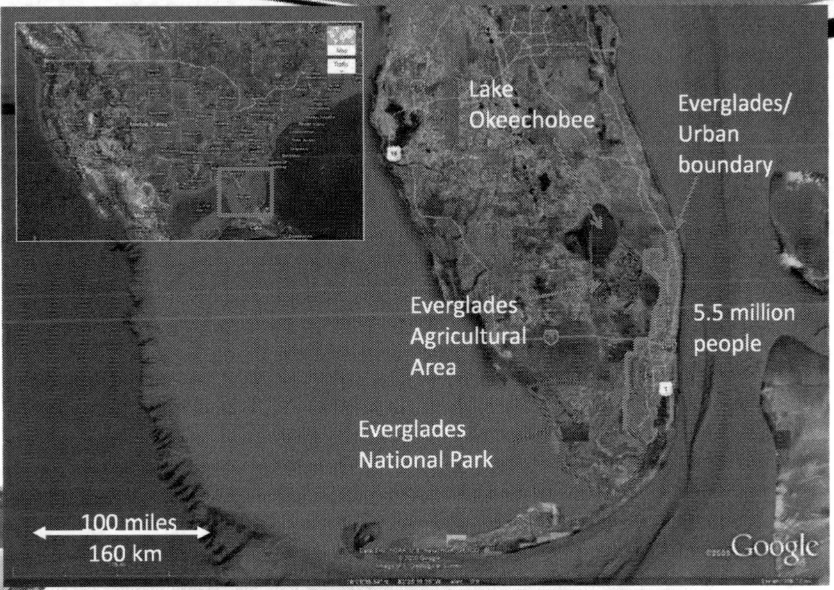

Hydro-economic Modeling

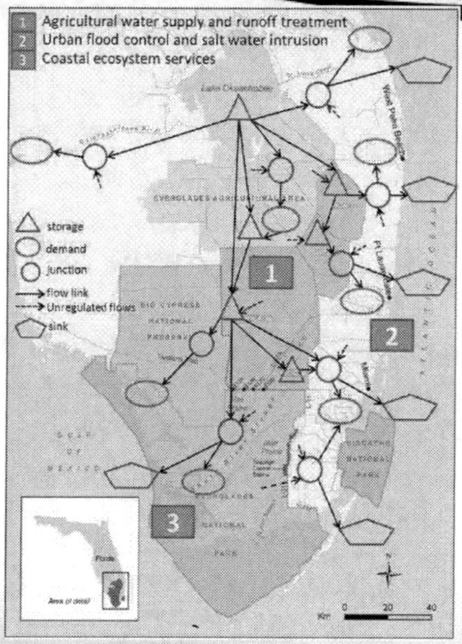

Stormwater Treatment Areas

http://sofia.usgs.gov/publications/ofr/2007-1374/

A Broader Systems Approach

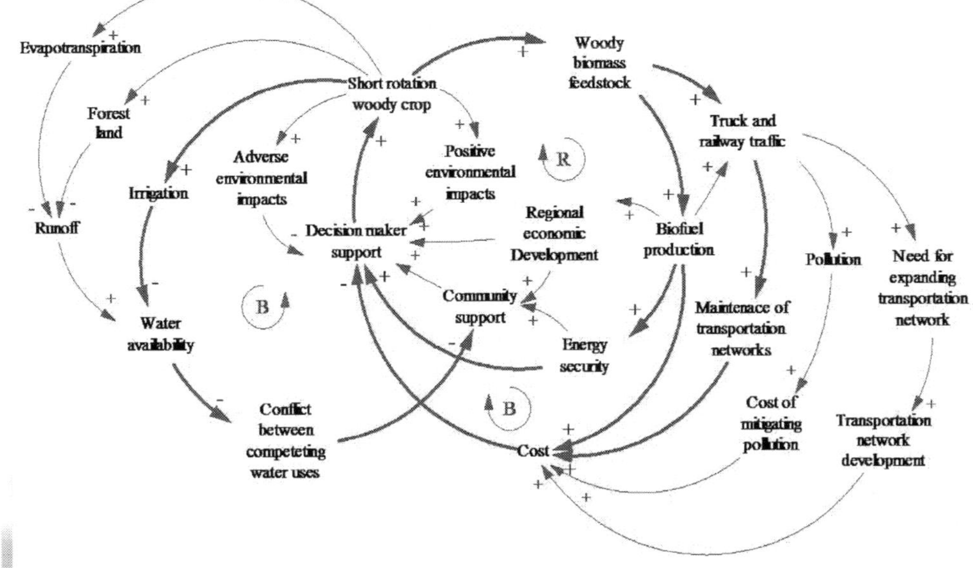

Conclusions and Recommendations

- Research is needed to better understand hydrologic impacts and trade-offs of many bioenergy systems, particularly monoculture and mixed forest-based systems.
- Water quality impacts of bioenergy systems are expected to be similar to those of other agricultural systems, where best management practices can greatly reduce impacts.
- Cumulative impacts of large-scale bioenergy development at watershed scales may be difficult to avoid, however, due to spatial variability in factors affecting pollutant transport to affected water bodies (e.g., soils, fertilizer application rates, drainage networks).

Conclusions and Recommendations

- Although integrated management approaches can significantly increase water use efficiency and reduce water quality impacts, bioenergy production in many locations is expected to be limited by water availability in the near future.
  - Country-level analysis are often not appropriate for policy making.
- Water use policies and instruments for bioenergy must be evaluated not only in relation to their effectiveness and efficiency, but also with respect to their socio-economic impacts.
  - Policies should be evaluated based on measures of direct and indirect economic impacts, impacts on all water users and uses, and lifecycle impacts.

Implications of Woody Bioenergy Feedstock Production for Water Supply and Hydrologic Regulation Services

Jose Antonio Gutierrez-Lopez, Julian Licata, Tom Pypker, Heidi Asbjornsen

Introduction: Why should we care about bioenergy and water resources?

- Water scarcity quickly becoming the world's most critical issue for human well-being and ecosystem sustainability: UN "International Decade Water for Life 2005-2015"
- Growing global pressure to increase bioenergy crop production: major consequences for water resources, often not considered!
- Current understanding? Important knowledge gaps?

WATER FOR LIFE
2005-2015

Background: The Water Cycle

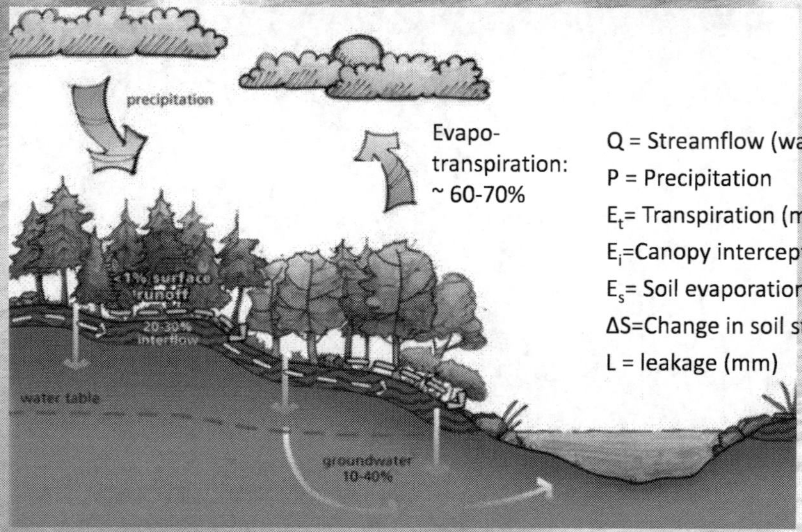

Q = Streamflow (water yield)
P = Precipitation
E_t= Transpiration (mm)
E_i=Canopy interception (mm)
E_s= Soil evaporation (mm)
ΔS=Change in soil storage
L = leakage (mm)

Streamflow (water yield) = Inputs - Outputs ± Storage

$$\mathbf{Q\ (WY)=P - E_t + E_i + E_s + \Delta S + L}$$

Land use change (LUC) impacts on vegetation water use

- Forests *almost always* use more water compared to shrubs, grasslands, or crops.
  - Deep roots
  - Higher leaf area
  - Longer growing season

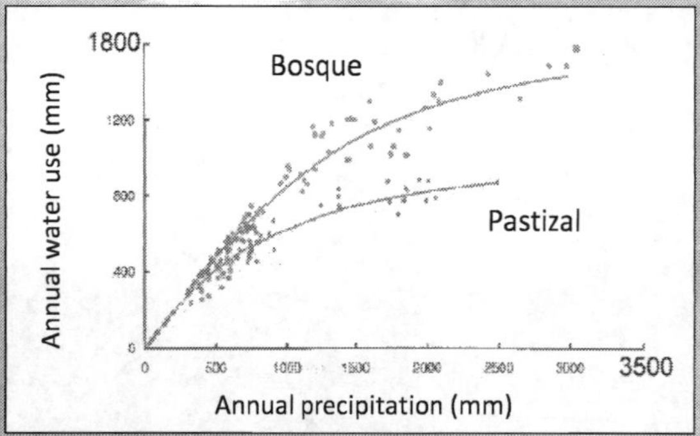

Zhang 2004

What's so unique about (woody) bioenergy plantations vs. other types of LUC?

- Exotic species
- Rapid growth rates
- High planting density
- High leaf area index
- Short rotation length

High Productivity →
- Water quantity
- Water "timing" (seasonal distribution)

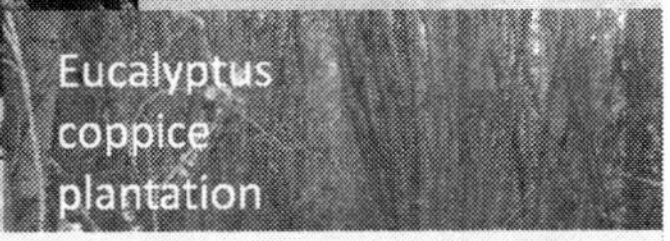

Few studies on woody bioenergy plantations and water use, but many similarities to other plantations

Water Quantity: Transpiration

Streamflow (water yield) = Inputs - Outputs ± Storage

$$Q\ (WY) = P - E_t + E_i + E_s + \Delta(S+G) + L$$

Transpiration rates for exotic species often greater than native forests: Douglas-fir plantations vs. native *Nothofagusantartica*forests: Patagonia, Argentina

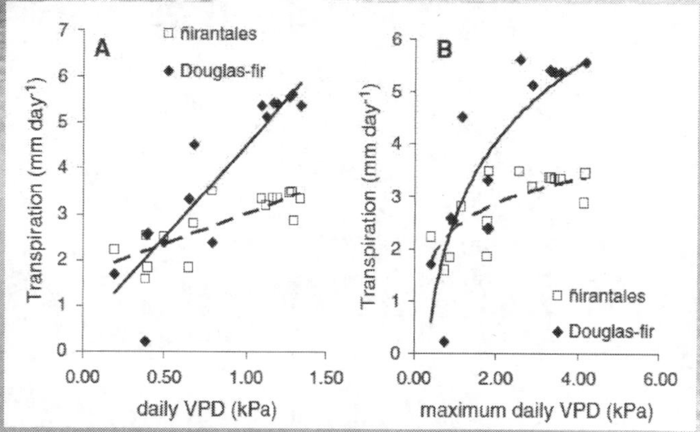

- Rooting depth and access to water
- Stomatal sensitivity to drought (conservative strategy)

Gyenge et al. 2008, 2009

Rotation time for woody biomass crops may affect total water use: Conceptual model

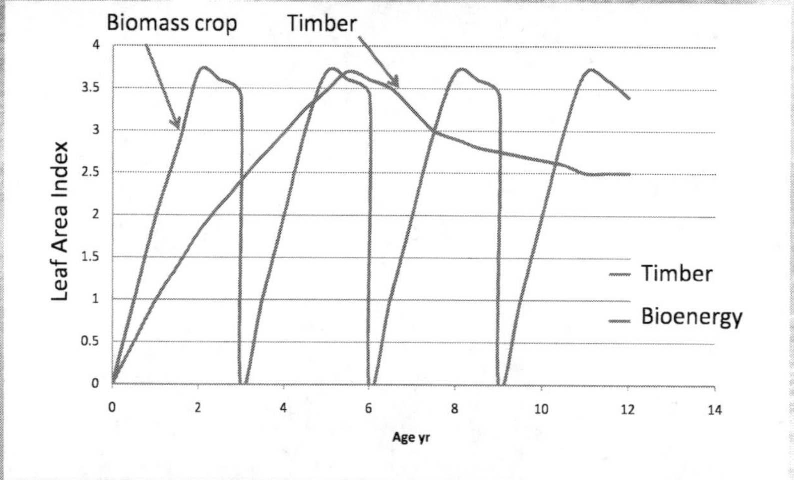

- Woody biomass crops often have higher growth rates and leaf area indices, **but shorter rotation times**
- Potential for higher total water use over time

Licata et al., unpubl.

Biomass crops often planted at high density: Modeled effects of high vs. low density on transpiration

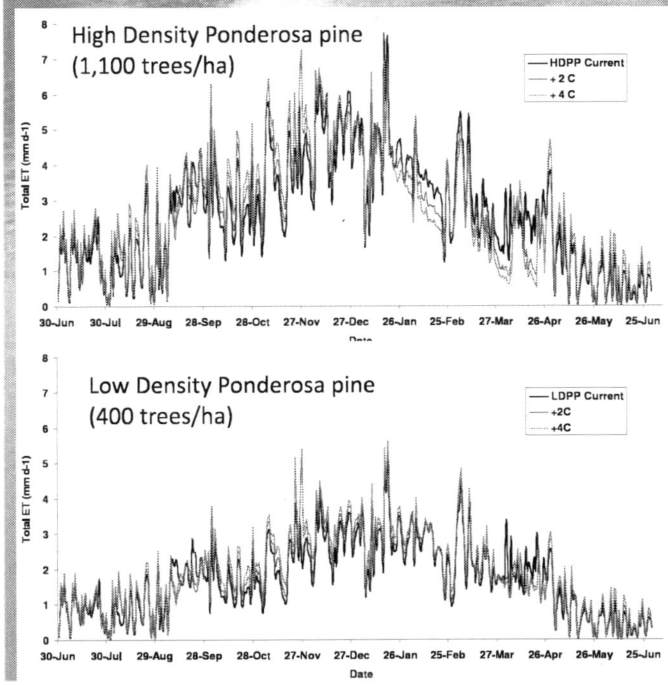

Temperature change current, +2 C and +4 C

Biomass crops may be less resilient to future climate change

Licata et al., unpubl.

Water Quantity: Canopy Interception

Streamflow (water yield) = Inputs - Outputs ± Storage

$$Q\ (WY) = P - E_t + E_i + E_s + \Delta(S+G) + L$$

Canopy Interception (E_i):
Productive vs. non-productive water use

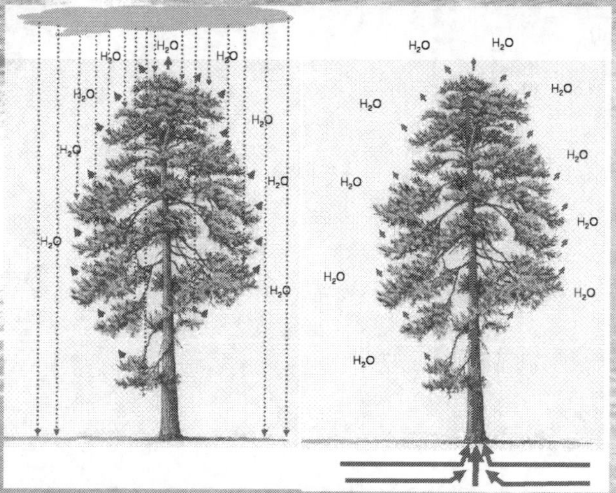

(Licata et al., 2010)

- E_i(usually) =non-productive water loss
- Throughfall = productive water use if infiltrated into the soil (if transpired by vegetation and not lost via runoff)

Canopy Interception (E_i): Patagonia
Morphological and structural traits of leaves and branches are critical determiningE_iLosses

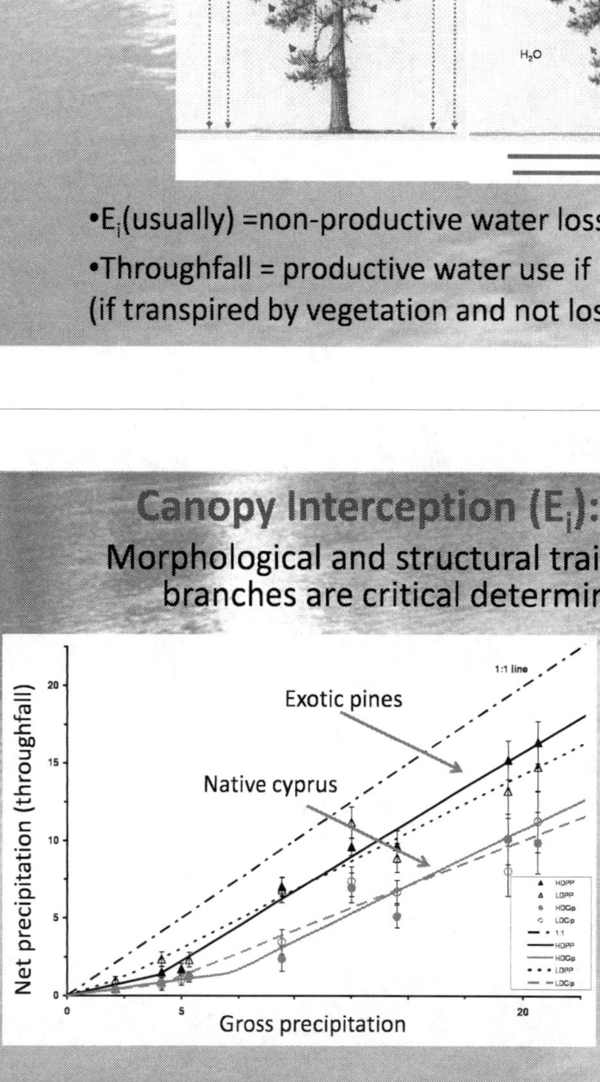

- Exotic pines had lower E_i losses and higher soil water recharge than native cypress forests.
- Exotic pines had higher E_t AND productivity.
- Species selection important!

Licata et al., 2010

Water Quantity: ET and Streamflow

Streamflow (water yield) = Inputs - Outputs ± Storage

$$Q\ (WY) = P - E_t + E_i + E_s + \Delta(S+G) + L$$

= ET

Establishing plantations on former (non-degraded) grasslands in South Africa:Effects on streamflow

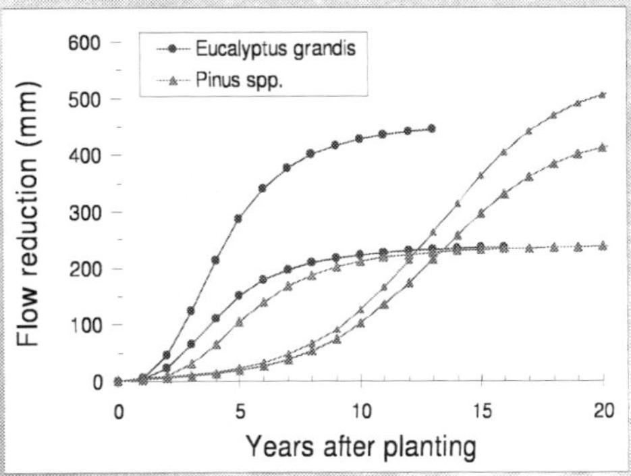

- Trees have higher water use than grasses
- Streamflow reduced
- Effects observed sooner for eucalyptus vs. pine plantations

Scott et al. 1997

Conversion of mature forest to eucalyptus plantations in New Zealand: Effects on streamflow

- Young rapidly growing plantations have higher transpiration rates than mature forest.
- Streamflow reduced under eucalyptus during early stand development.
- Streamflow returns to pre-conversion levels after 80-100 y

Age	Plottranspiration (mm/day)
14	2.2 mm
45	1.4 mm
160	0.8 mm

Roberts et al. 2001

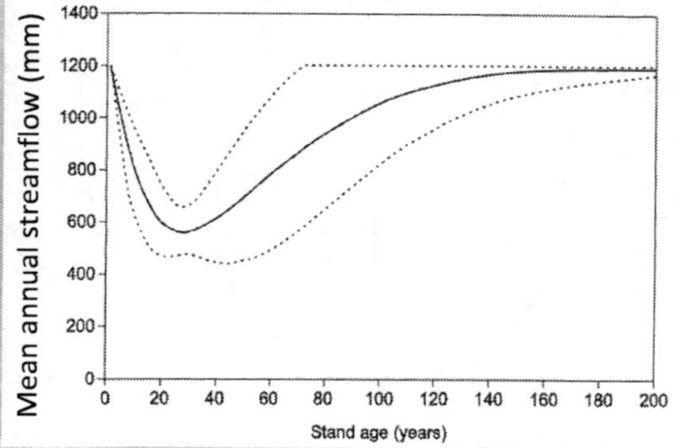

Kuczera1985

Water Timing (seasonal distribution)

Dry Season Flows: Critical for downstream hydrologic services!

Conversion of native forest to pine plantations in Chile: Effects on dry season streamflow

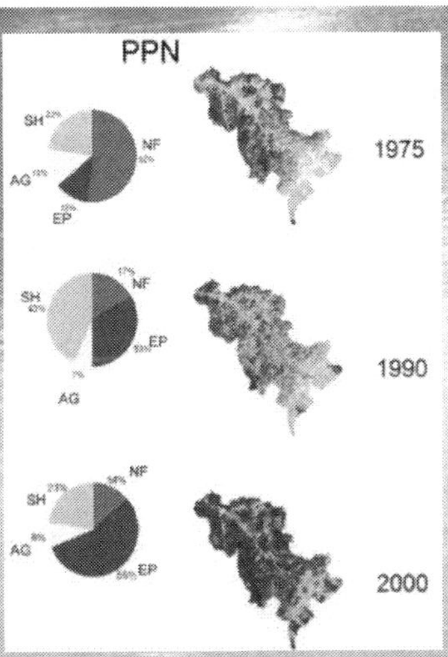

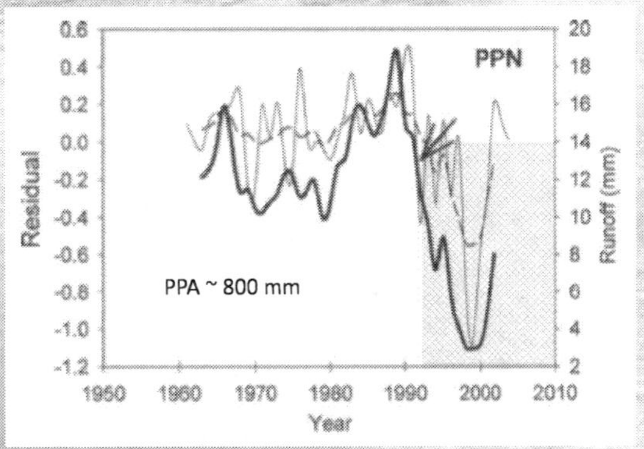

- Watershed in Chile: 250 km2
- Approximately 50% of native forest converted to radiata pine plantations between 1978 and 1997

- Dry season streamflow reduced by 40%
- Greater water use during wet season by pines.

Little et al. 2009

Water Timing (seasonal distribution)

Can planting trees *__on degraded soils__* increase dry season streamflow?

- Net balance between the amount of increased water loss (due to increased ET) and gains in groundwater water recharge (due to increased soil infiltration rates).
- Key: soil hydraulic properties & recharge

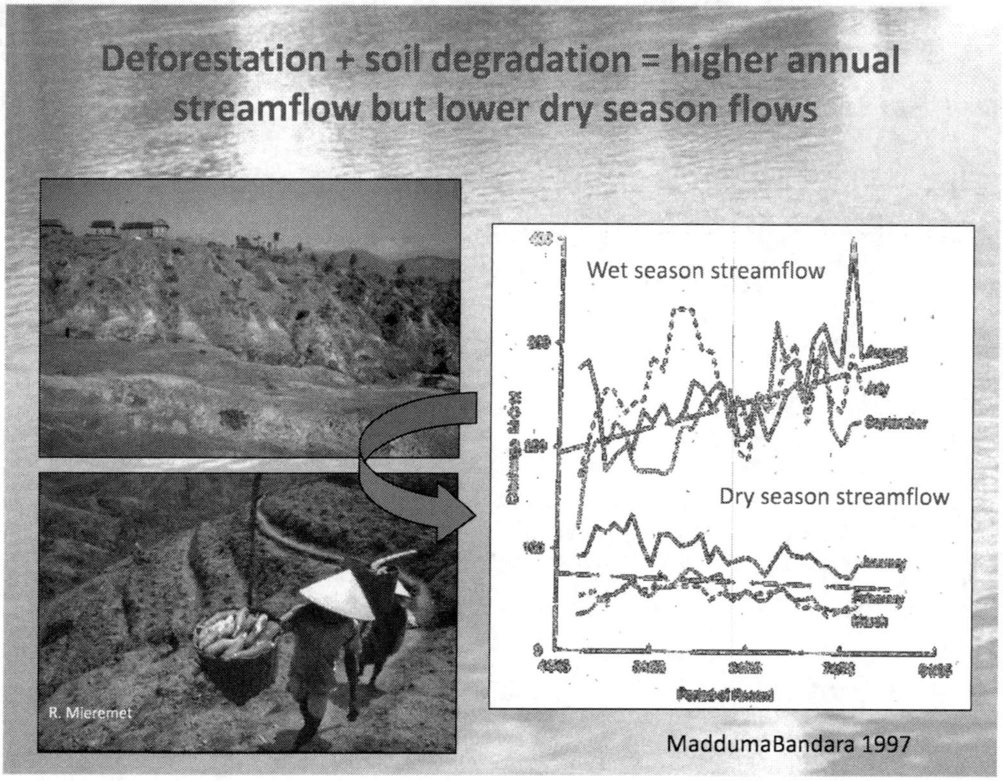
Deforestation + soil degradation = higher annual streamflow but lower dry season flows
Wet season streamflow
Dry season streamflow
R. Mieremet
MaddumaBandara 1997

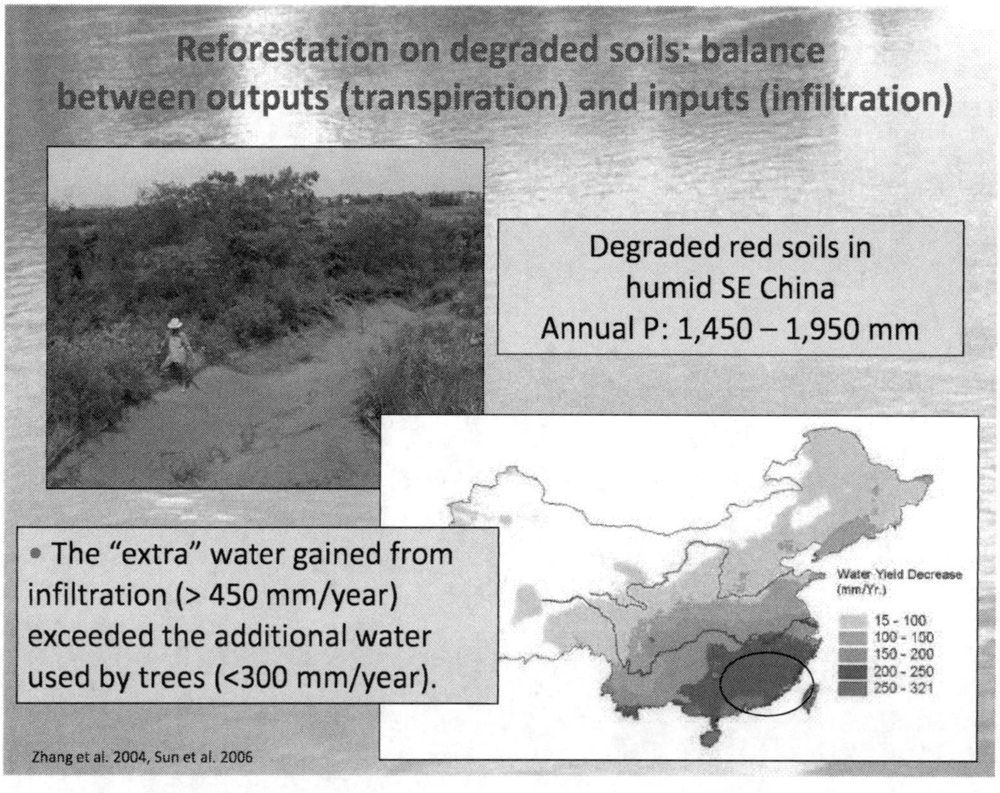
Reforestation on degraded soils: balance between outputs (transpiration) and inputs (infiltration)
Degraded red soils in humid SE China
Annual P: 1,450 – 1,950 mm
• The "extra" water gained from infiltration (> 450 mm/year) exceeded the additional water used by trees (<300 mm/year).
Water Yield Decrease (mm/Yr.)
15 - 100
100 - 150
150 - 200
200 - 250
250 - 321
Zhang et al. 2004, Sun et al. 2006

Pine Reforestation of Degraded Lands: Nepal

Field saturated soil hydraulic conductivity

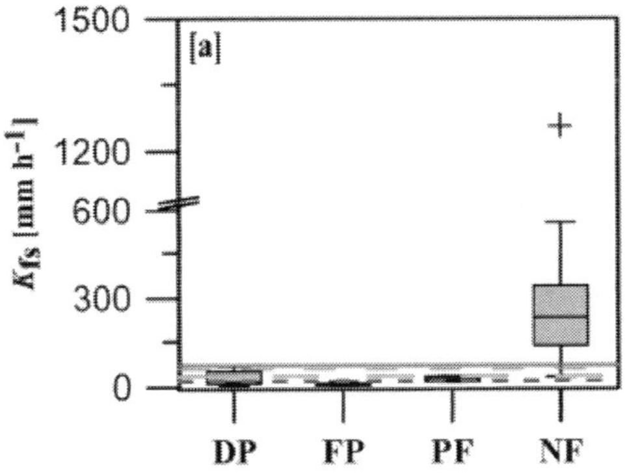

DP = Degraded Pasture (DP)
FP = Foot Path (FP)
PF = Pine Reforestation (PF)
NF = Native Forest (NF)

- Pine reforestation did not improve soil hydraulic properties after 25 years.

Ghimire et al. 2013

Opportunities for managing biomass plantations for hydrologic services

- Rotation length
- Stand density
  - planting
  - thinning
- Genetic improvement:
  - Productivity
  - Water use efficiency
- Species selection
- Species combinations

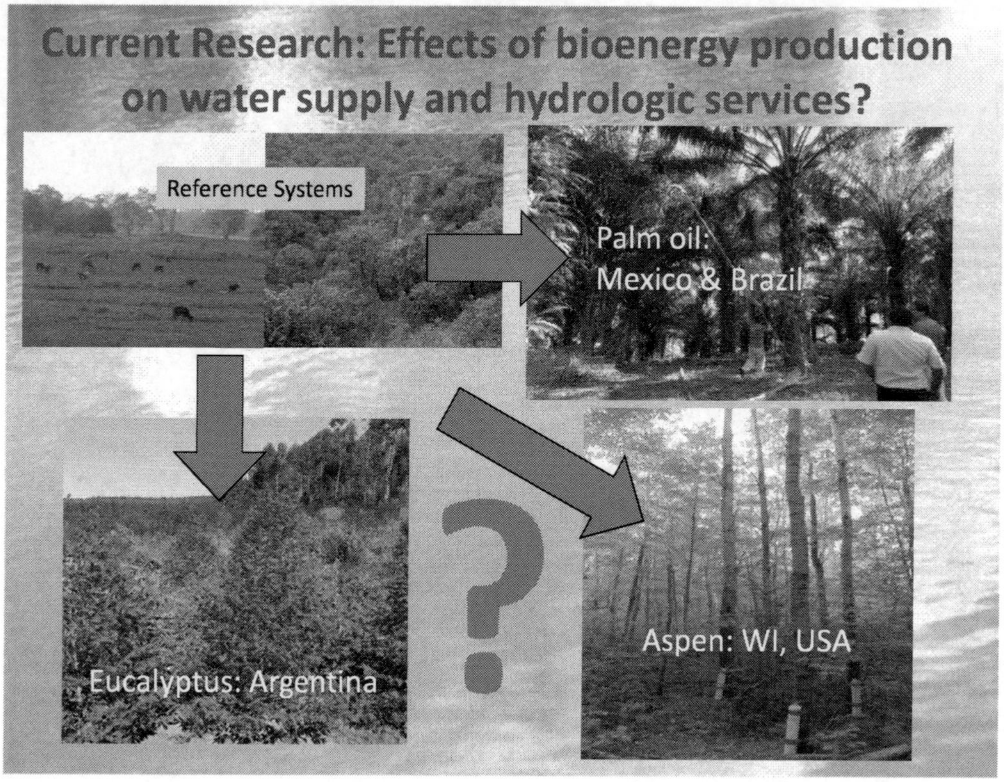
Current Research: Effects of bioenergy production on water supply and hydrologic services?
Reference Systems
Palm oil:
Mexico & Brazil
?
Eucalyptus: Argentina
Aspen: WI, USA

Acknowledgements:
Funding: NSF-PIRE, NSF-RCN

Obrigado! Obrigada!
Perguntas?

RCN Conference on
Pan American Biofuels & Bioenergy
Sustainability
Recife, Brazil July 22-25, 2014

PAN AMERICAN BIOFUELS AND BIOENERGY SUSTAINABILITY
AN NSF RESEARCH COORDINATION NETWORK

IMPACTS OF SUGARCANE EXPANSION ON WATER RESOURCES IN AREAS OF SÃO PAULO STATE

Marjorie M. Guarenghi
Arnaldo Walter

UNICAMP

University of Campinas - UNICAMP
Faculty of Mechanical Engineering - FEM

1

UNICAMP

Summary

1. Introduction
2. Database and areas of study
3. Methods of analysis
4. Results and discussions
5. Conclusion

2

Introduction

- The expansion of sugarcane cropping for bioethanol production can impact:

  - quantity of water resources
  - quality of water resources

- The consumption of water associated to sugarcane and ethanol production has decreased during the last years due to:

  - reduction of water withdraws for the industrial phase through the optimization of some processes and water reuse;
  - use of fertirrigation

3

Introduction

- However:

Land use changes Different management practices	can modify the runoff processes in the watersheds and the river flows
Vinasse (fertirrigation) Use of agrochemicals	contribute to the pollution of water bodies

4

Introduction

➢ The state of São Paulo:
→ main producer of sugarcane and ethanol in Brazil
→ is supposed to be the one with the largest water monitoring network

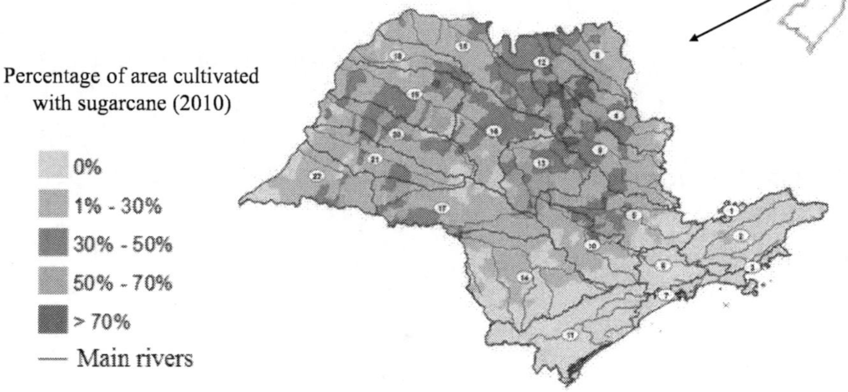

5

Objective

➢ To evaluate whether it is possible to observe **impacts of sugarcane** expansion both on the **quantity** and the **quality** of water bodies.

⇩

Using database publically available of water monitoring in the state of São Paulo

⇩

What can be understood from the data publically available?

6

UNICAMP

Database and areas of study

➢ **Database:**

Monitoring stations regarding **precipitation, discharge** and **quality parameters** provided by:

- Brazilian Water Agency (ANA)
- Department of Water and Energy of the State of São Paulo (DAEE)
- Environmental Agency of the State of São Paulo (CETESB).

7

UNICAMP

Database and areas of study

➢ **Selection of studied areas:**

<u>Discharge monitoring stations</u>:

- with the longest historical series (above 15 years);
- with less than 10% of failures;
- located in areas with significant expansion of sugarcane over the years;

* stations located near hydroelectric plants were avoided;
small scale rivers basins were prioritized

<u>Water quality stations</u>:

- availability of parameters related to agrochemical inputs and vinasse.

Potassium,
Total phosphorus,
Nitrogen series
(nitrate, nitrite, ammonium)

8

Database and areas of study

➢ **Limitations:**

→ unavailability of long time series of discharge and variables of water quality in areas with growing sugarcane cultivation.
→ many data of monitoring stations have several failures, mainly in small scale basins

➢ The methodology for evaluating the effects of land use change in river flow is more conclusive in small-scale basins

➢ However, the hydrological monitoring network in small-scale basins has constrains
→ data is available for just few years
→ the series have several failures.

➢ In Brazil, the implementation of a water monitoring network was motivated by the importance of hydroelectric plants
→ shortage of monitoring stations in basins with less than 500 km^2

9

Areas of study

➢ **Region of Jaú River**

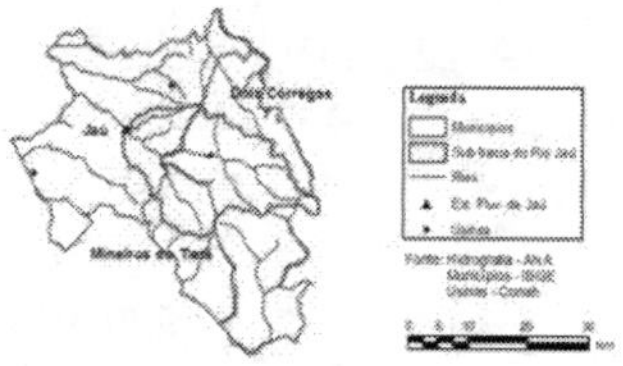

➢ **Region of Pontal and Ribeirão Preto**

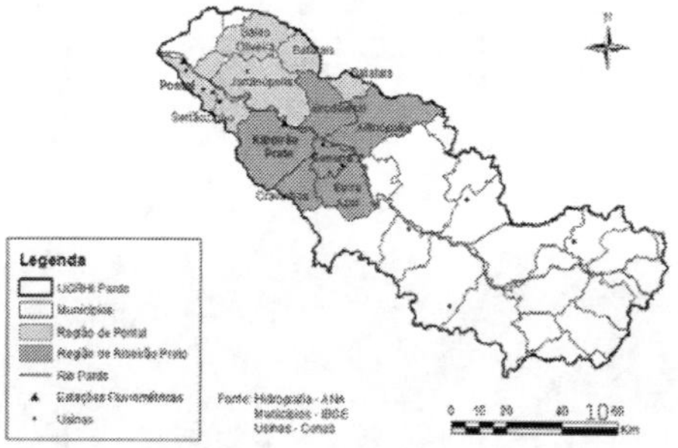

Areas of study

UNICAMP

➢ **Region of Jaú River**

- Jaú River
- Drainage watershed of 417 km²
- Period of evaluation:
  1982 – 1999 (discharge) → the area planted with sugarcane increased 20% (regarding the total area)

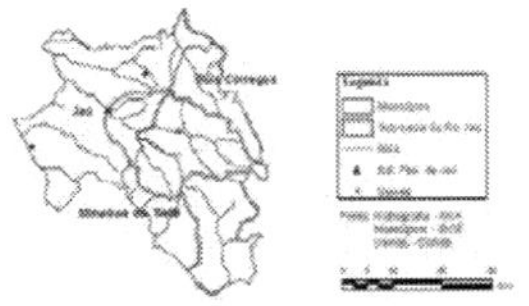

- Three mills of sugar/ethanol are placed close to the Jaú River.

➢ **Region of Pontal and Ribeirão Preto**

- Pardo River
- Drainage watershed of 12.445 km² (Pontal)
  10.679 km² (Ribeirão Preto)
- Period of evaluation:
  1974-2011 (discharge) →sugarcane increased 40%
  1989-2011 (water quality)

- Seven sugar/ethanol mills are located near the discharge and water quality stations analyzed.

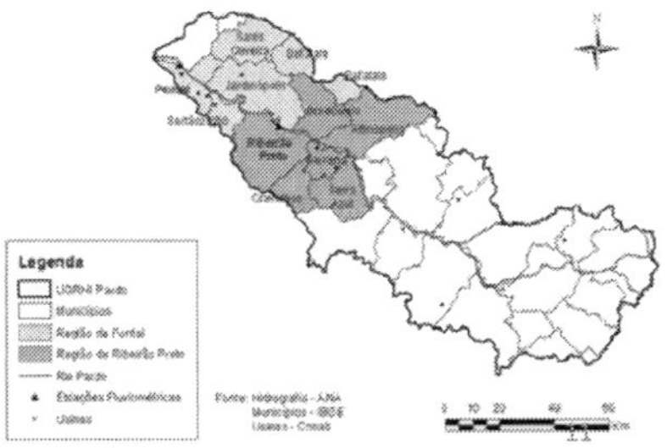

11

Methods of analysis

UNICAMP

➢ The historical time series provided by the monitoring stations were analyzed using nonparametric procedures.

Nonparametric tests } Mann-Kendall's test → to detect trends
Pettitt's test → to determine abrupt changes

➢ Correlation tests : Kendall's correlation

Discharge data —correlated with→ - sugarcane planted area

Concentrations of water quality variables —correlated with→ - sugarcane planted area
- population growth.

12

Results - Discharge analysis

Quantity analysis:

➢ Mann-Kendall's test → Pettitt's test: no trend and no changing point were detected in total annual discharge of the rivers analyzed.

➢ The discharge time series presented better correlation to precipitation series.

➢ No significant correlation was observed between sugarcane growth and the river flows.

13

Results – Quality analysis

Quality parameters series in Pardo River:

	Station of Pontal			Station of Ribeirão Preto		
	Trend test (MK/MKS)	**Correlation test** Population	Sugarcane	**Trend test** (MK/MKS)	**Correlation test** Population	Sugarcane
Potassium	-	-	-	-	-	-
Phosphorus	-	-	-	T(+)	sign	-
Nitrite	T(+)	sign	sign	T(+)	sign	sign
Nitrate	T(+)	sign	sign	T(+)	sign	sign
N_{amon}	T(+)	sign	sign	T(+)	sign	sign

T(+): significant increasing trend . Significant level at 5%

➢ The increase on concentrations can be associated to the
- → fertilizers leaching;
- → discharge of sewage and industrial effluents without appropriated treatment from urban areas

➢ Pettitt's test → 1998/1999 → not correspond to years of increase of sugarcane planted area

14

Conclusions

Quantity analysis:

- According to the tests, **no significant results were observed**
- The stations with the largest time series correspond to **meso- or large-scale basins**.
  → **several** anthropogenic, climatic and geographical **factors impact** on water resources
  → it is difficult to isolate the effects of sugarcane expansion on catchment hydrology

Quality analysis:

- Water quality **parameters related to sugarcane activity** showed **significant increasing** trends in general, but it is difficult to isolate the effects of sugarcane from the effects of residues from urban areas
- Impact assessments on **water quality** were also **constrained** by the **shortage** of important monitoring **parameters related to agriculture practices**
- The bulk of **quality monitoring** stations are in **large rivers**
- Several time series have **failures** that constrain the assessment along the time.

15

Conclusions

- Among the biggest sugarcane producers, the state of São Paulo is supposed to be the one with the largest water monitoring network.
  → time series are poor for analyzing the relationship between land use changes and its effects on water resources.
- For the proper evaluation of impacts on water resources caused by the expansion of sugarcane activity, using statistical procedures:
  → it would be necessary to monitor specific areas in small-scale watersheds.
- This would allow proper data gathering in regions where the influence of other factors could be reduced.
- Investments are required for the installation of new gauging stations to expand the network of water controlling stations.

16

Thank you!

Marjorie M. Guarenghi * and **Arnaldo Walter**

*marjorie@fem.unicamp.br

Faculty of Mechanical Engineering
University of Campinas – UNICAMP
Campinas, SP - Brazil

Laboratório Nacional de Ciência e Tecnologia
do Bioetanol - CTBE/CNPEM

17

Energy-Water Nexus: An Input-Output Dynamical MODEL

Geraldo Andrade de Oliveira
and
Fernando Menezes Campello de Souza

IFPE/UFPE

<gerandr@gmail.com>\<fmcs@hotlink.com.br>

20 de julho de 2014

Objectives

Geral

General

To establish a rational for better energy-water nexus management.

Specifics

- Make a broad review in order to frame the issues;
- Propose a model that could be adapted for local solutions (Regional Economy — Supply Chain Model).

Energy-Water Issues (1)

Highlights(1)

- There are tradeoffs: To generate energy, massive quantities of water are consumed, and to deliver clean water, massive quantities of energy are consumed;
- Both are essential to our health, quality of life, and economic growth, and demand for these resources continues to rise;
- People die, and one cannot grow food, if water is not available. Without energy, one cannot run computers, or power homes, schools, offices, farms, and industrial plants.
- As the world's population grows in number and affluence, the demands for both resources are increasing faster than ever.

Energy-Water Issues(2)

Highlights(2)

- Only about 2.5% of the world's water is freshwater. Unfortunately, less than 1% is accessible via surface sources and aquifers, and the rest is imprisoned in underground reservoirs and in permanent ice and snow cover; Also, the available water is often not clean or not located close to population centers.
- Desalination, a process that removes salt from water, is the most energy-intensive and expensive option for treating water and is used where alternatives are very limited;
- Many people are concerned about the perils of peak oil — running out of cheap oil, and only a few are voicing concerns about peak water. Almost no one is addressing the conflict between the two: water restrictions are hampering solutions for generating more energy, and energy problems, particularly rising prices, are curtailing efforts to supply more clean water;

Energy-Water Nexus (1)

Central Problem: Water Restrictions

- Water is ubiquitous in energy production: in power generation; the extraction, transport and processing of fossil fuels; and, increasingly, in irrigation for the cultivation of raw materials used for the production of biofuels. Similarly, energy is vital to the water supply needed to power that collect, transport, distribution and treat it systems;
- Each faces increasing demands and restrictions in many areas as a result of economic and population growth and climate change, which amplify the mutual vulnerability of energy and water;
- Globally, agriculture is the main user of water, accounting for 70% of water use, followed by industry (including mining and power generation) in 19% and municipal networks that meet the water needs of public and private users in 11%.

Energy-Water Nexus (2)

The Water and Energy Standoff Solutions

- The choices on the future mix of plants used to generate electricity can relieve tension between water and energy. Renewable energy technologies such as wind turbines and photovoltaic panels use little or no water;
- Even fossil fuel technologies provide opportunities to reduce water demand;
- New refrigeration technologies, such as cooling systems and hybrid dry, can also reduce pressure on water systems;
- If energy companies have difficulty finding enough water to cool it´s power plants, blackouts can force them to buy electricity from other sources, which can raise utility bills.

Energy-Water Nexus (3)

The Art of Mathematical Modeling

- This theme has attracted world-wide attention in various contexts like economy, ecology, environment, energy policies and to small, medium and large industries;
- The subject emerges many distortions of information, conflict of interest and undue desire and even contributions from many to the problem detected does not proceed properly there;
- One way to consolidate opinion and strengthen the arguments is through the effort of mathematical modeling of these contexts;

Energy-Water Nexus (4)

The Art of Mathematical Modeling

- Optimal Solution requires Mathematical Modeling;
- The mathematical tool has the power to, in a few lines of information; clarify irrefutable arguments, as a basis for informed decisions to be made.
- Energy-Water Nexus is a real problem that requires complex solution. It is necessary to analyze in the general context (Macro-Solution Optimization), that will serve to long term planning (05–10 years) and supply chain context (Local-Solution Optimization),that will serve to short term planning (01–05 years);
- This work proposes a mathematical supply chain model to analyze the Energy-Water Nexus Issue: An Input-Output Dynamical Model.

E-W Nexus: An Input-Output Dynamical Model

Energy-Water Nexus: An Input-Output Dynamical Model

E-W Nexus: An Input-Output Dynamical Model (1)

The Supply Chain Model

- The supply chain management is a central tool in business administration. The input-output matrix, which is part of an underlying mathematical model, summarizes the matching of supply and demand amongst the various sectors of the economy;
- Two important economic variables are supply, $\boldsymbol{s}$, and demand, $\boldsymbol{d}$. It is assumed that if demand grows, then supply will grow, or should grow, to match it;
- Since demand is always varying, there will be always a dynamic equilibrium. The system is, in general, always moving. It is the dynamics of this movement that one wants to study and control,and in the case of the energy-water interplay this is crucial, since these two variables are closely intertwined.

E-W Nexus: An Input-Output Dynamical Model (2)

Input-Output Dynamics

The equilibrium point is defined by:

$$\boldsymbol{s}(t) = \boldsymbol{d}(t). \tag{1}$$

The dynamical systems works:

$$\begin{aligned} \frac{dx}{dt} &= u \\ x &= Ax + B\frac{dx}{dt} + s \\ u &= K(d - s) \end{aligned} \tag{2}$$

- $\boldsymbol{x}$ is the system total production;
- $\boldsymbol{Ax}$ is part of the total production which is used in the production process;
- $\boldsymbol{B}\frac{d\boldsymbol{x}}{dt}$ is the part of the total production which is put in stock (energy and water need to be stocked);
- $\boldsymbol{s}$ is the supply to satisfy the external demand.

E-W Nexus: An Input-Output Dynamical Model

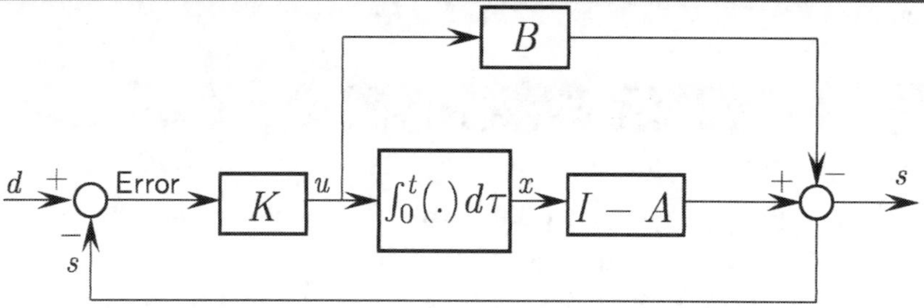

Figura: Block diagram of a supply chain dynamical model.

$$\begin{aligned} \frac{dx}{dt} &= u \\ x &= Ax + B\frac{dx}{dt} + s \\ u &= K(d - s) \end{aligned} \tag{3}$$

- if $\boldsymbol{x}$ is a straight, $\boldsymbol{B}\frac{d\boldsymbol{x}}{dt}$ is constant;
- if $\boldsymbol{x}$ is constant, $\boldsymbol{B}\frac{d\boldsymbol{x}}{dt}$ is zero;
- ...

$$\frac{dx}{dt} = -K(I - BK)^{-1}(I - A)x + K(I - BK)^{-1}d \tag{4}$$

$$s = (I - BK)^{-1}(I - A)x - (I - BK)^{-1}BKd \tag{5}$$

E-W Nexus: An Input-Output Dynamical Model

Simulations

The following values were used:

$$\boldsymbol{A} = \begin{bmatrix} 0.30 & 0.15 \\ 0.10 & 0.13 \end{bmatrix}; \quad \boldsymbol{B} = \begin{bmatrix} 0.020 & 0.010 \\ 0.010 & 0.025 \end{bmatrix}; \quad \boldsymbol{K} = \begin{bmatrix} 0.30 & k_{12} \\ k_{21} & 1.25 \end{bmatrix}$$

- Stability Condition — The matrix $-K(I - BK)^{-1}(I - A)$ characteristics polynomial roots should have negative real parts;
- The $\boldsymbol{B}$ Matrix — The problem with reducing the storage policy is the risk of go through periods of abrupt change in demand and the company be required to set up a matrix K with high values, destabilizing the productive system. In case of availability of energy, for example, the elimination of stocks of energy can cause an economic collapse.

E-W Nexus: An Input-Output Dynamical Model

Simulations

- The $\boldsymbol{K}$ Matrix — As already mentioned, it is responsible for the production planning. It was named control force;
- The Objective Functional — The model described leads the analysis of input-output matrix in the direction of modeling a system of optimal control.The quadratic sum of the error (difference between demand and supply) was used as a functional objective, but any other criterion could be used (including cost, for instance).

E-W Nexus: An Input-Output Dynamical Model

Separated Policies for Energy and Water

By changing the values of k_{11} and k_{22}, one could diminish the errors. This would have the effect of increasing the intensity of the control effort (cost). The eigenvalues of matrix $-K(I-BK)^{-1}(I-A)$ are:

$$K = \begin{bmatrix} 0.30 & 0.00 \\ 0.00 & 1.25 \end{bmatrix} \quad \rightarrow \quad \begin{cases} \lambda_1 = -1.13 \\ \lambda_2 = -0.21 \end{cases}$$

The error can be seen from trajectories shown in the graphics in figures 2 and 3. The sum of squared errors was 316.

E-W Nexus: An Input-Output Dynamical Model

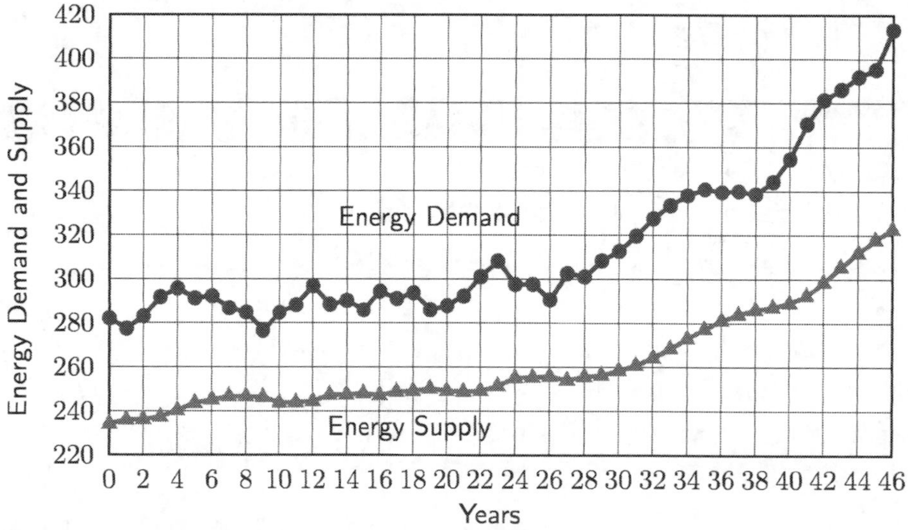

Figura: Energy demand and supply curves, when the K matrix off-diagonal elements are null, i.e., $k_{12}=0$ and $k_{21}=0$ (independent control of energy and water).

E-W Nexus: An Input-Output Dynamical Model

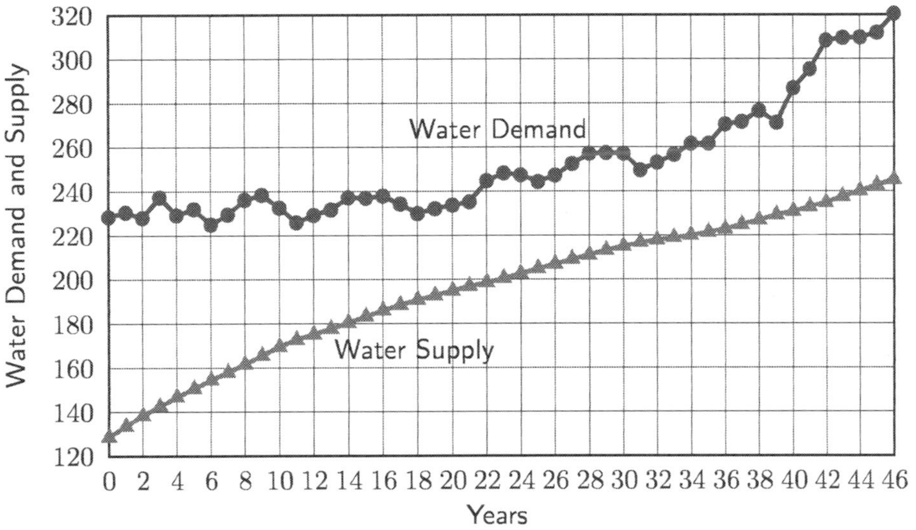

Figura: Water demand and supply curves, when the K matrix off-diagonal elements are null, i.e., $k_{12} = 0$ and $k_{21} = 0$ (independent control of energy and water).

E-W Nexus: An Input-Output Dynamical Model

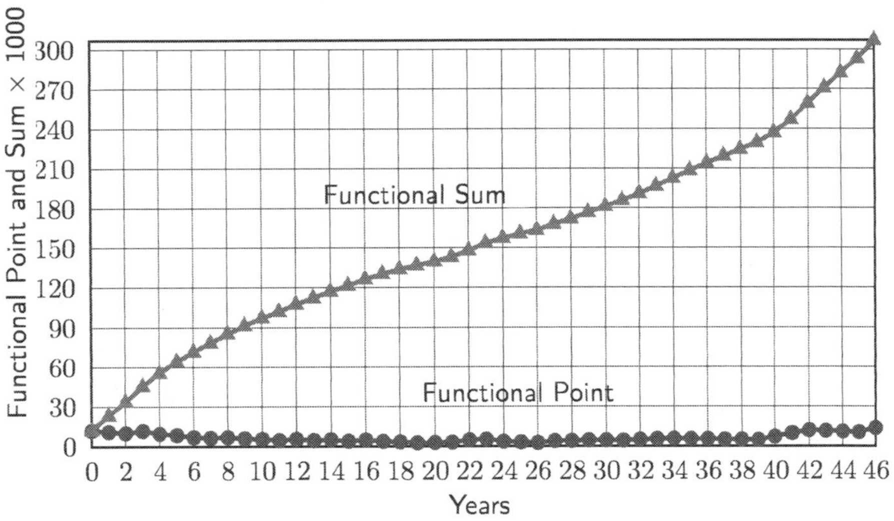

Figura: Functional curves, when the K matrix off-diagonal elements are null (independent control of energy and water).

E-W Nexus: An Input-Output Dynamical Model

Integrated Policies for Energy and Water

Instead of increasing the intensity of the separable controls (k_{11} and k_{12}), one introduces non-zero elements in the off-diagonal of the matrix K, namely, $k_{12} \neq 0$ and $k_{21} \neq 0$. The eigenvalues of matrix $-K(I - BK)^{-1}(I - A)$ are:

$$K = \begin{bmatrix} 0.30 & 0.50 \\ -0.80 & 1.25 \end{bmatrix} \quad \rightarrow \quad \begin{cases} \lambda_1 = -0.75 \\ \lambda_2 = -0.63 \end{cases}$$

The error can be seen from trajectories shown in the graphics in figures 5 and 6. It is visible that the error, in this case, is smaller than in the case of separated policies. The sum of squared errors was 126. The ratio was then 2.50.

The results presented here depended, of course, upon the matrices A and B that were used. The objective of the simulation is to highlight the flexibility that one can achieve.

E-W Nexus: An Input-Output Dynamical Model

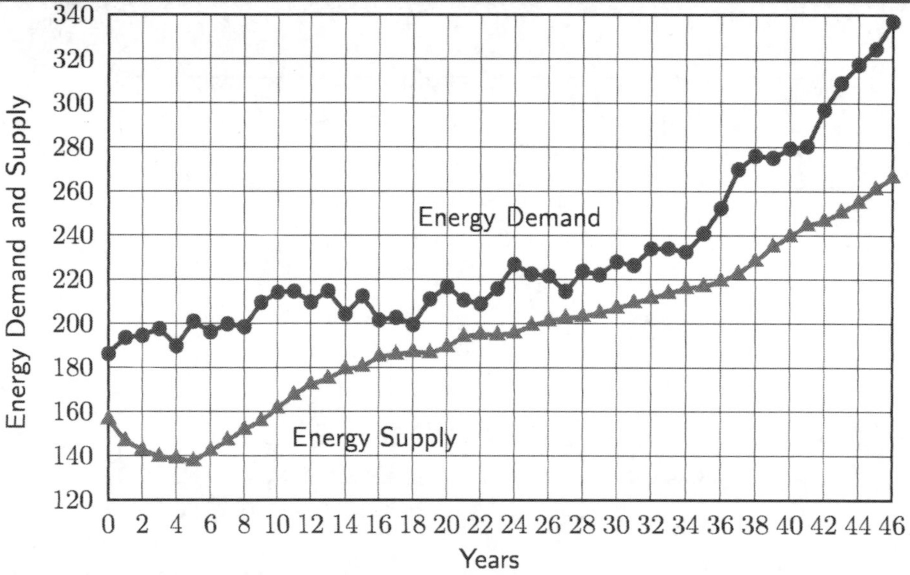

Figura: Energy demand and supply curves, when the K matrix off-diagonal elements are non-null, with values $k_{12} = 0.50$ and $k_{21} = -080$ (integrated control of energy and water).

E-W Nexus: An Input-Output Dynamical Model

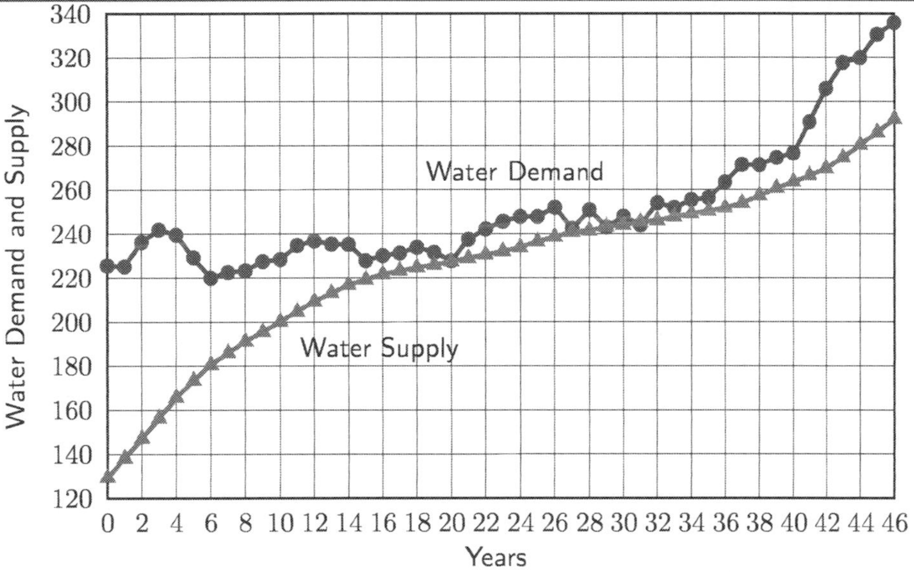

Figura: Water demand and supply curves, when the K matrix off-diagonal elements are non-null, with values $k_{12} = 0.50$ and $k_{21} = -080$ (integrated control of energy and water).

E-W Nexus: An Input-Output Dynamical Model

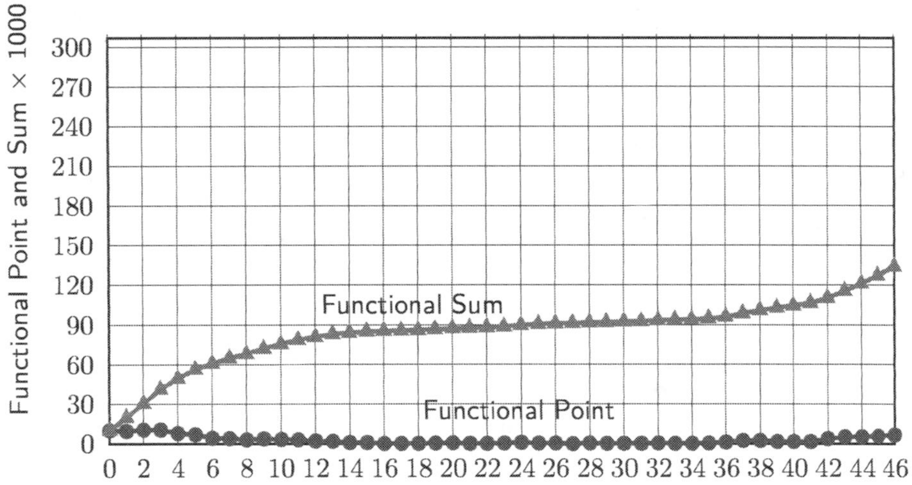

Figura: Functional curves, when the K matrix off-diagonal elements are not null (integrated control of energy and water).

E-W Nexus: An Input-Output Dynamical Model

Three Time Scales for Control

There are three time scales for control. The manager will act in the elements of matrix:

- A (technology development)
- B (stock policy), and
- K (production planning).

In the simulations presented here, the optimality criterion was the quadratic error (to be minimized), and the A and B matrix were fixed. Only the elements of matrix K were used as parametric controls.
A more general objective functional can be used, including, for instance, the control cost.
One can use probabilistic algorithms in order to compute the parametric controls.

Energy-Water Nexus: Conclusions (1)

- The time has come to consider both issues as one.
- As the population grows, water will be more and more scarce. And it cannot be substituted. It will be harder and harder to make water, a precious liquid, available to the population and its economy. More and more energy will be needed. The case of sea water desalination is emblematic; reverse osmosis requires lots of energy. Energy, for its turn, will need more and more water in order to be produced. The only way to escape from this trap is to improve the technology for producing energy (fuels, thermoelectric, for instance) using less water, on one hand, and use progressively more energy primary sources that do not need water in order to produce usable energy, on the other hand.

Energy-Water Nexus: Conclusions (2)

- There is a close connection, or nexus, between energy and water. It takes a significant amount of water to create energy. Water is used to cool steam electric power plants — fueled by coal, oil, natural gas and nuclear power — and is required to generate hydropower. Water is also used in great quantities during fuel extraction, refining and production. It takes a significant amount of energy to extract, move and treat water for drinking and irrigation. It is used in the collection, treatment and disposal of waste water. Energy is also consumed when water is used by households and industry, especially through heating and cooling. Water and energy policy, planning and management must be integrated to encourage conservation, motivate innovation and ensure sustainable use of water and energy.
- One may not realize it, but when one uses energy, one is also using water indirectly — lots of it!

Energy-Water Nexus: Suggestions For Future Studies(2)

A National Energy-Water Supply Chain Model

In the case of two or more country´s regions, it´s possible to make the local matrixes intertwined. One can insert more variables in the problem. This will increase the quantity of matrixes elements ...

With the local matrixes intertwined, the authorities could minimize the costs and risk of non energy-water availability.

Energy-Water Nexus: Questions???

???

Modelagem Hidro-Econômica na Bacia do Araguari para avaliação do impacto da produção de cana-de-açúcar e biocombustíveis no valor econômico da água

PROF. GUILHERME FERNANDES MARQUES, PHD
GUILHERME.MARQUES@UFRGS.BR

INSTITUTO DE PESQUISAS HIDRÁULICAS UFRGS

Núcleo de Planejamento e Gestão de Recursos Hídricos

Contexto atual e justificativa

- Expansão na demanda por fontes energéticas renováveis, com potencial incremento se o país retomar o crescimento econômico;

- Segundo projeções da Empresa Brasileira de Pesquisa Energética (EPE, 2008a), uma expansão sustentada anual média do PIB de 5% no próximo decênio irá resultar em um crescimento no consumo total de eletricidade a um taxa de 5,5% ao ano.

- Crescentes preocupações sobre o uso de combustíveis fósseis e seus impactos ambientais.

Contexto atual e justificativa

- Fontes primárias de energia empregadas no Brasil

	2007	2006
	%	
Energia Não Renovável	**53,6**	**55,1**
Petróleo e Derivados	36,7	37,8
Gás Natural	9,3	9,6
Carvão Mineral e Derivados	6,2	6,0
Urânio (U_3O_8) e Derivados	1,4	1,6
Energia Renovável	**46,4**	**44,9**
Produtos da Cana-de-açúcar	16,0	14,5
Energia Hidráulica e Eletricidade	14,7	14,8
Lenha e Carvão Vegetal	12,5	12,7
Outras Renováveis	3,1	2,9

Brasil, Empresa de Pesquisa Energética. (2008b). *Balanço Energético Nacional 2008: Ano base 2007: Resultados Preliminares.* Rio de Janeiro: EPE, 2008. 44p.

INSTITUTO DE PESQUISAS HIDRÁULICAS UFRGS *Núcleo de Planejamento e Gestão de Recursos Hídricos*

Contexto atual e justificativa

- O Brasil é um dos maiores produtores de cana-de-açúcar do mundo, com uma área cultivada de 5,34 milhões de hectares, dos quais 2,66 milhões para produção de alcool.
- Os maiores produtores mundiais de etanol são os Estados Unidos e o Brasil, que utilizam como matéria prima o milho e a cana-de-açúcar respectivamente.
- O Brasil produziu 22,5 milhões de m3 de alcool em 2007, um aumento de 26% em relação a 2006. No mercado mundial, o Brasil participa com 50% nas exportações de etanol, a maior parte para os países da Índia, Japão e Estados Unidos

INSTITUTO DE PESQUISAS HIDRÁULICAS UFRGS *Núcleo de Planejamento e Gestão de Recursos Hídricos*

Contexto atual e justificativa

- Devido às vantagens energéticas, econômicas e ambientais do etanol da cana-de-açúcar sobre o do milho, existe um grande potencial para o aumento da demanda por este combustível em nível mundial.

INSTITUTO DE PESQUISAS HIDRÁULICAS UFRGS *Núcleo de Planejamento e Gestão de Recursos Hídricos*

Contexto atual e justificativa

- Esta demanda trará impactos para as bacias hidrográficas das regiões produtivas:
  - Aumento na demanda pela água no caso dos sistemas irrigados.
  - Produção de efluentes e resíduos.
  - Expansão de fronteiras agricolas.
- Na ausência do planejamento adequado, estes impactos irão produzir vários efeitos negativos:
  - Aumento na escassez hídrica.
  - Agravamernto de conflitos pelo uso da água.
  - Poluição e outros impactos ambientais.

INSTITUTO DE PESQUISAS HIDRÁULICAS UFRGS *Núcleo de Planejamento e Gestão de Recursos Hídricos*

Foco e objetivos do projeto

- Estudar os sistemas hídricos quanto à sua capacidade de suporte em um cenário mundial cujo principal desafio é a produção de energia mais limpa.
- Objetivos:
  - Estudar o impacto da produção de etanol de cana-de-açucar irrigada no valor econômico da água, na medida em que esta torna-se um insumo para a produção de energia.
  - Identificar qual a produção "ótima" de etanol considerando demandas econômicas que competem por água e energia na bacia do Rio Araguari.
  - Identificar qual o nível de eficiência na produção de etanol necessário para sua viabilidade regional na bacia, reduzindo a demanda por combustíveis fósseis.

PESQUISAS HIDRÁULICAS UFRGS *Núcleo de Planejamento e Gestão de Recursos Hídricos*

Para que servirão os resultados produzidos ?

- Estudo de cenários futuros "e se" de abastecimento e escassez hídrica envolvendo diferentes demandas por energia e pela água;
- Auxílio na identificação da alocação otimizada da água entre demandas econômicas, especialmente aquelas relacionadas com a produção de biocombustíveis;
- Auxílio no desenvolvimento de políticas e planos de gestão da água visando sua conservação e aproveitamento, considerando aspectos econômicos.

PESQUISAS HIDRÁULICAS UFRGS *Núcleo de Planejamento e Gestão de Recursos Hídricos*

Metodologia

- Modelo de otimização, *economically driven.*
- Abordagem de PL, com função objetivo convexa *piece wise*.
- Funções econômicas para agricultura irrigada produzidas por modelo de produção quadrático, calibrado via PMP (*positive mathematical programming*).

Metodologia – funções econômicas

- As funções econômicas podem ser estimadas com o emprego de modelos que maximizam o lucro agrícola sujeito a restrições de disponibilidade de água, capital e outros *inputs*.

$$y = [\alpha_1 \quad \alpha_2 \quad \alpha_3] \cdot \begin{bmatrix} x_1 \\ x_2 \\ x_3 \end{bmatrix} - [x_1 \quad x_2 \quad x_3] \cdot \begin{bmatrix} \gamma_{11} & \gamma_{12} & \gamma_{13} \\ \gamma_{21} & \gamma_{22} & \gamma_{23} \\ \gamma_{31} & \gamma_{32} & \gamma_{33} \end{bmatrix} \cdot \begin{bmatrix} x_1 \\ x_2 \\ x_3 \end{bmatrix}$$

$$Max \sum_G \sum_i p_i f_{Gi}(x_1, x_2, x_3) - \omega_1 x_1 - \omega_2 x_2 - \omega_3 x_3$$

$$sujeito\ a \quad \sum_G \sum_i x_{1,G,i} \le X_{1,G}$$

$$\sum_G \sum_i x_{2,G,i} \le X_{2,G}$$

Funções econômicas e o custo da escassez da água

- O custo da escassez é a área sob a curva de demanda entre os pontos do abastecimento atual e a demanda máxima (ponto onde a disponibilidade a pagar pela água, ou valor marginal, é igual a zero).

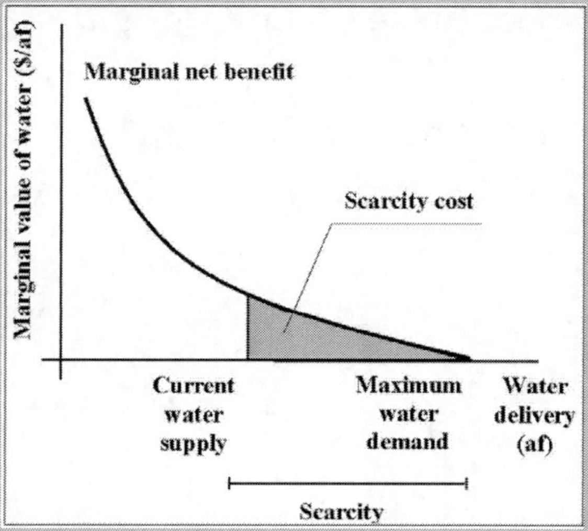

Custo da escassez = função de penalidade

- Função objetivo: minimiza o custo da escassez da água.
- Variáveis de decisão: volumes alocados a cada demanda e armazenados em reservatórios.
- Equações de restrição: balanço de massa, limitações físicas de infraestrutura e outros condicionantes como vazões ecológicas e restrições operativas hidráulicas.

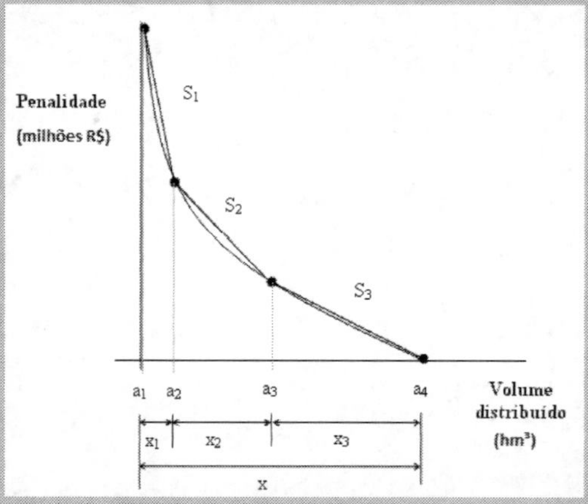

Metodologia - Otimização de um sistema hídrico

O sistema hídrico modelado pode incluir:

- Fontes de abastecimento;
- Reservatórios;
- Demandas;
- Retiradas e retorno de água.

•Sistema representado como um *grafo*

•Presença de arcos (links) e nós

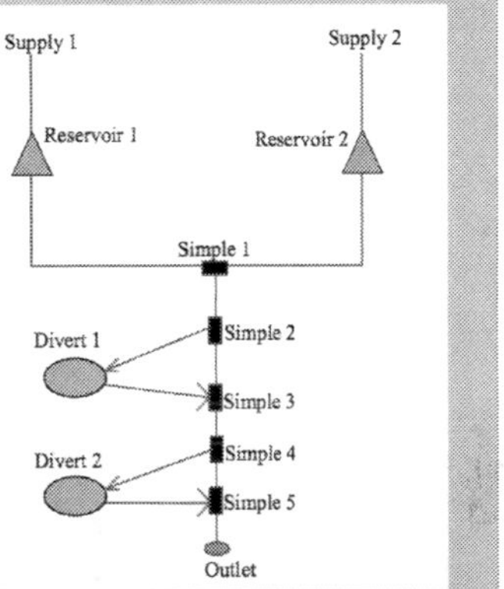

INSTITUTO DE PESQUISAS HIDRÁULICAS UFRGS

Núcleo de Planejamento e Gestão de Recursos Hídricos

Equações de restrição

Nós simples

$$\sum_{n \in out} R(n,t) = \sum_{n \in in} Q(n,t) + SUP(n,t)$$

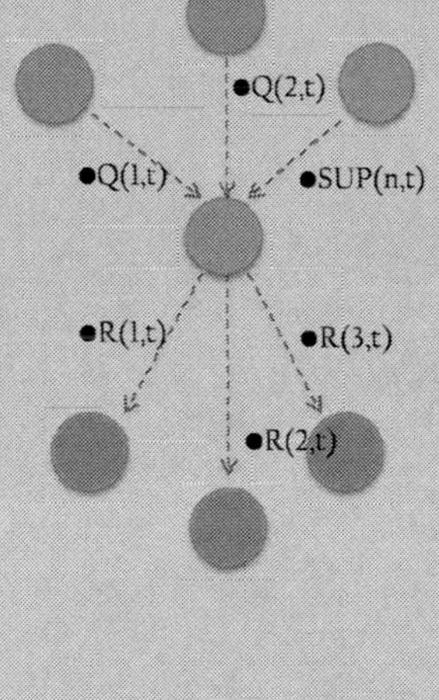

INSTITUTO DE PESQUISAS HIDRÁULICAS UFRGS

Núcleo de Planejamento e Gestão de Recursos Hídricos

Equações de restrição

Nós de demanda

$$R(n,t) = R_j * \sum_{n \in in} Q(n,t) \quad n \in nrr$$

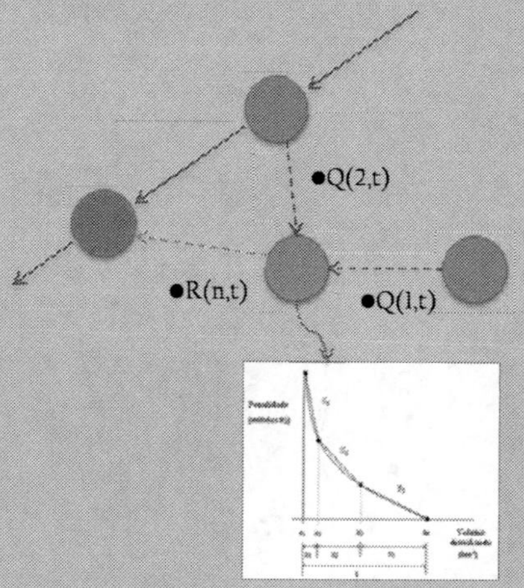

PESQUISAS HIDRÁULICAS UFRGS *Núcleo de Planejamento e Gestão de Recursos Hídricos*

Equações de restrição

Nós de reservatório

$$S(n,t) = S(n,t-1) + \sum_{n \in in} Q(n,t) - \sum_{n \in out} R(n,t) \quad n \in nl$$

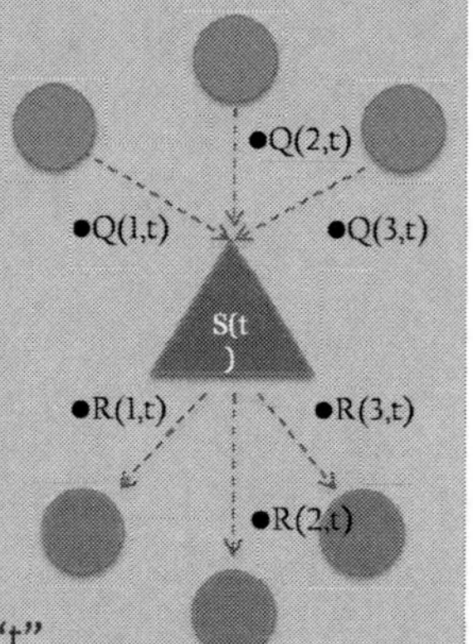

S(n,t) é o volume armazenado no nó "n" ao fim do intervalo "t"

PESQUISAS HIDRÁULICAS UFRGS *Núcleo de Planejamento e Gestão de Recursos Hídricos*

Formulação

$$CET = \sum_{t} \left(\sum_{PT} (CMAX_{PT} - \sum_{CST} X_{t,PT,CST} * CM_{CST,PT}) + \sum Ex_{t,PT} * CEX; \right)$$

$$\sum_{i} VOL_{t,PT,i} - \sum_{j} VOL_{t,j,PT} = \begin{pmatrix} S_{t+1,PT} - S_{t,PT} - P_{t,PT} + E_{t,PT} + Ex_{t,PT}, \text{ IF } PT \in RES_{PT} \\ CONT_{t,PT}, \text{ IF } PT \in DF_{PT} \\ DM_{t,PT}, \text{ IF } PT \in DEM_{PT} \\ Q_{t,PT}, \text{ IF } PT \in IN_{PT} \end{pmatrix} \quad (9)$$

$$X_{t,PT,CST} \leq REST_{CST,PT} \quad (10)$$

$$DM_{t,PT} = \sum_{CST} X_{t,PT,CST} \quad (11)$$

$$VOL_{t,PT,i} \leq UP_{PT,i} \quad (12)$$

$$VOL_{t,PT,i} \geq LOW_{PT,i} \quad (13)$$

$$S_{t,PT} \geq Smin_{PT} \quad (14)$$

$$S_{t,PT} \geq Smax_{PT} \quad (15)$$

$$CONT_{t,PT} \geq LBOUND_{t,PT}; \quad (16)$$

$$CONT_{t,PT} \leq UBOUND_{t,PT};$$

INSTITUTO DE PESQUISAS HIDRÁULICAS UFRGS

Núcleo de Planejamento e Gestão de Recursos Hídricos

Área de estudo

Cenários analisados

- **Situação atual**
  - Áreas plantadas de cana de açucar e outras culturas segundo SIDRA 2006
- **Situação sem subsídio na gasolina;**
  - Aumento no preço da gasolina: 29,05%
  - Percentual do preço do produto sobre o preço da bomba: 34,6%
  - Aumento de 29,05% no preço da gasolina, resulta em 29,05% * 34,6% = 10% no aumento do preço na bomba

Cenários analisados

- Média do preço considerando a elasticidade preço cruzada etanol gasolina (dados literatura);
- Aumento resultante da demanda de etanol com o incremento do preço da gasolina é de aproximadamente: 3,7 milhões m^3/ano

Cenários analisados

- **Produção de cana-de-açucar destinado ao etanol no Brasil, safra de 2013/2014: 339,8 milhões de ton,**
  - 27,2 milhões m3 de etanol (CONAB, 2013) e uma razão de 0,08 m3 de etanol/ton de cana;
- **Aumento potencial na demanda de etanol: 13,67%, (3,7 milhões m^3)**
  - 46,5 milhões de toneladas de cana / ano;
  - O triângulo mineiro produziu 23 milhões de toneladas (6,8% do total nacional);
  - Equivale a 3,14 milhões de toneladas de cana;
  - Equivale a 23.776 ha irrigados (a 132,2 TCH)

Demanda por água

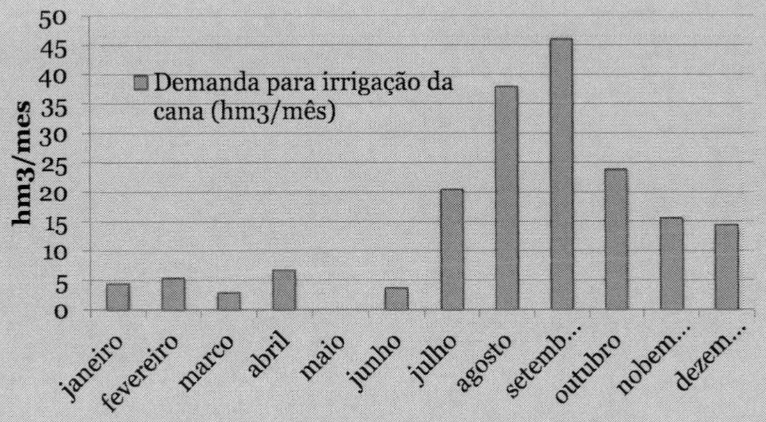

Sistema modelado

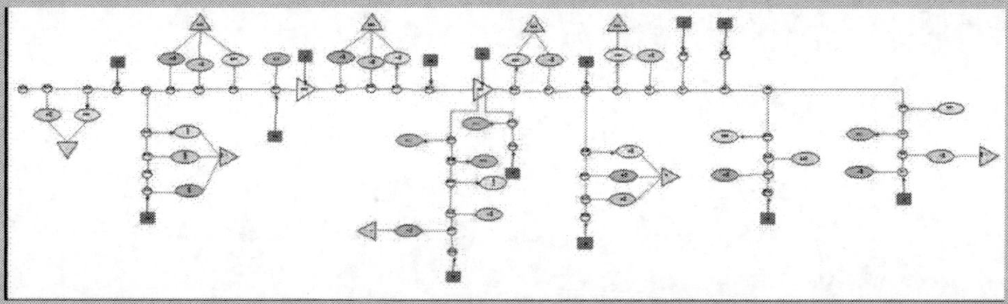

Land Use Changes and Biofuel Feedstock Production in Brazil: The Role of Irrigation Water

Ana C. G. Carneiro, Universidade Federal de Pernambuco
Hector M. Nuñez, Centro de Investigación y Docencia Económicas,
Márcia M.G.A. Moraes , Universidade Federal de Pernambuco
Hayri Önal, University of Illinois at Urbana-Champaign

PAN-AMERICAN BIOFUELS & BIOENERGY SUSTAINABILITY

July 22-25, 2014, Recife, Brazil

World Ethanol Production

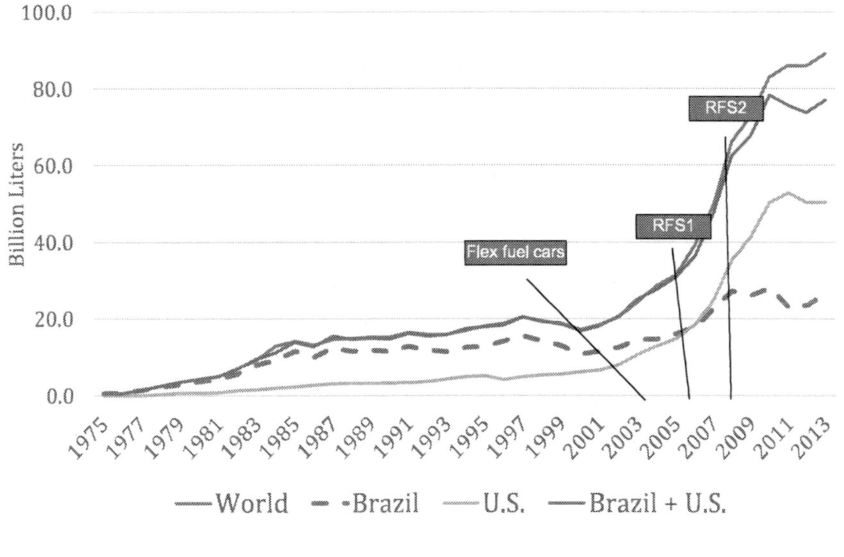

RCN,July 22-25, 2014, Recife, Brazil

RFS Biofuel Blending Mandates
(Bil.Gallons/Yr)

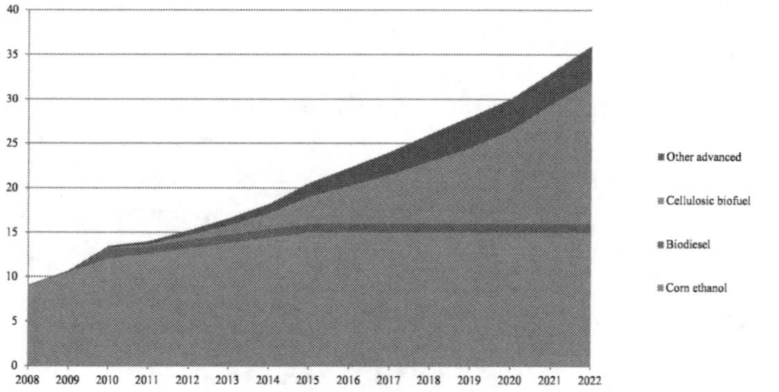

RCN,July 22-25, 2014, Recife, Brazil

Light Duty Vehicle Fleet Projections, Brazil

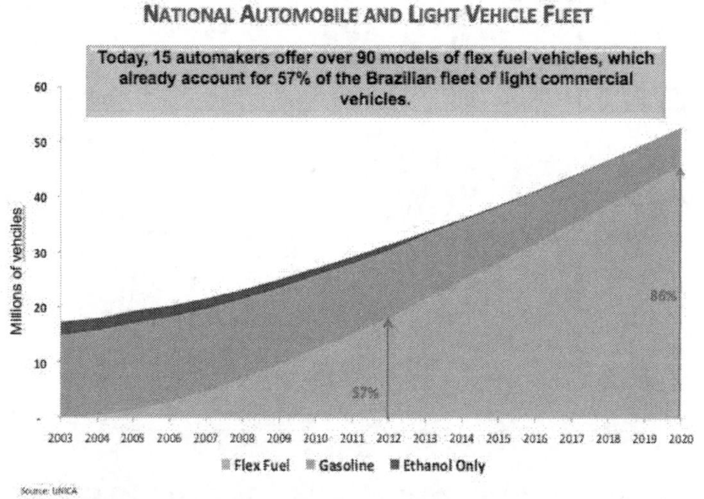

RCN,July 22-25, 2014, Recife, Brazil

Research Issues

- Can Brazil ethanol industry respond to the growing demand for biofuels in the domestic and global markets?
- Investigate the potential for expansion in sugarcane and ethanol production, impacts on resource utilization in agriculture and food/feed commodity prices considering the competition between sugarcane and other crops.

RCN,July 22-25, 2014, Recife, Brazil

Resource Availability

- Planted Area of Sugarcane in 2007= 7.1 million ha (Censo Agropecuário - IBGE), suitable area for expansion ZAE = 64.7 million ha (19 million ha with high potential)
- Expansion in biofuel production requires additional cropland for sugarcane production , can be within 1-5 million ha.
- Various modeling studies have looked at the expansion of sugarcane areas by intensification in the livestock sector and converting some of the pasturelands into crop production.
- **But, no comprehensive modeling study has looked at the other important resource: WATER**
- **Out of 24 billion m^3 of irrigation water use in 2007, 14.5 billion m^3 was used for sugarcane (60%) !**

RCN,July 22-25, 2014, Recife, Brazil

Methodology

- We developed a multi-region, multi-market spatial equilibrium model including the agricultural and fuel sectors of Brazil and US to determine the production and consumption of food and fuel products in both countries and trade between them.
- The model maximizes producers' + consumers' surplus subject to resource constraints, technical constraints and policy constraints.
- Explicit demand functions are used for food commodities, km driven by light duty vehicles (conventional, flex-fuel, and ethanol-dedicated).

Spatial Aspects

- The Brazil component includes 137 mesoregions producing eight annual crops and sugarcane, beef-cattle production under three different pastures types and production systems.
- Sugarcane expansion is limited to agro-ecological zones.
- Ethanol and livestock transportation costs between production and consumption regions are included.
- **Besides regional land constraints, the model also includes regional water use-supply constraints which incorporate irrigation water requirements and projected water supply for irrigated agriculture in 2030.**

Water Constraint

$$\sum_i w_{irt} \cdot \Delta_{ir} \boldsymbol{LAND_{ir}} \leq W_{rt}$$

w_{irt}= water consumption by crop *i* in region *r* in month *t* (m^3/ha/mo)
Δ_{ir}= irrigated/planted acreage of crop *i* in region *r*
W_{rt}= Total water supply in region *r* in month *t*
$\boldsymbol{LAND_{ir}}$=Land allocation for crop i in region r (endogenous)

$$\Delta_{i,r} = \frac{Area\ Ir_{i,r}}{Area\ Plant_{i,r}}$$

RCN,July 22-25, 2014, Recife, Brazil

Water Data

- Data from irrigated and harvested areas in the states and municipalities provided by the IBGE Agricultural Census of 2006.
- The technical coefficients of water use published by Ministry of Environment in a partnership with Federal University of Viçosa (2011).
  - These were based on:
    - Evapotranspiration data from the database of ONS 2005 (1950-2000).
    - Precipitation data available from the National Water Agency - ANA by month for the period from 2000 to 2008.
    - Coefficients of crop moisture and irrigation efficiency application parameters from ANA / GEF / UNEP / OAS (2002).

RCN,July 23-25, 2014

Derivation of the Water Parameters

$$K_{it,m} = \left[\frac{(ETO_{t,m} \cdot Kc_{it,m} \cdot Ks_{i,m}) - pef_{it,m}}{Ea_{i,m}}\right] \cdot 10$$

$$w_{irt} = \sum_{m \in r} [AreaIr_{im} / AreaIr_{ir}] . K_{itm}$$

$$W_{rt} = \sum_{i} w_{irt} \cdot AreaIr_{ir}$$

RCN,July 22-25, 2014, Recife, Brazil

Projected Water Coefficients (2030)

For the 2030 projections, we update the irrigation water requirements based on the temperature and precipitation predictions of global climate scenarios **(IPCC, 2013)**.

$$K'_{it,m} = K_{it,m} + dK_{it,m}$$

$$dK_{it,m} = \frac{\partial K_{it,m}}{\partial ETO_{t,m}} \cdot \frac{\partial ETO_{t,m}}{\partial T} \cdot dT_{t,m} + \frac{\partial K_{it,m}}{\partial Pef_{t,m}} \cdot dPef_{it,m}$$

$$w'_{it,r} = w_{it,r} + dw_{it,r}$$

$$dw_{it,r} = \left[\frac{(Kc_i \cdot Ks_{i,r})}{Ea_{i,r}}\right] \cdot 10 \cdot p_{t,r} \cdot 0{,}457 \cdot dT_{t,r} - \left[\frac{10}{Ea_{i,r}}\right] \cdot dPef_{it,r}$$

RCN,July 22-25, 2014, Recife, Brazil

Projected Water Coefficients (continued)

$$\Delta'_{i,r} = \frac{Area\ Ir_{i,r} + Area\ Ir'_{i,r}}{Area\ Plant_{i,r} + Area\ Ir'_{i,r}}$$

$$Area\ Ir'_{i,r} = Area\ Ir_{i,r} + \left(Area\ Projeto_r * \frac{Area\ Ir_{i,r}}{area\ Ir_r} \right)$$

$$W'_{rt} = \sum_i w'_{irt} \cdot AreaIr'_{ir}$$

Estimated Data Variation of Precipitation and Temperature from IPCC for region 12 in the months of Dec -Feb

Dec-Feb	Absolute Temperature				Percentage Change in Precipitation Media			
IPCC Models\Families	A2	B2	A1FI	B1	A2	B2	A1FI	B1
CCSRNIES	0.9981	1.0123	1.1420	0.8406	-1.15	-2.28	-2.27	-0.45
CSIRO	0.9188	0.9804	0.9923	1.0422	-2.18	-1.4	-2.35	-1.45
ECHAM4	0.9568	0.9610	1.0525	0.8265	4.35	5.2	4.79	4.47
HADCM3	1.0022	0.8445	1.0924	0.7347	-0.57	3.57	-0.62	3.11
NCAR PCM	0.7206	0.7354	0.7927	0.6325	1.61	2.22	1.77	1.91
CGCM2	1.0471	1.1868			1.92	3.07		
GFDL-R30	0.8799	0.9197			2.87	0.84		

Source: IPCC, 2013.

New Water supply for Irrigated Agriculture in Brazil in 2030

24 billion m^3 in 2007

IPCC Model	IPCC scenarios [using the] NEW PEF			
	A2	B2	A1	B1
	Millions M^3			
CCSRNIES	27022	27428	26579	26122
CSIRO	26294	26103	26302	26278
ECHAM4	26212	25899	26219	25931
HADCM3	27540	27119	27655	26996
NCAR_PCM	25951	25802	25934	25854
CGCM2	26138	26246	26137	26137
GFDL-R30	25619	25548	26137	26137
Average	26397	26306	26423	26208

Data

- The model is calibrated using 2007 as the base year
- The data inputs include the base year domestic and global commodity prices and quantities demanded, price-demand elasticities, historical crop mixes (2003-2009), crop yields, costs of crop and livestock production, processing costs, and costs of fuel transportation.
- Data sources: USDA, Department of Agriculture of Argentina, EIA, IBGE , AgraFNP, CONAB , EMBRAPA , CEPEA , PECEGE, EPE, ANFAVEA, FAO , COMTRADE, FUNARBE, elasticities from the literature; Water coefficients are calculated using the MMA data and the data obtained from FUNARBE (2011).

Model Validation - Crop Acreages (2007)

(in million ha)	Observed	Original Model w/o water restriction	Extended model (w/water restriction)
Total cropland	47.98	42.65	41.83
Sugarcane	7.09	6.19	6.04
Corn	14.01	11.91	11.86
Soybeans	20.57	22.97	22.43
Wheat	1.86	2.78	2.78
Cotton	1.13	0.98	0.91
Rice	2.92	2.07	2.04
Sorghum	0.67	0.77	0.75
Pasture Land	127.00	124.40	124.95

RCN,July 22-25, 2014, Recife, Brazil

Scenario Analysis (2030)

- *Scenario-1:* US, Brazil, EU and World ethanol mandates, disregarding the water constraints
- *Scenario-2:* Same as Scenario-1 except that the water constraints are included

RCN,July 22-25, 2014, Recife, Brazil

Simulated Land Use (2030)

	Base Run	Scenario-1 w/o water	Scenario-2 w/water
Sugarcane	6.04	15.90	6.34
Corn	11.86	15.13	16.49
Soybeans	22.43	30.06	30.97
Wheat	2.78	2.73	2.65
Cotton	0.91	1.44	1.54
Rice	2.04	2.46	2.45
Sorghum	0.75	0.61	0.54
Total Cropland	**41.83**	**63.52**	**56.29**
Pasture Land	**124.95**	**108.79**	**116.02**
Total Agr. Land	166.78	172.31	172.31

RCN,July 22-25, 2014, Recife, Brazil

Simulated Performance of the Sugarcane Industry in 2030

	Scenario-1 w/o water	Scenario-2 w/water	Percent Change
SCane Acreage (Mil. Ha)	15.90	6.35	-60
SCane Production (Mil. MT)	1,870.44	702.99	-62
for sugar (Mil. MT)	459.84	440.54	-4
for ethanol (Mil. MT)	1,410.60	262.45	-81
Ethanol production (Bil. lt)	115.93	21.56	-81
Hydrous ethanol price ($/lt)	0.41	0.71	73

RCN,July 22-25, 2014, Recife, Brazil

Simulated Sugarcane Areas- Scenarios 1&2

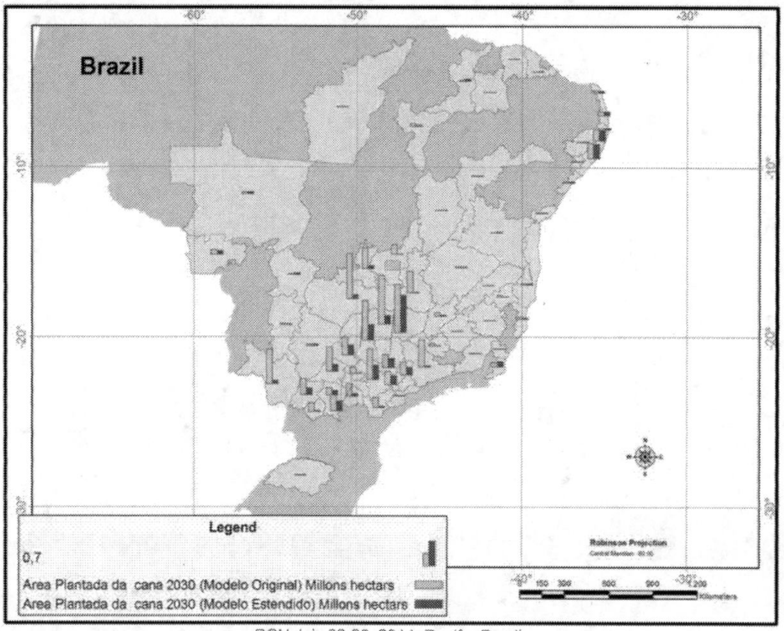

RCN,July 22-25, 2014, Recife, Brazil

Simulated Pasture Areas- Scenarios 1&2

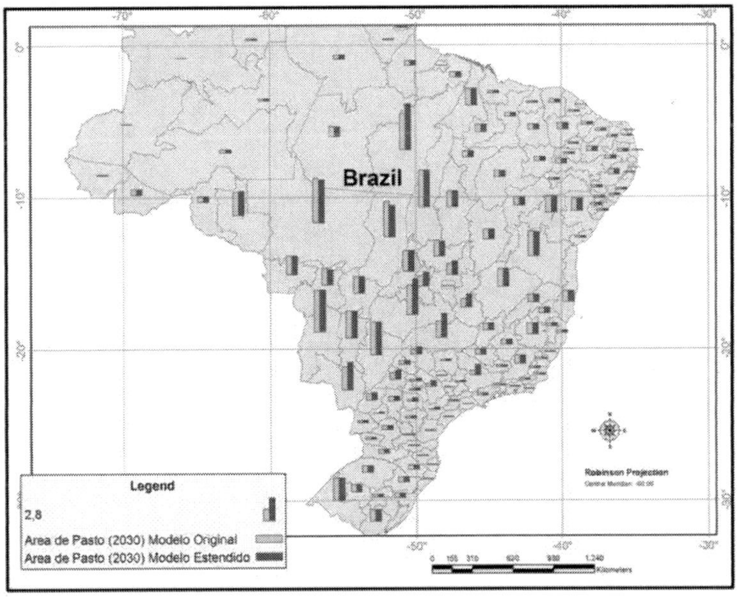

RCN,July 22-25, 2014, Recife, Brazil

Concluding Remarks

- The simulation results indicate that consideration of regional irrigation water constraints would have substantial impacts on land use, livestock intensification, and expansion of sugarcane production potential.
- Without infrastructure investments and improvement in efficiency, irrigation water requirements due to climate change may put significant pressure on other uses of water.
- Water scarcity may intensify conflicts in regions such as Northeastern Brazil where water availability is limited and irrigation water has important role for socioeconomic development.
- The results under various water availability scenarios may provide useful insight to agricultural producers and public policy makers.

RCN,July 22-25, 2014, Recife, Brazil

Concluding Remarks

- Investment in irrigation technologies, which can lead to an increase in use efficiency, should enhance the expansion of crops, even without an increase in the availability of water for the irrigated agriculture sector.
- To this end, public water allocation policies need to encourage the rational and efficient use of water through an effective management of demand.
- The possible impacts of the likely expansion of biofuel production in developing countries requires more attention internationally in approaches to public policies that help to secure property rights to land and water for the poorest.
- Results of economic- integrated models are able to support the decision in the design of management tools for effective demand, such as water pricing, and others that induce an economically optimal allocation

RCN,July 22-25, 2014, Recife, Brazil

Thank You!

Bioenergy and biodiversity: Key lessons from the Pan American Region

AL Mayer, Michigan Technological University USA
KL Kline, Oak Ridge National Laboratory USA
F Martinelli, R Medeiros, Conservation International BRAZIL
COFD Oliveira, A Walter, University of Campinas BRAZIL
G Sparovek, University of São Paulo BRAZIL
L Venier, Canadian Forest Service CANADA

To keep in mind for discussion:

What are priority research needs for biodiversity – that are relevant to Pan America bioenergy development?

Costs and benefits of bioenergy to biodiversity

Costs

- Land cover/management change
  - LC change -> loss of habitat: a major threat to biodiversity
  - Expansion into natural areas = less area for biodiversity support
  - Species-area curve predicts decline in species with less available habitat
- Decreased ecosystem services/ functioning
  - Nutrient cycling, water filtration
- Unsustainable biomass harvests

Benefits

- Land cover/management change
- Climate change mitigation
  - Climate change is a major threat to biodiversity
- Some bioenergy crops can provide habitat and improve ecosystem services/ functioning
  - Advanced or "2nd generation" biofuels
- Biomass can be managed in ways that are sustainable and help conserve habitat
  - Increased value of ecosystem

Bioenergy mandates and biodiversity protection laws

- Many countries have laws protecting biodiversity (e.g., Endangered Species Act in the US), nature reserves and parks, air and water quality (ecosystem services)
- 14 countries in Pan American region have legislation promoting biofuel blends or bioenergy production
  - USA: blending requirements by state, federal level production targets
  - http://www.biofuelsdigest.com/bdigest/2013/12/31/biofuels-mandates-around-the-world-2014/

Country	Blending targets
Argentina	B10, E5
Brazil	E20, B5
Canada	E5 (<E8.5), B2
Columbia	E8
Chile	E5, B5
Costa Rica	E7, B20
Ecuador	B5
Jamaica	E10
Mexico	E2
Panama	E5
Paraguay	E24, B1
Peru	E7.8, B2
Uruguay	E5 and B5 by 2015
USA	Varies by state

Biofuels production in Pan America

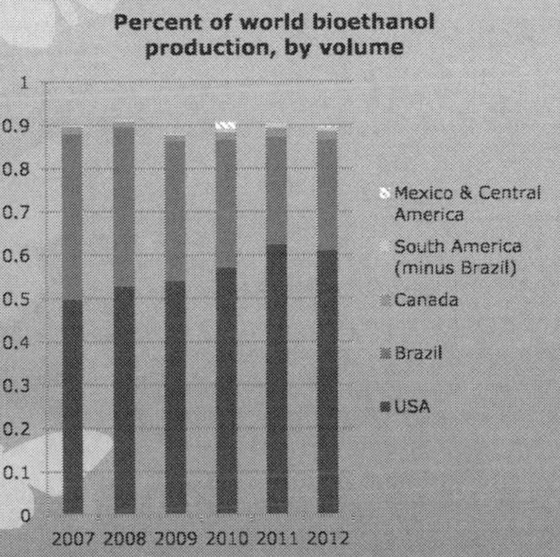

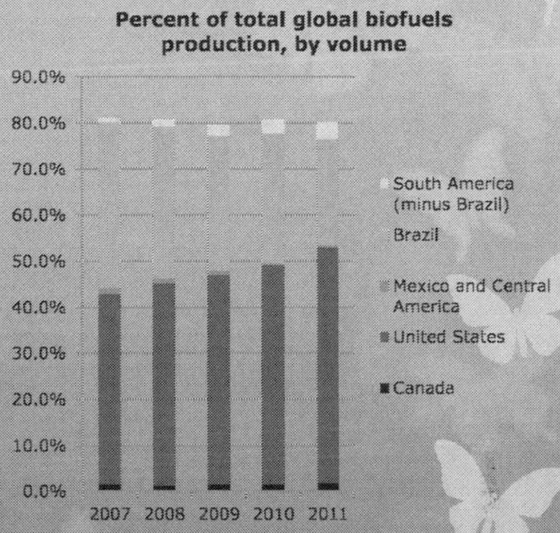

www.ethanolrfa.org/pages/annual-industry-outlook

http://www.eia.gov/cfapps/ipdbproject/IEDIndex3.cfm?tid=79&pid=79&aid=1

Wood Pellets in British Columbia, Canada

- Canada: produces 3 million tonnes per year
- British Columbia: 66% of Canadian pellet production
  - Feedstock from 22 million ha of public forests, 2 million ha private forests
    - 80-95% of feedstock from industrial residues, plus tops and branches post-harvest
  - 94% of pellets exported to Europe

Wood Pellets in British Columbia, Canada

- Benefits: reduction in debris and slash piles facilitates replanting after harvest and reduces fire risk
- Public forests required compliance with management guidelines; monitoring by interested parties
- Biodiversity concerns are primarily related to possible future changes in practices:
  - Deadwood collection
  - Excessive collection of coarse woody debris may negatively impact beetles, woodpeckers, and other deadwood obligates
  - Forests infested with mountain pine beetle could become targets for chips and pellets
    - At present this is not considered to be economically viable

Biofuels in the United States

- Prior research and initial implementation of renewable fuel mandates focused primarily on ethanol from corn (maize)
  - Rapid growth in production required little change in machinery, methodology, infrastructure
- New research to facilitate shift to advanced or 2nd gen feedstocks:
  - Perennial grasses (switchgrass, miscanthus), fast-growing deciduous trees (hybrid poplar, hybrid willow)
    - Many options for integration with existing systems
      - Buffer strips, riparian zones
      - Inter-cropping
      - Winter cover crops and biomass rotations
      - Mono-culture perennials...
  - Harvest residues (tree tops and branches, corn stover)

Biofuels in the United States

- Comparison of biodiversity impacts of grain-based biofuels versus 2nd generation biofuels:
  - Biodiversity supported is higher in and around second-generation
    - Landscape-level heterogeneity
    - Reduction in negative impact to aquatic systems
    - Still concern about impacts of conversion of natural cover or "marginal" land
  - Second generation plots provide ecosystem services
    - Water filtration
    - Soil stabilization
    - Habitat (breeding, feeding, stopover) for some species

U.S. Bioenergy supply model Billion Ton Update (USDOE 2011)

- Forecasts of potential biomass
  - POLYSYS partial equilibrium model of US agricultural and forestry sectors.
  - 20-year projections of economic availability of biomass (price, location, scenario)
- Forest resources
  - Logging residues
  - Forest thinnings (fuel treatments)
  - Conventional wood
  - Fuelwood
  - Primary mill residues
  - Secondary mill residues
  - Pulping liquors
  - Urban wood residues
- Agricultural resources
  - Crop residues
  - Grains to biofuels
  - Perennial grasses
  - Perennial woody crops
  - Animal manures
  - Food/feed processing residues
  - MSW and landfill gases
  - Annual energy crop (added for 2011 Update)

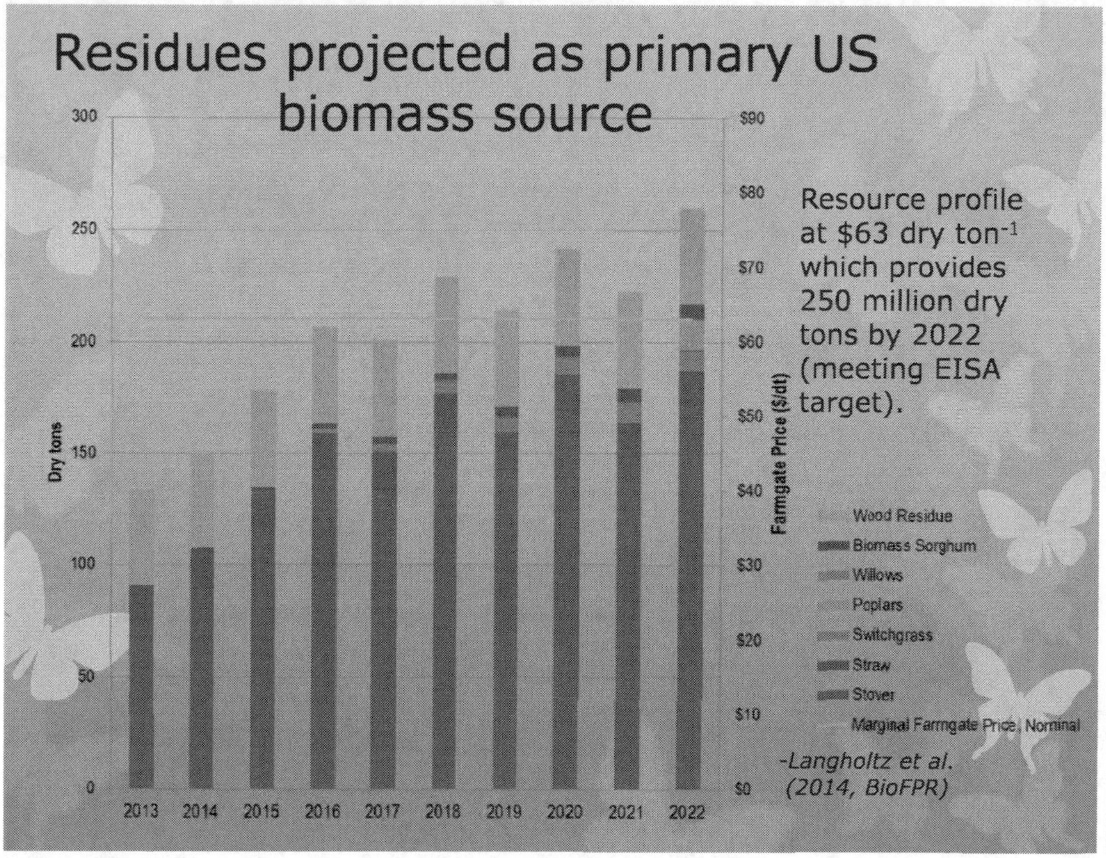
Residues projected as primary US biomass source
Resource profile at $63 dry ton-1 which provides 250 million dry tons by 2022 (meeting EISA target).
Dry tons
Farmgate Price ($/dt)
300
250
200
150
100
50
0
$90
$80
$70
$60
$50
$40
$30
$20
$10
$0
2013
2014
2015
2016
2017
2018
2019
2020
2021
2022
Wood Residue
Biomass Sorghum
Willows
Poplars
Switchgrass
Straw
Stover
Marginal Farmgate Price, Nominal
-Langholtz et al. (2014, BioFPR)

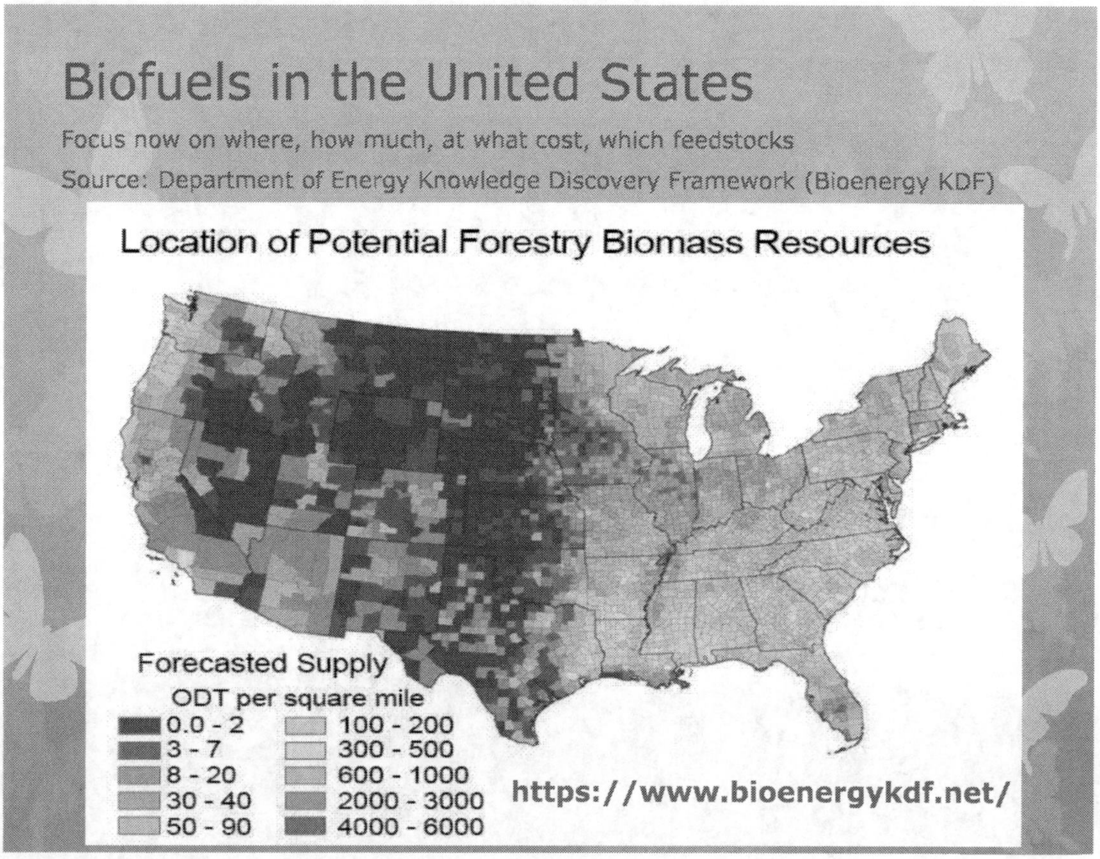
Biofuels in the United States
Focus now on where, how much, at what cost, which feedstocks
Source: Department of Energy Knowledge Discovery Framework (Bioenergy KDF)
Location of Potential Forestry Biomass Resources
Forecasted Supply
ODT per square mile
0.0 - 2
3 - 7
8 - 20
30 - 40
50 - 90
100 - 200
300 - 500
600 - 1000
2000 - 3000
4000 - 6000
https://www.bioenergykdf.net/

Biofuels in Brazil

Dynamics of land use governed by expansion and consolidation phases

- Expansion: extraction, deforestation, pasture, agriculture
  - Complicated land tenure, socioeconomics
  - Ag production secondary: less specialized; extensive
- Consolidation: increasingly specialized, intensive agriculture
  - Supported by infrastructure (roads and bridges)
  - Increased land values requires greater returns
  - Sugarcane, soy and palm oil for biofuels
- Policies needed to avoid further damage to key biodiversity/ high conservation value (HCV) areas
  - Cerrado (savannah region) under intense development pressure now

Biofuels in Brazil: sugarcane

- Sugarcane has been grown in Brazil for 500 years
- Most sugarcane is far from Amazon
- To address concerns about LUC, Brazil government developed Agro-Ecological Zoning for sugarcane
  - Focus on former pasture lands
  - Concerns remain about expansion into Cerrado

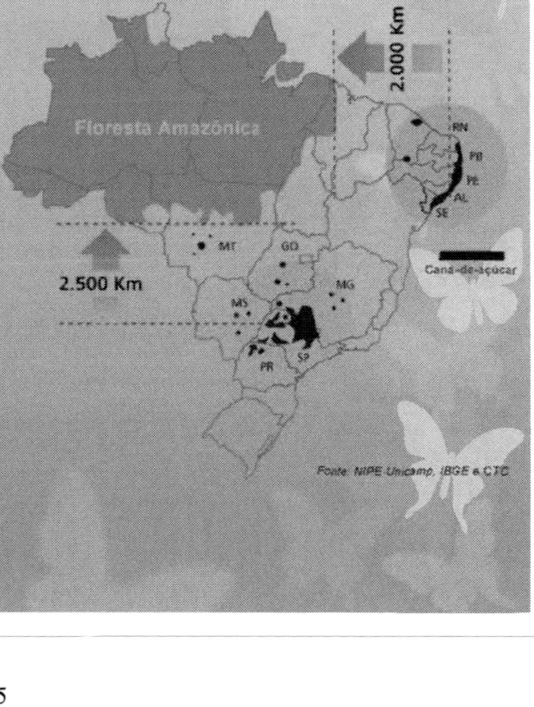

Biofuels in Brazil: Palm Oil

- Agro-Ecological Zoning aims to restrict new palm oil plantations to previously degrade lands
- Some concerns remain in Pará region – area of rapid growth – with high biodiversity and endemism, endangered
  - Potential effects on water quality
  - Other effects of development in rural areas

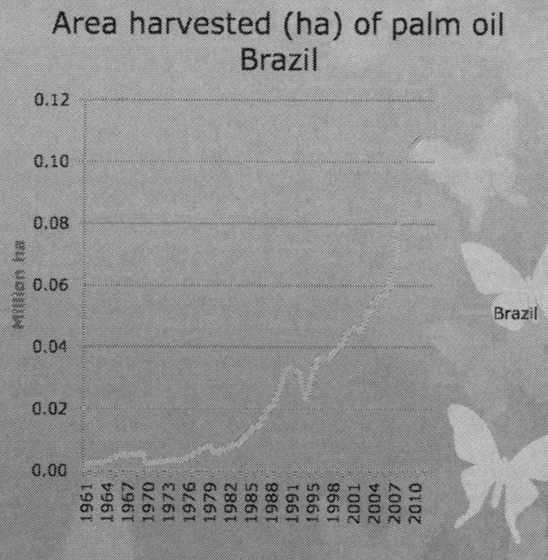

Biodiversity protection through Certification

	RSB	ISCC	Bonsucro
Biodiversity Principle	Avoid high biodiversity areas	Avoid high biodiversity areas	Actively manage for biodiversity
Issues addressed	Ecosystem services, corridors	biodiverse grasslands, carbon stocks, peatlands	Assessment of impacts, mitigation measures
No-go areas	Addressed	Indirectly addressed	Assumes existing plantations only
HCV	Yes	Yes	Yes
Degraded lands	Used to be	No	Yes
Invasive species	Yes	Yes	Yes
Threatened species	Yes	Yes	Yes

Pan-American Patterns

- Concern for biodiversity and ecosystem protection is explicit – legal and enforcement mechanisms improving
- Advances in identifying bioenergy production systems that are more harmonious with natural habitats and conservation goals
- Policies restricting bioenergy production to less diverse, already disturbed, and marginal lands
- Most growth and new development is NOT driven by bioenergy per se
  - Successful feedstocks have large non-energy markets
    - Bioenergy is one of several, minor, co-products
  - Bioenergy incentives – when they exist – tend to support large industrial interests
  - Difficult for new crops/technologies to penetrate fuel markets, especially if dependent on biofuel market alone.

To keep in mind for discussion:

What are priority research needs for biodiversity – that are relevant to Pan America bioenergy development?

Thank you

Center for Bioenergy Sustainability
http://www.ornl.gov/sci/ees/cbes/

CBES
Center for BioEnergy Sustainability

See the website for

- Reports
- Forums
- Other presentations
- Recent publications

Acknowledgements: paper under preparation with support from the NSF Research Collaboration Network (RCN NSF Award #1140152) Project – David Shonnard PI, and the PIRE Project for Environmental and Social Sustainability Assessment of Bioenergy in Pan America – Kathy Halvorsen PI. Other collaborators include co-authors (Title slide) and others

Kline's research was supported by the U.S. Department of Energy (DOE) Bio-Energy Technologies Office and performed at Oak Ridge National Laboratory (ORNL). Oak Ridge National Laboratory is managed by the UT-Battelle, LLC, for DOE under contract DE-AC05-00OR22725.

The views in this presentation are those of the author, Keith L. Kline, who is responsible for any errors or omissions.

LAND SPARING VERSUS LAND SHARING: HOW MIGHT BIODIVERSITY AND BIOENERGY COEXIST?

Audrey L. Mayer (almayer@mtu.edu)
Michigan Technological University
School of Forest Resources and Environmental Science
Department of Social Sciences
Houghton, MI USA

MODERN HUMAN LAND USE AND BIODIVERSITY LOSS

- Human land use does not automatically lead to less biodiversity
  - Terra preta soils in Amazon
  - Forest islands in savanna in Africa
- In this talk, bioenergy production ≈ agriculture
  - Food, fiber, fuel
  - Intensive industrial
  - High diversity, polyculture

BIOENERGY PRODUCTION IN DIVERSE AREAS

- Countries in western hemisphere producing much of the world's biofuels at present
  - Volume vs. hectare
- The hemisphere has many biodiversity hotspots, including high biodiversity biomes that have been substantially reduced by land use change
  - Tallgrass and shortgrass prairies of North America
  - Atlantic forest, cerrado in Brazil

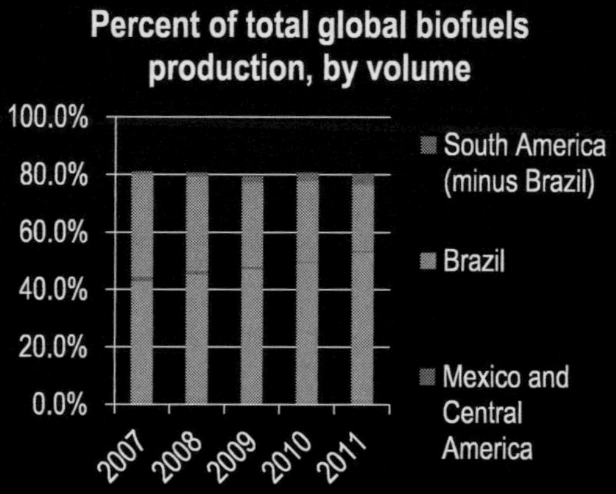

LAND SPARING (OR LAND INTENSIFICATION)

- Isolate production to as small of an area as possible
  - Allows for maximum land area for biodiversity and intact ecosystems
- Use all technology available to produce maximum yields
  - Fertilizers
  - Pesticides
  - Genetic modification
  - Antibiotics/hormones

Photo: Garth Lenz (www.garthlenz.com)

Photo: Wired Magazine/ Mishka Henner

LAND SHARING

- Agroforestry or agroecology
  - Planting food and medicinal crops interspersed to create a structure similar to native habitat
    - Shade-grown coffee
    - Intercropping of grains and seed crops in grassland/prairie
- Polyculture or permaculture
  - Using perennials instead of annuals to produce food

Photo: Arbor Day Foundation

Photo: The Land Institute

PRO-LAND SPARING

- Maximize diversity of ecosystem services
- Island Biogeography Theory
  - Species-Area curve: more area supports more species
- Landscape Matrix
  - Flip land use so that intensive agriculture are islands in a sea of natural land cover
    - Reduction in pests
    - Buffers for runoff

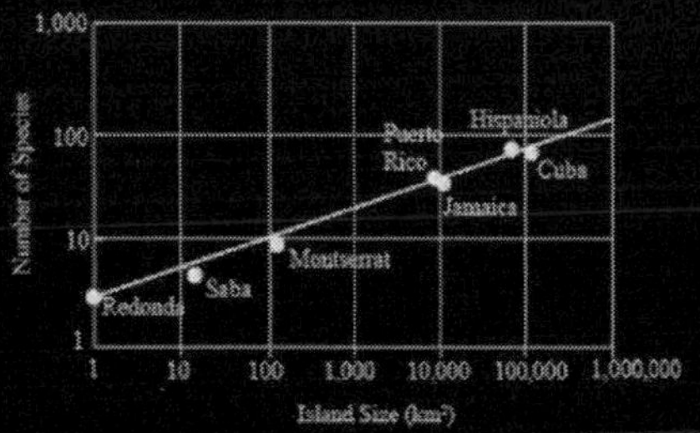

Amphibian and reptile species in Caribbean

LAND SPARING AND BIOENERGY PRODUCTION

Hybrid Poplar for cellulosic ethanol, bio-oil, wood chips

Photo: Michigan State University

Switchgrass for cellulosic ethanol, pellets

Photo: Paul Knox

ANTI-LAND SPARING

- Can impacts of high-intensity production be isolated to production fragments?
  - DNA drift from GMO varieties
- How much can we compress agriculture and still meet demand?
  - Food
  - Energy
  - Fiber
  - What are limits to land productivity?

Arable land per capita (ha in use/person)

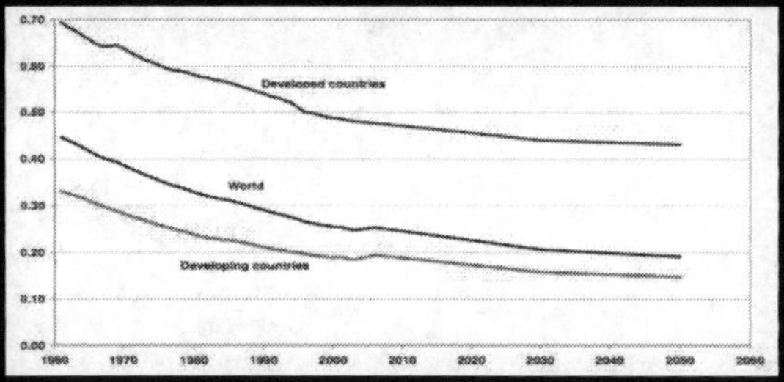

Jelle Bruinsma (2009) FAO

PRO-LAND SHARING

- Agriculture (and bioenergy) production can:
  - Mimic structure of natural habitats
  - Support some biodiversity
  - Provide some ecosystem services
- Patches of mixed production would have less sharp edges/boundaries with natural patches

Photo: Polyculture Design

LAND SHARING AND BIOENERGY PRODUCTION

Wood pellets from forest harvest residues of boreal forests (Sweden)

Photo: SCA Group

Cellulosic biofuels from tallgrass prairie (Wisconsin)

Photo: University of Wisconsin

ANTI-LAND SHARING

- Even diverse production areas may support fewer species, and support fewer rare, endemic, or specialist species
- Far more land is required to produce same amount of bioenergy
- May be difficult to regulate

Density of endemic birds, mammals and amphibians in Central and South America. Image: Clinton Jenkins

ANDERSON-TEIXEIRA ET AL. 2012, *ECOLOGICAL APPLICATIONS*

- Model compares land sharing vs. land sparing and impact on greenhouse gas regulation
- Advantages of land sparing vs. land sharing depend upon type of ecosystem displaced
- Average GHG regulation is maximized through land sparing scenarios
  - *"Whenever bioenergy crops have substantially low ecosystem services than the ecosystems with which they are competing for land, the most effective strategy for meeting bioenergy demand while maximizing ecosystem services on a landscape level is one of land sparing"*
  - Probably a similar case for biodiversity as well
- Caveats: GHG regulation not dependent upon landscape spatial configuration of land use types

HEATON ET AL. 2013, *BIOFUELS BIOPRODUCTS & BIOREFINING*

- Room for intensification in developing countries
  - Higher yields on existing ag land possible
  - Prevent conversion of high diversity natural areas
- Could increase diversity in production areas in developed countries
  - Yield already probably at maximum
  - Increase ecosystem services in existing ag lands

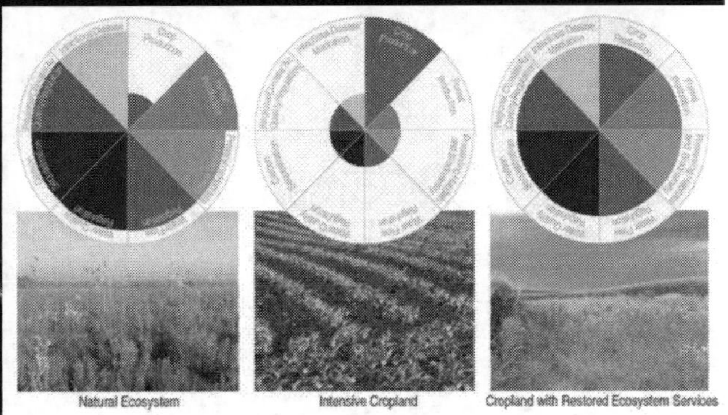

IMMERZEEL ET AL. 2014, *GLOBAL CHANGE BIOLOGY BIOENERGY*

- Reviewed 53 publications on bioenergy and biodiversity
- Lots of discussion on land sparing vs. land sharing, but not much data
- Understanding how bioenergy and biodiversity can coexist must include indicators at both the field and landscape scale
  - Impact of bioenergy production on landscape heterogeneity is key

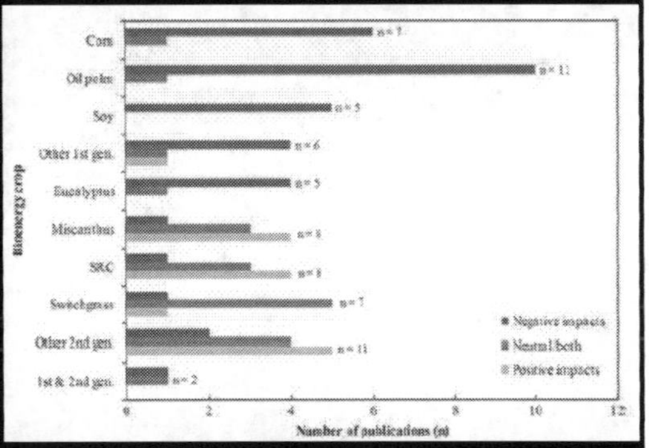

CONCLUSIONS?

- Land sparing vs. land sharing decisions likely to be highly context-specific:
  - Type of biomass
  - Method of harvest
  - Temperate vs. tropical
  - Existing landscape heterogeneity
  - Rare/endemic species
- How do existing policies (e.g., renewable fuels standards) drive land use decisions?
- As always, more research needed

THANK YOU! QUESTIONS?

Integrating Biodiversity Metrics in Life-Cycle Assessment of Intensive Biomass Production in North American Forests:

Challenges and Considerations

Jake Verschuyl (National Council for Air and Stream Improvement; NCASI)
T. Bently Wigley (National Council for Air and Stream Improvement; NCASI)
Caroline Gaudreault (NCASI; Canada Operations)
Manuele Margni (Polytechnique Montreal)
Darren Miller (Weyerhaeuser)
Sam Riffell (MS State)
Brian Titus (Nat. Res. Canada)
Kirsten Vice (NCASI; Canada Operations)

ncasi Environmental research for the forest products industry since 1943

Natural Resources Canada / Ressources naturelles Canada
Canadian Forest Service / Service canadien des forêts

POLYTECHNIQUE MONTRÉAL LE GÉNIE EN PREMIÈRE CLASSE

ATE MISSISSIPPI Forest & Wildlife

Weyerhaeuser

Life Cycle Assessment

Life Cycle Assessment: "compilation and evaluation of the inputs, outputs and the potential environmental impacts of a product system throughout its life cycle" (ISO 2006a, b)

Life Cycle Impact Assessment (LCIA) is one of the phases in the LCA methodology

LCIA is used to assess the relative environmental significance of Life Cycle Inventory results (e.g. resources extracted, releases to air, water, etc.)

LCIA requires quantification of meaningful indicators

Areas of Protection

- Human health
- Natural environment
- Natural resources - (sustainability considerations)
- Man-made structures and ecosystems

Udo de Haes et al. (1999) – SEATAC Europe

Impact Categories for LCIA

Nine commonly applied Impact Categories:

1. Climate change
2. Stratospheric ozone depletion
3. Photo-oxidant formation
4. Acidification
5. Nitrification
6. Human toxicity
7. Ecotoxicity
8. Depletion of abiotic resources (fossil fuels, minerals, etc.)
9. Unsustainable use of biotic resources (wood, fish, etc.)

Others? Land Use?

Levels of Biodiversity

- Ecological diversity (ecosystems and landscapes; abiotic & biotic)

Direct vs. Indirect Biodiversity Indicators

Direct indicators - biological or taxon based

- Indicator species
- Guilds
- Richness
- Presence of functional groups

Indirect indicators - vegetation structure or habitat based

- Often selected in the absence of information on direct indicators
- Usefulness depends on:
  - Whether the relationship is between organisms and their habitat is linear
  - Taxonomic groups being considered, and the correlation of their responses to habitat manipulations

Commonly Used Indicators of Biodiversity Response to Land Use

• Species Richness	Geyer, (2010) Goedkoop and Spriemsma (2001) Koellner and Scholz (2008) Kollner (2000) Lindeijer (2000) Schmidt (2008)
• Species-area relationship	DeSchryver et al. (2010) Koellner and Scholz (2008) Kollner (2003) Schmidt (2008)
• Number of threatened species	Koellner and Scholz (2008) Schmidt (2008) Lenzen (2009)
• Fragmentation	Jordaan et al. (2009) Schenk (2001)

Challenges Associated with Intrinsic Complexities of Biodiversity

- Consideration of scale largely absent
- Most data are geographically or taxonomically restricted
- Disproportionate focus on indicators that reflect changes in species richness; ignoring the other levels of biodiversity
- Indicators often miss functional and structural attributes of biodiversity
- Invasive species and overexploitation (drivers of biodiversity) are often overlooked

(Curran et al. 2011)

Challenges Associated with Assessment of Biodiversity Effects in LCA Framework

- Existing LCIA methods developed for Europe often using plant species richness as the primary indicator
- Reaction of plants to land use change may not be indicative of the other 98% of terrestrial species
- May approximate some land use impacts on biodiversity but are not intended to provide site specific guidance (Baan et al. 2012)

Commonly Selected Indicators of Forest Biodiversity

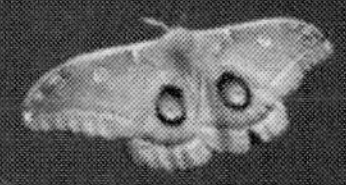
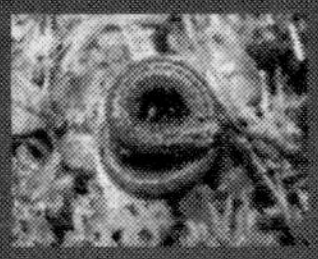

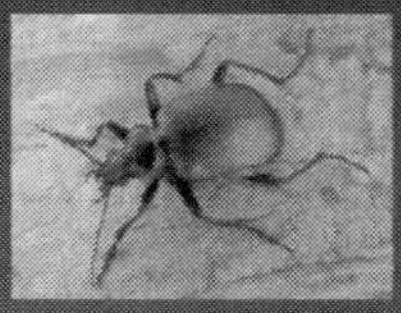

- Direct (taxonomic groups)
  - Birds
  - Cavity nesting birds
  - Small mammals
  - Lepidoptera
  - Carabid beetles
  - Terrestrial salamanders
- Indirect
  - Habitat Connectivity/Fragmentation
  - Stand age - structure
  - Selected or "rare" habitat conditions (e.g. late seral)
  - Climax (undisturbed) plant communities

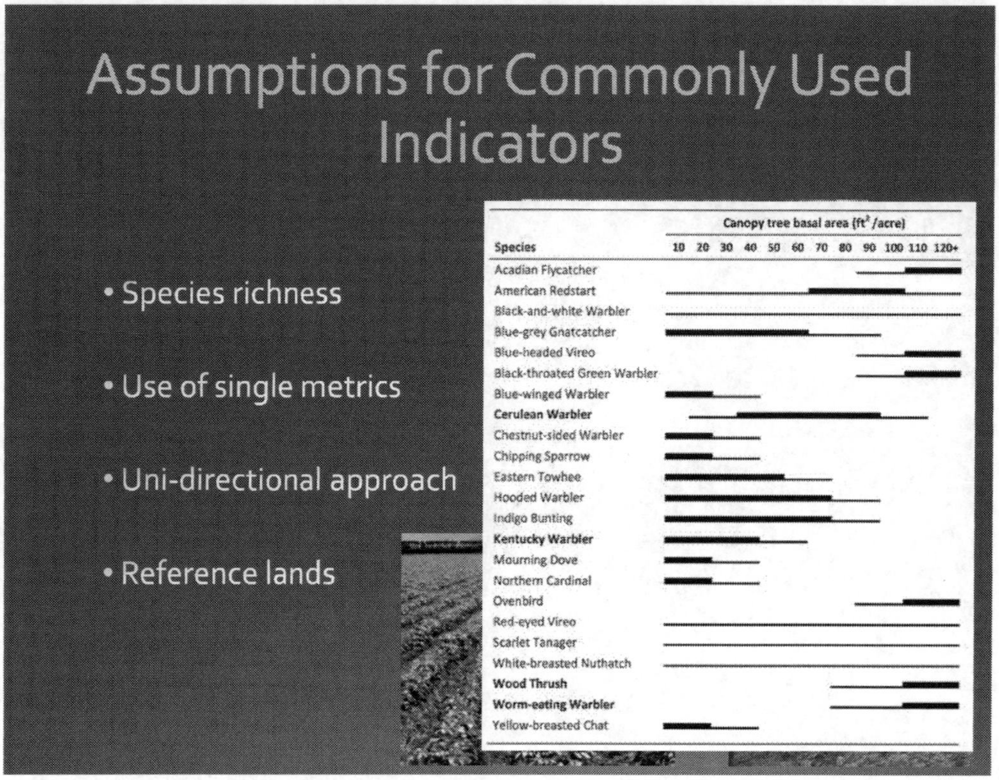
Assumptions for Commonly Used Indicators
• Species richness
• Use of single metrics
• Uni-directional approach
• Reference lands
Canopy tree basal area (ft²/acre)
Species
10 20 30 40 50 60 70 80 90 100 110 120+
Acadian Flycatcher
American Redstart
Black-and-white Warbler
Blue-grey Gnatcatcher
Blue-headed Vireo
Black-throated Green Warbler
Blue-winged Warbler
Cerulean Warbler
Chestnut-sided Warbler
Chipping Sparrow
Eastern Towhee
Hooded Warbler
Indigo Bunting
Kentucky Warbler
Mourning Dove
Northern Cardinal
Ovenbird
Red-eyed Vireo
Scarlet Tanager
White-breasted Nuthatch
Wood Thrush
Worm-eating Warbler
Yellow-breasted Chat

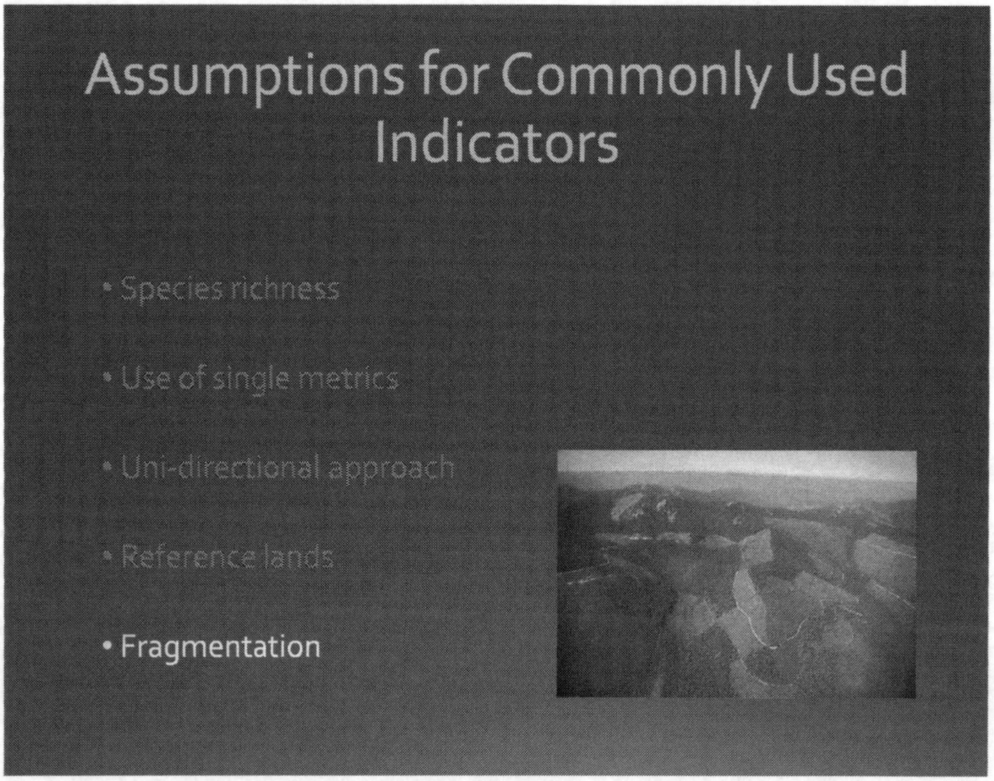
Assumptions for Commonly Used Indicators
• Species richness
• Use of single metrics
• Uni-directional approach
• Reference lands
• Fragmentation

Fragmentation is a Complex Concept

- "Fragmentation" is defined and measured in different ways
- Habitat area may mean more than habitat configuration
- "Habitat" is a species-specific concept
- In managed forests, habitat change is not permanent

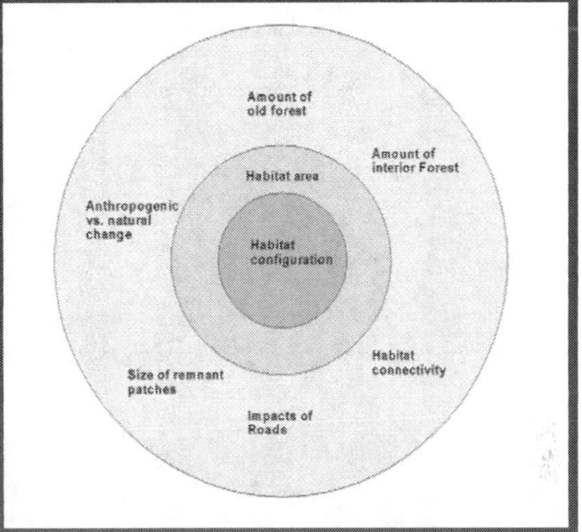

Source: Wedeles and Sleep. 2008. NCASI TB 959

Fragmentation is a Complex Concept

- Site productivity can influence effects of disturbance and habitat change on species
- Species mobility will also play a role in response

Implications for Biomass Production Systems

Biodiversity response to forest biomass harvest will be dependent on:

- Biomass production system
- Landscape context
- Geographic scale of harvest
- Frequency and intensity of harvest
- Structure of wildlife communities present
- Species' life history traits

Meta-analyses of Biodiversity Response to Biomass Harvesting

Removal of woody debris

Thinning and fuels treatments

Short-rotation woody crops

Intercropping biomass crops on existing forest lands

Biodiversity Response to Removal of Harvest Residues

Existing studies
- Have not been conducted in all geographies
- Some taxa are under-represented
- Many have not have involved actual biomass harvests

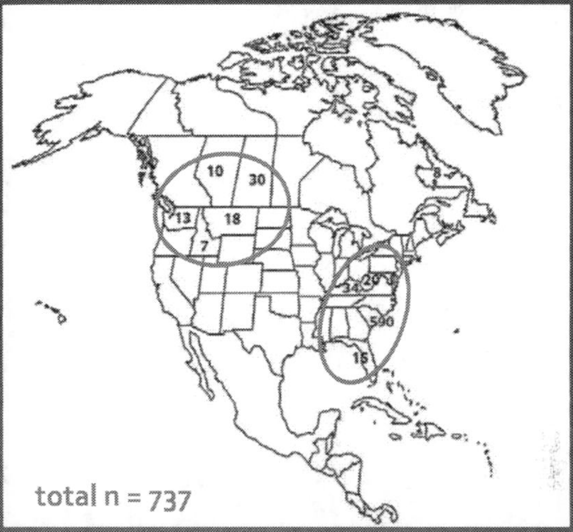

Source: Riffell et al. 2011 Forest Ecology and Management

Removal of Harvest Residues

Key findings
- Birds are likely to respond negatively
- Response may be driven by lower invertebrate biomass
- Little evidence of strong responses by other taxa

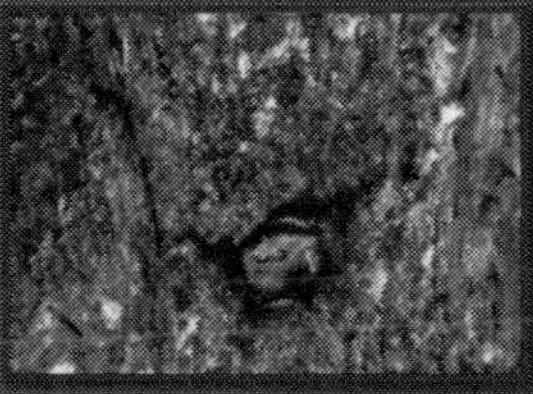

Information Needs
- Better geographic coverage of research
- How much woody debris is left after biomass harvests?
- How do DCWD, FWD, and litter interact?
- How does response vary with landscape context?

18

Biodiversity Response to Forest Thinning

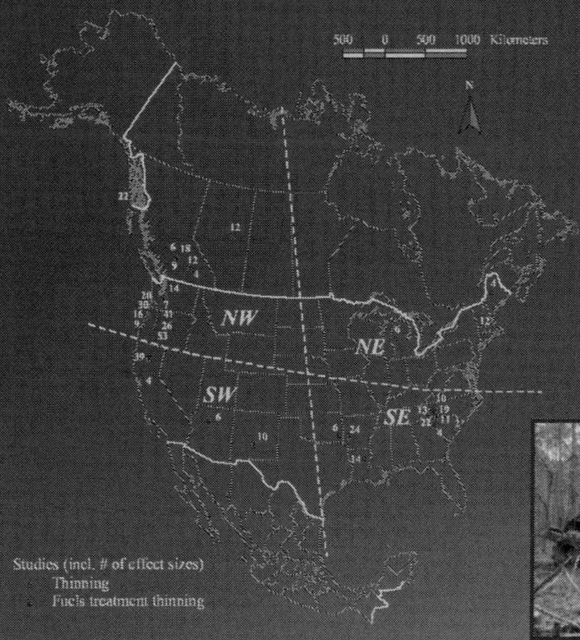

- Precommercial
- Commercial
- Fuels-treatments

- 33 studies
- 505 individual effect sizes
- 54% of effect sizes for birds
- 17 of 33 studies on mammals

Forest Thinning

Key findings

- Most taxa appear to respond positively
- Species of high conservation priority may warrant special attention
- Biodiversity response likely depends on:
  - Thinning intensity
  - Time since thinning
  - Landscape factors

Information Needs

- Only 3 studies were related to reptiles, all from SE
- Response of reproductive success
- More data about community-level response and response by high-priority species

20

Intercropping Biomass Crops

Limited research suggests

- Diversity in switchgrass may be
  - Higher than for row crop agriculture
  - Lower than for mixed native warm season grasses
- Weak positive correlations between row spacing and diversity measures
- Positive responses by some early-seral species

Information Needs

- Influences on biodiversity of
  - Stand and landscape characteristics
  - Timing of harvests
  - Other intercropped species
- Can intercropping extend period of early-seral habitat conditions?

21

Short-rotation Woody Crops

Key findings

- Studies typically compare poplar / cottonwood plantations to mature hardwood forests
- Reported fauna characteristic of these two seral classes
- Few studies or regions

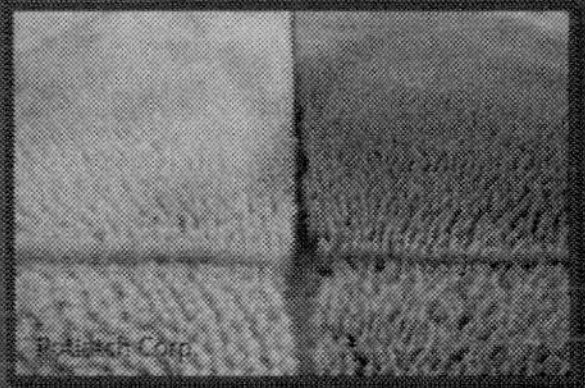

Information Needs

- Better taxonomic and geographic coverage
- Data on responses throughout the entire rotation and across multiple spatial scales
- Comparisons with multiple reference types
- Bird nesting success – not just abundance

22

Summary

Important to consider indicator quality and assumptions when assessing biodiversity response to land use within LCA

Existing literature indicates:

- Negative effects of forest residue harvest on birds; neutral for other taxonomic groups
- Positive effects of thinning for most taxonomic groups
- Response to SRWC and intercropping is uncertain

Improved information about biodiversity response will inform

- Decisions about ongoing management
- Discussions about the need for and nature of incremental improvements in sustainability guidance

23

- Riffell, S., J. P. Verschuyl, D. Miller, T. B. Wigley. 2011. Biofuel harvests, coarse woody debris, and biodiversity – A meta-analysis. Forest Ecology and Management 261:878-887.
- Verschuyl, J. P., S. Riffell, D. Miller, T. B. Wigley. 2011. Biodiversity Response to intensive biomass production from forest thinning in North American forests – A meta-analysis. Forest Ecology and Management 261:221–232
- Riffell, S., J. P. Verschuyl, D. Miller, T. B. Wigley. 2011. Biodiversity response to short-rotation woody crops. Global Change Biology Bioenergy 3:313–321.
- Riffell, S., J. P. Verschuyl, D. Miller, T. B. Wigley. 2012. Potential biodiversity response to intercropping herbaceous biomass crops on forest land. Journal of Forestry 110(1): 42-47

Questions?

1

The Sustainability of Forest Residue for Bioenergy in Canada: What can biodiversity tell us?

Venier, L.A., Aubin, I., Webster, K. Fleming, R., Hazlett, P., Titus, B.

Natural Resources Canada   Ressources naturelles Canada

Canada

2

Forest Residue

Tree-length Harvest

Residue Roadside Piles

Natural Resources Canada   Ressources naturelles Canada

Canada

3

Interest in using forest biomass is high

- Reduction in greenhouse gas emissions
- New jobs in energy and forestry sector
- Diversification of Canadian energy portfolio

Natural Resources Canada Ressources naturelles Canada

Canada

4

Context

- **A National Scan of Regulations & Practices Relevant to Biomass Harvesting** 2010. World Wildlife Fund- Forest Products Association of Canada.
- **eNGO and Conservation Group Outreach on Biomass: Position and rationale regarding the use of biomass for electricity/heat production.** Dagg et al. 2011 Pembina institute.
- **Fuelling a BioMess: Why burning trees for energy will harm people, the climate and forests.** 2011. Greenpeace.

Natural Resources Canada Ressources naturelles Canada

Canada

Dagg et al. 2011

"It is the opinion of many organizations that forest residue is not an acceptable biomass resource.
This material is essential for biodiversity, wildlife habitat, soil fertility and forest productivity"

Estimates of CWD volume

Stand	Managed (m^3/ha)	Unmanaged (m^3/ha)
Ontario (upland conifer) (Parton 2013)	15.21 (21-50%)	30.3, 34.1, 73.4
Ontario (Pj) (Parton 2013)	9.81 (47%)	20.8
BC (Densmore 2011)	78.8-167% of unmanaged	
BC (Densmore 2011)	4.4-80 % of CWD large pieces of unmanaged	
Sweden (Fridman and Walheim 2000)	0.5-11 (2-15%)	11-91
Fennoscandia(1-140 yrs) (Siitonen 2001)	2-10 (2-17%)	60-90
Fennoscadina >140 yrs (Siitonen 2001)	15.9 (18-27%)	60-90

Biodiversity and deadwood

- Northern Europe has hundreds of red listed species associated with deadwood while Canada does not
- Quantity and quality is important
- Examples
  - High bryophyte and rove beetle diversity in late decay logs (Cole et al. 2008, Buddle et al. 2006)
  - Insect species composition changes with decay class (Vanderwel et al. 2006)
  - Decay class and log diameter are important predictors of suitable foraging substrate for Black-backed Woodpeckers (Tremblay et al. 2009)

Review of some Canadian Literature…

Mammals

JM

Journal of Mammalogy, 93(6):1456–1468, 2012

If we build habitat, will they come? Woody debris structures and conservation of forest mammals

Thomas P. Sullivan,* Druscilla S. Sullivan, Pontus M. F. Lindgren, and Douglas B. Ransome

- Red-backed voles and weasels were more abundant in piles.
- Removal of deadwood limits management options for mammals.

Birds

Research Papers

Habitat Requirements of Breeding Black-Backed Woodpeckers (*Picoides arcticus*) in Managed, Unburned Boreal Forest

Besoins en termes d'habitat chez le Pic à dos noir (*Picoides arcticus*) nichant en forêt boréale non brûlée et sous aménagement

Junior A. Tremblay [1], Jacques Ibarzabal [2], Christian Dussault [1], and Jean-Pierre L. Savard [3]

Avian Conservation and Ecology 4(1): 2 online http://www.ace-eco.org/vol4/iss1/art2/

- Threshold of $35m^3$/ha of deadwood
- $15m^3$/ha early decay class
- Majority of foraging on recently dead snags

Natural Resources Canada Ressources naturelles Canada

Canada

Macro-arthropods

Stand composition and structure of the boreal mixedwood and epigaeic arthropods of the Ecosystem Management Emulating Natural Disturbance (EMEND) landbase in northwestern Alberta[1]

Timothy T. Work, David P. Shorthouse, John R. Spence, W. Jan A. Volney, and David Langor

Canadian Journal of Forest Research 34: 417-430 (2004)

Forest Ecology and Management 321 (2014) 19-28

Contents lists available at ScienceDirect

Forest Ecology and Management

journal homepage: www.elsevier.com/locate/foreco

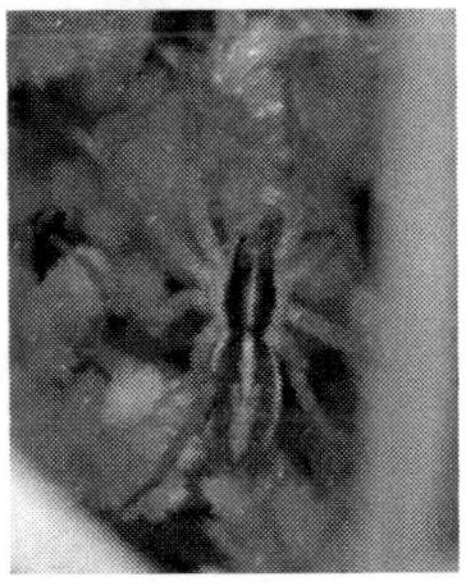

Reductions in downed deadwood from biomass harvesting alter composition of spiders and ground beetle assemblages in jack-pine forests of Western Quebec

Timothy T. Work [a,c,*], Suzanne Brais [b,c], Brian D. Harvey [b,c]

Natural Resources Canada Ressources naturelles Canada

Canada

Macro-arthropods

ZooKeys 258: 31–52 (2013)
doi: 10.3897/zookeys.258.4174
www.zookeys.org

RESEARCH ARTICLE

ZooKeys

Initial responses of rove and ground beetles (Coleoptera, Staphylinidae, Carabidae) to removal of logging residues following clearcut harvesting in the boreal forest of Quebec, Canada

Timothy T. Work[1], Jan Klimaszewski[2], Evelyne Thiffault[2], Caroline Bourdon[2], David Paré[2], Yves Bousquet[3], Lisa Venier[4], Brian Titus[5]

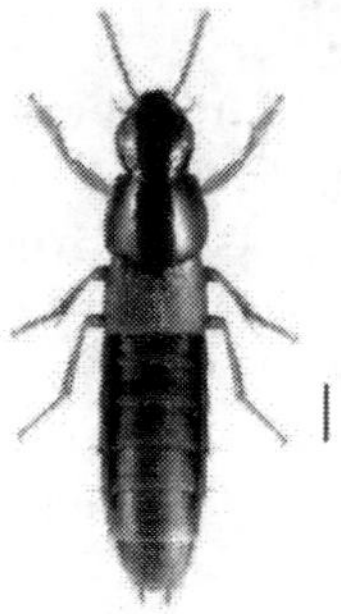

Quedius labradorensis

ELSEVIER

Available online at www.sciencedirect.com

SCIENCE DIRECT

Forest Ecology and Management 225 (2006) 190–199

Forest Ecology and Management

www.elsevier.com/locate/foreco

Insect community composition and trophic guild structure in decaying logs from eastern Canadian pine-dominated forests

Mark C. Vanderwel *, Jay R. Malcolm, Sandy M. Smith, Nurul Islam

Natural Resources Canada / Ressources naturelles Canada

Canada

Macro-arthropods

esa

ECOSPHERE

Estimating species loss of saproxylic insects under scenarios of reduced coarse woody material in eastern boreal forests

Timothy T. Work[1,2,†] and Annie Hibbert[2]

Volume 2(4) Article 41

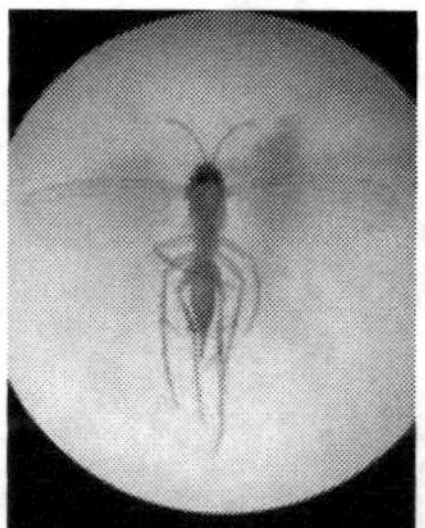

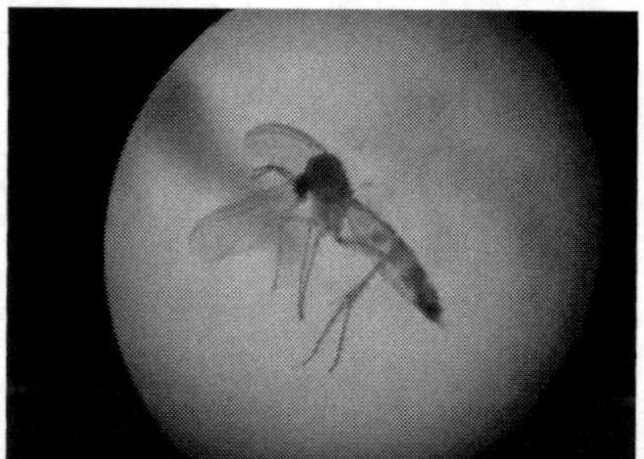

Natural Resources Canada / Ressources naturelles Canada

Canada

13

Microbes

The ISME Journal (2012) 6, 2199–2218
© 2012 International Society for Microbial Ecology All rights reserved 1751-7362/12
www.nature.com/ismej

ORIGINAL ARTICLE

Significant and persistent impact of timber harvesting on soil microbial communities in Northern coniferous forests

This article has been corrected since Advance Online Publication and a corrigendum is also printed in this issue.

Martin Hartmann[1,7], Charles G Howes[1], David VanInsberghe[1], Hang Yu[1,8], Dipankar Bachar[2,3], Richard Christen[2,3], Rolf Henrik Nilsson[4,5], Steven J Hallam[1,6] and William W Mohn[1]

More than a decade after harvesting, diversity and structure of soil fungal communities remained significantly altered by different levels of organic material removal but differences were small.

Natural Resources Canada Ressources naturelles Canada

Canada

14

LTSP design and intensive forest biomass removals

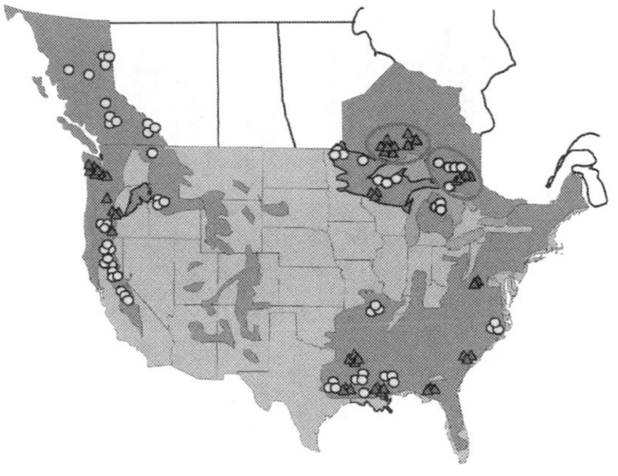

Powers 2006. Long-term soil productivity: genesis of the concept and principles behind the program. CJFR

- Conceptual model - organic matter a major variable regulating soil processes affecting productivity
- Modification of site organic matter is a main effect treatment
- Long-term forest growth measurements
- Pre- and post-harvest measurements of site C and nutrient pools that enable accurate determination of site removals and retention

Natural Resources Canada Ressources naturelles Canada

Canada

15

LTSP treatments

- Targeted infertile coarse textured and shallow soils
- 14 sites, mature 60-125 year-old boreal stands – Pj and Sb
- Harvest treatments tree-length (TL), full-tree harvest (FT), full-tree harvest and forest floor removal (FFR) – 30m X 30m plots X 3 replicates/site
- Sites replanted – NE Pj, NW Sb

Natural Resources Canada Ressources naturelles Canada

Canada

Mature Control

Tree Length Harvest

Whole Tree Harvest

Forest Floor Removal (Bladed)

Natural Resources Canada Ressources naturelles Canada

Canada

17

LTSP Biodiversity work

- Mohn et al. Soil microbial communities; University of British Columbia
- Berch et al. Soil fauna (collembola and mites) BC Ministry of Forests
- Venier, Rousseau et al. Soil fauna (collembola and mites) Canadian Forest Service and University of Quebec at Montreal

Natural Resources Canada Ressources naturelles Canada

Canada

18

Dead wood as a surrogate for biodiversity

- British Columbia Forest and Range Evaluation Program
- Data collected from 2006-2009
- Comparison of dead wood in uncut retention patches and harvested areas in 3 forest regions (18 biogeoclimatic subzones)
- Volume of CWD left on harvested sites was similar or higher than within retention patches
- Density of large pieces of CWD (>=20 cm diameter and >=10 m long) is significantly lower
- No direct measures of biodiversity

Natural Resources Canada Ressources naturelles Canada

Canada

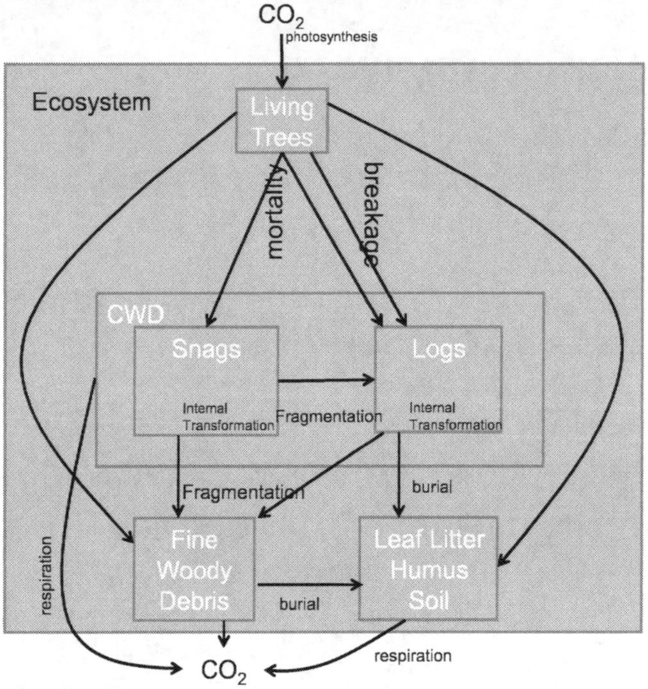

Natural Resources Canada   Ressources naturelles Canada

Harmon et al 1986

Canada

Collaborative science: multi partnership; Island Lake and Forêt Montmorency

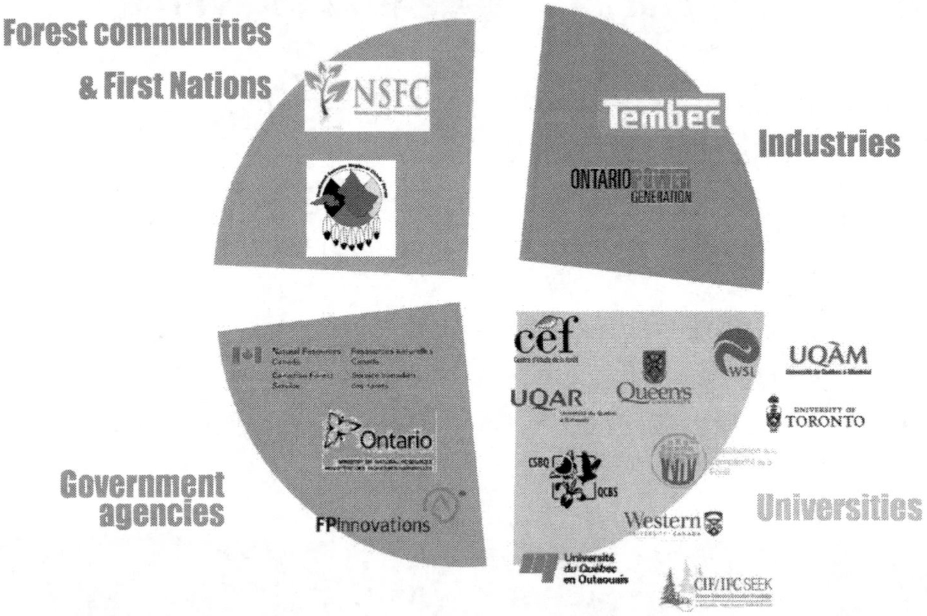

Natural Resources Canada   Ressources naturelles Canada

Canada

21

Natural Resources Canada   Ressources naturelles Canada   Canada

22

Natural Resources Canada   Ressources naturelles Canada   Canada

23

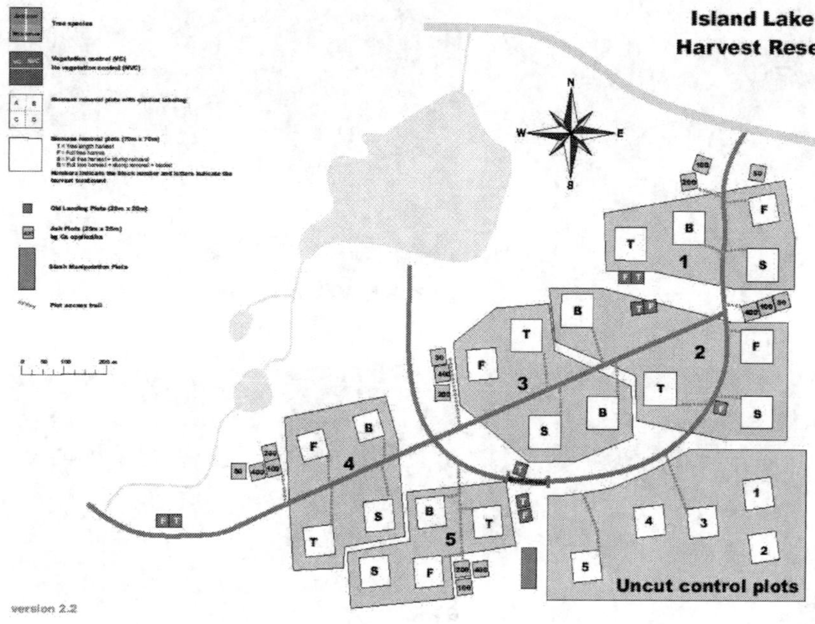

Natural Resources Canada / Ressources naturelles Canada

Canada

Island Lake Biomass Research and Demonstration Area

Island Lake post-harvest coarse wood (all aboveground > 5 cm diameter)

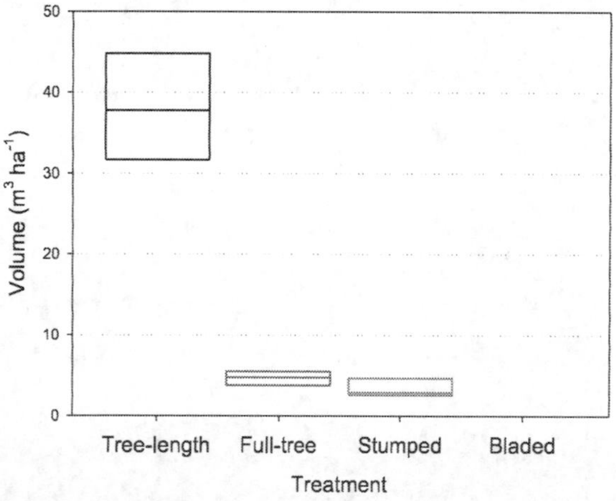

Natural Resources Canada / Ressources naturelles Canada

Canada

Natural Resources Canada
Ressources naturelles Canada
Canada

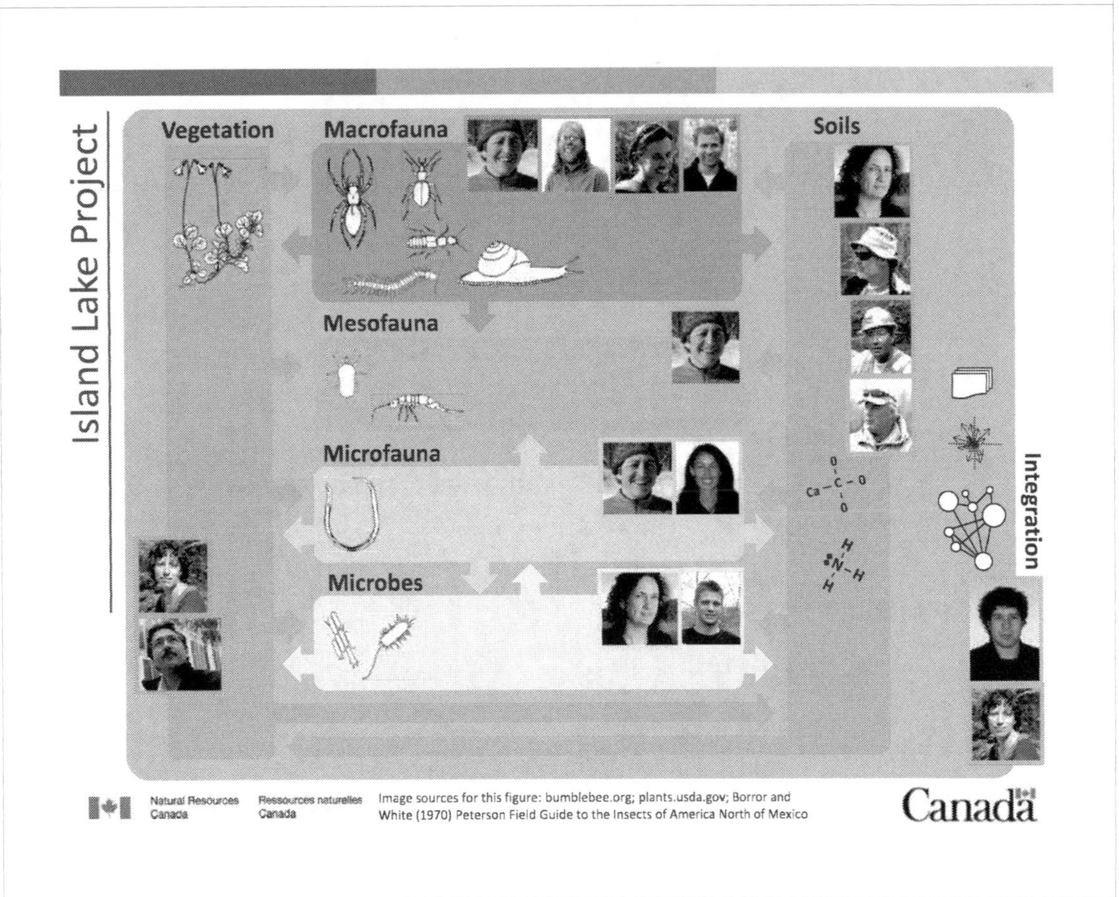
Island Lake Project
Vegetation
Macrofauna
Soils
Mesofauna
Microfauna
Microbes
Integration
Natural Resources Canada
Ressources naturelles Canada
Image sources for this figure: bumblebee.org; plants.usda.gov; Borror and White (1970) Peterson Field Guide to the Insects of America North of Mexico
Canada

Multi-trophic scale project

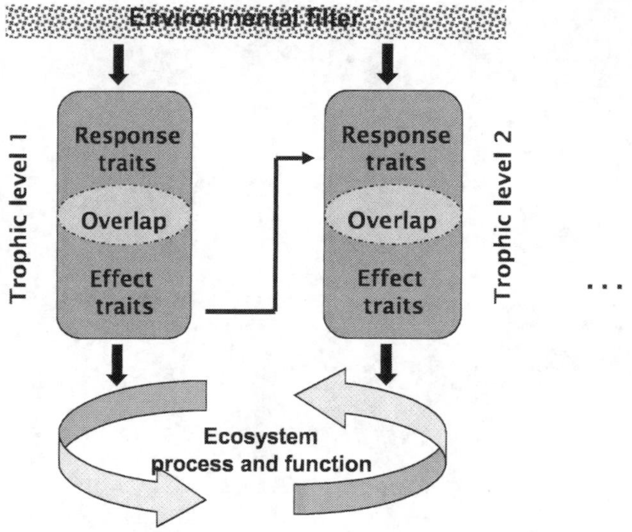

Lavorel & Garnier 2002. Funct. Ecol.
Moretti et al. submitted

Natural Resources Canada Ressources naturelles Canada

Canada

Conclusions

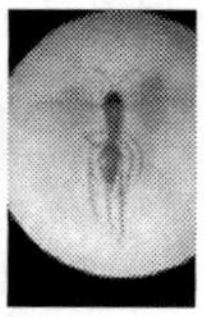

- Work on sustainability of biomass removal based on biodiversity is in early stages in Canada especially relative to work on soil and tree productivity.
- There is much interest in thresholds of biomass removal or the corollary of how much biomass needs to be left on site to be sustainable. Studies to address this question are preliminary to date but ongoing.

Natural Resources Canada Ressources naturelles Canada

Canada

Conclusions

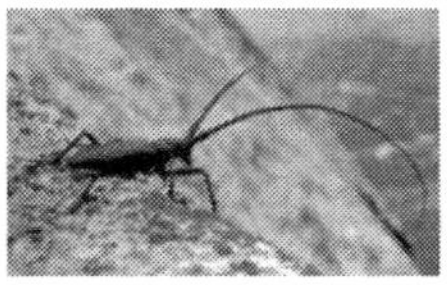

29

- Most work to date has looked at short-term impacts. There are some opportunities with the LTSP network to examine longer – term impacts but only for species that operate at very small scales.
- Biomass inventories are being suggested and used as surrogates for biodiversity because of the difficulty of monitoring biodiversity. A related idea is the modelling of deadwood to provide long-term predictions of biomass availability relative to forest type and disturbance or harvesting approaches.
  - Direct links between biodiversity and deadwood quantity and quality will still need to be made using experimental approaches

Natural Resources Canada Ressources naturelles Canada

Canada

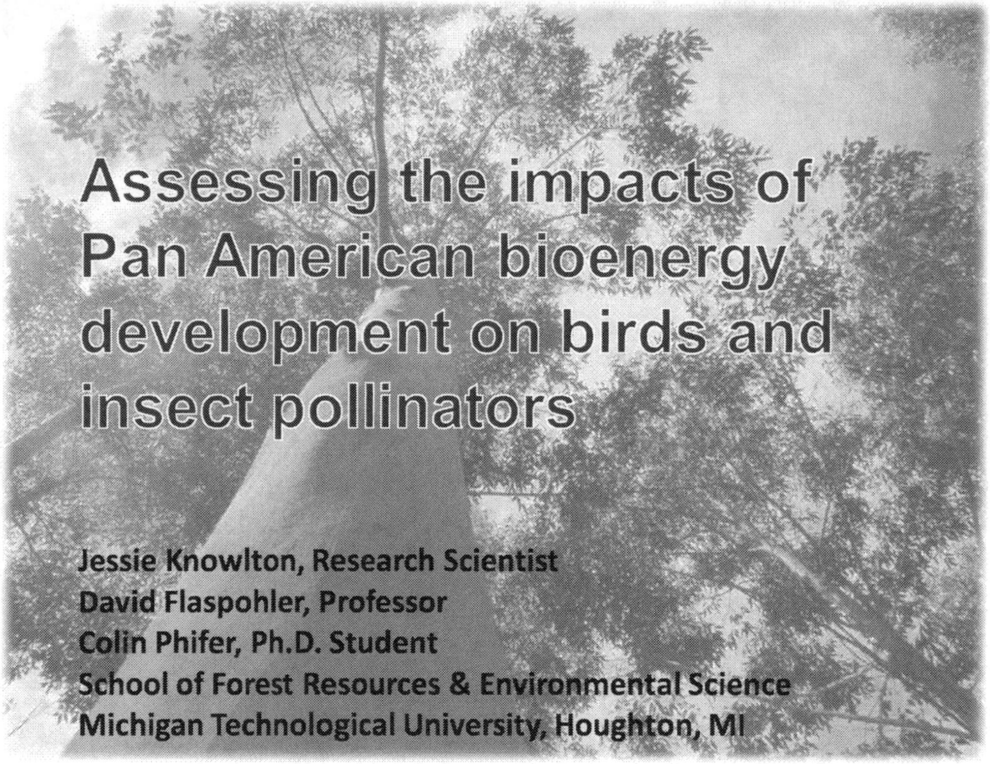

Assessing the impacts of Pan American bioenergy development on birds and insect pollinators

Jessie Knowlton, Research Scientist
David Flaspohler, Professor
Colin Phifer, Ph.D. Student
School of Forest Resources & Environmental Science
Michigan Technological University, Houghton, MI

Partnerships in International Research & Education (PIRE) Sustainability, Ecosystem Services, and Bioenergy Development across the Americas

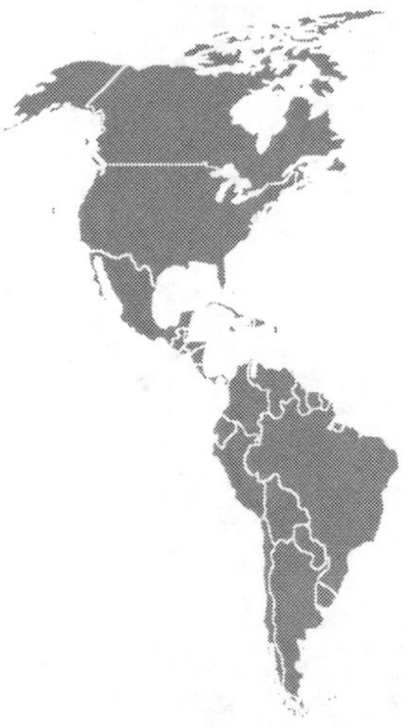

Research Question:

How will Pan American bioenergy development impact socio-ecological systems, and associated ecosystem services, and how can those impacts best be measured, modeled, and mitigated?

**Partnerships in International Research & Education (PIRE)
Sustainability, Ecosystem Services, and
Bioenergy Development across the Americas**

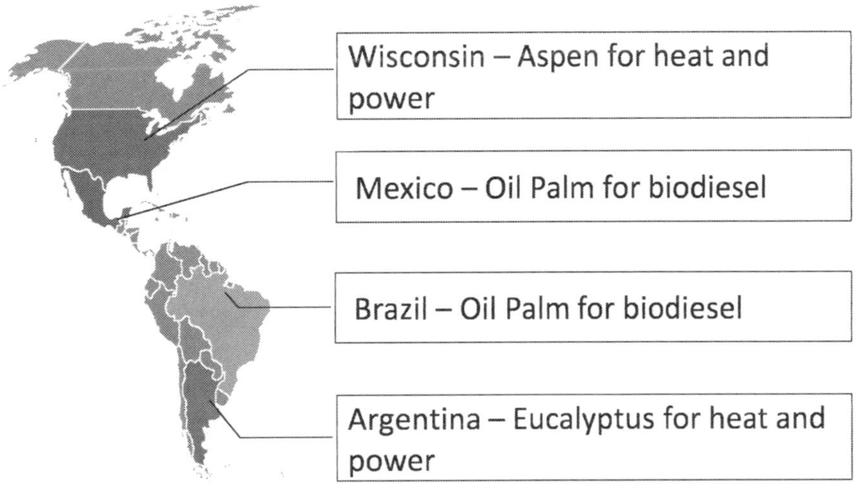

**Partnerships in International Research & Education (PIRE)
Sustainability, Ecosystem Services, and
Bioenergy Development across the Americas**

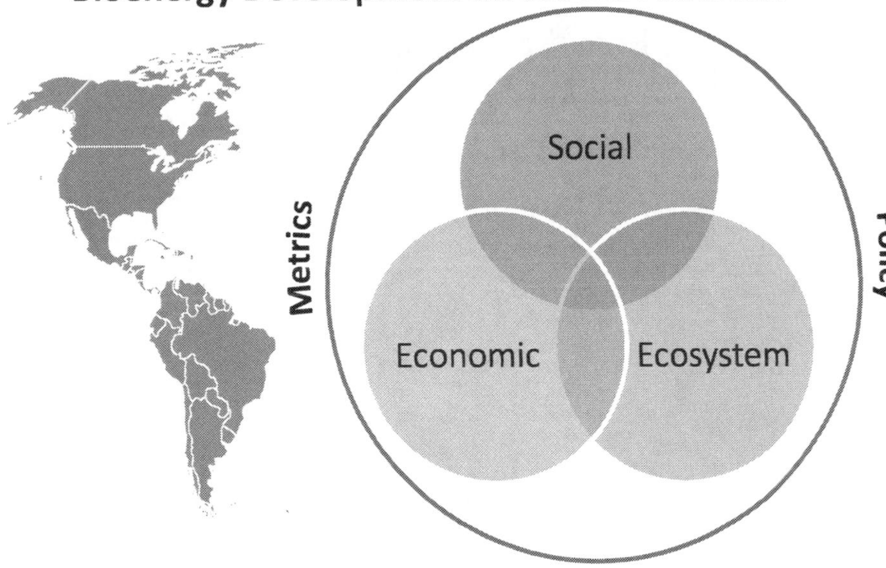

Partnerships in International Research & Education (PIRE)
Sustainability, Ecosystem Services, and
Bioenergy Development across the Americas

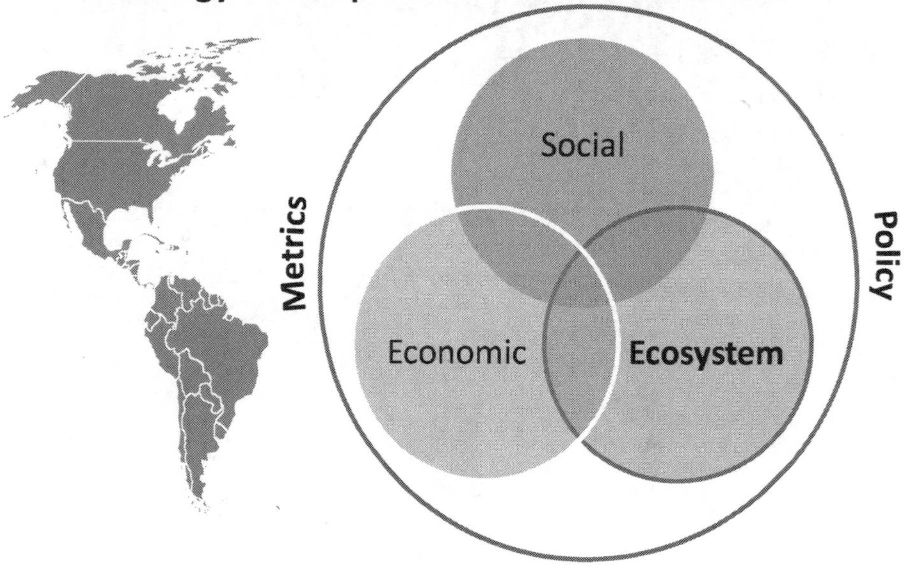

How will bioenergy development affect ecological systems?

1) Identify broad patterns of ecosystem response to land use change associated with bioenergy production systems
2) Identify how local conditions (e.g. climate, soils) affect ecosystem responses to bioenergy production
3) Assess tradeoffs between ecosystem services provisioned by alternative land use systems

How will bioenergy development affect ecological systems?

1) Identify broad patterns of ecosystem response to land use change associated with bioenergy production systems
2) Identify how local conditions (e.g. climate, soils) affect ecosystem responses to bioenergy production
3) Assess tradeoffs between ecosystem services provisioned by alternative land use systems

How will bioenergy development affect ecological systems?

1) Identify broad patterns of ecosystem response to land use change associated with bioenergy production systems
2) Identify how local conditions (e.g. climate, soils) affect ecosystem responses to bioenergy production
3) Assess tradeoffs between ecosystem services provisioned by alternative land use systems

How will bioenergy development affect ecological systems?

Quantify how bioenergy production alters:

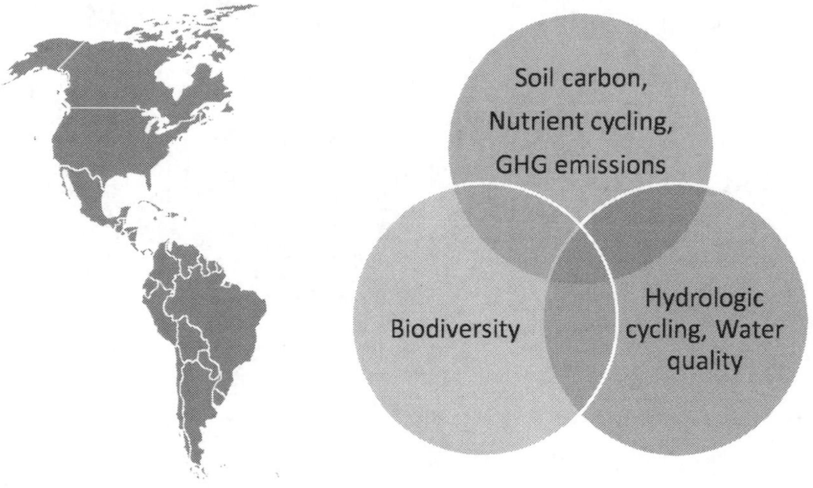

Experimental Design

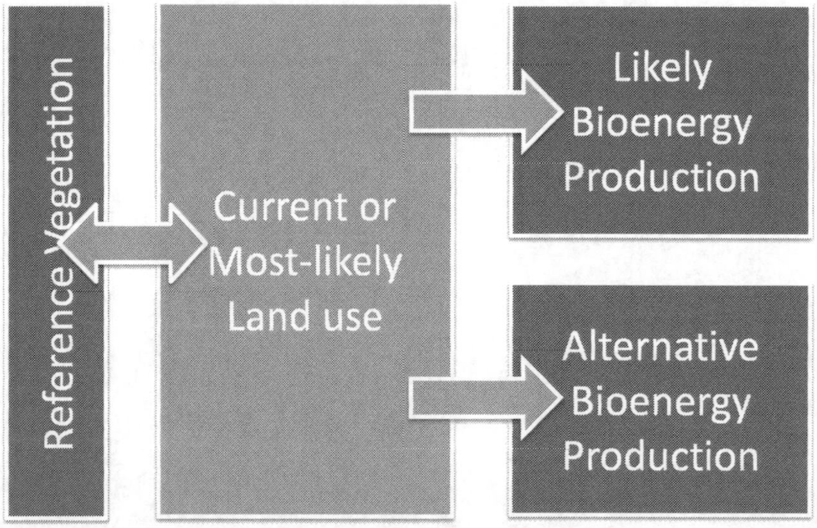

Argentina: Large plantations

Citrus plantation
Eucalyptus plantation
Blueberries
Grazing
Argentina: Mixed Use

Argentina: Pasture

Argentina: El Palmar National Park (Reference)

Villahermosa, Tabasco

Brazil

- Vegetation type, land use and deforestation
- Availability of areas according to agro-ecological zoning
- Land ownership structure

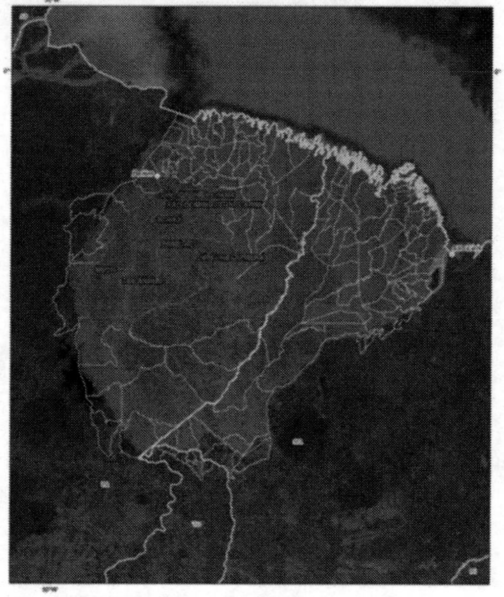

How will bioenergy development affect ecological systems?

Quantify how bioenergy production alters:

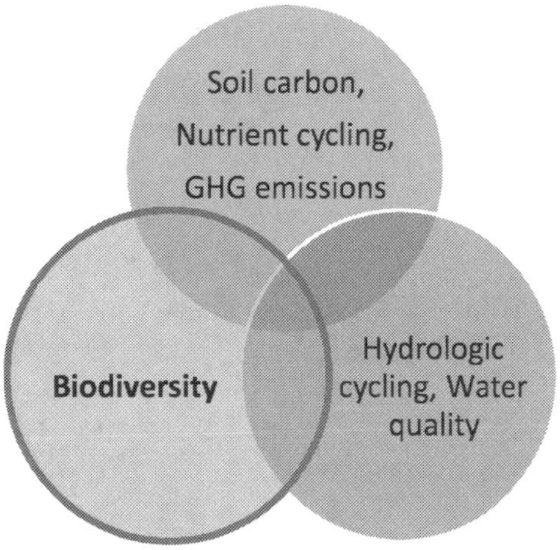

Biodiversity: Birds and Insects

- Cost and time effective
- Sensitive to habitat change
- Indicator species
- Widely studied
- Important ecosystem service providers

Biodiversity Research Objective

Examine the impacts of both **local** and **landscape** scale factors in production systems on **birds** and **insects** and the **ecosystem services** they provide

Argentina: Local scale

Landscape Scale

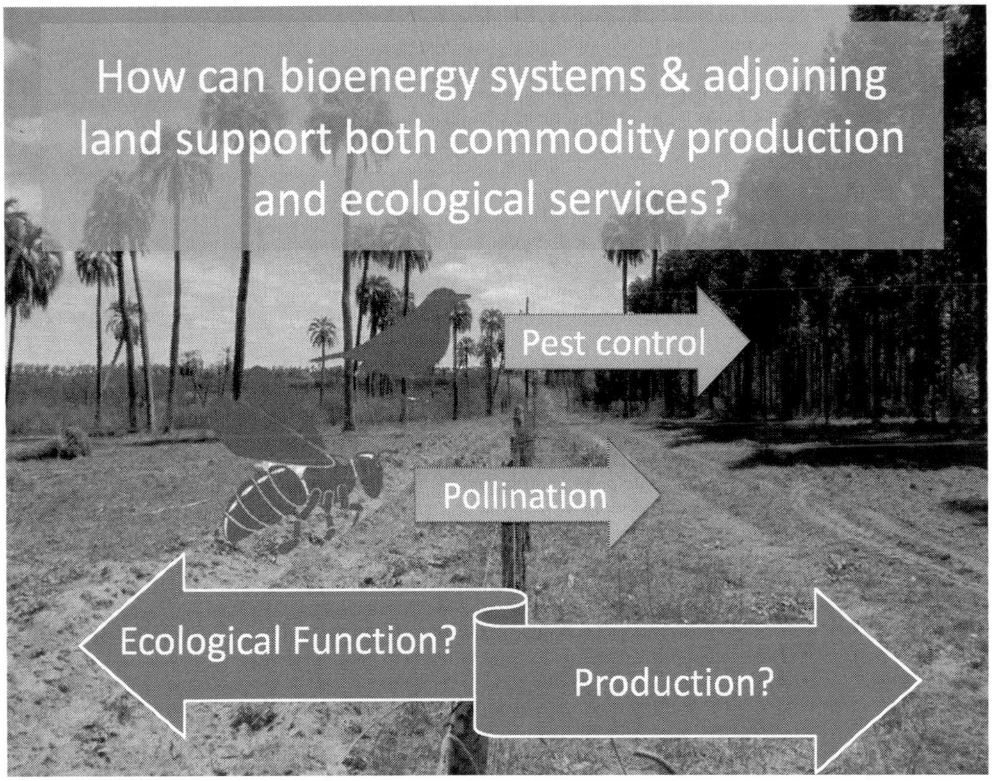

Biodiversity Research Objective

Examine the impacts of both **local** and **landscape** scale factors in production systems on **birds** and **insects** and the **ecosystem services** they provide

In each country and land-use type:

1) Which species are present?
2) How do species use modified landscapes?
3) What ecosystem services are provided?

Biodiversity Design

- Replicated 5 km^2 landscapes, each consisting of >80% of one of the following:
  - 1) Plantation
  - 2) Cattle pasture (most likely alternative)
  - 3) Mixed cropland/small plantation
  - 4) Natural vegetation

Three replicates each, for a total of 12 study landscapes/country

Which species are present?

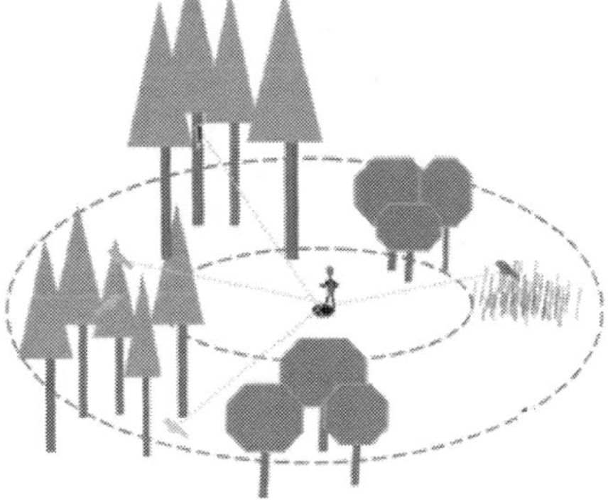

Graphic: USGS PWRC

Which species are present?

- Bee bowl sampling
- Blue vane traps
- Active netting
- Canopy traps

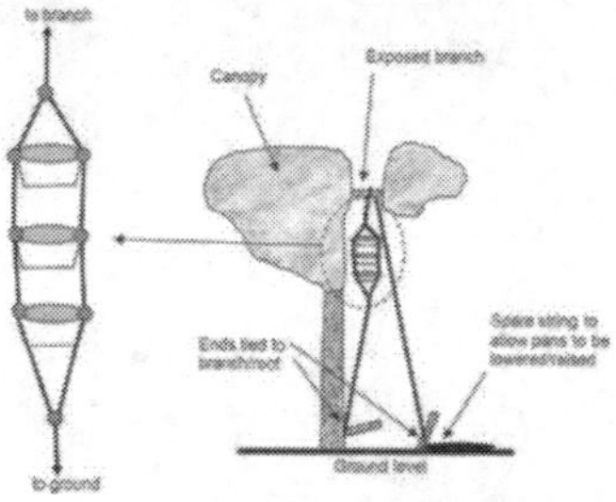

Nuttman et al. 2011

Expected Results

- Species richness
- Density and/or Abundances of individuals
- Functional group diversity
- Rare and threatened species presence
- Habitat specialists vs. generalists

How do species use these modified landscapes?

Gilles et al. 2011

Hagen et al 2011.

Expected Results

- Home range habitat composition and sizes
- Measure of habitat use and preference
- Barriers to movement

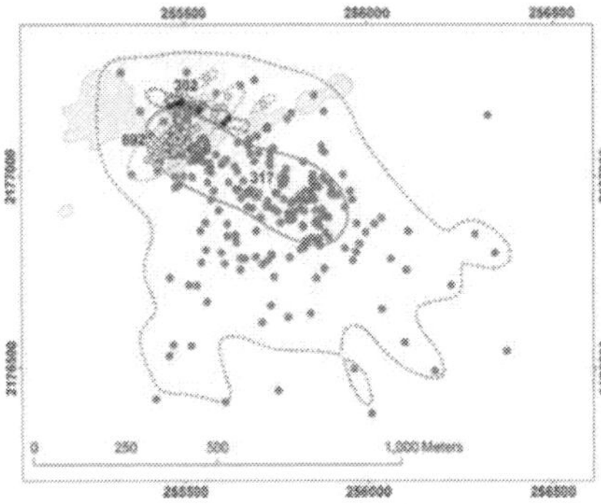

Wu et al. 2014

What ecosystem services are provided?

- Experimental approachs (local level)
  - Adjacent crop pollen limitation
  - Bird pest removal

(exclosures)

D. Karp

Expected Results

- Are the other crops in the landscape pollination limited?
- Do birds provide a pest removal service to crops in the landscape?

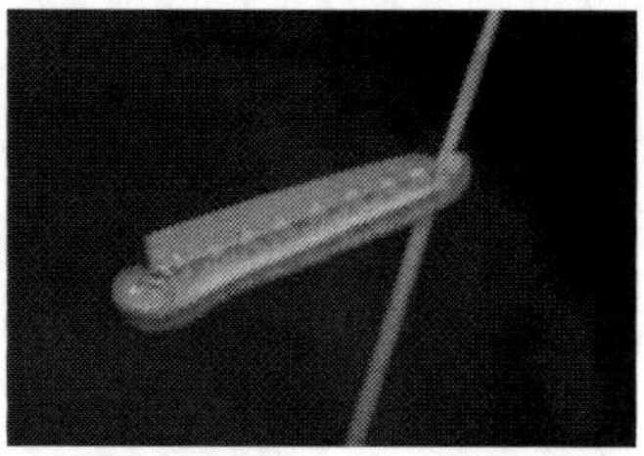

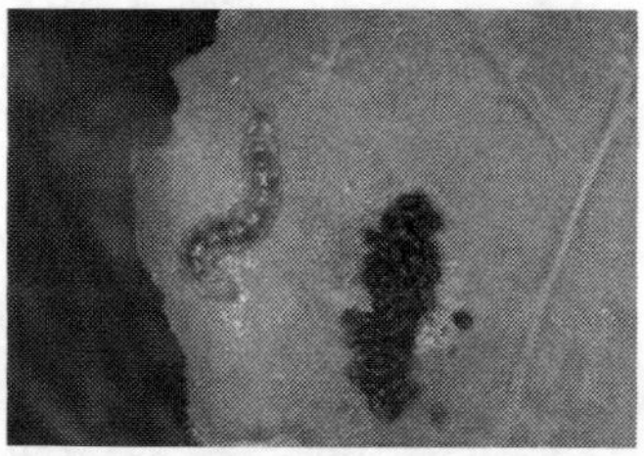

What ecosystem services are provided?

- Modeling approach (landscape level): InVEST
  - How changes in ecosystems are likely to lead to changes in benefits that flow to people
- Inputs:
  - Habitat quality and quantity
  - Pollinator model
    - Species and abundances of pollinators
    - Flight ranges

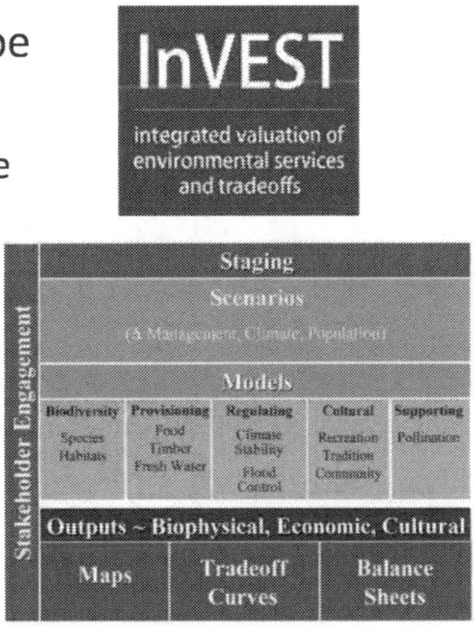

The Natural Capital Project, Stanford

Expected Results

- Under future land use change scenarios:
  - Levels of biodiversity protection
  - Ecosystem services
- Integrate with the carbon and water ecosystem PIRE subteams and the socio-economic and policy teams

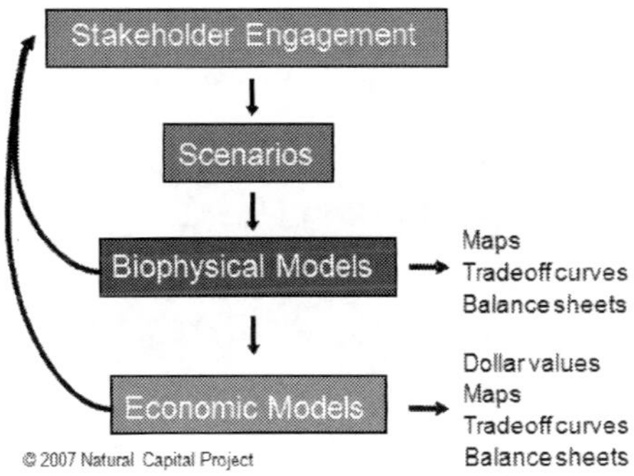

Integration

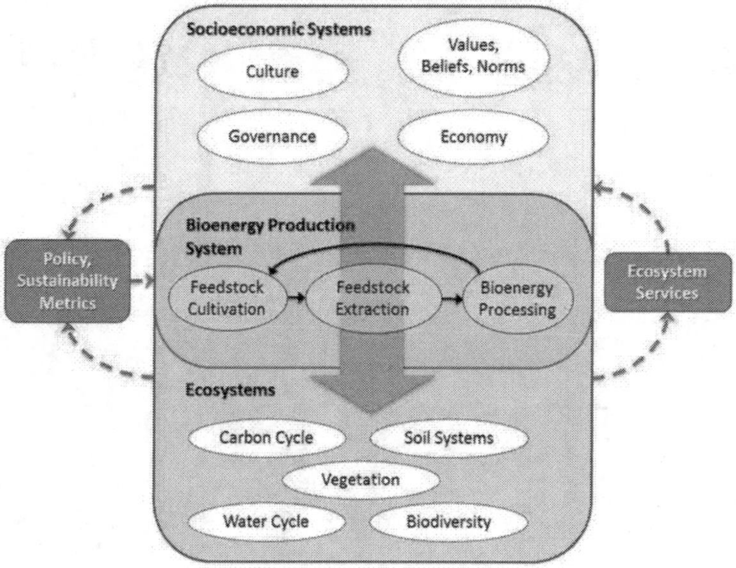

Contact Information:

- **Biodiversity:**
  - Jessie Knowlton (MTU) jlknowlt@mtu.edu
  - David Flaspohler (MTU) djflaspo@mtu.edu
  - Colin Phifer (MTU) ccphifer@mtu.edu

NSF SFI Forest Resources & Environmental Science Ecosystem Science Center

POLICIES FOR THE SUSTAINABLE DEVELOPMENT OF BIOFUELS IN PAN AMERICA

Barry Solomon, Aparajita Banerjee,
Kathy Halvorsen
Michigan Technological University, USA;
Alberto Acevedo, INTA, Buenos Aires;
Amarella Eastmond, Universidad Autonoma de Yucatan, Merida, Mexico

Prepared for the Conference on Pan American Biofuels and Bioenergy Sustainability, Recife, Brazil, 22-25 July 2014

Introduction

- Biofuel production greatly expanded during 2000-10, though output more stagnant since

  - U.S. & Brazil the world leaders, mostly w/Ethanol; less so Biodiesel (followed by Germany, Indonesia, France; Argentina)

- many concerns over Sustainability of Biofuels raised: use of Food (ie Maize) for Fuel, Land Use Changes, Greenhouse Gas Emissions, Water Use, Land Tenure, Labor & Human Rights, etc.

Introduction - cont.

- Why increased interest in Biofuels in Pan America?

- Increase Economic Returns from Agriculture
- Alternative to Imported Oil
- Lower Greenhouse Gas Emissions
- Sustainable Development

- We'll review 5 National case studies of Biofuel development, Policy Goals, Results

Biofuels Production in Pan America

- U.S. & Brazil began modern Ethanol industries in 1970s, today account for ~95% of Ethanol production in W. Hemisphere (~85% worldwide)
- Canada doubled Ethanol production in last few years; output elsewhere in region small
- Biodiesel Production → more recent & more competitive – Brazil, US, Argentina all close

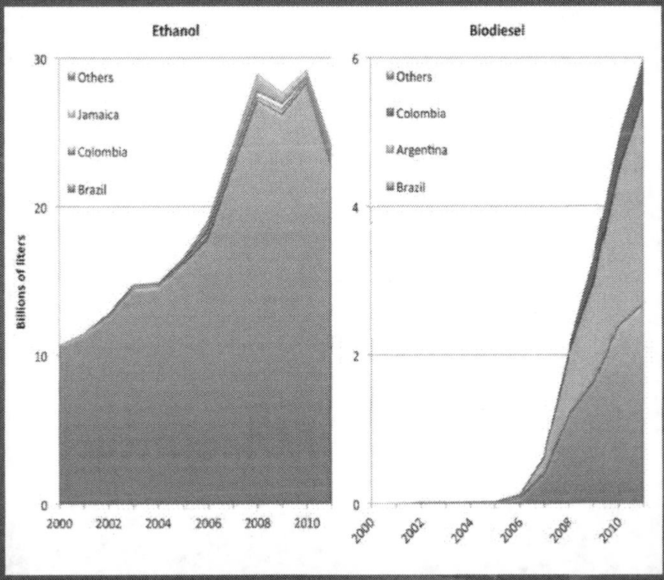

Biofuel Sustainability Governance

- proliferation of International Sustainability Standards for Biofuels & their Feedstocks → 6 adopted in Latin America, but many more ..

- most Certification Schemes done by non state actors, voluntary, performance self-reported (& thus not 100% credible), and:
- with so many standards the Schemes somewhat inconsistent, confusing, and:
- sustainability certification doesn't = sustainability!

Food vs. Fuel, or Both? Regional Production, Cooperation & Trade

- Agricultural Sectors very important in Region

- Maize – US & Brazil major producers; Mexico & Colombia large importers (Argentina also exporter)
- Soybeans – US, Brazil & Argentina dominate
- Sugarcane – Brazil dominates
- Canola – Canada dominates
- among these, only Maize a Staple; Soy Oil for Biodiesel doesn't conflict with Soybean Meal use

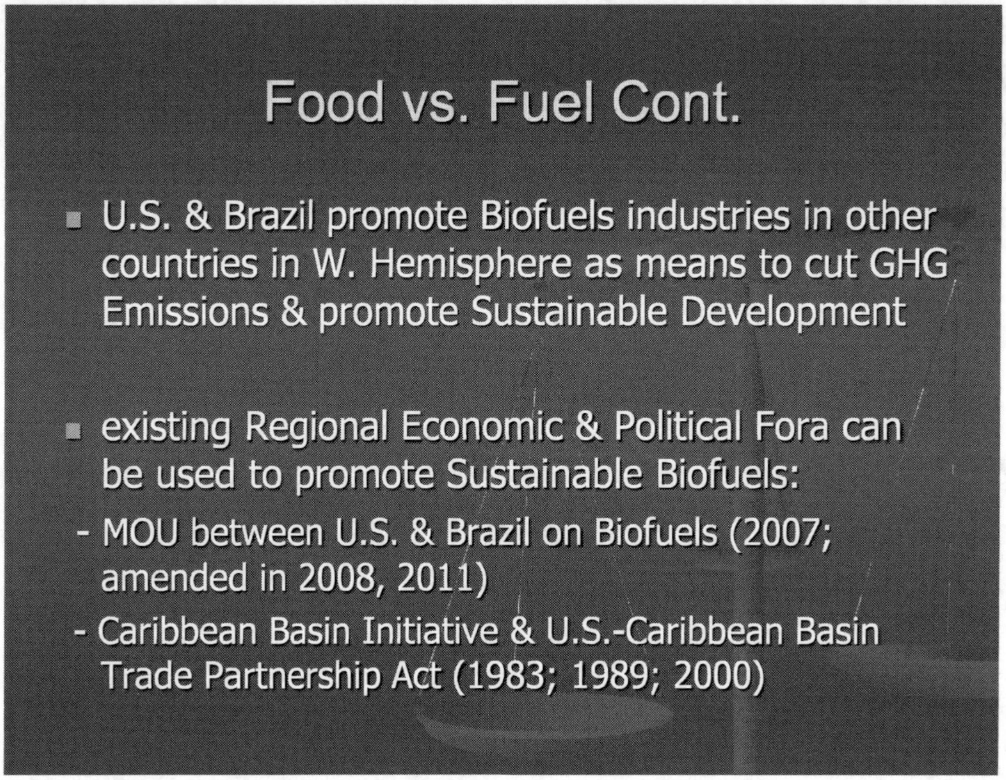
Food vs. Fuel Cont.
U.S. & Brazil promote Biofuels industries in other countries in W. Hemisphere as means to cut GHG Emissions & promote Sustainable Development
existing Regional Economic & Political Fora can be used to promote Sustainable Biofuels:
- MOU between U.S. & Brazil on Biofuels (2007; amended in 2008, 2011)
- Caribbean Basin Initiative & U.S.-Caribbean Basin Trade Partnership Act (1983; 1989; 2000)

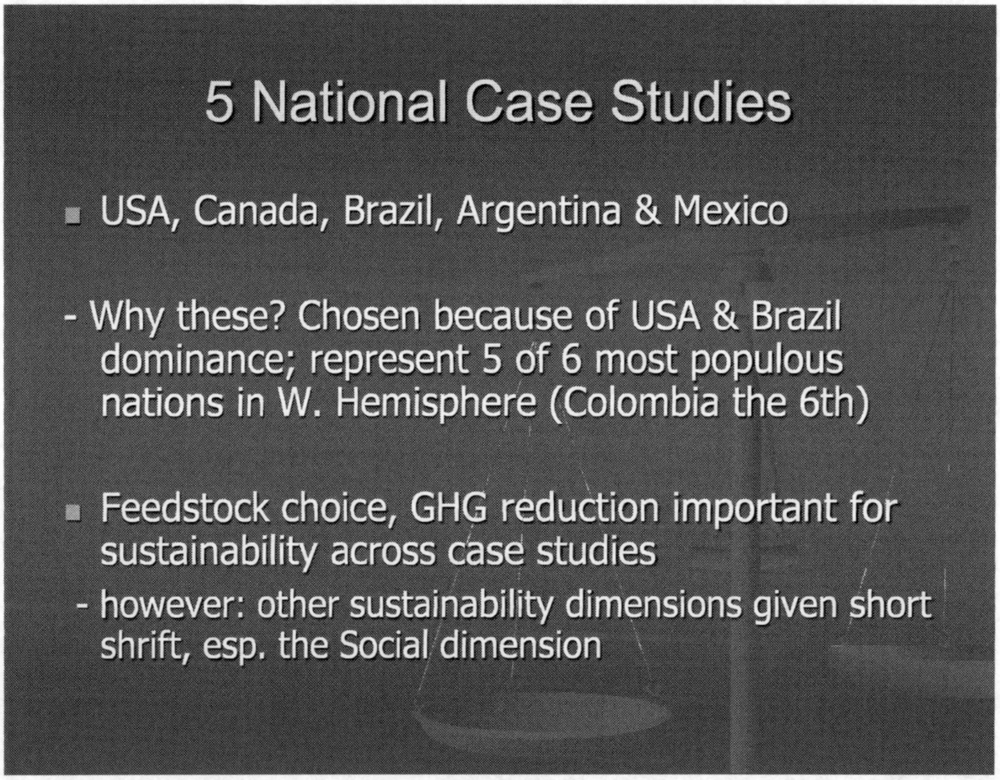
5 National Case Studies
USA, Canada, Brazil, Argentina & Mexico
- Why these? Chosen because of USA & Brazil dominance; represent 5 of 6 most populous nations in W. Hemisphere (Colombia the 6th)
Feedstock choice, GHG reduction important for sustainability across case studies
- however: other sustainability dimensions given short shrift, esp. the Social dimension

5 National Case Studies

- 3 feedstocks dominate in Pan America: Maize (Corn), Sugarcane, Soybean Oil
- Maize presents largest conflict with staple Food needs, Sugars & Oils less so
- U.S. EPA has estimated GHG reductions: Ethanol → Maize -21%, Sugarcane -61%; Biodiesel → Soy -57%, & Palm Oil – 11-17%

United States Case

- **Biofuel Policy Goals**:
  - Increase Energy Security
  - Help achieve Oil Independence
  - Benefit farmers (+ GHG cuts a focus since 2007)
- Ethanol initial focus; 98% dependent on Maize, mostly from 6 farm states, historically subsidized (main subsidy since 1979 ended in 2012)
- Ethanol mandated varying levels since 2005 Renewable Fuel Standard, amended in 2007, greatly increasing standard through 2022

U.S. – cont.

- Biodiesel: little production before 2005
- Feedstock mostly from Soy Oil & Yellow Grease, most from 4 states (TX, IA, MO, IL) & subsidized

  - Biodiesel Mandated & shift toward Cellulosic Biofuels mandated by U.S. 2007 law but occurring slowly!

- **Biofuel Policy Results**: Mixed; Large Production since 2007-08, but some down years; other factors matter too, e.g. MTBE Phase-out, Market & Financial Conditions

U.S. – cont.

- Ethanol – Maize is Staple food crop & large use as fuel negatively impacts Mexico; other problems: historically large Subsidies, Water Use, Pollution from crop growth, less Biodiversity ➔ not Sustainable long-term

- Biodiesel – smaller scale, less crop pollution, in theory more Sustainable, but Industry crashed in 2008-10, thus volatile thus far

Canada Case

- Canada a late comer to Biofuels
- **Biofuel Policy Goals**:

- reduce GHG emissions
- rural economic development

- Federal RFS Mandate from 2008 Law, for 2015: 5% Ethanol blend required, 2% Biodiesel; plus 5% to 8.5% ethanol mandates in 5 provinces; also excise tax exemptions

Canada – cont.

- Similar feedstock profile to U.S., although no Soy use; Ethanol mostly from Maize, some Wheat; Biodiesel from Grease & Canola Oil
- **Biofuel Policy Results**: rapid Ethanol production growth last few years (mostly Ontario), mandate not met yet w/large U.S. imports; Biodiesel production minimal; No conflict with Food Production or water use, some environmental concerns
- NOT Sustainable yet, but better potential than U.S.

Brazil Case

- **Biofuel Policy Goals:**

  - since 1975: goal of independence from foreign oil
  - (GHG emissions cut a focus since 2008)
- Old industry, revived during '70s Oil Crisis
- Ethanol was initial focus as well:

  - historical subsidies, but mostly eliminated since late 1990s
  - blending mandate from Proalcool program since 1975; currently 18-25%, reduced from 25% in 2011

Brazil – cont.

- Ethanol ~100% Sugarcane

  - Ethanol production - efficient & low cost
- Ethanol largely Sustainable when considering GHG, Feedstock, Land Use Change, Socioeconomics

  - *However*: some social concerns raised, e.g. some violations of workers' rights in North & NE

- Biodiesel: ~80% Soy, rest from fats, various oils

  - Blend Mandates - 2% from 2003-08, 5% 2010 onward

Brazil – cont.

- Biodiesel has more Deforestation & GHG emissions concerns, since:

- Soy expansion in Mato Grosso correlated with Deforestation of Amazon Rainforest, destruction of Cerrado forests (tropical savanna) & other Ecosystems, thus much higher GHG Emissions (?)
- But: Biofuels only a *small* part of Soy production;
- Task Force & Round Table on Responsible Soy created to address these problems

Brazil – cont.

- **Biofuel Policy Results:**

- Ethanol: largely Successful, since (flexible?) Blend Mandates met & Brazilian oil imports minimal since 2007); *However* : Ethanol imports from U.S. have been large since 2011
- Biodiesel: 5% Blend Mandate being met, though Sustainability Concerns persist;

  300+ Sustainability Certifications issued (mostly soy oil, less so sugarcane ethanol)

Argentina Case

- **Biofuel Policy Goals**: diversify fuel supply, promote economic development, energy self-sufficiency & energy security
- Biodiesel dominates (ethanol production, largely from sugarcane, very small, + corn ethanol plants since 2013)
- 2006 Biofuel Law mandated 5% blend, raised to 7%, 8%, then 10% this year (with greater domestic consumption vs. exports)

Argentina – cont.

- Biodiesel feedstock ~99+% soy oil (the rest from used vegetable oil)
- **Biofuel Policy Results**:

- Biodiesel has met goals, though production way down in 2013 as market adjusts to cutoff from EU, but up again in 2014; Ethanol production a little short of mandate

Argentina – cont.

- Sustainability? Biodiesel production from Soy Oil a Co-Product so doesn't interfere with Food
- 80%+ of Soybean production uses No Till Agriculture & has Low GHG Emissions

~88 Sustainability Certificates issued, yet still some Sustainability questions, *since*: Land Use Change may causes Deforestation (poor enforcement); high use of Glyphosphate Pesticide; production on fragile areas; some problems w / Land Rights & Rural Development

Mexico Case

- **Biofuel Policy Goals:**

- addressed by 2008 Biofuels Law - sets E2 production goal for Guadalajara, Mexico City, Monterrey, but NOT mandated; sustainability addressed
- domestic Maize use restricted/banned
- biofuels industry just starting & very minimal production so far (based on sorghum, jatropha, palm & cooking oils) & not really commercial

Mexico – cont.

- **Biofuels Policy Results**:

- given lack of commercial production Mexico failing on Economic & Social dimensions of Sustainability
- a sustainability certificate issued for jatropha production, then company fired 100s Workers!
- too much emphasis placed on jatropha so other crops needed (eg sugarcane, sorghum, palm oil?)

Conclusions

- National Biofuel Policies in Pan America focus on cutting foreign oil imports, rural economic development, GHG reduction
  - Subsidies & Mandates dominate policy instruments
- Programs achieving policy goals: Brazil biofuels; also Argentina biodiesel, US biofuel somewhat; Canada improving
- Program achieving least: Mexican biofuels

Conclusions

- **Sustainability?**: Feedstock Choice, GHG Emissions, Deforestation, Soil Loss, Pollution, Land/Labor Rights, Subsidies all matter!
- Some Pan America nations have "potentially" Sustainable Biofuels programs: Brazil (ethanol) & Argentina (biodiesel) but improvements needed
- **Future Research Q:** How well are Sustainability Certification Programs working in practice? Statistical analysis of policies & certification on production/trade of biofuels

UNIVERSITY OF OREGON

Federal woody biomass policy– strategies for building regional business capacity in Oregon?

Cassandra Moseley, University of Oregon

1

Co-authors

- Emily Jane Davis, Oregon State Univ.
- Yeon-Su Kim, Northern Arizona Univ.
- Anne Mottek Lucas, Mottek Consulting
- Max Nielsen-Pincus, Portland State Univ.
- Ted Bilek, U.S. Forest Service

Funding: U.S. Forest Service

2

Outline

- Context
- Woody Biomass Utilization Grant Program (WoodyBUG)
- Findings
- Implications for Policy
- Implications for Research

© 2012 Google

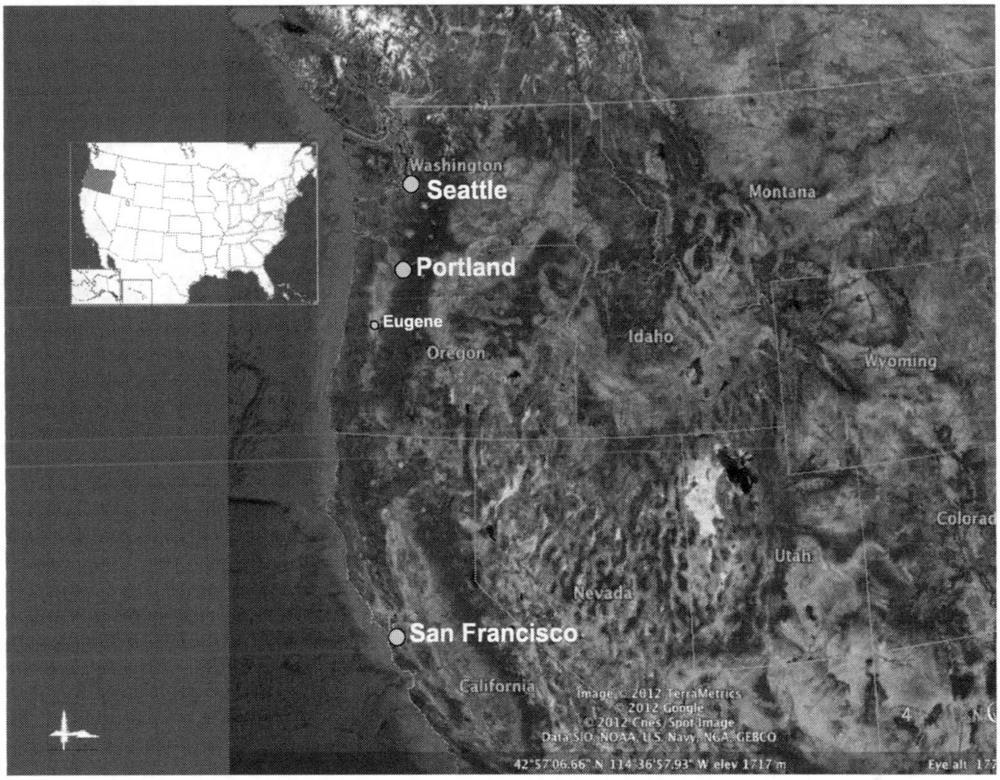

Electricity

POWER GENERATION MAP

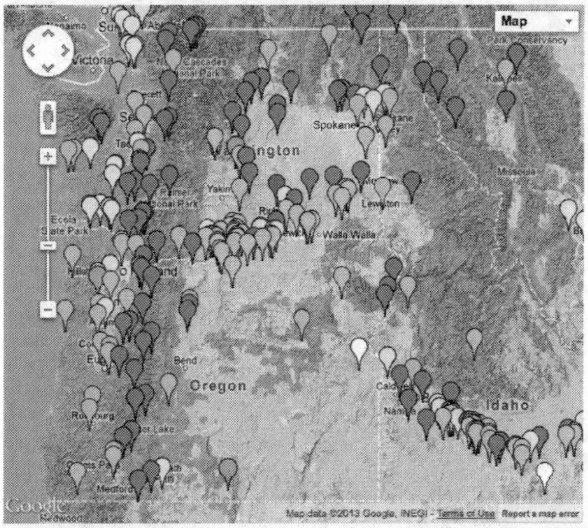

These are the top power producers in each category, or those producing more than 25mw.

- For details, click any marker or choose from the boxes below.
- Other formats: Google Earth and Excel

☑ - Select a Hydropower project - ☐ only treaty dams
☑ - Select a Natural gas project -
☑ - Select a Coal project (see MT, WY, NV) -
☑ - Select a Biomass project -
☑ - Select a Wind project - wind flow map
☑ - Select a Geothermal project -
☑ - Select a Solar project -
☑ - Select a Nuclear project -

More information and your comments.
Last updated 5/6/2013

5

Woody Bioenergy

- Feed stocks: Native species, softwood
- Space heat
- Process heat (esp. kilns for drying wood)
- Pellets, biobricks, kiln-dried fire wood
- Combined heat and power
- Stand alone electric (very rare)

6

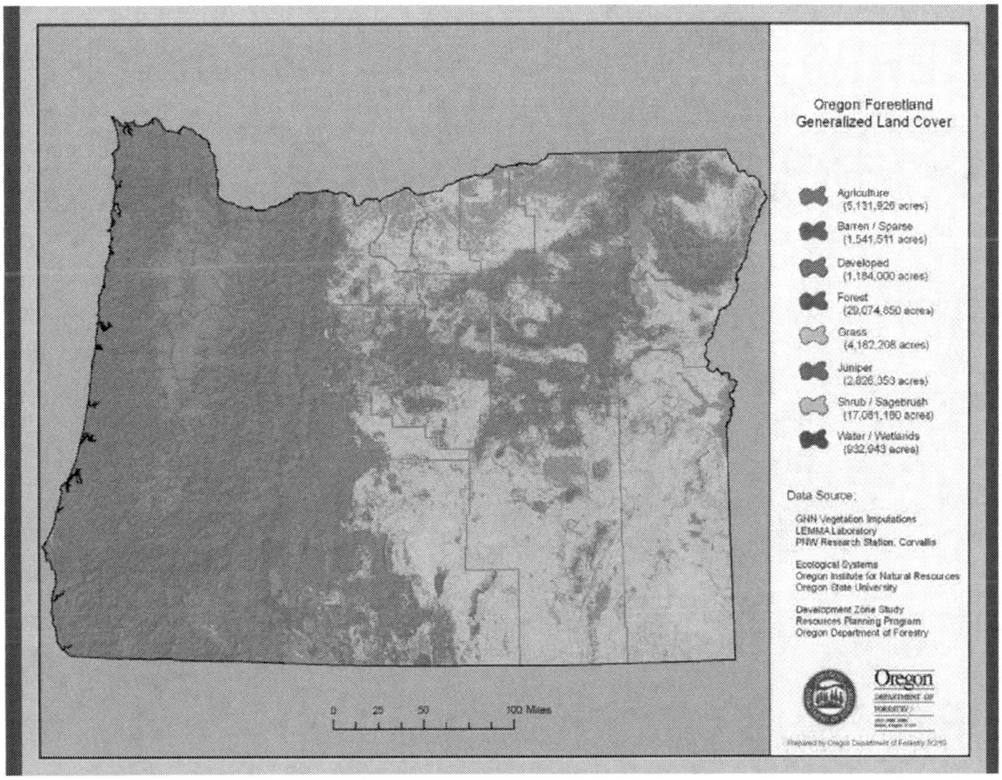
Oregon Forestland
Generalized Land Cover
Agriculture
(5,131,926 acres)
Barren / Sparse
(1,541,511 acres)
Developed
(1,184,000 acres)
Forest
(29,074,650 acres)
Grass
(4,182,208 acres)
Juniper
(2,826,353 acres)
Shrub / Sagebrush
(17,081,180 acres)
Water / Wetlands
(932,943 acres)
Data Source:
GNN Vegetation Imputations
LEMMA Laboratory
PNW Research Station, Corvallis
Ecological Systems
Oregon Institute for Natural Resources
Oregon State University
Development Zone Study
Resources Planning Program
Oregon Department of Forestry
0 25 50 100 Miles
Oregon

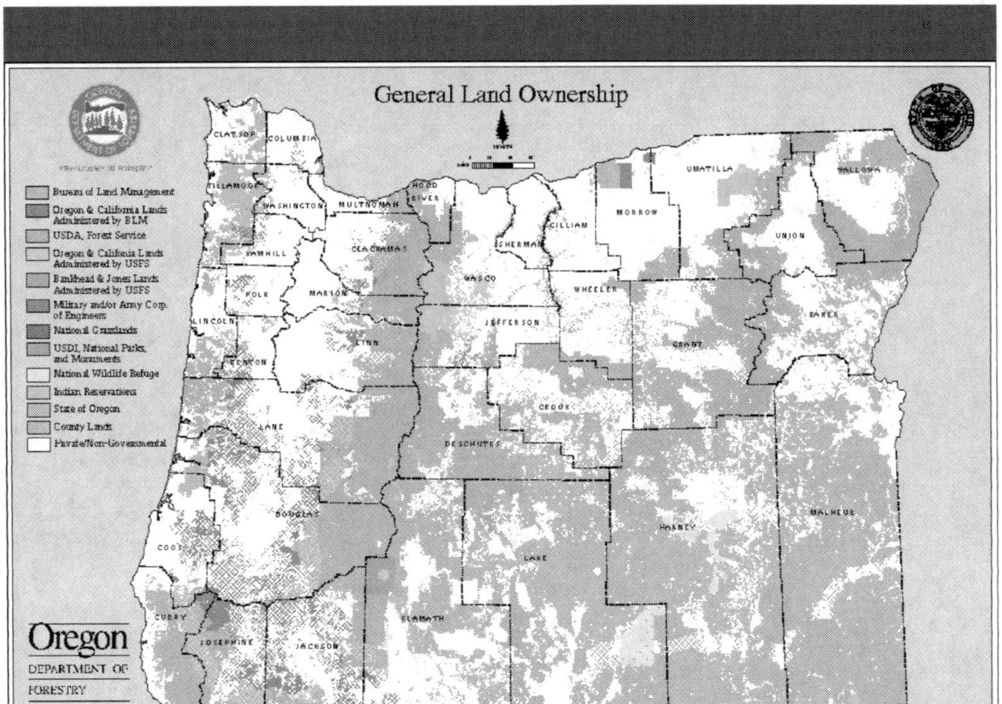
General Land Ownership
Bureau of Land Management
Oregon & California Lands Administered by BLM
USDA, Forest Service
Oregon & California Lands Administered by USFS
Bankhead & Jones Lands Administered by USFS
Military and/or Army Corp. of Engineers
National Grasslands
USDI, National Parks, and Monuments
National Wildlife Refuge
Indian Reservations
State of Oregon
County Lands
Private/Non-Governmental
CLATSOP
COLUMBIA
TILLAMOOK
WASHINGTON
MULTNOMAH
HOOD RIVER
MORROW
UMATILLA
WALLOWA
GILLIAM
SHERMAN
UNION
YAMHILL
CLACKAMAS
WASCO
POLK
MARION
WHEELER
LINCOLN
JEFFERSON
BAKER
GRANT
LINN
BENTON
CROOK
LANE
DESCHUTES
DOUGLAS
HARNEY
MALHEUR
COOS
LAKE
CURRY
KLAMATH
JOSEPHINE
JACKSON
Oregon
DEPARTMENT OF FORESTRY

Merry Christmas
INSTRUCTIONS
National Forest
Sample Permit
Sample Permit
CHRISTMAS TREE PERMIT
3
10

11

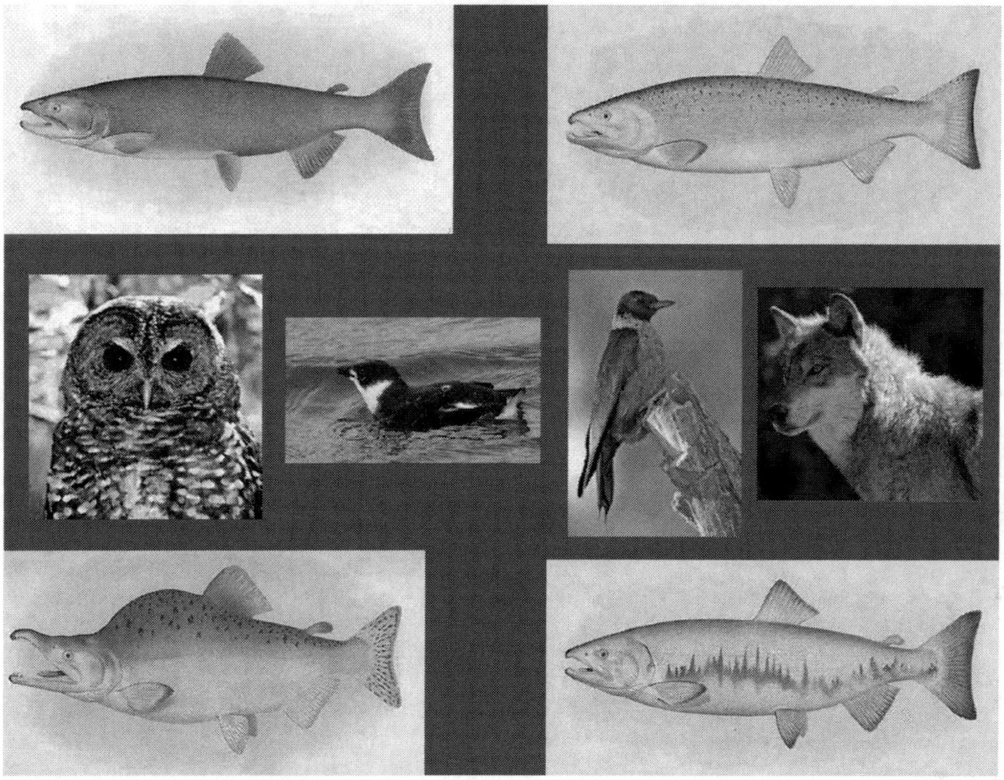

NIKE
KEEN
Columbia
Sportswear Company
STARBUCKS
COFFEE
BOEING
REI
www.rei.com
14

Policy Environment

- As much about forest policy as it is about energy policy
  - Community safety from wildfire
  - Endangered species recovery
  - Landscape-scale restoration
  - Community economic opportunity
  - Declining federal budgets for forest mgt.
  - Social conflict over forest management

15

16

Woody Biomass Utilization

- Advocates hope that it can:
  - Contribute to forest restoration by reducing treatment costs for restoration
  - Create local economic opportunity
    - Contribute to wood products viability
    - Jobs, businesses
    - Lower energy costs
    - Reduce leakage of energy dollars in local economy

17

Federal Woody Biomass Utilization Grant Program (Woody BUG)

- Grants to businesses increase utilization of biomass from hazardous fuels reduction
  - Harvesting and processing equipment, market development, technical assistance.
  - US$5 million in grants annually (nationally)
- Question: How did this program change change regional business capacity?

© 2012 Google
Image © 2012 DigitalGlobe
18
Google

Methods

- Comparative case study (2005-2010)
  - Eastern Oregon and Eastern Arizona
  - In-depth interviews with key informants (Forest Service, NGOs, and grant recipients)
  - Economic impact analysis

19

Figure 1 Eastern Oregon Woody Biomass Utilization grantees

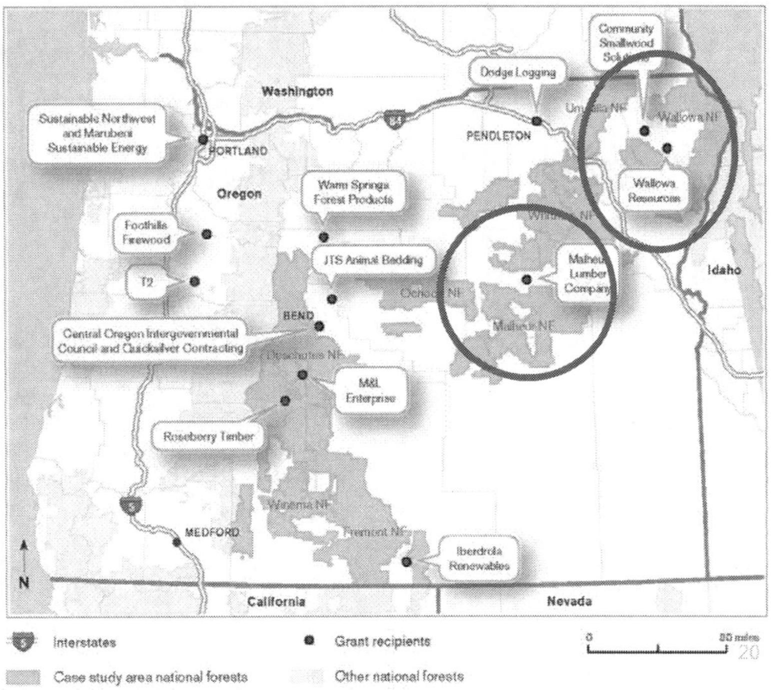

20

Malheur Lumber Company

- Dimensional lumber
  - Small log line, "large" log line
  - Mill residuals– animal bedding, pulp
- Arid landscape, far from transportation
- Struggling with supply access
  - Partially dependent on federal forests
- Market volatility on demand side

21

Malheur Lumber Company

- WoodyBUG Grant: Focus on product diversification
  - $250,000 for log shaver, drumb dryer, and shaving bagging (animal bedding)
  - Mix of federal and private funding
  - A couple of years later, got addition $ to add new pellet/biobrick facility

22

Wallowa County

- All saw milling capacity gone
  - Culturally still important
  - Need for jobs
  - Need for forest restoration/fuels reduction
- Local NGO
  - Rebuilding small scale wood products capacity

23

Woody BUG grants

- $250,000 Expand existing post and pole plant to include clean chips, hog fule, and fire wood
- $250,000 horizontal wood processer for slash and wood waste

24

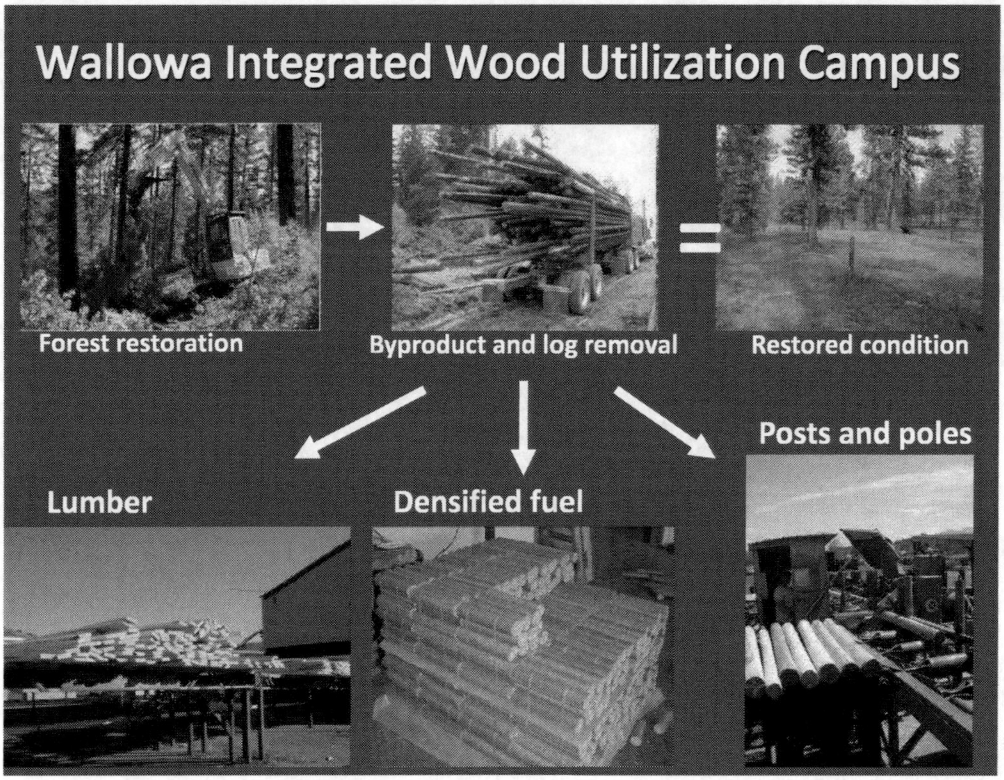

Findings – Business Capacity

- Program filled key gaps in supply chains in subregions by investing in harvesting and processing equipment creating new capacity, leveraging further investments
- But, many grantees combined this with other assistance to succeed
- And, filling these equipment gaps revealed other challenges

26

Findings – NGO's

- Non-governmental organizations played key roles in regional efforts
  - Market development
  - Technical assistance to smaller businesses
  - Risk sharing with businesses
  - Grant writing, paperwork

27

Findings – In Sum

- Small investments can make a big difference, if used to fill key gaps in supply chain
- NGO's are important for assisting small businesses and building regional strategy
- Narrow programs combined with other programs to make viable businesses
- Didn't address supply or demand issues

28

Implications for policy

- Critical to pay attention to all parts of the supply chain simultaneously
- Given policy fragmentation, cluster development strategy may be necessary to succeed
  - Regional intermediaries to coordinate strategy and align policies

29

Implications for research

- Because policy environment is extremely fragmented
  - Studying one policy at a time limited utility
- Biomass policy is often not "energy" policy, but social, natural resource, agriculture and environmental policy – need to look well beyond energy policy to understand policy drivers.

30

Thank you
Cassandra Moseley
Ecosystem Workforce Program
Institute for a Sustainable Environment
University of Oregon
cmoseley@uoregon.edu
http://ewp.uoregon.edu
31

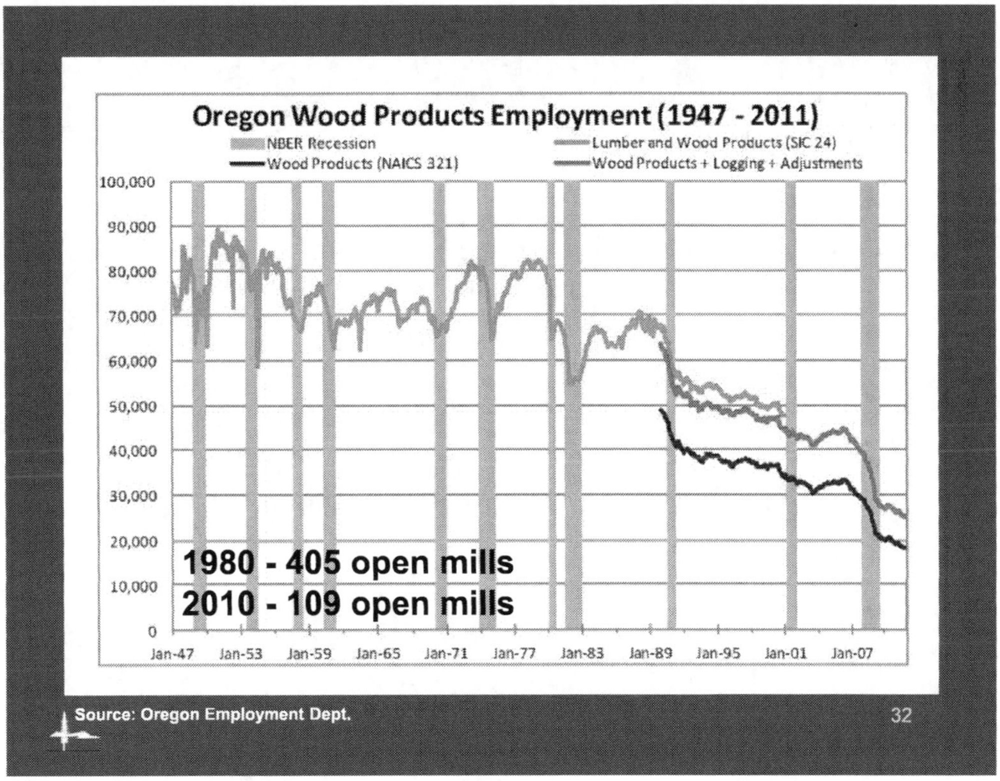
Oregon Wood Products Employment (1947 - 2011)
NBER Recession
Lumber and Wood Products (SIC 24)
Wood Products (NAICS 321)
Wood Products + Logging + Adjustments
100,000
90,000
80,000
70,000
60,000
50,000
40,000
30,000
20,000
10,000
0
Jan-47
Jan-53
Jan-59
Jan-65
Jan-71
Jan-77
Jan-83
Jan-89
Jan-95
Jan-01
Jan-07
1980 - 405 open mills
2010 - 109 open mills
Source: Oregon Employment Dept.
32

Indirect effects of bioenergy: international standards and science

"Policies for the Sustainable Development of Biofuels in Pan America" session

PAN AMERICAN BIOFUELS AND BIOENERGY SUSTAINABILITY
AN NSF RESEARCH COORDINATION NETWORK

Recife, Brazil
July 23, 2014

Maggie Davis
Oak Ridge National Laboratory

In collaboration with Keith Kline (ORNL) and conveners of the working group, Fred Ghatala & Diego Goldin, as well as Chuck Corr and the US Technical Advisory Group for ISO 13065 , & other ORNL staff from Center for BioEnergy Sustainability

Special thanks to Kristen Johnson and Alicia Lindauer, Alison Goss Eng, among others at the Department of Energy (DOE), BioEnergy Technologies Office (BETO)

Views expressed are the author's.

CBES
Center for BioEnergy Sustainability

Climate Change
SCIENCE INSTITUTE

OAK RIDGE
National Laboratory

Outline

- Relevance
- Background:
  - International Organization for Standardization- ISO 13065 "Sustainability Criteria for Bioenergy"
  - Group to assess the <u>state of the science</u> of "indirect effects"
- Conclusions:
  - Diverse results represented in literature
  - Consensus statement
- Resolutions
- Concluding remarks

OAK RIDGE
National Laboratory

Relevance – bioenergy trade, bio-economy

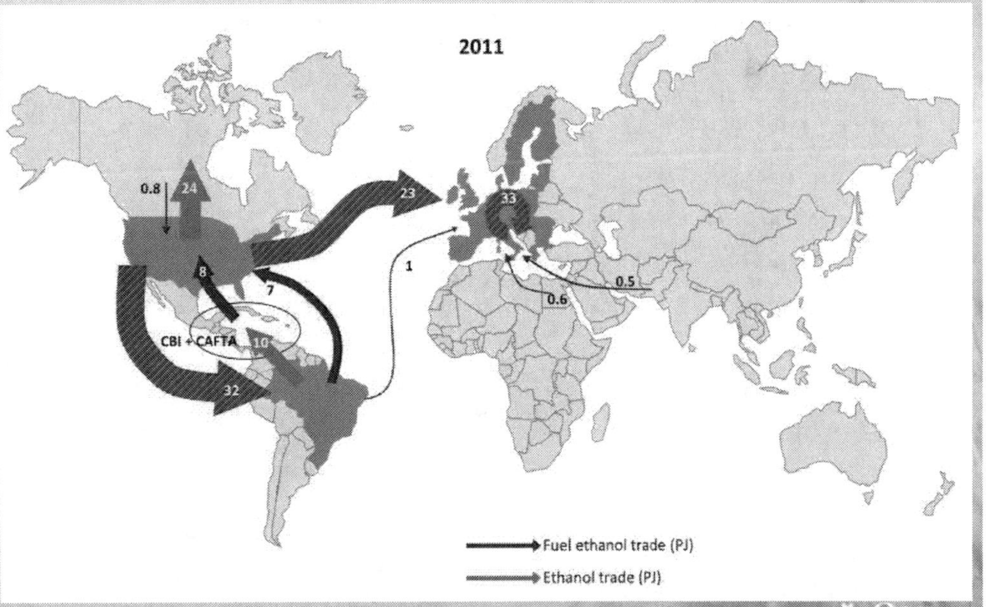

Source: Lamers et al., RSER, 15 (2011) 2655–2676

Relevance: Standards & Industry

- Barriers and inefficiencies are emerging from requirements for sustainable bioenergy
- Diverse ... inc... tent methods and numerous indicators with va...le def...ons ...
  - incre... t... act...
  - Un... k... co... investm...
  - Create barriers to growth in production and trade of clea...

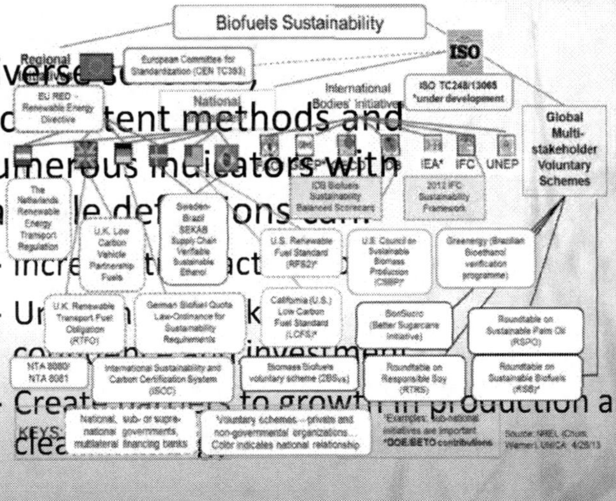

Background: International Standards

What is a standard?

- A standard is a document that
  - Provides consistency
    - In requirements
    - In guidelines
    - In specifications
  - Can be used to ensure consistent and appropriate
    - Materials
    - Products
    - Processes
    - Services

Why develop standards?

- Comparable assessment
- Help ensure products and services are "fit for purpose"
- Reduce costs by minimizing waste and errors; increasing productivity
- ***Facilitate free and fair global trade***
  - **Access to new markets**
  - **Level the playing field for new entrants**

Source: adapted from www.ios.org

Background: Approach & Objectives

- Guide efforts toward science-based approaches
  - Help to determine state of science on controversial and politicized issues
- Enable informed decisions about energy choices that support continual improvements in energy systems
- Develop consensus on common terms and methods

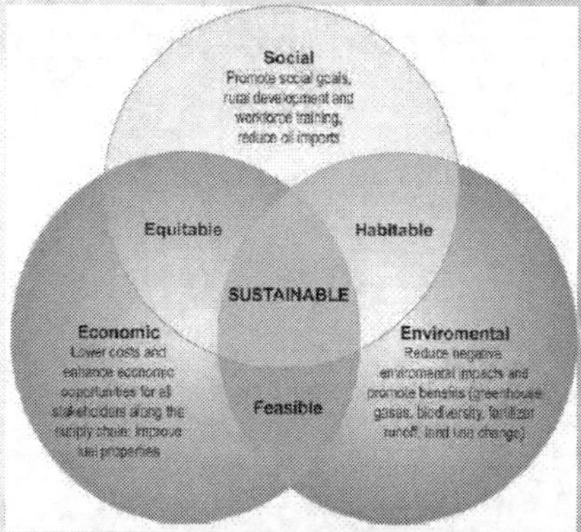

FIG from BETO MPP, Slide adapted from KL Kline presentation for DOE EERE webinar, "Global Solutions for Global Challenges: International Collaborations to Advance Bioenergy Research"

Sustainability is a trajectory
How we define and measure is fundamental to understanding
Photo credit: Orbiter News Dec. 22, 2010

Measurement is challenging...
"Not everything that can be counted counts, and not everything that counts can be counted."
-William Bruce Cameron
NASA Images

Background: Why ISO?

- Global reach and impact
  - 163 member countries
  - 19,500 published International Standards
- Rio 1992: Series of Environmental Standards (ISO 14000)
  - 250,000 users
  - Applied in 155 countries
- Social Responsibility (ISO 26000:2010)
- ISO 14064:2006 and ISO 14065:2007 standards to provide
  - An internationally agreed framework for measuring GHGs
  - so that " a tonne of carbon is always a tonne of carbon"

International Organization for Standardization

Source: http://www.iso.org/iso/rio_20_forging_action_with_agreement.pdf

ISO 13065 "Sustainability Criteria for Bioenergy"

"Standardization in the field of sustainability criteria for **production, supply chain and application of bioenergy**."

Indirect effects work group mandate: "critically assess and present results on indirect effects" (e.g., LUC and food security)

Foundation:

ISO guidelines for drafting of standards states that:

> "Test methods should be clearly identified and be consistent with the purpose of the standard. **They should be objective, concise and accurate, and produce unambiguous, repeatable and reproducible results,** so that results of tests made under defined conditions are comparable" (ISO 1994).

Process: 161 publications reviewed; 4 reports submitted (2012-13)

Indirect effects work group - Results

Literature Review/Statements on State of Science Applicable to Indirect Effects:

United Nations (UN)-Energy (2000). Sustainable Bioenergy: A framework for decision makers. p. 1-61.

"If biofuel feedstock production competes for water supplies, it could make water less readily available for household use, threatening the health status and thus the food security status of affected individuals. On the other hand, if modern bioenergy replaces more polluting sources or expands the availability of energy services, it could make cooking both cheaper and cleaner, with positive implications for food utilization." (p. 31)

OAK RIDGE National Laboratory

Indirect effects work group - Results

Literature Review/Statements on State of Science Applicable to Indirect Effects:

Babcock, B. (2011). The Impacts of US Biofuel Policies on Agricultural Price Levels and Volatility. Issue Paper No. 35. International Centre for Trade and Sustainable Development.

"These results indicate that US ethanol subsidies during this period had little impact on consumer prices and quite modest impacts on crop prices." (pg. vii)

OAK RIDGE National Laboratory

Indirect effects work group - Results

Literature Review/Statements on State of Science Applicable to Indirect Effects:

Berndes, G., Bird, N., & Cowie, A. (2010). Bioenergy, land use change and climate change mitigation. IEA Bioenergy, ExCo: 2010:03.

"The GHG effects of LUC are difficult to quantify with precision in relation to a specific bioenergy project, particularly for iLUC where the causes are often multiple, complex, interlinked and change over time." (p. 6)

OAK RIDGE National Laboratory

Indirect effects work group - Results

Literature Review/Statements - Continued:

Baffes, J., & Haniotis, T. (2010). Placing the 2006/08 commodity price boom into perspective. Policy Research Working Paper 5371. The World Bank. July, 2010.

"This paper concludes that **a stronger link between energy and nonenergy commodity prices is likely to be the dominant influence on developments in commodity, and especially food, markets**. Demand by emerging economies is unlikely to put additional pressure on the prices of food commodities. The paper also argues that the **effect of biofuels on food prices has not been as large as originally thought, but that the use of commodities by financial investors (the so-called "financialization of commodities") may have been partly responsible for the 2007/08 spike.** ..." (p. i)

OAK RIDGE National Laboratory

Indirect effects work group - Results:

WG4 summary consensus from report submitted to PC on Feb 17, 2012:

The science on indirect effects is nascent and rapidly evolving. This makes it difficult to reach consensus on the state of the current science.

The conclusion, based on the expertise of, and literature reviewed by, the work group, is that **the state of the science in terms of evidence based research is inconclusive or contradictory regarding indirect effects of bioenergy.**

Indirect effects work group - Results:

- **WG4 summary consensus continued:**

There has been more emphasis on sustainability and indirect effects of bioenergy than on baseline (often fossil fuel) scenarios.

Recommendations:

An economic operator should not be held responsible for indirect effects

Model input assumptions require more corroboration from evidence-based research

There needs to be equitable treatment of direct and indirect effects for any energy options being analyzed including baseline fuel(s) that would be replaced by proposed bioenergy sources.

OAK RIDGE National Laboratory

Indirect effects work group- Resolutions

Definitions:

Direct effects can vary widely and their measurement will depend on the defined system boundaries for analysis.

Estimation and attribution of indirect effects depend on quantitative and qualitative methods and assumptions used.

The Standard considers the measurable effects that are under the control of the economic operator and caused by the process being analyzed

Concluding remarks

Successful sustainability assessment requires:

- Agreement on goals and purpose (priorities)
- Valid representation of conditions based on clear, consistent definitions for variables of concern; e.g.,:
  - land attributes (primary forest, grassland, high-carbon...)
  - management practices
  - baseline trends and change dynamics
- Understanding causes and effects (empirical data to test models and hypotheses)
- Effective incentives for compliance and continual improvement

Successful standard requires low transaction costs relative to value-added

Thank you

davismr@ornl.gov

OAK RIDGE National Laboratory

Thank you

CBES
Center for BioEnergy Sustainability

http://www.ornl.gov/sci/ees/cbes/

See the website for: Reports, Forums, presentations, publications

Acknowledgements:
Fred Ghatala (Waterfall group-Canada), Diego Goldin (INTA-Argentina), Michael Wang (ANL), Chuck Corr, Paul Trupo, Chris Farley, Fahran Robb, US TAG, and ISO PC248; Helena Chum; Tom Wilbanks, Virginia Dale and ORNL CBES colleagues, B Davison, LM Eaton, RA Efroymson, PN Leiby, M Langholtz, LR Lynd, A McBride, GA Oladosu, ES Parish, RD Perlack, N Singh, S Kang, SB Wright, LL Wright; and DOE OBP staff Zia Haq, Kristen Johnson, Alicia Lindauer-Thompson, P Grabowski, A Goss-Eng.

Department of Energy (DOE), BioEnergy Technologies Office (BETO)

Argonne NATIONAL LABORATORY

This research was supported by the U.S. Department of Energy (DOE) Bio-Energy Technologies Office and performed at Oak Ridge National Laboratory (ORNL). Oak Ridge National Laboratory is managed by the UT-Battelle, LLC, for DOE under contract DE-AC05-00OR22725. The views in this presentation are those of the authors, who are responsible for any errors or omissions.

OAK RIDGE National Laboratory

References

- Dale VH, KL Kline, LL Wright, RD Perlack, M Downing, RL Graham. 2011. Interactions among bioenergy feedstock choices, landscape dynamics and land use. *Ecological Applications* 21(4):1039-1054.
- Efroymson, R. A., V. H. Dale, K. L. Kline, A. C. McBride, J. M. Bielicki, R. L. Smith, E. S. Parish, P. E. Schweizer, D. M. Shaw. 2012. Environmental indicators of biofuel sustainability: What about context? Environmental Management DOI 10.1007/s00267-012-9907-5
- Giglio L., J. T. Randerson, G. R. van derWerf, P. S. Kasibhatla, G. J. Collatz, D. C. Morton, and R. S. DeFries. Assessing variability and long-term trends in burned area by merging multiple satellite fire products. Biogeosciences, 7, 1171–1186, 2010.
- Kline KL, et al. 2011. Scientific analysis is essential to assess biofuel policy effects [published as a peer-reviewed article] in response to the paper by Kim and Dale on "Indirect land use change for biofuels: Testing predictions and improving analytical methodologies."] *Biomass and Bioenergy*
- Kline KL, Dale VH, Grainger A. 2010. Challenges for Bioenergy Emission Accounting. *Science e-letter.* (2 March 2010)
- Kline, KL and Dale, VH. 2008. Biofuels, causes of land-use change, and the role of fire in greenhouse gas emissions. Science 321:199 http://www.sciencemag.org/cgi/reprint/321/5886/199.pdf

- Kline KL, Dale VH, Grainger A. 2010. Challenges for Bioenergy Emission Accounting. *Science e-letter.* (2 March 2010)
- Kline, KL and Dale, VH. 2008. Biofuels, causes of land-use change, and the role of fire in greenhouse gas emissions. Science 321:199 http://www.sciencemag.org/cgi/reprint/321/5886/199.pdf

- Lamers, P, C Hamelinck, M Junginger, and A Faaij, 2011. Renewable and Sustainable Energy Reviews 15: 2655–2676
- International Energy Agency (IEA), Technology Roadmap: Biofuels for transport, 2011. Accessed from http://www.iea.org/publications/freepublications/publication/Biofuels_Roadmap_WEB.pdf

- **References included in indirect effects literature review and quoted here are available by request. Please email me at davismr@ornl.gov**

Parting thoughts– why 'sustainable' bioenergy?

Ethical and scientific considerations

- Conserve resources for future generations
- Efficient alternatives to fossil fuels needed
- Improve land use efficiency
  - Incentives to invest in better soil management
  - Over 400 million hectares burn *each year*
- Sustainable development goals
  - "Living within our means"
  - Integrated land-use planning
  - More sustainable rural livelihoods
- Climate change, resilience and adaptation
  - Incentives to manage landscapes for multiple benefits including climate change mitigation and improved system resilience

Parting thoughts– standards

Can certification ensure sustainability?

No, *because n*othing can **ensure** sustainability and…

1. There are too many opportunities for substitution in biomass markets
2. Transaction costs for certification, monitoring and verification are too high relative to value of products
3. *Uncertainty:* is there political will and sufficient market premium to justify certification?
4. "Setting a bar" does not necessarily improve anything (e.g., wastes)
5. Even well-designed schemes can be too easily "gamed" and it only takes a few well-publicized cases to undermine credibility

Photo: José Luis Gómez; Fondo Acción, Colombia

Slide adapted from Kline presentation for IEA Joint Task 38-40-43 presentation on LUC: http://ieabioenergy-task38.org/workshops/campinas2011

OAK RIDGE National Laboratory

Parting thoughts– standards

Can a standard support more sustainable outcomes?

Yes, *if* it –

1. Is developed with users to meet their needs (context specific)
2. Provides science-based tools that promote learning
3. Creates incentives that shift production toward more sustainable paths
4. Is adaptable to changing contexts and priorities
5. Encourages all to participate
6. Can be implemented on a "level playing field"
7. Is transparent and easily adopted.

Photo: José Luis Gómez; Fondo Acción, Colombia

Slide adapted from Kline presentation for IEA Joint Task 38-40-43 presentation on LUC: http://ieabioenergy-task38.org/workshops/campinas2011 also available on CBES website

OAK RIDGE National Laboratory

Supplementary information on ISO 13065, Process to date:

- Proposed by Germany, Brazil (2008)
- Mandate approved (2009)
- Work groups finalized (2011)
- Over 40 entities on the Project Committee (PC-248)
- Committee reviewing 830 comments received on 2nd Draft
  - An intermediate draft was approved to continue to a draft international standard ballot
- Target completion and publication date: 2016

OAK RIDGE National Laboratory

ISO 13065, GHG Methodology Work Group

References TS 14067 and explains applicability to bioenergy

TECHNICAL SPECIFICATION **ISO/TS 14067**

First edition 2013-05-15

Greenhouse gases — Carbon footprint of products — Requirements and guidelines for quantification and communication

Specifies principles, requirements and guidelines for the quantification of the carbon footprint of a product (CFP), based on ISO 14040 and ISO 14044 and on environmental labels and declarations (ISO 14020, ISO 14024 and ISO 14025 for communications)

OAK RIDGE National Laboratory

BIOFUEL SUSTAINABILITY IN LATIN AMERICA & THE CARIBBEAN – A REVIEW OF RECENT EXPERIENCES

Robert Bailis[1], Barry Solomon[2],
Chrstine Moser[3], Tina Hildebrandt[4]
[1]Yale University, USA
[2]Michigan Technological University, USA
[3]Leuphana Universtat, Germany
[4]EDEKA - Germany

Prepared for the Conference on Pan American Biofuels and Bioenergy Sustainability, Recife, Brazil, 22-25 July 2014

Introduction

- Vast majority of biofuel production in Latin America & Caribbean (LAC) derived from well established farm sectors with inequitable land distribution & land use
- Doubts raised about biofuel sustainability in LAC for these reasons, plus many other issues e.g. food insecurity, water pollution, GHG emission, labor rights violations, etc.

Introduction - cont.

- Globally there's been proliferation of biofuel mandates & blending targets, tax exemptions, other subsidies, etc. encouraging more ethanol & biodiesel production
- Also encouraged are use of new feedstocks, & use of "marginal" or "degraded" lands"
- We'll review these trends + governance mechanisms for sustainability, implementation, how to address challenges to sustainable production, + directions for future research

National Biofuels Policy Objectives

- In LAC today there are several **policy objectives** for biofuels development; 10 are widespread plus an 11th in 2 countries only:

- energy security/rural energy access;
- gain competitive advantage/new markets;
- stimulate economic growth;
- promote rural development;
- strengthen or diversify farm sector

Policy Objectives

- job creation
- reduce poverty/social inclusion/improve rural life (less countries)
- mitigate climate change/join carbon markets
- protect the environment more generally
- avoid impacts on food security (less countries)
- create alternatives to illicit drug cultivation (Colombia & Peru only)

Governing Biofuel Sustainability in LAC & Role of Export Markets

- Governance mechanisms introduced by wide range of actors → national & sub-national governments, intra & supra govt. orgs, private corporations, civil society actors

- **Mechanisms**: policies, regulations, voluntary certification schemes, sustainability standards, meta standards, codes of conduct
- may be single issue or comprehensive

Governance cont.

- US & EU major biofuel export targets from LAC
  - US requires greater GHG reduction from biofuel (2007) imports, and transportation fuel providers in Calif.

- EU passed Renewable Energy Directive (RED) in 2009 to mandate 10% renewable transport fuel; must reduce lifecycle GHG emissions, avoid high bio-diverse lands, lands with *high carbon stocks*, peatlands ➔incentives to produce on severely degraded or heavily contaminated lands
  - European Commission (EC) monitors implementation

Governance cont.

- 14 international biofuel sustainabiity standards accepted by EC; 6 adopted in LAC:
  - Roundtable on Sustainable Biomaterials (RSB)
  - Int. Sustainability & Carbon Certification (ISCC)
  - Biomass biofuel, sustainability voluntary scheme (2BSvs)
  - Round Table on Responsible Soy (RTRS)
  - Roundtable on Sustainable Palm Oil (RSPO)
  - Bonsucro (sugarcane feedstock)

BIOMASS
BIOFUELS
SUSTAINABILITY
2BS®vs
voluntary scheme

RTRS
Certified Soy
RTRS
Certified Soy
RTRS
Certified Soy

Governance cont.

- also: Inter-American Development Bank – *Biofuels Scorecard*; Global Bioenergy Partnership – 24 sustainability indicators, pilot tested in Colombia; ISO developing voluntary bioenergy sustainability criteria (Brazil & Germany cooperation – ref. Maggie's talk)

- Schemes usually have multiple criteria & indicators to demonstrate compliance, and have involved multiple stakeholders across supply chain
- - social issues often downplayed or are weak

Status of Sustainability Standard Adoption in LAC

- Voluntary biofuel project certification gaining traction in LAC, incl. qualifying standards for RED

- based on nation by nation review, as of July 2014: 315 certificates in Brazil, 88 in Argentina, 9 in Paraguay, 6 in Guatemala, 5 each in Colombia & Uruguay, 4 in Costa Rica, 3 each in Peru & Nicaragua; 2 each in Chile & in Mexico

- most popular schemes: RTRS, 2BSvs, ISCC, Bonsucro; only a few have adopted RSP, RSB

Standard Adoption cont.

- Biggest surprise → few takers of RSB standards, which are most detailed & comprehensive, had very widespread stakeholder input worldwide, thus high awareness among practitioners
  - 12 principles, 35 criteria, 200 indicators!
- the popular 2BSv (industry led), in contrast, only requires applicants meet RED minimum criteria in order to access profitable EU biofuel markets

Standard Adoption cont.

- ISCC scheme covers similar range of criteria as RSB but has seen much higher uptake rate
- main difference between ISCC & RSB: ISCC had less stakeholder involvement in standard setting & governance, & has fewer criteria & indicators to comply with, thus probably less burdensome
  - overall: cost of achieving certification gives strategic advantage to large-scale industrial firms vs. small-scale producers

Key Challenges cont.

- **GHG Emissions & Land Use Change**
  - GHG reduction targets set in USA & EU; incorporated into many certification schemes, though specific reduction targets not set in LAC
  - GHG reduction estimates vary widely due to different assumptions & methodologies
  - sugarcane can reduce emissions 80-90%, palm 35-90%, jatropha 30-60%, soy 50%, or much less
  - however actual emissions reduction may be less due to dLUC (direct land use change) & iLUC (indirect)

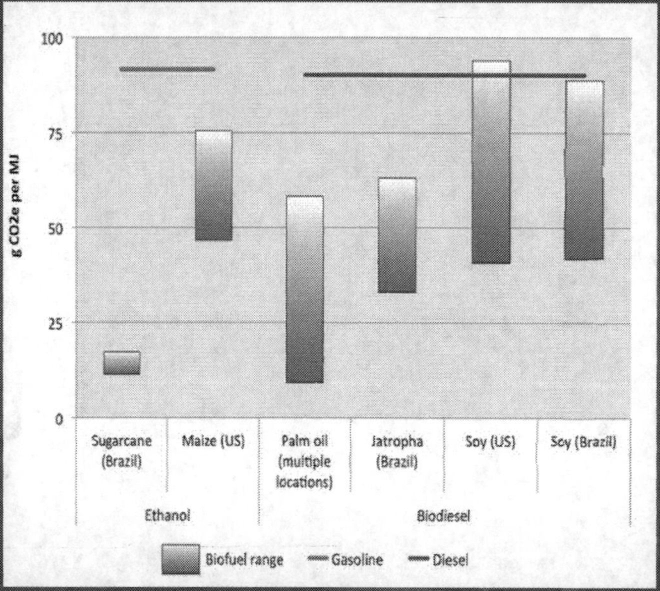

Key Challenges cont.

- **GHG Emissions & Land Use Change**

  - dLUC results from feedstock cultivation displacing natural vegetation
  - iLUC results if biofuel feedstock affects commodity markets in ways that induce LUC elsewhere
  - however: dLUC can be positive, eg perennial crops planted on degraded pasture lands may increase C
  - most sustainability schemes *acknowledge* iLUC, but only the US EPA & CA incorporate it, and have been subjected to legal challenges

Key Challenges cont.

- **Water Accessibility & Quality**

  - large scale farming production can have large impacts on nearby communities (incl. evapo-transpiration & irrigation, eg Colombia), from water demand for feedstock production & processing, plus runoff or pollution from refining
  - fertilizer & pesticide use high in LAC
  - salts from cooling towers & boilers must be discharged
  - ethanol also discharges vinasse waste, though Brazil increasingly using it for ferti-irrigation in cane fields

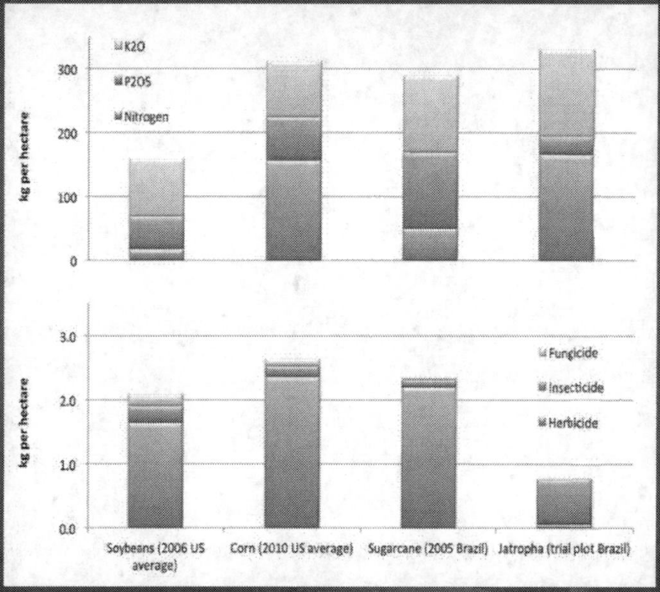

Key Challenges cont.

- **Food Security**
  - occurs through *indirect* impact of biofuels on global markets, effecting prices of staple grains & oils
  - in 2007-08, e.g. biofuel expansion may have added 20-40% to increased price of staple foods, *but*: many other factors account for more
  - ethanol production from maize? Little such production in Latin America, but Mexico & Colombia are major importers, thus backlash (Mexican 'tortilla crisis' & backlash in 2007)

Key Challenges cont.

- **Food Security**
  - factors within LAC countries are relevant in even less direct ways
  - while soy & sugar are not really staple foods, expanded cultivation for biofuel could displace land used for food crops, mainly affecting local communities
  - among the 6 biofuel certification schemes adopted in LAC, only ISCC & RSB have food security criteria, and for ISCC isn't a core criterion

Key Challenges cont.

- **Land Tenure Issues**
  - generations of unequal land ownership in the region have resulted in some of the highest such land & wealth disparities in the world, + conflict/violence
  - most biofuel production highly concentrated, *not* helping the rural poor
  - sugarcane & soybean production highly concentrated, e.g. in Brazil ~70% of cultivation done by 340 large-scale mills, w/avg. holding of 30,000 ha

Key Challenges cont.

- **Land Tenure Issues**

  - result of inequities has been social reform movements, via *Via Campesina* (global movement of peasants & small farmers w/roots in Central America, & Brazil's Landless Rural Workers' Movement (MST in Portuguese) ➔ some opposition to biofuels
  - most sustainability standards have land rights provisions, but rely largely on existing law & international treaties protecting indigenous rights; 2BSv doesn't mention land tenure & not a *core* criterion for Bonsucro; RSB goes further but few use it

Conclusions/Future Research

- many social/environmental concern accompanied rapid biofuel development in LAC
- 400+ feedstock cultivators in region certified as "sustainable" operations
- unclear if this means real sustainability in LAC, or simply market savvy businesses "green-washing" their operations in order to access foreign markets in EU & USA
- active monitoring & impact evaluation needed to determine true sustainability of biofuel operations

Conclusions/Future Research cont.

- too early to say if sustainable biofuel certification schemes in LAC (esp. ISCC, Bonsucro, RTRS) are strong enough to safeguard against adverse environmental & social effects
- additional mechanisms besides these voluntary certification schemes needed, e.g. make more schemes RED eligible & make it more rigorous?

 - regional cooperation could facilitate this

- also need for greater incorporation of social movements & their concerns into biofuel sector

Trajectory of Wood Energy Policies in the US: Sustainability and Governance in Global Markets

Dennis Becker, Associate Professor
University of Minnesota

Department of FOREST RESOURCES
University of Minnesota

Drivers of Wood Energy in the US

- Restore forests—reduce risk of wildfire
- Maintain and enhance biological diversity
- Mitigate GHG impacts
- Enhance forest productivity
- Increase economic development
- Transition to renewable energy economy

Projected Renewable Energy

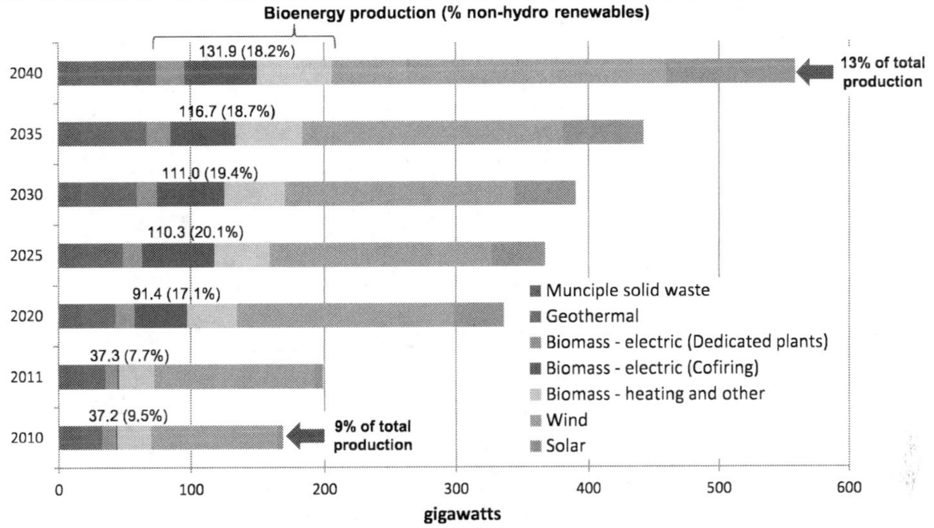

Source: EIA, Annual Energy Outlook 2013 (Table A16)

Projected Biomass Feedstock

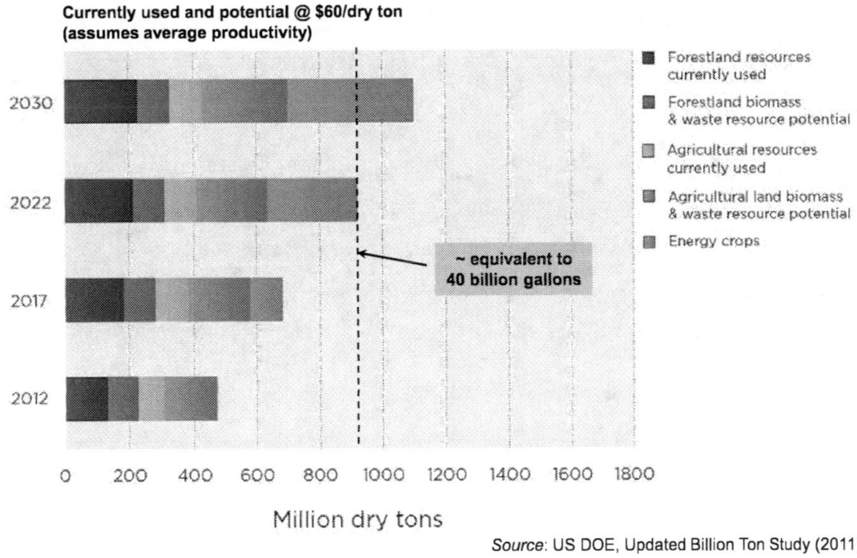

Source: US DOE, Updated Billion Ton Study (2011)

Euphoria or Fear??

Forest Management

- **Federal Policy:**
  - Clean Water Act, Endangered Species …
- **State Policy:**
  - Best Management Practices (nearly all forested states)
    - Supplemental biomass harvest guidelines (15 states)
  - Agriculture biomass harvest guidelines in development
  - California AB32: forestry offsets; land conversion rules
- **Governance:**
  - FSC/SFI/PEFC third-party forest certified (49.4 million ha)
- **Gaps:**
  - Inconsistent state practices, and BMP monitoring
  - Market recognition

Best Management Practices

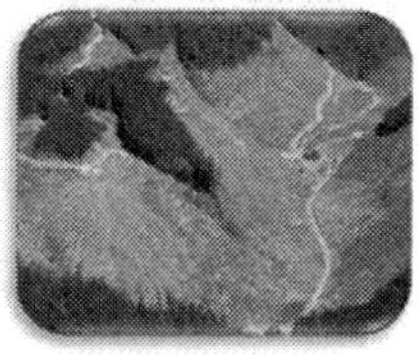

- **Three levels of BMPs:**
  - Required by law or regulation
  - Necessary to comply with law or regulation
  - Recommended practices
- **Guidelines for:**
  - Streamside management zones
  - Forest road construction
  - Site preparation
  - Reforestation
  - Prescribed burning
  - Minor drainage
  - Wildlife management
  - Stream crossings
  - Timber harvesting
  - Reforestation
  - Prescribed burning
  - Pesticides and fertilization
  - Endangered species management

Feedstock Source

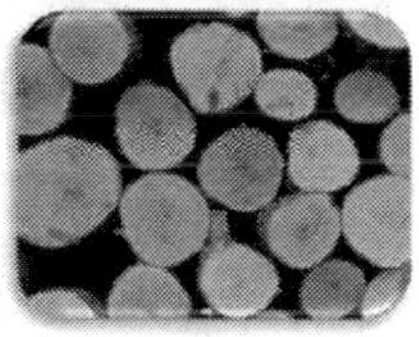

- **Federal Policy:**
  - Farm Bill Energy Title – BCAP (certified energy crops)
  - Federal Production and Investment Tax Credits looped: 1.1¢/kWh; closed-looped (2.2¢/kWh)
- **State Policy:**
  - Massachusetts RPS procurement standards to protect ecological functions (residuals vs. roundwood)
  - Oregon Producer Tax Credit for biomass ($10/green ton)
- **Governance:**
  - Pending EPA rules for biogenic carbon ("tailoring rule"); residuals/roundwood
- **Gaps:**
  - Inconsistent policy definitions of biomass
  - Non-durable policies

Tiered System

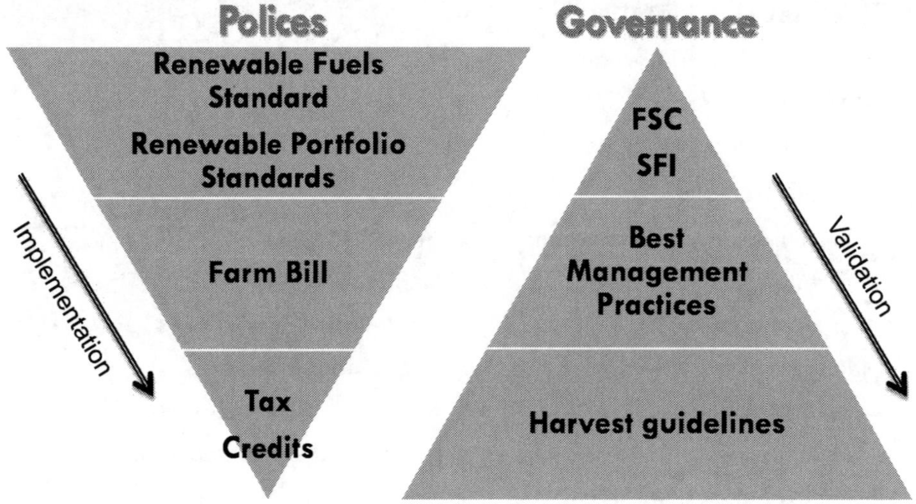

Global Governance—Local Standards

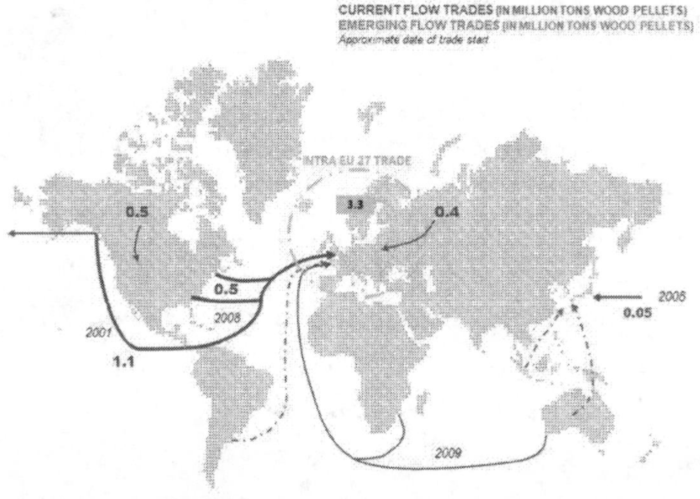

Adapted from Poyry 2011, Sikkema 2009

Map of wood pellets trade flows worldwide (2010 status).

For more information contact:

Dennis R. Becker
Associate Professor
Department of Forest Resources
University of Minnesota
drbecker@umn.edu
612.624.7286

Research Webpage:
http://www.forestry.umn.edu/People/Becker/index.htm

Center for Environment & Natural Resource Policy:
http://enrpolicy.forestry.umn.edu/Research/index.htm

MEEM 3502 - Product Realization 2 Final Project

Analysis of the Pogo Stick

Brian Haupt
Jack Babcock
Garrett Dubie

Instructor, Radheshyam Tewari

Submitted on December 12, 2014

Gimbal Engineering

Executive Summary

Gimbal Engineering has chosen to analyze the design and manufacturing of a Pogo Stick. Pogo sticks are traditionally comprised of a spring, shaft and foot pegs, designed for the fun and enjoyment of children. Gimbal Engineering composed of Brian Haupt, Garrett Dubie and Jack Babcock, who are all mechanical engineering undergraduate students at Michigan Technological University. Gimbal Engineering has gained its credibility and experience through a series of rigorous engineering classes and projects. Pogo sticks are designed to accommodate a leaf spring, rubber bands in tension, or simple coil springs to generate a response appropriate for the given age group of the intended audience. Main design practices include testing for forced input to determine frequency of system response and to determine the failure modes associated with pogo stick use. The main goal of analysis for this report will focus on the types of material used to manufacture pogo sticks, the vibration response of the system and the cycle life of an average pogo stick.

Table of Contents

Background

The pogo stick is most commonly known as a toy used to jump up and down off the ground through the use of a shaft, spring, and pedals to stand on. The idea for the pogo stick originated from a patent filed by George H. Herrington for stilts that utilized compression springs. Figure 1 is an image of the idea from 1881 [1].

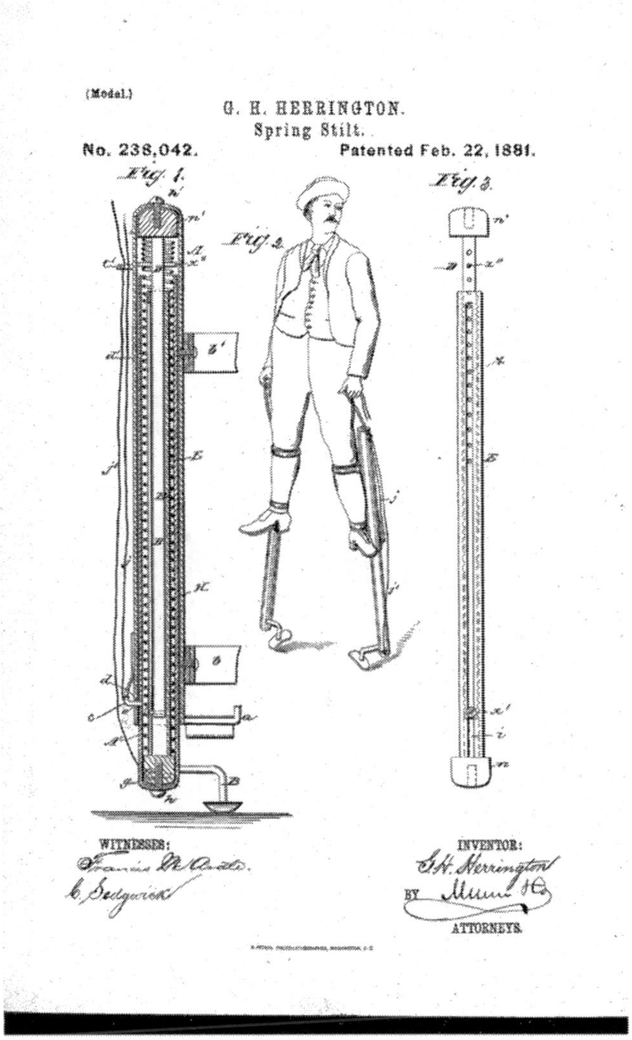

Figure 1: The illustration shows the compression springs that surround the two hollow shafts that are able to slide in and out of each other and also how the user would operate the stilts.

The purpose of this patented invention was "for leaping great distances and heights," according to Herrington. However, there were a few more innovations to the spring stilt that made it become more like the pogo stick most people know of today. The name "pogo" is thought by some to come from the names of the inventors of the more modern-looking pogo stick from Germany. Hans Pohlig and Ernst Gottschall filed a patent in Hanover in 1920 [2]. This pogo stick was made specifically for entertainment purposes and its German name, literally translated, means "spring end hopping stilt." The final, most common, design was patented by George B. Hansburg in 1957. This design is the two-handle design seen on nearly every iteration of the pogo stick since. Figure 2 is the patent illustrations from 1955 [3].

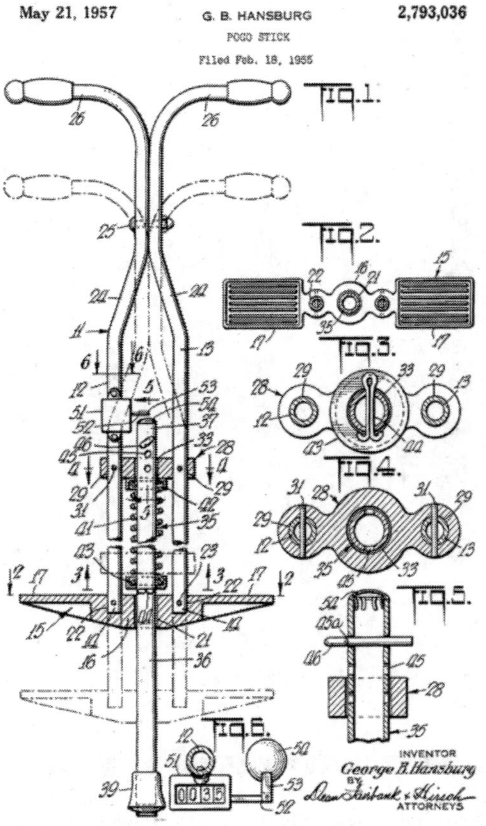

Figure 2: Common, "two-handle," pogo-stick design.

The group decided to analyze and research the pogo stick because of its simple design (that offers many different variations to analyze) and the potential for improvement on the latest trends. The technology and design choices that goes into the latest models of pogo sticks is a great opportunity to see different systems that essentially do the same thing. Also, since this product is also a toy to potentially be marketed towards children it is interesting to see the safety precautions that are implemented in the design.

Componentry/Functionality of Main Components

There are many different kinds of pogo sticks which will be discussed later. However, they all consist of a few main components.

1. The handlebars.
2. Pegs, or pedals to stand on.
3. The internal system that transfers the kinetic energy from falling to potential, and then back again to kinetic energy making it possible to jump up and down.

The handle bars serve as a source of balance for the user of the pogo stick. Since the handlebars are not a part of the critical component design, it is up to the designer and manufacturer to make sure that they do not get in the way of the user or other components yet still provide a reliable grip so that the user can maintain their balance. The internal system that converts the energy is the most important part of the pogo stick. Without it, it would only be a shaft with handles and pegs. The most common is similar to the initial invention seen in figure 1. Others consist of multiple, high-strength rubber and rely on its elasticity to spring back, providing transfer of energy and motion. The last common design is a leaf spring or a bending rod that is put into compression, bends and flexes back providing the transfer of energy. These designs will be discussed later and in greater detail.

Manufacturing Aspects

The original pogo sticks, crafted of wood, can be seen on a manufacturing scale as more of a prototype than a manufactured product because of the very low production volumes. As the pogo stick became a full scale product and was being shipped around the world, the wood was replaced with bent steel tubing to avoid rotting of the sticks during shipping [4]. During this era the sticks were composed of this bent steel tubing and held together by bolts. The spring was more than likely a simple spring steel coiled /forged on a mandrel and heat treated and coated (most likely painted in this era) to avoid heavy corrosion. As production volumes and manufacturing technologies jumped in the late 20th century the pogo stick designs shifted away from the bent steel tubing into a multi-part steel or aluminum extrusion frame which was often welded together. This change was likely a strength and packaging optimization, and the improved automation of the manufacturing allowed to maintain a relatively low cost with the increased number of processes. In this era many parts were probably still painted, however some parts in more recent years may have been powder coated for improved corrosion resistance and available manufacturing technologies. In this era many plastic trim and foam pieces were also added, which will be further discussed in the Latest Trends/Advanced Technologies section of this report. These plastic portions are likely injection molded or vacuum-thermo formed depending on the dimensions of the part because this is much cheaper than machining pieces and can reduce waste materials. The 21st century gave way to cheaper and more accessible materials and processes such as higher strength aluminums, a larger variety of steels, cheaper automation, and better coating technology. These factors again allowed for more complex pogo stick designs (as discussed in the Latest Trends section) without increased manufacturing costs. Many of the pieces, such as the main body of the pogo stick, could be composed completely of aluminum extrusion which may be pre-purchased or prepared on site. Springs can still be made by bending hot spring steel extrusion around a mandrel, but often times in large production settings springs are made using an automated forming machine to decrease the cost and cycle time per part. A smaller scale machine is shown below [5]. The machine operates by uncoiling spring steel from a reel and feeding it through the machine. a series of levers, arms, and dies, are used to form the spring into the desired shape. In this case, the spring is a simple compression spring so the only processes which the machine must perform is the coiling and cutting of the spring. For use in a pogo stick, the springs may be pre-set to avoid deformation but would probably not be ground square.

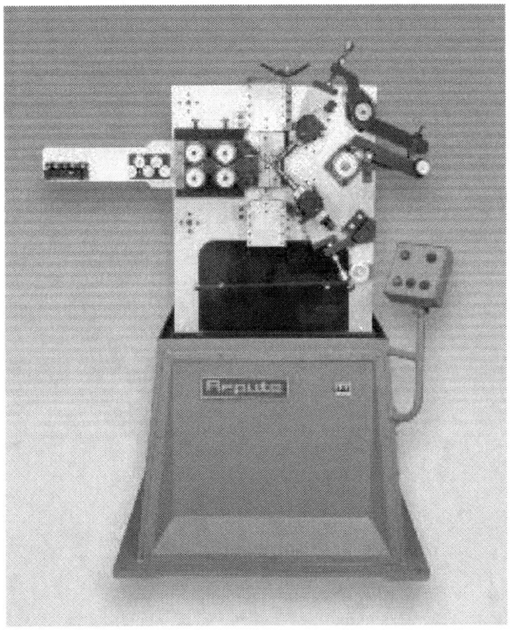

Figure 3: Automated forming machine for the manufacturing of small springs

With an increase in technology, the 21st century also came with a wave of lawsuits. From dining to transportation, companies are pushing safety more than ever to avoid liabilities, as must be true for companies manufacturing pogo sticks. Often times these safety considerations can add complexity to a design, and can make manufacturing more difficult which could increase the cost of a product. In addition to these safety considerations, there are more stringent policies relating to the waste and manufacturing of goods. One large difference between manufacturing in current times and in the past is the use of scrap materials. Historically, it may have been possible to purchase a product with very little scrap material in it. The reality of the 21st century is that almost any product a consumer would purchase has some scrap material in it. From a manufacturing standpoint this increases the need for quality control of both raw materials and product. As mentioned previously, the extrusions used in manufacturing pogo sticks may be prepared off or on-site. Regardless of where the extrusion is prepared, the metal content can vary significantly based on the scrap that was used in preparing the raw materials, and hence the strength of the material can vary from batch to batch. Luckily, with the increased process automation that was discussed previously it is possible to sample material and product more frequently to ensure low-liability products.

Design Analysis

For vibration analysis, every pogo stick is designed as a single degree of freedom (SDOF) mass-spring-damper system, with a damper optional, depending on child safety versus performance. The entire system is simplified in figure 4, where the spring is the total stiffness of every spring on the pogo stick, and the damper is the total damping in the system. The main design focus, however, is the spring since most pogo sticks are only equipped with rubber stoppers to provide very minor damping in comparison to spring stiffness.

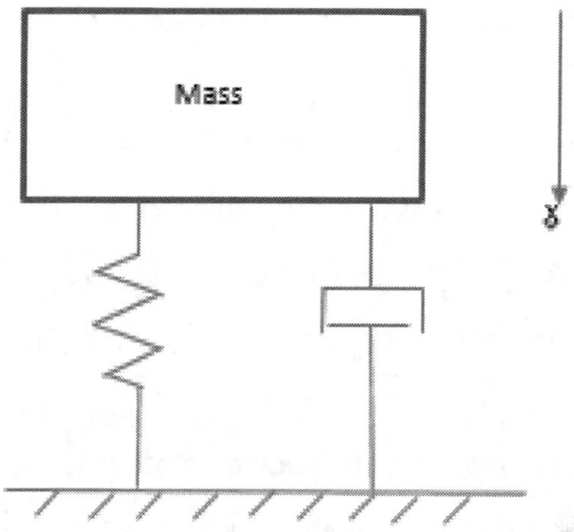

Figure 4: Mass, spring and damper pogo system simplification

By simplifying the pogo stick into a mass, spring and damper system, a free body diagram can be drawn. The forces which act on a pogo stick are the weight, the spring forces caused by initial weight deflection and the response to the input force from the user, and the damping force acting against the input.

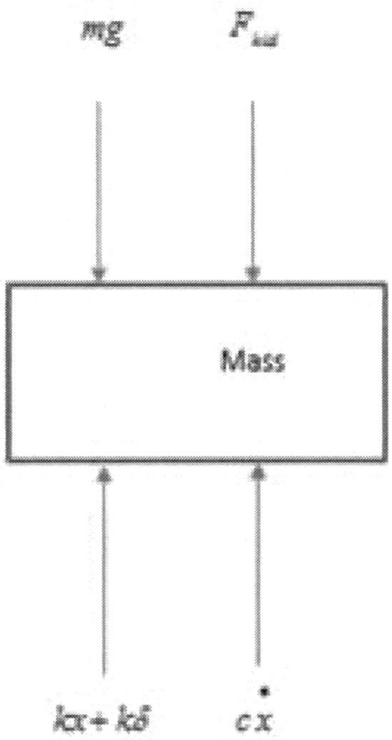

Figure 5: Free body diagram of simplified pogo system

These forces can thus be summed and reorganized to represent the response of the system to a given input. This representation is shown in equation 1 below. The input force will take the form of a step function since the input is the iterative jumping onto the pogo stick.

$$m\ddot{x} + kx + c\dot{x} + = \dot{F}_{user}$$

The main purpose of a pogo stick is to manipulate the amount of excitation is produce per given input, that is, how to regulate the amount of jump produced by a child's input. [6] In order to produce such an intricate analysis, several parameters need to be established through experimentation. The total mass, equivalent stiffness, equivalent damping, and frequency of input are manipulated to achieve the perfect pogo stick design [7].

The total mass of the system is a simple parameter to test for. Apart from simply measuring the pogo stick on a scale, the mass of the user must be taken into account. [8] Average mass values for given age groups is well researched in the health field, with numerous tabulated results for the masses of different age groups. The manufacturer must first establish a desired age group, retrieve the correlating mass, and design around that mass. The most effective procedure to follow, however, would be to test for the upper, lower, and mid range values of mass when testing for stiffness and damping.

Equivalent stiffness is another non trivial experiment. The pogo stick will stay at rest and measured to length. A known mass, preferably within the intended user's weight range, will be placed on the spring, and the deflection is measured. [9] In experiments to follow the full compression forces for a pogo stick spring will be tested for to determine the failure modes.

The amount of damping the system undergoes is also a well tabulated data source. Usual amounts of damping is centered on the rubber stopper against concrete or pavement. [10] The associated damping coefficient is about 250 n-m/s. Other forms of damping is to pressurize the spring housing allowing the spring to act against a pocket of air. This air pocket will act as a damper against the input force and would most likely be installed in children's pogo sticks focused on ultra safety. For the sake of this report, such a system will not be considered, since ultra high performance or super jumping pogo sticks will also not be discussed.

An example for nominal values used to test for pogo stick response will be used to show a typical pogo stick's vibration pattern. Using matlab simulink to reconstruct a mass spring damper , the forced response of the pogo stick X can be determined as a function of time. The model will be for a 30 kg user, which is a rough average for a twelve year old child, operating at a period of one second, a damping value of 256 n-m/s, which is estimated damping between a rubber stopper and concrete, and a stiffness of 1000 n-m, which is within usable pogo stick range. This test resulted in the following response, described in figure 6.

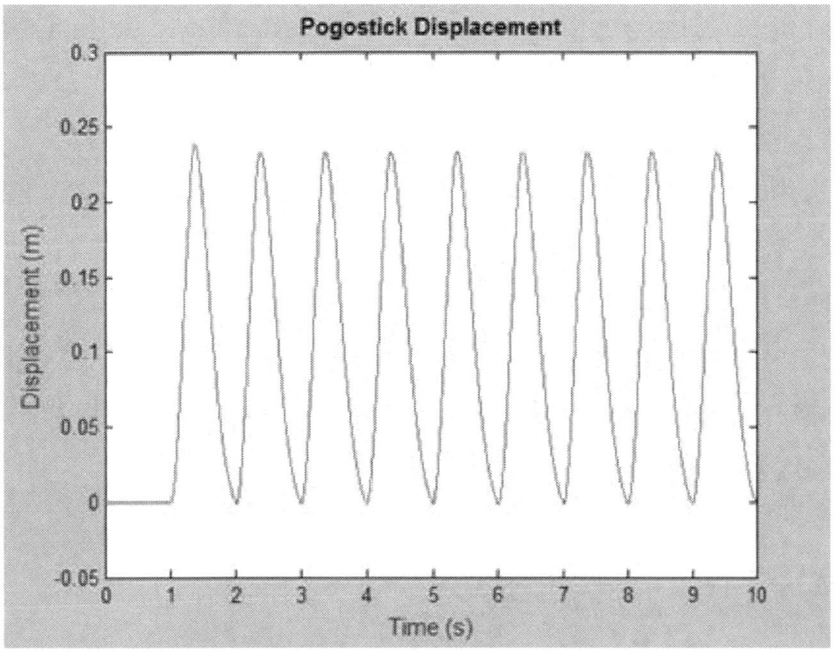

Figure 6: Pogo stick system response

Figure six describes the level of displacement the pogo stick will produce given a twelve year old child's input. The greatest output is about a quarter of a meter, which is expected, on average, on a pogo stick designed for a child. Also, the applied frequency outputs a frequency that matches the initial expected value of one jump per second, as the steady state response of the system reaches a peak at each second.

The only realistic approach at manipulating this response, however, is to change the equivalent stiffness . Since the damping is fixed for a rubber stopper against concrete, the force input frequency is fixed given a child's input, and the mass is an externally controlled variable, the stiffness is the single manipulated design feature to produce different jumps. This reality is desirable, since there is not any complex research and development manufacturer must pay into to produce a better model pogo stick, that is, there is only one variable to the displacement output. This means the response changes linearly as stiffness changes.

For product development analysis, designing for pogo failure becomes the center of analysis. The two types of springs for analysis will be the leaf spring and coil spring design. The third design, comprised of several rubber bands in tension, is a high performance model and beyond the scope of this report. Also to be tested for failure will be the welds on the footpegs.

For coil springs, the ends will be considered grounded and squared to comply with shaft dimensions, as shown in figure 7. An educated guess will suggest the major diameter will fall within a few millimeters short of the shaft inner diameter, and the pitch will vary depending on desired spring stiffness. This type of problem is most closely associated with cam and spring design, where a spring is not statically loaded, but rather succomed to a step input, much like a revolving cam against a spring.

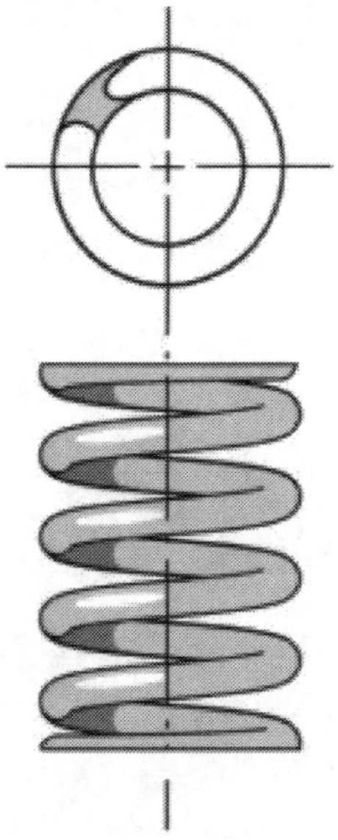

Figure 7: Physical representation of a spring with squared and ground ends

The parameters required to determine the appropriate coil spring design includes the frequency of applied force, the weight of the user, the applied force of the user not including weight and allowable spring length. Spring length can be iteratively found using a base example and changing for different desirable spring stiffness. The clash allowance ought to be assumed as 10% since the weight of the user will vary. The diameter will need to comply with the shaft inner diameter.

With the aforementioned parameters defined, the spring length, coil number, and solid length can be determined. Using equation 2: below

$$Ck_w = \frac{\pi \tau_{max} d^2}{8F_{max}}$$

where tau max is the allowable shear stress the spring can undergo, F max is the maximum assumed force, D is the spring diameter, and CK wahl is the spring ratio multiplied by the Wahl factor for cyclic loading, can the spring ratio and pitch diameter can be found. Ck wahl is used in a stress correction factor chart for helical springs. The spring stiffness will be found experimentally at a known applied force. The number of coils required is known through equation 3 below.

$$N = \frac{dG}{8C^3k}$$

Where G is the shear modulus of desired spring. Usually this number will be associated with steel or aluminum shear modulus'. The solid length is found using equation 4.

$$l_f = l_s + \frac{F_{Solid}}{k}$$

Buckling is determined using the free length and equation 5. This value will be used to determine if the spring will undergo suntable movement and fail.

$$\frac{l_f}{D}$$

Spring surge does not need to be calculated since the frequency of spring excitation is very small at one Hz compared to the natural frequency.

The next type of spring commonly used is the leaf spring. For iterative loading, the analysis is very similar, but determining the spring rate changes slightly. The type of spring used in leaf spring pogo sticks is the semi elliptic style leaf spring, shown .in figure 8 There is only one leaf to design.

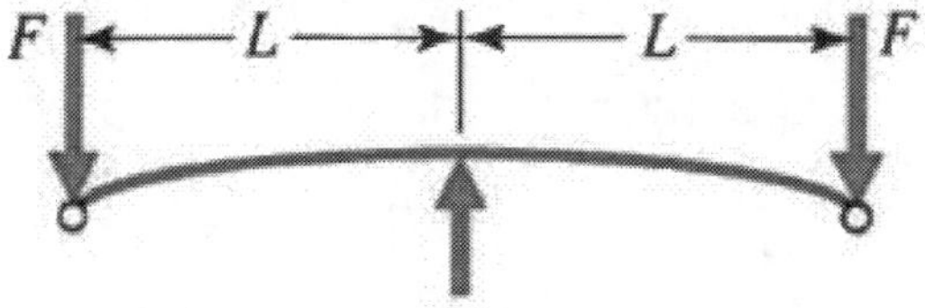

Figure 8: Semi-elliptic leaf spring

The stress of the leaf spring can be determined using equation 6, which can be used for maximum and minimum applied stresses. The length is dependent on the user intended, and the force will be half of the input force on each end.

$$\sigma = \frac{6FL}{bh^2}$$

The only welded component to a pogo stick are the footpegs, and in some cases the handlebars. For the sake of this design it will be assumed that the handlebars are simply bent horizontally at 90 degrees, limiting weld design to only the footpegs. Each footpeg will receive half of the downforce produced by the user. The footpeg welds must be able to withstand iterative down force to avoid failure. The welds will be butt joint welds.

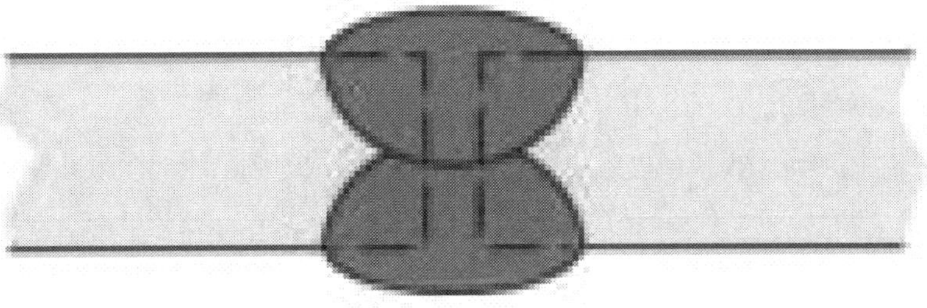

Figure 9: Butt weld for foot pegs

The design of a weld is determined by the safety factor desired by the manufacturer. With a known applied input force and a known yield stress of the desired material, a throat area can be designed for the safety factor desired by the manufacturer. The Safety Factor value will most likely fall between 10 and 15. This relation is given by equation 7 below.

$$F = \frac{.58 s_y A}{SF}$$

Latest trends/Advanced technologies

For most people the first thought that comes to mind when we hear the term "pogo stick" is either an old rusted bit of steel tubing with a rubber foot and a coil spring, or for some it may be even more traditional, such as a wooden framed pogo stick with a coil spring. Over the past century technologies which were once military class have been applied to everyone's favorite backyard toy, the pogo stick.

In the last few decades improvements to the pogo stick have involved different plastics and foams. From an engineering viewpoint, these changes have been identified as efforts to make a more aesthetically desirable product, to make a product which is lighter (and therefore easier to control for a child) and to reduce the vibrations experienced by the operator during common use. The 90's came with a demand for vibrant, exciting colors. In order to sell more product, pogo sticks were more than likely given plastic components. Adding these components was likely cheaper than the extra manufacturing step to add paints and stickers and allowed for easy structural reinforcement. As cheap plastics and foams became more readily available it became easy to apply these foams to the handles and bodies of the pogo stick to help avoid injury when falling during use. Additionally, the damping that was added by applying a foam coating would theoretically reduce the vibrations that the rider experienced due to the impact with ground and the ringing of the spring.

Several attempts have been made in recent years to move away from the traditional coil spring and toward a more elegant solution without such harsh noise and vibration performance. Additionally, there has been a push for both a safer stick, and one that will allow the user to jump higher. Perhaps the safest configuration that has been observed is the ball style pogo stick, as it reduces the number of moving parts that could make it a dangerous toy for small children and increases the contact area with the ground (hence reducing the stress that would be experienced if a user hopped on someone's hand or foot).The ball style pogo stick, example shown below, uses a rigid body with a ball of compressed air on the lower and costs around $40 [11].

Figure 10: A safe version of a pogo stick designed for children

Pogo sticks geared toward an older audience come with more sophisticated and sometimes more dangerous designs. There are three designs which can summarize the more advanced configurations for the pogo stick. These three designs are the BOWGO, the VURTEGO, and the FLYBAR as termed by various companies and are shown in the image below [12].

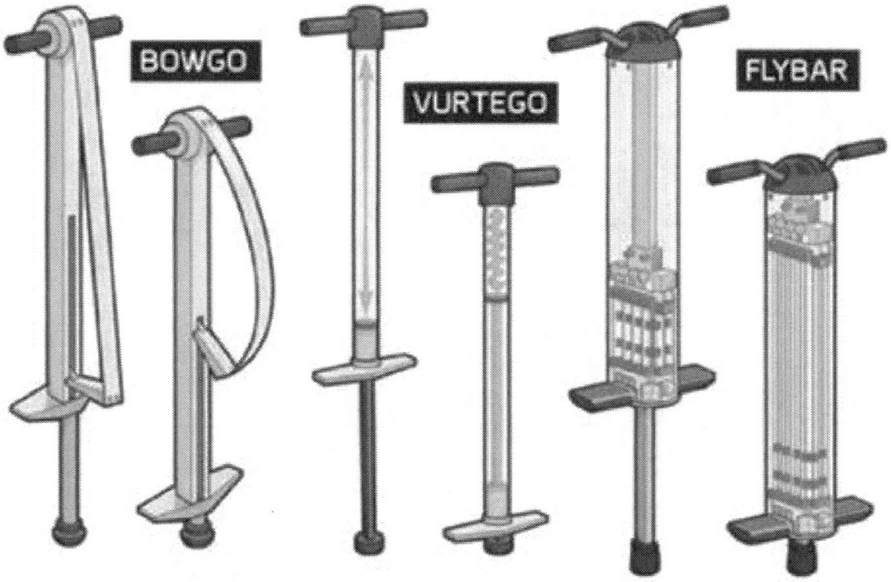

Figure 11: Illustrations of several modern pogo sticks [12**]**

The BOWGO is currently mass produced and sold by Razor (the company that brought us the Razor Scooter a few years back) and operates using a design similar to a leaf spring. It is unclear as to whether the device has a linear or non-linear spring rate, however a non-linear spring rate could allow for a wider weight range of user. As shown in the image below the leaf spring essentially acts in the system in a similar manner to the coil spring, but undergoes a bending loading when compressed. While this version of the pogo stick sells for around $200, there have been many complaints of injuries due to the leaf spring contacting the operator's legs during use [12].

The VURTEGO was designed to be geared toward adult operators, as it allows for a much heavier weight, however it could also be scaled for a child. The VURTEGO operates using a plunger system which creates an air spring when the foot of the pogo receives a load. These pogo sticks have been positively accepted on the market so far, as there is little possibility for injury due to moving parts and allows the user to jump higher than other sticks but comes at the hefty price of up to $400 [13].

The last of the major designs in pogo sticks today is the FLYBAR. With a price tag of around $200, the FLYBAR hit the market receiving mixed reviews. While most users were satisfied with the operation of the device when it was in operating condition, many users complained that the rubber bands, used as springs, "dry-rotted" or snapped within reasonable use [13]. Additionally users said that the device was bulky and heavier than other pogo sticks. An image of the bulky spring system is shown below.

Figure 12: The elastic mechanism within a Flybar pogo stick.

Materials Selection

Each brand's pogo stick uses different materials due to their main component design. A commonality between most pogo stick designs is the aluminum housing and frame. Aluminum provides a cost effective and strong frame for the internal workings of common, "department store," pogo sticks. However, some manufacturers of higher-end, extreme sports pogo stick have used a carbon-based polymer to encase the inner-workings of the pogo stick. Unlike aluminum, carbon polymers are tough to dent and require a lot of force to fracture the housing. If a user were to jump as high as 10 feet (as some marketers claim is possible), and then fall, the user should be more worried about their own body rather than their pogo stick.

As for materials used in the main components, the many design choices have just as many materials choices. The BOWGO stick uses a fiberglass bending rod to provide the springback. Depending on the type of fiberglass used, it can have a compressive strength from 20 to 50 ksi. This is more than enough to withstand an average adult bouncing up and down for extended periods of time.

The FLYBAR brand pogo stick uses elastomers that determine the force of the bounce. Their website doesn't go into detail, however, these thrusters act like hyper-strength rubber bands and undergo high tensile loads as the user hits the ground then the forces in the elastomers will spring back launching the user high into the air. The FLYBAR pogo stick also employs plastic bearings in the design to provide smooth movement across all of the inner workings. As with any bearing, they recommend keeping them free from sand and other grit so that the fatigue life will be as long as possible.

Finally, the VURTEGO pogo stick utilizes both aluminum and stainless steel designs. Both designs are built for a pneumatic energy transfer system in which the air inside the tube compresses as the user bounces down and that pressure is what springs the user back up. Again, whether it be aluminum or steel, these materials have the strength required to withstand the pressure of the air inside the tube as well as the tensile force the user places on the apparatus every time they slam onto the ground.

Conclusion

In conclusion, pogo sticks are mass spring damper systems designed and manufactured for the enjoyment of children. Pogo sticks come in great variety, and with each different iteration, the amount of jump achieved is manipulated to accommodate the user of the modern day. Pogo sticks have become more advanced, using lighter technology as reminiscent of the space age. Pogo stick competitions are held, which greatly relies on high performance pogo stick designs.

Challenges faced by the team was determining the real history on pogo sticks. Pogo sticks are fairly simple, and do not allow for too much variation. It required a lot of digging to see any real change in how pogo sticks changed. The only remedy for this challenge would be time and the availability of pogo stick history. Another challenge the team faced was determining failure design. Pogo sticks are not fixed to the ground, rather there is an unknown impact force associated with every jump. Although this is outside the scope of the class, having the ability to test for this value would have been useful to produce a more wholesome report.

Comments and Recommendations

Our Team felt as though the work being requested of us for this report was fair and allowed for us to analytically use what we learned in our product realization courses here at Michigan Technological University. While we were not required to perform extensive calculations or generate generic cad models of any kind, we were able to use what we learned without wasting valuable time. We appreciate that we were not asked to create another theoretical product, as this process becomes quite old after Engineering Fundamentals, Engineering Design Process, and Product Realization I. That being said, it initially seemed as though the requested page limit is somewhat extensive for the scale of the project, and could be shortened by 25%. Additionally, it may be beneficial to reduce the group sizes to 2-3 members with these shorter reports, again due to the nature and scope of the project. Our team struggled in selecting what product to analyze and it would be nice if you were able to provide future classes with a list of possible products to help the brainstorming process. We firmly believe that as a first iteration, this project was successfully achieved its intended goals and in the future iterations it will continue to improve.

Acknowledgements

Our team would like to thank Radheshyam Tewari and the other professors of Michigan Technological University for their contributions to the education of each team member in the field of Product Realization and Mechanical Engineering. The knowledge that was gained through these courses will stay with the team in all that we do and we look forward to continued relations with the university.

References

[1] "Patent US238042 - Spring-stilt." Google Books. Accessed December 10, 2014. http://www.google.com/patents/US238042#v=onepage&q&f=false.

[2] "Patentschrift Nr 352704" DEPATISnet. Accessed December 10, 2014. https://depatisnet.dpma.de/DepatisNet/depatisnet?window=1&space=menu&content=treffer&action=pdf&docid=DE000000352704A

[3] "Patent US2793036 - Pogo Stick." Google Books. Accessed December 10, 2014. http://www.google.com/patents/US2793036?dq=2793036.

[4] "History of Pogo Sticks." History of Pogo Sticks. Accessed December 11, 2014. http://www.pogostickusa.com/history.htm.

[5] "Image." http://img1.tradeget.com/repute_industries/ALDCJYM41rsm20.jpg. Accessed December 10, 2014.

[6] G. B. Hansburg, "Pogo Stick," U.S. Patent 2793036 A, 02/18/1955

[7] J. P. Laird, "Analog Simulation of Mechanical Impact," unpublished

[8] Disabled World (August 7th, 2014). Average Height to Weight Chart-Babies to Teenagers [Online]. Available http://www.disabled-world.com/artman/publish/height-weight-teens.shtml

[9] R.W. Chabay, B. A. Sherwood, "The Stiffness of an Interatomic Bond," in Matter and Interactions, 3rd edition, Hoboken, John Wiley &sons, 2012.

[10] S. H. Sawant, "Experimental Verification of Damping Coefficient by Half Power Band Width Method," in International Journal of research in Aeronautical and Mechanical Engineering, Vol. 2, Issue. 7, pp. 8-13, Kolhapur, India, 2014.

[11] "How the pogo stick lept from classic toy to extreme sport." http://www.smithsonianmag.com/science-nature/how-the-pogo-stick-leapt-from-classic-toy-to-extreme-sport-18123864/?no-ist=&utm_campaign=Feed%3A+smithsonianmag%2Fscience-nature%2Ftechnology-space+%28Technology+&+Space+|+Smithsonian.com%29=&page=1. Accessed December 11, 2014

[12] "Amazon product page." http://www.amazon.com/Razor-10067112-BoGo-Pogo-Stick/dp/B0047ERP8K. Accessed December 10, 2014

[13] "Amazon Product Page." http://www.amazon.com/Vurtego-V3-Pogo-Stick-Medium/dp/B00IIX412Y/ref=sr_1_1?s=sporting-goods&ie=UTF8&qid=1418265312&sr=1-1&keywords=vurtego+pogo. Accessed December 10, 2014

Biogeochemical Research Priorities for Sustainable Bioenergy Feedstock Development in Pan-America

H. Gollany[1], B. Titus[2], D.A. Scott[3], H. Asbjornsen[4], S. Resh[5], R. Chimner[5], A., D. Kaczmarek[6], L. Leite[7], A. Ferreira[7], K. Rod[8], J. Hilbert[9], M. Galdos[7] and M. Cisz[5]

[1] USDA Agricultural Research Service, OR, USA
[2] Natural Resources Canada, Canadian Forest Service
[3] USDA Forest Service, AL, USA
[4] Natural Resources & Environment, University of New Hampshire, NH, USA
[5] Forest Resources & Environmental Sciences, Michigan Technological University, MI, USA
[6] USDA Forest Service (ret'd), Aiken, SC, USA
[7] EMBRAPA, Brazil
[8] Washington State University, USA
[9] INTA, CIA, Buenos Aires, Argentina

RCN Conference on Pan American Biofuels & Bioenergy Sustainability
Recife, Brazil, 22-25 July 2014

Natural Resources Canada Ressources naturelles Canada

Canada

Soil and nutrient cycling principles are universal but operate at local (site) level within the landscape

- Local soils, sites/ecosystems, climate
- Local feedstock type (demand on soil reserves; quality of OM input)
- Local management practices (including intensity)

Pedro Martin Ureta's farm, nr General Levalle , Cordoba, Argentina; whudat.de

Importance of OM

Physical properties

- Reduced OM → reduced soil aggregate size

bulk density ↑ → rooting ↓ ↘ nutrient uptake ↓

water repellency ↑ → sorption ↓ ↗

↘ erosion ↑

- Few studies on aggregates in forestry *(exception:* Lupi et al. 2007.
  - Lupi AM, Conti M, Fernández R, Cosentino D, López G. 2007. Efecto de las prácticas de repoblación forestal sobre el carbono orgánico del suelo y la estabilidad de los agregados en el noreste de Argentina. *[Effect of reforestation practices on soil organic carbon and aggregate stability in northeastern Argentina. In Spanish with English abstract.]* Investigación Agraria: Sistemas y Recursos Forestales 16:230–240)

Importance of OM

Physical properties

- Reduced OM → reduced soil aggregate size
- Roadbed for equipment (mostly forestry)
  - Reduces rutting (erosion) and compaction

Importance of OM

Physical properties

- Reduced OM → reduced soil aggregate size
- Roadbed for equipment (mostly forestry)
  - Reduces rutting (erosion) and compaction
  - But compaction in forestry can sometimes increase productivity (soil water-related?)
    - *Long-Term Soil Productivity (LTSP) network in Canada & US: benefit of collaborative networks with common research designs (see Eric Vance, on Steering Committee)*
    - *(See also CIFOR network in tropics and sub-tropics)*

Importance of OM

Physical properties

Biological properties

- Reduced OM

  →reduced earthworms, fungal growth, microbial community structure and function, below-and above-ground saproxylic organisms (and larger above-ground organisms)

  →affects decomposition and biogeochemical cycling processes in soil

7

Importance of OM

Physical properties

Biological properties

Chemical properties

- OM removal → reduced nutrient availability
  - Loss of nutrient capital in removed OM
  - Changed dynamics (pH, CEC, electrical conductivity)

8

Importance of OM

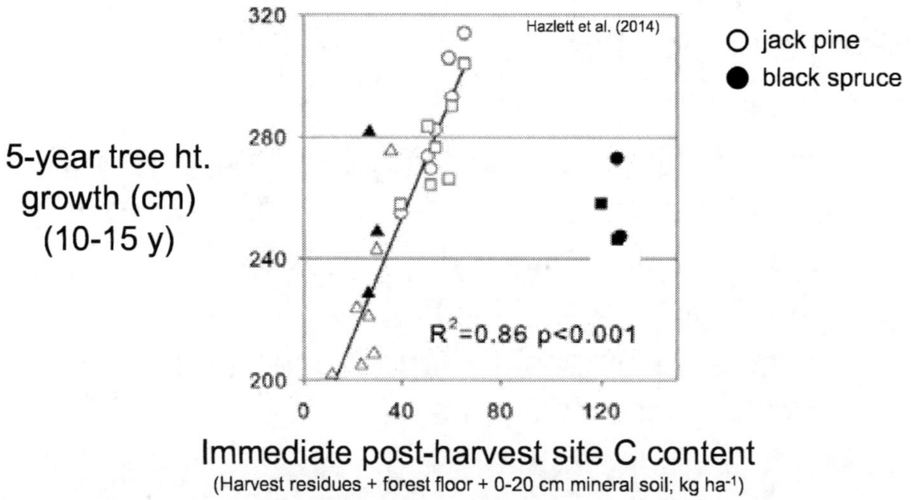

In forestry, OM can be "friend or foe" (Prescott et al. 2000)

Importance of OM

Hazlett, P.; Morris, D.; Fleming, R.L. 2014. Effects of biomass removals on site carbon and nutrient retention and jack pine tree growth across a site productivity gradient in upland boreal forests of Ontario. 2014. Soil Science Society of America Journal.

Prescott CE, Maynard DG, Laiho R (2000) Humus in northern forests: friend or foe? For. Ecol. Manage. 133:23–36

Importance of OM

- *What are the biomass retention thresholds that sustain soil properties and functions for different combinations of crops, management systems, sites and soils? (this threshold interacts with biodiversity threshold)*

Nutrient management

- Chemical fertilization
  - Fertilization ≠ OM retention
  - Fertilization + OM retention → improve both yield & soil quality

Pendleton Plots, OR

60- to 70-year data for winter wheat, with fallow

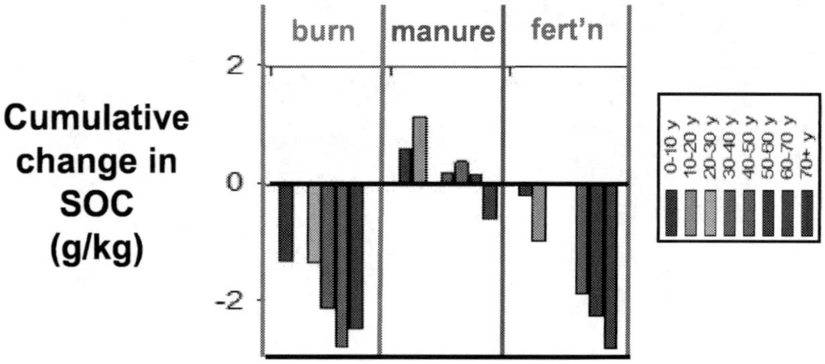

Nutrient management

- Chemical fertilization
  - Fertilization ≠ OM retention
  - Fertilization + OM retention → improve both yield & soil quality
  - Forest Stewardship Council (FSC): discourages forest fertilization in some countries/regions
- Conserve P ("peak phosphorus")
- Ash from bioenergy production
  - If addresses limitations; lacks N and P; can be toxic
- Biochar
  - Outcomes variable; not all biochars are created equal
  - Need strategic research approach to elucidate mechanisms (Jeffery et al. 2011)

Nutrient management

- *What nutrient management/amelioration practices are economically and environmentally sustainable?*
  - How do we better define "precision" agriculture/ agroforestry/forestry to minimize fertilizer use?
  - How do we optimize nutrient availability and plant uptake while conserving OM (and P)?
  - On which sites does addition of ash or biochar increase soil productivity/long-term soil C sequestration?

Process-based models

In the absence of long-term data, reliable models would help us predict the future (within limits)

- Long-term results can be very different from short-term results
- Field trials for generating long-term data can be expensive to maintain
- Current models have varying degrees of predictive capability

How can we improve models for predicting long-term soil impacts of biomass removals?

Off-site impacts

- GHGs: CO_2, N_20 (nitrous oxide) and NO_x
  - Decomposition → CO_2 (soil respiration)
  - Increase biomass growth (including fertilization) → increase CO_2 sequestration
  - Fertilization → 69% of U.S. N_20 emissions
    - N_20 is 200-300x more potent (per-molecule) than CO_2
    - (Much less of a problem in forestry)

Off-site impacts

- GHGs: CO_2, N_20 (nitrous oxide) and NO_x
- Eutrophication
  - >400 eutrophication-associated hypoxic "dead zones" in oceans (245,000 km^2)
    - (But 1 million km^2 natural hypoxic zones; Helly & Levin 2004)
  - Optimal fertilization ("precision" management)
  - Reduce leaching/run-off with SRWC, perennial grasses in boundary areas

Off-site impacts

- GHGs: CO_2, N_20 (nitrous oxide) and NO_x
- Eutrophication
- Sedimentation
  - Erosion (OM, topography, soil disturbance)

Off-site impacts

- GHGs: CO_2, N_20 (nitrous oxide) and NO_x
- Eutrophication
- Sedimentation
- Waste disposal
  - Ash
  - Biochar

19

Off-site impacts

- GHGs: CO_2, N_20 (nitrous oxide) and NO_x
- Eutrophication
- Sedimentation
- Waste disposal

- *How can we reduce off-site impacts?*
- *How can we use "waste" products to increase growth (if this does not cause other problems)?*

20

Soil sustainability indicators

Qualitative (±) and quantitative indicators of soil "quality" and function

- Land management
  - Site-level prescriptions
  - Landscape-level planning
- LCA
  - If metric can be scaled spatially and temporally
- Governance
  - Guidelines, BMPs, regulations, policies
- Markets
  - Certification
  - Trade (import regulations)

Soil sustainability indicators

Suggested indicators include (McBride et al. 2011):

- Total organic C and N
- Extractable P
- Bulk density
- Stream concentrations of NO_3, P, suspended sediments
- GHG emissions
- productivity

Soil sustainability indicators

- *Which soil indicators are*
  - Scientifically valid, operationally feasible, and affordable?
  - Universal vs. crop- or site-specific?
- *How can we aggregate multiple indicators?*
  - e.g., C *vs.* C + N *vs.* C + N + base cations
- *How can we scale up spatially?*
  - feedstock supply chains → investor confidence, policy
  - LCA → soil impacts in context of entire supply chain

Suggested biogeochemistry research strategies

- Long-term field trial remeasurements
  - Comparative studies of old field trials
  - Money cannot buy time
- Collaboration increases value: "multi-" principle
  - Multi-feedstocks, -disciplinary, -agency, -national
- Build networks of new trials addressing emerging issues with coordinated common designs and measurements
  - Shared insight, more efficient, better comparisons, statistically stronger meta-analyses
  - E.g., CIFOR, LTSP, etc.
- Strategic syntheses, reviews, meta-analyses, "thought" articles

Suggested biogeochemistry research strategies

- Link research networks to bioenergy operations
  - Operators install plots as they go
- Scale up from site to landscape level
  - (Weyerhaeuser; Eric Sucre)
- Comparative case studies integrating biogeochemical research with other disciplines and industry
  - (e.g., PIRE)
- Balance research funding opportunities within bioenergy sector
  - Technological research with (not at cost of) sustainability research

25

Obrigado!
Gracias!
Merci!
Thank you!

Soil Organic Carbon Measurement Protocols: A USA and Brazil comparison and recommendation

"Biogeochemical Cycles" session

PAN AMERICAN BIOFUELS AND BIOENERGY SUSTAINABILITY
AN NSF RESEARCH COORDINATION NETWORK

Recife, Brazil
July 23, 2014

Maggie Davis
Oak Ridge National Laboratory

In collaboration with Keith Kline (ORNL), Bruno Alves (EMBRAPA), Marcelo Galdos (CTBE), Douglas L. Karlen (USDA), Dana Abulebdeh (UNC-Charlotte), as well as staff from Center for BioEnergy Sustainability (CBES) and the Climate Change Science Institute (CCSI)

Special thanks to Kristen Johnson and Alicia Lindauer, Alison Goss Eng, among others at the Department of Energy (DOE), BioEnergy Technologies Office (BETO)

This research was supported by the U.S. Department of Energy (DOE) Bio-Energy Technologies Office and performed at Oak Ridge National Laboratory (ORNL). Oak Ridge National Laboratory is managed by the UT-Battelle, LLC, for DOE under contract DE-AC05-00OR22725. The views in this presentation are those of the authors, who are responsible for any errors or omissions.

CBES Center for BioEnergy Sustainability
Climate Change Science Institute
Oak Ridge National Laboratory

Outline

- Background: SOC importance
- Relevance
- Methods:
  - Literature review
  - Comparison
  - Analysis with co-authors
- Emerging recommendations
- Concluding remarks

Oak Ridge National Laboratory

Background: SOC importance

SOC:

- is a reservoir containing about two to three times more carbon than the atmosphere or vegetation.
- can significantly impact the global carbon cycle.
- reserves can be lost through inappropriate management practices such as
  - excessive tillage, drainage, or removal of plant biomass, or other disturbances.

A simplified version of the carbon cycle in vegetation and soil.
Source: Steiner, Christopher 2008. Accessed from: http://www.biochar.org/joomla/index.php?option=com_content&task=view&id=67&Itemid=7&limit=1&limitstart=4

Conversely, improvements in land management (e.g., no-tillage, cover crops, forage-based rotations, etc.) that promote sequestration of SOC can simultaneously sustain or even increase SOC and contribute to the drawdown of global carbon dioxide (CO_2) emissions

OAK RIDGE National Laboratory

The performance paradox: You can't manage what you can't (or don't) measure.

Source: NASA Images

Background: SOC importance

- SOC is often seen as the most important indicator of soil quality
- Measuring and understanding changes in SOC has thus become an essential activity for land managers around the world

Category	Indicator	Units
Soil quality	1. Total organic carbon (TOC)	Mg/ha
	2. Total nitrogen (N)	Mg/ha
	3. Extractable phosphorus (P)	Mg/ha
	4. Bulk density	g/cm^3
Water quality and quantity	5. Nitrate concentration in streams (and export)	concentration: mg/L; export: kg/ha/yr
	6. Total phosphorus (P) concentration in streams (and export)	concentration: mg/L; export: kg/ha/yr
	7. Suspended sediment concentration in streams (and export)	concentration: mg/L; export: kg/ha/yr
	8. Herbicide concentration in streams (and export)	concentration: mg/L; export: kg/ha/yr
	9. storm flow	L/s
	10. Minimum base flow	L/s
	11. Consumptive water use (incorporates base flow)	feedstock production: m^3/ha/day; biorefinery: m^3/day

McBride A, et al. 2011. Indicators to support environmental sustainability of bioenergy systems. Ecological Indicators 11(5) 1277-1289.

OAK RIDGE National Laboratory

Relevance

- SOC has received significant attention from researchers and policy makers: climate change, bioenergy and bio-product industries
- Lack of consistency in measurement and verification methods

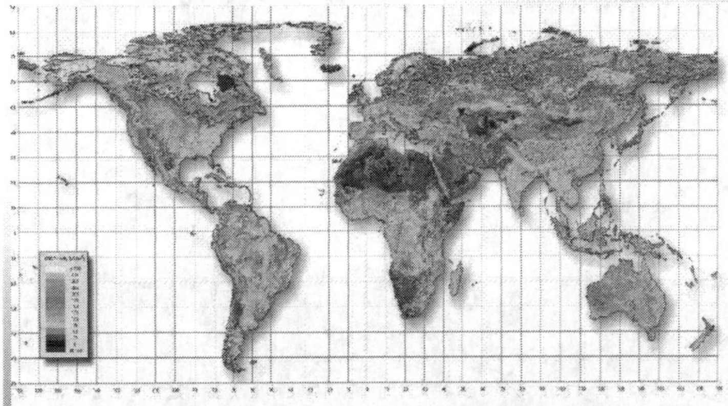

Estimates of Global Soil Organic Carbon Density from amended Harmonized World Soil Database (t C ha^{-1})

HWSD data sources:
"...differences in landform, parent material, ..., and time of sampling were often not described explicitly in the source materials."

"...variables have been determined in many laboratories according to various methods and these methods are not necessarily comparable..."

Source: The European Commission, 2014; FAO HWSD, 2009

Relevance

US regulations (RFS2 and LCFS) require a lifecycle analysis of emissions for nearly all fuel options.

- GREET model for GHG emissions, and for estimating GHGs from land use change these include SOC emission factor:
  - estimates of emissions from a loss of SOC resulting from a change in land cover
- Bioenergy producers that can show a GHG offset for their fuel could have an advantage in the increasingly competitive global bioenergy market
- Results of lifecycle analyses are extremely important to these producers and the end users seeking the most sustainable bioenergy option

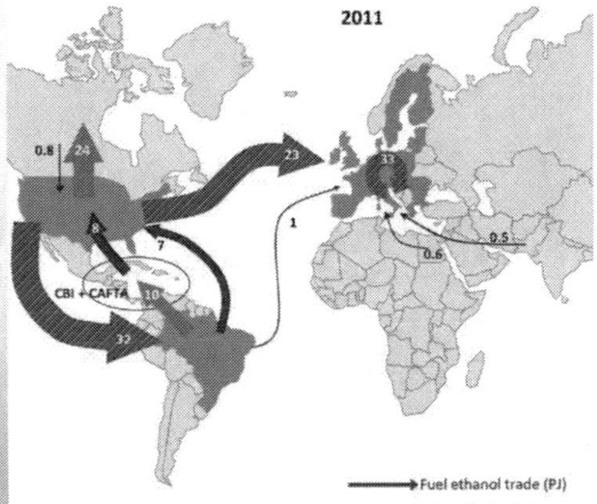

Source: Lamers et al., RSER, 15 (2011) 2655–2676

Methods: Literature review

Bioethanol: corn in the US and sugarcane in Brazil

- focus for many sustainability assessments
- Aim is for standardized measurement protocols

Methods: Comparison, and Analysis

Table 1. Summary of reviewed studies measuring SOC for bioenergy crops, US

Author	Crop[a]	Depth (cm)	Sample Collection Method	Sample Design
Bolinder et al. (1999)	C	30	Cores	Random
Clapp et al. (2000)	C	30	Cores	Random
Wilts et al. (2004)	C/W/O	45	Cores	Random
Duiker and Lal (1999)	C/W	30	Cores	Systematic
Halvorson et al. (2005)	C/W	15.2	Cores	Random
Hooker et al. (2005)	C/W/B	15	Cores	Systematic
Motta et al. (2000)	C/S/Co/So	30	Cores	Random
Olson et al. (2005)	C/S	75	Cores	Systematic
Al-Kaisa et al. (2005)	C-S	15	Cores	Random
Varvel and Wilhelm (2010)	C/C-S/S-C	30	Cores	Random
Aziz et al. (2013)	C/C-S/W	30	Cores	Random
Evers et al. (2013)	C/S/SG	15	Cores	Random
Lee et al. (2007)	SG	90	Cores	Random
Yang and Wander (1998)	C/S	30	Cores	Random
Yang and Wander (1999)	C/S	90	Cores	Random
Rhoton et al. (2002)	C/Co	15.2	Cores	Random

Methods: Comparison, and Analysis

Table 2. Summary of reviewed studies measuring SOC for bioenergy crops, Brazil

Author	Crop[a]	Depth (cm)	Sample Collection Method	Sample Design
Cerri et al. (2004)	Sc	20	Pits	Systematic
Souza et al. (2005)	Sc	30	Cores	Random
Razafimbelo et al. (2006)	Sc	10		Systematic
Resende et al. (2006)	Sc	60	Pits/auger*	Systematic
Szacks et al. (2007)	Sc	30	Pits	Systematic
Galdos et al. (2009)	Sc	100	Pits	Systematic
Pinheiro et al. (2010)	Sc	100	Pits/ auger*	Systematic
Tivet et al. (2013)	Sc	100	Pits	Random
Rossi et al. (2013)	Sc	60	Pits/ auger*	Systematic
Calegari et al. (2008)	C/S	60	Pits	Systematic
Sisti et al (2004)	S	100	Pits/ auger*	Systematic
Zotarelli et al. (2013)	C/S	100	Pits/ auger*	Systematic
Diekow et al. (2006)	C	100	Pits	Systematic

*Pits were opened for quantification of soil bulk density and samples for C analysis were from composite samples taken by auguring

Emerging results and recommendations

Position and density of sampling points

- Sampling design is crucial step & the design of an efficient, cost-effective sampling and estimation scheme is the main challenge researchers face
- Systematic and random sampling are the two types of designs used to estimate carbon stocks
  - Methodologies for carrying out these two designs are well documented (e.g., see IPCC, 2006).
- Stratified sampling not commonly used in the reviewed studies
  - Stratified random sampling and systematic random sampling are useful for within-field variability
  - high variability in soil types and within-field conditions common in many bioenergy producing areas in both the US and Brazil
  - may warrant use of the SRS technique

OAK RIDGE National Laboratory

Emerging results and recommendations

Sampling procedures- Number of samples

- Sample design and statistically reliable results
  - Cerri et al. recommends nine trenches, covering an area of one hectare, be evaluated
  - GRACEnet are more vague
- based on available resources

Sampling procedures- Frequency and timing

- time taken for soil carbon sink (i.e. a new equilibrium) to occur is highly variable
- IPCC guidelines suggest new steady state $\geq$ 20 yr
- Sampling once restricts recognizing how SOC varies temporally
- based on available resources, other external factors

OAK RIDGE National Laboratory

Emerging results and recommendations

Sampling procedures- pits and cores

- Brazil used pits
  - For some investigations, pits were used only to take undisturbed samples for bulk density;
  - complement an increased number of samples collected using an auger
- U.S. studies generally collected soil samples with a coring device
  - GRACEnet protocol
- Emerging technology (e.g., rotary core device)

More research needed

Emerging results and recommendations

Sampling procedures- Sampling depth

- SOC is generally unevenly distributed over varying soil depths
- several studies show that long-term accumulation of carbon also takes place in deeper soil layers (e.g., 30-100cm)
- Variable depths in US and BR
  - Cerri et al. depth recommendations ≥ 30cm
  - GRACEnet ≥100cm
- budget or time limitations; management practices are focus

More research needed

Author	Crop[a]	Depth (cm)
Cerri et al. (2004)	Sc	20
Souza et al. (2005)	Sc	30
Razafimbelo et al. (2006)	Sc	10
Resende et al. (2006)	Sc	60
Szacks et al. (2007)	Sc	30
Galdos et al. (2009)	Sc	100
Pinheiro et al. (2010)	Sc	100
Tivet et al. (2013)	Sc	100
Rossi et al. (2013)	Sc	60
Calegari et al. (2008)	C/S	60
Sisti et al (2004)	S	100
Zotarelli et al. (2013)	C/S	100
Diekow et al. (2006)	C	100
Bolinder et al. (1999)	C	30
Clapp et al. (2000)	C	30
Wilts et al. (2004)	C/W/O	45
Duiker and Lal (1999)	C/W	30
Halvorson et al. (2005)	C/W	15.2
Hooker et al. (2005)	C/W/B	15
Motta et al. (2000)	C/S/Co/So	30
Olson et al. (2005)	C/S	75
Al-Kaisa et al. (2005)	C-S	15
Varvel and Wilhelm (2010)	C/C-S/S-C	30
Aziz et al. (2013)	C/C-S/W	30
Evers et al. (2013)	C/S/SG	15
Lee et al. (2007)	SG	90
Yang and Wander (1998)	C/S	30
Yang and Wander (1999)	C/S	90
Rhoton et al. (2002)	C/Co	15.2

Emerging results and recommendations

Sampling procedures- Sampling depth increments

- All sampling protocols emphasize the importance of sampling soil proportionately (i.e., taking samples from a constant horizontal area)
- Cerri et al. & GRACEnet protocols recommend samples to be taken in 10 cm increments throughout the soil profile
- Most of the reviewed literature used depth increments of 5 or 10 cm for the upper soil profile (30 cm)
  - depth increments >30 cm varied substantially
  - 10 to 20 cm increments in BR (in general)
  - 20 to 30 cm depth increments (in general)

More research needed

Concluding remarks

More research needed in key areas

- SOC laboratory processing protocols - Bulk Density & Soil C fractionation
- Long term soil experiments
  - Management practices, baseline data and comparisons
  - Chronosequence
  - Estimating SOC over space and time

Input welcomed as we continue to develop recommendations

OAK RIDGE National Laboratory

Thank you

Research and paper development with financial support provided by: U.S. Department of Energy, BioEnergy Technologies Office (BETO); U.S. Department of Agriculture - Agricultural Research Service; Brazilian Center of Science and Technology of Bioenergy (CTBE), Grant 2012/06933-6 from the São Paulo Research Foundation (FAPESP), NLAE, ORNL, UT-Battelle LLC,

USDA CTBE Laboratório Nacional de Ciência e Tecnologia do Bioetanol Embrapa

Special thanks to the National Science Foundation RCN project and conference organizers from Michigan Tech University.

PAN AMERICAN BIOFUELS AND BIOENERGY SUSTAINABILITY
AN NSF RESEARCH COORDINATION NETWORK

Michigan Tech
Michigan Technological University

Acknowledgements:
Co-authors from Brazil and US (Keith Kline, Dana Abulebdeh, Doug Karlen; Bruno Alves; Marcelo Galdos) as well as ORNL staff that offered advice and input; Helena Chum; Tom Wilbanks, Virginia Dale and ORNL CBES colleagues, B Davison, LM Eaton, RA Efroymson, PN Leiby, M Langholtz, LR Lynd, A McBride, GA Oladosu, ES Parish, RD Perlack, N Singh, S Kang, SB Wright, LL Wright; and DOE OBP staff Zia Haq, Kristen Johnson, Alicia Lindauer-Thompson, P Grabowski, A Goss-Eng.

CBES Center for BioEnergy Sustainability

http://www.ornl.gov/sci/ees/cbes/

OAK RIDGE National Laboratory

Selected references:

- Cerri, C.E.P., Galdos, M.V., Carvalho, J.L.N.C., Feigl, B.J., Cerri, C.C., 2013. Quanitfying soil carbon stocks and greenhouse gas fluxes in the sugarcane agrosystem: point of view. Sci. Agric. **70**. 361-368.
- Doran, J.W., Jones, A.J, 1996 Editors, Methods for Assessing Soil Quality: Soil Science Society of America Special Publication **49,** SSSA, Madison, WI.
- Hansen J, Kharecha P, Sato M, Masson-Delmotte V, Ackerman F, Beerling DJ, Hearty PJ,..., Zachos JC, 2013. Assessing "Dangerous Climate Change": Required Reduction of Carbon Emissions to Protect Young People, Future Generations and Nature. DOI: 10.1371/journal.pone.0081648
- Houghton, R.A., 2007. Balancing the Global Carbon Budget, Annu. Rev. Earth Planet. Sci. 35:313–47. Downloaded from arjournals.annualreviews.org
- Jandl, R., Rodeghier, M., Martinex, C., Cotrufo, M.F., Bampa, F., Wesemael, B., Harrison, R.B., Guerrini, I.A., Richter jr, D.D., Rustad, L., Lorenz, K., Chabbi, A., Miglietta, F., 2014. Current status, uncertainty, and future needs in SOC monitoring. Science of the total environment **468.** 376-383.
- Kwon, H-Y, S. Mueller, J.B. Dunn, & M.M. Wander, 2013. Modeling state-level soil carbon emission factors under various scenarios for direct land use change associated with United States biofuel feedstock production. Biomass and Bioenergy 55 (2013) 299-310.
- McBride, A.C., Dale, V.H., Baskaran., L.M., Downing, M.E., Eaton, L.M., Efroymson, R.A., Garten, C.T., Kline, K.L., Jager, H.L., Mulholland, P.J., 2011. Indicators to support environmental sustainability of bioenergy systems. Ecol. Indicators **11**. 1277-1289
- Meki NM., Snider JL., Kiniry JR., Raper RL., Rocateli AC., 2013. Energy sorghum biomass harvest thresholds and tillage effects on soil organic carbon and bulk density. Industrial Crops and Products 43: 172-182.
- Olson, K.R., 2013. SOC sequestration, storage, retention and loss in U.S. croplands: Issues paper for protocol development. Geoderma **195.** 201-206.
- Panagos P, Van Liedekerke M, Jones A, Montanarella L. European Soil Data Centre: response to European policy support and public data requirements. Land Use Policy 2012;29(2):329–38.
- Reeves, PG., 1997. The roles of soil organic matter in maintain soil quality in continuous cropping systems. Soil & tillage research **43**. 131-167.
- Wang, M., Jeongwoo, H., Dunn, J.B., 2012. Well-to-wheels energy use and greenhouse gas emission of ethanol from corn, sugarcane, and cellulosic biomass for US use. Env. Research Letters **7**. DOI: 10.1088/1748-9326/7/4/045905.
- U.S Department of Agriculture, Economic Research Service using data from the USDA Production, Supply, and Distribution data base and Global Agricultural Trade System
- U.S Department of Agriculture, 2006. Model Simulation of Soil Loss, Nutrient Loss, and Change in SOC. www.nrcs.usda.gov/internet/FSE_DOCUMENTS/nrcs143_013137.pdf
- US Department of Agriculture, Natural Resources Conservation Service. NRCS, RaCA. http://www.nrcs.usda.gov/wps/portal/nrcs/detail/soils/survey/?cid=nrcs142p2_054164 Accessed 3/17/2014. Data citation: Soil Survey Staff. Rapid Assessment of U.S. Soil Carbon (RaCA) project. Available online. June 1, 2013 (FY2013 official release).

RCN Conference on Pan American Biofuels & Bioenergy Sustainability
Recife, Brazil July 22-25, 2014

Advances in the knowledge on the impacts of slash management on soil quality after forest harvesting in the North East of Argentina

Ana Maria Lupi
Instituto de Suelos- INTA Castelar
Buenos Aires- Argentina
Lupi.ana@inta.gov.ar

RCN Conference on Pan American Biofuels & Bioenergy Sustainability
Recife, Brazil July 22-25, 2014

Distribution of forest plantations in Argentina.
source: http://www.minagri.gob.ar/new/0-0/forestacion/index.php
Accessed July / 2014

90% of the forest area in Argentina is concentrated in the NE region (928,067 ha)

Misiones Province
Corrientes Province
Entre Ríos Province
and Delta of Buenos Aires province

91% of the area corresponds to Pinus and Eucalyptus species

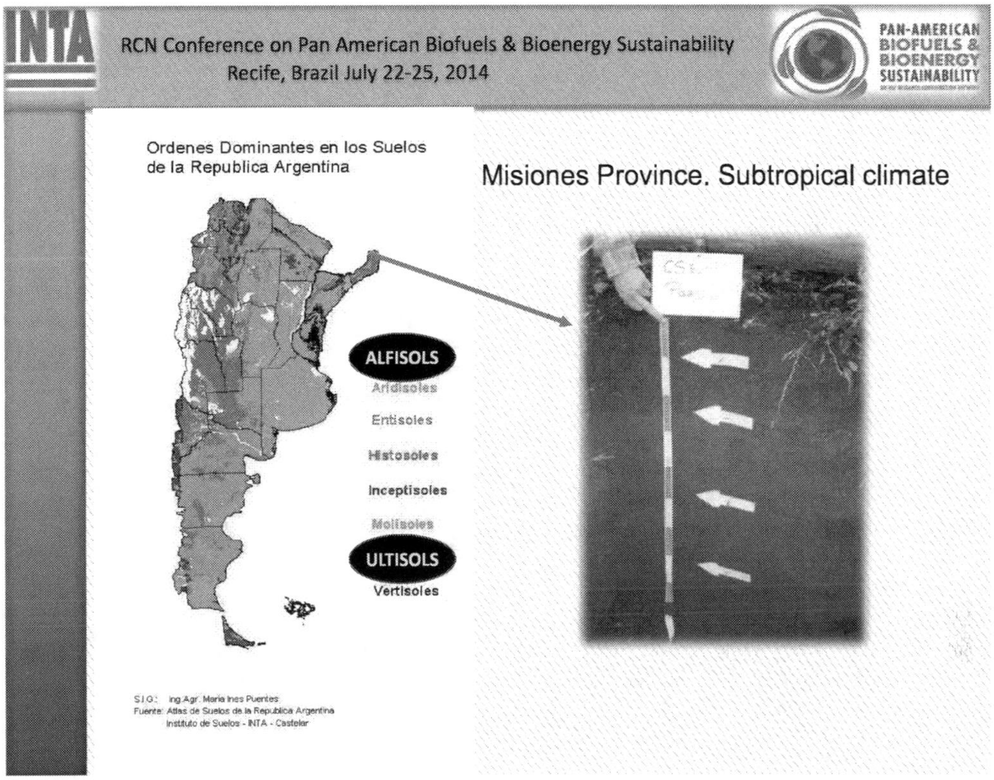
INTA
RCN Conference on Pan American Biofuels & Bioenergy Sustainability
Recife, Brazil July 22-25, 2014
PAN-AMERICAN BIOFUELS & BIOENERGY SUSTAINABILITY
Ordenes Dominantes en los Suelos de la Republica Argentina
Misiones Province. Subtropical climate
ALFISOLS
Aridisoles
Entisoles
Histosoles
Inceptisoles
Molisoles
ULTISOLS
Vertisoles

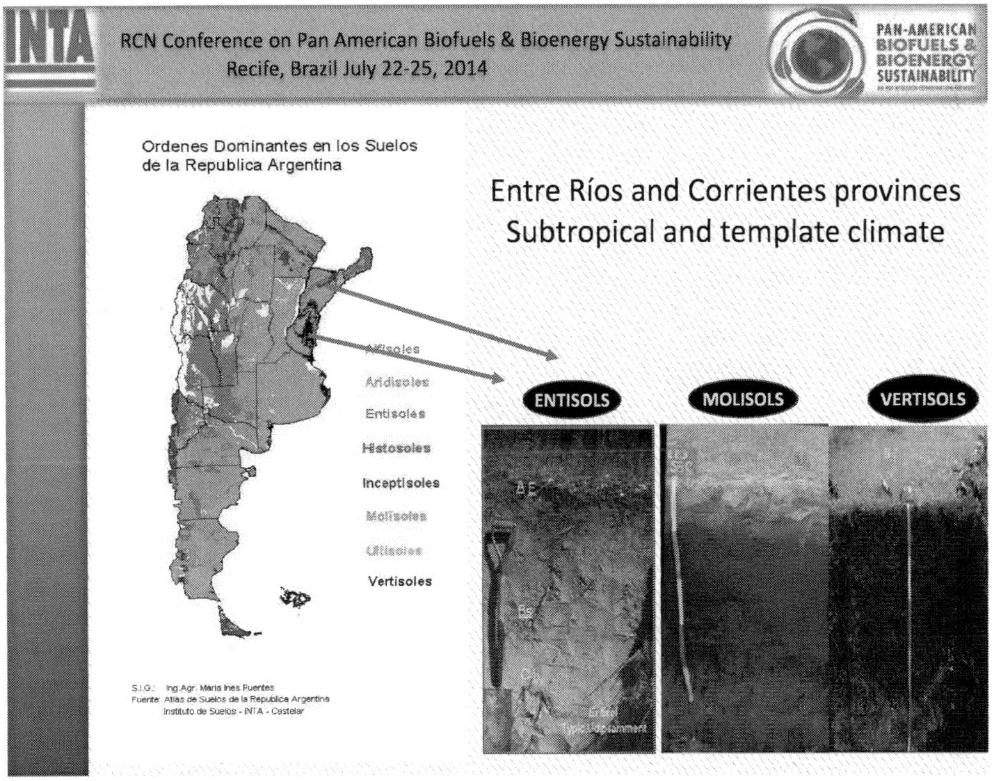
INTA
RCN Conference on Pan American Biofuels & Bioenergy Sustainability
Recife, Brazil July 22-25, 2014
PAN-AMERICAN BIOFUELS & BIOENERGY SUSTAINABILITY
Ordenes Dominantes en los Suelos de la Republica Argentina
Entre Ríos and Corrientes provinces
Subtropical and template climate
Aridisoles
Entisoles
Histosoles
Inceptisoles
Molisoles
Ultisoles
Vertisoles
ENTISOLS
MOLISOLS
VERTISOLS

INTA

RCN Conference on Pan American Biofuels & Bioenergy Sustainability
Recife, Brazil July 22-25, 2014

PAN-AMERICAN BIOFUELS & BIOENERGY SUSTAINABILITY

In the past 15 years there has been considerable discussion about the environmental impact and sustainability of some harvesting operations and slash management practices.

- Forest plantation are in good productivity soil
- We need to preserve productive capacity
-We are looking for management practices that maintain long-term soil quality .

Forest harvesting and site preparation are practices that can have a negative impact on soil quality.

INTA

RCN Conference on Pan American Biofuels & Bioenergy Sustainability
Recife, Brazil July 22-25, 2014

PAN-AMERICAN BIOFUELS & BIOENERGY SUSTAINABILITY

Burning of residues

mid 1990s

Retention of residues

Removal of residues

RCN Conference on Pan American Biofuels & Bioenergy Sustainability
Recife, Brazil. July 22-25, 2014

TOTAL BIOMASS AND NUTRIENTS CONTENT

1996 – Harvest residue in 21-year-old *Pinus elliottii* plantation Misiones Province

Type of residue	Biomass ($Mg.ha^{-1}$)	N ($kg.ha^{-1}$)	P ($kg.ha^{-1}$)	K ($kg.ha^{-1}$)
Fine debris: <1 cm diameter	20,6	230	28	160
Large debris: (1-15 cm dia.) + bark	20,5	67	15	30

UREA (46-0-0)	593,3 (kg ha^{-1})
TSP	210 (kg ha^{-1})
KCl	372 (kg ha^{-1})

RCN Conference on Pan American Biofuels & Bioenergy Sustainability
Recife, Brazil July 22-25, 2014

2005 - Harvest residues ($Mg.ha^{-1}$) of *Eucalyptus grandis* after 15 years, Entre Rios Province. *Garcia, M de los A. y equipo. INTA EEA Concordia.*

	Peluderte	Hapludol
Residue retention (with bark)	62,7	64,0
Residue retention (no bark)	67,5	59,7
Burning residue	51,9	73,0

	N (kg ha^{-1})	P (kg ha^{-1})	K (kg ha^{-1})
Peluderte (Vertisol)	506,0	16	35
Hapludol (Molisol)	214,2	12	60

UREA (46-0-0)	476/1124 (kg ha^{-1})
TSP	60/95 (kg ha^{-1})
KCl	60/115 (kg ha^{-1})

INTA

RCN Conference on Pan American Biofuels & Bioenergy Sustainability
Recife, Brazil July 22-25, 2014

PAN-AMERICAN BIOFUELS & BIOENERGY SUSTAINABILITY

2010- Residual biomass and nitrogen, phosphorus and potassium content, by harvesting methods in *Pinus elliottii* plantation. 24 years-old. **Entisol.** Aparicio J. INTA EEA Bella Vista.

Residual biomass ($Mg.ha^{-1}$)	Bole only	85,8	Aditional 60,3 $Mg. ha^{-1}$
	Whole Tree	25,5	
N ($kg.ha^{-1}$)	Bole only	513	Aditional 344 $kg\ ha^{-1}$
	Whole Tree	169	
P ($kg.ha^{-1}$)	Bole only	37	Aditional 25 $kg\ ha^{-1}$
	Whole Tree	12	
K ($kg.ha^{-1}$)	Bole only	78	Aditional 59 $kg\ ha^{-1}$
	Whole Tree	19	

UREA (46-0-0)	764,4 ($kg.ha^{-1}$)
TSP	125,0($kg.ha^{-1}$)
KCl	112,5($kg.ha^{-1}$)

INTA

RCN Conference on Pan American Biofuels & Bioenergy Sustainability
Recife, Brazil July 22-25, 2014

PAN-AMERICAN BIOFUELS & BIOENERGY SUSTAINABILITY

Residual biomass and nitrogen, phosphorus and potassium content, by harvesting methods, in *Pinus taeda,* 20 years old. **Ultisol.** Fernandez et al., (2014)

Residual biomass ($Mg.ha^{-1}$)	Bole only	74,0	Aditional 25,2 $Mg. ha^{-1}$
	Whole Tree	48,8	
N ($kg.ha^{-1}$)	Bole only	423,3	Aditional 129,5 $kg\ ha^{-1}$
	Whole Tree	305,6	
P ($kg.ha^{-1}$)	Bole only	27,8	Aditional 7,6 $kg\ ha^{-1}$
	Whole Tree	20,2	
K ($kg.ha^{-1}$)	Bole only	117,9	Aditional 33,3 $kg\ ha^{-1}$
	Whole Tree	84,7	

UREA (46-0-0)	287,8 ($kg.ha^{-1}$)
TSP	38,0($kg.ha^{-1}$)
KCl	63,5($kg.ha^{-1}$)

EFFECTS ON THE SOIL.....

Soil organic carbon, nitrogen concentration and C/N ratio in different slash management treatments of *P. taeda*.

Lupi et al., (2014) CACS.

Statistical analysis of soil samples 2, 5, 7 and 12 years after treatment application

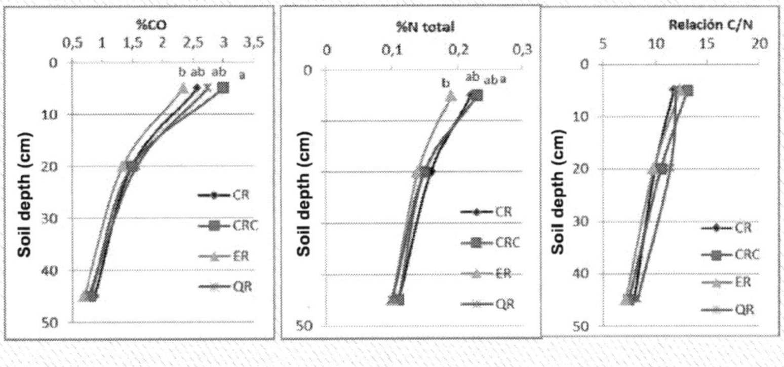

CR: retention of residue; CRC: retention of secondary forest residue; ER: residue removal; QR; burning of residue

INTA
RCN Conference on Pan American Biofuels & Bioenergy Sustainability
Recife, Brazil July 22-25, 2014
PAN-AMERICAN BIOFUELS & BIOENERGY SUSTAINABILITY

Available phosphorus and pH after differents slash management treatments for *P. taeda*.

Lupi et al., (2014) CACS.

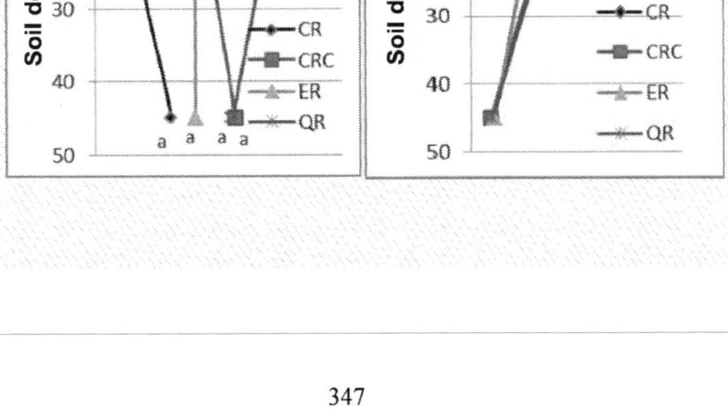

Soil organic carbon concentration in two soil after different slash management treatments for *Eucalyptus grandis*

	0-5 cm	5-10 cm
	Peluderte (Vertisol) 18 month	
Residue retention	21,2 (7,8)a	15,75,6)a
Residue removal	20,4(4,8)a	15,5(1,6)a
Residue burning	22,4(6,5)a	13,6(2,2)a
	Hapludol (Molisol) 22 month	
Residue retention	28,1(5,3)a	15,7(0,7)a
Residue removal	27,9(3,1)a	15,6(0,9)a
Residue burning	22,3(7,2)a	14,1(0,6)a

Lupi et al., in review, Ciencia Florestal. SM

RCN Conference on Pan American Biofuels & Bioenergy Sustainability
Recife, Brazil July 22-25, 2014

Phosphorus in soils under forests

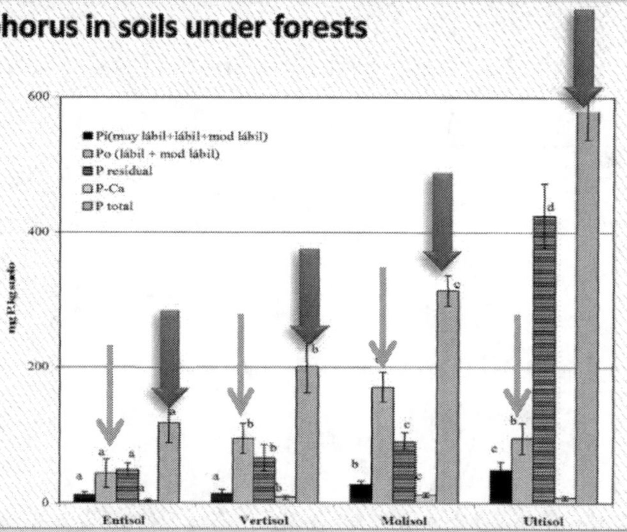

Major differences in soil origin, mineralogy, and climatic environments.
Contrasting levels of total phosphorus (red arrows).

Lupi et al., (2012). Changes in soil phosphorus in different forest residue management. Agrociencia Uruguay, Special Issue. Striving for Sustainable High Productivity through Improved. Soil and Crop Management – September 2012

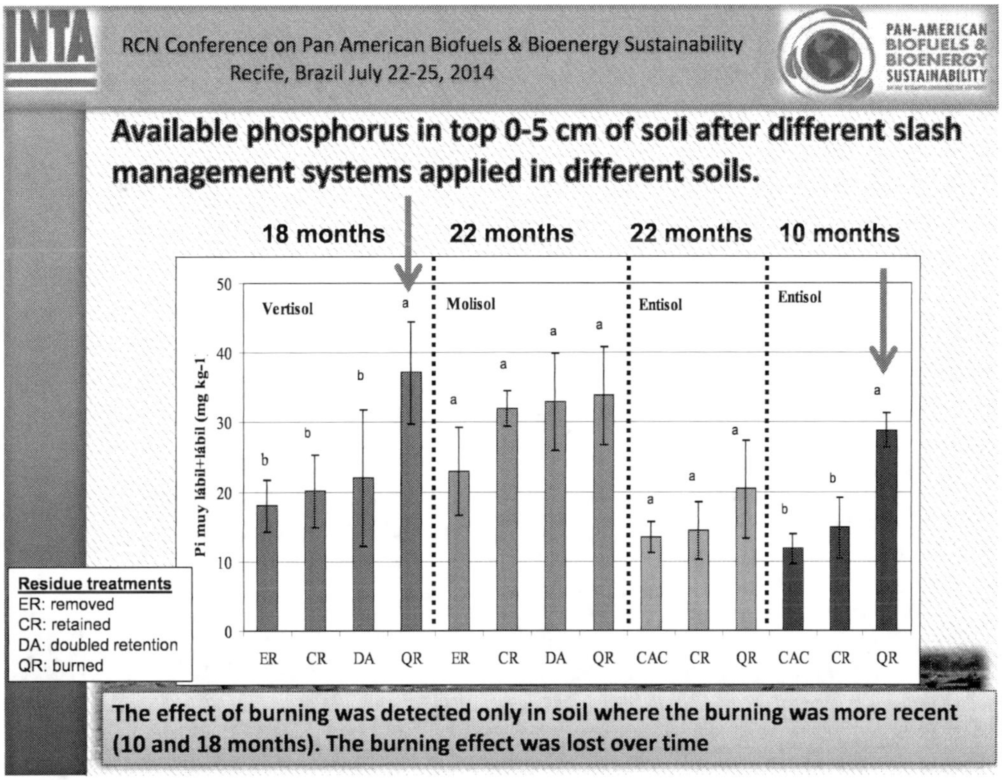
INTA
RCN Conference on Pan American Biofuels & Bioenergy Sustainability
Recife, Brazil July 22-25, 2014
PAN-AMERICAN BIOFUELS & BIOENERGY SUSTAINABILITY
Available phosphorus in top 0-5 cm of soil after different slash management systems applied in different soils.
18 months
22 months
22 months
10 months
Vertisol
Molisol
Entisol
Entisol
Pi muy lábil+lábil (mg kg-1
0
10
20
30
40
50
ER CR DA QR ER CR DA QR CAC CR QR CAC CR QR
Residue treatments
ER: removed
CR: retained
DA: doubled retention
QR: burned
The effect of burning was detected only in soil where the burning was more recent (10 and 18 months). The burning effect was lost over time

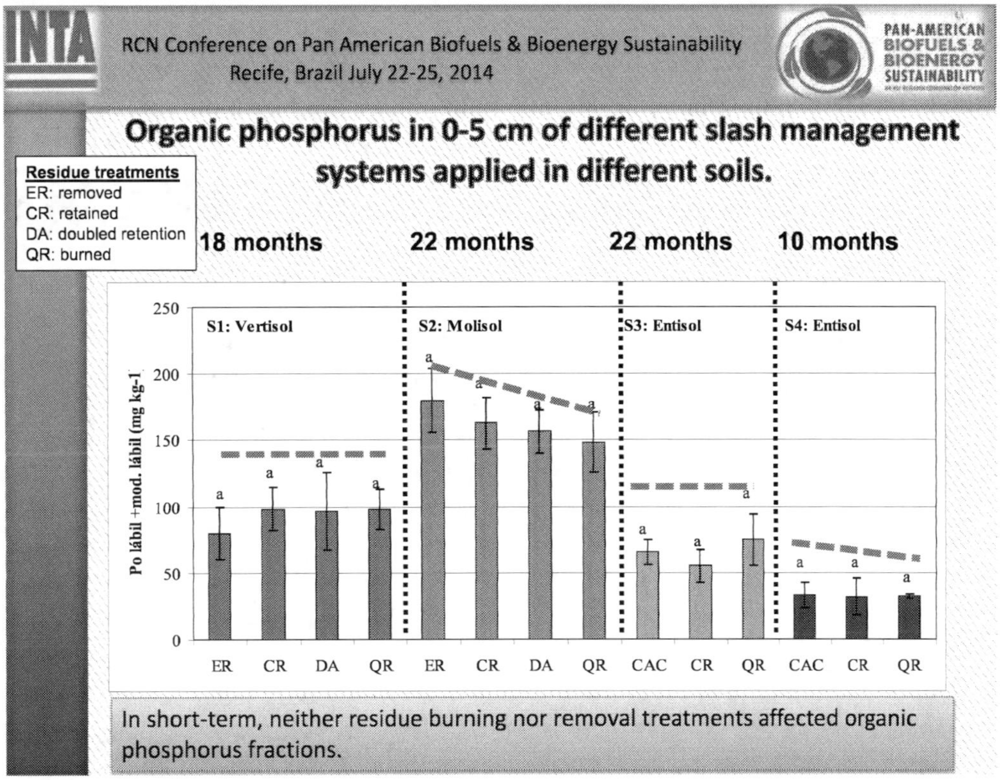
INTA
RCN Conference on Pan American Biofuels & Bioenergy Sustainability
Recife, Brazil July 22-25, 2014
PAN-AMERICAN BIOFUELS & BIOENERGY SUSTAINABILITY
Organic phosphorus in 0-5 cm of different slash management systems applied in different soils.
Residue treatments
ER: removed
CR: retained
DA: doubled retention
QR: burned
18 months
22 months
22 months
10 months
S1: Vertisol
S2: Molisol
S3: Entisol
S4: Entisol
Po lábil +mod. lábil (mg kg-1
0
50
100
150
200
250
ER CR DA QR ER CR DA QR CAC CR QR CAC CR QR
In short-term, neither residue burning nor removal treatments affected organic phosphorus fractions.

RCN Conference on Pan American Biofuels & Bioenergy Sustainability
Recife, Brazil July 22-25, 2014

Soil aggregate size after sieving different soils (0- 5 cm) in water

ER: Residue removed
CR-PC: Residue retained
QR-PI: Residue burned
DAR: Double residue retention
MN: native forest

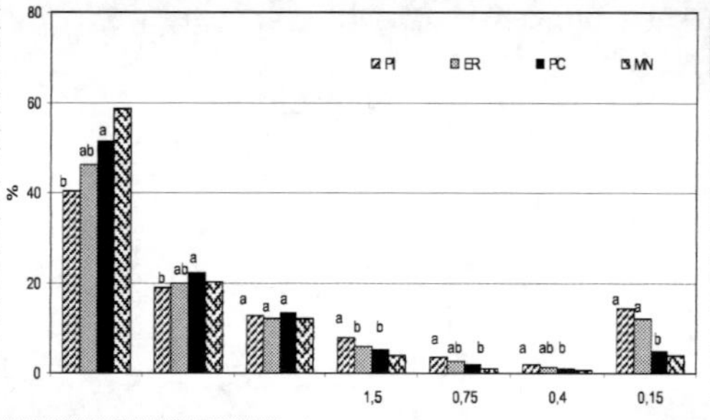

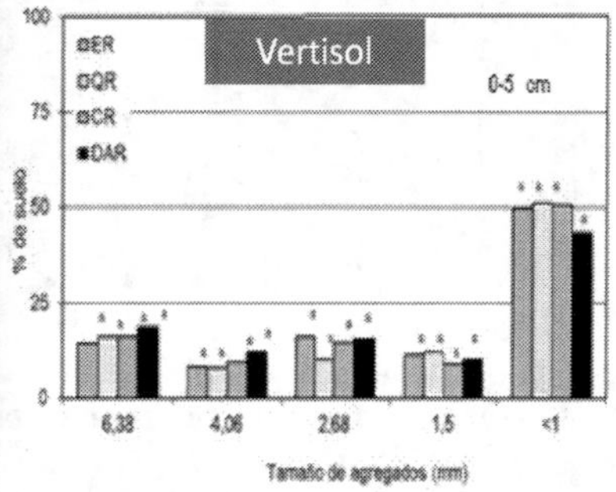

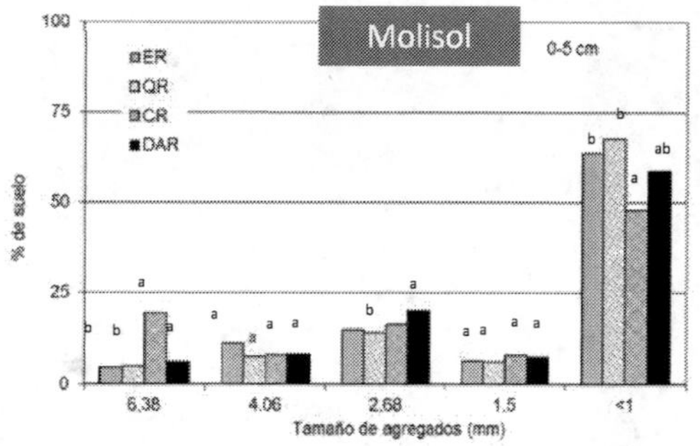

RCN Conference on Pan American Biofuels & Bioenergy Sustainability
Recife, Brazil July 22-25, 2014

Saturated hydraulic conductivity (Ksat: mm h^{-1})

	Vertisol	Molisol
Residue retained	8,22 (7,81)a	2,38 (1,52)a
Residue removed	3,56 (2,9)b	1,30 (1,68)a
Residue burned	2,52 (1,92)b	2,15 (1,56)a

Vertisol is naturally is more stable than Molisol. Aggregate size was greater in Vertisols.

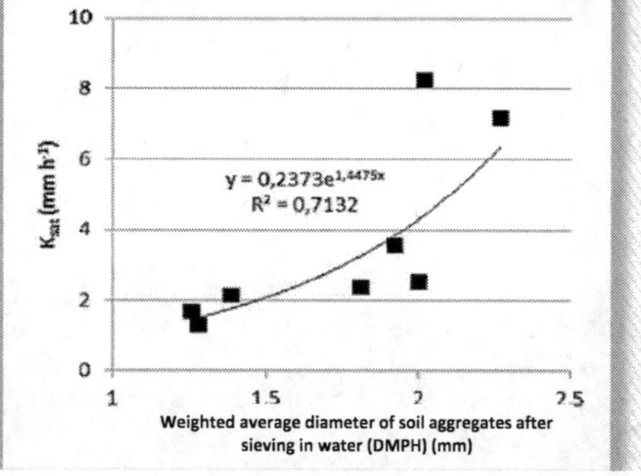

INTA RCN Conference on Pan American Biofuels & Bioenergy Sustainability
Recife, Brazil July 22-25, 2014
PAN-AMERICAN BIOFUELS & BIOENERGY SUSTAINABILITY

Humic acids in natural grassland and after different slash management treatments in *E. grandis*. Entisol.

	% C	% N	% H	% S	%O	O/C	C/N
Gassland	51,8	3,8	4,7	0,53	39,1	0,75	13,7
Residue removed	52,8	3,3	5,1	0,44	38,3	0,73	15,9
Residue retained	54,3	3,3	5,3	0,42	36,6	0,67	16,5
Residue burned	55,0	3,6	5,9	0,44	34,9	0,64	15,4

- Humic acid grassland is richer in nutrients.
- Fire creates organic structures enriched with C and N.
- Organic structures lose oxygen because of burning temperature.

INTA RCN Conference on Pan American Biofuels & Bioenergy Sustainability
Recife, Brazil July 22-25, 2014
PAN-AMERICAN BIOFUELS & BIOENERGY SUSTAINABILITY

Relative intensities (%), aromaticity index and decomposition ratio derived from ^{13}C-NMR spectra of humic acids from soil.

	C=O	Lignine	carbohydrates	polypeptides	Poly péptides, Lipids	Descomp. rate	Aromatocity index
Gassland	16	30	7	25	22	0.69	36
Residue removed	16	37	3	14	30	1.76	44
Residue retained	15	33	4	14	34	1.89	39
Residue burned	16	41	5	12	26	1.5	49

No change.
Memory soil.

+ Aromatic Compounds
- cellulose

INTA RCN Conference on Pan American Biofuels & Bioenergy Sustainability
Recife, Brazil July 22-25, 2014
PAN-AMERICAN BIOFUELS & BIOENERGY SUSTAINABILITY

In summary ...

1- Forest soils are relatively fertile. We just evaluated a second-rotation forest.

The wide variability of environments relates to different climates and soils that differ in their formation and evolution. Soils have special functional features.

2- Removal of harvesting residue generates additional outputs of nutrients and C. These outputs are not replenished.

3- In the short term the soils studied show different behaviors. Some soils are non-responsive. We have little information on long-term impacts.

4- We have evidence of changes in the quality of soil organic matter by burning or removal of harvesting residue. We do not know the implications for nutrient cycling.

(See also Abstract on RCN conference website for detailed text)

INTA RCN Conference on Pan American Biofuels & Bioenergy Sustainability
Recife, Brazil July 22-25, 2014
PAN-AMERICAN BIOFUELS & BIOENERGY SUSTAINABILITY

What do not we know?

1- Long-term studies. Networks of experiments in different environments in which changes in soil processes are evaluated.

- Balance of soil organic matter and changes in soil physical properties related to the entry and movement of water and oxygen.
- Soil microbiology has not yet been addressed (changes in diversity / quantity).

2- Know the potential and the limitations of each environment/site type and manage each one with specific site-level guidance for sustainable production (BMPs).

3- Teach about possible impacts of extractive management systems.

4- Promote nutrient replenishment and field debarking as mitigation measures.

.... and much more

REFERENCES *(from Abstract on RCN conference web site)*

Fernandez, R.; Von Wallis, A.; Pahr, N.; Friedl, A., Martiarena, R.; Lupi, A. 2014. Implicancias de la modalidad de cosecha forestal sobre el contenido de nutrientes. XXIV Congreso Argentino de la Ciencia del Suelo II Reunión Nacional "Materia Orgánica y Sustancias Húmicas" *Producción sustentable en ambientes frágiles* Bahía Blanca, 5 al 9 de mayo de 2014

Fernández, R.; Martiarena, R.; Goya, J.; Frangi, J.; Von Wallis, A.; Lupi, A.; Pahr, N. 2012a. Simulación del contenido de nitrógeno y fósforo remanentes según escenarios de cosecha de rodales de *Araucaria angustifolia*(Bert.) O. Ktze de 40 años. XIX Congreso Latinoamericano y XXIII Congreso Argentino de la Ciencia del Suelo. Mar del Plata, Argentina – 16 al 20 de abril de 2012.

Fernández, R.; Martiarena, R; Goya, J.; Frangi, J.; Von Wallis, A.; Lupi, A.; Pahr, N. 2012b. Simulation of remaining potassium, calcium and magnesium content in relation to harvest regimes in a *Araucaria angustifolia* stand of 40 years. Aceptado para presentación oral. 19th ISTRO Conference. 24-28 de Setiembre de 2012, Montevideo, Uruguay.

Giuffré, L., Fernández, R.; Lupi, A.M.; Heredia O.; Pascale, C. 2002. Efecto de diferentes técnicas de manejo de residuos de la cosecha forestal en algunas propiedades de un Kandiudult en el NE Argentino. Agricultura Técnica, 62 (1): 143 -149.

Goya J.; Pérez C.; Frangi J.; Fernández R. 2003. Impacto de la cosecha y destino de los residuos sobre la estabilidad del capital de nutrientes en plantaciones de *Pinus taeda* L. Ecología Austral 13: 139-150.

Fernández, R. A.; Martiarena, R. A.; Lupi, A. M.; Von Wallis, A.; Pahr N. 2010. Crecimiento del Pinus taeda a los 12 años y condición química del suelo e el NE de Argentina: Efecto del Manejo de residuos de la cosecha en el establecimiento. Congreso ISTRO 12 al 14 de julio, 2010. Colonia, Uruguay

Fernández, R.; Lupi, A.; Pahr, N.; Reis, H.; O´Lery, H.; Gelid, M.; Martínez, S. Técnicas de manejo de residuos de cosecha para el establecimiento forestal y su impacto sobre la condición química de los suelos rojos del Noreste de Argentina. En: Avances en Ingenieria Agrícola. Pp: 243-248. ISBN 950-29-0593-8. Ed. Facultad Agronomía (UBA). 2000.

Frangi, J.; Goya, J.; Larocca F. 1999.Ciclo de nutrientes en plantaciones de E. grandis de distintas edades en la provincia de Entre Rios.Informe Final PIA 11/96 Proyecto Forestal de Desarrollo. SAGPyA-BIRF.

Lupi, A. M; Conti, M.; Fernández, R.; Cosentino, D.; López G. 2007. Efecto de las prácticas de repoblación forestal sobre el carbono orgánico del suelo y la estabilidad de los agregados en el noreste de Argentina Investigación Agraria: Sistemas y Recursos Forestales 16(3): 230-240

Lupi, A.M.; Conti, M.; Fernandez; R. 2012a. Calidad del carbono organico del suelo e diferentes técnicas de manejo de residuos forestales. Ciência Florestal, Santa Maria, 22 (2): 295-303

Lupi, A. M.; Quintero, C.; Boca, T.; Garcia, M D Los A.; Diaz, D.; Boschetti G.; Aparicio J.; Ingaramo L. 2012b. Changes in soil phosphorus in different forest residue management. Agrociencia Uruguay, Special Issue Eds: Oswaldo Ernst , Mario Pérez Bidegain, José Terra, Mónica Barbazán. 128-134

Lupi, A. M.; Garcia, M D Los A.; Diaz, D.; Ingaramo, L.; Boca, T. Manejo de residuos de *Eucalyptus grandis* Hill ex Maiden en sitios contrastantes. *en evaluación.* Martiarena R.; Fernández, R.; Lupi, A.; Alegranza, D.; Bischoff, D.; Muñoz, D.. 2004. Efecto de la preparación de terreno sobre el crecimiento de Araucaria angustifolia. XIX Congreso Argentino de la Ciencia del Suelo. II Simposio Nacional sobre suelos Vertisólicos. Paraná, Entre Ríos. 22 al 25 de Junio de 2004.

Martiarena, R.; Goya, J.; Fernández, R.; Frangi, J.; Lupi A.M. 2010. Exportación de nitrógeno y fósforo durante la cosecha de una cronosecuencia de *Araucaria angustifolia* (BERT) O. KTZE. Jornadas Forestales Eldorado, Misiones. Junio 2010. 10 p

Martiarena, R; Frangi, J; Pinazo, M.; Von Wallis, A.; Fernández, R. 2011a. Effect of Thinning and Harvest Type on Storage and Losses of Phosphorous in *Pinus taeda* L. Plantations in Subtropical Argentina. International Journal of Forestry Research. Article ID 761532.10 p.

Martiarena, R; Pinazo, M; Von Wallis, A; Knebel, O; Pahr, N. 2011b. Alternativas de manejo silvícola para la conservación de nutrientes en sistemas forestales en Misiones, Argentina. Ci. Suelo (Argentina) 29(1): 39-48.

Mortola, N. 2013. Comportamiento de las formas de fósforo en un Ultisol con diferentes manejos de implantación forestal. Tesis de Magister. FAUBA. 99 p

Von Wallis, A. 2013. Contenidos de materia orgánica y condición física de un Kandiudult de Misiones bajo diferentes sistemas de preparación del terreno forestal y bosque nativo. Tesis de Maestria FAUBA. 68 p

INTA

RCN Conference on Pan American Biofuels & Bioenergy Sustainability
Recife, Brazil July 22-25, 2014

PAN-AMERICAN BIOFUELS & BIOENERGY SUSTAINABILITY

MANEJO DE RESIDUOS DE COSECHA FORESTAL, SU EFECTO SOBRE LA CALIDAD DE LOS SUELOS DEL NORESTE DE ARGENTINA

AVANCES EN EL CONOCIMIENTO

Ana Maria Lupi
Instituto de Suelos- INTA Castelar
Buenos Aires- Argentina
Lupi.ana@inta.gov.ar

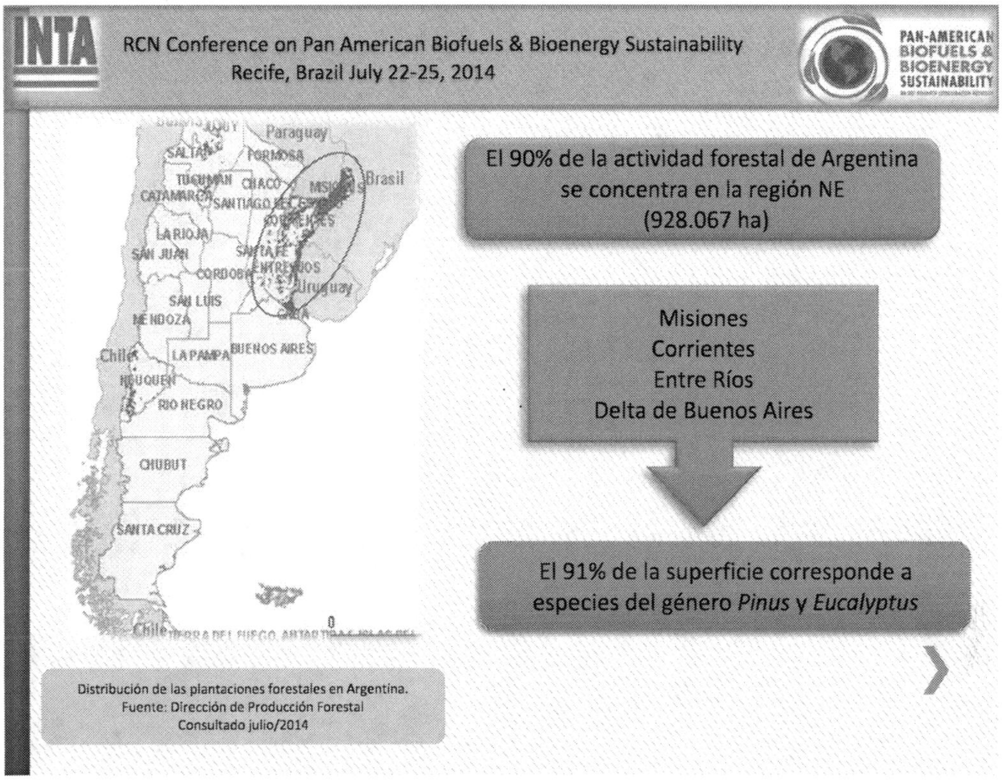
INTA
RCN Conference on Pan American Biofuels & Bioenergy Sustainability
Recife, Brazil July 22-25, 2014
PAN-AMERICAN BIOFUELS & BIOENERGY SUSTAINABILITY
Paraguay
Brasil
Uruguay
BUENOS AIRES
CHUBUT
SANTA CRUZ
El 90% de la actividad forestal de Argentina se concentra en la región NE (928.067 ha)
Misiones
Corrientes
Entre Ríos
Delta de Buenos Aires
El 91% de la superficie corresponde a especies del género *Pinus* y *Eucalyptus*
Distribución de las plantaciones forestales en Argentina.
Fuente: Dirección de Producción Forestal
Consultado julio/2014

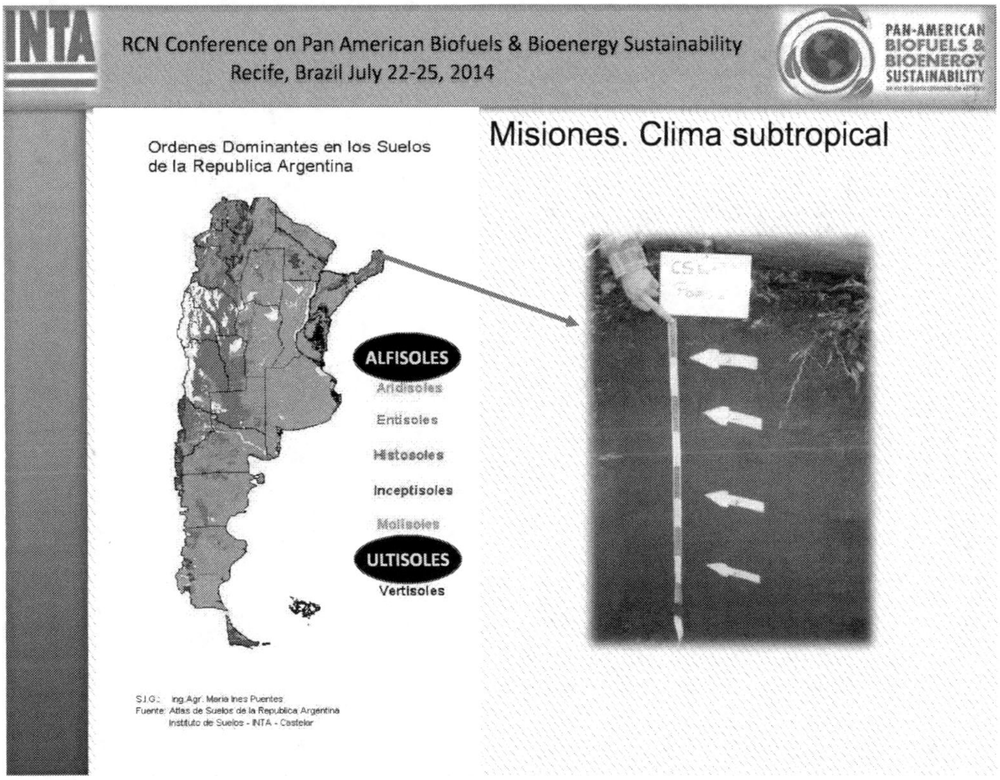
INTA
RCN Conference on Pan American Biofuels & Bioenergy Sustainability
Recife, Brazil July 22-25, 2014
PAN-AMERICAN BIOFUELS & BIOENERGY SUSTAINABILITY
Ordenes Dominantes en los Suelos de la Republica Argentina
Misiones. Clima subtropical
ALFISOLES
Aridisoles
Entisoles
Histosoles
Inceptisoles
Molisoles
ULTISOLES
Vertisoles

INTA

RCN Conference on Pan American Biofuels & Bioenergy Sustainability
Recife, Brazil July 22-25, 2014

PAN-AMERICAN BIOFUELS & BIOENERGY SUSTAINABILITY

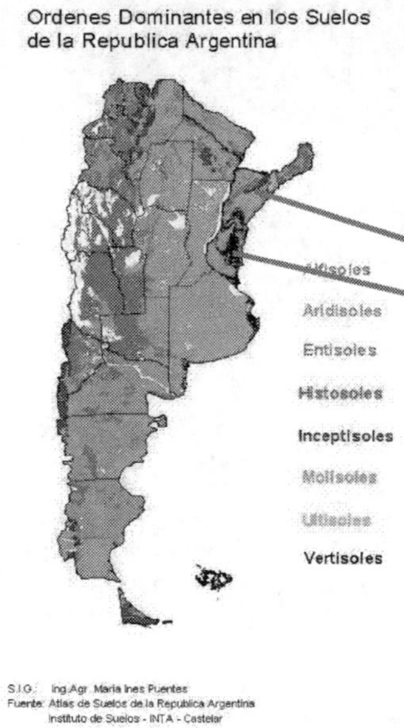

Entre Ríos y Corrientes
Subtropical y templado

ENTISOLES · MOLISOLES · VERTISOLES

INTA

RCN Conference on Pan American Biofuels & Bioenergy Sustainability
Recife, Brazil July 22-25, 2014

PAN-AMERICAN BIOFUELS & BIOENERGY SUSTAINABILITY

En los últimos 15 años existe un debate importante sobre el impacto ambiental y la sostenibilidad de algunas prácticas o sistemas de manejo

- Suelos de buena productividad
- Necesitamos conservar su capacidad productiva
- Buscar prácticas de manejo que mantengan la calidad del suelo a largo plazo

La cosecha forestal y la preparación del sitio son las prácticas que pueden tener impacto negativo sobre la calidad del suelo.

RCN Conference on Pan American Biofuels & Bioenergy Sustainability
Recife, Brazil July 22-25, 2014

PAN-AMERICAN BIOFUELS & BIOENERGY SUSTAINABILITY

Quema de residuos de la cosecha

Mediados años 1990

Conservación de los residuos de la cosecha

Extracción de residuos de la cosecha

RCN Conference on Pan American Biofuels & Bioenergy Sustainability
Recife, Brazil. July 22-25, 2014

PAN-AMERICAN BIOFUELS & BIOENERGY SUSTAINABILITY

CANTIDAD DE RESIDUOS Y NUTRIENTES

1996- Residuos de cosecha de *Pinus elliottii*. Misiones

Tipo de residuos	Biomasa ($Mg.ha^{-1}$)	N ($kg.ha^{-1}$)	P ($kg.ha^{-1}$)	K ($kg.ha^{-1}$)
Finos. < 1 cm diámetro	20,6	230	28	160
Gruesos + corteza. 1-15 cm di.	20,5	67	15	30

UREA	593,3 (kg ha^{-1})
SFT	210 (kg ha^{-1})
KCl	372 (kg ha^{-1})

2005- Residuos de cosecha ($Mg.ha^{-1}$)de *Eucalyptus grandis de 15 años, Entre Ríos.* *Garcia, M de los A. y equipo. INTA EEA Concordia.*

	Peluderte	Hapludol
Conservación de residuos CC	62,7	64,0
Conservación de residuos SC	67,5	59,7
Quema de residuos	51,9	73,0

	N ($kg\ ha^{-1}$)	P ($kg\ ha^{-1}$)	K ($kg\ ha^{-1}$)
Peluderte (Vertisol)	506,0	16	35
Hapludol (Molisol)	214,2	12	60

UREA	476/1124 ($kg\ ha^{-1}$)
SFT	60/95 ($kg\ ha^{-1}$)
KCl	60/115 ($kg\ ha^{-1}$)

RCN Conference on Pan American Biofuels & Bioenergy Sustainability
Recife, Brazil July 22-25, 2014
PAN-AMERICAN BIOFUELS & BIOENERGY SUSTAINABILITY

2010-Biomasa remanente y contenido de Nitrógeno, Fósforo y Potasio, por tipo de cosecha, en un rodal de *Pinus elliotti* de 24 años. **Entisol.** Aparicio J. INTA EEA Bella Vista.

Biomasa residual ($Mg.ha^{-1}$)	FE	85,8	adicional de 60,3 $Mg.\ ha^{-1}$
	AE	25,5	
N ($kg.ha^{-1}$)	FE	513	adicional de 344 $kg\ ha^{-1}$
	AE	169	
P ($kg.ha^{-1}$)	FE	37	adicional de 25 $kg\ ha^{-1}$
	AE	12	
K ($kg.ha^{-1}$)	FE	78	adicional de 59 $kg\ ha^{-1}$
	AE	19	

FE: Fuste entero. AE: árbol entero

UREA	764,4 ($kg.ha^{-1}$)
SFT	125,0($kg.ha^{-1}$)
KCl	112,5($kg.ha^{-1}$)

RCN Conference on Pan American Biofuels & Bioenergy Sustainability
Recife, Brazil July 22-25, 2014

PAN-AMERICAN BIOFUELS & BIOENERGY SUSTAINABILITY

Biomasa y contenido de Nitrógeno, Fósforo y Potasio, por tipo de cosecha, en un rodal de *Pinus taeda* de 20 años. **Ultisol.** Fernandez et al., (2014)

Biomasa residual (Mg.ha^{-1})	FE	74,0	adicional de 25,2 Mg. ha^{-1}
	AE	48,8	
N (kg.ha^{-1})	FE	423,3	adicional de 129,5 kg ha^{-1}
	AE	305,6	
P (kg.ha^{-1})	FE	27,8	adicional de 7,6 kg ha^{-1}
	AE	20,2	
K (kg.ha^{-1})	FE	117,9	adicional de 33,3 kg ha^{-1}
	AE	84,7	

FE: Fuste entero. AE: árbol entero

UREA	287,8 (kg.ha^{-1})
SFT	38,0(kg.ha^{-1})
KCl	63,5(kg.ha^{-1})

RCN Conference on Pan American Biofuels & Bioenergy Sustainability
Recife, Brazil July 22-25, 2014

PAN-AMERICAN BIOFUELS & BIOENERGY SUSTAINABILITY

ALGUNOS EFECTOS SOBRE EL SUELO.....

Concentración de CO ; N total y relación C/N, en diferentes sistemas de manejo de residuos de cosecha de *P. taeda*.

Lupi et al., (2014) CACS.

Integrando muestreos de suelos a los 2, 5, 7 y 12 años de aplicar los tratamientos

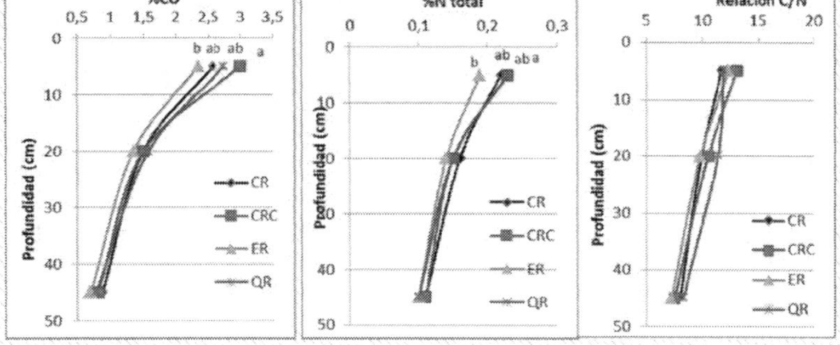

CR: conservación de residuos .CRC: Conservación de residuos-bosque secundario ER: Extracción de residuos. QR; quema de residuos.

pH y fósforo disponible en diferentes sistemas de manejo de residuos de cosecha de *P. taeda*.

Lupi et al., (2014) CACS.

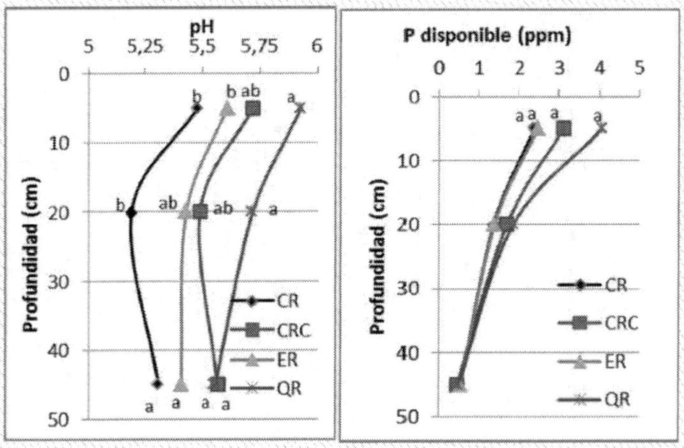

INTA
RCN Conference on Pan American Biofuels & Bioenergy Sustainability
Recife, Brazil July 22-25, 2014
PAN-AMERICAN BIOFUELS & BIOENERGY SUSTAINABILITY

Carbono orgánico total en dos suelos con similares sistemas de manejo de residuos de *Eucalyptus grandis*

	0-5 cm	5-10 cm
	Peluderte (Vertisol) 18 meses	
Conservación de residuos	21,2 (7,8)a	15,75,6)a
Extracción de residuos	20,4(4,8)a	15,5(1,6)a
Quema de residuos	22,4(6,5)a	13,6(2,2)a
	Hapludol (Molisol) 22 meses	
Conservación de residuos	28,1(5,3)a	15,7(0,7)a
Extracción de residuos	27,9(3,1)a	15,6(0,9)a
Quema de residuos	22,3(7,2)a	14,1(0,6)a

Lupi et al., en evaluación Ciencia Florestal. SM

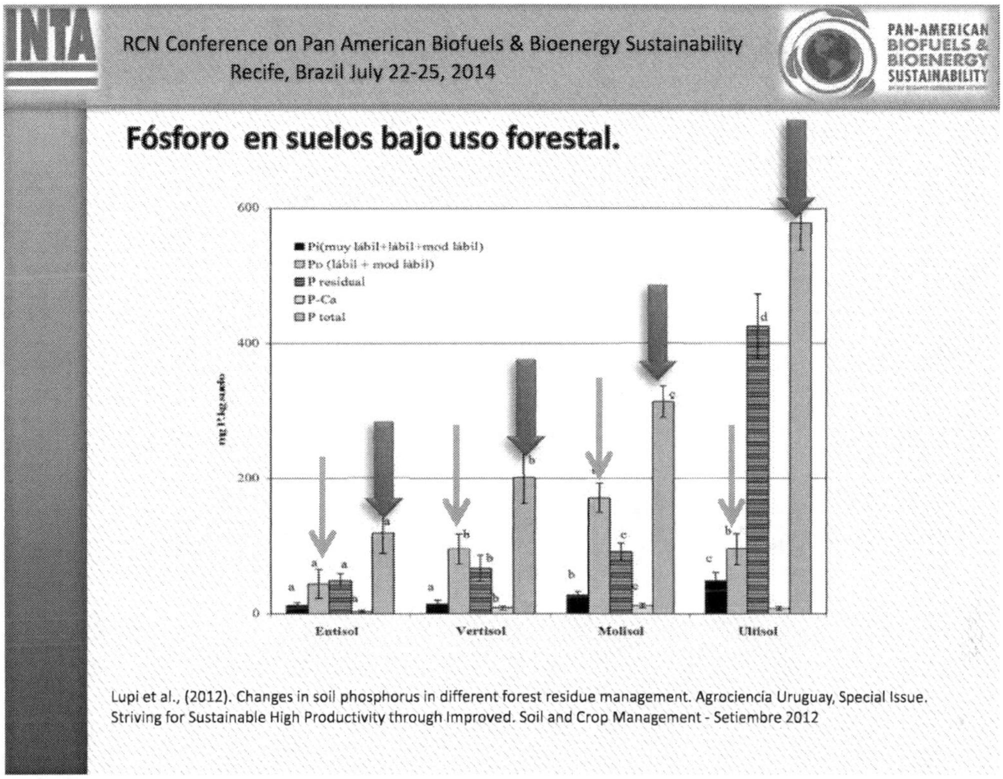
INTA
RCN Conference on Pan American Biofuels & Bioenergy Sustainability
Recife, Brazil July 22-25, 2014
PAN-AMERICAN BIOFUELS & BIOENERGY SUSTAINABILITY
Fósforo en suelos bajo uso forestal.
Pi(muy lábil+lábil+mod lábil)
Po (lábil + mod lábil)
P residual
P-Ca
P total
600
400
200
0
Entisol
Vertisol
Molisol
Ultisol
Lupi et al., (2012). Changes in soil phosphorus in different forest residue management. Agrociencia Uruguay, Special Issue. Striving for Sustainable High Productivity through Improved. Soil and Crop Management - Setiembre 2012

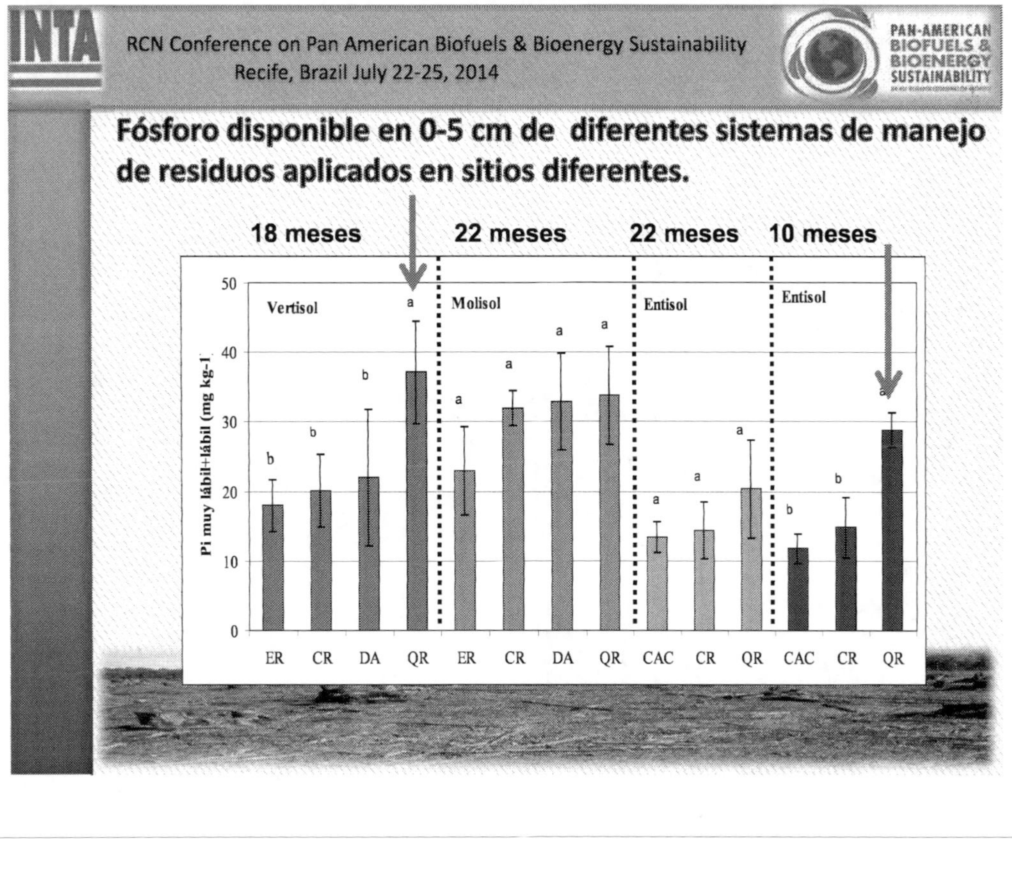
INTA
RCN Conference on Pan American Biofuels & Bioenergy Sustainability
Recife, Brazil July 22-25, 2014
PAN-AMERICAN BIOFUELS & BIOENERGY SUSTAINABILITY
Fósforo disponible en 0-5 cm de diferentes sistemas de manejo de residuos aplicados en sitios diferentes.
18 meses
22 meses
22 meses
10 meses
Vertisol
Molisol
Entisol
Entisol
Pi muy lábil+lábil (mg kg-1)
50
40
30
20
10
0
ER CR DA QR ER CR DA QR CAC CR QR CAC CR QR

INTA

RCN Conference on Pan American Biofuels & Bioenergy Sustainability
Recife, Brazil July 22-25, 2014

PAN-AMERICAN BIOFUELS & BIOENERGY SUSTAINABILITY

Fósforo orgánico total en 0-5 cm de diferentes sistemas de manejo de residuos forestales.

18 meses 22 meses 22 meses 10 meses

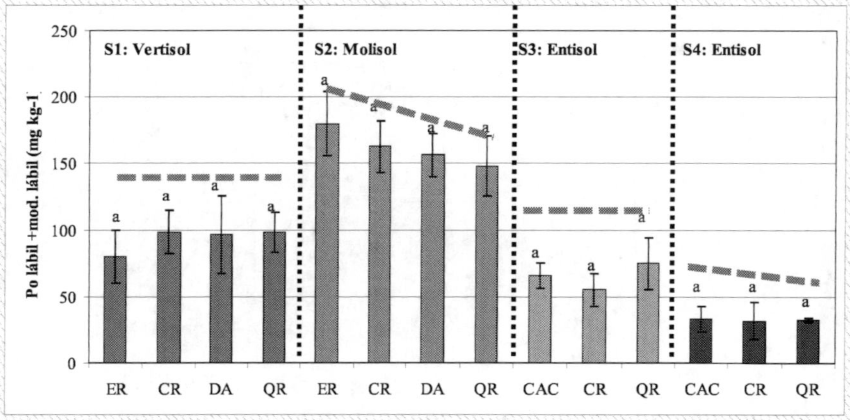

INTA

RCN Conference on Pan American Biofuels & Bioenergy Sustainability
Recife, Brazil July 22-25, 2014

PAN-AMERICAN BIOFUELS & BIOENERGY SUSTAINABILITY

Tamaño de los agregados luego del tamizado en agua

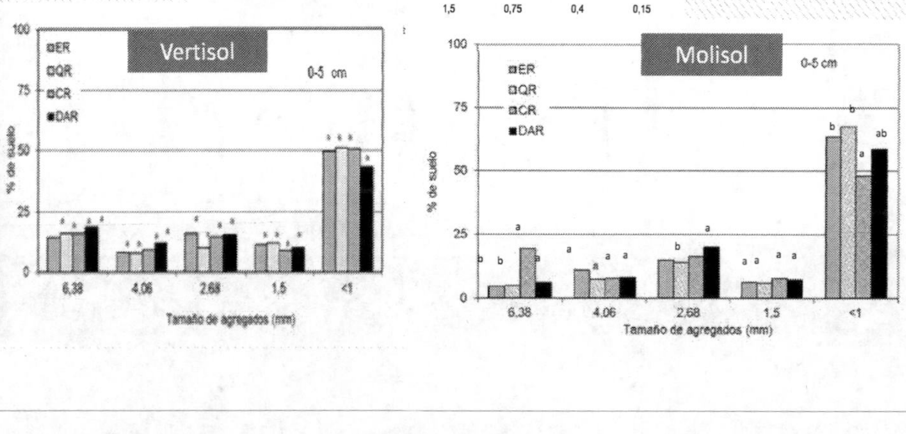

RCN Conference on Pan American Biofuels & Bioenergy Sustainability
Recife, Brazil July 22-25, 2014

Conductividad hidráulica saturada superficial (mm h^{-1})

	Vertisol	Molisol
Conservación de Residuos	8,22 (7,81)a	2,38 (1,52)a
Extracción de Residuos	3,56 (2,9)b	1,30 (1,68)a
Quema de Residuos	2,52 (1,92)b	2,15 (1,56)a

Vertisol, naturalmente es mas estable que el Molisol. DMPA mayores en Vertisol.

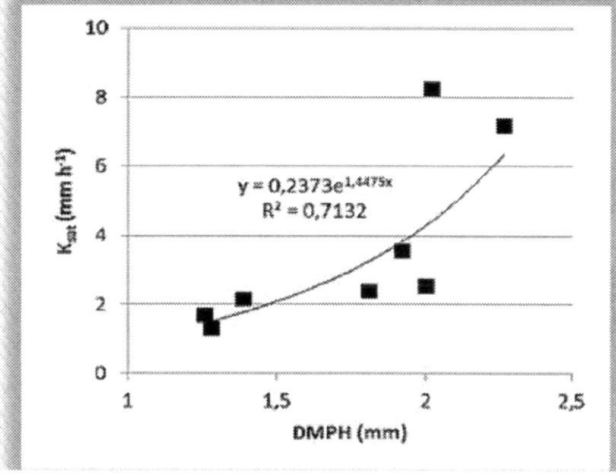

RCN Conference on Pan American Biofuels & Bioenergy Sustainability
Recife, Brazil July 22-25, 2014

Ácidos húmicos de pastizal natural y de diferentes sistemas de manejo de residuos de *E. grandis. Entisol.*

	% C	% N	% H	% S	%O	O/C	C/N
Pastizal Natural	51,8	3,8	4,7	0,53	39,1	0,75	13,7
Extracción de residuos	52,8	3,3	5,1	0,44	38,3	0,73	15,9
Conservación de residuos	54,3	3,3	5,3	0,42	36,6	0,67	16,5
Quema de residuos	55.0	3.6	5,9	0,44	34,9	0,64	15,4

Estructuras orgánicas más ricas en C y N y pérdida de oxígeno por temperatura de la quema.

RCN Conference on Pan American Biofuels & Bioenergy Sustainability
Recife, Brazil July 22-25, 2014

Intensidades relativas (%), índice de aromaticidad y relación de descomposición derivados de los espectros de ^{13}C-RMN de ácidos húmicos de un suelo.

	C=O	Lignina	Carbohidratos	polipéptidos	Poli péptidos, Lípidos	Relación descomp.	Índice de aromaticidad
Pastizal	**16**	30	7	25	**22**	**0.69**	**36**
Extracción de resduos	16	37	3	14	30	1.76	44
Cons. Residuos	15	33	4	14	34	1.89	39
Quema	16	41	5	12	26	1.5	49

memoria del suelo

+ Comp. Aromáticos
- celulosa

RCN Conference on Pan American Biofuels & Bioenergy Sustainability
Recife, Brazil July 22-25, 2014

EN SINTESIS

1- Los suelos forestales son relativamente fértiles y recién evaluamos un segunda rotación forestal.

La gran variabilidad de ambientes se presenta con suelos que difieren en su formación y evolución, con características de funcionamiento particulares.

2- La extracción de residuos genera salidas adicionales de nutrientes y de C a nivel de lote. No se reponen.

3- En el corto plazo lo suelos muestran comportamientos diferentes. Algunos suelos son indiferentes.

4- Tenemos evidencias de cambios en la calidad de la materia orgánica por quema o exportación de residuos. No sabemos las implicancias en cuanto al ciclo de los nutrientes.

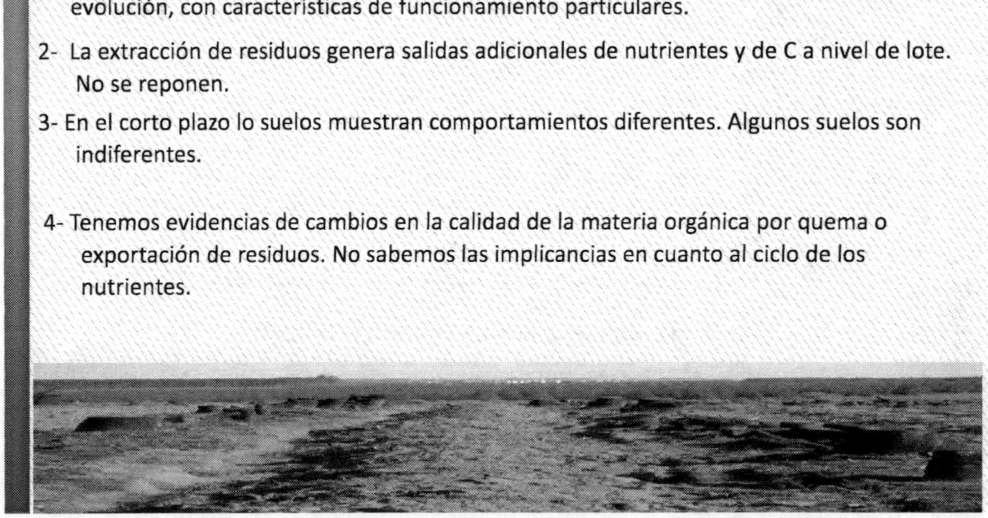

INTA

RCN Conference on Pan American Biofuels & Bioenergy Sustainability
Recife, Brazil July 22-25, 2014

PAN-AMERICAN BIOFUELS & BIOENERGY SUSTAINABILITY

¿Qué nos falta?

1- Estudios de largo plazo- red de experimentos en ambientes diferentes donde se evalúen cambios de procesos edáficos.

- Balance del MO y cambios en propiedades físicas del suelo relacionadas con el ingreso y movimiento del agua y aire.
- Microbiología de suelos es un aspecto no abordado (cambios en la diversidad/ cantidad).

2- Conocer las potencialidades y las limitantes de cada ambiente y manejar cada uno con una silvicultura sitio especifica para una producción sostenible (buenas prácticas de manejo).

3- Alertar sobre posibles impactos de sistemas de manejo extractivos .

4- Promover la reposición de nutrientes y el descortezado a campo como medida de mitigación

Lipid Accumulation and Nitrogen Removal for Chlorella vulgaris and Scenedesmus Obliquus Using Wastewater

Ruiz Marin, A. Sr.,
Canedo Lopez, Y. Sr.,
Originally presented on: 7/23/2014 14:15:00 - 14:40:00

Lipid accumulation and nitrogen removal for *Chlorella vulgaris* and *Scenedesmus obliquus* using wastewater

[a*]Ruiz-Marín Alejandro, Canedo-López Yunuen [a]

[*a] Facultad de Química, Universidad Autónoma del Carmen. C. 56 #4, Av. Concordia, Col. Benito Juárez, C.P. 24180. Ciudad del Carmen, Campeche, México. Email: aruiz@pampano.unacar.mx; ycanedo@pampano.unacar.mx

Nitrogen limitation has been widely proposed as a method to increase lipid content of microalgae in biodiesel-oriented processes. However, this is typically accompanied by a reduction on the growth rate, and as a result, the overall lipid productivity does not necessarily increase. In this study a novel multi-stage nitrogen-reduction process is proposed, in order to promote a balance between growth rate and lipid accumulation which could result in a net increase of lipid-productivity in microalgae, while simultaneously reducing nitrogen concentrations in wastewater. *Chlorella vulgaris* and *Scenedesmus obliquus* were grown initially in nitrogen-rich (90 mg L^{-1}) artificial-wastewater medium, followed by sequential dilutions (50% v/v) in fresh medium with $N\text{-}NH_4$ concentrations of 60, 40, and 20 mg L^{-1}, respectively. The overall lipid productivity was compared to those obtained in various two-stage nitrogen reduction processes, wherein the nitrogen-rich culture was followed by a 50% v/v dilution in fresh medium containing 30, 20, or 10 mg L^{-1} $N\text{-}NH_4$ in the second stage. Increased net lipid-productivity was observed for both species in the two-stage mode, although nitrogen depletion was not achieved in these cases. On the other hand, in the sequential mode only *C. vulgaris* exhibited a net lipid-productivity increment. The highest lipid productivities occurred in the two-stage mode for both *S. obliquus* and *C. vulgaris* (194.9 and 133.5 mg L^{-1} d^{-1}, respectively). The lipid productivities achieved in this study are among the highest reported in the open literature to date, and the fatty-acid profiles are adequate for biodiesel production.

Biomass Harvesting and Site Productivity: Is Policy ahead of Science?

Eric D. Vance, National Council for Air and Stream Improvement (NCASI)
W. Michael Aust, Virginia Tech
Robert E. Froese, Michigan Technological University
Robert B. Harrison, University of Washington
Larry A. Morris, University of Georgia
Brian D. Strahm, Virginia Tech

RCN Conference on Pan American Biofuels and Bioenergy
July 22-25, 2014
Recife, Brazil

Why Harvest Residues?

- Renewable energy
  - Electrical, thermal (pellets)
  - Biofuels
- Particle board, composite materials
- Forest health (fire, insects, disease)
- Forest operations (site preparation)

Environmental Questions

- Removal of biomass traditionally left on site
- Shorter rotations, more frequent site entries
- Site productivity
- Water
- Wildlife

Common U.S. Biomass Harvesting Guideline Provisions

- Retain 15-30% or "as much as possible"
- Restrictions on shallow, coarse-textured, nutrient-poor soils
- Minimize area in roads, landings, skid trails

Forest Guild Biomass Retention and Harvesting Guidelines

(February, 2012)

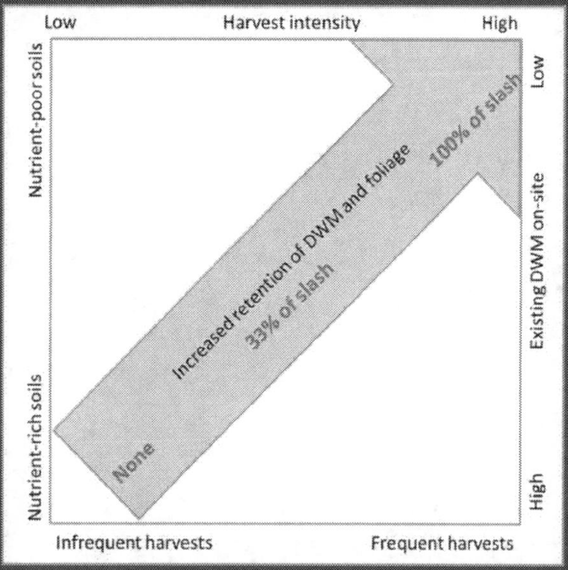

Forest Guild Biomass Retention and Harvesting Guidelines

- "Where science remains inconclusive, we relied on field observation and professional experience"

Frames of Reference

- "Natural" is optimal, more retention is better
- Conventional harvests retain residues
- Biomass harvesting removes all residues
- No beneficial practices
- Changes in State = Changes in Function

Perceived Benefits of Traditional Residue Management

25 to 80% of Non-Bole Residues Left after Operational Biomass Harvesting Due to Technical Limitations, Incidental Breakage

(Nurmi 2007, Ralevic et al. 2010, Klockow et al. 2013)

Sites in Minnesota following biomass harvesting
(The Forestry Forum)

Ecological Effects of Harvest Residue Management in Southern Pine Plantations

(USDA-AFRI, NC State Univ., Univ. of Georgia, Weyerhaeuser, Plum Creek, Georgia Pacific, NCASI)

- 8 sites in NC and GA
- Residues retained
- Residues removed
- 15% or 30% retained
- Clumped or dispersed

Harvest Residue Management in Southern Pine Plantations

- Small mammals, reptiles, amphibians, arthropods
- Plant community composition
- Soil properties, variability
- Technical input on BHGs
- Logger/manager survey

Residues Retained After Operational Biomass Harvesting in North Carolina, USA

S. Fritts, C. Moorman, D. Hazel, B. Jackson
NC State Univ., Univ. of Georgia

NO BHG, 15CLUS, 15DISP, 30CLUS, 30DISP, BOHarv

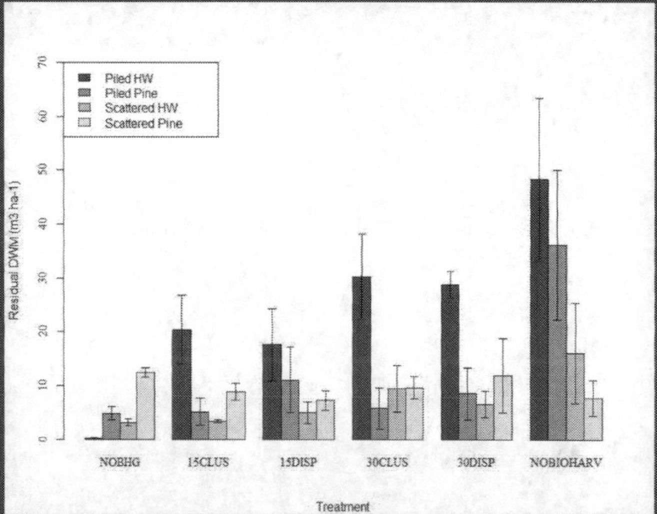

New Technologies will Increase Harvesting Efficiency

Analysis of 53 Temperate and Boreal Forest Studies

(Thiffault et al. 2011)

- Biomass harvesting can reduce soil nutrient pools, particularly in the forest floor
- “No clear impact of whole-tree harvest on soil C”
- “No consistent, unequivocal and universal effects of biomass harvesting on soil productivity”

Forest Service Long-Term Soil Productivity Project

(Ponder et al. 2012)

- 10 years, 45 installations across North America
- Whole-tree harvest: "little consistent effect on any response variable, including tree growth"
- Vegetation control consistently increased tree biomass
- Soil nutrient demands may increase

Ecological Effects of Logging Residue Harvest in Great Lakes Aspen Stands

(Michigan Tech, Plum Creek, NCASI)

("Residues removed" treatment; Michael Premer, Michigan Tech)

- 40 year chronosequence of harvested stands
- With or without residue removal

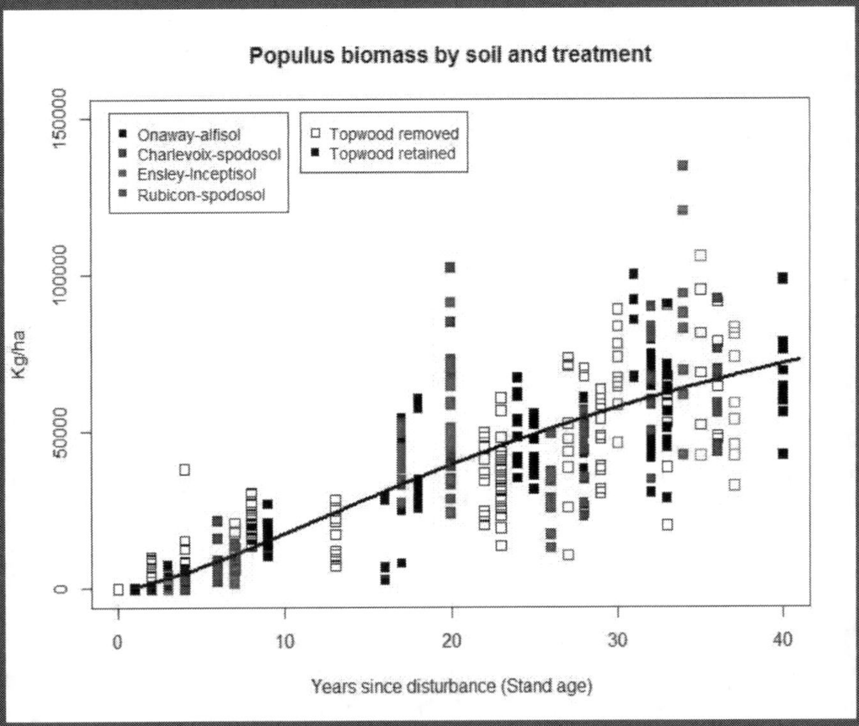

Fall River Long-Term Soil Productivity Project

(Univ. of Washington, Weyerhaeuser, USFS, NCASI)

- Conventional bole-only removal
- Total stem (bole-only to a 5cm top)
- Total-tree removal
- Total-tree + legacy-wood removal

Fall River: Effects of Harvesting, Compaction and Vegetation Control on Age 10 Tree Volume
(Holub et al. 2013)

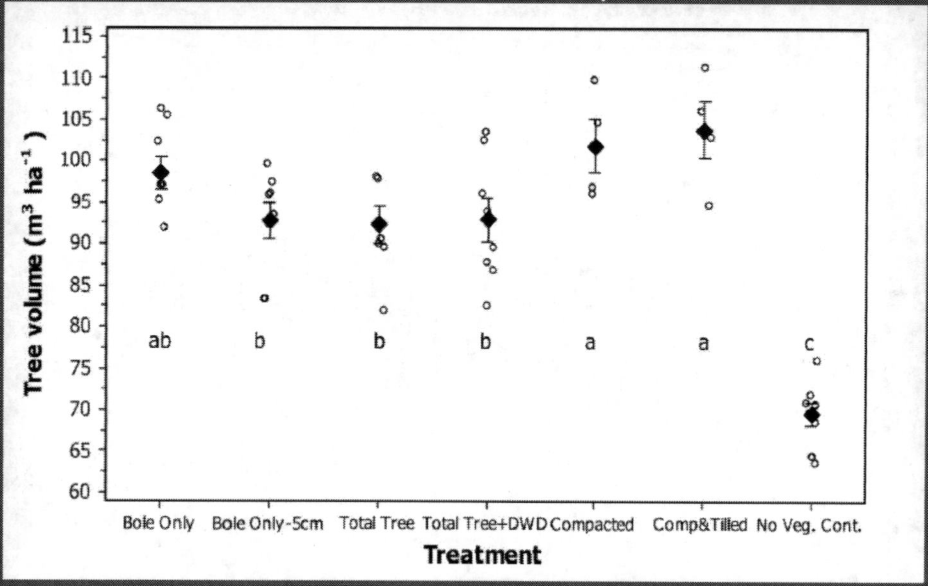

Can Nutrient Deficiencies be Prevented or Corrected?

- Evidence suggests yes
  (e.g., Helmisaari et al. 2011, Jacobson et al. 2000, Smith et al. 2000)
- Lack of evidence for uncorrectable declines
- Many studies do not include fertilization
- Fertilization not an option on some lands
  - Management restrictions
  - Economics

Soil Erosion May Impact Soil Productivity and Water Quality

- Decades of Research: BMPs for water quality effective with 90% implementation
  - (Ice et al. 2011, Schilling et al. 2009)
- Biomass Harvesting Can Exacerbate Disturbance
  - (Curran and Howes 2011)
- Extra precautions
  - Minimize bare soil
  - Retain residues where needed
  - Monitor

Moving Forward in the Face of Uncertainty

- Default to operational expertise, best management practices
- Monitor Function rather than State
- Assess economic and bureaucratic deterrents
- Identify sensitive sites
- Mitigate, modify, or suspend practices if problems occur

Discussion

Standardization of Environmental LCA of Biofuels in the Pan American Region

David R. Shonnard[1,4], Bethany Klemestrud[1], Julio Sacramento-Rivero[2], Freddy Navarro-Pineda[2], Jorge Hilbert[3], Robert Handler[4], Nydia Suppen[5], Richard P. Donovan[6]

[1] Department of Chemical Engineering, Michigan Technological University, Houghton, MI USA
[2] Faculty of Chemical Engineering, Universidad Autónoma de Yucatán, Merída, Mexico
[3] INTA, National Agricultural Technology Institute, Buenos Aires, Argentina
[4] Sustainable Futures Institute, Michigan Technological University, Houghton, MI USA
[5] CADIS - Center for LCA and Sustainable Design, Mexico City, Mexico
[6] Assistant Director for Research Development, University of California, Irvine, CA

RCN Conference on Pan American Biofuel and Bioenergy Sustainability
July 23, 2014
Golden Tulip Recife Palace
Recife, Pernambuco, Brazil

Outline

- ❑ Pan American Focus for LCA
- ❑ LCA is a Promising Tool for Biofuel Sustainability
- ❑ Challenges in Biofuel LCA
- ❑ Study Scope, Method, and Results
- ❑ Summary and Acknowledgements
- ❑ Recommendations for Future Research

Why a Pan American Focus?

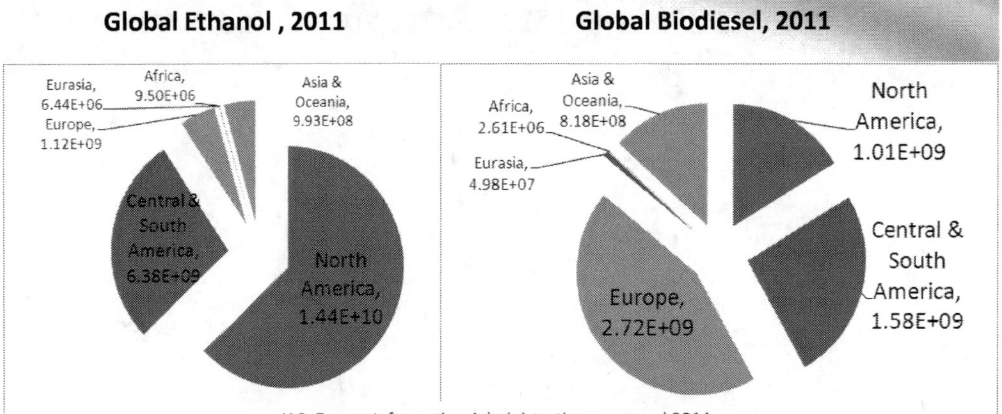

U.S. Energy Information Administration, accessed 2014

- ❑ **The vast majority of transportation biofuel production**
- ❑ **No prior studies on standardization of LCA in Pan American region**

Biofuel LCA – a Promising Tool

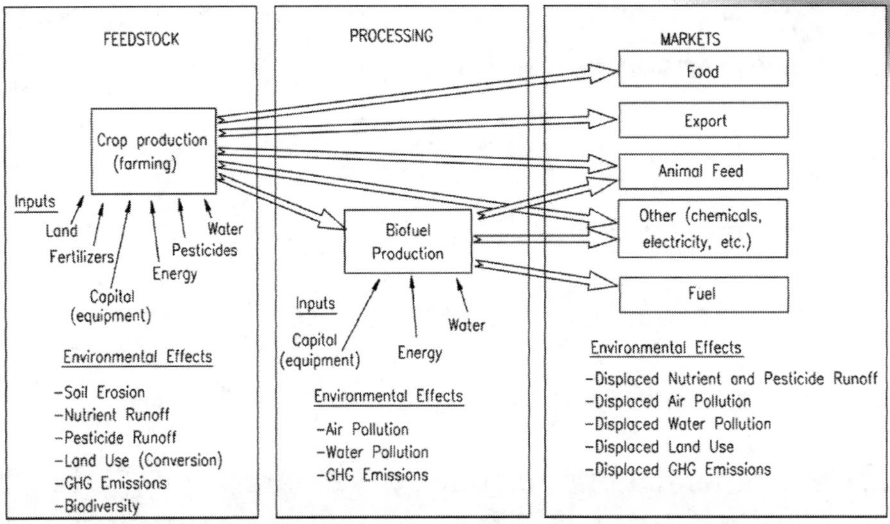

KAMMEN et al (2007) Energy and greenhouse impacts of biofuels: a framework for analysis. *Organization for Economic Co-Operation and Development (OECD)*, Berkeley, USA: International Transport Forum

Methodology of biofuels' LCA

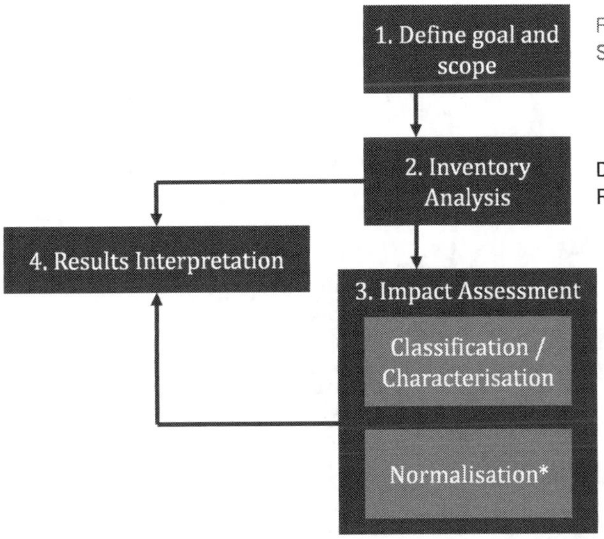

Functional unit
System boundaries
(space and time)

Data quality
Representativeness

Methodology selection
Categories Selection
Allocation method

Analysis of biofuels' LCAs in LA

Functional unit
(depends on the goal of the study)

System boundaries

Overview of biofuels' LCA

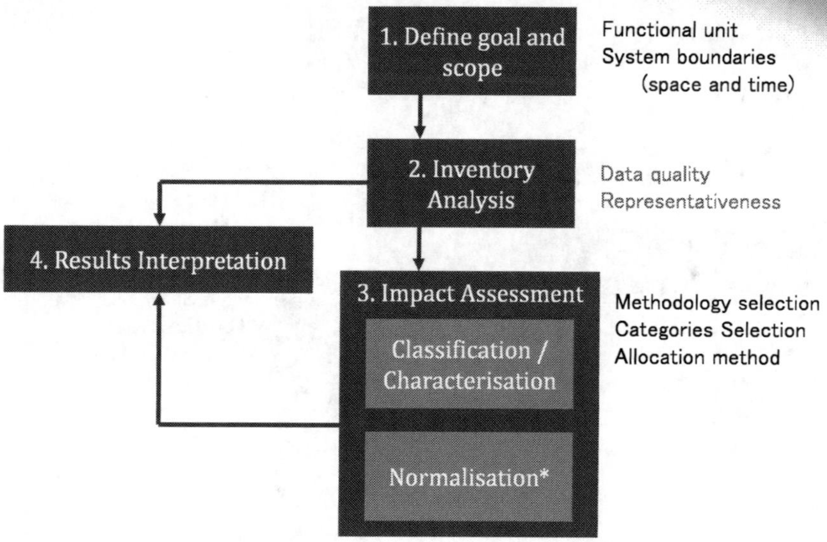

NSF Research Coordination Network Conference: 22-25 July 2014, Recife, Brazil 7

Analysis of biofuels' LCAs in LA

LCI database

NSF Research Coordination Network Conference: 22-25 July 2014, Recife, Brazil 8

Overview of biofuels' LCA

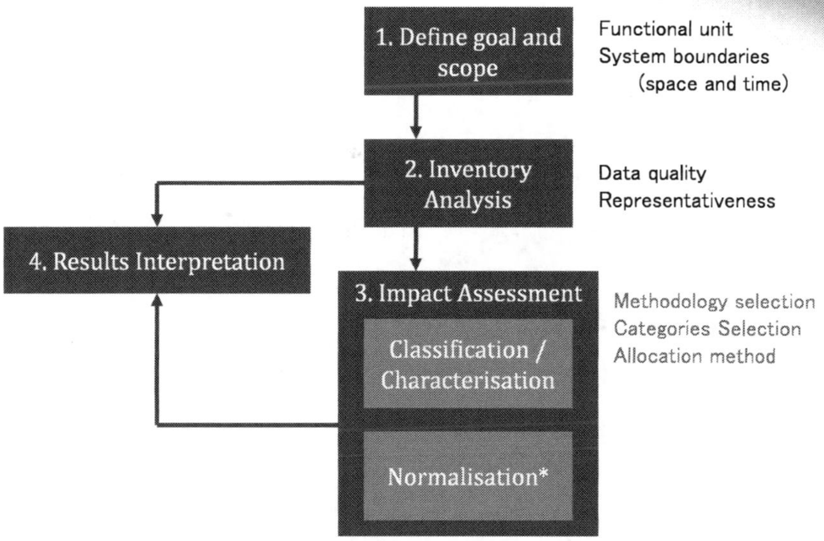

Analysis of biofuels' LCAs in LA

Impact Assessment Methodology

Analysis of biofuels' LCAs in LA

CML Impact Categories

NSF Research Coordination Network Conference: 22-25 July 2014, Recife, Brazil 11

Regulatory requirements

	Study scope	Functional unit	Allocation	Selected time period	GHG emission reduction required
EU-RED	Attributional	gCO_{2eq}/MJ	Energy	Annualized emissions over 20 years.	35%[a]
US-RFS2	Consequential	GHG reduction compared to fossil fuel, (MJ)	System expansion	100 years with 2% discount rate or 30 year with 0% discount rate.	Conventional biofuel: 20%[b] Advanced biofuels: 50% Biomass-based diesel: 50% Cellulosic biofuel: 60%

[a] This value will rise to 50% on January 2017 and will be 60% on 2018 for those facilities which production starts on or after January 2017.
[b] Below gasoline.
[c] RSB (2011).

NSF Research Coordination Network Conference: 22-25 July 2014, Recife, Brazil 12

Attributional vs Consequential LCA

Consequential: biofuel pathway plus external consequences

Attributional: biofuel pathway only

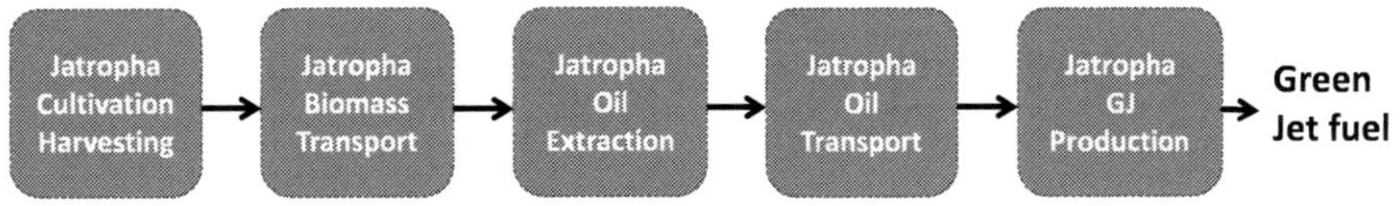

Consequences:
- indirect land use change emissions
- biofuel co–product displacement of market products

Study Scope

- Shed some light on biofuels LCA' s common practices in the Americas
- Aiming to provide standardised guidance of LCA methodology, for greater:
  - Accuracy
  - Transparency
  - Comparability

Analysis of biofuels' LCAs in LA

- Critical evaluation of biofuel LCAs
  - Several conversion pathways
- Papers from the last 10 years
- Total of 74 relevant LCAs selected for the analysis
  - Available USA and EU studies are far more, mainly on 1st generation biofuels
- Recommendations for standardization of LCA methodology

NSF Research Coordination Network Conference: 22-25 July 2014, Recife, Brazil 15

Results: Country location

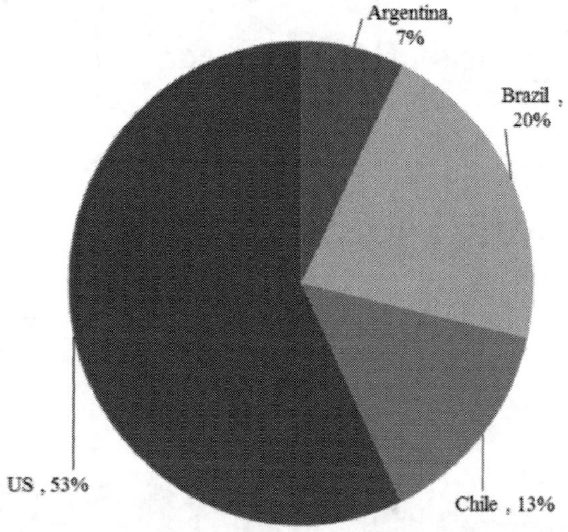

NSF Research Coordination Network Conference: 22-25 July 2014, Recife, Brazil 16

Results: LCA Framework

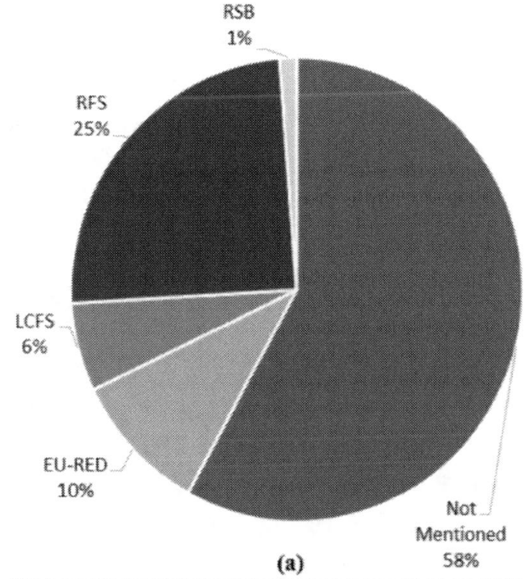

(a)

NSF Research Coordination Network Conference: 22-25 July 2014, Recife, Brazil 17

Results: Functional Unit

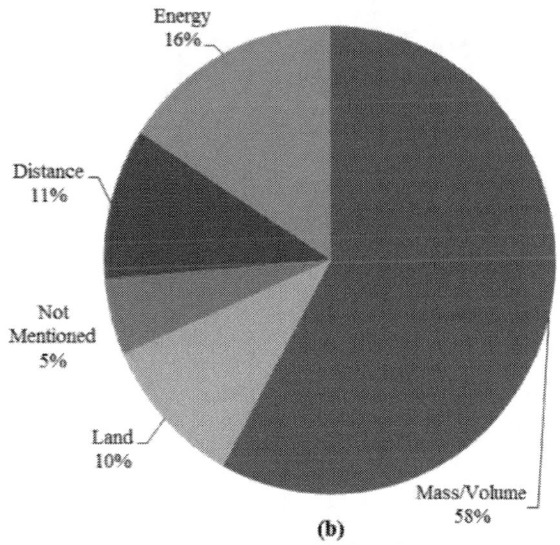

(b)

NSF Research Coordination Network Conference: 22-25 July 2014, Recife, Brazil 18

Results: Allocation method

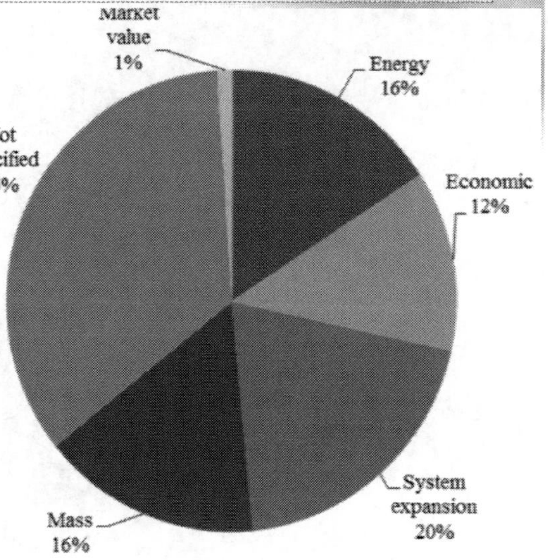

NSF Research Coordination Network Conference: 22-25 July 2014, Recife, Brazil 19

Results: Inputs and inventory data

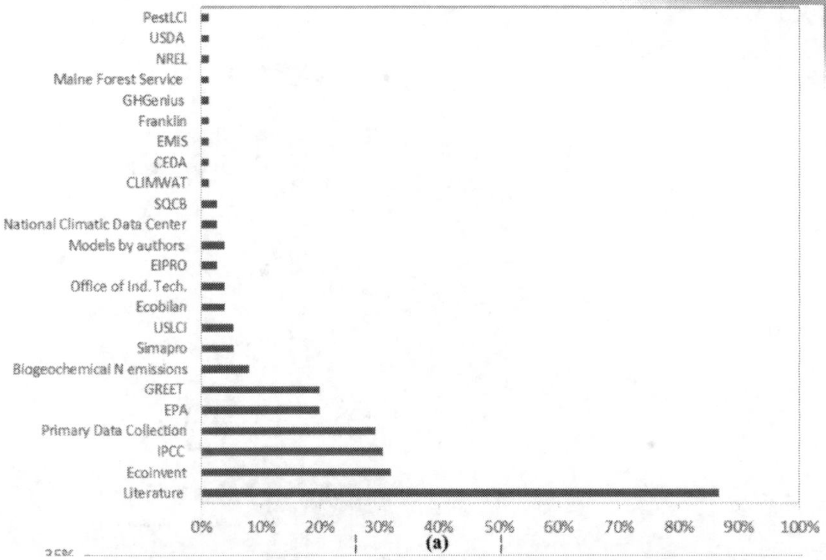

NSF Research Coordination Network Conference: 22-25 July 2014, Recife, Brazil 20

Results: N_2O emissions

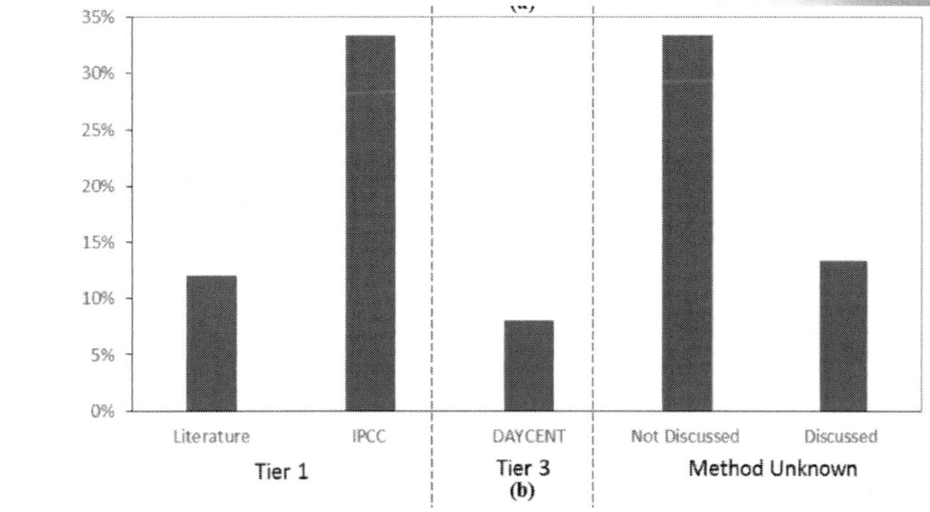

Results: Impacts assessment

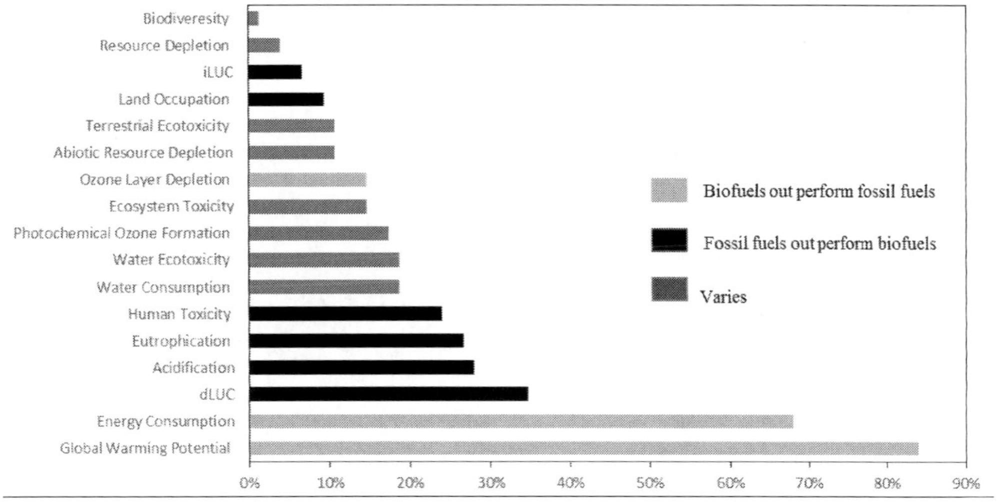

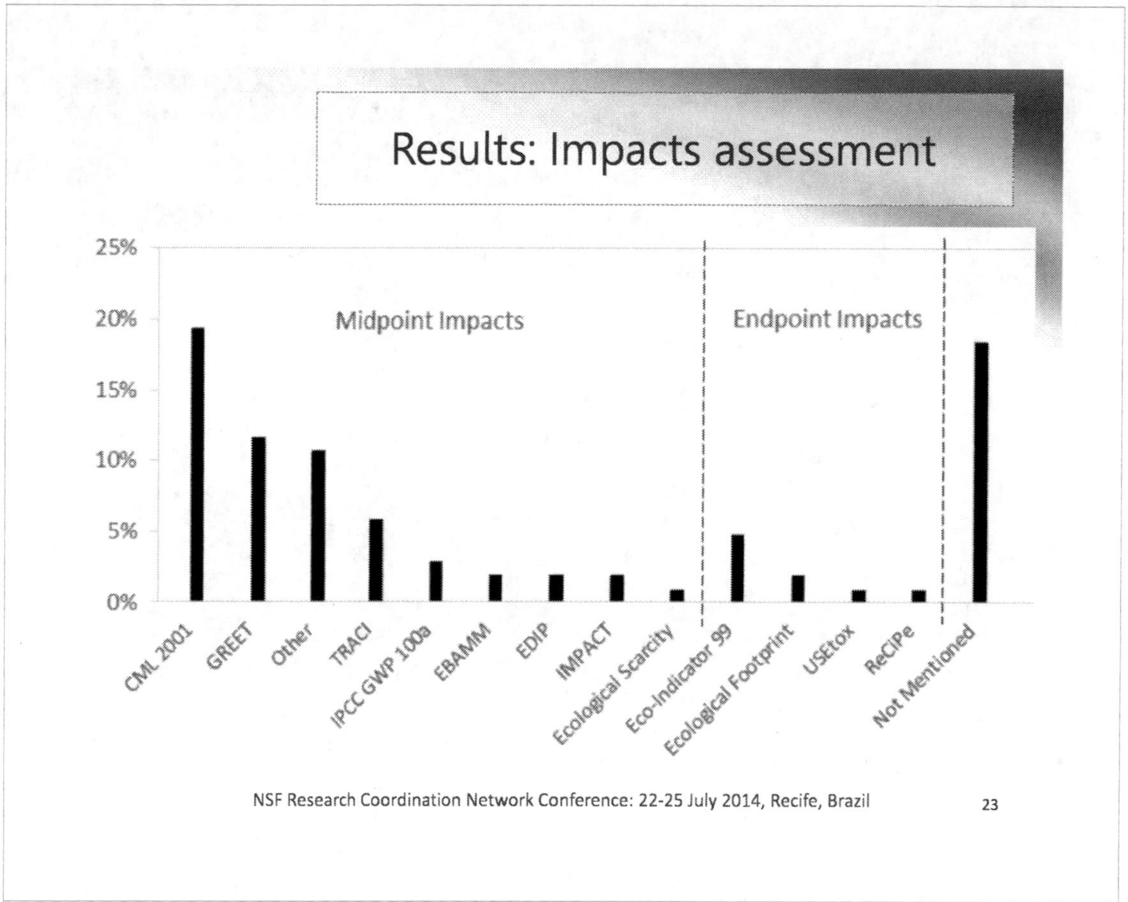

Results: Impacts assessment
25%
20%
15%
10%
5%
0%
Midpoint Impacts
Endpoint Impacts
CML 2001
GREET
Other
TRACI
IPCC GWP 100a
EBAMM
EDIP
IMPACT
Ecological Scarcity
Eco-Indicator 99
Ecological Footprint
USEtox
ReCiPe
Not Mentioned
NSF Research Coordination Network Conference: 22-25 July 2014, Recife, Brazil
23

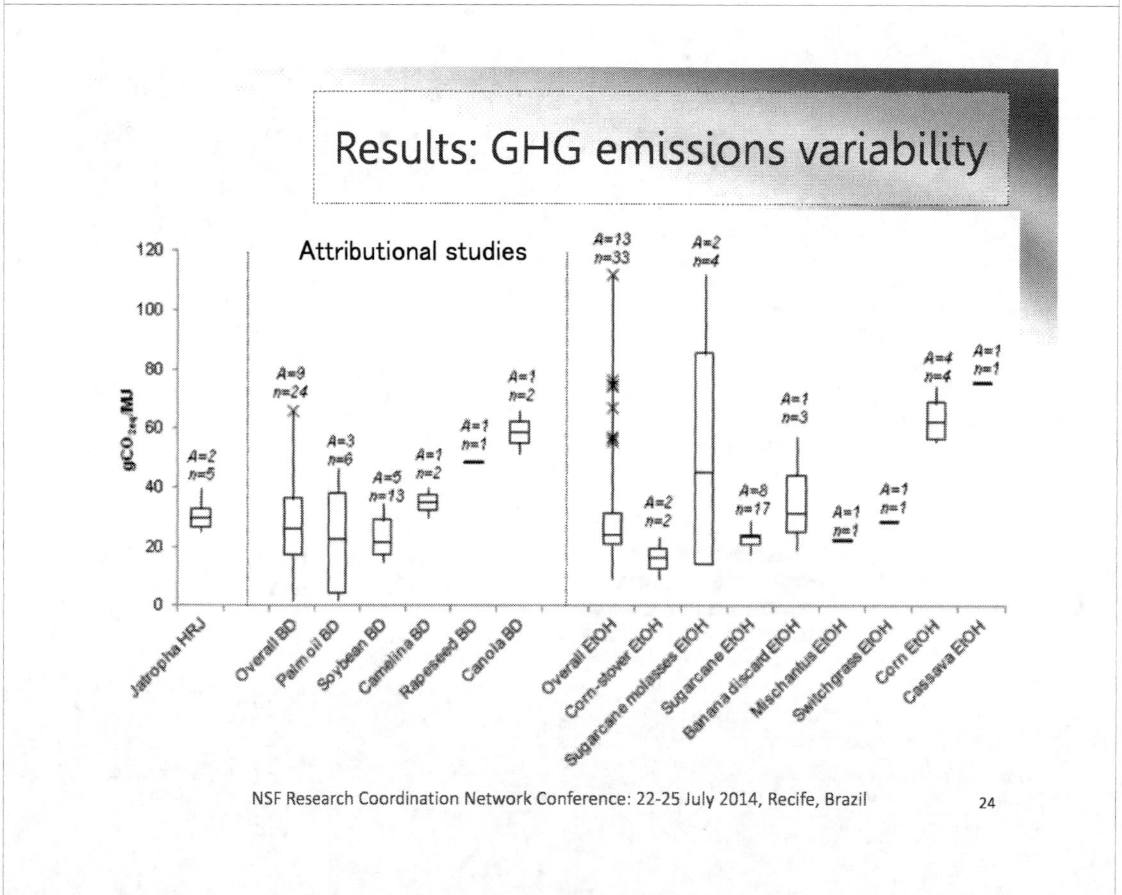

Results: GHG emissions variability
120
100
80
60
40
20
0
Attributional studies
gCO2eq/MJ
A=2 n=5
A=9 n=24
A=3 n=6
A=5 n=13
A=1 n=2
A=1 n=1
A=1 n=2
A=13 n=33
A=2 n=2
A=2 n=4
A=8 n=17
A=1 n=3
A=1 n=1
A=1 n=1
A=4 n=4
A=1 n=1
Jatropha HRJ
Overall BD
Palm oil BD
Soybean BD
Camelina BD
Rapeseed BD
Canola BD
Overall EtOH
Corn-stover EtOH
Sugarcane molasses EtOH
Sugarcane EtOH
Banana discard EtOH
Mischantus EtOH
Switchgrass EtOH
Corn EtOH
Cassava EtOH
NSF Research Coordination Network Conference: 22-25 July 2014, Recife, Brazil
24

Results: GHG emissions variability

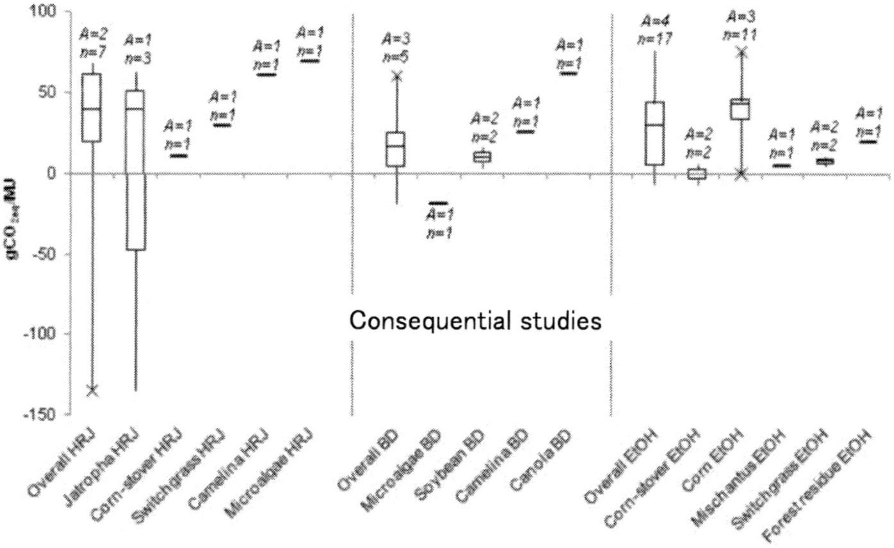

Summary

- Data Sources: Inventory databases each have advantages and limitations, but detecting and eliminating errors remains a high priority.
- Accounting: Different regulatory environments will force biofuel LCA practitioners to meet evolving methodology constraints
- System Boundary: The path to biofuel sustainability will involve ever broader system scope and boundaries in biofuel LCA.

Recommendations

- Biofuel LCA Guidance Frameworks: Future research should quantify the impact of these frameworks and schemes for the same pathways in order to isolate this single study variable.
- Life Cycle Inventory Data Quality: Research is needed to understand inventory data variability and to incorporate it on a statistical basis for use in uncertainty analysis in LCA
- Cultivation Systems: Research is needed in both experimental and modeling aspects on soil properties such as organic matter, carbon, cycling of nutrients (N, P, K), management of water, erosion, emissions, and yields. This research should be conducted in a coordinated way on multiple cropping systems, various climates and in multiple locations throughout Pan America.

NSF Research Coordination Network Conference: 22-25 July 2014, Recife, Brazil 27

Recommendations

- Life Cycle Impact Categories: Future research should continue with GHG emissions, but also expand with increasing momentum into water availability and degradation issues (nutrient runoff and management), biodiversity, criteria air pollutants (PM, H_2S, Hg, etc.), and human/ecosystem toxicity.
- Systems Analysis for Sustainability: Beyond traditional LCA, future research should address the data, framework, and methodology issues in systems analysis for sustainability
- Outreach: A funding mechanisms should be considered for translating LCA research out to the professional communities who are impacted by study results, to policy makers, and also to the general public.

NSF Research Coordination Network Conference: 22-25 July 2014, Recife, Brazil 28

Acknowledgements

This material is based upon work supported in part by the U.S. National Science Foundation grant CBET-1140152"RCN-SEES: A Research Coordination Network on Pan American Biofuels and Bioenergy Sustainability".

We would like to thank the U.S. National Science Foundation for partial support under Award Number 1105039, "OISE-PIRE Sustainability, Ecosystem Services, and Bioenergy Development Across the Americas".

Questions?

Argonne NATIONAL LABORATORY

Biofuel Life-Cycle Analysis: Possible for Standardization?

Jeongwoo Han, Michael Q. Wang, Jennifer B. Dunn, and Hao Cai
Center for Transportation Research
Argonne National Laboratory

2014 RCN Conference on Pan American Biofuels & Bioenergy Sustainability
July 22-25, 2014
Recife, Brazil

U.S. DEPARTMENT OF ENERGY

US Transportation Sector in 2013

Energy Consumption
Total Energy: 28 Trillion MJ (28% of Total US)
Petroleum: 27 Trillion MJ (72% of Total US)

CO_2 Emissions
1.8 Billion Tonnes
(34% of Total US)

Crude Source
Domestic: 43%
Import: 57%

Source: U.S. Energy Information Administration. Annual Energy Outlook 2014 Early Release (http://www.eia.gov/forecasts/aeo/)

- To reduce the transportation energy consumption, Corporate Average Fuel Economy (CAFE) standard was implemented.
- However, more dramatic reductions in crude import and GHG emissions require the development of **alternative fuels**, such as natural gas-based fuels, electricity and biofuels

2

Life cycle Analysis (LCA) is used in US and EU policies targeting expanded use of biofuels

Renewable Fuel Standard (US)

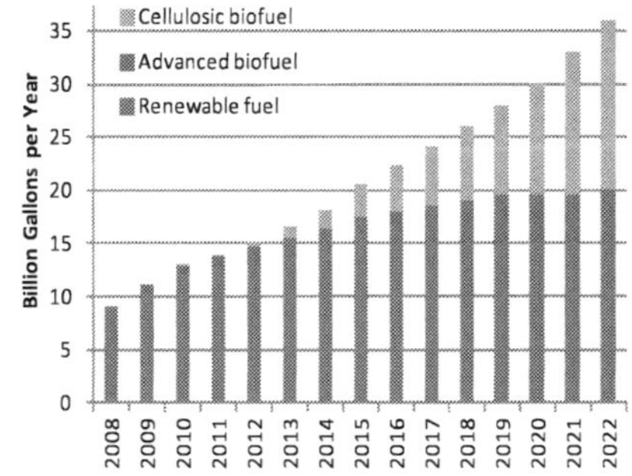

Life Cycle GHG Threshold

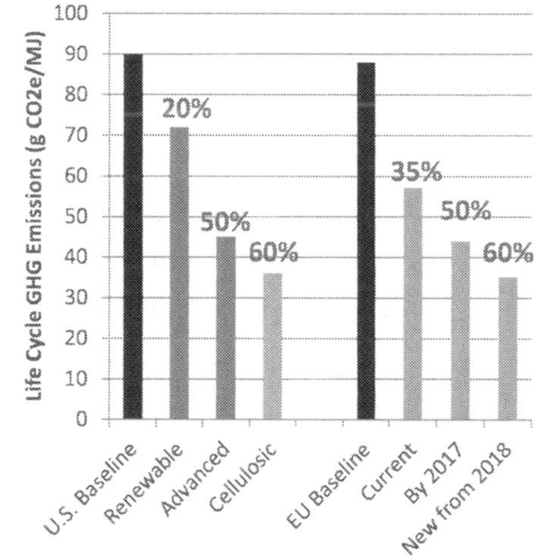

Renewable Energy Directive & Fuel Quality Directive (EU)

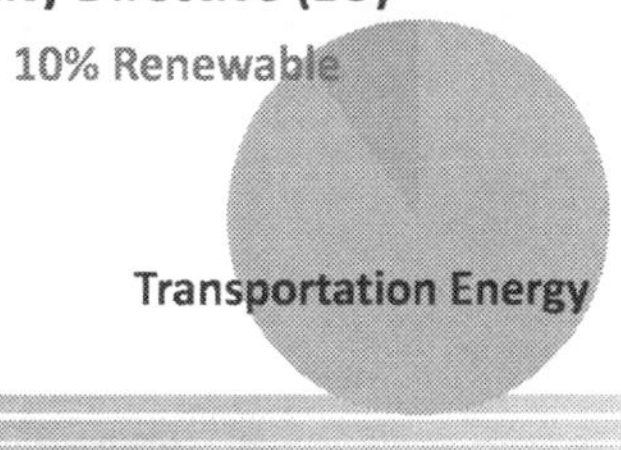

Under Low-Carbon Fuel Standard (LCFS) by California Air Resource Board, credit is accounted for the amount of GHG reductions (Performance-based)

3

Life Cycle Analysis of Transportation Fuels

- Life cycle analysis (LCA): Systematic accounting of the energy use and emissions at every stage of the production, use, disposal and recycle of a product
- Well-To-Wheel (WTW) analysis: Specific to transportation fuels
- WTW analysis takes into account the direct fuel use and its upstream energy use and associated emissions

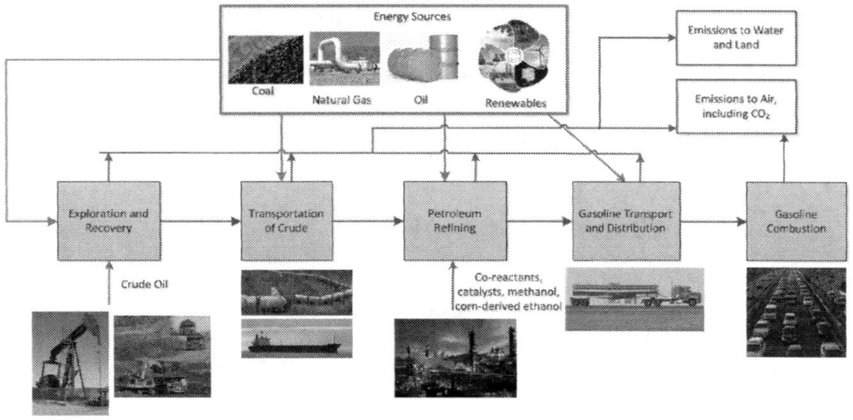

4

Life Cycle Analysis (LCA) Models Available Worldwide for Transportation Fuel Examination

- The GREET model at Argonne National Laboratory
  - CARB LCFS
  - One of a suite of models of EPA RFS2 (along with DAYCENT, GREET, FASOM & FAPRI-CARD)
- The lifecycle emission model (LEM) at University of California at Davis
- Canadian GHGenius model
- BioGrace database in Europe
  - EU Renewable Energy Directive (RED) and Fuel Quality Directive (FQD)
- Other generic LCA models (e.g. SimaPro and Gabi) that can be applied to examine transportation fuels and vehicle technologies
- Newly emerging consequential LCA methods based on economic interactions within a country and/or in the world

5

Technical Issues with LCA

- Modeling framework
  - Consequential vs. attributional
- Modeling decisions
  - System boundary definition (including land-use change)
  - Allocation methodology
- Data inputs
  - Regional/geo-spatial assumptions
  - References & databases used
  - Technology development over time
  - Sensitivity of LCA parameters and uncertainty analysis

6

Land-Use Change Overview

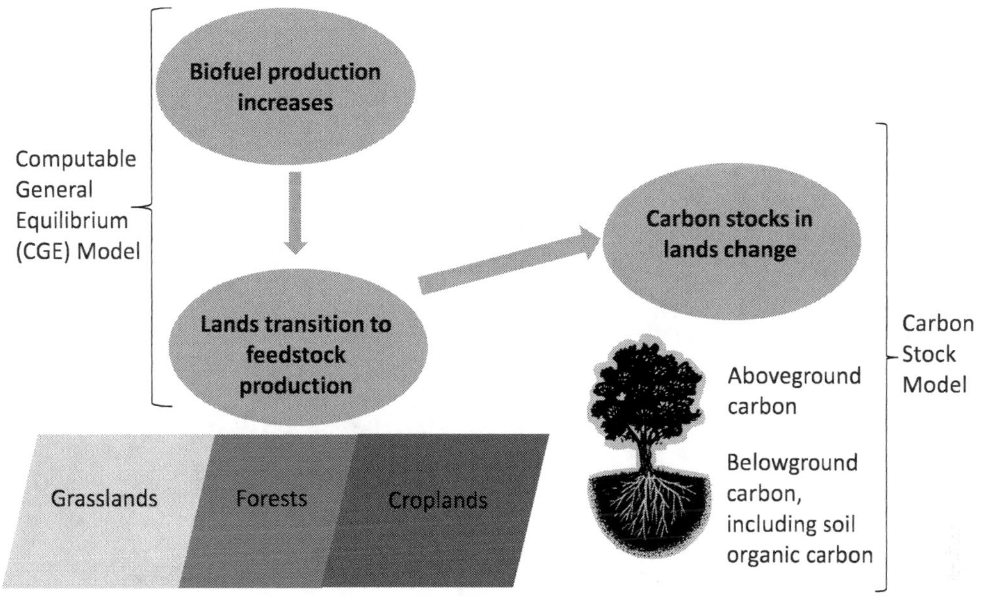

7

Estimates of LUC GHG emissions for the corn ethanol

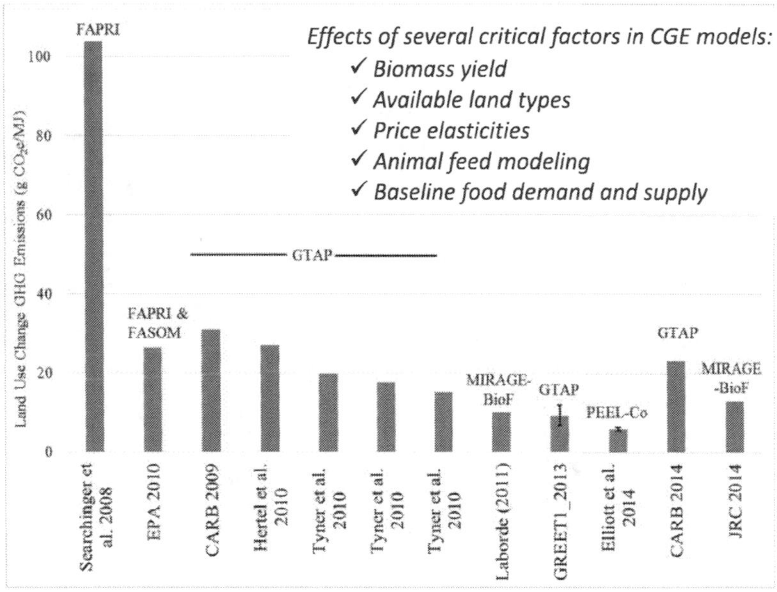

EU's proposed ILUC emissions for cereals and other starch rich crops: 12 gCO_{2e}/MJ

8

Estimates of LUC GHG emissions for sugarcane ethanol and soybean diesel

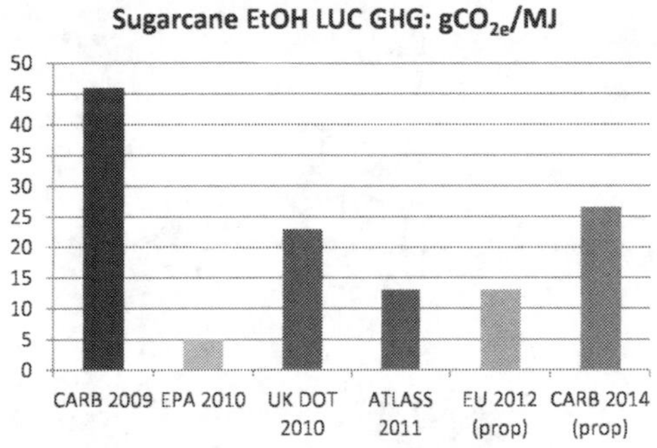

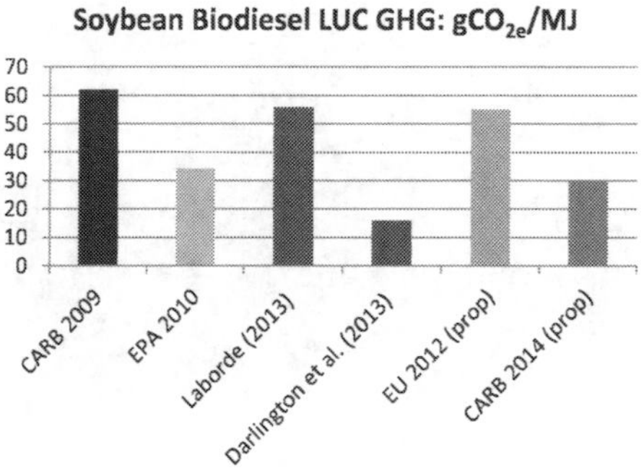

9

Choice of co-product methods can have significant LCA effects for biofuels

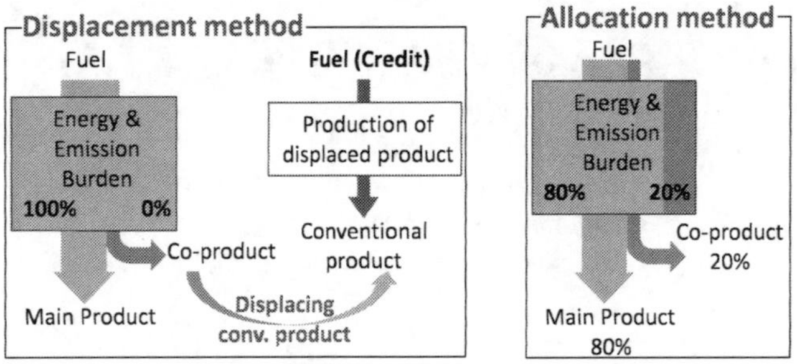

EPA generally uses displacement method while EU uses energy allocation methods

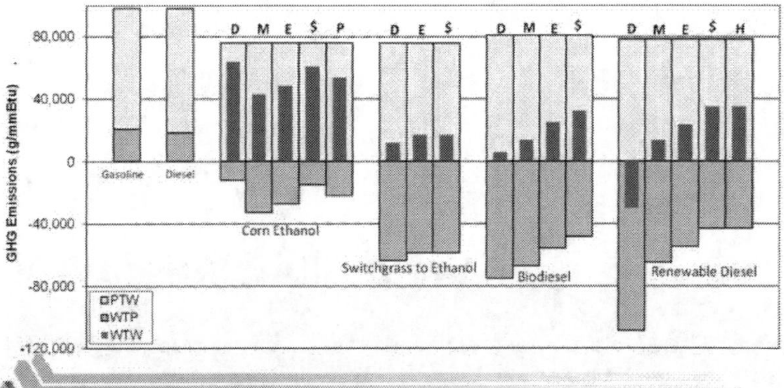

D: Displacement
M: Mass based
E: Energy Based
$: Market Value
P: Process Purpose
H: Hybrid Allocation

10

Harmonization of Rapeseed HEFA/HVO Pathway

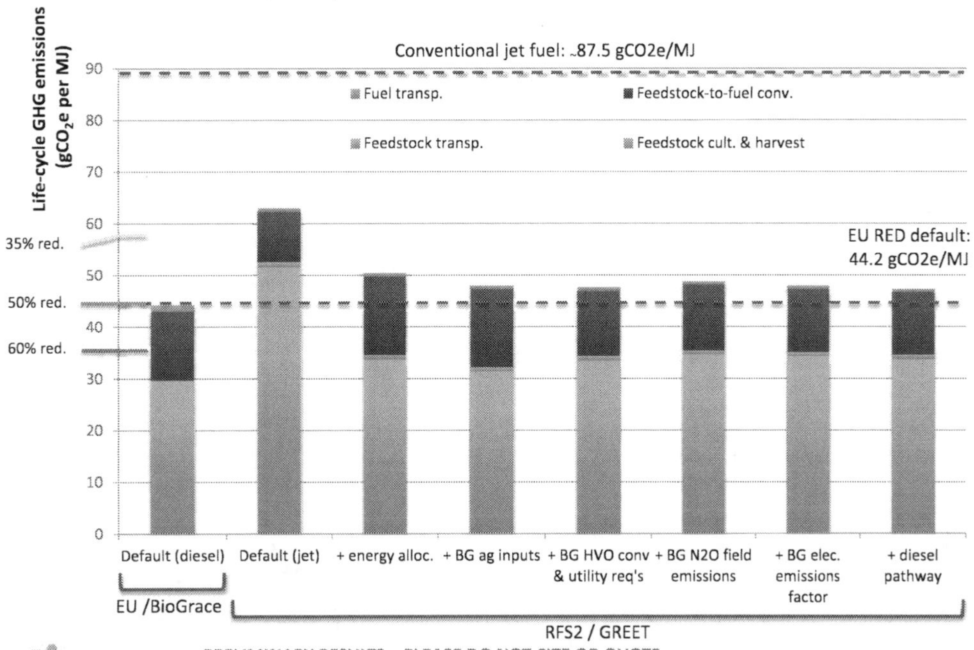

PRELIMINARY RESULTS – PLEASE DO NOT CITE OR QUOTE

Courtesy of Robert Malina

Harmonization of Soybean HEFA/HVO Pathways

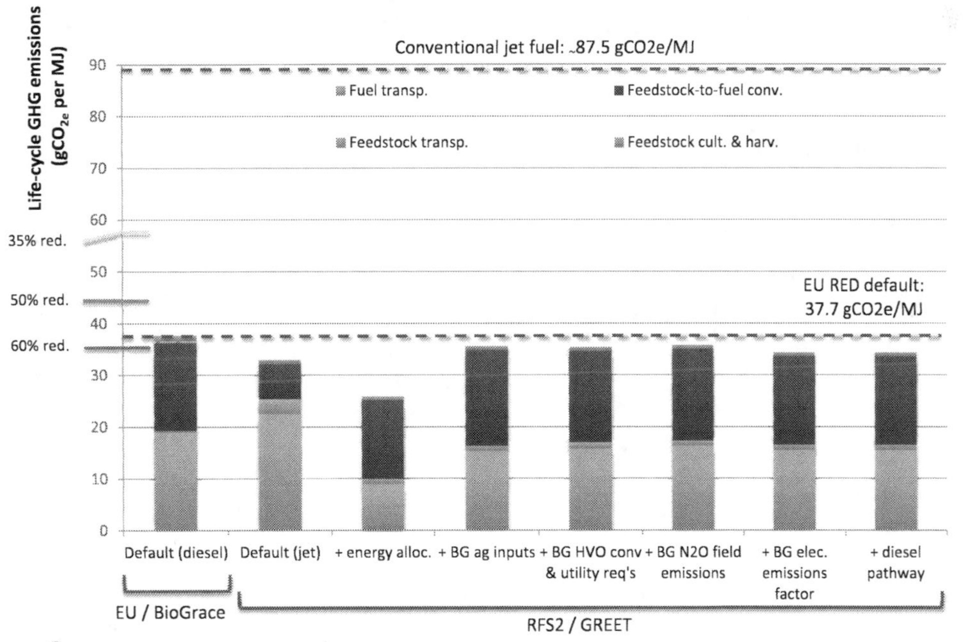

PRELIMINARY RESULTS – PLEASE DO NOT CITE OR QUOTE

Courtesy of Robert Malina

System Boundary Issue for Waste-based Feedstock

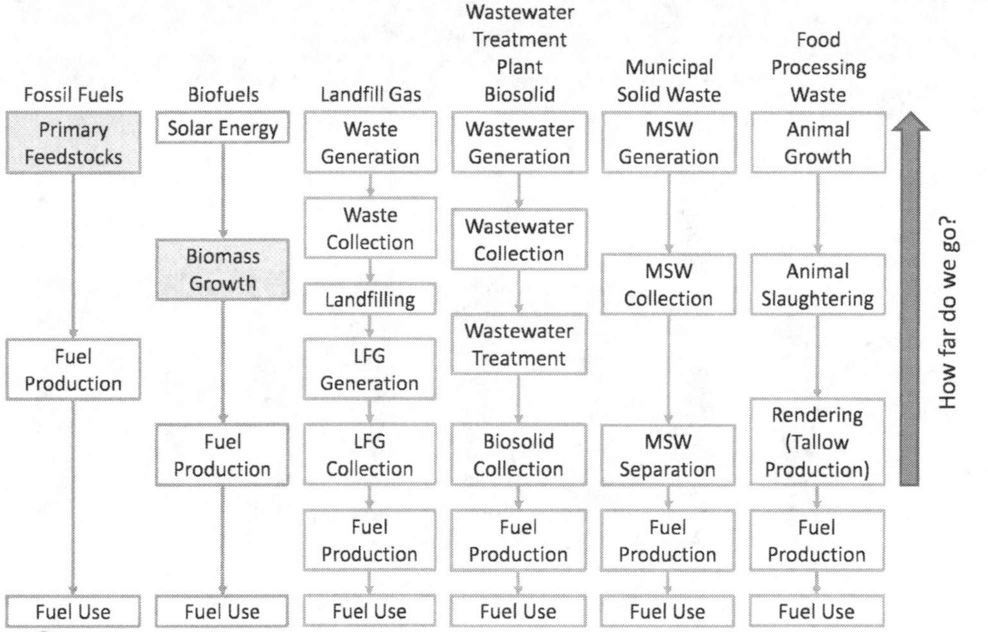

13

System boundary of waste-based feedstocks affects LCA results significantly

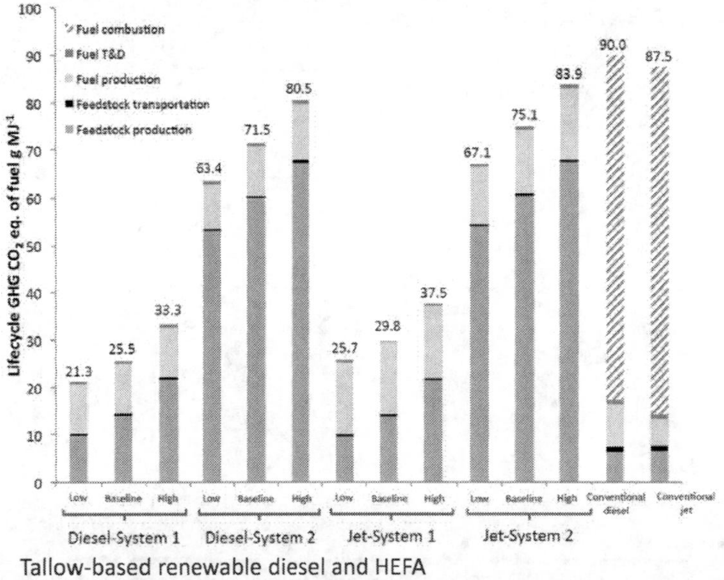

Seber et al. 2014

Tallow-based renewable diesel and HEFA

System 1 starts at rendering while System 2 starts at animal growth

14

Concluding Remarks

- Differences in lifecycle results for pathways assessed due to:
  - Allocation rules (Energy vs. Displacement):
    Δ 7-12 gCO_{2e}/MJ for HEFA pathways
  - System boundaries, including land-use change:
    Up to 55 gCO_{2e}/MJ for oil seed
  - Agricultural inputs
    Δ 2-9 gCO_{2e}/MJ for HEFA pathways
- Not all differences are indicative of a need for harmonization (systematic vs. parametric differences)
- In addition to technical discrepancy in LCA, regulatory framework is also an important factor
  - Baseline fuels' GHG intensity
  - Threshold- vs. performance-based

15

Questions/Comments

jhan@anl.gov

16

To reduce the transportation energy consumption, Corporate Average Fuel Economy (CAFE) standard was implemented

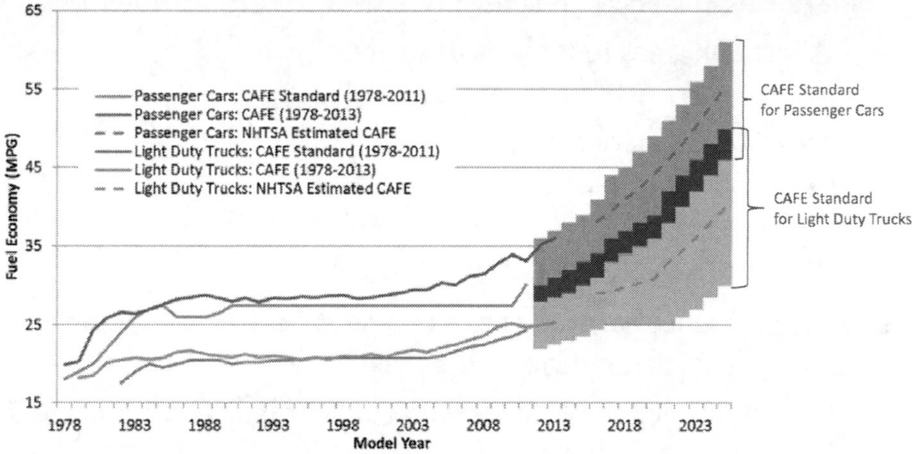

- Cumulative reduction in fuel consumptions is expected to be 16~31% during 2017-2060
- Cumulative reduction in GHG emissions is expected to be 22~34% during 2017-2100
- However, more dramatic reductions in crude import and GHG emissions require the development of **alternative fuels**, such as natural gas-based fuels, electricity and biofuels

Source: U.S. National Highway Traffic Safety Administration (NHTSA). CAFE – Fuel Economy (http://www.nhtsa.gov/fuel-economy)

17

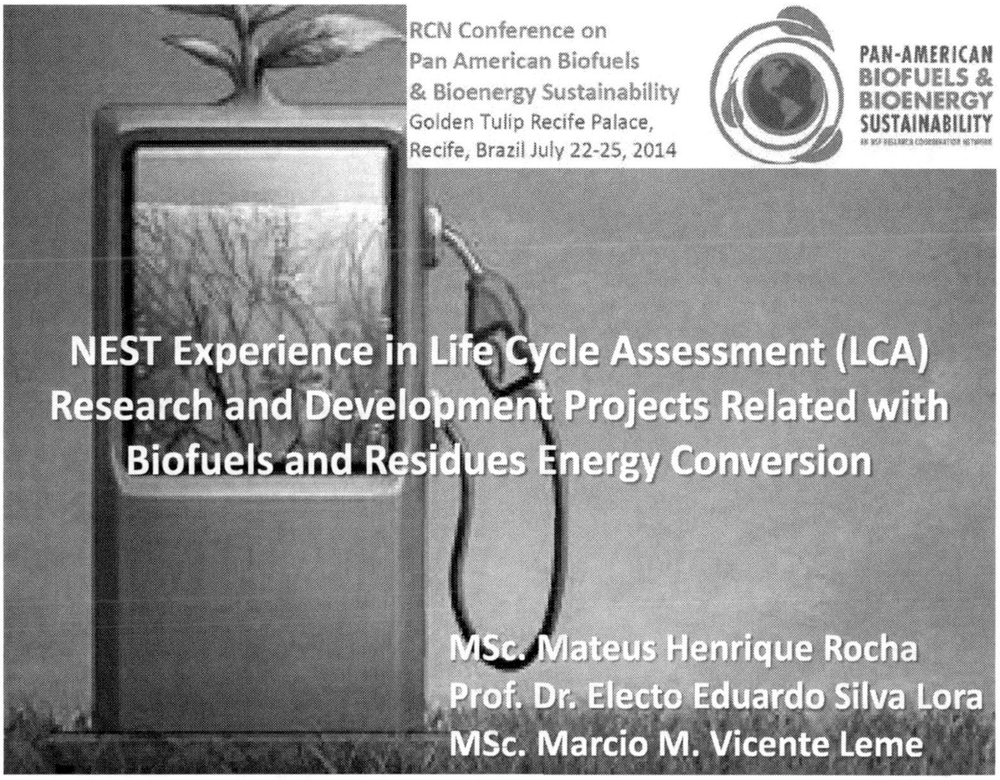

UNIFEI Universidade Federal de Itajubá

NEST

Excellence Group in Thermal and Distributed Generation

Mechanical Engineering Institute

EXCELLENCE GROUP IN THERMAL POWER AND DISTRIBUTED GENERATION - NEST

Professor Dr. Electo S. Lora | Professor Dr. Osvaldo J. Venturini | Professor Dr. Vladimir M. Cobas | Professor Dr. Ricardo D. M. Carvalho | Professor Dr Rubenildo Vieira Andrade | Professor DrJosé Carlos Escobar Palacio

++++ PhD and Master degree students

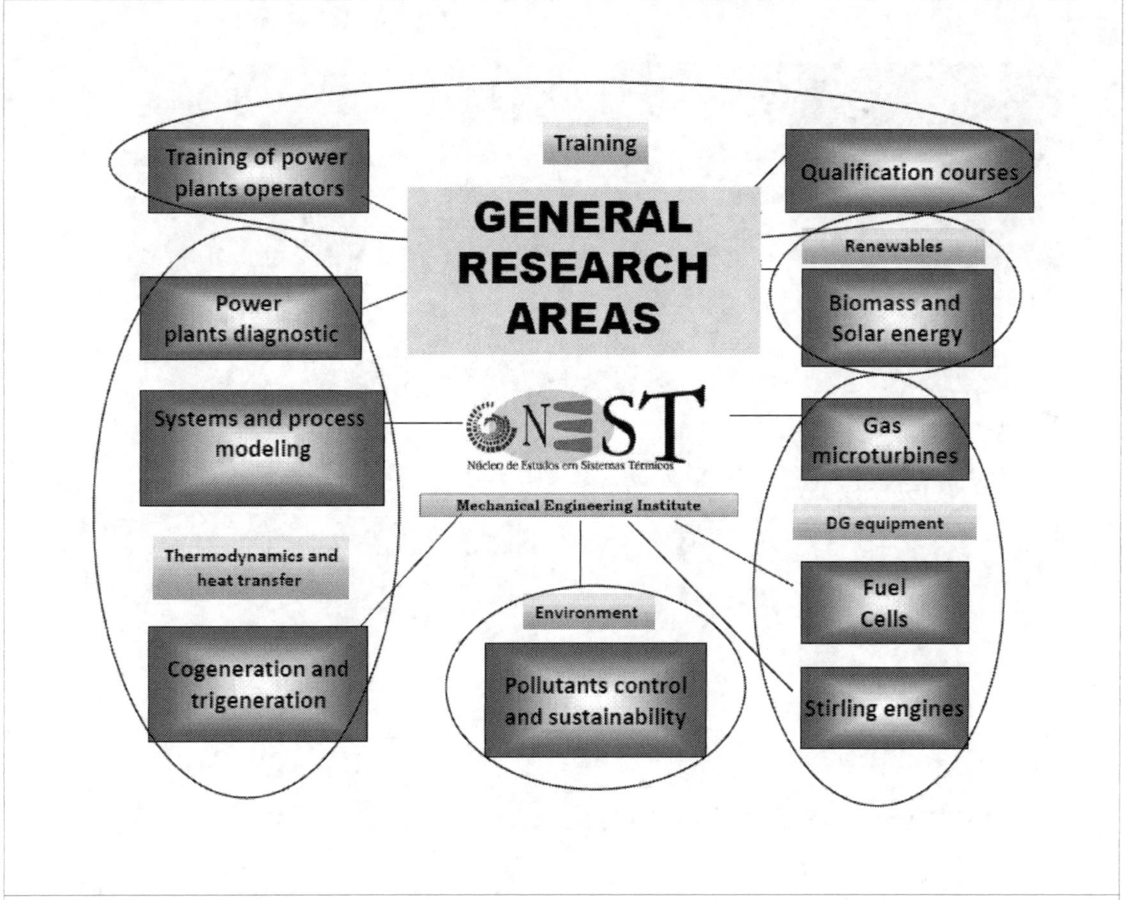
Training
Training of power plants operators
Qualification courses
GENERAL RESEARCH AREAS
Renewables
Biomass and Solar energy
Power plants diagnostic
Systems and process modeling
NEST
Núcleo de Estudos em Sistemas Térmicos
Mechanical Engineering Institute
Gas microturbines
DG equipment
Thermodynamics and heat transfer
Fuel Cells
Environment
Cogeneration and trigeneration
Pollutants control and sustainability
Stirling engines

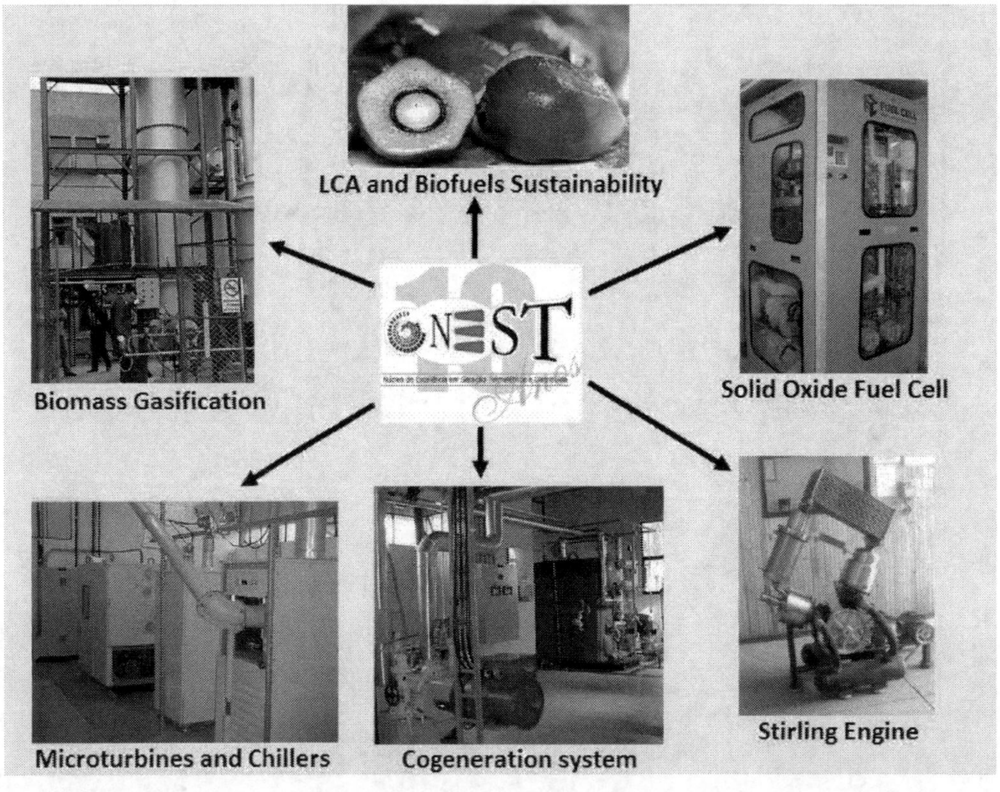
LCA and Biofuels Sustainability
Biomass Gasification
NEST
Solid Oxide Fuel Cell
Stirling Engine
Microturbines and Chillers
Cogeneration system

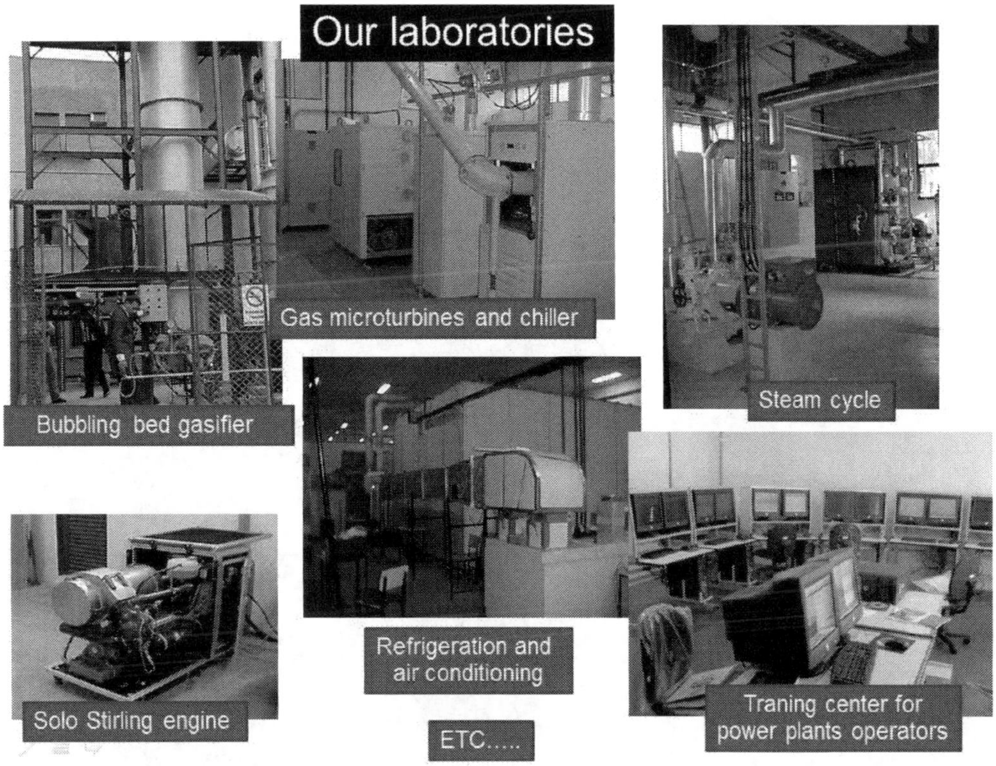
Our laboratories
Gas microturbines and chiller
Steam cycle
Bubbling bed gasifier
Refrigeration and air conditioning
Solo Stirling engine
Traning center for power plants operators
ETC.....

PUBLISHED BOOKS AT NEST/UNIFEI

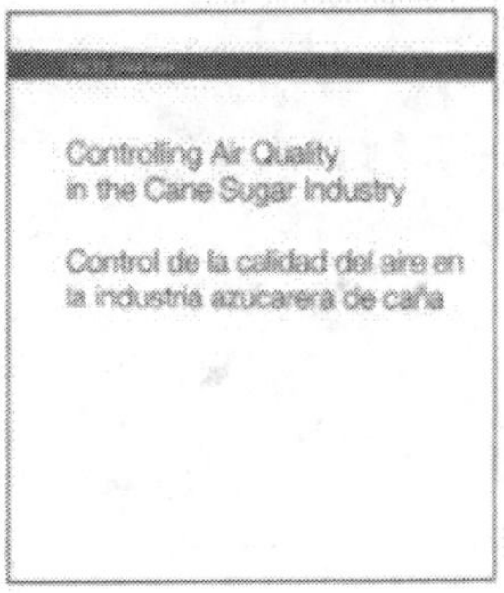

Chapter 1 – Biofuels: fundamentals
Chapter 2 – Combustion of Biomass
Chapter 3 – Biodiesel
Chapter 4 – Biogas
Chapter 5 – First Generation Bioethanol
Chapter 6 – Pyrolysis and Gasification of Biomass
Chapter 7 – Biomass to Liquid
Chapter 8 – Lignocellulosic Bioethanol
Chapter 9 – Utilization of $BioH_2$ in Fuel Cells
Chapter 10 – Utilization of Biofuels in IC Engines
Chapter 11 – Utilization of Biofuels in Microturbines
Chapter 12 – Residues Stillage and Glycerin
Chapter 13 – Cogeneration in Biofuels Production
Chapter 14 – Planning and Projects Management in Biofuels
Chapter 15 – Environmental Integrated Evaluation
Chapter 16 – The future of Biofuels: Biorefineries, Biodiesel from Algae and Microbial Fuel Cells

New installations: Office and Laboratories

Funding
ANP/PETROBRAS
R$ 4.151.430,00

To be completed in 2015

NEST Experience in Research and Development Projects Related With Biomass Energy Conversion

Main NEST Published Papers in Biomass Energy Conversion

BIOMASS AND BIOENERGY 61 (2014) 236–244

Available online at www.sciencedirect.com

ScienceDirect

ELSEVIER

http://www.elsevier.com/locate/biombioe

BIOMASS & BIOENERGY

Biomass gasification in a downdraft gasifier with a two-stage air supply: Effect of operating conditions on gas quality

CrossMark

Ana Lisbeth Galindo[a], *Electo Silva Lora*[a,*], *Rubenildo Viera Andrade*[a], *Sandra Yamile Giraldo*[a], *Rene Lesme Jaén*[b], *Vladimir Melian Cobas*[a]

[a] *Excellence Group in Thermal Power and Distributed Generation-NEST, Institute of Mechanical Engineering, Federal University of Itajubá, Brazil*

[b] *Faculty of Mechanical Engineering, Center for Energy Efficiency Studies, University of Oriente, Cuba*

Evaluation of the quality of the producer gas in a two-stage, air supply downdraft gasifier. The gas composition and its lower heating value were also determined. Experimental tests were performed varying the operating conditions of the gasifier: the air flow between 18 Nm^3/h and 22 Nm^3/h. The results show that a fuel gas, with tar and particulate matter content of 54.25±0.66 mg/Nm^3 and 102.4±1.09 mg/Nm^3, respectively, was obtained, for a total air flow rate of 20±0.45 Nm^3/h and an air ratio, between the two stages, of 80%. For these conditions, the lower heating value of the gas was 4.74±0.5 MJ/Nm^3.

Main NEST Published Papers in Biomass Energy Conversion

Renewable Energy 60 (2013) 427–432

Contents lists available at SciVerse ScienceDirect

ELSEVIER

Renewable Energy

journal homepage: www.elsevier.com/locate/renene

Prediction by mathematical modeling of the behavior of an internal combustion engine to be fed with gas from biomass, in comparison to the same engine fueled with gasoline or methane

CrossMark

Felipe O. Centeno González[a,*], Khamid Mahkamov[b], Electo E. Silva Lora[a], Rubenildo V. Andrade[a], René Lesme Jaen[c]

[a] The Centre for Excellency in Thermoelectric and Distributed Generation (NEST), The Federal University of Itajuba, Av. BPS 1303 Pinheirinho, Itajuba, MG, Brazil
[b] School of Computing, Engineering and Information Sciences, Northumbria University, Ellison Building, Newcastle upon Tyne NE1 8ST, UK
[c] University of Oriente, Sede J.A. Mella. Ave. de las Américas s/n y calle I. C.P. 90900 Santiago de Cuba, Cuba

The performance of a spark ignition internal combustion engine fueled with synthesis gas (syngas) from biomass gasification was evaluated using an analytical mathematical model. The model predicted of the internal temperatures profiles, heat flow, as well as the work and pressure in relation to crank angle. It was used also to evaluate the influence of the rotation speed, the air ratio and the ignition timing on the engine indicated power. It was found that when feeding the engine with syngas, a power output between 59 and 65% could be obtained, in comparison it's powered by gasoline.

Main NEST Published Papers in Biomass Energy Conversion

Applied Thermal Engineering 52 (2013) 109–119

Contents lists available at SciVerse ScienceDirect

ELSEVIER

Applied Thermal Engineering

journal homepage: www.elsevier.com/locate/apthermeng

Exergetic and economic comparison of ORC and Kalina cycle for low temperature enhanced geothermal system in Brazil

CrossMark

Carlos Eymel Campos Rodríguez[a,*], José Carlos Escobar Palacio[a], Osvaldo J. Venturini[a], Electo E. Silva Lora[a], Vladimir Melián Cobas[a], Daniel Marques dos Santos[b], Fábio R. Lofrano Dotto[c], Vernei Gialluca[d]

[a] Federal University of Itajuba (UNIFEI), Mechanical Engineering Institute – IEM, Excellence Group in Thermal Power and Distributed Generation (NEST), Minas Gerais, Brazil
[b] AES Tietê, Bauru, São Paulo, Brazil
[c] FAROL Pesquisa, Desenvolvimento e Consultoria, Brazil
[d] Gênera Serviços e Comércio LTDA, Brazil

Thermodynamic analysis, of both the first and second law of thermodynamic of two different technologies (ORC and Kalina cycle) for power production through an enhanced geothermal system. In order to find a better performance of both thermal cycles it were evaluated 15 different working fluids for ORC and three different composition of the ammonia-water mixture for the Kalina cycle. At the end the two cycles was compared using an economic analysis with the fluid that offers the best performance for each thermal cycle which are R-290 for ORC and for Kalina cycle a composition of the mixture of 84% of ammonia mass fraction and 16% of water mass fraction.

Main NEST Published Papers in Biomass Energy Conversion

BIOMASS AND BIOENERGY 58 (2013) 76–86

Available online at www.sciencedirect.com
SciVerse ScienceDirect
http://www.elsevier.com/locate/biombioe
ELSEVIER
BIOMASS & BIOENERGY

Assaí – An energy view on an Amazon residue

Marcos Alexandre Teixeira[c,*,2], *José Carlos Escobar Palacio*[b], *César Rodriguez Sotomonte*[b], *Electo Eduardo Silva Lora*[b], *Osvaldo José Venturini*[b], *Dirk Aßmann*[a,1]

[a] GIZ – Deutsche Gesellschaft für Internationale Zusammenarbeit GmbH, Energy Program, Brazil. c/o Edifício Rodolpho de Paoli, Av. Nilo Peçanha, 50 - 30° andar, grupo 3009, Centro - Rio de Janeiro, RJ 20020-906, Brazil

[b] Núcleo de Excelência em Geração Termelétrica e Distribuida – NEST, Instituto de Engenharia Mecânica – IEM, Universidade Federal de Itajubá – UNIFEI, Av. BS 1303, CP 50, Itajubá, MG 97500-093, Brazil

[c] Federal Fluminense University, Department of Agricultural Engineering and the Environment, Rua Passo da Pátria 156, CEP 24.210-240 Niterói, RJ, Brazil

The paper analyzed the economic feasibility of electricity generation using the residues from the exploitation of Assaí. The electricity generation cost for a 1 MW conversion systems, considering 5.5 US$/ton biomass price, were evaluated: conventional steam cycle with backpressure turbine (66.97 US$/MWh), steam cycle with extraction condensation turbine (92.11 US$/MWh), Organic Rankine cycle (122 US$/MWh) and a gasifier/ICE set (102 US$/MWh). Based on financial performance, backpressure steam turbine was the best option, and gasifier/ICE should be further considered due its operation flexibility. For any system, minimal electricity commercialization price for economical feasibility found was 150 US$/MWh.

Main NEST Published Papers in Biomass Energy Conversion

Renewable Energy 37 (2012) 97–108

Contents lists available at ScienceDirect
Renewable Energy
journal homepage: www.elsevier.com/locate/renene
ELSEVIER

Theoretical and experimental investigations of a downdraft biomass gasifier-spark ignition engine power system

Felipe Centeno[a], Khamid Mahkamov[b], Electo E. Silva Lora[a,*], Rubenildo V. Andrade[a]

[a] The Centre for Excellency in Thermoelectric and Distributed Generation (NEST), The Federal University of Itajuba, Av. BPS 1303 Pinheirinho, Itajuba, MG, Brazil

[b] School of Computing, Engineering and Information Sciences, Northumbria University, Ellison Building, Newcastle upon Tyne NE1 8ST, UK

A mathematical model was developed to predict steady state performance of a biomass downdraft gasifier/spark ignition engine power system. A mathematical model of the integrated system consists of two parts: the fixed bed downdraft gasifier and spark ignition internal combustion engine models. The numerical results obtained using the proposed model are in a good agreement with data produced with the use of other theoretical models and experimental data published in literature. The proposed model is applicable for modelling integrated downdraft gasifier/engine biomass energy systems and can be used for more accurate adjustment of design parameters of the gasifier and the engine in order to provide the higher overall efficiency of the system.

Main NEST Published Papers in Biomass Energy Conversion

BIOMASS AND BIOENERGY 35 (2011) 3465–3480

Available at www.sciencedirect.com

ScienceDirect

http://www.elsevier.com/locate/biombioe

ELSEVIER

BIOMASS & BIOENERGY

Experimental study on biomass gasification in a double air stage downdraft reactor

Juan Daniel Martínez [a,b,*], *Electo Eduardo Silva Lora* [a,**], *Rubenildo Viera Andrade* [a], *René Lesme Jaén* [c]

[a] Núcleo de Excelência em Geração Termelétrica e Distribuída, Instituto de Engenharia Mecânica, Universidade Federal de Itajubá, Av. BPS 1303, Itajubá, Minas Gerais, Brazil

[b] Grupo de Investigaciones Ambientales, Instituto de Energía, Materiales y Medio Ambiente, Universidad Pontificia Bolivariana, Circular 1ra N° 70 – 01, Bloque 11, Medellín, Colombia

[c] Facultad de Ingeniería Mecánica, Centro de Estudios de Eficiencia Energética, Universidad de Oriente, Av. Las Américas s/n C.P. 90900, Santiago de Cuba, Cuba

Experimental study of the gasification of a wood biomass in a moving bed downdraft reactor with two-air supply stages. The gasifier produces a combustible gas with a CO, CH_4 and H_2 concentrations of 19.04%, 0.89% and 16.78% volume respectively, at a total flow of air of 20 Nm^3/h and an air between stages of 80%. For these conditions, the LHV of the gas was 4539 kJ/Nm^3. Results from the calculation model show a useful gas power and cold efficiency around 40 kW and 68%, respectively. The resulting ER under the referred operation condition is around 0.40.

Main NEST Published Papers in Biomass Energy Conversion

BIOMASS AND BIOENERGY 33 (2009) 1101–1107

Available at www.sciencedirect.com

ScienceDirect

http://www.elsevier.com/locate/biombioe

ELSEVIER

BIOMASS & BIOENERGY

Review

Estimate of the electric energy generating potential for different sources of biogas in Brazil

Karina Ribeiro Salomon, *Electo Eduardo Silva Lora*

Federal University of Itajubá – UNIFEI, Excellence Group in Thermal Power and Distributed Generation – NEST/IEM, Av BPS 1303, CP 50, Itajubá, MG, CEP 37.500-903, Brazil

It was carried out an evaluation of the quantities of organic residues coming out from the sugar and alcohol industry (stillage), Municipal Solid Waste (MSW) and liquid wastes (sewage) and livestock residues (bovine and swine manure) in Brazil. Finally the electricity generation potential of biogas out of the evaluated sources of organic residues in Brazil was estimated. The results of this study indicate that the potential regarding the production of biogas out of the aforementioned organic residues of electricity production using could meet an energy demand of about 1.05 to 1.13 %. Constraints for biogas energy utilization was identified and discussed.

NEST Experience in Life Cycle Assessment (LCA) Related With Biofuels

Renewable and Sustainable Energy Reviews 37 (2014) 435–459

Contents lists available at ScienceDirect

Renewable and Sustainable Energy Reviews

journal homepage: www.elsevier.com/locate/rser

Life cycle assessment (LCA) for biofuels in Brazilian conditions: A meta-analysis

Mateus Henrique Rocha[a,*], Rafael Silva Capaz[b], Electo Eduardo Silva Lora[a], Luiz Augusto Horta Nogueira[b], Marcio Montagnana Vicente Leme[a], Maria Luiza Grillo Renó[a], Oscar Almazán del Olmo[c]

[a] NEST – Excellence Group in Thermal Power and Distributed Generation, Institute of Mechanical Engineering, Federal University of Itajubá, Av. BPS 1303, Itajubá, Minas Gerais State, CEP: 37500-903, Brazil
[b] IRN – Institute of Natural Resources, Federal University of Itajubá, Av. BPS 1303, Itajubá, Minas Gerais State, CEP: 37500-903, Brazil
[c] ICIDCA – Instituto Cubano de Investigaciones de los Derivados de la Caña de Azúcar, Via Blanca y Carretera Central 804, San Miguel Del Padrón, A.P. 4036, La Habana, Cuba

Biofuel	Reference	Country	LCA environmental impacts	LCEA (energy balance)
Ethanol – sugarcane	Capaz [93] – Base case	Brazil	No	Yes
	Macedo et al. [94]	Brazil	GHG emissions	Yes
Biodiesel – Soybean oil	Capaz [43] – Base case	Brazil	No	Yes
	Cavalett and Ortega [95]	Brazil	Emergy Analysis and Embodied Energy Analysis	Yes
	Tsoutsos et al. [96]	Greece	Yes	No
	Pradhan et al. [97]	USA	No	Yes
	Carraretto et al. [98]	Italy	Yes	Yes
Biodiesel – Palm oil	Costa [99] – Base case	Brazil	Yes	Yes
	Kamahara et al. [100]	Indonesia	No	Yes
	Papong et al. [101]	Thailand	No	Yes
	Souza et al. [102]	Brazil	GHG emissions	Yes
	Pleanjai and Gheewala [103]	Thailand	No	Yes

Scheme of the system boundary of the ethanol production from sugarcane.

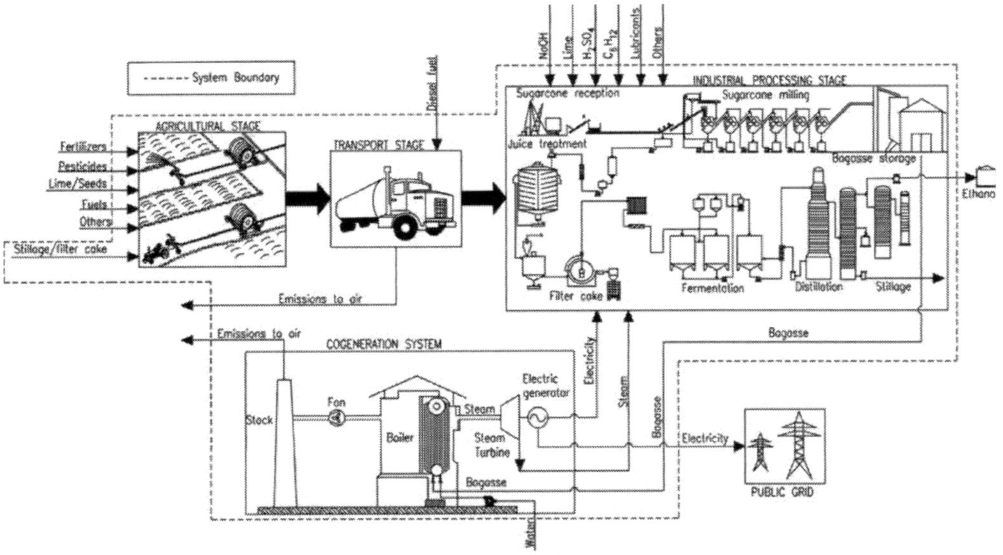

Scheme of the system boundary of the biodiesel production from soybean

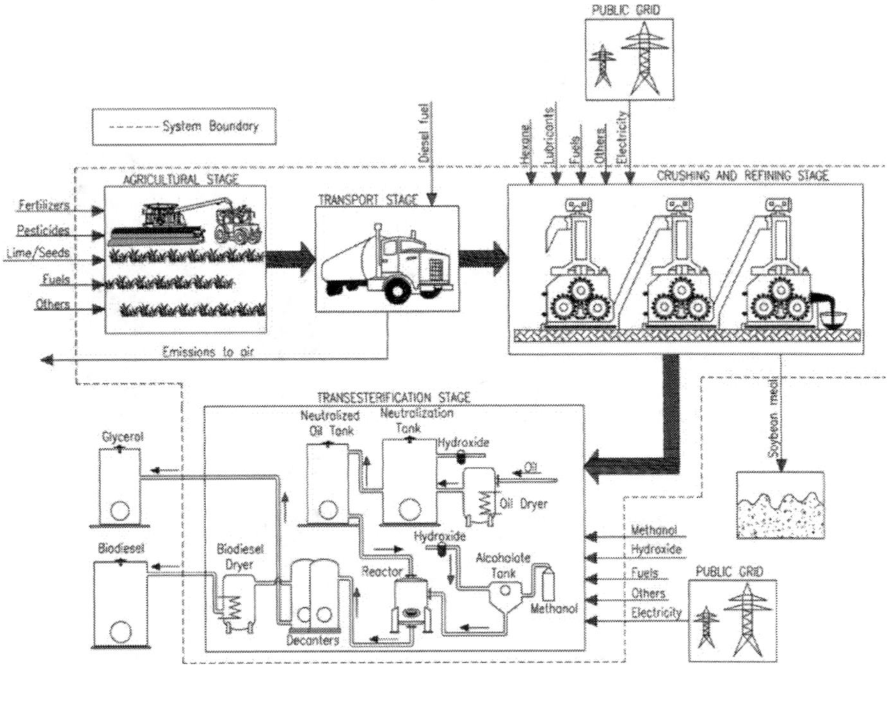

Scheme of the system boundary of the biodiesel production from palm oil

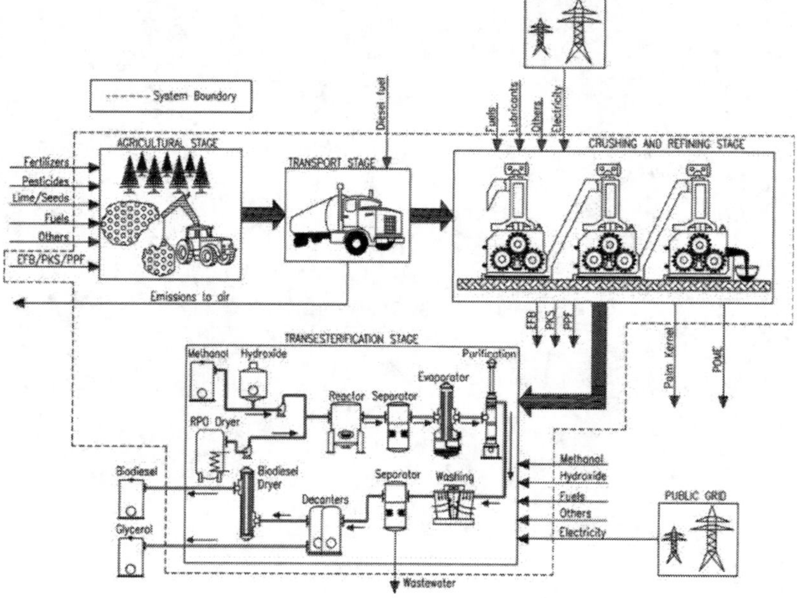

Global Warming Potential in life cycle of ethanol production

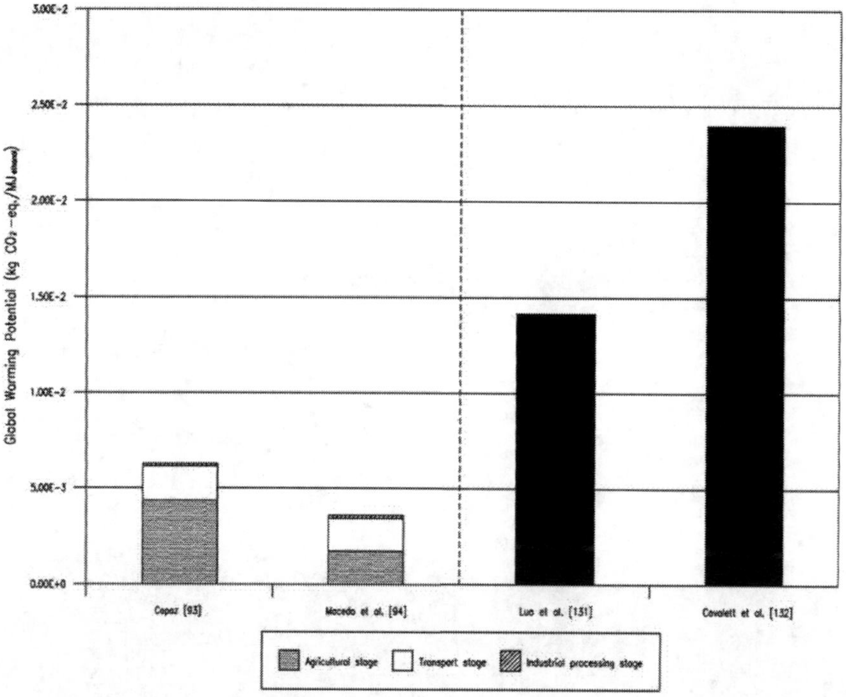

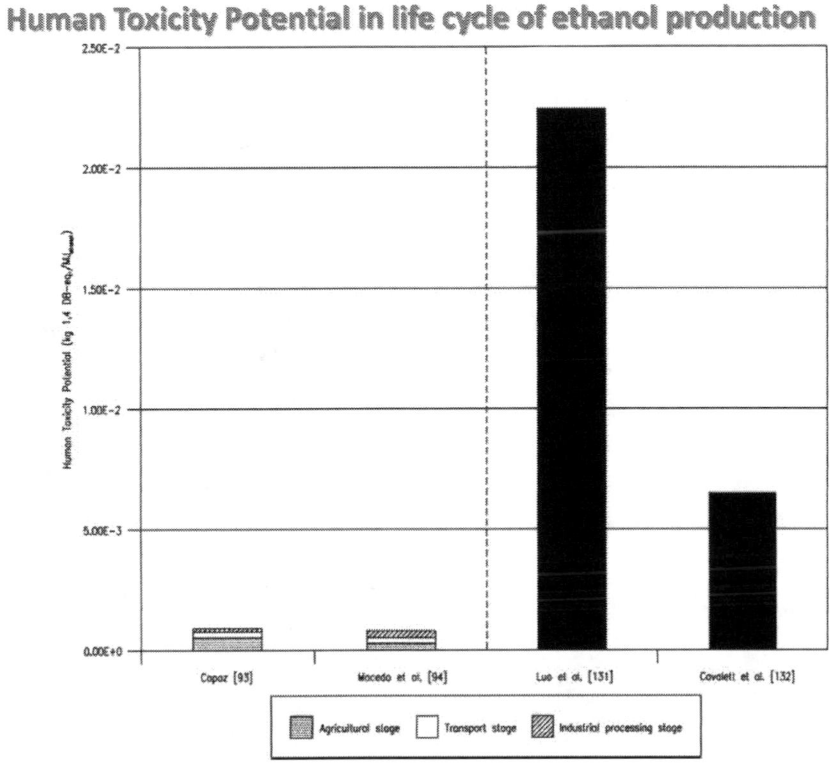
Human Toxicity Potential in life cycle of ethanol production
Human Toxicity Potential (kg 1,4 DB-eq./MJ)
2.50E-2
2.00E-2
1.50E-2
1.00E-2
5.00E-3
0.00E+0
Capaz [93]
Macedo et al. [94]
Luo et al. [131]
Cavalett et al. [132]
Agricultural stage
Transport stage
Industrial processing stage

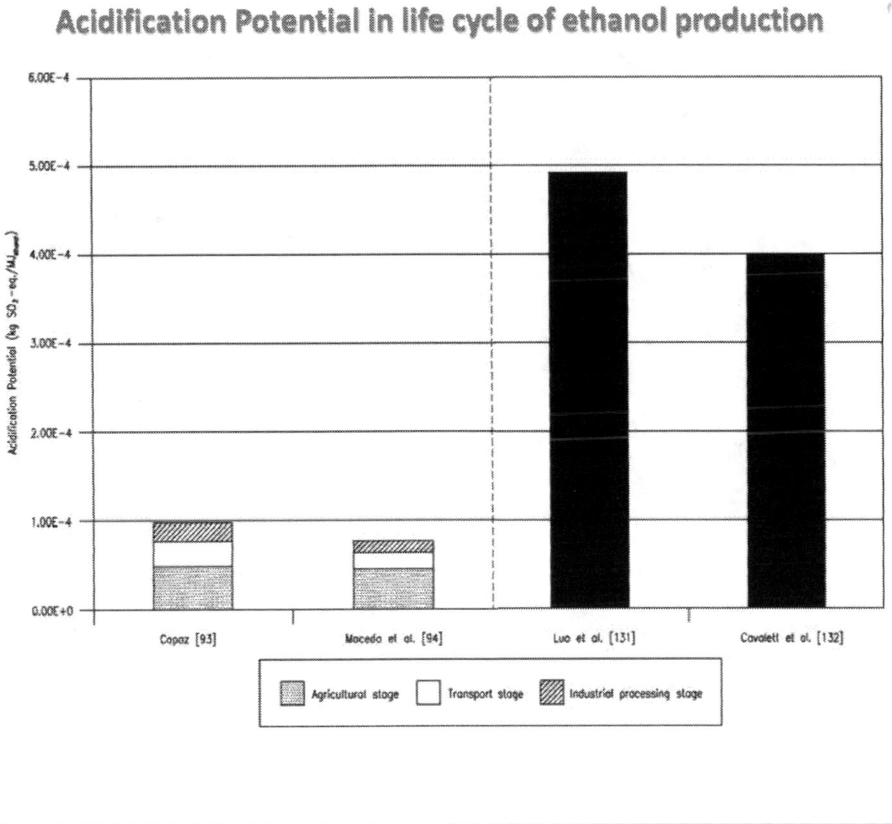
Acidification Potential in life cycle of ethanol production
Acidification Potential (kg SO2-eq./MJ)
6.00E-4
5.00E-4
4.00E-4
3.00E-4
2.00E-4
1.00E-4
0.00E+0
Capaz [93]
Macedo et al. [94]
Luo et al. [131]
Cavalett et al. [132]
Agricultural stage
Transport stage
Industrial processing stage

Eutrophication Potential in life cycle of ethanol production

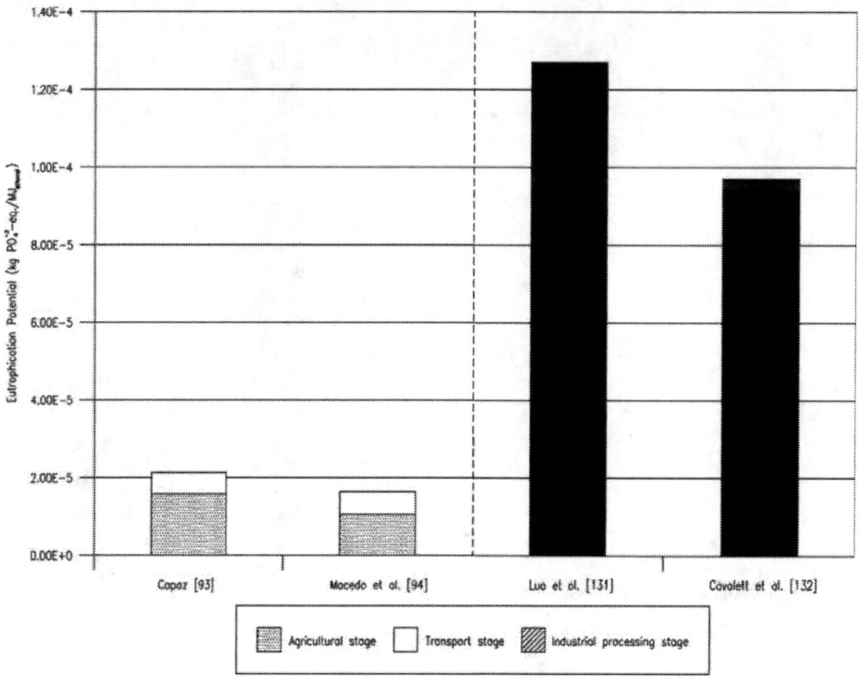

Global Warming Potential in life cycle of biodiesel production

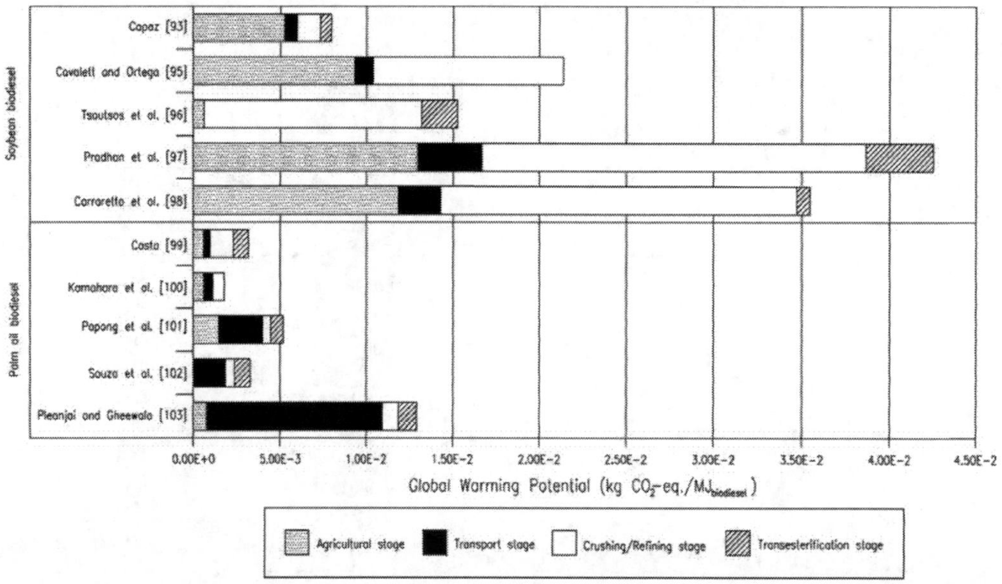

Human Toxicity Potential in life cycle of biodiesel production

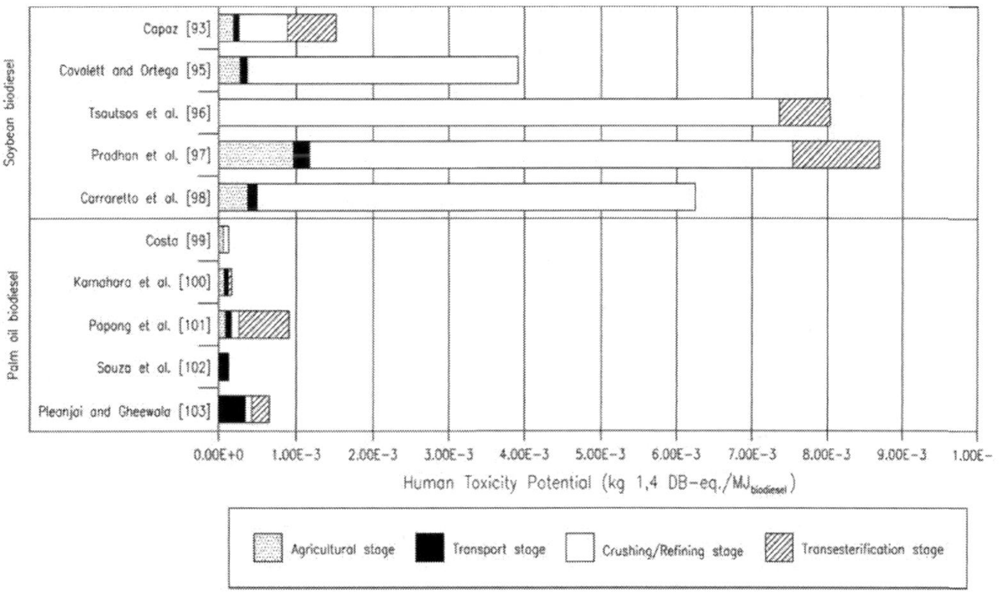

Acidification Potential in life cycle of biodiesel production

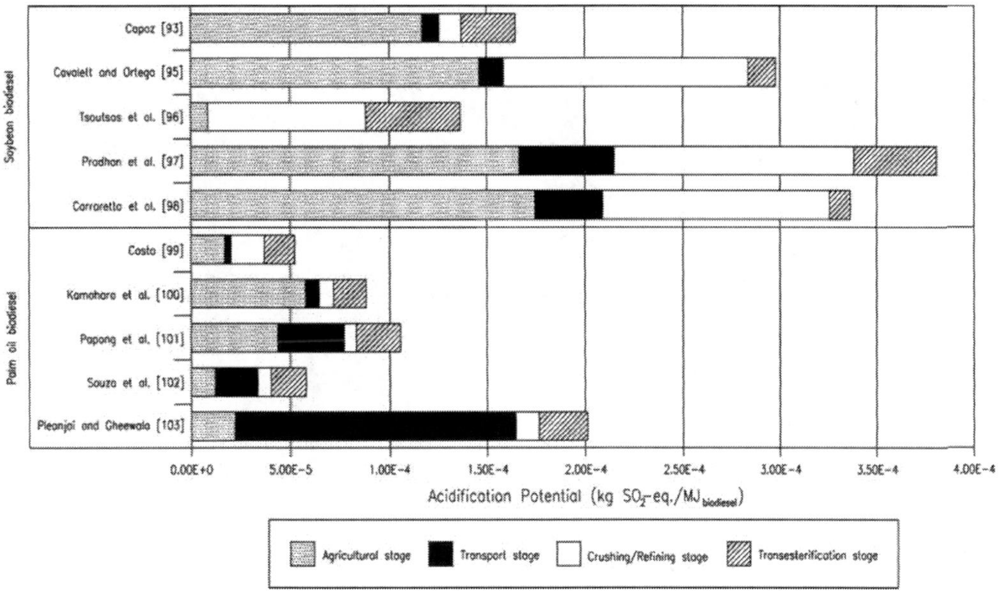

Eutrophication Potential in life cycle of biodiesel production

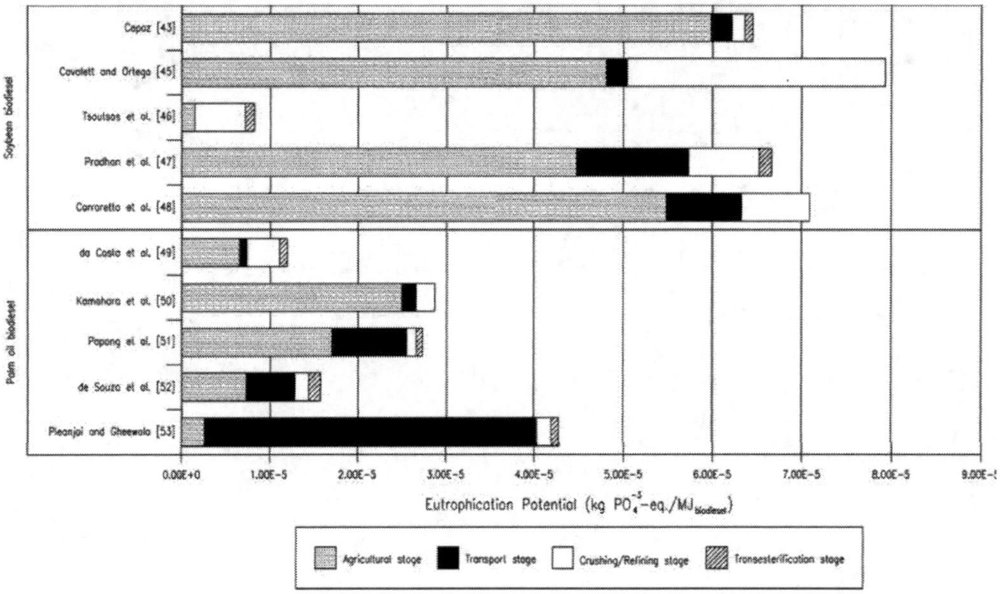

Net Energy Ratio of Biofuel

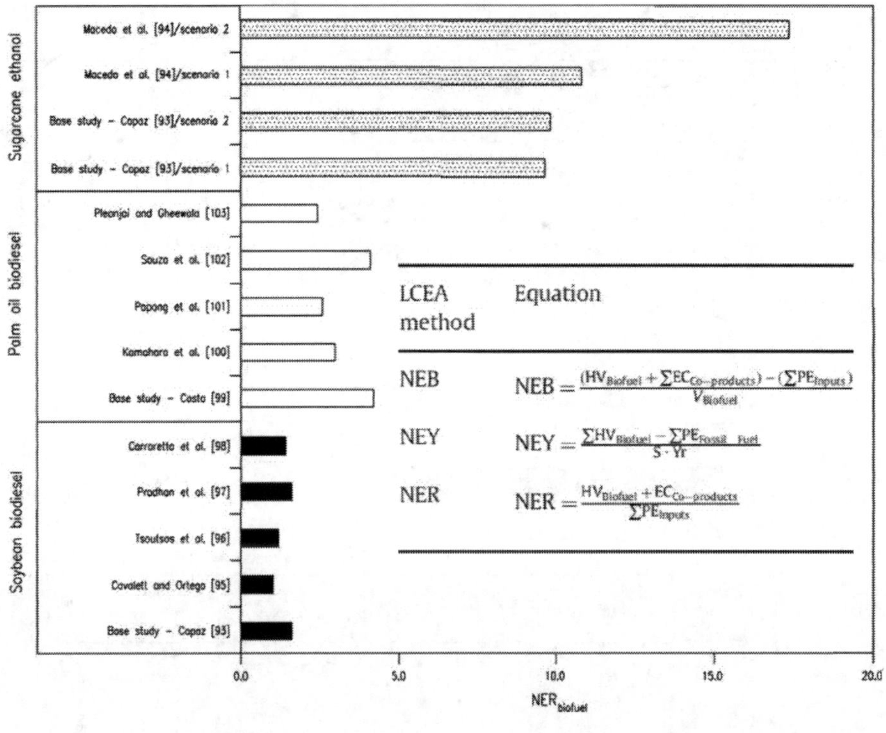

LCEA method	Equation
NEB	$NEB = \frac{(HV_{Biofuel} + \sum EC_{Co-products}) - (\sum PE_{Inputs})}{V_{Biofuel}}$
NEY	$NEY = \frac{\sum HV_{Biofuel} - \sum PE_{Fossil\ Fuel}}{S \cdot Yr}$
NER	$NER = \frac{HV_{Biofuel} + EC_{Co-products}}{\sum PE_{Inputs}}$

Net Energy Ratio Total

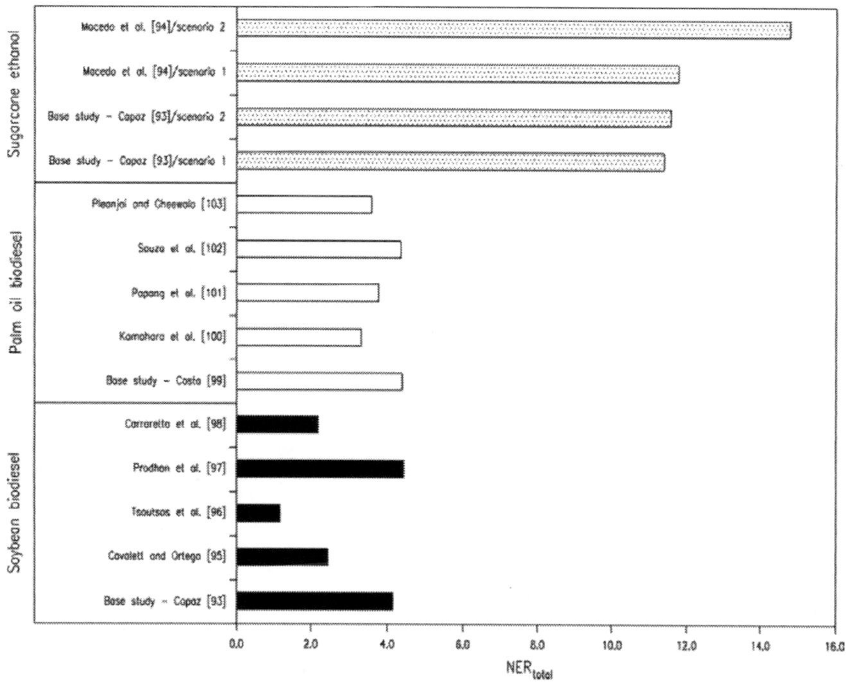

Existing Tools and Constraints in Biofuels Analysis

Energy 36 (2011) 2097–2110

Contents lists available at ScienceDirect

Energy

journal homepage: www.elsevier.com/locate/energy

ELSEVIER

Issues to consider, existing tools and constraints in biofuels sustainability assessments

Electo E. Silva Lora[a], José C. Escobar Palacio[a], Mateus H. Rocha[a], Maria L. Grillo Renó[a], Osvaldo J. Venturini[a,*], Oscar Almazán del Olmo[b]

[a] NEST – Excellence Group in Thermal Power and Distributed Generation, Federal University of Itajubá, Instituto de Eng. Mecanica – IEM, Av. BPS 1303, CP 50, Itajubá-MG, Brazil
[b] ICIDCA – Instituto Cubano de Investigaciones de los Derivados de la Caña de Azúcar, Via Blanca y Carretera Central 804, San Miguel del Padrón, A.P. 4036, La Habana, Cuba

Claims related to the negative consequences of biofuel programs are frequent; mainly those related to the biofuels/food competition and sustainability. This work contributed for the development of a framework for sustainability indicators as a tool for performance assessment. The most used indicators to measure the biofuels sustainability are: Life Cycle Energy Balance, quantity of fossil energy substituted per hectare, co-product energy allocation, life cycle carbon balance and changes in soil utilization. LCA's of biofuels was compared emphasizing their advantages and disadvantages. Main constraints related to the studied frontiers, as well as the lack of reliable data and their effects was also discussed. Suggestions and recommendations were made to improve existing methodologies for biofuels sustainability evaluation.

Sustainability of biofuels is a multicriterial approach

- **Productivity:** depends on the type of biomass crop, related with the efficiency of the soil utilization. The specific productivity (kg/ha) of the first-generation biofuels vary in a wide range.
- **GHG emissions:** considering that the main justification for biofuel production expansion is the global warming mitigation, it is necessary to verify to which extent the production of a biofuel really reduces the GHG emissions.
- **Land use and carbon stocks changes:** soil organic carbon is of global importance because of its role in the global carbon cycle and, therefore, in the mitigation or worsening of atmospheric levels of GHG.
- **Fertilizer efficiency:** a fraction of the applied fertilizers can be transferred by volatilization to the atmosphere, some of their constituents having a high GWP such as N_2O.
- **Co-product and residues utilization:** as fertilizer, energy, animal feeding, production of chemicals, etc.

Sustainability of biofuels is a multicriterial approach

- **Potential for cropland expansion (marginal and low productivity soils):** assessment of the world agro-ecological productivity shows a potential of 700–800 million of hectares available for bioenergy productions.
- **Impacts on water resources (water depletion):** the production of biofuel feedstock and its industrial processing consumes considerable amounts of water.
- **Soil:** soil degradation by erosion and salt increase.
- **Impacts on biodiversity:** this is related with habitat degradation of many species due to deforestation and other land use changes.
- **Costs:** balance between biofuels cost of production and market prices.
- **Land use:** the need of land for other human activities besides energy production, especially for food production.
- **Social issues:** working conditions, rural development, food prices, impact on communities.

Renewable and Sustainable Energy Reviews 13 (2009) 1275–1287

Contents lists available at ScienceDirect

Renewable and Sustainable Energy Reviews

journal homepage: www.elsevier.com/locate/rser

Biofuels: Environment, technology and food security

José C. Escobar [a], Electo S. Lora [a,*], Osvaldo J. Venturini [a], Edgar E. Yáñez [b], Edgar F. Castillo [c], Oscar Almazan [d]

[a] NEST - Excellence Group in Thermal Power and Distributed Generation, Mechanical Engineering Institute, Universidade Federal de Itajubá, Brazil
[b] CENIPALMA, Oil Palm Research Center - Cenipalma, Calle 21 # 42-C-47, Bogotá, Colombia
[c] CENICAÑA - Sugarcane Research Center of Colombia, Calle 58 N, # 3BN-110, A.A., 9138 - Cali, Colombia
[d] ICIDCA - Instituto Cubano de Investigaciones de los Derivados de la Caña de Azúcar, Via Blanca y Carretera Central 804, San Miguel del Padrón, A.P. 4036, La Habana, Cuba

This work carried out a assessment of the causes of the rise in the demand and production of biofuels. It was discussed different vegetable raw materials sources and technological paths to produce biofuels, as well as issues regarding production cost and the relation of their economic feasibility with oil international prices. The environmental impacts of programs that encourage biofuel production, farmland land requirements and the impacts on food production are also discussed, considering the LCA as a tool. It was concluded that the rise in the use of biofuels is inevitable and that international cooperation, regulations and certification mechanisms must be established regarding the use of land, the mitigation of environmental and social impacts caused by biofuel production. It is also mandatory to establish appropriate working conditions and decent remuneration for workers of the biofuels production chain

Life Cycle Energy Analysis of Biodiesel Production

Renewable Energy 34 (2009) 2905–2913

Contents lists available at ScienceDirect

Renewable Energy

journal homepage: www.elsevier.com/locate/renene

Technical note

The energy balance in the Palm Oil-Derived Methyl Ester (PME) life cycle for the cases in Brazil and Colombia

Edgar Eduardo Yáñez Angarita [a,*], Electo Eduardo Silva Lora [b], Rosélis Ester da Costa [b], Ednildo Andrade Torres [c]

[a] Oil Palm Research Center - CENIPALMA Cll 20A # 43 A 50, Piso 4, Bogotá D.C, Colombia
[b] Federal University of Itajubá/Excellence Group in Thermal and Distributed Generation NEST (IEM/UNIFEI), Brazil
[c] Bahia Federal University - UFBA, Brazil

The Output/Input energy relation in the biodiesel production life cycle can be an important indicator of the techno-economic and environmental feasibility evaluation of production of biodiesel from different oleaginous plants. Due to increasing environmental concerns about the emissions from fuel-derived atmospheric pollutants, alternative sources of energy have been receiving greater attention. This work does not look to carry out a complete LCA but rather just to focus on the energy balance in the Palm Oil-Derived Methyl Ester life cycle, taking into account practices in Brazil and Colombia. This work will show the differences between the results attained for the two cases. The Output/Input energy relation for the evaluated case studies ranged from 3.8 to 5.7, with an average value of 4.8.

PME System Boundaries

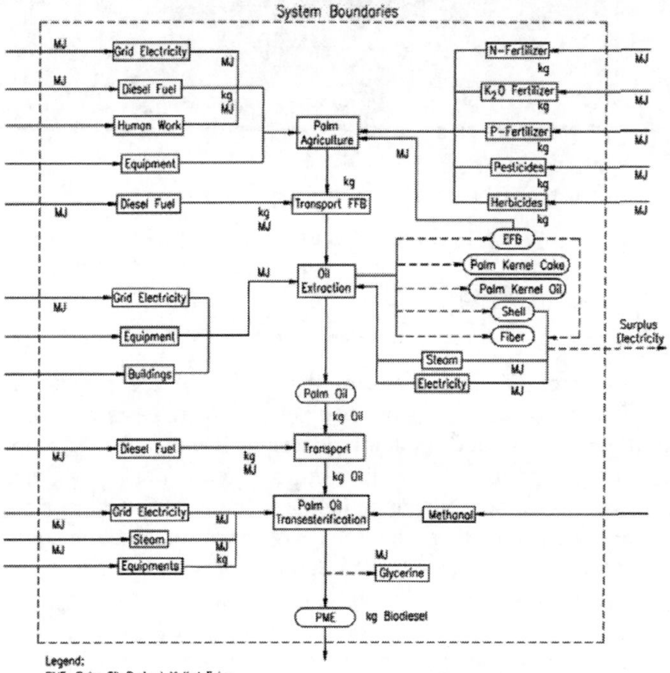

Legend:
PME: Palm Oil-Derived Methyl Ester
EFB: Empty Fruit Brunches
FFB: Fresh Fruit Brunches

Energy Consumption related to Fuels and Agrochemicals for Agricultural Stage Production of Biodiesel

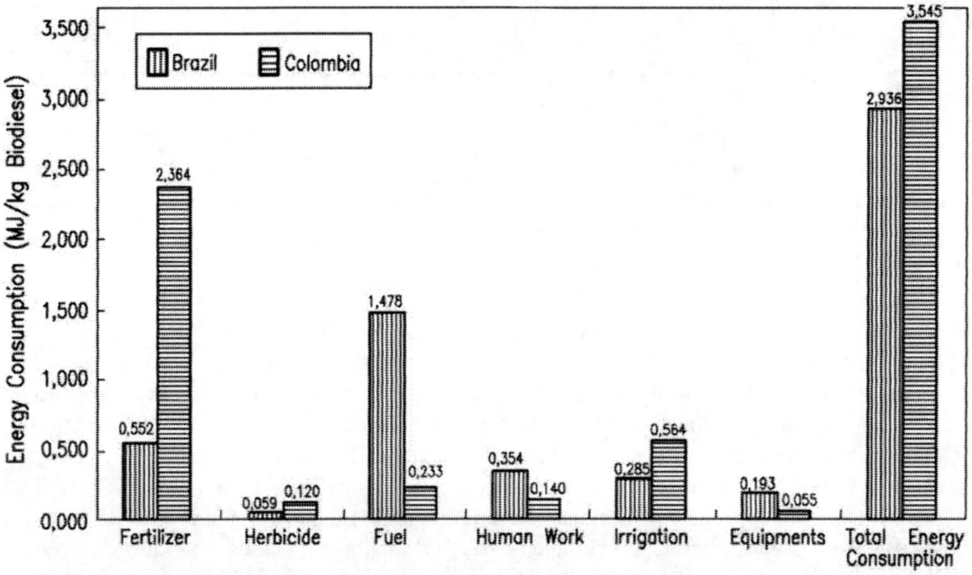

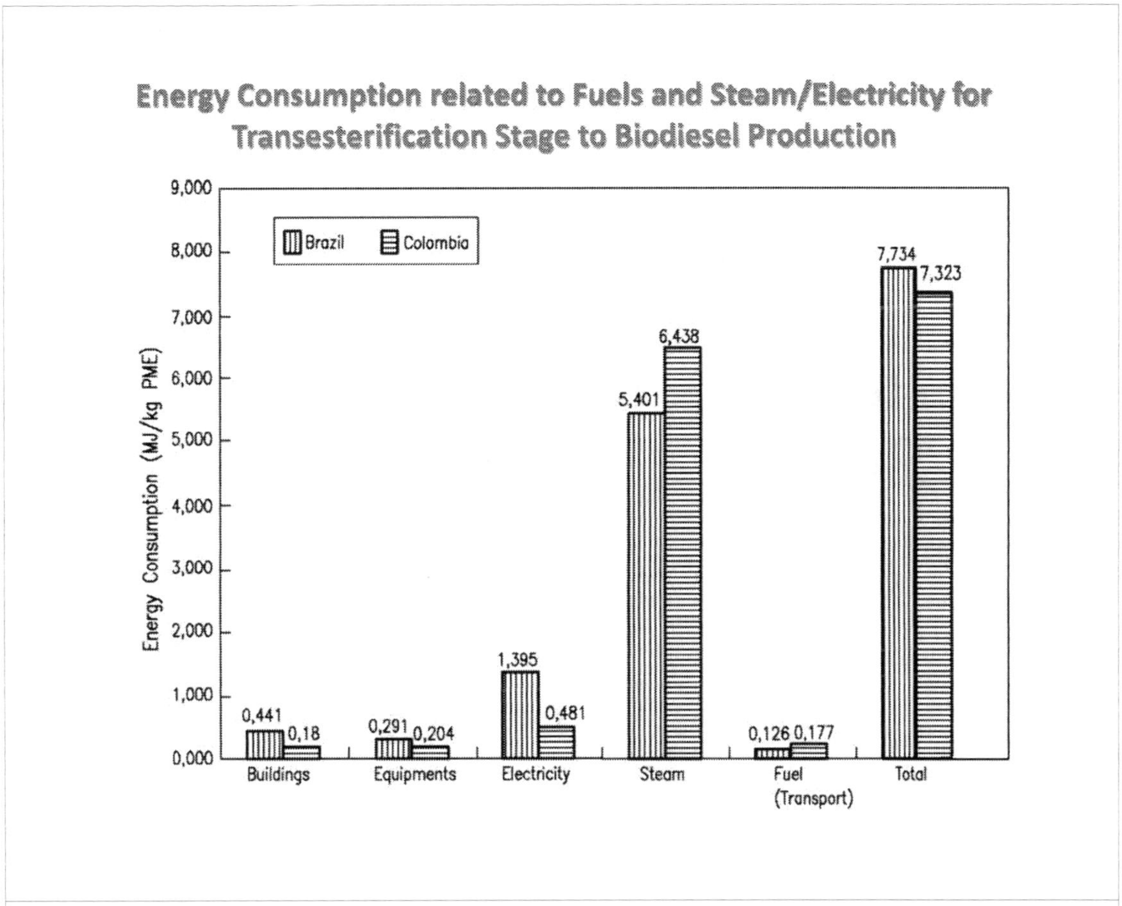
Energy Consumption related to Fuels and Steam/Electricity for Transesterification Stage to Biodiesel Production
Brazil
Colombia
Energy Consumption (MJ/kg PME)
9,000
8,000
7,000
6,000
5,000
4,000
3,000
2,000
1,000
0,000
0,441
0,18
0,291
0,204
1,395
0,481
5,401
6,438
0,126
0,177
7,734
7,323
Buildings
Equipments
Electricity
Steam
Fuel (Transport)
Total

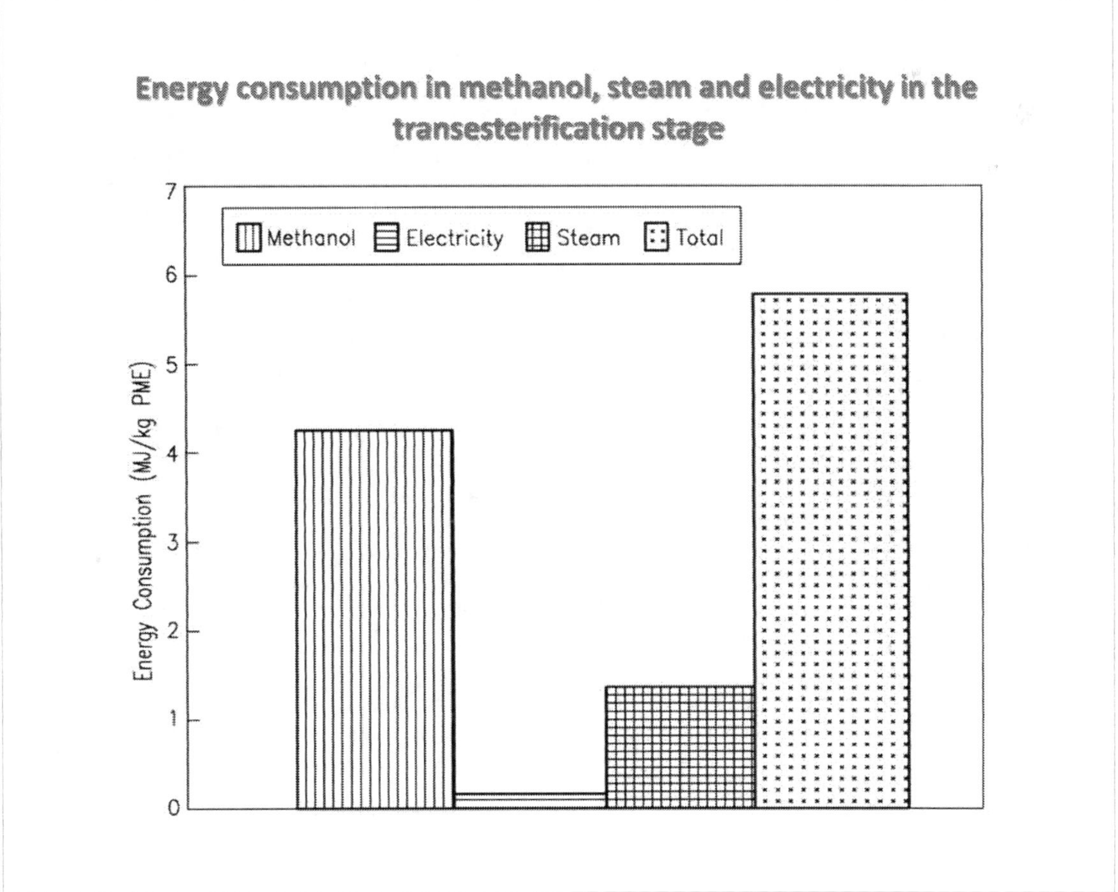
Energy consumption in methanol, steam and electricity in the transesterification stage
Methanol
Electricity
Steam
Total
Energy Consumption (MJ/kg PME)
7
6
5
4
3
2
1
0

NEST Experience in Life Cycle Assessment (LCA) Related With Residues Energy Conversion

Resources, Conservation and Recycling 87 (2014) 8–20

Contents lists available at ScienceDirect

Resources, Conservation and Recycling

journal homepage: www.elsevier.com/locate/resconrec

Techno-economic analysis and environmental impact assessment of energy recovery from Municipal Solid Waste (MSW) in Brazil

Marcio Montagnana Vicente Leme[a], Mateus Henrique Rocha[a,*], Electo Eduardo Silva Lora[a], Osvaldo José Venturini[a], Bruno Marciano Lopes[b], Cláudio Homero Ferreira[b]

[a] NEST – Excellence Group in Thermal Power and Distributed Generation, Institute of Mechanical Engineering, Federal University of Itajubá, Av. BPS 1303, Itajubá, Minas Gerais State CEP: 37500-903, Brazil

[b] CEMIG – Electric Company of Minas Gerais State, TE/AE, Av. Barbacena 1200 – 16° andar – B1 Belo Horizonte, MG CEP: 30190-131, Brazil

This work compared from a techno-economic and environmental point of view, different alternatives to the energy recovery from the MSW generated in Brazilian cities. The environmental analysis was carried out using current data collected in Betim, a 450,000 inhabitants city that currently produces 200 tones of MSW/day. Four scenarios were designed, whose environmental behavior were studied applying the LCA methodology. The results show the landfill systems as the worst waste management option and that a significant environmental savings is achieved when a wasted energy recovery is done. The best option, which presented the best performance based on considered indicators, is the direct combustion of waste as fuel for electricity generation.

Energy and economic performance for the 100,000 inhabitants case (landfill option)

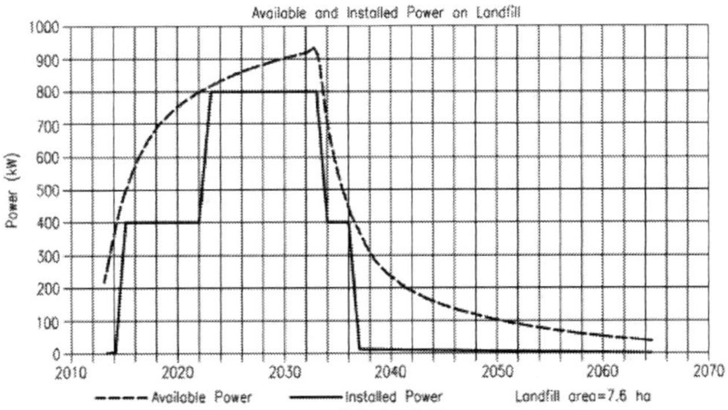

Schedule for generator modules

Modules	Power (kW)	Start of operation	End of operation
Module 1	400	2015	2025
Module 2	400	2023	2033
Module 3	400	2026	2036

Total energy produced: 103,154 MWh
Biogas generator sets: Caterpillar CG132-8
supplied by biogas/Electric Power: 400 kW_e
Electrical efficiency: 42.8%

Landfill Biogas Utilization Specification:

Collection system efficiency: 75.0%
Fugitive biogas emissions: 25.0%
Biogas used for energy generation: 42.5%
Biogas send to flares: 32.5%

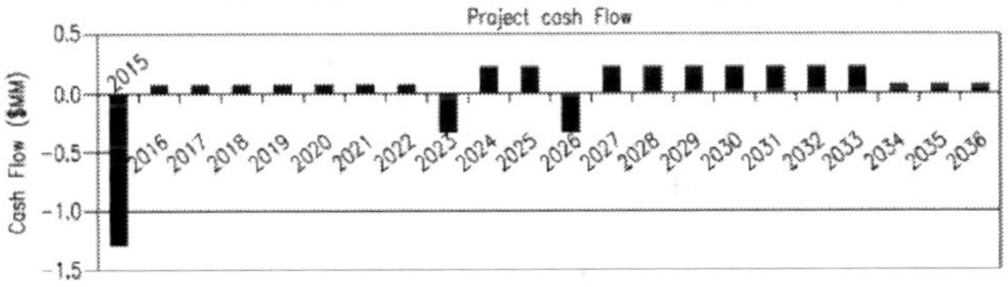

Economic Results:

NPV: $ −659,204.13	
IRR: 0.4%	
Capital costs	$ 1,828,329.86
O&M costs	$ 2,059,040.76
Energy sales	$ 3,130,335.93
CER's sales	$ 97,830.56
CER's costs	$ 694,783.67
LCOE	$ 82.60/MWh

Energy and economic performance for the 500,000-inhabitant scenario (landfill option)

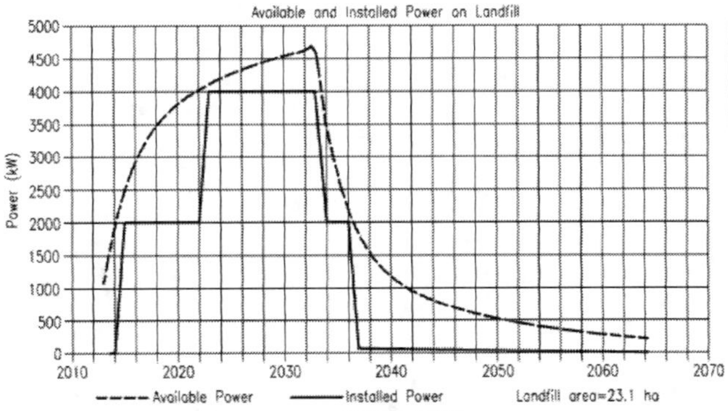

Schedule for generator modules

Modules	Power (kW)	Start of operation	End of operation
Module 1	2000	2015	2025
Module 2	2000	2023	2033
Module 3	2000	2026	2036

Total energy produced: 515,771 MWh
Biogas generator sets: Caterpillar CG170-20
supplied by biogas/Electric Power: 2000 kW_e
Electrical efficiency: 42.9%

Landfill Biogas Utilization Specification:

Collection system efficiency: 75.0%
Fugitive biogas emissions: 25.0%
Biogas used for energy generation: 42.4%
Biogas send to flares: 32.6%

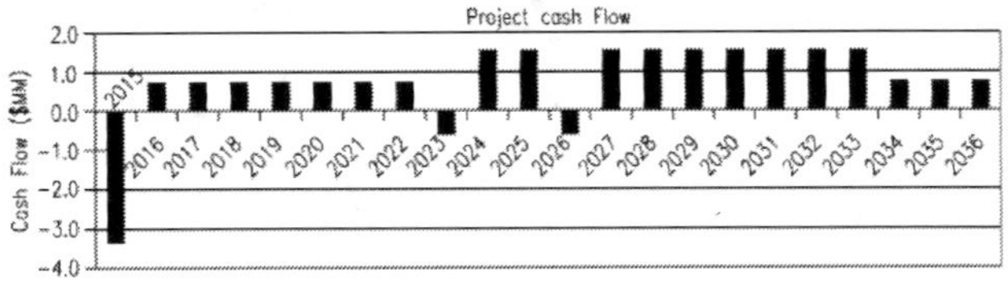

Economic Results:

NPV: $ 3,004,678.30	
IRR: 15.6%	
Capital costs	$ 5,937,432.82
O&M costs	$ 7,198,721.32
Energy sales	$ 15,651,679.66
CER's sales	$ 489,152.78
CER's costs	$ 1,457,208.34
LCOE	$ 49.00/MWh

Energy and economic performance for the 1,000,000 inhabitant's scenario (landfill option)

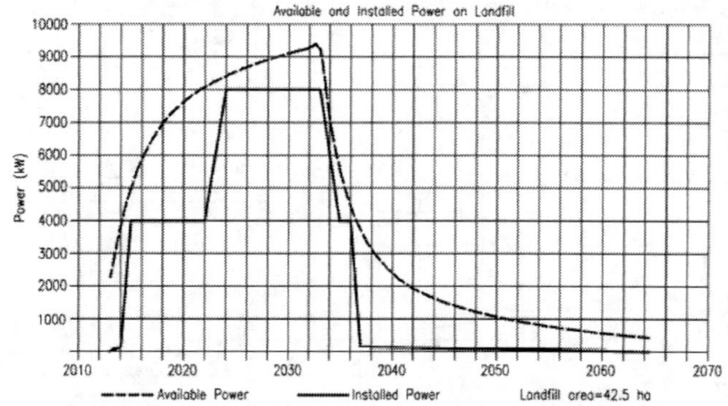

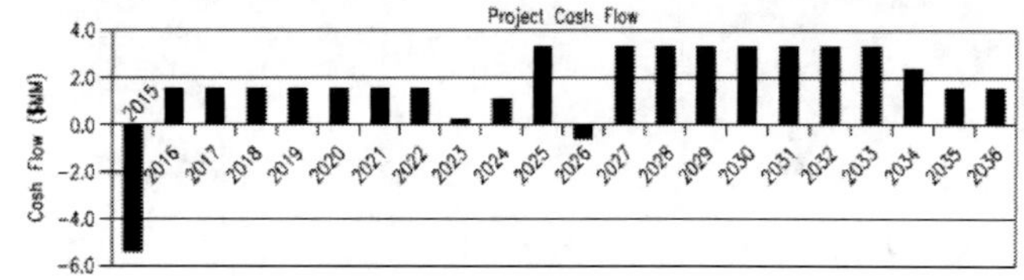

Schedule for generator modules

Modules	Power (kW)	Start of operation	End of operation
Module 1	2000	2015	2025
Module 2	2000	2015	2025
Module 3	2000	2023	2033
Module 4	2000	2024	2034
Module 5	2000	2026	2036
Module 6	2000	2026	2036

Total energy produced: 1,031,542 MWh
Biogas generator sets: Caterpillar CG170-20
supplied by biogas/Electric Power: 2000 kW_e
Electrical efficiency: 42.9%

Landfill Biogas Utilization Specification:

Collection system efficiency: 75.0%
Fugitive biogas emissions: 25.0%
Biogas used for energy generation: 42.4%
Biogas send to flares: 32.6%

Economic Results:

NPV: $ 8,793,264.25	
IRR: 21.5%	
Capital costs	$ 10,504,505.26
O&M costs	$ 12,686,389.02
Energy sales	$ 31,006,423.99
CER's sales	$ 977,734.54
CER's costs	$ 2,165,501.11
LCOE	$ 41.15/MWh

Results of the Waste-to-Energy facility cases

100,000 Inhabitants Scenario		500,000 Inhabitants Scenario		1,000,000 Inhabitants Scenario	
Power output	2350 kW	Power output	11,930 kW	Power output	23,880 kW
Energy produced	488,800 MWh	Energy produced	2,481,440 MWh	Energy produced	4,967,040 MWh
Economic Results		Economic Results		Economic Results	
NPV	$ −73,857,512	NPV	$ −189,861,280	NPV	$ −269,667,068
Total COI	$ 28,952,471	Total COI	$ 104,609,442	Total COI	$ 182,181,518
O&M costs	$ 71,733,149	O&M costs	$ 220,452,589	O&M Costs	$ 358,107,897
Energy sales	$ 14,451,102	Energy sales	$ 73,362,401	Energy Sales	$ 146,847,789
WTB	$ 12,285,580	WTB	$ 61,379,339	WTB	$ 122,855,797
CER' sales	$ 91,427	CER' sales	$ 459,011	CER' sales	$ 918,761
CER' cost	$ 457,516	CER' cost	$468,544	CER' cost	$ 482,336
LCOE [$/MWh]	397.00	LCOE [$/MWh]	233.40	LCOE [$/MWh]	184.40

Life Cycle Assessment Boundaries of the Scenario 1

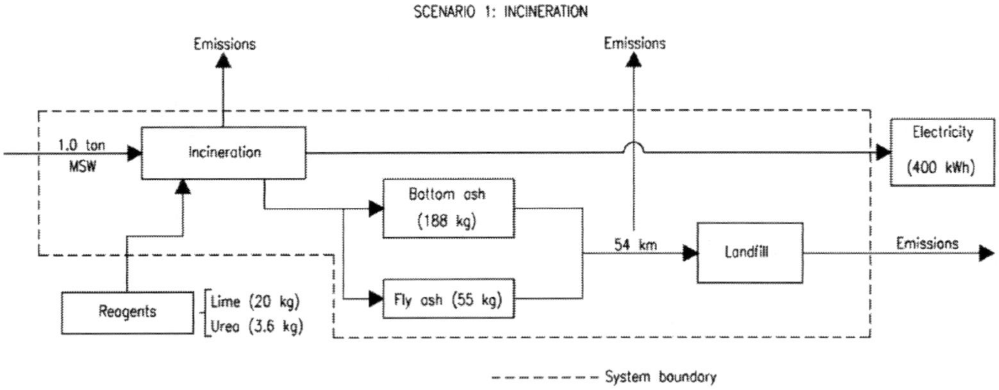

Life Cycle Assessment Boundaries of the Scenario 2, 3 and 4

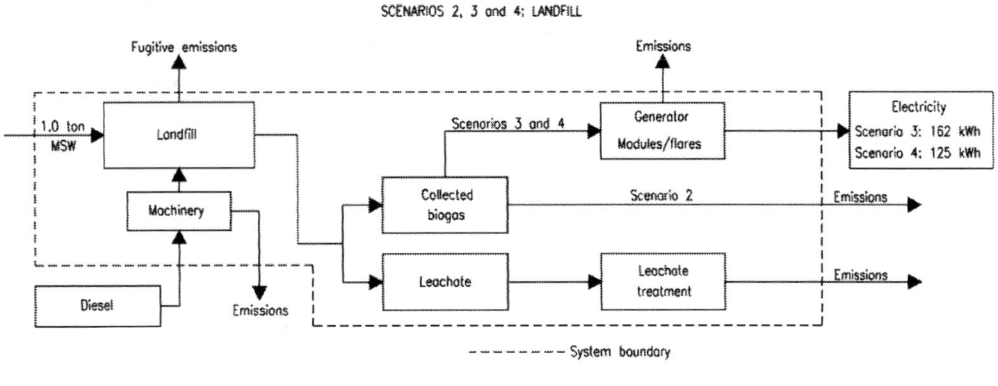

Emissions characteristics of biogas combustion in different systems

	Biogas Burning Systems emissions and removal efficiency		
	Pollutant	Value	Unit
Flare	NO_x	19.3	µg/kJ
	CO	22.54	µg/kJ
	Particulates	7.28	µg/kJ
	Dioxins/Furans	0.205	pg TEQ/kJ
	Pollutants removal efficiency	99.7	%
IC Engine	NO_x	1077	µg/kJ
	CO	784.2	µg/kJ
	Particulates	21.5	µg/kJ
	Pollutants removal efficiency	97.2	%
Gas Turbines	NO_x	125.9	µg/kJ
	CO	393.2	µg/kJ
	Particulates	38.23	µg/kJ
	Pollutants removal efficiency	94.4	%

LCA characterization for evaluated scenarios

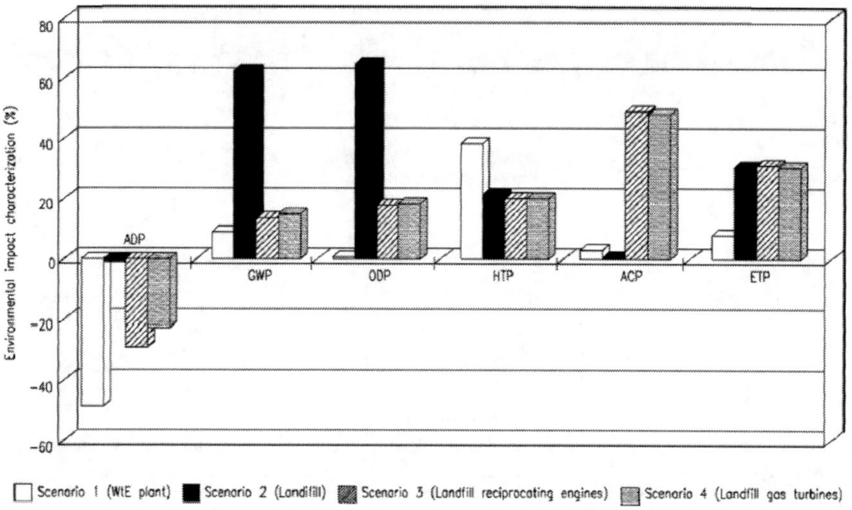

Scenario	Energy recovery		LCA characterization					
			ADP	GWP	ODP	HTP	ACP	ETP
Units	kWh/tonne MSW	%	kg Sb-eq.	kg CO_2-eq.	kg CFC-11-eq.	kg 1,4 DB-eq	kg SO_2-eq.	kg PO_4^{-3}-eq
Scenario 1 – WtE	400.0	18.0	−0.21	285.0	0.00	331.9	0.68	0.66
Scenario 2 – Landfill	0.0	0.0	0.00	2052	0.0132	182.4	0.00	2.53
Scenario 3 – Landfill reciprocating ICE	162.0	7.4	−0.12	464.0	0.0036	175.7	11.87	2.57
Scenario 4 – gas turbines	125.0	5.7	−0.10	478.0	0.0037	175.4	11.65	2.51

Energy Conversion and Management 86 (2014) 981–991

Contents lists available at ScienceDirect

Energy Conversion and Management

journal homepage: www.elsevier.com/locate/enconman

Sugarcane biorefineries: Case studies applied to the Brazilian sugar–alcohol industry

Maria Luiza Grillo Renó[a,*], Oscar Almazán del Olmo[b], José Carlos Escobar Palacio[a], Electo Eduardo Silva Lora[a], Osvaldo José Venturini[a]

[a] Federal University of Itajubá, Av. BPS 1303, CP 50 Itajubá, MG, Brazil
[b] ICIDCA – Instituto Cubano de Investigaciones de los Derivados de la Caña de Azúcar, Vía Blanca y Carretera Central 804, San Miguel del Padrón, C.P. 4036 La Habana, Cuba

The biorefineries can reach the higher overall yield from the raw materials, with the lowest environmental impact, at minimum energy input and giving the maximum of the energy output. The biorefinery is the true valuable option of a wide diversification, with by-products like the single cell protein and biogas from the distillery stillage, new oxidants like methanol, second generation biofuels, biobutanol, etc. In this context this paper presents a study of five different configurations of biorefineries. Each case study being a system based on an autonomous distillery or sugar mill with an annexed distillery and coproduction of methanol from bagasse. The paper includes the use of sugarcane harvest residues (mainly sugarcane trash) and a BIG–GT plant as alternatives to fulfill the energy demands of the complex

Environmental impacts of the products by impact category

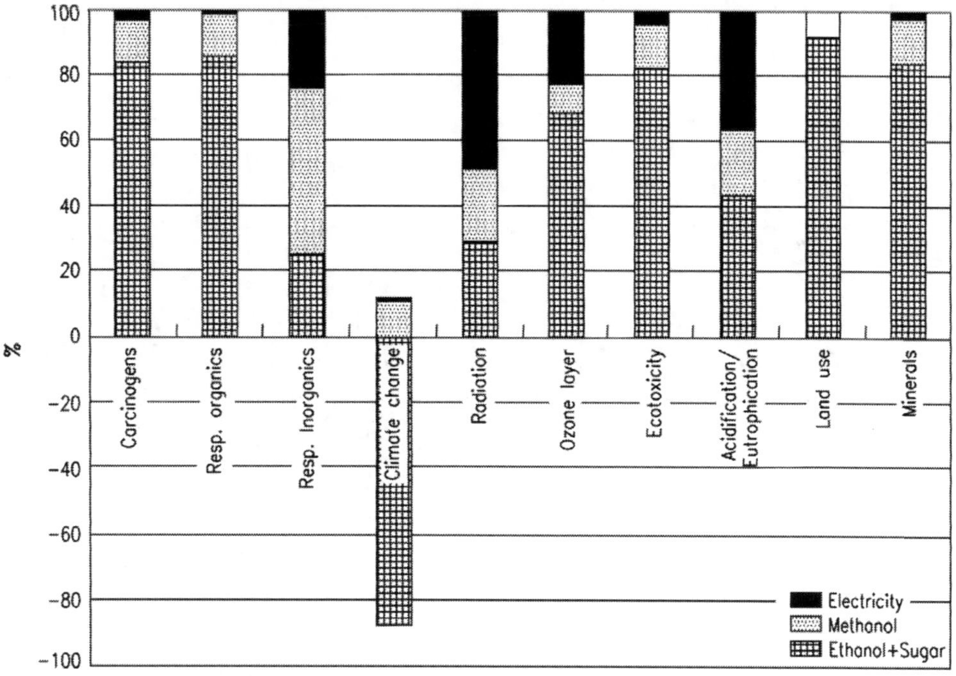

International Sugar Journal, Vol 112, Issue 1343, 2010.

Use of the life cycle assessment (LCA) for comparison of the environmental performance of four alternatives for the treatment and disposal of bioethanol stillage †

By M.H. Rocha [1*], E.E.S. Lora [1], O.J. Venturini [1], J.C.P. Escobar [1], J.J.C.S. Santos [1] and A.G. Moura [2]

[1] Excellence Group in Thermal Power and Distributed Generation (NEST), Federal University of Itajubá (UNIFEI) Av. BPS, 1303, Itajubá, Minas Gerais, CEP 37500-903, Brazil.
[2] Dedini Indústria de Base S.A. Rod. Rio Claro/Piracicaba, km 26,3, Bairro Cruz Caiada, Piracicaba, São Paulo, CEP 13412-900, Brazil.
* Contact author: Email: mateus.rocha@unifei.edu.br

This work shows the results of the application of the LCA methodology for the analysis and comparison of four alternatives for stillage treatment and disposal: conventional *"in natura"* fertirrigation, anaerobic digestion, concentration until 40% for fertirrigation and concentration until 65% for combustion in boilers using fuel oil as supplementary fuel. The milling capacity of 1.99 million tonnes of sugarcane per crop, producing 154,000 tonnes of sugar and 81,000 m³ of ethanol. The mill is located near the city of Sertãozinho, Brazil and local soil characteristics were also considered. The Simapro software and the CML 2 baseline 2000 are used as support tools. Conventional and concentrated stillage fertirrigation alternatives have the best environmental performance. In the combustion of stillage, we considered the installation of pollution control devices for SO_x and NO_x with 95% efficiency. From the point of view of climate change, based on the life cycle greenhouse gases balance, the best alternative was anaerobic digestion.

Scheme of the Fertirrigation Scenario

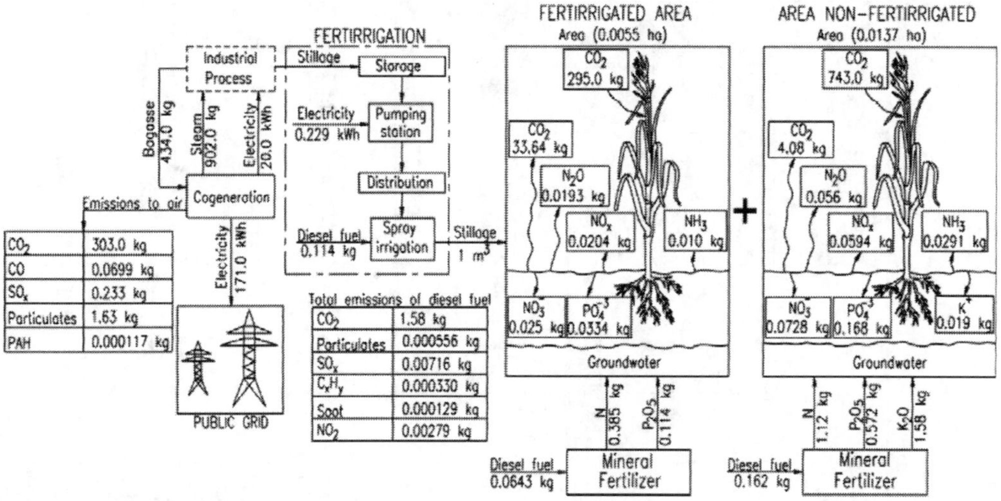

Scheme of the Anaerobic Digestion Scenario

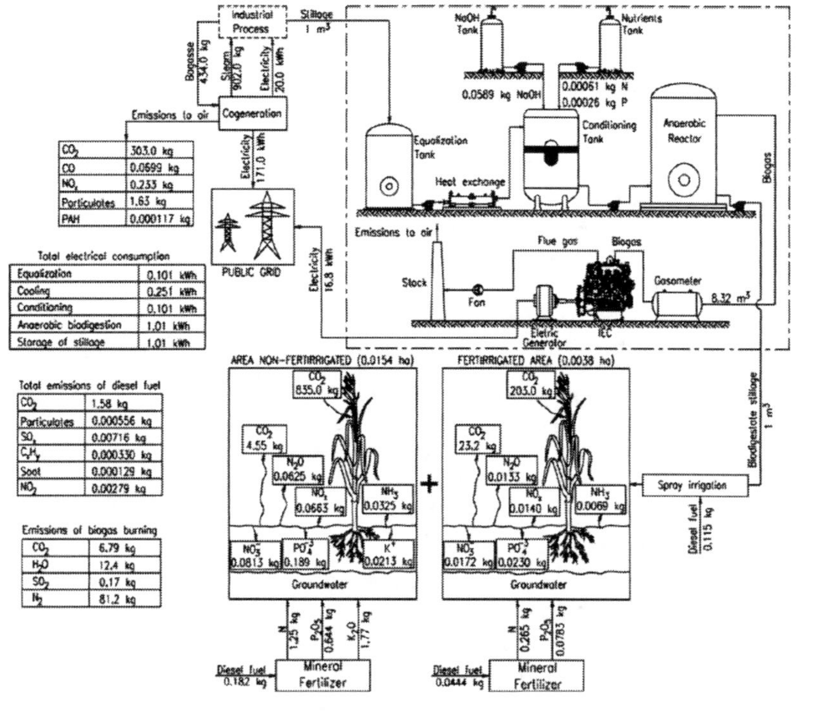

Scheme of the Concentration up to 40% Scenario

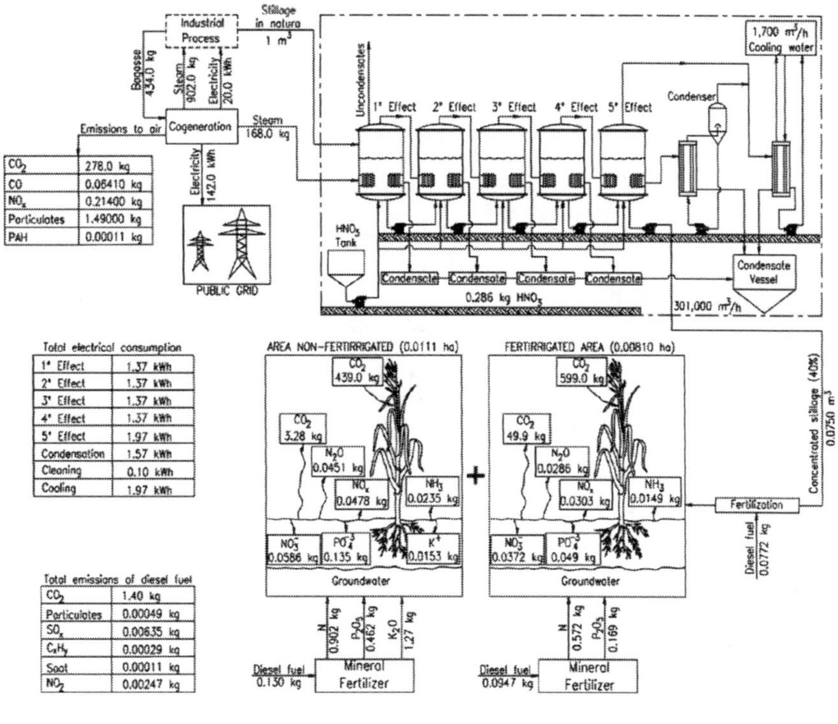

Scheme of the Concentration up to 65% and Combustion With Auxiliary Fuel Scenario

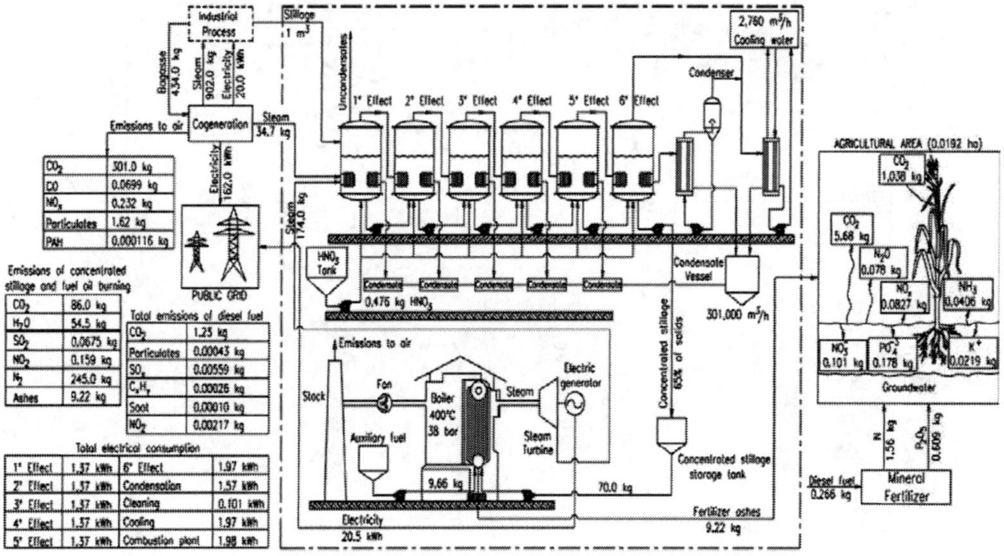

Rate of emission of CO_2–equivalent in function of the electricity exported for the public grid

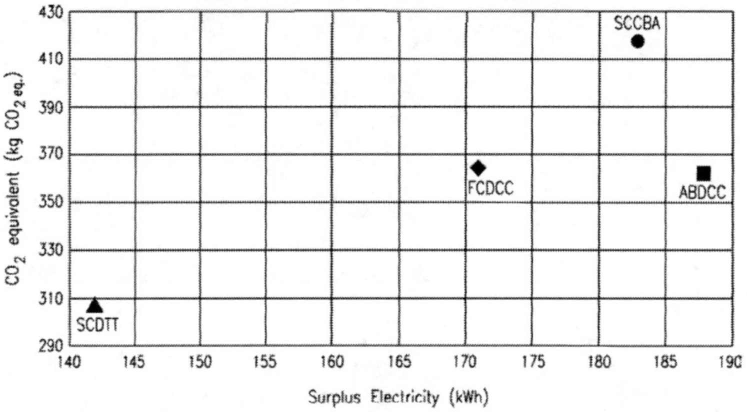

Energy Balance of Evaluated Stillage Disposal Scenarios

Scenario	Renewable energy output (MJ/m³ stillage)	Fossil fuel energy input (MJ/m³ stillage)	ER
FCDCC	118.45	2477.36	20.9
ABDCC	133.17	2538.56	19.1
SCDTT	152.56	2372.96	15.6
SCCBA	544.16	2564.36	4.7

Characterization of the Environmental Impacts of the Considered Scenarios

Impacts categories	Unit	FCDCC	ABDCC	SCDTT	SCCBA
ADP	kg Sb eq.	−0.0455	−0.0551	−0.0171	0.1430
GWP	kg CO_2 eq.	−698.0	−703.0	−749.0	−648.0
ODP	kg CFC-11 eq.	0	0	0	0
HTP	kg 1,4-DB eq.	66.9	66.9	61.3	66.7
FEP	kg 1,4-DB eq.	0.0203	0.0204	0.0185	0.0234
MEP	kg 1,4-DB eq.	36.0	38.1	33.1	43.2
TEP	kg 1,4-DB eq.	0.00031	0.00523	0.00027	0.00373
POP	kg C_2H_2	−0.00419	0.00337	−0.00264	0.00283
ACP	kg SO_2 eq.	0.0103	0.1950	0.0556	0.1670
ETP	kg PO_4^{-3} eq.	0.6430	0.6700	0.5970	0.5940

Percentile Comparison of the Contribution of Each Evaluated Scenario in Each Impact Category

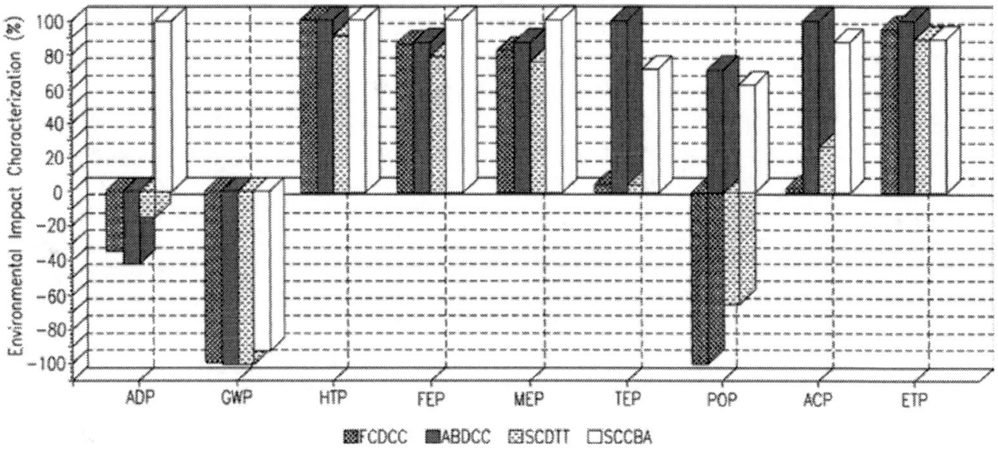

Conclusions

- The world population, requirements for food and feed permanently increases. This creates growing demands for production and consumption of biofuels in the last years. While high oil prices favored this growth, it was also supported and drove by policies such as mandates, targets and subsidies catering to energy security and climate change considerations.

- To evaluate the biofuels sustainability it must be done serious, multidirectional and holistic evaluation and comparison of the environmental life cycle impacts, complemented with an energy balance of biofuels.

- It must be identified the intensity and interrelations of aggressions, the environment has suffered several detrimental effects, as a consequence of the production of biofuels. The results are a useful reference for future scientific works related to the development of new technologies and/or an effective adequateness of the existing ones, for the production of biofuels, with higher efficiency, effectiveness and safety, looking for a true sustainability.

- The role of the transportation of the crops, from the field to the factory, in the total environmental impact, making the distance a limiting factor and showing clearly the need to give priority to the development of ways and means to reduce it.

Conclusions

- In the cradle-to-gate LCA it is observed that the consumption of fossil fuels is associated with several impacts, such as, Abiotic Depletion and Global Warming Potential that indicates the necessity to reduce it in logistic and industrial stages, adopting other alternatives.

- The use of the co-products and residues of the production of the biofuels, for electrical and thermal energy, through cogeneration, to be used in the processes, improve the ecological and economic benefits, increasing the appeal of biofuels. The use of co-products in process, to generate steam and electricity, results in good indicators because it decreases or cancels the external dependence of fossil fuels and electricity, and allows the supply of renewable electricity or fuels to use in other process. This practice could, in some cases, represent the satisfaction of 85 to 95% of the total energy demand.

- A wide range of high added value products such as enzymes, organic acids, biopolymers, electricity, and molecules for food and pharmaceutical industries could be obtained upgrading of industrial co-products of biofuels production (BIOREFINERIES). Further works are required to have a more realistic consideration of the co-products and biomass valorization contribution to LCA indicators. The LHV approach for co-products has been used in real industrial applications process on animal feed and surplus electricity generation.

Thank you very much for your attention!
Itajubá/MG between the mountains
MSc. Mateus Henrique Rocha
mateus0@yahoo.com.br
The Aristotelian holistic principle: "The whole is grater that the addition of its parts"

Environmental Implications of Jatropha Biofuel from a Silvi-Pastoral Production System in Central-West Brazil

Presented at RCN-SEES for Pan American Biofuels and Bioenergy Sustainability

"Standardization of Environmental Life Cycle Assessments of Biofuels"

22 - 25 July 2014

Rob Bailis - Associate Professor; Yale School of Forestry & Environmental Studies

Outline

- Some background on Jatropha LCA
- The silvi-pastoral system
  - Co-production with cattle
  - Detoxification
  - Other innovations
- Results
- Future directions

Previous work…

Jatropha WTW GHGs using generic Brazilian conditions in 2009 (no LUC)

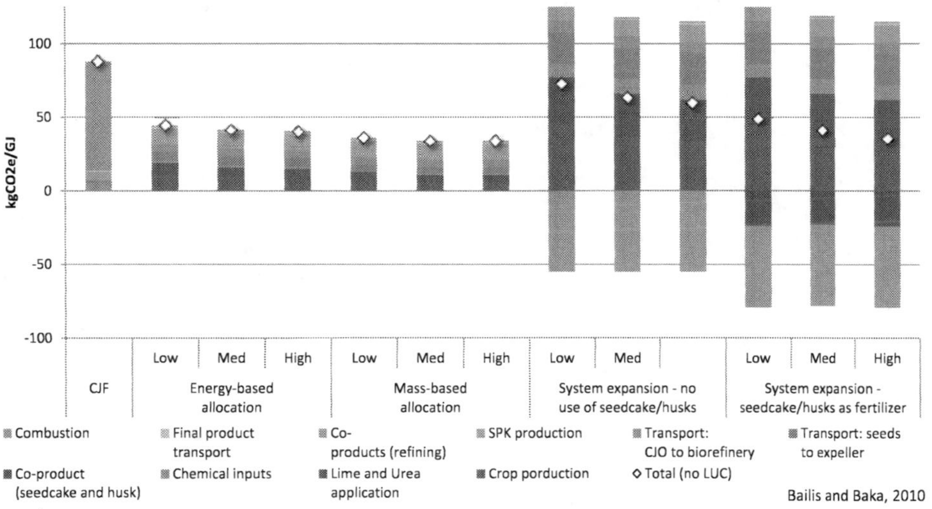

Bailis and Baka, 2010

What about LUC?

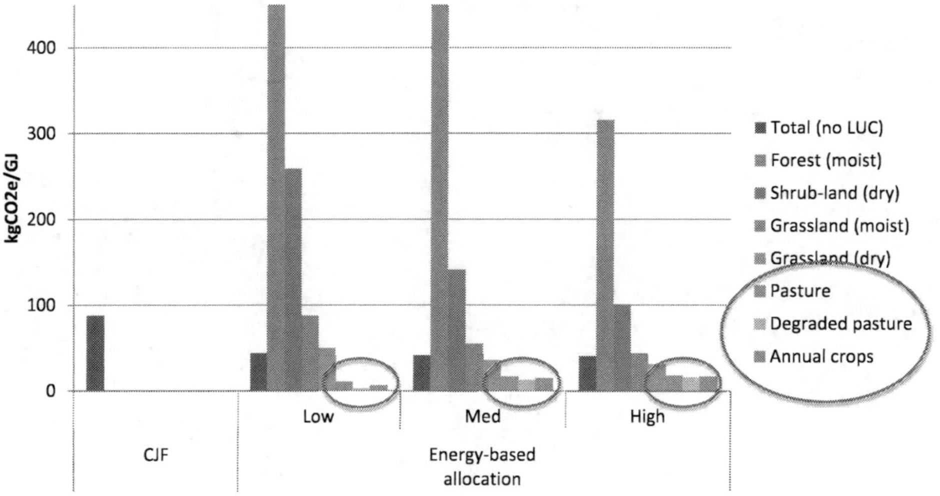

Bailis and Baka, 2010

Project-based LCA

- Site-specific silvi-pastoral production
- Utilizing existing pasture
  - 160 Mha of pasture in Brazil
  - 10 Mha is "degraded"
- Multiple innovations
- Generalizable to other perennial species?

Innovations

1. Silvipastoral production
   - Co-produce oilseeds and cattle
2. Detoxify press-cake and use animal feed
3. Use hybrid seeds
4. Oil extraction with locally produced ethanol
5. Cogeneration using husks and shells
6. Alternative refining pathway

Goal and Scope

Compare environmental impacts of conventional jet fuel to Jatropha-SPK produced in a novel silvi-pastoral system*

- System boundaries

  Farm + Oil Extraction + Refining
- Functional unit
  - 1 GJ of fuel
- Treatment of co-products
  - Mass, energy, and economic allocation
  - System expansion

* SPK - synthetic paraffinic kerosene

Innovations

1. Co-produce oilseeds and cattle
2. Detoxify press-cake and use animal feed
3. Use hybrid seeds
4. Oil extraction with locally produced ethanol
5. Cogeneration using husks and shells
6. Alternative refining pathway

1. Co-production with cattle

- Stocking 2 AU* per ha
- Inputs nourish pasture and Jatropha

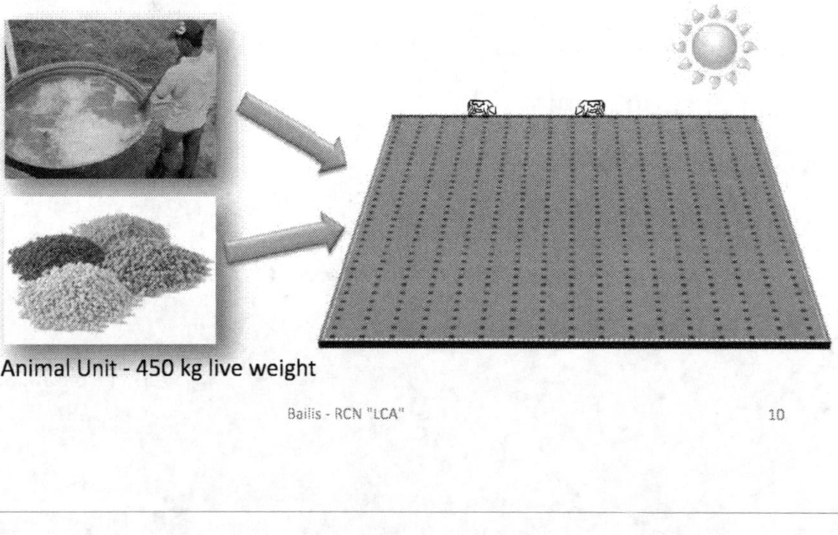

* AU = Animal Unit - 450 kg live weight

1. Co-production with cattle

- Annual productivity per ha:
  - Jatropha 1.95 tons dry seed plus co-products
  - Pasture 10-17 t/ha
  - Cattle 2 UA (900 kg)

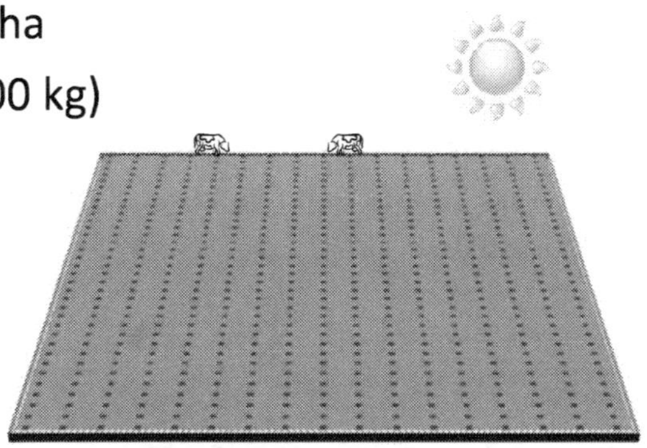

* UA = Unidade Animal = 450 kg live weight

1. Co-production with cattle

- Cattle produced here displace cattle produced elsewhere

...where and by how much?

* UA = Unidade Animal = 450 kg live weight

1. Co-production with cattle

- Co-produced cattle might displace:
  - Cattle in managed, natural, or degraded pasture
  - Cattle grazing in recently deforested land

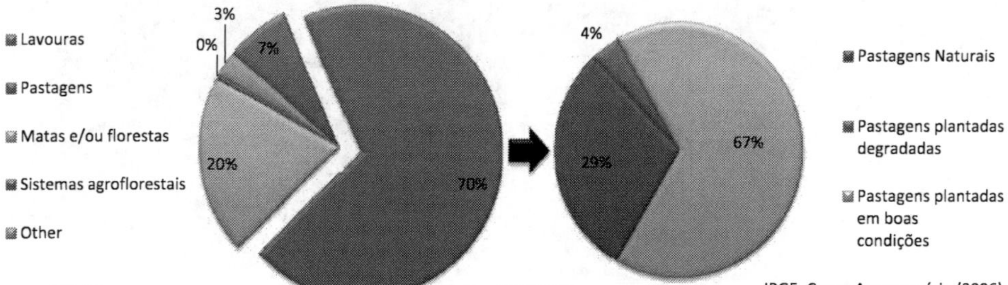

IBGE, Censo Agropecuário (2006)

1. Co-production with cattle

- Emissions from 1 AU:
  - Enteric fermentation: 56 kg CH_4 per year
  - Manure: 1 kg CH_4 and 1.2 kg N_2O per year
  - Inputs: NPK, diesel...
  - Land cover change:
    - Degraded – small increase in C
    - Managed pasture – no change C
    - Forest – large C emissions avoided

      Stocking rate in forest regions 0.9 AU/ha

      Carbon stocks ~ 150 t-C/ha (IPCC)

1. Co-production with cattle

Cederberg et al (2011):

- ~6% of Brazil's cattle were produced on recently cut forest in LAR
- 44 kg CO2e/kg cattle averaged over all regions
- 352 $kgCO_2e$ per FU

2. Press-cake detox

- Press-cake detoxification
  - 1 dry ton of seed yields 250kg of Jatropha seedcake (JSC)
    - 60% protein
  - Assume product displaces soymeal
    - 19% domestic animal feed market in Brazil (USDA)
    - Corn has bigger market share
    - maybe more realistic to displace maize?

2. Press-cake detox

- Press-cake detoxification
  - Substitution on a protein basis
    - 1 kg Jatropha presscake ≈ 1.3 kg soymeal
  - 0.7 kg CO_2e/kg soymeal (EcoInvent, 2010)
  - 16-18 kgCO_2e per FU via soymeal displacement

3. Hybrid seeds

- Yield
  - Originally expected 3.5 kg/tree
  - Revised down to 2.4 kg/tree
    - Tested sensitivity @ 1.2 kg/tree
- Spacing
  - 8x1 with large alleys for pasture
  - Considered monoculture @ 4x1

SG biofuels

- Improved yields
- Vigor and uniformity
- Reduced input costs
- Enhanced logistics & processing

4. Ethanol as solvent

- Replace hexane or pentane
  - Saves energy and reduces emissions

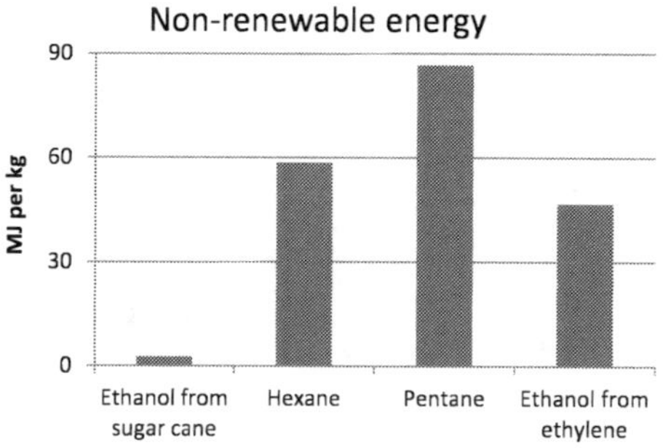

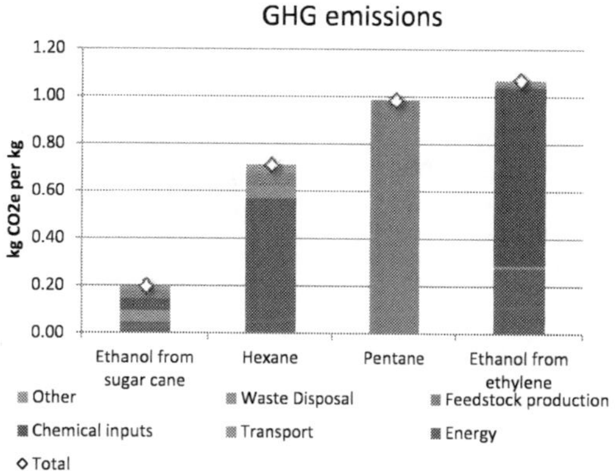

5. Cogeneration

- Cogen using seedcake and husks

1 FU yields

~ 100 kg husk and shell

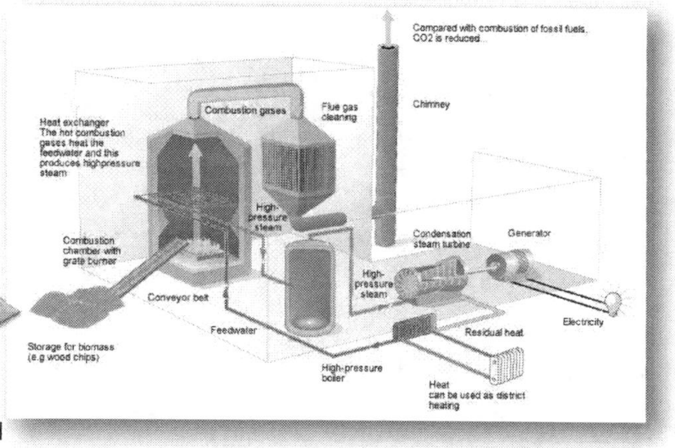

94 kWh electricity

- 11 kWh used internally
- 83 kWh exported to grid

6. Alternative refining pathways

Catalytic hydro-treatment (UOP)

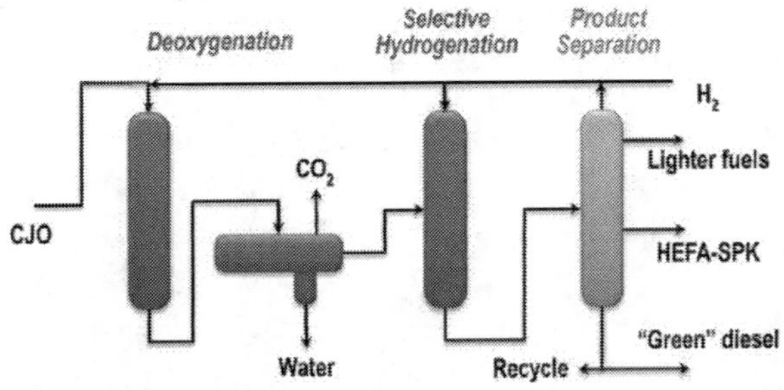

Catalytic decarboxylation (AliphaJet)

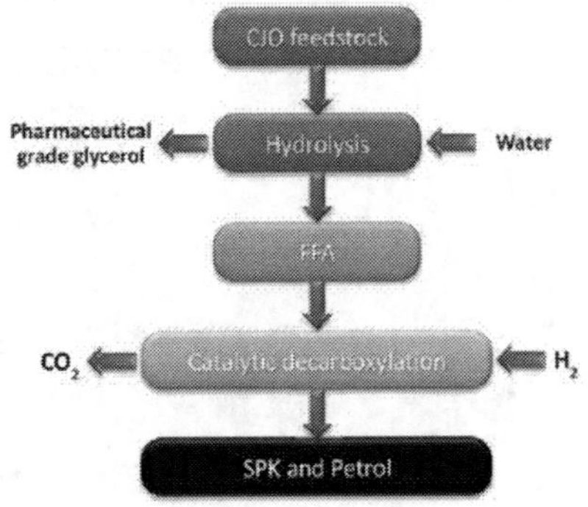

From http://www.uop.com/green-jet-fuel/

From http://www.syngest.com/AliphaJet/images/img_flow_graphic.jpg

Results (system expansion not shown)

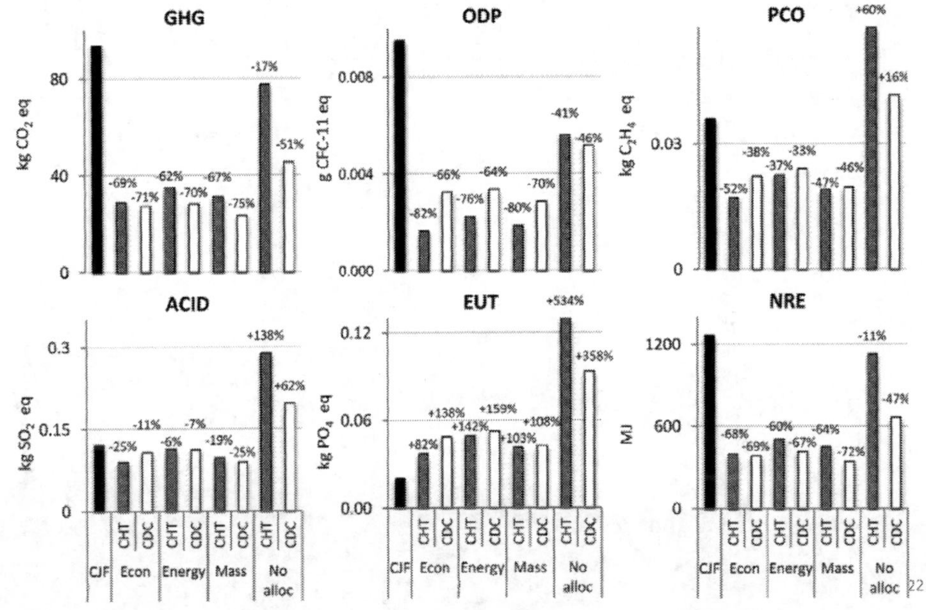

Results

- Reduced impacts in most indicators
- Improvements are robust across allocation methodologies
- Improvements persist if yields are 50% lower than expected

Remaining questions

- Carbon dynamics:
  - trees, pasture, and soil?
- Reality check on results:
  - Soy and cattle displacement
- The project failed and SG pulled out of Brazil…why?
- What are future prospects for silvipastoral-based production of oilseed feedstocks?

"A Gol aircraft is fueled with bio-kerosene in Rio; in particular, macaúba fruit and cane sugar, used to produce biofuel."
From p. 141 of *this month's* GOL in-flight magazine

Thanks! Obrigado! Gracias!

- AIRBUS for funding the Jatropha research

  Collaboration with Goksin Kavlak (Yale), SG Biofuels, JetBio and Rio Pardo Bioenergia

Land use change

- Little impact shifting from managed/degraded pasture to Jatropha:
  - 1 t-C per ha lost from managed pasture
    - ~11.6 kg CO2e/GJ lost over 20 years
  - 3 t-C gained in degraded pasture
    - ~5.0 kg CO2e/GJ gained over 20 years

GHG reduction requirements in a sample of existing sustainability initiatives					
Initiative	**Allocation**	**GHG Reduction requirement**	**ILUC**	**LUC**	**Time frame**
US-RFS	Displacement method	Conventional biofuels: **20% lifecycle GHG threshold** (below gasoline) Advanced biofuels: 50% lifecycle GHG threshold Biomass-based diesel: 50% lifecycle GHG threshold Cellulosic biofuel: 60% lifecycle GHG threshold	Yes	Yes	100 year with 2% discount rate OR 30 year with 0% discount rate
CA-LCFS	GREET methodology	**10%** reduction in GHG emissions across fleet	Yes	Yes	30 year project horizon
UK RTFO	Substitution approach preferred but economic also permitted	Targets to overall level of GHG saving achieved by the biofuel supplied in each obligation period: **2008–2009, 40%, 2009–2010, 45%, 2010–2011, 50%, etc.** The level of GHG saving is an overall target for all fuels and feedstock reported by a fuel supplier Will follow the EC-RED Directive	X	Yes	No dLUC with a carbon payback time over 10 years
Dutch NTA 8080	Based on energy content	**For heat and power**: at least 70% if reference case is Dutch mixture of electricity or coal, or at least 50% if reference case is natural gas. **For transportation fuels:** at least 50%; for flows of biomass for which in the EC-RED "a typical GHG emission saving of less than 50% is included as transition period till 2012, a minimum of 35%	X	Yes	Annualized emissions based on 20 years
EC-RED	Based on energy content	At least **35% GHG emission** reduction compared to reference fuel Rising to **50% on January 2017 and 60% in 2018** for biofuels and bioliquids produced in installations in which production started on or after January 2017.	X	Yes	Annualized emissions with 20 yrs timeframe
RSB	Guidelines under development	Biofuel shall have lower GHG emissions than the fossil fuel baseline and shall contribute to the minimization of overall GHG emissions. The threshold **(10, 40 and 70% reductions are under discussion)** will be set at the conclusion of the test period	?	Yes	Based on IPCC methodology

Scenarios

Scenario	Description
Fossil fuel reference	Average US kerosene-based jet fuel
Baseline Jatropha SPK	Jatropha in Brazilian conditions with no innovations
CJO co-production with cattle	Energy Allocation Mass Allocation Economic Allocation No accounting for cattle Accounting for cattle but no avoided LUC Accounting for cattle and avoided LUC

LCA methodologies

- Our approach:
  - Collect primary data
  - Follow accepted LCA protocols
    - Allocate with system expansion when possible
  - Use LCA software common to industry and academia
  - Undergo extensive peer review

Allocation

Milling Products	Mass (kg per ton of dry seed)	mass %	energy (MJ /kg)	energy %	Value (R$/ton)	value (%)
shells	390	28%	19	22%	100	5%
husks	380	28%	19	22%	100	4%
CJO	360	26%	39.6	43%	1681	71%
Cake (or meal)	250	18%	18.3	14%	700	20%
TOTAL (fruit)	1380		24		621	

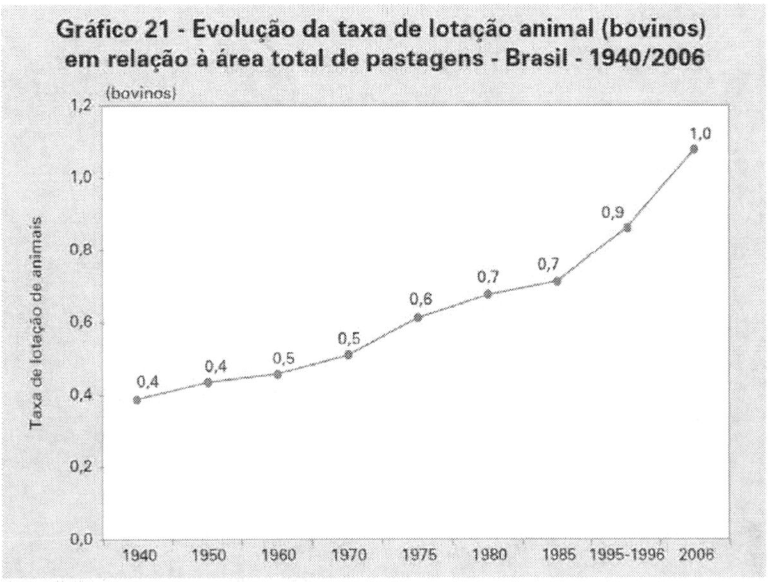

Fonte: IBGE, Censo Agropecuário 1940/2006.

Allocation factors

Table S1: Factors used to allocate impacts among co-products by energy, mass or economic value (8x1 spacing) [a]

	Mass		Energy		Economic value	
1. Crop production	**Units produced per ton of dry seed**	**%**	**Calorific value (MJ/unit)**	**%**	**Economic value (R$/unit)** [b]	**%**
Jatropha fruit (inc. husk)	1380 kg	88%	12.2	96%	0.39	64%
Cattle (carcass wt) [c]	111 kg	12%	7.8	4%	2.66	36%
2. Oil Extraction [d]						
CJO	360 kg	59%	39.6	67%	1.70	71%
JSC	250 kg	41%	18.3	22%	0.55	20%
Electricity to grid	640 kWh	NA	3.6	11%	0.12	9%
3 a. Oil refining: HEFA-CHT [e]						
J-SPK	174 kg	56%	44.3	58%	10	53%
Other hydrocarbons	133 kg	43%	42.0	42%	18	47%
3 b. Oil refining: HEFA-CDC [f]						
J-SPK	238	80%	44.3	86%	10	97%
Other hydrocarbons	58	20%	28.7	14%	2.1	3%

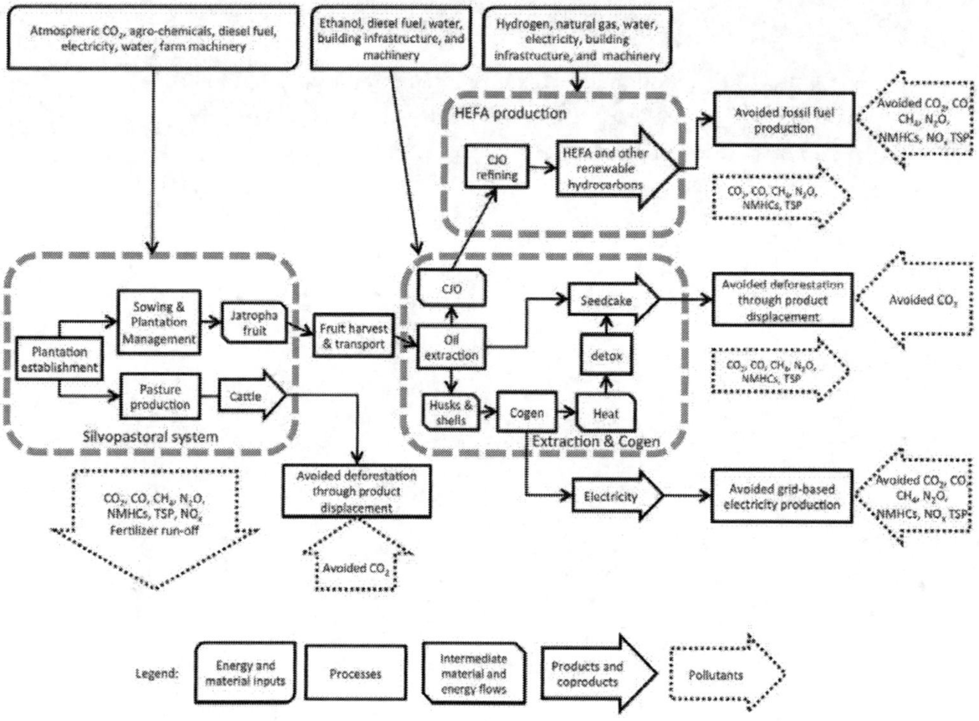

Reference scenario

- Fossil jet fuel

Figure 7-11. GHG Emissions for Liquid Fuels Produced Domestically

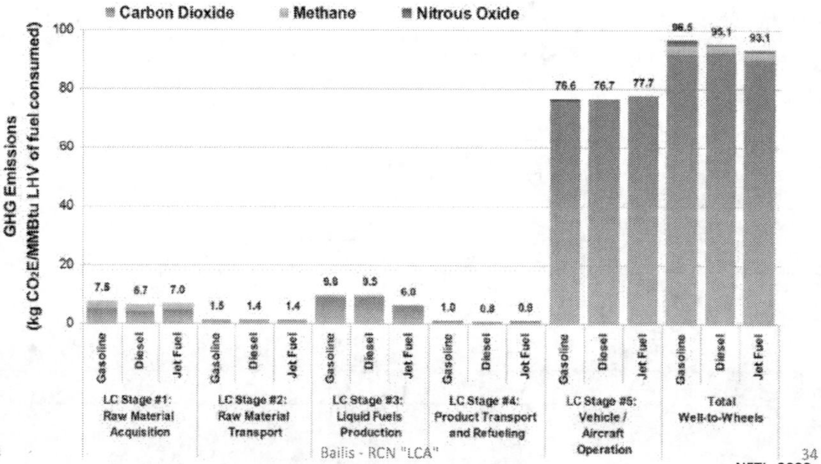

NETL, 2008

Sensitivity to spacing and yield

Table 2. Differences in Impacts between 8 × 1 Intercropping and 4 × 1 Monoculture[a]

	allocation methodology									
	economic		energy		mass		no allocation		system expansion	
impact category[b]	CHT	CDC	CHT	CDC	CHT	CDC	CHT	CDC	CHT	CDC
GHG	−24%	−34%	−33%	−45%	−30%	−42%	−41%	−51%	82%	78%
ODP	8%	5%	−19%	−14%	−15%	−12%	−23%	−18%	−12%	−38%
PCO	6%	6%	−19%	−20%	−17%	−18%	−22%	−23%	−8%	−7%
ACID	8%	9%	−14%	−16%	−13%	−14%	−17%	−18%	−161%	−192%
EUT	12%	12%	−16%	−16%	−14%	−14%	−19%	−19%	−19%	−14%
NRE	5%	7%	−13%	−17%	−11%	−15%	−17%	−21%	−22%	−51%

[a]Negative entries (shaded) indicate the 4 × 1 monoculture system performs better than the 8 × 1 intercropping system. [b]Impact categories are defined in Figure 2.

Table 3. Relative Increase in Each Category of Impacts Resulting if Yield Is 50% Lower than Expected with Inputs Held Constant in the 8 × 1 Intercropping Scenario

	econ		energy		mass		no alloc		syst exp	
impact[a]	CHT	CDC	CHT	CDC	CHT	CDC	CHT	CDC	CHT	CDC
GHG	15%	21%	35%	48%	31%	42%	47%	59%	9%	9%
ODP	30%	21%	63%	46%	59%	40%	75%	59%	37%	122%
PCO	27%	28%	59%	60%	54%	54%	68%	69%	23%	21%
ACID	22%	25%	51%	56%	45%	51%	59%	63%	554%	659%
EUT	31%	32%	65%	66%	59%	61%	73%	74%	75%	56%
NRE	19%	27%	43%	57%	37%	50%	57%	70%	71%	168%

[a]Impact categories are defined in Figure 2.

Results – system expansion

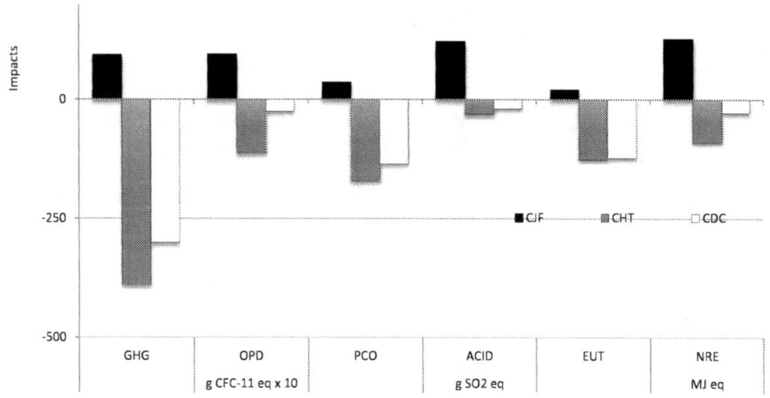

Sustainability criteria (Article 17 – EU Directive) (1)

Emition reduction

a) Feedstock production
b) Transformation (Industry).
c) Transport

January 2010	January 2017	January 2018
35%	50 %	60 %

To avoid default values there is a need to calculate real values

Default values have not been modified

Rules for calculating the greenhouse gas impact of biofuels, bioliquids and their fossil fuel comparators

A. *Typical and default values for biofuels if produced with no net carbon emissions from land-use change*

Biofuel production pathway	Typical greenhouse gas emission saving	Default greenhouse ga emission saving
sugar beet ethanol	61 %	52 %
wheat ethanol (process fuel not specified)	32 %	16 %
rape seed biodiesel	45 %	38 %
sunflower biodiesel	58 %	51 %
soybean biodiesel	40 %	31 %
palm oil biodiesel (process not specified)	36 %	19 %

Green House emitions Frias Plant Santiago del Estero

AG-ENERGY
VILUCO

Objetive of the work

MAIN OBJETIVE
Perform a complete inventory of GHG ivolved in the production and conversion of soybeans in Frias Plant
Follow up this study over three years in order to capture interanual variations

First research core team

Technical seminars

Starting Implementation

Methodological tools used

2006 IPCC directives for national GHG inventories

EB 50 – Executive MDL board "Guidelines on apportioning emissions from production processes between main product and co-and by-products ".

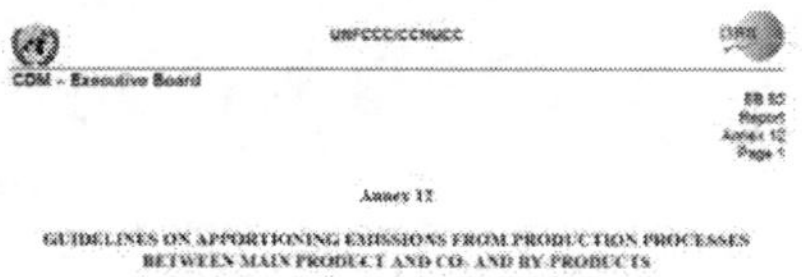

ACM0017 Methodology "Approved consolidated baseline and monitoring methodology Production of biodiesel for use as fuel".

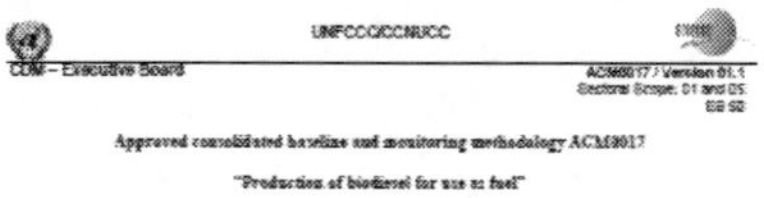

DIRECTIVE 2009/28/CE
European Union Parliament and council
April 23 2009

Anexo V
Directive EU-RED

Industry Transport

$$E = e_{ec} + e_l + e_p + e_{td} + e_u - e_{sca} - e_{ccs} - e_{ccr} - e_{ee},$$

Agriculture Agriculture

E	Total emissions from the use of the fuel (soybean based biodiesel)
E_{ec}	Emissions from the cultivation raw material (soybean)
e_l	Annualised emissions from carbon stock changes caused by land-use change
e_p	Emissions from processing
e_{td}	Emissions from transport and distribution
e_u	Emissions from the fuel in use
e_{sca}	Emission savings from soil carbon accumulation via improved agricultural management
e_{ccs}	Emission saving from carbon capture and geological storage
e_{ccr}	Emission saving from carbon capture and replacement;
e_{ee}	Emission saving from excess electricity from cogeneration

Anexo V concepts included in the study

	Concepto	Incluido
e_{ec} =	Emissions from the cultivation raw material (soybean);,	Yes
e_{l} =	Annualised emissions from carbon stock changes caused by land-use change,	No Assumption of no change in carbon stoks in soils since January 2008.
e_{p} =	Emissions from processing	Yes
e_{td} =	Emissions from transport and distribution emissions	Yes
e_{u} =	Emissions from the fuel in use	No Europen Directive - Anex V - Páragraph 13:Emissions from the fuel in use, eu, shall be taken to be zero for biofuels and bioliquids
e_{sca} =	Emission savings from soil carbon accumulation via improved agricultural management,	No No changes in carbon stoks due to agricultural practces
e_{ccs} =	Emission saving from carbon capture and geological storage,	No There are not any geological storage in place.
e_{ccr} =	Emission saving from carbon capture and replacement	No No biomass is used for fossile fuel replacement.
e_{ee} =	Emission saving from excess electricity from cogeneration .	No No electricity is generated.

Emition calculator

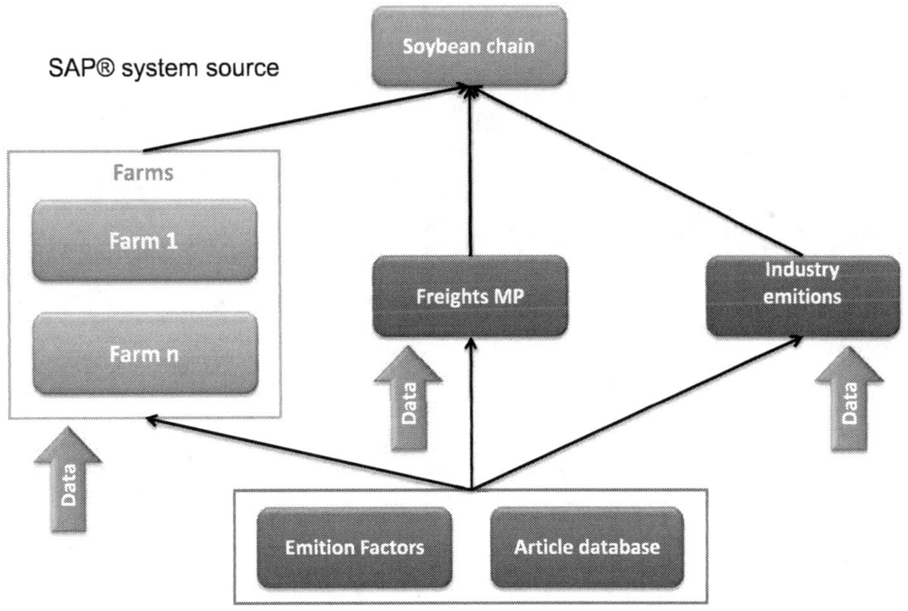

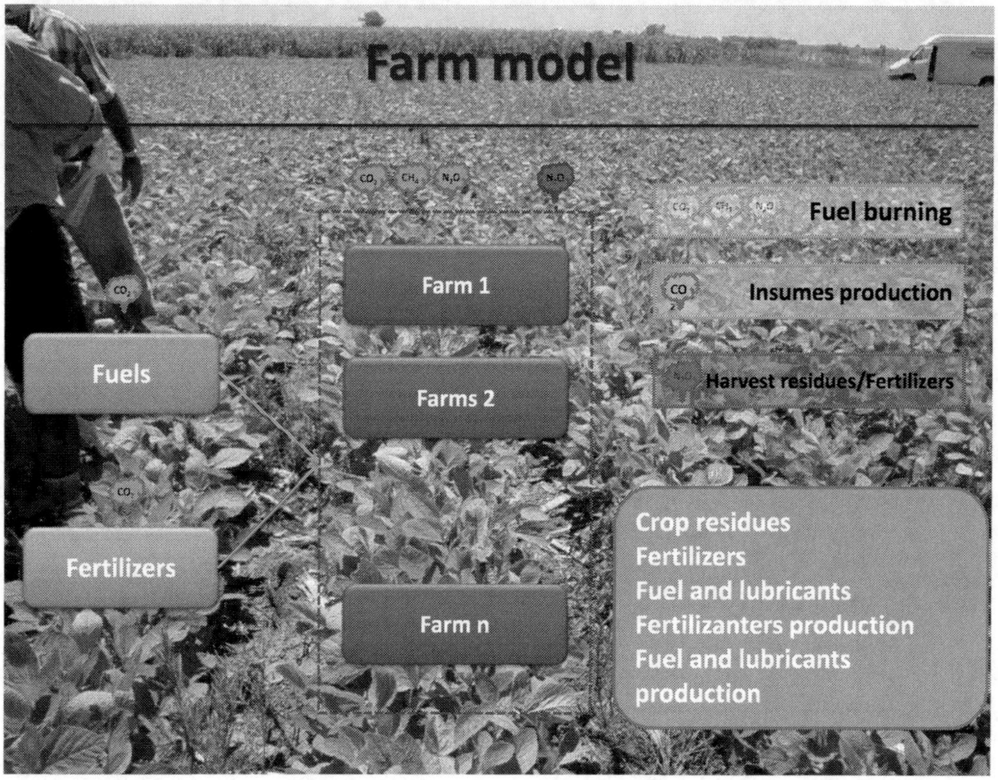

1. Emissions from the extraction or cultivation of raw materials (e_{ec})

Art. 6:, eec, shall include emissions from the extraction or cultivation process itself; from the collection of raw materials; from waste and leakages; and from the production of chemicals or products used in extraction or cultivation. Capture of CO2 in the cultivation of raw materials shall be excluded. Certified reductions of greenhouse gas emissions from flaring at oil production sites anywhere in the world shall be deducted. Estimates of emissions ......

Agriculture module

- N in agriculture residues, including N fixing crops & return of forage/pastures to the soil
- N suplied to the soil artificially.
- CO_2 coming from Urea use.
- CO_2 coming from fossile fuel employed
- Emissions associated to the life cycle of fertilizers and fossile fuels

- Raw material transport (CO_2 generated by fossile fuel use)

1.1 - N from agriculture residues

DATA SOURCE
Crop type
Production field
Field surface

Method Chapter 11 - Volume 4 IPCC 2006 Guides Level 1 Direct and indirect sources

CALCULATION FLOW

- Step 1: Yield of the crop Kgs/hectare.
- Step 2: Use of equation 11.7 to calculate N of Agricultural residues Cálculo del N de residuos agrícolas, (F_{CR})
- Step 3: Calculate direct emissions by using equation 11.1 of the table 11.1.
- Step 4: Calculate the indirect emissions by lixiviation using equation 11.10 of the table 11.3.

1.2 – Synthetic Fertilizers

DATOS USED
Quantity
Type of fertilizer
Composition

Methodology according to chapter 11 volume 4 of the IPCC 2006 guide - Level 1.Emition sources "Direct" & "Indirect x Deposition in the atmosphere and lixiviation" & CO_2 by the use of urea and derivates

PROCEDURE

- Step 1: Calculate the ammount of synthetic fertilizers applied (FSN) according to type and composition.
- Step 2: Calculate the direct emissions by using equations 11.1 from the table 11.1
- Step 3: Cálculate the indirect emitions by atmosphere deposition by using equation 11.09 and table 11.3.
- Step 4: Calculate the indirect emissions through lixiviation using equation 11.10 & table 11.3.
- Step 5: Calculate the equivalent urea quantity applied (FUREA).
- Step 6: Calculate the CO2 emissions by urea use using equation 11.3

1.3 – Fuel & lubricants

DATA USED
Type of field work
Quantity

Emisions (CO_2-N_2O-CH_4) asociated to the fuel burning for all the field operations preparation, planting, fertilizer and herbicides and harvesting

PROCEDURE

- Step 1: Estimation of fuel and lubricant consumption by converting each field operation for surface to liters of fuel and lubricants. All operations in Viluco are outsorced with specific contractors. Mean numbers are used..
- Step 2: Calculate the direct emissions multiplying the liters by the emission factor

1.4 – Fertilizantes production

DATA USED
Quantity
Type of fertilizer

"A Review of Greenhouse Gas Emission Factors for Fertiliser Production. Sam Wood and Annette Cowie Research and Development Division, State Forests of New South Wales. Cooperative Research Centre for Greenhouse Accounting - For IEA Bioenergy Task 38 - June 2004"

Product	Country	Composition	g CO_{2e}			Reference
		N:P:K	per kg N	per kg Product	CO_2:N_2O:CH_4	
Urea	Europe Average	46:0:0	*4018.9*	1848.7	97.5:0.1:2.3	Davis and Haglund (1999)
Urea	Europe Average	46:0:0	*1326.1*	610.0	-	Kongshaug (1998)
Urea	Europe: Modern Tech.	46:0:0	*913.0*	420.0	-	Kongshaug (1998)
Urea	Europe	*46:0:0*[a]	1707.3	*785.4*	-	Kuesters and Jenssen (1998)
UAN	Europe	*32:0:0*[a]	3668.0	*1173.8*	36.6:63.4:0.0	Kuesters and Jenssen (1998)
UAN	Europe Average	32:0:0	*5762.9*	1844.1	59.1:39.5:1.4	Davis and Haglund (1998)
UAN	Europe Average	32:0:0	*4093.8*	1310.0	-	Kongshaug (1998)
UAN	Europe Modern Tech.	32:0:0	*2000.0*	640.0	-	Kongshaug (1998)

(a) Composition from Kongshaug (1998).
Note: Figures in *italics* are derived values, based on % N composition.

PROCEDURE

- Step 1: Multiply the quantity of fertilizers by the corresponding emission factor.

1.5 – Production of fuels and lubricants

DATA SOURCE
Quantity

"Approved consolidated baseline and monitoring methodology ACM0017 "Production of biodiesel for use as fuel" - v.01.1 - UNFCCC - CDM Executive Board".

ID Number:	9
Parameter:	EF_{PROD}
Data unit:	tCO_2e/t petrodiesel
Description:	Emission factor for production of crude oil
Source of data:	-
Value to be applied:	The emission factor for the production of crude oil (EF_{PROD}) to be used in equation 16 is **0.073 tCO2e/t petrodiesel**[10]. A global value was calculated with the assumption that that upstream emissions with respect to crude oil production in Annex I countries is zero
Any comment:	-

ID Number:	10
Parameter:	EF_{REF}
Data unit:	tCO_2e/t petrodiesel
Description:	Emission factor related to oil refinery
Source of data:	-
Value to be applied:	The emission factor related to oil refinery (EF_{REF}) shall be one of the following: a) In the absence of a country-specific data, a global average figure of **0.233t-CO2/t-petrodiesel** can be used.[11] b) If refining occurs in the host country, reliable local emission factors from an official information source (e.g. national communications) may be used instead of the default emission factor
Any comment:	-

CALCULATION

- Step 1: Multiply the quantity of fuels and lubricants by the corresponding emission factor. .

1.6 – Farm structural costs / no productive fields

DATA SOURCE
Energy
Operations
Fertilizers

Common Energy emitions for all the farm operations in order to maintain all plots without production including fertilizer and agrochemical use.

DISTRIBUTION

- Total emitions are assigned according to the physical production of each crop, in order to assure a fair distribution between all the farm emitions associated with production.

Summary per field emissions

RESUMEN EMISIONES DE GASES DE EFECTO INVERNADERO
Resumen Abastecimiento de Materias Primas de Campos Propios - SOJA
2010-2011

		1280	Arbolito	Cabeza de Caballo	Las Marias	Los Guayacan	Pozo de la Espuela	Rancho Grande	San Jorge	Sara	Total
Producción Soja	Superficie (Hectáreas)	380	1.834	1.203	2.697	1.840	5.168	4.696	682	987	**33.382**
	Produccion (Toneladas)	1.481	6.086	3.063	7.986	3.972	18.255	15.771	2.529	3.408	**104.234**
	Rendimiento (Tn/Ha)	3,90	3,32	2,55	2,96	2,16	3,53	3,36	3,71	3,45	**3,12**
Emisiones Soja (Kgs CO_{2eq})	Residuos de cosecha	107.489	461.665	252.852	627.062	348.376	1.360.374	1.192.197	185.957	255.592	**8.051.244**
	Fertilizantes	-	-	2.810	264.029	-	-	-	-	-	**408.128**
	Combustibles y Lubricantes	47.857	143.736	115.745	213.008	160.793	415.467	356.956	85.606	124.533	**2.838.723**
	Produccion Fertilizantes	-	4.607	1.178	110.670	-	19.720	46.810	-	4.930	**269.398**
	Produccion Combustibles y Lubricantes	4.625	13.900	11.192	20.596	15.536	40.166	34.525	8.282	12.037	**274.486**
	Estructura campo+Lotes sin produccion [1]	36.679	38.338	36.679	40.086	38.772	38.013	38.338	36.679	37.627	**681.594**
	Total x cultivo	**196.650**	**662.247**	**420.456**	**1.275.451**	**563.477**	**1.873.741**	**1.668.826**	**316.524**	**434.719**	**12.523.573**
Emisiones x Tn Soja (Kgs CO_{2eq}/TN)	Residuos de cosecha	73	76	83	79	88	75	76	74	75	**77**
	Fertilizantes	-	-	1	33	-	-	-	-	-	**4**
	Combustibles y Lubricantes	32	24	38	27	40	23	23	34	37	**27**
	Produccion Fertilizantes	-	1	0	14	-	1	3	-	1	**3**
	Produccion Combustibles y Lubricantes	3	2	4	3	4	2	2	3	4	**3**
	Estructura campo+Lotes sin produccion	25	6	12	5	10	2	2	15	11	**7**
	Total x Tonelada	**133**	**109**	**137**	**160**	**142**	**103**	**106**	**125**	**128**	**120**

[1] Asignado por Tonelada producida

2 – Feedstock transport module

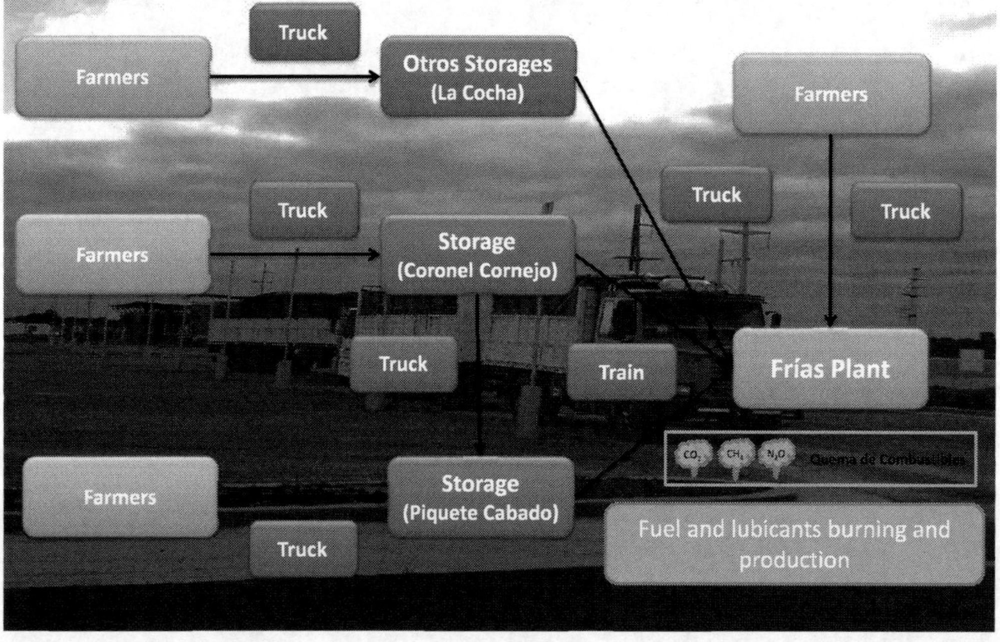

2.1 – Transport by Truck

PROFIT® system source

DATA SOURCE
TRIPS
Kilómeters per trip

"Approved consolidated baseline and monitoring methodology ACM0017 "Production of biodiesel for use as fuel" - v.01.1 - UNFCCC - CDM Executive Board".

Emisiones por km recorrido		Unidades	Ecuacion	Valor
	Consumo específico de Gas-Oil	Lt/ 100 Km	Tabla 1.39 - IPCC 1996 - Heavy Duty	29,90
	Consumo ajustado por IDA y Vuelta	Lt/ 100 Km	Según ACM0017 / Version 01.1	59,80
FECO2 LTS	Factor de emision de CO2	KgsCO2/Lts	Ver Hoja Factores de emision Incluye LCA	2,92
CO2	Emisiones CO2 por Transporte por Km	KgsCO2/Km	Ajustado por ida y vuelta	1,74
FEN2O LTS	Factor de emision de N2O	mg N2O/Km	IPCC 2006 - Cuadro 3.2.5 - Pre-Euro Diesel - Autobus - Rural	30,00
N2O	Emisiones N2O por Gas-Oil Transporte	KgN2O/Km	Ajustado por ida y vuelta	0,00
FECH4 Lts	Factor de emision de CH4	mg CH4/ km	IPCC 2006 - Cuadro 3.2.5 - Pre-Euro Diesel - Autobus - Rural >	80,00
CH4	Emisiones CH4 por Gas-Oil Transporte	KgCH4/km	Ajustado por ida y vuelta	0,00
$FE_{CO2eq\ Unidad}$	Factor de emision x KM recorrido	$KgsCO_{2eq}$ /Km	FE total x Km	1,76

2.2 – Transport by Train

PROFIT® system source

DATA SOURCE
Tons
Wagons
Wagons per train

"Approved consolidated baseline and monitoring methodology ACM0017 "Production of biodiesel for use as fuel" - v.01.1 - UNFCCC - CDM Executive Board".

Emisiones por km recorrido		Unidades	Ecuacion	Valor
	Consumo específico de Gas-Oil	Lt/ Km	Locomotora General Motors GT 22 CW [1]	3,40
	Consumo ajustado por IDA y Vuelta	Lt/ Km	Según ACM0017 / Version 01.1	6,80
FECO2 LTS	Factor de emision de CO2	KgsCO2/Lts	Ver Hoja Factores de emision Incluye LCA	2,92
CO2	Emisiones CO2 por Transporte por Km	KgsCO2/Km	Ajustado por ida y vuelta	19,82
FEN2O LTS	Factor de emision de N2O	mg N2O/Km	No hay dato	-
N2O	Emisiones N2O por Gas-Oil Transporte	KgN2O/Km	Ajustado por ida y vuelta	-
FECH4 Lts	Factor de emision de CH4	mg CH4/ km	No hay dato	-
CH4	Emisiones CH4 por Gas-Oil Transporte	KgCH4/km	Ajustado por ida y vuelta	-
$FE_{CO2eq\ Unidad}$	Factor de emision x KM recorrido	$KgsCO_{2eq}$ /Km	FE total x Km	19,82

[1] Fuente: www.forotransportes.com/showthread.php?t=4357

Transport summary MP

RESUMEN EMISIONES DE GASES DE EFECTO INVERNADERO
FLETES MATERIAS PRIMAS - SOJA
2010-2011

Desde:		Recibido en:	Frias	Alimenta	Cereales del Sur	Otros	Total
Productores	Toneladas	Tn	194.618	14.106	4.846	684	**214.254**
	Distancia Prom.	Km/Viaje	270	30	190	30	**252**
	Emisiones	KgsCO2Eq	3.153.121	23.548	57.713	1.249	**3.235.631**
	Emisiones x Tn	KgsCO2Eq/Tn	**16,2**	**1,7**	**11,9**	**1,8**	**15,1**
Interplantas (Camion)	Toneladas	Tn	-	-	-	2.492	**2.492**
	Distancia Prom.	Km/Viaje	-	-	-	129	**129**
	Emisiones	KgsCO2Eq	-	-	-	19.067	**19.067**
	Emisiones x Tn	KgsCO2Eq/Tn	-	-	-	**7,7**	**7,7**
Interplantas (Tren)	Toneladas	Tn	4.070	-	-	-	**4.070**
	Distancia Prom.	Km/Viaje	555	-	-	-	**555**
	Emisiones	KgsCO2Eq	78.717	-	-	-	**78.717**
	Emisiones x Tn	KgsCO2Eq/Tn	**19,3**	-	-	-	**19,3**
	Total emisiones	KgsCO2Eq	**3.231.838**	**23.548**	**57.713**	**20.317**	**3.333.415**

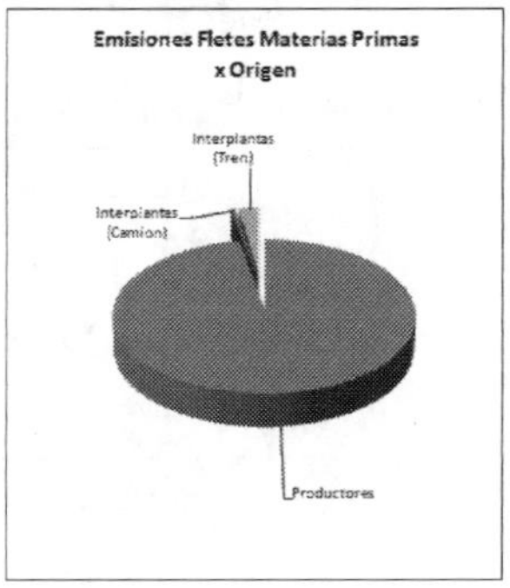

Resumen Ingresado a Frias [1]	Toneladas recibidas en Frias	Tn	198.688
	Total Emisiones por Fletes	KgsCO2Eq/Tn	3.333.415
	Promedio Emisiones x Tn	KgsCO2Eq/Tn	**16,8**

[1] Para utilizar en el calculo de emisiones del biodiesel se toman el total de las emisiones por flete pero tomando en cuenta las Tn recibidas en frias sin tener en cuenta la diferencia entre lo recibido de productores y lo ingresado efectivam

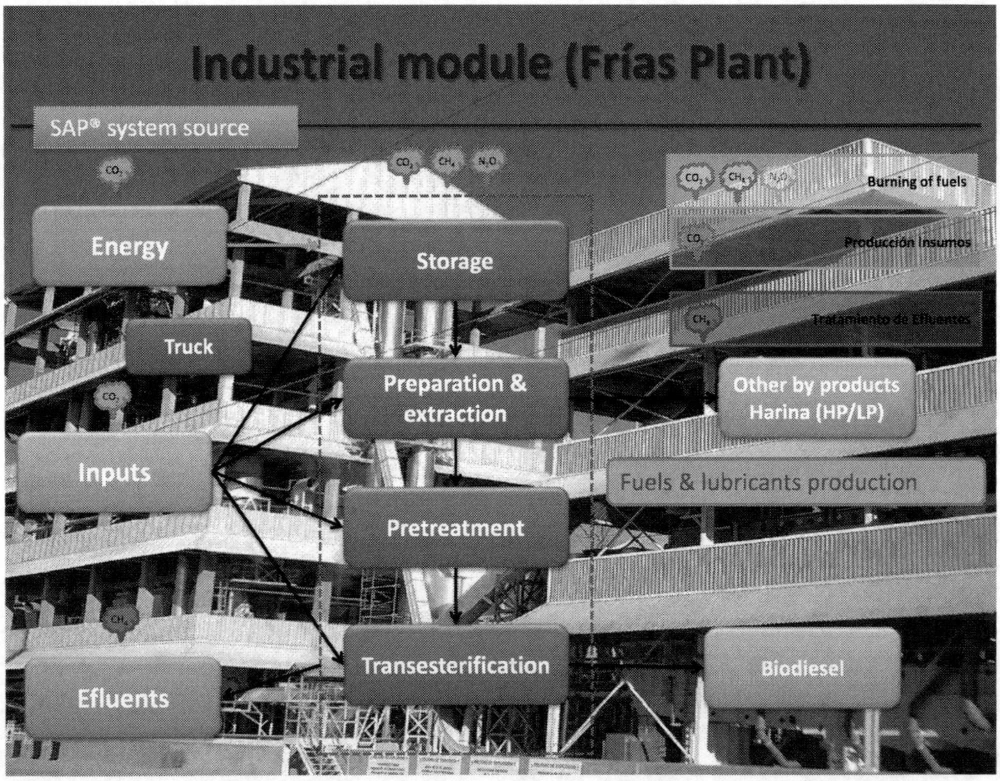

3. Transformation emission estimation (e_p)

Art. 11: The accounting of tranformation emitions e_p, shall include the emitions produced during the proper transformation of the feedstock, the residues and the emission produced by the manufacture of the different input materials

- CO_2 from the use of fossile fuels.
- Freigh of input fuels (CO_2 from the use of fossile fuels).
- Emissions related with the life cycle of input materials.
- CH_4 coming from the efluent treatment

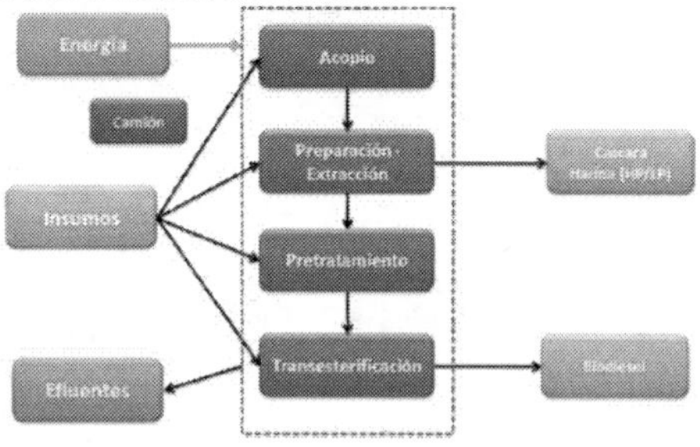

3.1 - Energy

EMPLOYED DATA
Consumption
Type of fuel
Sector distribution

Source of emissions associated with fuel consumption

Tipo de Combustible/Energético				Gas-Oil	Nafta	Lubricantes	G.L.P.	Gas Natural	Leña	Energia Electrica	Fuel-Oil
Variable	Descripcion	Unidades	Fuente	Lts	Lts	Lts	Kg	M³	Kg	KwH	Kg
PCI	Poder Calorifico Inferior	Kcal/unidad	Pagina 197 - Tabla 3.1-14. Factores de Emisión de CO2	8.619	7.607	8.497	10.980	8.300	2.300		9.800
D	Densidad	Kg/unidad	Pagina 197 - Tabla 3.1-16. Factores de Emisión de CO2	0,8450	0,7330	0,8850	0,5370	0,7190			0,9450
Frac Ox	Fraccion de Carbono Oxidado	%	Modulo Energia - Hoja 1/5 - Método de Referencia	0,990	0,990	0,990	0,990	0,995	0,870		0,990
C_c	Contenido de Carbono	tC/TJ	Pagina 197 - Tabla 3.1-16. Factores de Emisión de CO2	20,28	18,90	20,00	17,20	15,31	29,90		21,25
$FE_{CO2\ Kcal}$	Factor de emision de CO_2	$KgCO_2$/Kcal	FE_{CO2} = C_c*Frac Ox*44/12	0,0003022	0,0002872	0,0003040	0,0002614	0,0002339	0,0003993		0,0003230
$FE_{CO2\ unidad}$	Factor de emision de CO_2	$KgCO_2$/Unidad	FE x Unidad de consumo	2,66	2,19	2,58	2,87	1,94	0,92	0,351	3,17
FE_{N2O}	Factor de emision de N_2O	KgN_2O/TJ	Modulo Energia - Hoja 1-3	2,0	1,00	1,0	0,6	0,325	4,00		0,3
$FE_{N2O\ Kcal}$	Factor de emision de N_2O	KgN_2O/Kcal	Cambio de Unidades	0,000	0,000	0,000	0,000	0,000	0,000		0,000
$FE_{N2O\ unidad}$	Factor de emision de N_2O	$KgCO_2$/Unidad		0,000	0,000	0,000	0,000	0,000	0,000		0,000
FE_{CH4}	Factor de emision de CH_4	$KgCH_4$/TJ	Modulo Energia - Hoja 1-4	11,00	20,00	0,60	1,10	3,125	30,00		2,00
$FE_{CH4\ Kcal}$	Factor de emision de CH_4	$KgCH_4$/Kcal	Cambio de Unidades	0,000	0,000	0,000	0,000	0,000	0,000		0,000
$FE_{CH4\ unidad}$	Factor de emision de CH_4	$KgCO_2$/Unidad		0,008	0,001	0,000	0,000	0,000	0,000		0,000
$FE_{CO2eq\ unidad}$	Factor de emision de CO_{2eq}	$KgCO_{2eq}$/Unidad	FE total x Unidad	2,68	2,21	2,59	2,87	1,95	0,94	0,351	3,17
$FE_{CO2eq\ Kcal}$	Factor de emision de CO_{2eq}	$KgCO_{2eq}$/Kcal	FE total x Kcal	0,0003118	0,0002903	0,0003053	0,0002623	0,0002348	0,0004072		0,0003236
Emisiones LCA											
	Emisiones Extracción	$KgCO_{2eq}$/Kg producto	Metodologia MDL: ACM0017 / Version 01.1	0,073	0,073						
	Emisiones Refinado	$KgCO_{2eq}$/Kg producto	Metodologia MDL: ACM0017 / Version 01.1	0,233	0,233						
	Emisiones totales por Kg	$KgCO_{2eq}$/Kg producto		0,306	0,306						
$FE_{CO2eq\ unidad}$	Factor de emision de CO_{2eq}	$KgCO_{2eq}$/Unidad	FE total x Unidad	0,259	0,225	0,259	0,287	0,195			0,317
$FE_{CO2eq\ Kcal}$	Factor de emision de CO_{2eq}	$KgCO_{2eq}$/Kcal	FE total x Kcal	0,0000300	0,0000296	0,0000305	0,0000262	0,0000235			0,0000324
$FEtotal_{CO2eq\ Kcal}$	Factor de emision de CO_{2eq}	$FE_{CO2eq\ Kcal}$ + LCA	FE total x Unidad	2,946	2,434	2,853	3,162	2,144	0,937	0,351	3,488

3.2 – Input material transport
3.3 – Input material production

EMPLOYED DATA
Cuantity
Truck weight
Distance

Similar model to the one used for raw materials per truck

EMPLOYED DATA
Quantity of methanol

"Approved consolidated baseline and monitoring methodology ACM0017 "Production of biodiesel for use as fuel" - v.01.1 - UNFCCC - CDM Executive Board".

ID Number:	5
Parameter:	EF_{MeOH_PC}
Data unit:	tCO_2/t methanol
Description:	Specific emission per tonne of produced methanol
Source of data:	Apple 1998: <http://edj.net/sinor/SFR4-99art7.html> and 2006 IPCC Guidelines.
Measurement procedures (if any):	1.95 tCO_2/tonne produced methanol
Any comment:	Based on 30 GJ/tonne energy requirement and average of IPCC emissions factors for natural gas and diesel oil

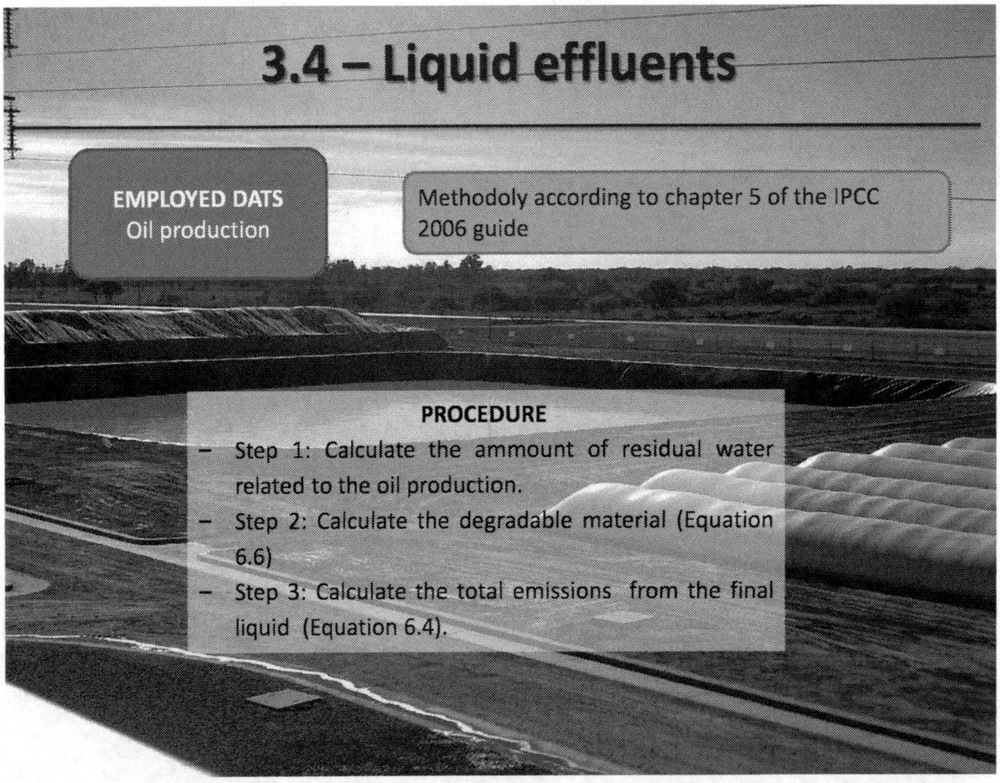

3.4 – Liquid effluents

EMPLOYED DATS
Oil production

Methodoly according to chapter 5 of the IPCC 2006 guide

PROCEDURE

- Step 1: Calculate the ammount of residual water related to the oil production.
- Step 2: Calculate the degradable material (Equation 6.6)
- Step 3: Calculate the total emissions from the final liquid (Equation 6.4).

Apropiation of the emitions Co-Productos

Mass balance: Emitions are appropiate according to real yields and mass balance (% weight) in each step.

Energy content: According to the European Union Directive
Where a fuel production process produces, in combination, the fuel for which emissions are being calculated and one or more other products (co-products), greenhouse gas emissions shall be divided between the fuel or its intermediate product and the co-products in proportion to their energy content (determined by lower heating value in the case of co-products other than electricity). ". Annex V – Point 17.

Market price: According to EB 50 – the executive board of the Clean Development Mechanism, for assigning of co-products.This methodology is being used for projects that generate cetrtiified emition green bonus.

Industry Emissions Summary

RESUMEN EMISIONES DE GASES DE EFECTO INVERNADERO
Planta FRIAS - Santiago del Estero
2.011

Emisiones por Etapa (kgsCO2eq)	Energía	Transporte insumos	Produccion Insumos	Efluentes	Total
Acopio	813.370	2	10	-	**813.382**
Preparacion - Extraccion	10.652.179	18.362	9.374	368.322	**11.048.237**
Pretratamiento	452.970	27.124	259	-	**480.354**
Transesterificacion	1.604.535	427.394	6.659.167	-	**8.691.096**
Total	**13.523.053**	**472.884**	**6.668.810**	**368.322**	**21.033.069**

			Emisiones por tonelada (KgsCO2eq/Tn)		
Etapa	Produccion	Rendimiento	x Balance de masas	x Contenido Energético	x Precio Mercado
Acopio					
Ingreso Total	170.139				
Soja (Seca+Humeda)	167.460	98%	5	5	5
Preparacion - Extraccion					
Ingreso Soja	157.169				
Harina (HP/LP)	113.318	72%	73	62	46
Cascara	8.029	5%	73	55	26
Aceite crudo	30.346	19%	73	118	185
Pretratamiento					
Aceite (crudo/desgomado)	30.387				
Aceite tratado	28.966	95%	17	17	17
Transesterificacion					
Ingreso Aceite	28.728				
Biodiesel	28.873	101%	301	301	301

Resumen Emisiones Industria por Tonelada de Biodiesel			
Acopio	26	26	26
Preparacion - Extraccion	76	123	193
Pretratamiento	17	17	17
Transesterificacion	301	301	301
Total (KgsCO2eq/Tn)	**420**	**467**	**536**

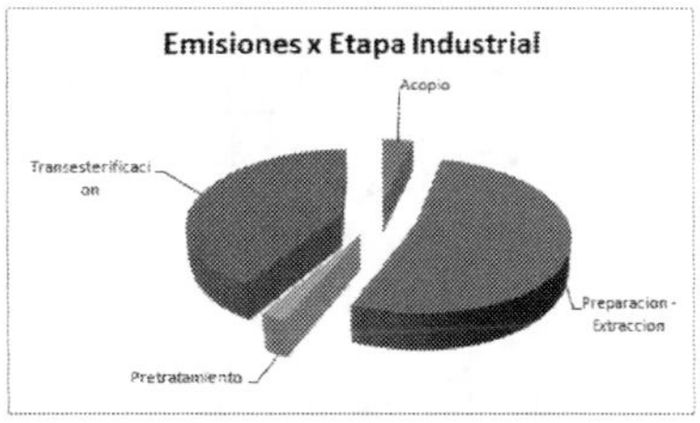

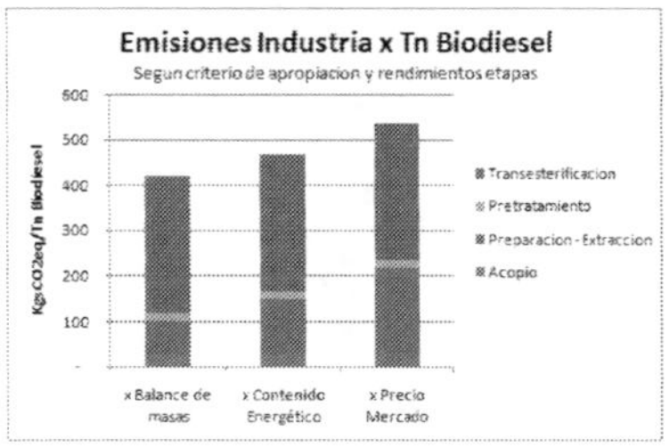

4. Oversea transport San Lorenzo - Rotterdam

Transporte terrrestre Frias-Puerto (Camión)

Distancia a Puerto	Km	750	Distancia a Puerto San Lorenzo
Aforo x Viaje	Tn/viaje	28,71	Promedio Octubre-Diciembre 2010
Emisiones a Puerto	KgsCO2eq /Tn	46,10	

Transporte martítimo San Lorenzo-Rotterdam (Buque)

Distancia a Puerto	Km	11.357	Distancia lineal corregida por un factor del 42%
Factor de emision x TN Km	$KgsCO_{2eq}$ /Tn Km	0,00418	Buque HANDY MAX (40.000 Tn Año 1980) - Bilan Carbone - V 5.0.
Emisiones a Puerto Destino	KgsCO2eq /Tn	94,94	Se considera el viaje ida y vuelta
Total Transporte a Europa	KgsCO2eq /Tn	141,04	

Summary per Ton

INTA

Resumen Emisiones Produccion de Biodiesel

de acuerdo a Directiva Europea de Biocombustibles - EU 2009/28/CE
2010-2011

AG-ENERGY VILUCO

	Emisiones por tonelada (KgsCO2eq/Tn Biodiesel)	x Balance de masas	x Contenido Energético	x Precio Mercado	
e_{ec}	Extracción o del cultivo de las materias primas	150	243	381	Promedio de los campos propios. Incluye Flete a Planta y Rendimiento Industria de acuerdo a Directiva Europea de Biocombustibles - EU 2009/28/CE. Incluye produccion de insumos.
e_p	Transformación (Industria)	420	467	536	Incluye produccion y transporte de insumos.
e_{td}	Transporte y distribución (Estimado)	141	141	141	Estimado Flete camión hasta San Lorenzo y Barco a Rotterdam.
E_B	**Emisiones procedentes de la produccion y uso de biodiesel (Kgs CO2eq/TN)**	**711**	**851**	**1.059**	

	Emisiones por tonelada (gCO2eq/Mj)	x Balance de masas	x Contenido Energético	x Precio Mercado	Directiva Europea de Biocombustibles - EU 2009/28/CE Valores Tipicos	Directiva Europea de Biocombustibles - EU 2009/28/CE Valores por Default	
e_{ec}	Etapa Agrícola	4	7	10	19	19	Valores directiva: Anexo V - Articulo D
e_p	Industria	11	13	14	18	26	Valores directiva: Anexo V - Articulo D
e_{td}	Transporte y distribución (Estimado)	4	4	4	13	13	Valores directiva: Anexo V - Articulo D
E_B	**Emisiones procedentes de la produccion y uso de biodiesel (g CO2eq/Mj)**	**19**	**23**	**29**	**50**	**58**	
E_F	Emisiones	83,8	83,8	83,8	83,8	83,8	Directiva Europea - Anexo V - Art. 19
RED	**Reduccion =$(E_F-E_B)/E_F$**	**77%**	**73%**	**66%**	**40%**	**31%**	

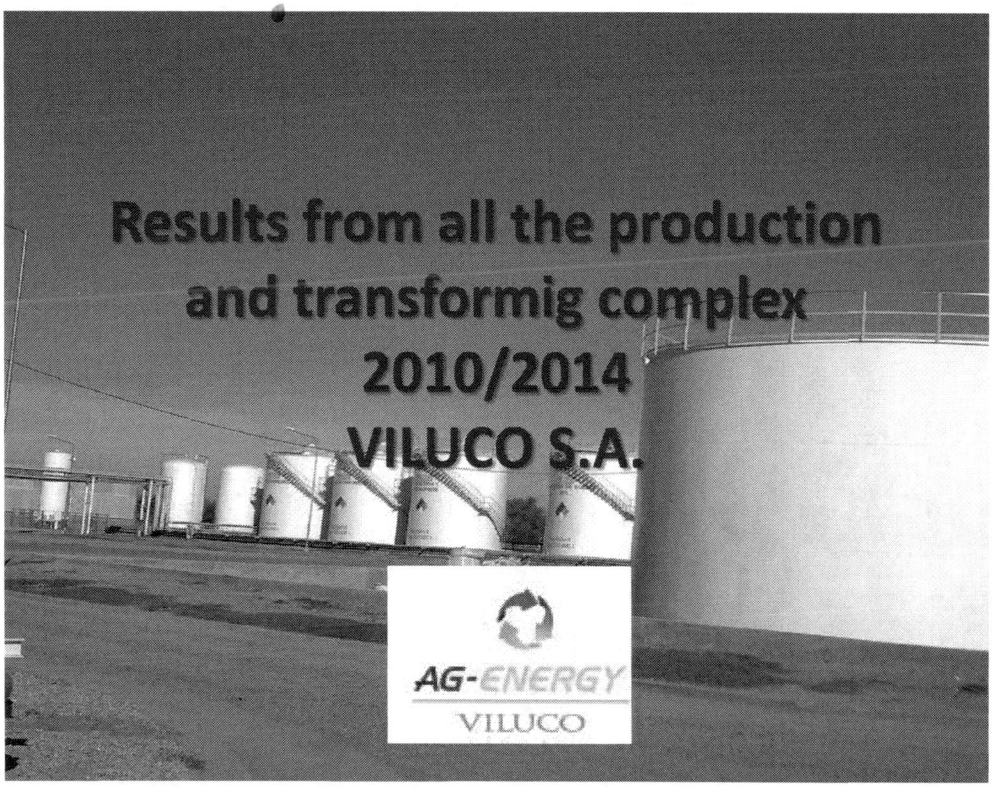

Interanual variation

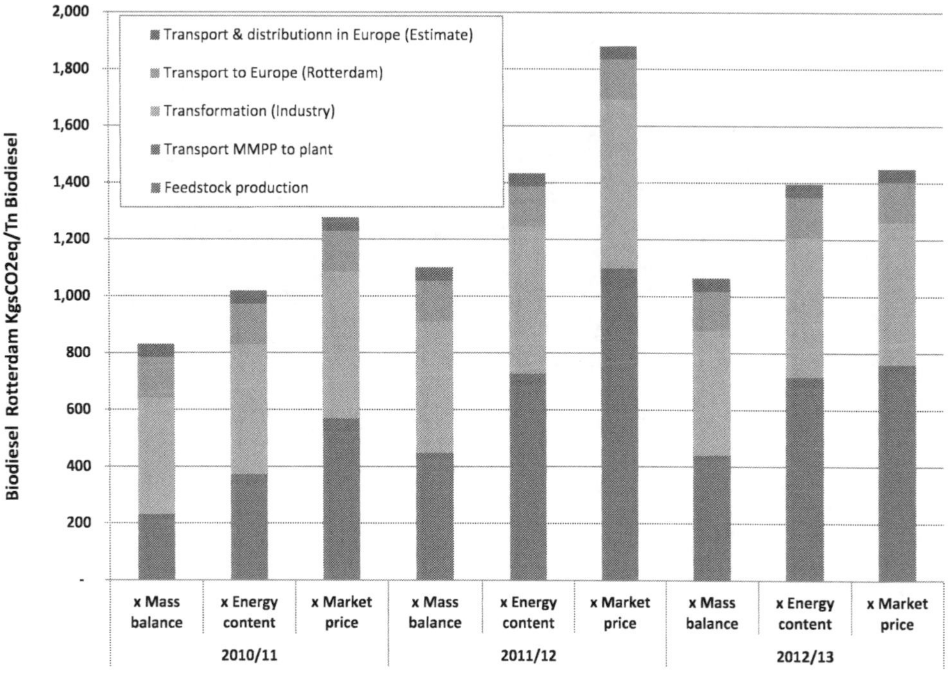

Yield effect over relative contribution

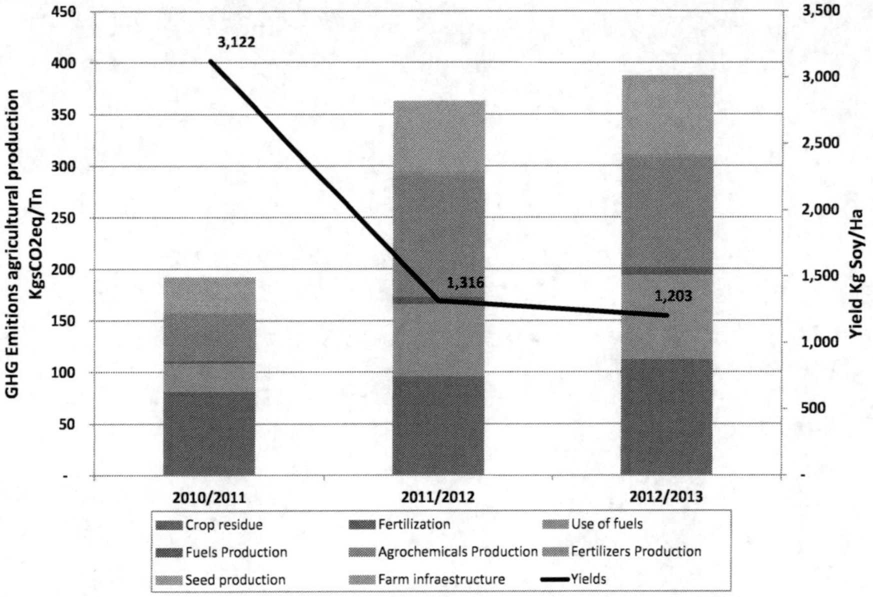

GHG REDUCTIONS

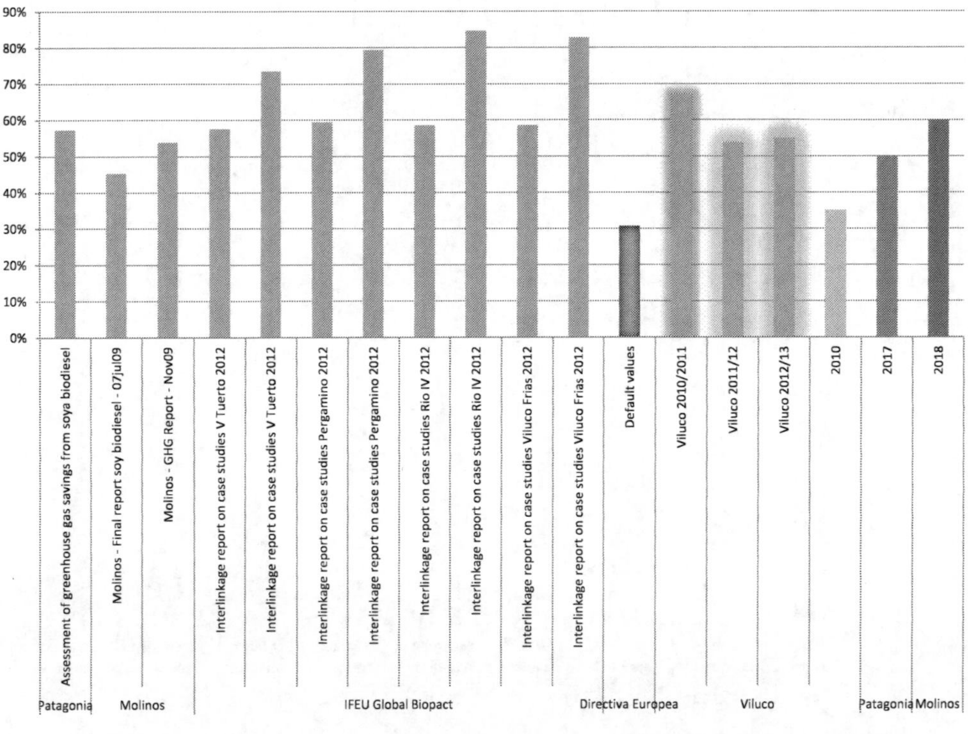

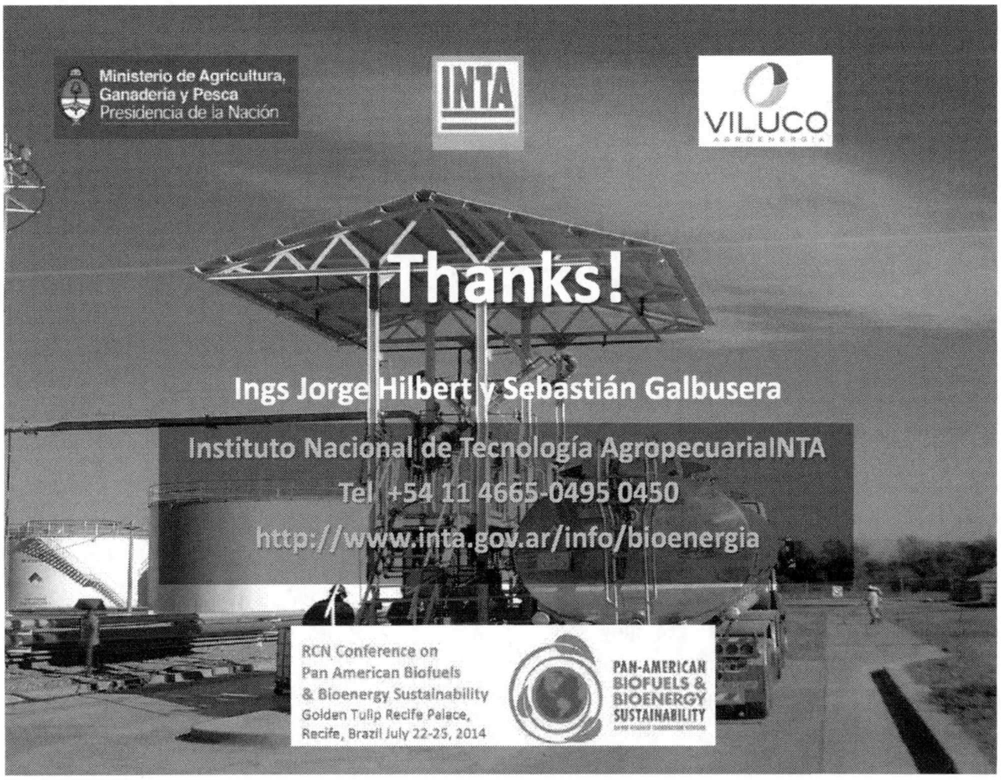
Ministerio de Agricultura, Ganadería y Pesca
Presidencia de la Nación
INTA
VILUCO
AGROENERGIA
Thanks!
Ings Jorge Hilbert y Sebastián Galbusera
Instituto Nacional de Tecnología AgropecuariaINTA
Tel +54 11 4665-0495 0450
http://www.inta.gov.ar/info/bioenergia
RCN Conference on
Pan American Biofuels
& Bioenergy Sustainability
Golden Tulip Recife Palace,
Recife, Brazil July 22-25, 2014
PAN-AMERICAN
BIOFUELS &
BIOENERGY
SUSTAINABILITY

Incorporating Bioenergy into Sustainable Landscape Designs

***Virginia H. Dale*[1,2], *Keith L. Kline*[2], *Jennifer Costanza*[3], *C. Tat Smith*[4], *Inge Stupak*[5], *Tim Volk*[6], *Arnaldo Cesar da Silva Walter*[7] and Camila Ortolan Fernandes de Oliveira[7]**

[1]University of Tennessee, Knoxville, TN, USA
[2]Oak Ridge National Laboratory, Oak Ridge, TN, USA
[3]North Carolina State University, Raleigh, NC, USA
[4]University of Toronto, Toronto, Canada
[5]Univeristy of Copenhagen, Copenhagen, Denmark
[6]State University of New York, USA
[7]UNICAMP/University of Campinas, Campinas, Brazil

July 2014 for RCN meeting

Center for BioEnergy Sustainability
Oak Ridge National Laboratory
http://www.ornl.gov/sci/ees/cbes/

U.S. DEPARTMENT OF ENERGY

OAK RIDGE NATIONAL LABORATORY
MANAGED BY UT-BATTELLE FOR THE DEPARTMENT OF ENERGY

Consider bioenergy within system as an opportunity to design landscapes that add value

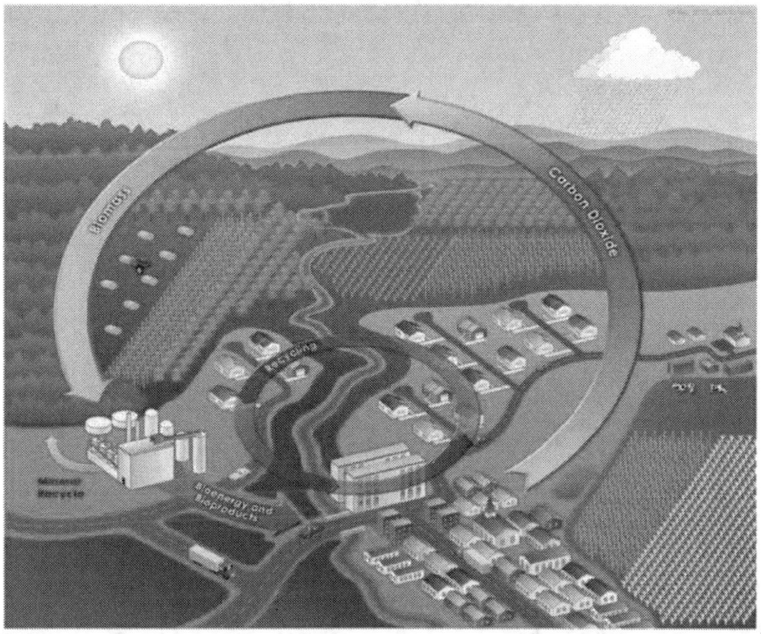

3 Managed by UT-Battelle for the U.S. Department of Energy

Landscape design is a plan for resource allocation.

- Suggest a way to manage for more sustainable provisions of bioenergy and other services
- Takes context, trends and current conditions into consideration

4 Managed by UT-Battelle for the U.S. Department of Energy

Negative impacts of bioenergy can be avoided or reduced by attention to three principles:

1. Conserve priority ecosystem and social services
2. Consider local context
3. Monitor effects of concern and adjust plans to improve performance over time

OAK RIDGE National Laboratory

Landscape design approach for bioenergy should be applied to particular contexts

- Set goals
  - Involve key stakeholders
  - Develop consensus approach
- Consider constraints
- Address wastes and other opportunities
- Evaluate and apply solutions
- Monitor for adaptive management

OAK RIDGE National Laboratory

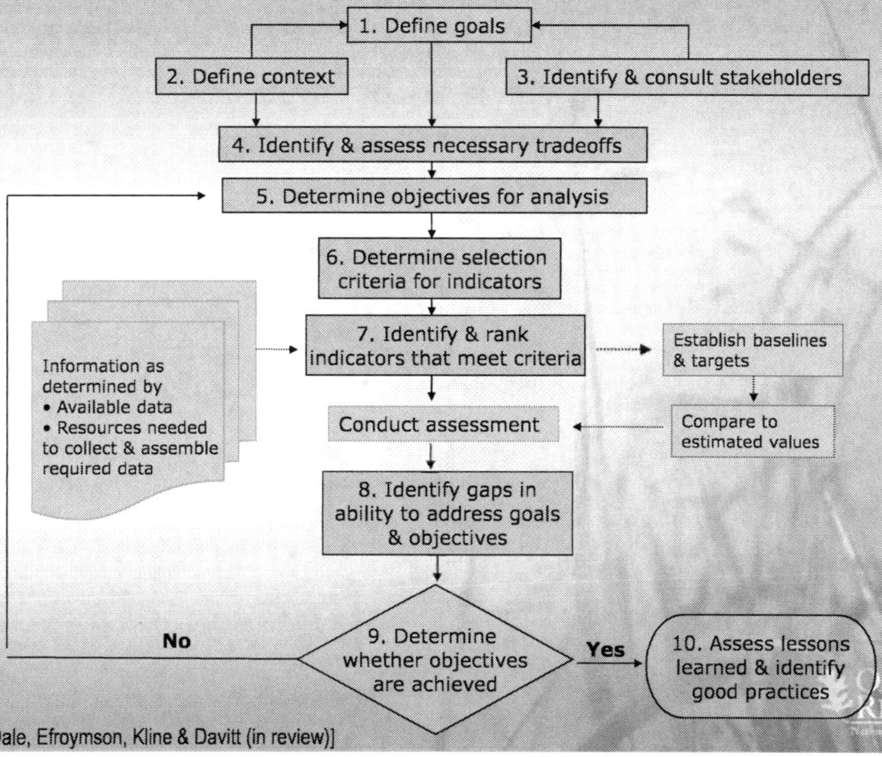

[Dale, Efroymson, Kline & Davitt (in review)]

Pressures and incentives for landscape design

- Legal demands or regulations
- Customer requirements or specifications
- Stakeholder concerns
- Competitive advantage
- Environmental and social pressure groups
- Reputation loss

[Building from Seuring and Muller (2008) Journal of Cleaner Production 16:1699-1710]

Obstacles to developing and deploying landscape design

- Landowner rights
- Traditional practices
- Up front planning required
- Coordination complexity/effort
- Higher initial costs

[Building from Seuring and Muller (2008) Journal of Cleaner Production 16:1699-1710]

OAK RIDGE National Laboratory

Recommended practices

- Stakeholder engagement throughout process
- Consider management options within the broader context
- Attention to site selection and environmental effects in the
  - location and selection of the feedstock
  - transport of feedstock to the refinery
  - refinery processing
  - final transport and dissemination of bioenergy.
- Monitoring and reporting of key measures of sustainability
- Attention to what is "doable"

OAK RIDGE National Laboratory

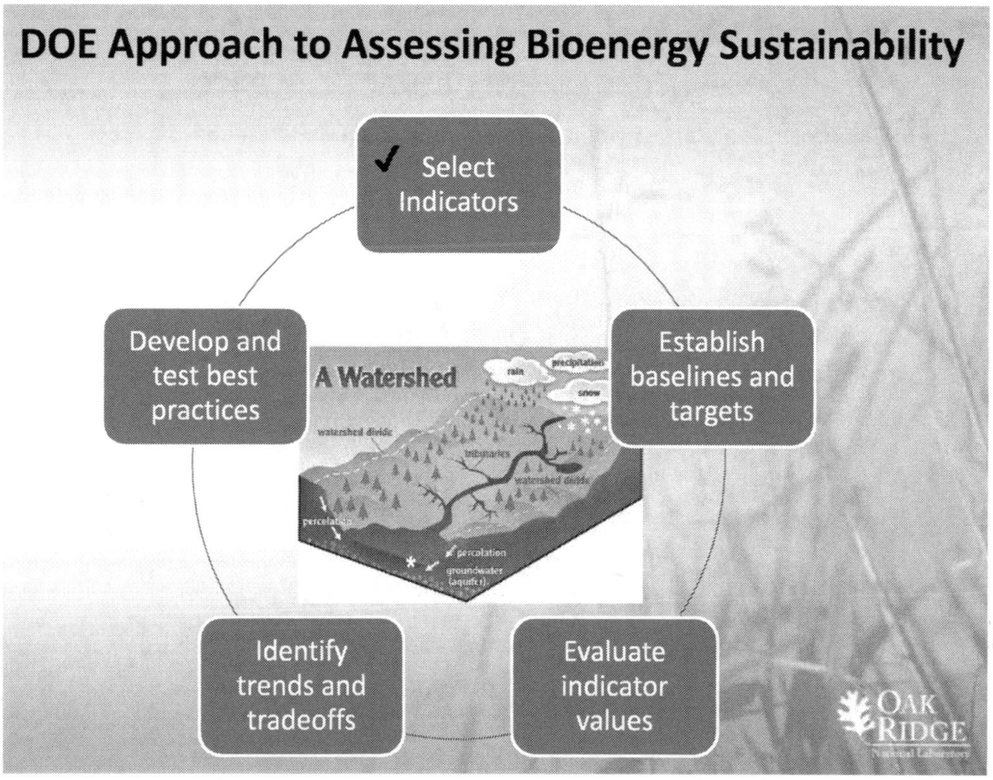
DOE Approach to Assessing Bioenergy Sustainability
✓ Select Indicators
Establish baselines and targets
Evaluate indicator values
Identify trends and tradeoffs
Develop and test best practices
A Watershed
precipitation
rain
snow
watershed divide
tributaries
percolation
groundwater (aquifer)
OAK RIDGE National Laboratory

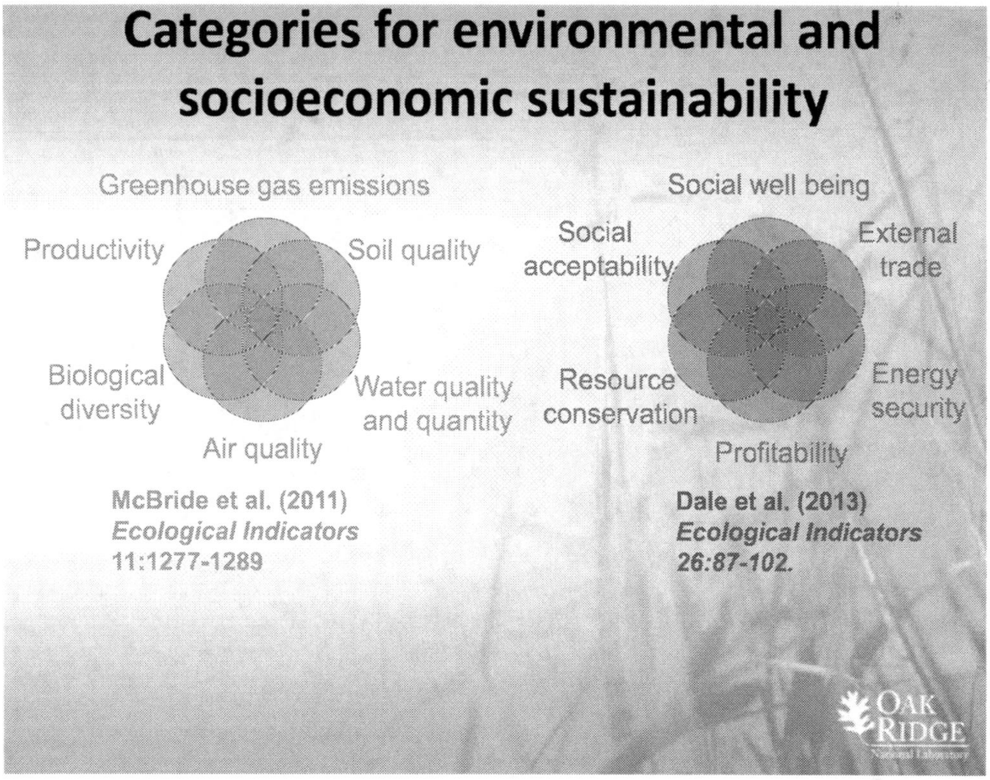
Categories for environmental and socioeconomic sustainability
Greenhouse gas emissions
Productivity
Soil quality
Biological diversity
Water quality and quantity
Air quality
McBride et al. (2011) Ecological Indicators 11:1277-1289
Social well being
Social acceptability
External trade
Resource conservation
Energy security
Profitability
Dale et al. (2013) Ecological Indicators 26:87-102.
OAK RIDGE National Laboratory

Landscape Design Involves Adapting Indicators to Particular Contexts

- **Indicator set is a starting point for sake of efficiency and standardization**
  - Particular systems may require addition of other indicators
  - Budget may require subtraction of some indicators
  - Some indicators more important for different supply chain steps
- **Protocols must be context-specific**

13 Managed by UT-Battelle for the U.S. Department of Energy

Efroymson et al. (2013) *Environmental Management* 51:291-306.

Landscape design for growing switchgrass in east Tennessee (USA)

An optimization model identified "ideal" locations for planting switchgrass for bioenergy in east Tennessee

Spatial optimization model

- **Considers**
  - **Farm profit**
  - **Water quality constraints**
- **Finds**
  - **"Business as usual" (profit only) compromises water quality**
  - **"Balanced" scenario offers farmer good price while enhancing water quality**

14 Managed by UT-Battelle for the U.S. Department of Energy

[Parish et al., *Biofuels, Bioprod. Bioref.* 6,58–72 (2012)]

Data available from Vonore for most indicators of socioeconomic sustainability

* Information not available from Vonore
\# not an issue for switchgrass

Category	Indicator	Units
Social well-being	Employment	Number of full time equivalent (FTE) jobs
	Household income	Dollars per day
	Work days lost due to injury	Average number of work days lost per worker per year
	Food security #	Percent change in food price volatility
Energy security*	Energy security premium	Dollars /gallon biofuel
	Fuel price volatility	Standard deviation of monthly percentage price changes over one year
External trade	Terms of trade	Ratio (price of exports/price of imports)
	Trade volume	Dollars (net exports or balance of payments)
Profitability	Return on investment (ROI)	Percent (net investment/ initial investment)
	Net present value (NPV)[2]	Dollars (present value of benefits minus present value of costs)

Category	Indicator	Units
Resource conservation	Depletion of non-renewable energy resources	MT (amount of petroleum extracted per year)
	Fossil Energy Return on Investment (fossil EROI)	MJ (ratio of amount of fossil energy inputs to amount of useful energy outputt
Social acceptability	Public opinion	Percent favorable opinion
	Transparency	Percent of indicators for which timely and relevant performance data are reported
	Effective stakeholder participation	Number of documented responses to stakeholder concerns and suggestions reported on an annual basis
	Risk of catastrophe	Annual probability of catastrophic event

Dale et al. (2013) *Ecological Indicators* 26:87-102.

Data available from Vonore for all indicators of environmental sustainability

Environment	Indicator	Units
Soil quality	1. Total organic carbon (TOC)	Mg/ha
	2. Total nitrogen (N)	Mg/ha
	3. Extractable phosphorus (P)	Mg/ha
	4. Bulk density	g/cm^3
Water quality and quantity	5. Nitrate concentration in streams (and export)	concentration: mg/L; export: kg/ha/yr
	6. Total phosphorus (P) concentration in streams (and export)	concentration: mg/L; export: kg/ha/yr
	7. Suspended sediment concentration in streams (and export)	concentration: mg/L; export: kg/ha/yr
	8. Herbicide concentration in streams (and export)	concentration: mg/L; export: kg/ha/yr
	9. storm flow	L/s
	10. Minimum base flow	L/s
	11. Consumptive water use (incorporates base flow)	feedstock production: m^3/ha/day; biorefinery: m^3/day

Environment	Indicator	Units
Greenhouse gases	12. CO_2 equivalent emissions (CO_2 and N_2O)	kgC_{eq}/GJ
Biodiversity	13. Presence of taxa of special concern	Presence
	14. Habitat area of taxa of special concern	ha
Air quality	15. Tropospheric ozone	ppb
	16. Carbon monoxide	ppm
	17. Total particulate matter less than 2.5μm diameter ($PM_{2.5}$)	μg/m^3
	18. Total particulate matter less than 10μm diameter (PM_{10})	μg/m^3
Productivity	19. Aboveground net primary productivity (ANPP) / Yield	gC/m^2/year

McBride et al. (2011) *Ecological Indicators* 11:1277-1289

While switchgrass has great sustainability benefits in east Tennessee, the low cost of natural gas translates to little demand for ethanol

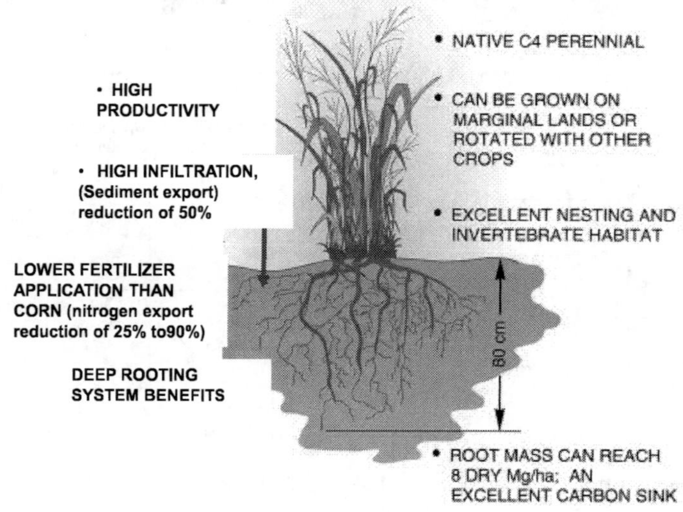

Dale et al. (2011) *Ecological Applications* 21(4):1039-1054.

Case Study of Eucalyptus in Brazil

Arnaldo Walter and Camila de Oliveira

- Legal and regulatory framework
  - Land use regulated by Forestry Code (amended in 2012).
  - "Permanent Preservation Areas" & "Legal Reserve Areas" defined
  - Identified appropriate areas for specific uses (e.g., eucalyptus and pines)
  - Foster good practices to reduce environmental impacts
- Institutional framework
  - Forestry Science and Research Institute (IPEF) calls for
    - "Landscape sustainable practices"
    - "Use of degraded areas".
  - Annual reports on Forestry Management by the industries highlight
    - Improving yield
    - Preserving water resources
    - Reducing & monitoring impacts on biodiversity
    - Adopting social programs
    - Reducing fragmentations
- Design
  - Integrating livestock into plantations
  - Integrating soy into planted forests.
  - Preserving natural vegetation
- Challenge in using pellets: logistics

Remediation Case Study: New York

Tim Volk

- Community Drivers: use of former industrial land and provision of renewable energy
  - Growing shrub willows on settling basins as alternative to standard geomembrane cap
  - Environmental monitoring willow fields for soils and water quality
  - Starting assessment of social factors in driving biomass use in the region
- Multifunctional systems
  - Sustainable Reuse Remedy
    - ✓ Use organic waste stream from local brewery to create favorable growing conditions
    - ✓ Manage water to minimize leaching to surface and ground water
    - ✓ Produce biomass
  - Shrub willow in highway rights of way for snow drift control and potential biomass production
  - Willow incorporated into riparian buffers
  - Potential for recreation uses

19 Managed by UT-Battelle for the U.S. Department of Energy

SE woody biomass case: By–products of tree harvest for saw timber and pulp used for bioenergy

20 Managed by UT-Battelle for the U.S. Department of Energy

Universiteit Utrecht

Global wood pellet production 2000 - 2010

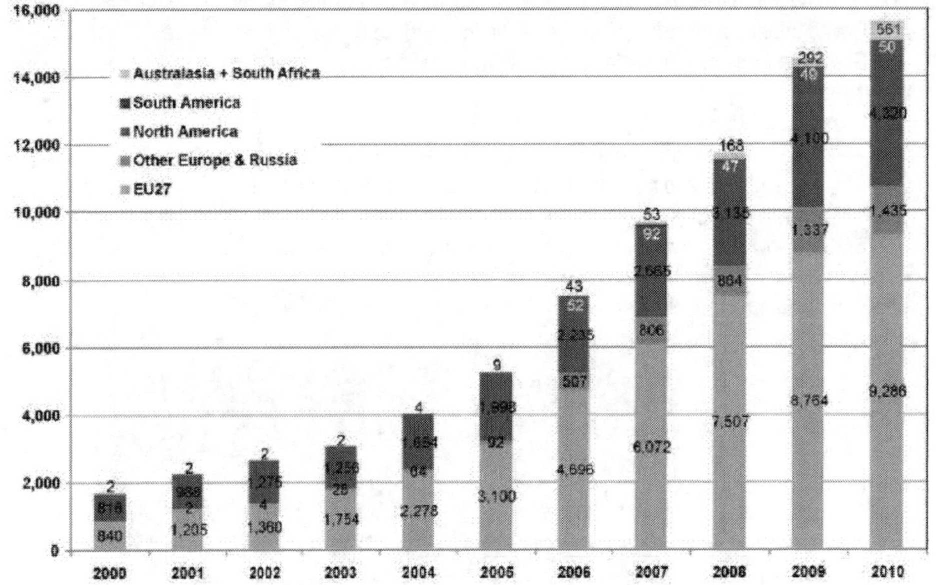

(Source: Lamers et al. RSER, 16(2012) 3176-3199

Trees are cut and sorted by size.

- Harvest meets standards of Forest Sustainability Initiative (FSI)
- Branches returned to forest
- Protection of places providing unique ecosystem services
- Largest trees shipped to China

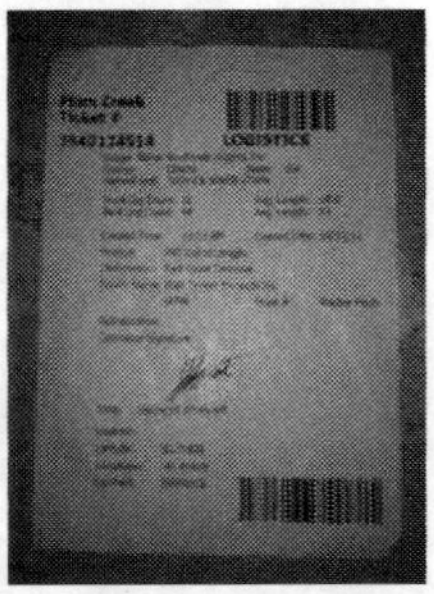

OAK RIDGE

IEA Task 43: Biomass feedstocks for energy markets

- **Overall approach**
  - Empirical case studies dealing with environmental, economic and social changes over time*
  - Looking for where can methodology* be coordinated to improve consistency and comparability among the individual case studies (to the extent it is possible and useful)
  - Policy messages: Barriers and opportunities to overcome them
- **Case studies**
  - Mobilization of forest* bioenergy supply chains in boreal and temperate forests (Canada, US and N Europe & Australia)
  - Mobilizing agricultural residues for bioenergy and biorefinieries
  - Regional biogas production from organic residues
  - Cultivation of grasslands and pastures – the sugarcane ethanol case
  - Integration of bioenergy crops into agricultural landscapes*

* ORNL is in discussion with IEA Task 43

Thank you!

CBES
Center for BioEnergy Sustainability

http://www.ornl.gov/sci/ees/cbes/

Opportunities and Challenges in the Design and Analysis of Biomass Supply Chain

SANGPIL KO,
Graduate Research Assistant
Michigan Technological University

July 23rd 2014. Recife

Article Authors:
- Pasi T Lautala
- Tuulli Laitinen
- Roger Bittencourt
- Amir Valente
- Michael R Hilliard
- Erin Webb
- Ingrid Busch
- Robert M Handler
- Richard Hess
- Mohammad S Roni
- Jorge Hilbert

Contents

I. Introduction of Biomass Supply Chain

II. Advanced Biomass Supply Chain System

III. Biomass Supply Chain Models (USA, Argentina)

IV. Biomass Transportation Models (USA, Brazil)

V. Challenges and Research Recommendations

I. Introduction of Biomass Supply Chain

I. Introduction

- The projection of replacing 30% of gasoline consumption by 2030 in the U.S. will require significant infrastructure investments on biomass over the next decades, and effective supply chain management.

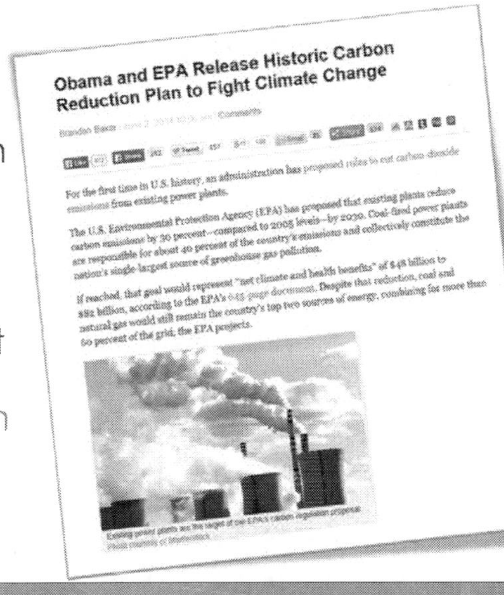

Obama and EPA Release Historic Carbon Reduction Plan to Fight Climate Change

For the first time in U.S. history, an administration has proposed rules to cut carbon dioxide emissions from existing power plants.

The U.S. Environmental Protection Agency (EPA) has proposed that existing plants reduce carbon emissions by 30 percent—compared to 2005 levels—by 2030. Coal-fired power plants are responsible for about 40 percent of the country's emissions and collectively constitute the nation's single-largest source of greenhouse gas pollution.

If reached, that goal would represent "net climate and health benefits" of $48 billion to $82 billion, according to the EPA's 645-page document. Despite that reduction, coal and natural gas would still remain the country's top two sources of energy, combining for more than 60 percent of the grid, the EPA projects.

I. Introduction

- The Goal of this paper is…

1) To **Introduce key components** of Biomass supply chains.

2) To **Provide examples of modeling applications** available for Biomass supply chain investigations.

I. Introduction

- Biomass Supply Chain Components
  - Biofuel industry is in progress to expand from a primarily sugar-based system to a cellulosic-based system.
  - Most of the feedstock of biomass include a common set of supply chain components.

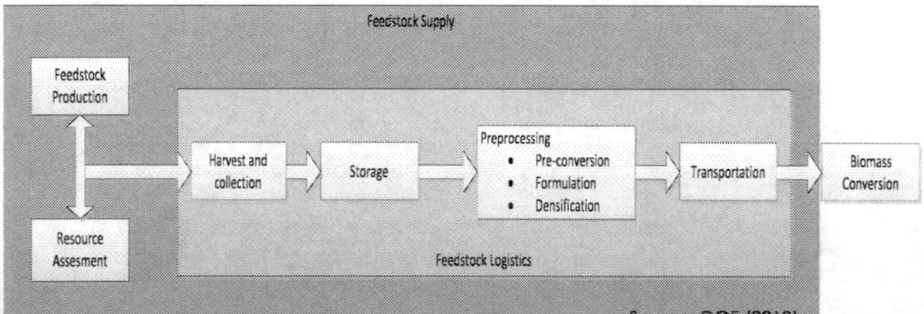

Source: DOE (2012).

I. Introduction

- Biomass Supply Chain Components

Biomass Cultivation	Naturally regenerated feedstock vs. intense cultivation procedures.
Biomass Harvesting	Removal of biomass from the growing landscape and consolidation for future use.
Biomass Transportation	Logistics of moving feedstock. Often significant cost component.
Biomass Storage	May be necessary at various points of supply chain.
Biomass Processing	Conversion into heat, power, fuel after size reduction or physical processing.
Storage and Final Transport	Direct use (for heat), or distribution for final use via electrical grid or transportation system for liquid products.

II. Advanced Biomass Supply Chain System

- Integrated Landscape management
- Advanced Biomass Feedstock Supply Chain

II. Advanced Biomass supply chain system

- Integrated Landscape Management
  - A New Concept to improve the biomass supply-demand dynamics and make more biomass available at lower access costs.
  - Traditional cropping system **vs.** Integrated landscape mgmt.

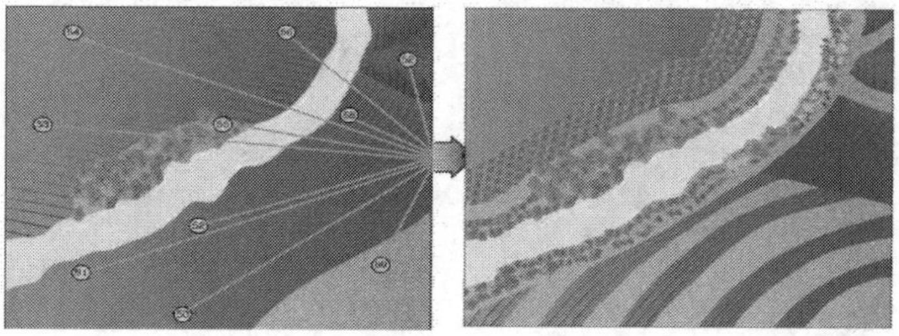

II. Advanced Biomass supply chain system

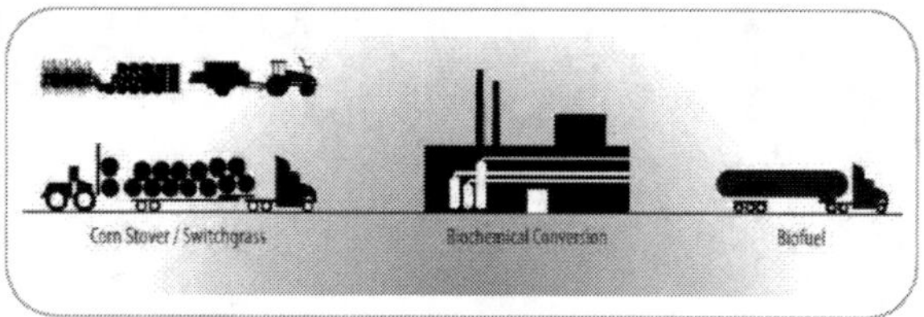

- Existing Biomass Supply Chain

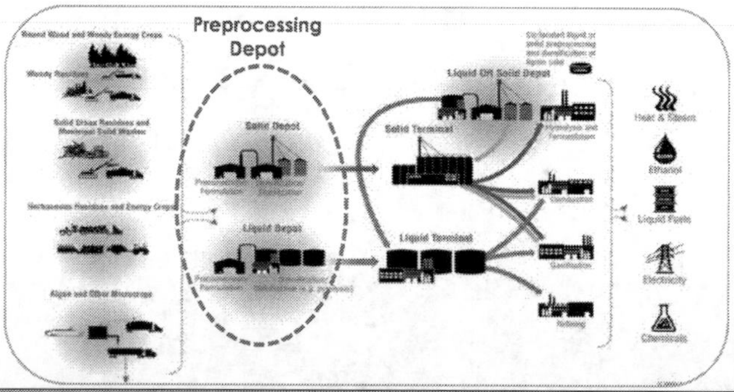

- Advanced Biomass Supply Chain included Preprocessing phase

II. Advanced Biomass supply chain system

Advanced biomass feedstock supply chain

- Advanced feedstock supply chain system is needed to enable sustainable integrated landscape management.

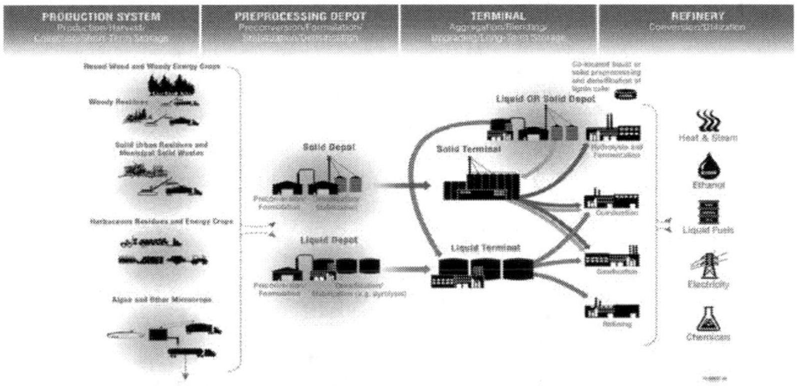

II. Advanced Biomass supply chain system

Advanced biomass feedstock supply chain

- Advanced supply chain system has the ability to utilize small quantity of multiple types of biomass produced from the integrated landscape mgmt.

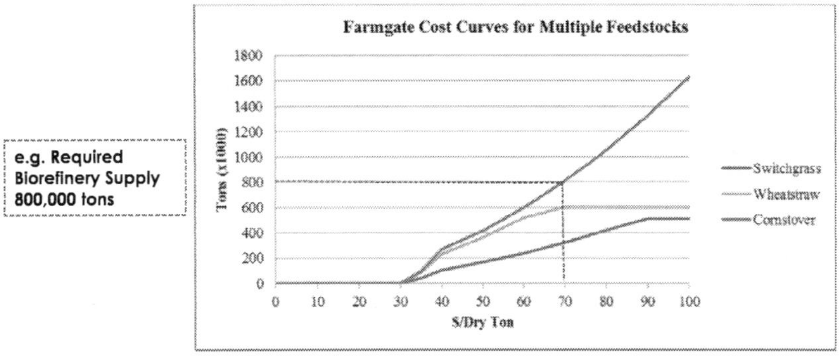

III. Biomass Supply Chain Models

- BILT model (USA)
- IBSAL model (USA)
- WISDOM model: Woodshed Analysis (Argentina)

BILT (Biofuel Infrastructure, Logistics and Transportation) Model

- Given a region and a level of biofuel demand, the model creates a biofuel supply chain for the region that minimizes costs under given set of assumptions. Information included in the model include:

1. **Feedstock Production**
   - e.g. Which feedstock?
2. **Feedstock Logistics**
   - e.g. How will the biomass be handled, processed and stored on the farm, in transit and before it is used at the refinery?
3. **Ethanol Production**
   - e.g. Which bio-refinery technologies will be employed?
4. **Ethanol Distribution**
   - e.g. How will ethanol be moved to blending facilities, and how will it be distributed between the facilities?

III. Biomass Supply Chain Models

- **BILT** (**B**iofuel **I**nfrastructure, **L**ogistics and **T**ransportation) Model

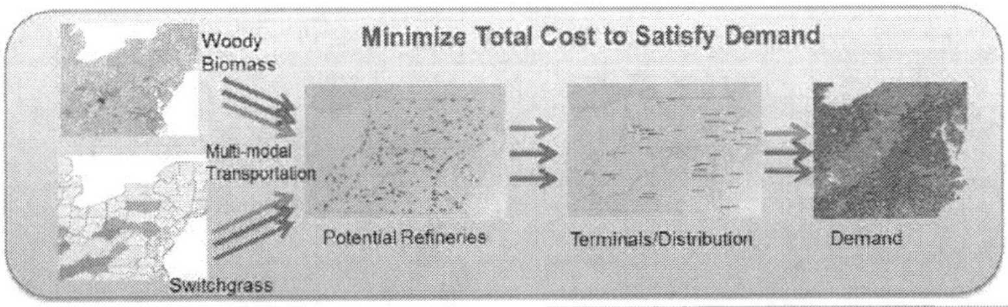

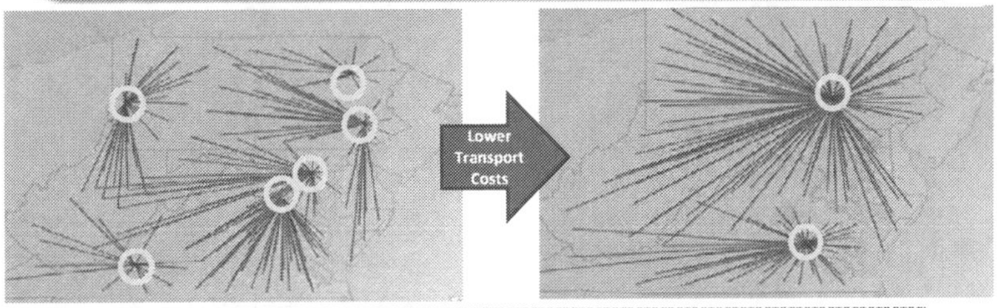

e.g. The six refineries are replaced by two large facilities when transportation costs are decreased.

III. Biomass Supply Chain Models

- **IBSAL** (**I**ntegrated **B**iomass **S**upply **A**nalysis and **L**ogistics) Model
  - The objective of this model to simulate biomass movement through the supply chain.

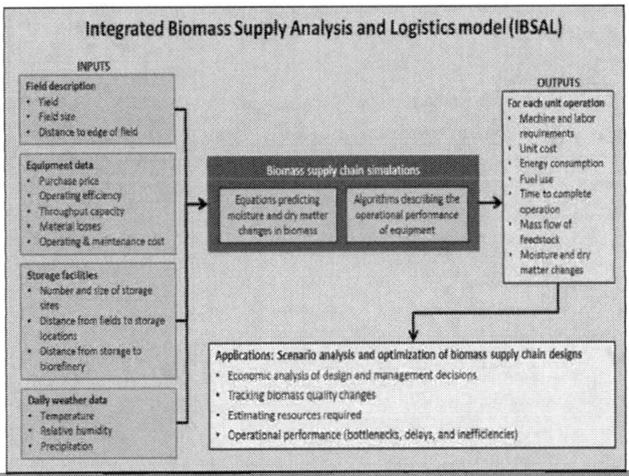

III. Biomass Supply Chain Models

IBSAL (Integrated Biomass Supply Analysis and Logistics) Model

- Simulation results are highly dependent on the quality and accuracy of data. For example, the cost of dry matter loss varies based on quantity lost and unit cost

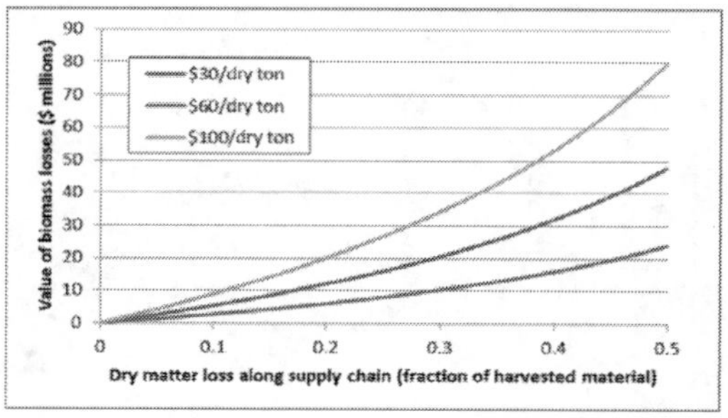

- Not properly accounting for losses along the supply chain results in inaccurate cost estimates.

III. Biomass Supply Chain Models

WISDOM (Woodfuel Integrated Supply/Demand Overview Mapping)

- Woodshed Analysis: the second phase analysis of WISDOM
- Mapping of sustainable and probable supply for woodshed delineation.

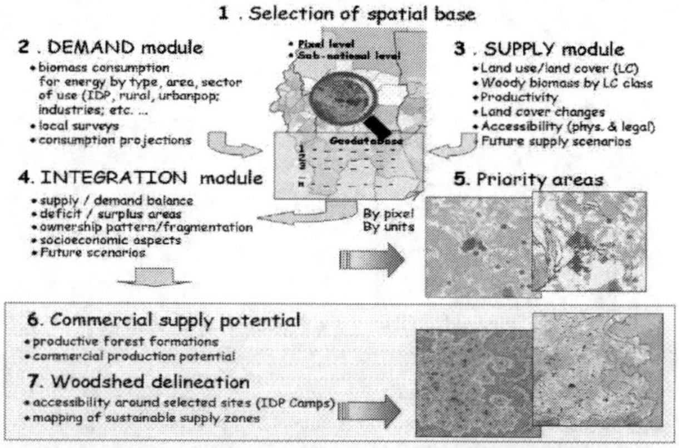

III. Biomass Supply Chain Models

- **WISDOM** (**W**oodfuel **I**ntegrated **S**upply/**D**emand **O**verview **M**apping)
  - Based on commercial balance surplus, an example of bio-shed analysis was carried out for the city of Cordoba considering minimum, medium and maximum productivity levels as follow maps:

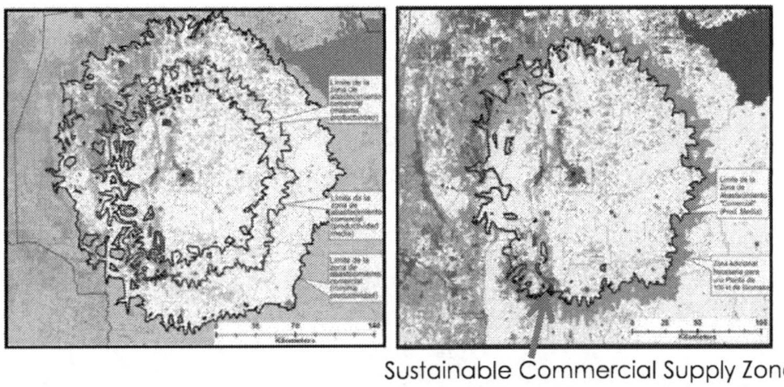

IV. Biomass Transportation Models

- MMLT model (USA)
- PrevFretes model (Brazil)

IV. Biomass Transportation Models

Transportation Options

- Most biomass transported by trucks
- Main Decision Criteria for Multi-Modal Selection: Cost Efficiency

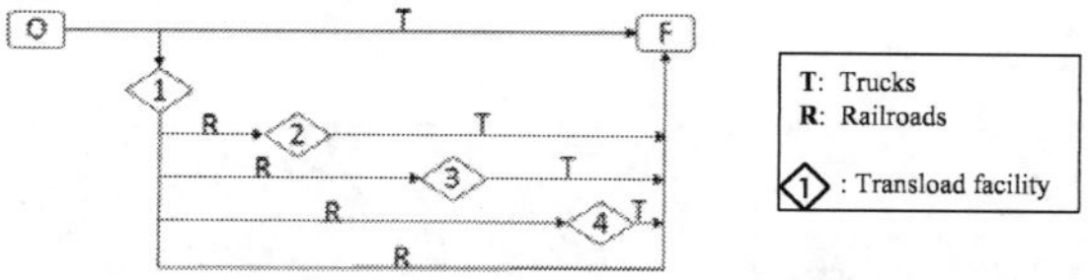

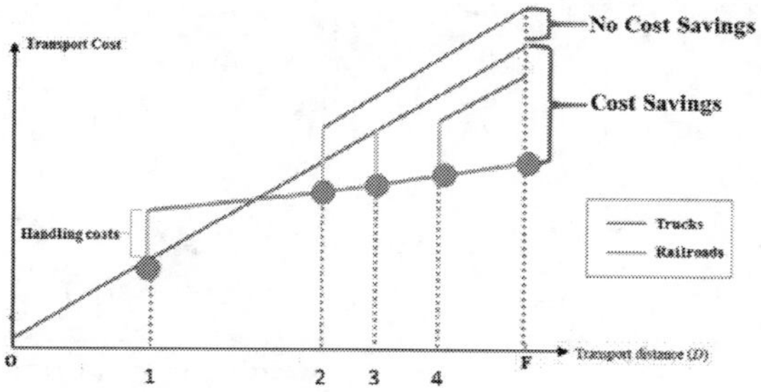

IV. Biomass Transportation Models

MMLT (Multi-Modal Log Transportation) Model

- Regional modeling tool to compare the cost of truck transportation with multi-modal (Truck/Rail) option.
- This model used loop optimization logic to investigate a data set of 100,000 actual log industry truck trips.
- Included evaluation of fuel price sensitivity

Optimization Results - Fuel Price $2.89 (Average 2007)					
Mode	Ton-Miles	Tons	Avg Trip Miles	Total Cost	Avg Cost / Ton
Non - Optimized					
Single Mode Trip: Truck	282,887,505.33	3,171,611.15	89.19	$ 37,912,725.50	$ 11.95
Optimized					
Single Mode Trip: Truck	206,206,174.60	2,675,989.03	77.06	$ 29,004,106.67	
Bi-Modal Trip: Rail Segment	62,532,744.09	495,622.12	126.17	$ 3,199,767.83	
Bi-Modal Trip: Truck Segment	14,148,586.65	495,622.12	28.55	$ 4,287,556.68	
Total Cost				$ 36,491,431.17	$ 11.51
Optimized Truck Totals	220,354,761.24	3,171,611.15	69.48		
Savings				$ 1,421,294.33	$ 0.45
Savings Percent				3.75%	

IV. Biomass Transportation Models

MMLT (**M**ulti-**M**odal **L**og **T**ransportation) Model

- The same model was used to develop 'cost gradient maps' to represent the cost efficient supply radius for a planned cellulosic ethanol plant.

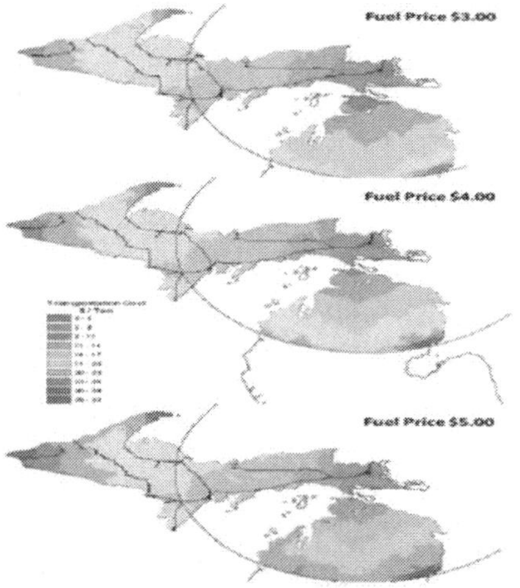

- Blue circle indicates 150mile distance from proposed facility.

IV. Biomass Transportation Models

PrevFretes, Freight Forecasting Model

- Freight Projection Cost for fuel and Oil byproducts. (Petrobras)
- Up to 5 Year Forecasting (Road, Rail, and Water way)
- Correlation Update every 4 month: Market value by Phone survey
- Decision Making Tool: Carrier, Logistics Operators, and Strategic Planners

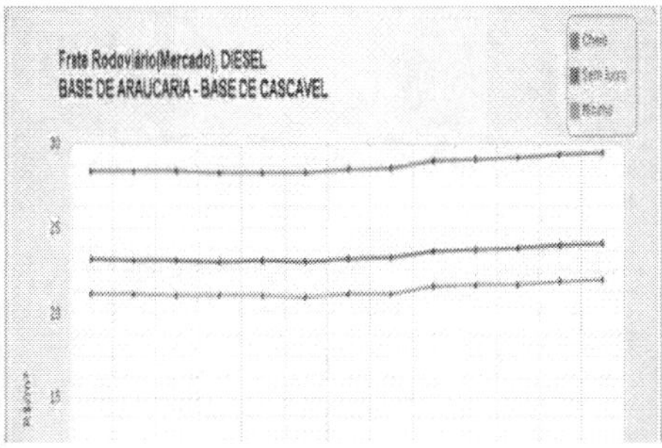

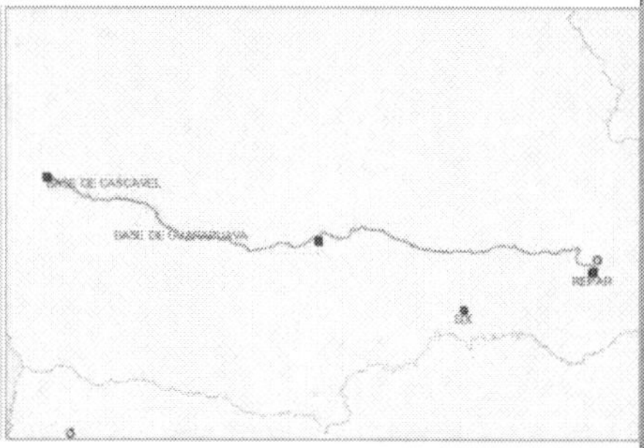

IV. Biomass Transportation Models

PrevFretes, Freight Forecasting Model

- Parameters
  - Different Modal Scenarios
  - Type of Service: Transfer or Distribution of Goods
  - Correlation Update every 4 month: Market Value by Phone Survey
  - Type / Material of Trailer or Boat

↓

- Results
  - Ideal vs. Market Value
  - Net Profit
  - Fees and Direct Taxes
  - Net and Gross Freight Fee

V. Challenges and Research Recommendations

- Challenges
- Recommendations

V. Challenges and Research Recommendations

- Some of the most prevalent Challenges include:
  - **Availability and quality of data** to analyze the various components of the supply chain.
  - Lack of a common **framework for sustainability indicators**.
  - Deficiency in integrated analysis when developing the supply chains and individual components.
  - Lack of **national or international policies** and standardization.
  - Lack of understanding the **effect of international energy markets** on sustainable bioenergy production.
  - Need to find shortcomings of current cellulosic conversion processes.

V. Challenges and Research Recommendations

- Recommendations **for Future Research:**
  - A **common framework** should be established that would allow comparison of potential implementation, or the impact of actual implementation.
  - More realistic representation of the biofuel supply chain requires **expanding the analysis of different logistics frameworks** and incorporating the temporal sequencing of infrastructure construction.
  - Advanced supply systems are essential to implement more sustainable biomass production schemes.
  - **The economics of the international energy markets** should be incorporated in the analysis.

Q & A

Thank you!

PAN-AMERICAN BIOFUELS & BIOENERGY SUSTAINABILITY
AN NSF RESEARCH COORDINATION NETWORK

Opportunities and Challenges for Biomass Supply Chains

PAN-AMERICAN BIOFUELS & BIOENERGY SUSTAINABILITY

Biomass Transportation Systems in Brazil: The Cases of the Ethanol Industry and the Constrains for Exporting Pellets

Arnaldo Walter
University of Campinas – Dept. Energy – FEM & NIPE
awalter@fem.unicamp.br

UNICAMP

PAN-AMERICAN BIOFUELS & BIOENERGY SUSTAINABILITY

UNICAMP

Summary

- Logistics in the sugarcane sector: the agricultural stage and ethanol distribution.
- Constraints for pellets production and commercialization.
- Conclusions.

PAN-AMERICAN BIOFUELS & BIOENERGY SUSTAINABILITY

UNICAMP

Logistics in the sugarcane sector

From the field to the mill

Source: Caixeta Filho (2012)

- The stage that is called CCT (in Portuguese) (Harvesting, Loading and Transportation) represents about 35% of the total costs of sugarcane (i.e., about 20% of ethanol production costs).
- The current tendency is mechanical harvesting, with use of large trucks to transport sugarcane to the mill.
- The current practice leads to high level of soil compression, impacting productivity.
- Mechanization has lead to a new paradigm on CCT.
- How do handle sugarcane straw is still an open issue.

From the field to the mill

DESCRIÇÃO	ILUSTRAÇÃO	NOME POPULAR	CAP. (MÉDIA) DE TRANSP.
Caminhão plataforma com um reboque acoplado		Romeu e Julieta	25 t/viagem
Caminhão plataforma com dois reboques acoplados		Treminhão	45 t/viagem
Cavalo mecânico com dois semi-reboques acoplados		Rodotrem	65 t/viagem

- Efforts for optimizing the operation: reducing idle time, losses of sugarcane, reducing shipment.

PAN-AMERICAN BIOFUELS & BIOENERGY SUSTAINABILITY

Traditional operation

UNICAMP

- Traditional loading after manual harvesting

PAN-AMERICAN BIOFUELS & BIOENERGY SUSTAINABILITY

Sugarcane harvesting

UNICAMP

- Mechanical harvesting with some degree of separation between sugarcane and trash

PAN-AMERICAN BIOFUELS & BIOENERGY SUSTAINABILITY

After mechanical harvesting

UNICAMP

PAN-AMERICAN BIOFUELS & BIOENERGY SUSTAINABILITY

At the mill

UNICAMP

Logistics for ethanol

Location of sugarcane mills

Source: Walter et al. (2014)

Railways in Brazil

Source: ANTF (2014)

Ports in Brazil

Source: MT (2014)

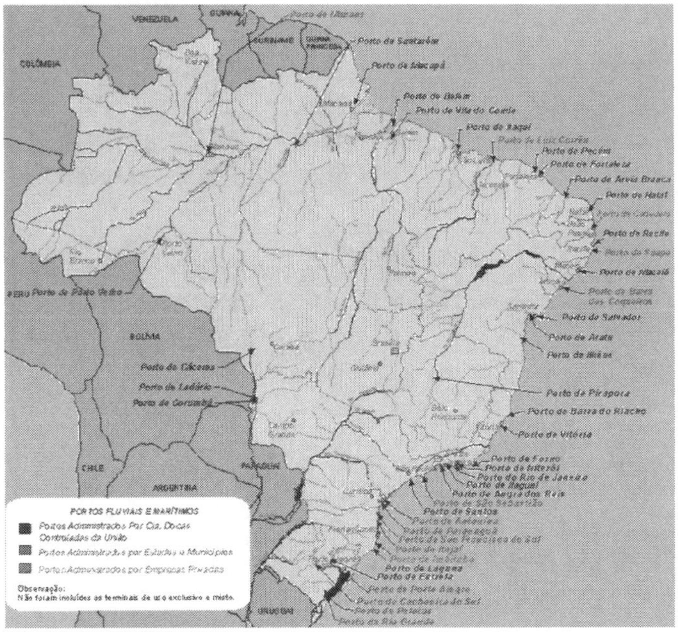

- In 2013, 87% of the total export (about 2.6 BL) were shipped in Santos and 10% in Paranaguá.

From the mill to the consumer

Source: Milanez et al. (2010)

- Due to the short distances and the relatively small volumes, in general ethanol flows by trucks from the mills to the distribution centers.
- About 80% of the ethanol produced moves from the distribution centers to the gas stations. About 20% of the production is transferred among distribution centers. Trucks are used in a large extent.
- The same (i.e., by far the transport by trucks) regarding the displacement of ethanol to the ports.
- The same picture in December 2013, according to Grupo de Pesquisa e Extensão em Logística Agroindustrial (Esalq-Log).

- In 2007 is was estimated that the transportation cost from the mill to the ports was 45-70 US$/m^3, i.e., 10-16% of the FOB price of exported ethanol (CGEE, 2010).

Typical trucks used

Source: Caixeta Filho (2012)

Carreta
5 a 6 eixos

Até 2001 - 30 a 35 m^3

Bi-trem
7 eixos

2001 - 45 m^3

Rodo-trem
9 eixos

2004 - 60 m^3

Transport by railways and pipelines

Source: Milanez et al. (2010)

- The best infrastructure for transporting ethanol by railways is in the South region (it is not the main producer region). 1.6 BL were transport by railways in 2008 and about 2.7 BL in 2009.
- In state of São Paulo (the largest producer), the average distance from the mills to the consumer market is 230-410 km. However, the use of railways has been extended.

- The feasibility of a pipeline with 500 km requires a minimum volume of about 4.5 BL/y. The investment would be about 1 million US$/km.
- In 2009, about 1 BL were transported by pipelines (basically, from storage units to the ports).

Focus on pipelines

Source: Logun (2013)

- Project to be developed from 2011 to 2020, with about 1,300 km of pipelines and capacity for 20 BL/y.

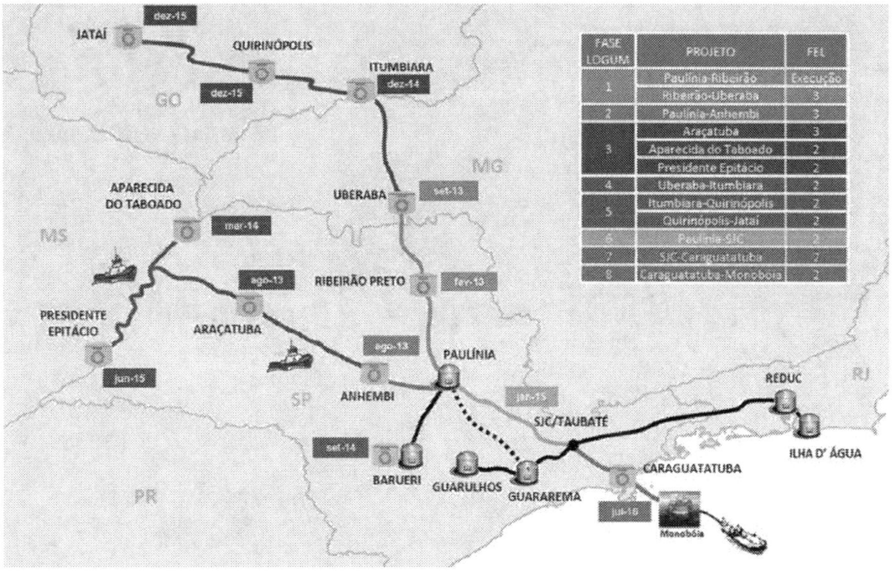

Actions by Petrobras (1)

Source: Cao (2012)

- The combined use of pipelines and boats. Nominal capacity in top right-side.

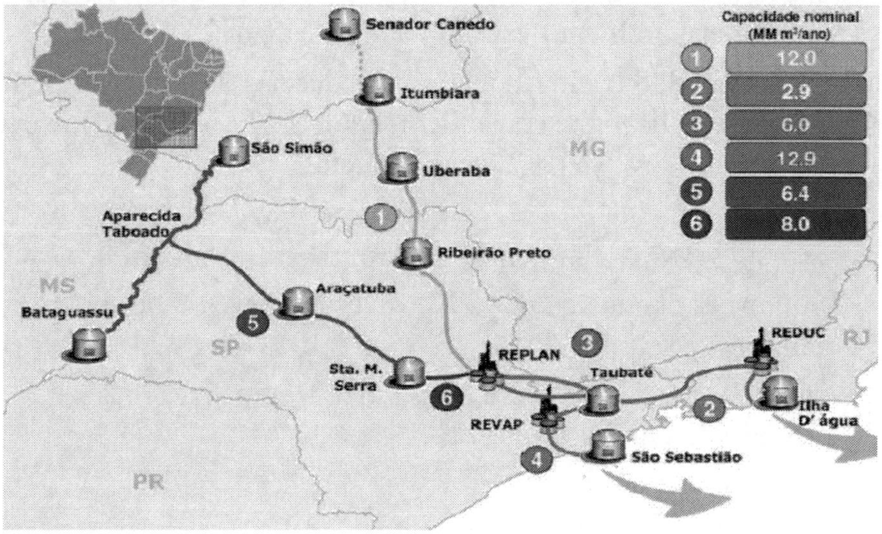

Actions by Petrobras (2)

Source: Cao (2012)

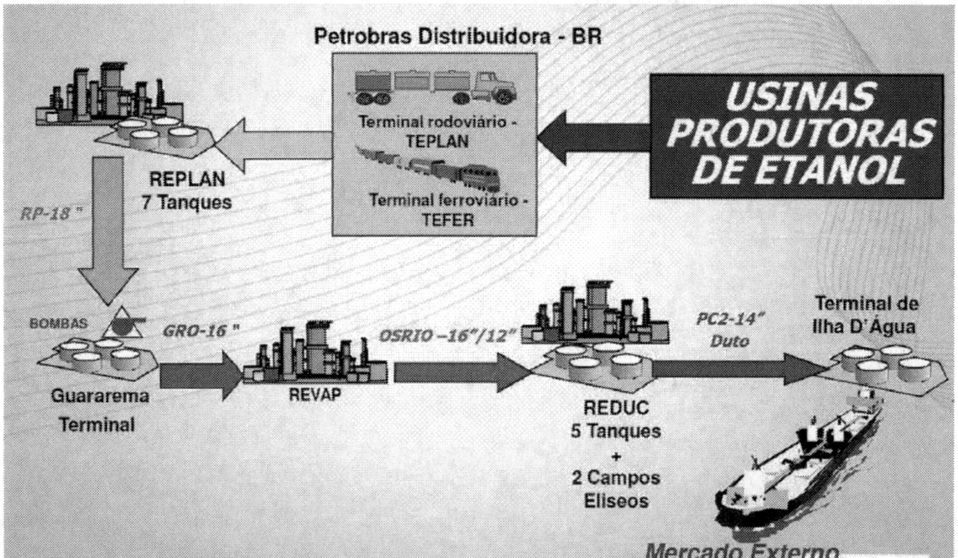

Constrains for producing and exporting pellets

Sources: Serrano (2009)

- In a study considering production of pellets from wood residues in Brazil it was estimated that the logistic cost in Brazil represented about 10% of the CIF price in Europe (in the reference case).
- However, transporting pellets by trucks (due to the small level of production), the break-even would be a distance of about 200 km; for 300 km the production wouldn't be feasible at all.
- Feasibility analysis by different investors, considering the use of industrial and agricultural residues, showed that the logistic costs is the main constrain.

Constrains for producing and exporting pellets (2)

Sources: Suzano (2012)

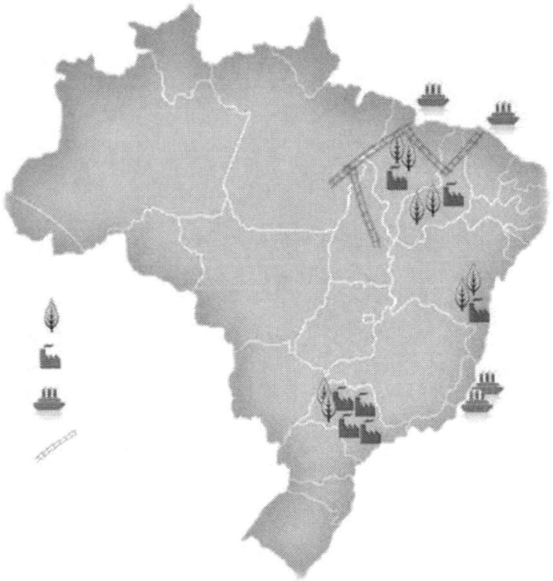

- Suzano is one of the majors producers of pulp and paper in Brazil.
- The company considered the production of pellets (2 Mt/y) from a dedicated eucalyptus plantation in NE Brazil.
- Even with a existing infrastructure, the project failed.

Conclusions

- Mechanical harvesting is a strong driver for changes on logistics in the sugarcane industry. Straw recovery is still a challenge.

- Logistics is an important drawback for large-scale production of bioenergy in Brazil (even for the well-established sector ethanol production).
- The logistic costs would be a serious barrier for exporting ethanol from more remote regions.
- For pellets, the lack of good railways is one of the most constrains. Investments should be in specific regions.

- It is necessary to (urgently) invest on railways, pipelines and on improving the ports!

Please, keep in touch

- awalter@fem.unicamp.br

Thank you!

Tropical Maize and Lipid Cane As Sustainable Bioenergy Crops

Vijay Singh
University of Illinois

RCN Conference on Pan American Biofuels & Bioenergy Sustainability
July 22-25, 2014
Golden Tulip Recife Palace
Recife, Brazil

What is Tropical Maize?

- Tropical Maize is Corn
  - Reproductive Asynchrony
  - Tropical Maize = Tropical x Temperate Maize Hybrid
  - High Biomass
  - High Stalk Sugars
  - Less Nitrogen requirement

History of tropical maize

From Stoskopf, 1981

CURRENT ANTHROPOLOGY Volume 44, Number 5, December 2003

Sweet Beginnings

Stalk Sugar and the Domestication of Maize[1]

by John Smalley and Michael Blake

Teosinte (*Zea mays* ssp. *parviglumis*) appears now to be the most widely agreed-upon candidate for the ancestor of domesticated maize (*Z. mays* spp. *mays*), but there are, at best, only partial answers to questions of how, when, and where this process took place (Wilkes 1967, 1985; Beadle 1980; Iltis 1972, 2000; Matsuoka et al. 2002). In a break with conventional wisdom, Hugh H. Iltis (2000:36 and quoted in Crosswhite 1982) has recently suggested that the direct ancestor of maize "was initially domesticated not for its grain but for its sugary pith or other edible parts." We elaborate on his suggestion by proposing that during *Zea*'s initial period of domestication the stalk provided a key source of sugar for many uses, including the making of alcoholic beverages. Furthermore, we suggest that the social importance of

Nutrient Use Efficiency

- **TM produces more biomass and more sugar than commercial corn hybrids with < 50% N fertilizer requirement**

US Sweetener Industry

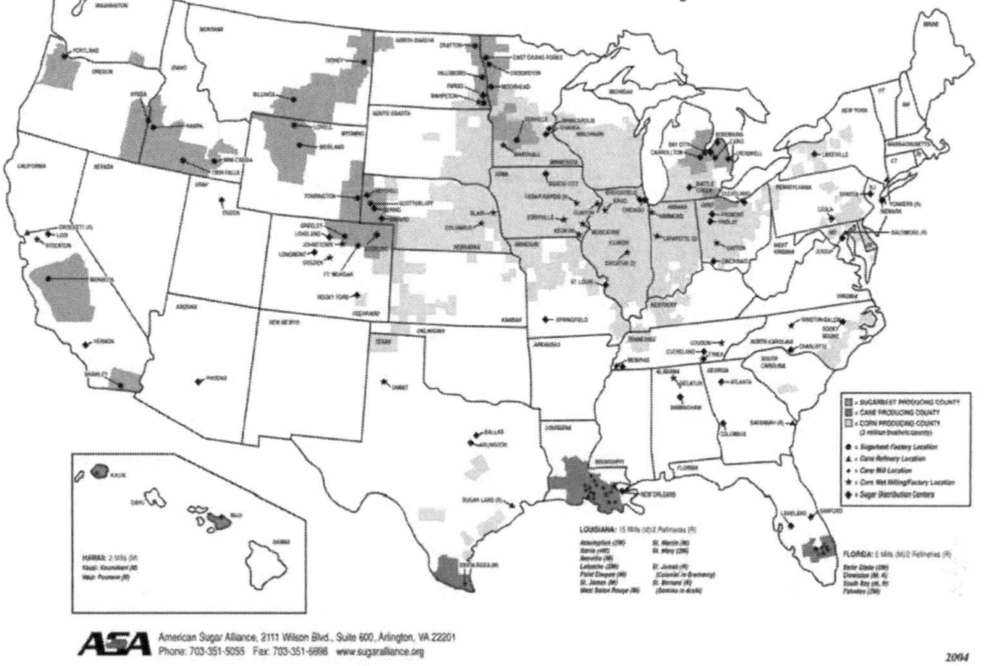

Tropical Maize

Preparing the syrup by using a press

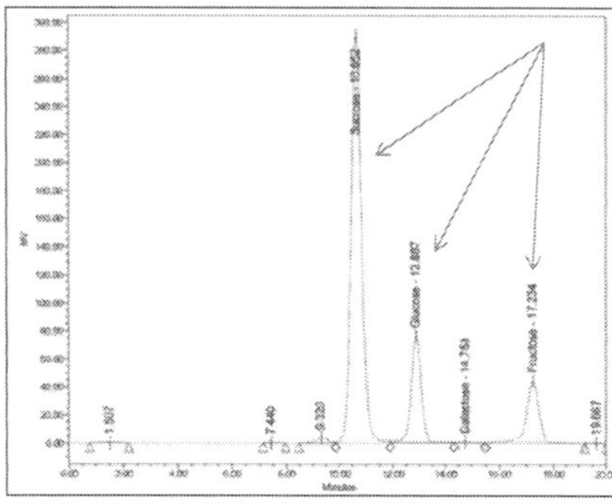

HPLC profile of the tropical maize syrup. Three main sugars content—Sucrose, Glucose and Fructose

Chen, M., Kaur, P., Dien, B.S., Below, F.E., Vincent, M.L. and Singh, V. 2013. Accumulation of fermentation sugars during tropical maize development. World Journal of Microbiology and Biotechnology 29:1509-1515.

Tropical Maize

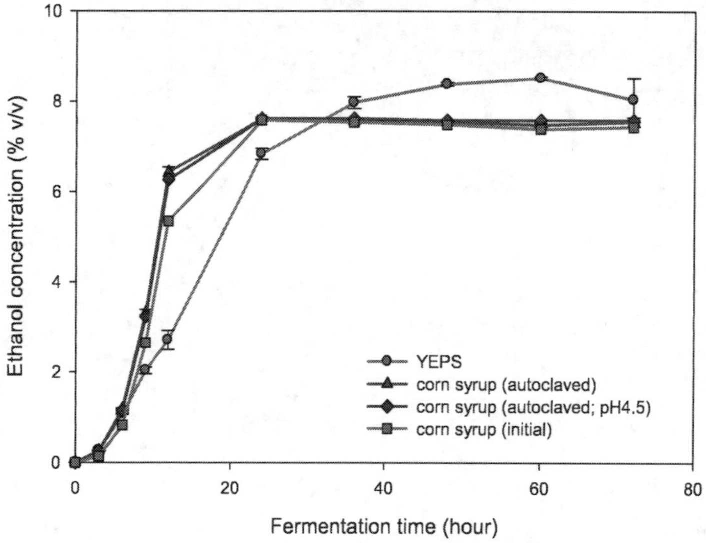

Chen, M., Kaur, P., Dien, B.S., Below, F.E., Vincent, M.L. and Singh, V. 2013. Accumulation of fermentation sugars during tropical maize development. World Journal of Microbiology and Biotechnology 29:1509-1515.

Tropical Maize

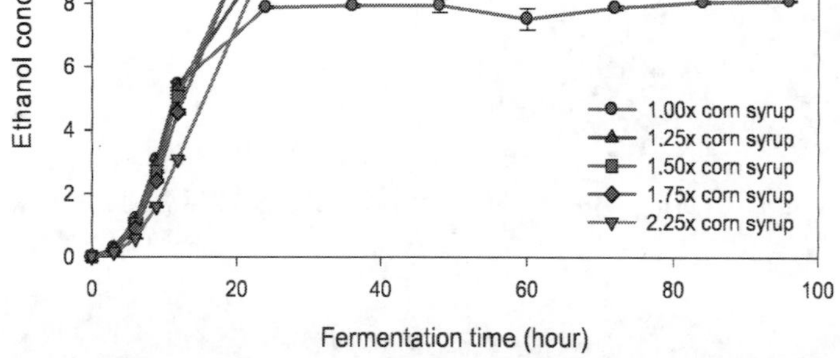

Chen, M., Kaur, P., Dien, B.S., Below, F.E., Vincent, M.L. and Singh, V. 2013. Accumulation of fermentation sugars during tropical maize development. World Journal of Microbiology and Biotechnology 29:1509-1515.

Conventional Sugarcane Ethanol Process

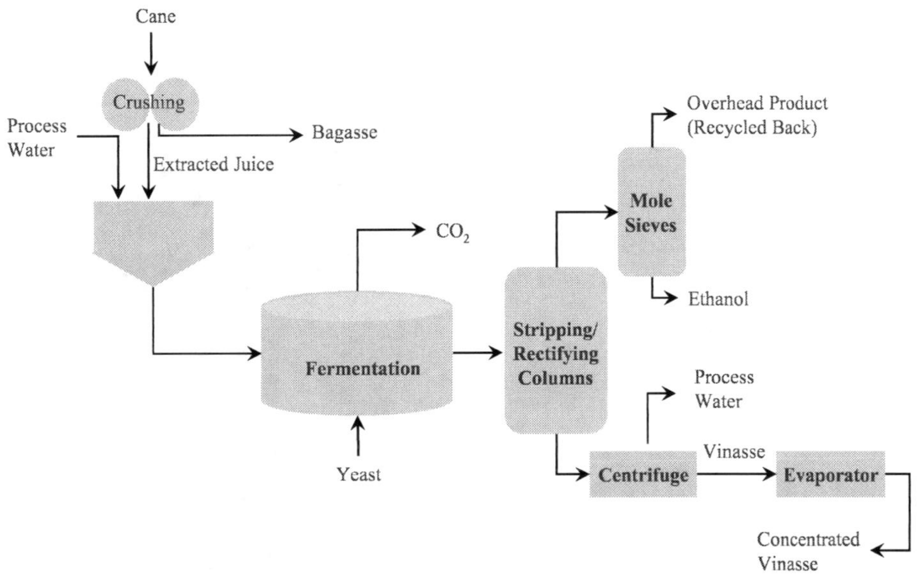

Tropical Maize

- Tropical maize requires few crop inputs such as nitrogen fertilizer, chiefly because it does not produce any ears.
- Require less processing than corn grain, corn stover, switchgrass, miscanthus.
- 25 percent or more sugar -- mostly sucrose, fructose and glucose
  - Easily fermented to ethanol
- Can be used in sugarcane to ethanol plant during the Inter-harvest

Lipid Producing Sugarcane (Lipid Cane)

Make World's Most Productive Sugar Crop as the Oil Crop

- Plants already naturally produce oil
  - Use it as part of their metabolism
- By up-regulating the genes that make oil and down-regulate the genes that use it
  - Oil is stored in stems
- Currently we have achieved lipid cane with 2% oil (dry basis)
- Goal is to achieve 20% oil (dry basis) in Lipid cane

Accumulate Oil in Sugarcane

- Increase Photosynthesis
- Increase Cold Tolerance

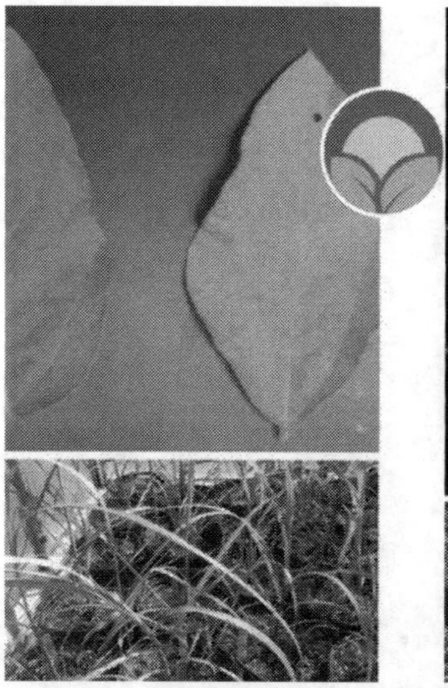

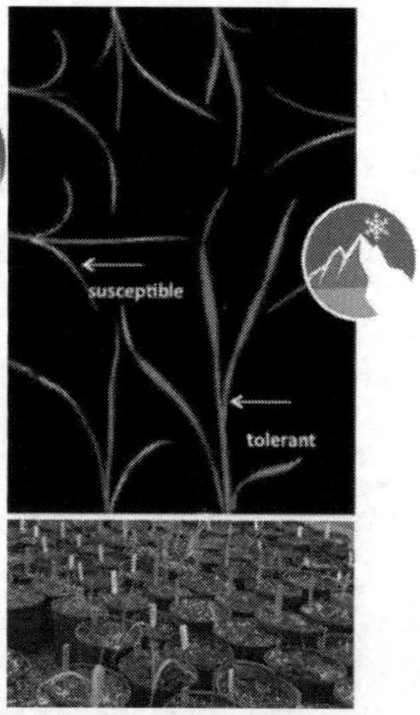

Techno-Economic Analysis of Lipid Cane with Different Lipid Content

- Lipid cane techno-economic models were developed and compared with the current existing soybean and corn processing plants
- Lipid cane process model with lipid content of 0%, 2%, 5%, 10% and 20%
- Soybean-biodiesel process model with solvent (hexane) exaction
- Corn-ethanol process model with oil extraction from thin stillage

Lipid Cane

- Sugarcane producing lipids
- 2% lipid in stems, higher photosynthesis efficiency and improved cold tolerance
  - Goal – 20% oil in stems, 50% improvement in phytosynthesis
- Grown in land not used for food or other crops and can produce 40 billion gallons of biodiesel in the US

Thanks!
ILLINOIS

Lignocellulosic Biomass Residues Identification and Selection in Five Countries of Latin America, As a Feedstock for Second Generation Ethanol Production

Ruz, E., PROCISUR
Hilbert, J., INTA
Carballo, S. M., INTA
Sundfeld, E.,
Coutinho, A.,
Perez, C.,
Mayeregger, M.,
Vazquez, D.,
Resquin, F.,
Ganduglia, F.,
Originally presented on: 7/23/2014 14:15:00 - 14:40:00

Lignocellulosic biomass residues identification and selection in five countries of Latin America, as a feedstock for second generation ethanol production i

i Regional research by IICA – PROCISUR

ii BABETHANOL leading researchers in PROCISUR member institutions:

Jorge A. Hilbert & Stella Maris Carballo INTA Argentina

Esdras Sundfeld & Alexander Coutinho EMBRAPA Brazil

Claudio Pérez INIA Chile

Manuel J. Mayeregger MAG/IPTA Paraguay

Daniel Vázquez & Fernando Resquín INIA Uruguay

Federico Ganduglia IICA Argentina Office

Emilio Ruz Jerez PROCISUR Executive Secretary

Emilio Ruz

Executive Secretary PROCISUR

www.procisur.org.uy

This presentation is part of a recently completed project: "New feedstock and innovative transformation process for a more sustainable development and production of lignocellulosic ethanol" (BABETHANOL) www.babethanol.com , a collaborative project for specific international cooperation actions SICA. Grant agreement N° 227498 of the

7th framework program of the European Commission.

The main objective of the BABETHANOL project was to develop new solutions for a more sustainable approach of second generation ethanol production, based on a "moderate, environmental-friendly and integrated" pre-treatment process of lignocellulosic biomasses. This pre-treatment of the biomass has been the main barrier to overcome in second generation ethanol. The new process, called CES (Combined Extrusion-Saccharification), is an alternative to the current pre-treatments of the state-of-the-art, requiring much energy, water, chemical products, detoxification and waste treatments. It has been developed and tested up to TRL 5 (Technology Readiness Level) from laboratory to pilot scales with seven selected biomasses covering a large range of diversified feedstock.

In parallel to the process development, a feedstock catalogue of diversified crop and agroindustrial wastes was built along the project from investigations run in Argentina, Brazil, Chile, France, Germany, Italy, Paraguay, Spain, United Kingdom and Uruguay, by one partner in each continent. The main idea was to identify concentrated amounts of indigenous lignocellulosic wastes currently available at local/regional scales for the prompt deployment of small, medium or large size second generation ethanol plants near the biomass production sites once the new process will be ready for industrialization.

We are hereby presenting the results of IICA-PROCISUR research ii in five countries of South America: Argentina, Brazil, Chile, Paraguay and Uruguay.

The final selection of the feedstock was performed after investigating: 1) the biomass availability taking into account current situation with competition for other uses (soil cover, animal feeding and/or bedding and energy); 2) the chemical compositions of the preselected biomasses; and 3) the concentration of feedstock at local level (within 100 km radius) to supply at least 30,000 t dry matter/year.

Argentina

Although Argentina is one of the largest producers and exporters of agricultural goods generating large amounts of residues, originated by the agricultural and agro-industrial sector, only a few were selected according to the required characteristics, availability and geographical dispersion. Precautious measures were taken into account for agricultural production systems since there are most under no tillage, and therefore, cover and organic material and nutrients must be reserved in the soils in order to avoid a rapid deterioration. In the case of agro-industrial products, competitiveness with other uses, physical and chemical characteristics and availability in volume and dispersion during the year were the main concerns. The biomasses found as the most suitable feedstock for the production of ethanol with the CES process were: **corn cobs, vineyard pruning, sugar cane field residues (tops and leaves) and bagasse, and eucalyptus field and industrial residues.** Less important was wheat straw because of its relatively low amount of cellulose.

For corn cob, three departments in the province of Cordoba: Marco Juarez, Unión and Río Cuarto, producing each of them over 100,000 t/year corn cob, were highlighted since they could match with minimum volume availability requirements. For vineyard pruning, the provinces with higher volumes of residues were San Juan and Mendoza. Two supply basins in a 25 km radius were detected at San Martin (100,000 t/year) and Maipu (35,000 t/year) departments in the province of Mendoza. For sugar cane field residues, two supply areas for ethanol plants were identified: the north, with 260,000 t/year and the south, with 220,000 t/year, both in a 20 km radius. For sugar cane bagasse, a main supply basin was identified around Concepción sugar cane mill in the province of Tucuman with 100,000 t/year bagasse available in 5 km radius and almost 200,000 t/year in a 30 km radius. Eucalyptus residues are concentrated in the provinces of Corrientes and Entre Rios. Forest residues amount to 170,000 and 230,000 t/year respectively.

Brazil

With regards to chemical composition and availability, **sugarcane residues (bagasse and trash) are the only feedstock convenient and currently available for ethanol production.** The residues from the production chains of soybeans, maize, banana, wheat, coffee and pineapple, with suitable chemical composition which are not available now-a-days under current uses, may offer a second list of additional feedstock that have the potential of being used in the future if the production of second generation ethanol picks up. In Brazil, the most logical alternative would be to locate a production unit of second generation ethanol, in an already existing or close to a processing unit of sugarcane. In this respect, there is a high concentration of industrial plants in the following states: Alagoas and Pernambuco (Northeast region), Goiás and Mato Grosso do Sul (Centre-West region), Sao Paulo and Minas Gerais (Southeast region) and Paraná (South region).

The minimum amount of residues required for the operation of a 30,000 tons of dry material per year processing capacity plant is guaranteed in the following situations: if the trash is not used (or if it is unavailable), in industrial plants with processing capacity of over 2 M tons of sugarcane per season and with the use of trash, in industrial plants with processing capacity over 0.5 and 2.0 million tons of sugarcane per season. There are several industrial plants of sugarcane in Brazil with processing capacities in these ranges, located in the CO/SE/S and N/NE regions. Moreover, there are areas with high concentration of industries in these regions.

Chile

From all the materials originally considered and assessed **wheat straw and corn stover field residues** as well as **vineyard and orchard pruning residues** were selected as potential feedstock suitable for the production of ethanol under the CES process. Best residues supplying basins were identified in the following areas: wheat straw in the Araucania region with over 500,000 t/year; corn stover in the O'Higgins region with over 500,000 t/year; apple and vineyard pruning residues in five regions (Coquimbo, Valparaíso, Metropolitana, O'Higgins and Maule) with amounts varying from 150,000 to

280,000 t/year in each region.

Paraguay

In general, the climate of Paraguay characterized for high temperatures and humidity all year round, influences the dynamic of the degradation of organic matter left in the fields, accelerating the decomposition process which requires that most of the crop residues should not be removed from the soil for sustainability reasons. For this reason, field residues from major cultures like sugar cane, soya, wheat and corn are not collected and therefore not available for other uses. **Sugarcane bagasse** is the only residue produced in large quantity which is available and suitable for ethanol production and the CES process. Although it is already used to produce energy, there are seasons when there would be sufficient surplus to supply 2nd generation ethanol plants. The province of Guaira appears as the best location for the supply of bagasse with 54,000 t/year in a 30 km radius.

Uruguay

Among a wide variety of residues of the Uruguayan agro-industry, the ones that apparently were most suitable as feedstock for the CES process were wheat, rice and forestry field residues, as well as rice and forestry industrial residues. During the chemical screening both rice residues presented a chemical composition that was not suitable for the production of ethanol with the CES process: for straw the cellulose was too low, for the hull the lignin was too high and for both of them the mineral content was too high. Although wheat straw had better chemical composition, cellulose content was slightly under the limit. So wheat straw was considered as a second choice feedstock. Among the forestry residues, Eucalyptus presented very high cellulose content and good content for all the other compounds, except for forestry industrial residues with lignin slightly above the limit. Several country departments produce sufficient biomass to supply ethanol production plant, especially Rivera, Paysandú, Río Negro and Rocha with amount available ranging from 90,000 to 140,000 t/year within a 30 km radius or less in some zones. So **Eucalyptus forestry residues** are considered first choice feedstock for the production of ethanol and for the CES process in Uruguay.

Uncovering System Behaviors in Biofuels Production and Use

A Multi-Actor System Approach

Datu B. Agusdinata, PhD

RCN Conference on Pan American Biofuels & Bioenergy Sustainability, Recife, Brazil, 2014

Actors' decisions and (in)actions matter

Understanding of:

- Motives, interests and decision options
- Rational decision criteria
- Irrational behaviors
- Dynamics and Feedback loop

EUROPEAN COMMISSION Directorate-General for Energy

U.S. DEPARTMENT OF ENERGY

VIRENT

cosan

uop A Honeywell Company

Troubles at "Multi-Actor Land"

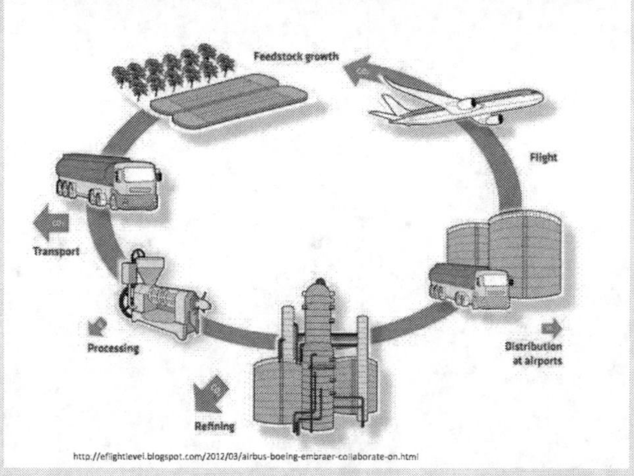

3. Asymmetry of cost-benefit distribution = Equity Issue

1. Individual actor vs. Supply Chain Network performance

2. "Achievable" Biofuels Life Cycle Impact Assessment

3

Observation 1: Ethanol Industry Boom and Bust

- **The US Energy Independent and Security Act of 2007 :**
  - 1.6 billion gallons in 2000 to 13.2 billion gallons in 2010
- **Biorefineries will likely face large financial uncertainties**
  - **Dried up capital** due to the credit crunch and the recent expiration of a federal subsidy for ethanol blenders
  - Recent **drought** has caused a spike in corn price leading to temporary shut down or scale back production of many ethanol refineries
  - Increased **price of corn** has squeezed refineries' profit margin below the sustainable level

Biofuels supply chain network is a complex system

Observation 2: Aviation Biofuels Potentials

- In the United States, the **aviation** sector is responsible for about **11%** of the total **transportation GHG emissions.**
- Aviation emission reduction goal:
  - carbon neutral growth by 2020
  - GHG emissions reduction by 50% compared to the 2005 baseline level by 2050

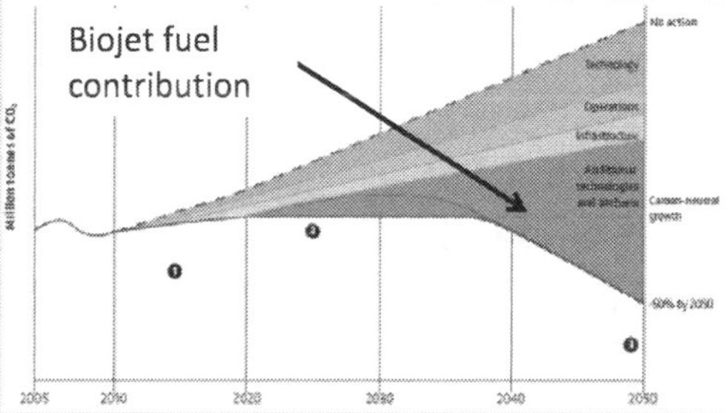

Source: Air Transport Action Group

For biojet fuels to achieve its GHG emission reduction potential, **technical** and **economic** hurdles must be overcome

Observation 3:
Many policymaking efforts on System Change involving multi-actors end up in Delay, Deadlock, or Court

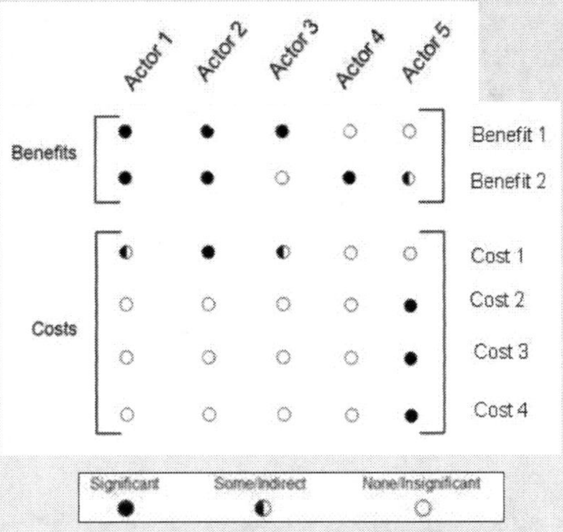

Credit: Mozdzanowska, A., & Hansman, R. J., 2008

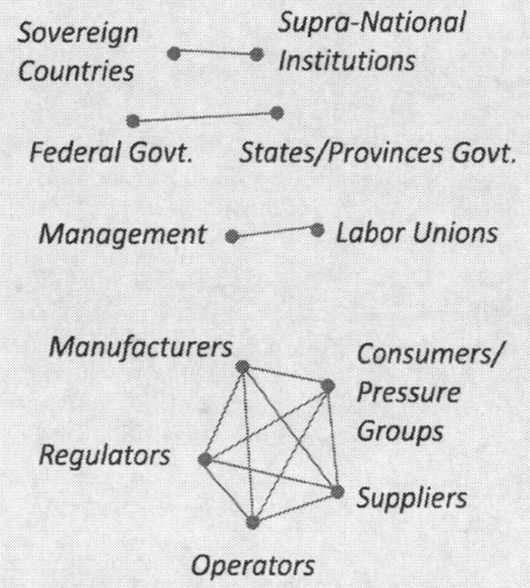

Need for a System Design for Equity/Fairness

Biofuels production and use = an industrial ecosystems

Features:

- an ecosystem consists of multiple agents or actors
- multiple resources
- decision is based on perceived payoff
- imperfect and obsolete knowledge to base their decisions
- lacks a central controller
- decisions are made in an uncoordinated fashion and asynchronously

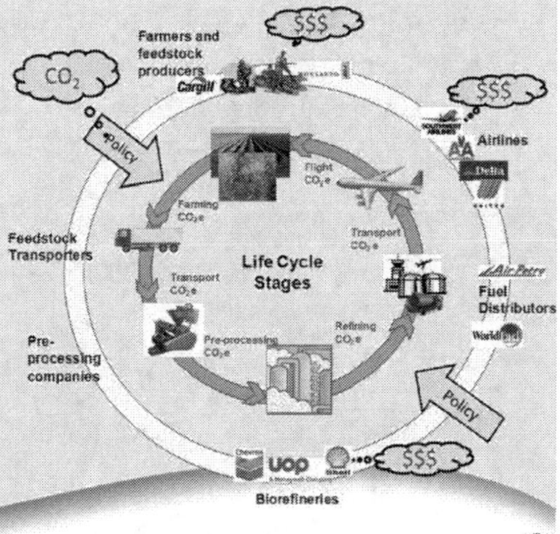

Dynamics of Biofuels Supply Chain Network

Farmers' Options	Biorefineries' Options	Users' Options
Decision 1: Energy Crops	Decision 1: Produce Biofuels	Decision 1: Biofuels
Decision 2: Food Crops	Decision 2: Not to Produce Biofuels	Decision 2: Fossil Fuels

$$\frac{df}{dt} = \alpha\,[\rho(f(t-\tau) - f(t)]$$

f : fraction of agents taking *Decision 1*
α : evaluation rate
ρ : **preference probability** function
G_1 : payoff from taking *Decision 1*
G_2 : payoff from taking *Decision 2*
σ : payoff uncertainty
τ : time delay

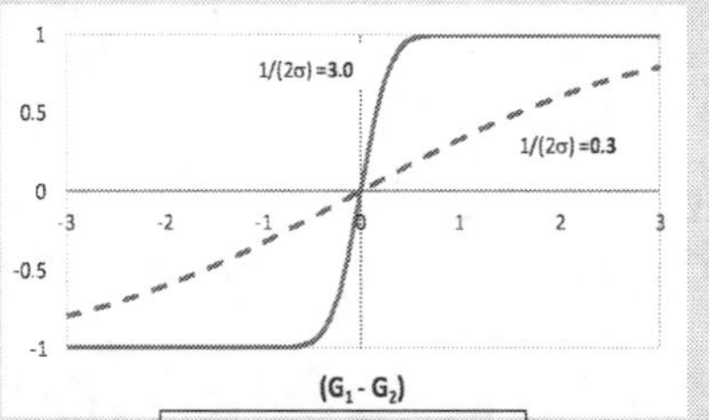

Error Function, erf

$$\rho = \frac{1}{2}\left[1 + \text{erf}\left(\frac{G_1(f) - G_2(f)}{2\sigma}\right)\right]$$

Behavior of Supply Chain Network

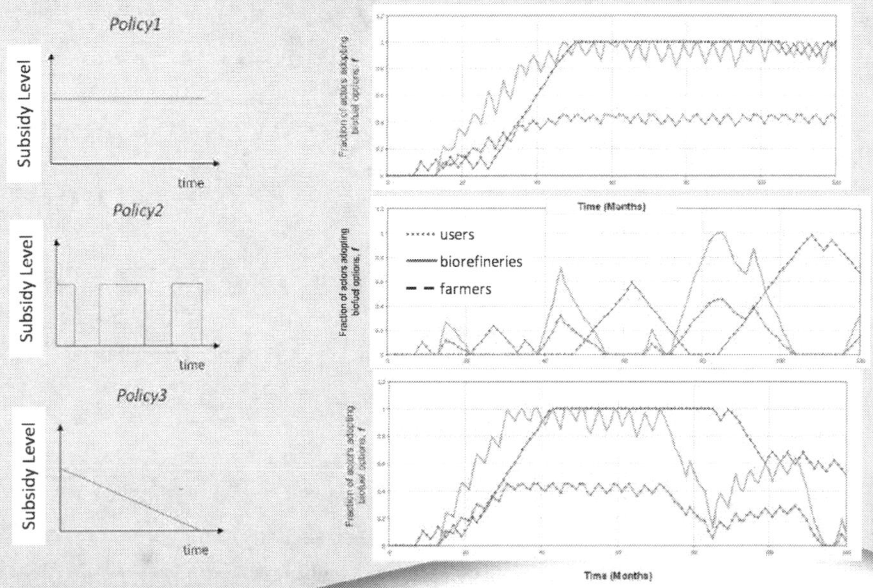

Agusdinata, D.B., Lee, S., and Zhao, F., Thissen, W., "Agent-based Simulation Modeling Framework for Uncovering System Behaviors in Biofuels Supply Chain Network," *Simulation: Transactions of the Society for Modeling and Simulation International*, (forthcoming)

Multi-Actor Life Cycle Assessment

- **Traditional LCA studies focus:**
  - mostly on the environmental performance of technology options
  - largely left out the economic aspect of the system in question
  - or at most include economic performance as a separate part
- **Multi-Actor LCA:**
  - Explicit consideration of supply chain actors
- **Expected Results:**
  - Better estimate of feedstock penetration level and life cycle emissions impact

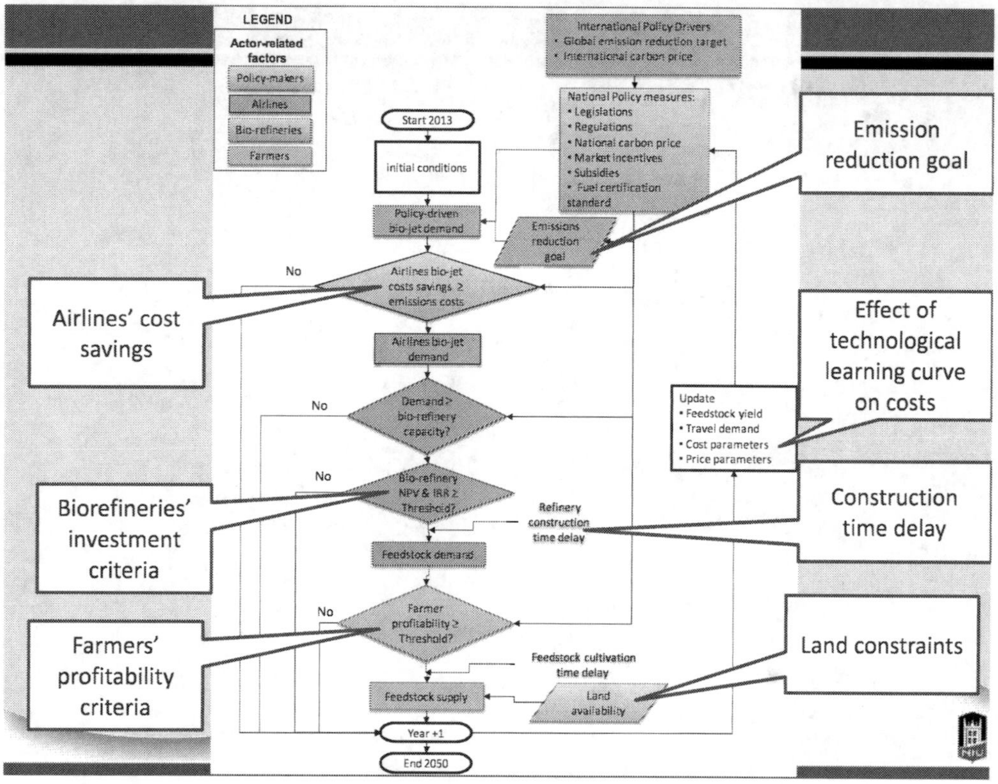

Result: production cost, NPV, IRR

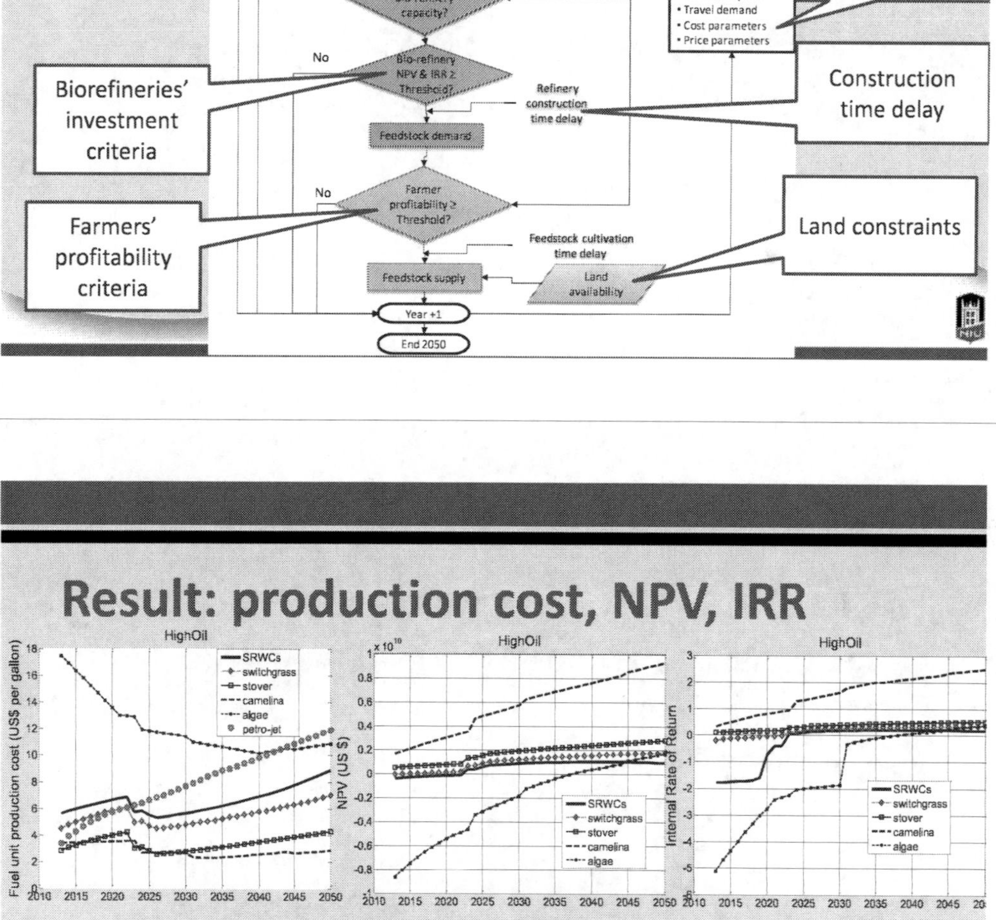

Reference: Agusdinata, D.B., Zhao, F., Ileleji, K.E. and DeLaurentis, D.A. "Life Cycle Assessment of Potential Bio-jet Fuel Production in the United States", *Environmental Science and Technology*, vol. 45 (21), pp. 9133–9143, 2011

Evolution of US Aviation Life Cycle Emissions

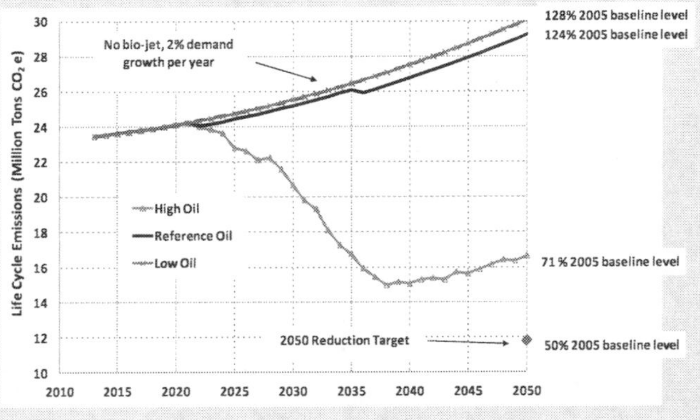

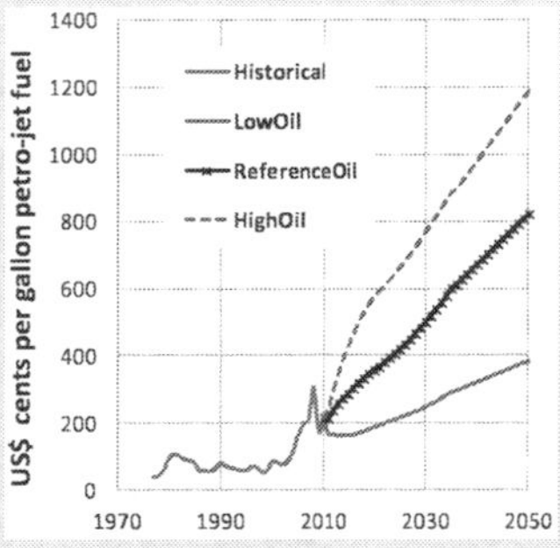

Source:

Lignocellulosic derived liquid fuels can only become competitive under high oil price scenario

Taking uncertainties into account

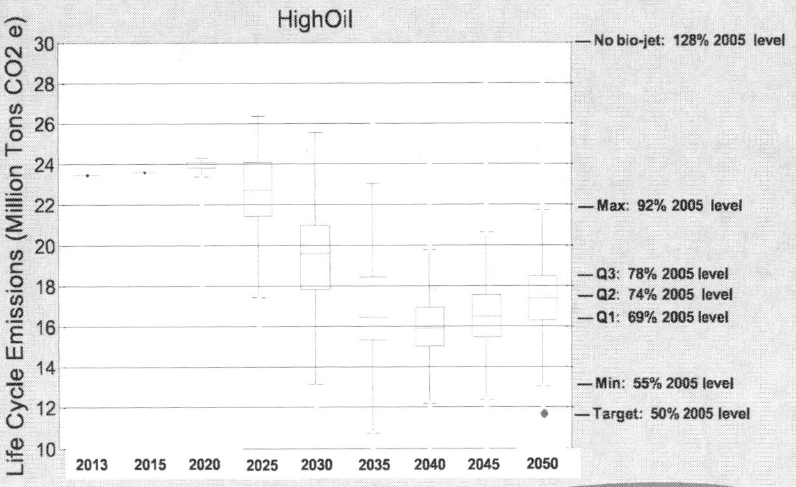

RESULTS: BIOFUELS EQUITY ISSUE

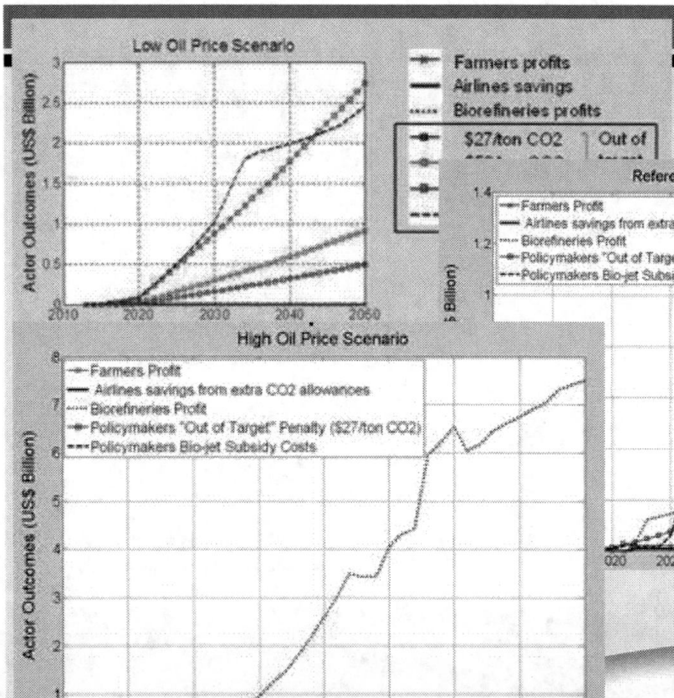

Preliminary Result (2)

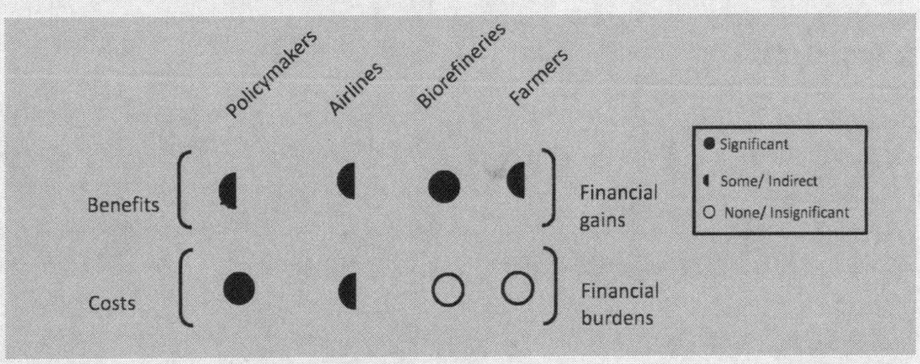

- **Biorefineries will be main beneficiaries**
  - But are exposed to financial risks
- **Policymakers (incl. consumers and taxpayers) will bear most the financial burdens**

Agusdinata, D.B., "System Design Framework for Equity/ Fairness among Actors, " Procedia Computer Science, Volume 16, , Pages 1122-1131, 2013

Concluding Remarks

- **Added values of the multi-actor system approach:**
  - Insights:
    - Understanding of **collective behavior** resulting from **decisions of individual actors**
    - A decision may perfectly be **rational** from an actor perspective but could be **detrimental** from the view of overall system
  - Policy implications:
    - **Policy design** more aligned to actors' interests
    - **Quantitative** basis for negotiation and compromise
- **Challenges:**
  - **Calibration** of the parameter space
  - **Validation** of the overall system behavior with empirical data or expert opinion

EXTRA SLIDES

Exploring the Meaning of Fairness

Equity Principles
Egalitarianism
Spoiler of equilibrium pays
Ability to pay
Grandfathering
Merit

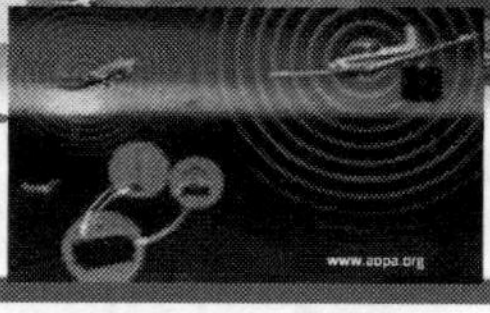

Approach

- most biofuel supply chain studies to date take the bio-refinery as the default customer under the assumption that the bio-refinery is in control of the entire supply chain.
- Most of these studies adopted the classical production/distribution MIP approach in order to design a network that maximizes bio-refinery profits

- As the SC network consist of a large number of firms from multiple interrelated industries, a complex adaptive system (CAS) perspective allows a supply network manager to make local decisions while considering the complexity of the overall system.
- Furthermore, it is argued that due to the prevalence of the use of information technology, supply chains have greatly increased in complexity almost to the level of biological system.
- In this environment there is a need for coordination strategies among supply chain actors to achieve an adaptive collective behavior.

SYSTEM DESIGN FOR EQUITY/FAIRNESS AMONG ACTORS

A System-of-Systems Definition for the Sustainability of Aviation Biofuels

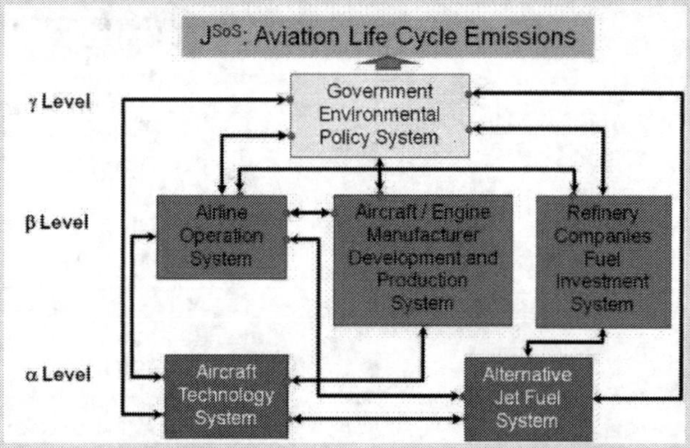

The common system objective, J_z^{SOS} transcends all individual objectives of actors, $J_z^{Actor\ i}$.

Agusdinata, D.B. and DeLaurentis, D.A.,"Addressing Equity Issues in Multi-Actor Policymaking via a System-of-Systems Approach: A Case Study in Aviation Emissions Reduction", Journal of Systems Science and Systems Engineering, vol. 20(1), pp. 1-24, 2011

IOWA STATE UNIVERSITY
Center for Sustainable Environmental Technologies

Smaller Is Better
Fostering Growth in the Biofuels Industry with Energy Manufacturing

Mark Mba Wright
Robert C. Brown
Mechanical Engineering
Iowa State University

Paul J. Componation
Industrial & Manufacturing Systems Engineering
Iowa State University

IOWA STATE UNIVERSITY
Center for Sustainable Environmental Technologies

Advanced manufacturing and clean energy technologies are pillars of future economic development

--President Obama 2013 State of the Union

Manufacture of energy conversion modules for distributed generation of renewable power

*--NSF Definition of **Energy Manufacturing***

2

IOWA STATE UNIVERSITY
Center for Sustainable Environmental Technologies

Energy Manufacturing Applied to Biorefining

- Concept for Distributed Biobased Manufacturing:
  - Manufacture of biomass processing modules for distributed conversion of biomass to biofuels and biobased products
- Attributes:
  - Mass production of processing modules
  - Procurement in quantity
  - Enhanced learning rates during manufacturing
  - Highly automated processing operations

3

IOWA STATE UNIVERSITY
Center for Sustainable Environmental Technologies

Advantages of Distributed Biobased Manufacturing

Conventional Biorefineries	Distributed Biobased Manufacturing
Large capital investments	Smaller capital investment
Long (years) deployment periods	Shorter deployment periods (months)
High technical and market risks	Lower project risks
Low learning rate – each facility is unique	High learning rate – accelerated by building multiple units
Complex feedstock supply logistics	Simpler supply logistics
Concentrated economic opportunity	Distributed economic opportunity

4

IOWA STATE UNIVERSITY
Center for Sustainable Environmental Technologies

Mass Production of Processing Modules Could Achieve Similar "Economies of Scale" Of Single Large Plants

Mass Production Example from Manufacture of Automobiles

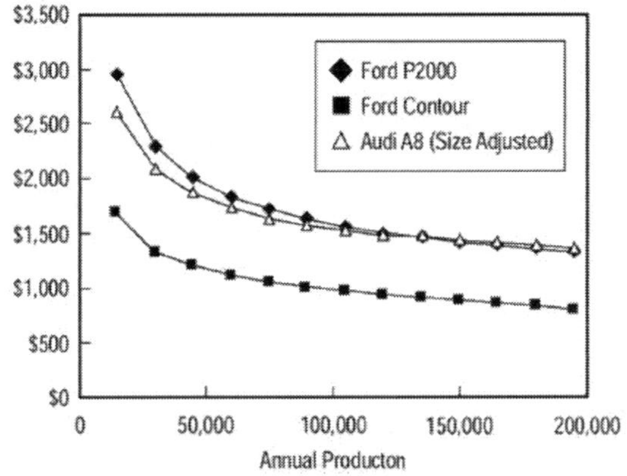

Large processing plant economies of scale:
Capital Costs proportional to (Plant Output)$^{0.66}$

Mass production economies of scale:
Capital Costs proportional to (Aggregate Output)$^{0.70}$

Source: A Kelkar, R Roth, J Clark. Automobile Bodies: Can Aluminum Be an Economical Alternative to Steel? JOM 53(8):28-32.2001

5

IOWA STATE UNIVERSITY
Center for Sustainable Environmental Technologies

Example from PV Solar Industry

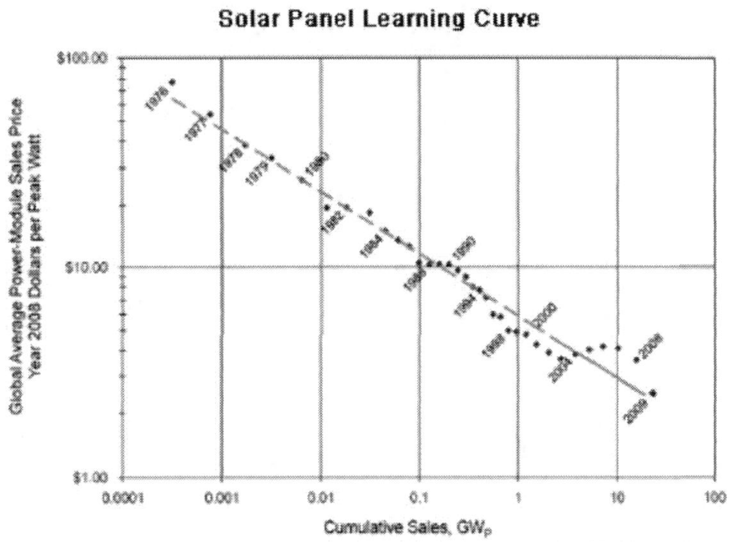

Ken Zweibel, GWU

6

IOWA STATE UNIVERSITY
Center for Sustainable Environmental Technologies

Examples Across Industries

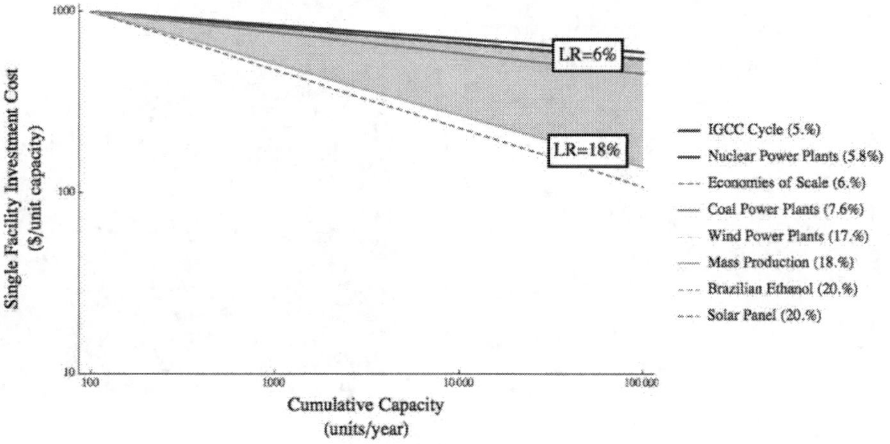

Source: Adapted from McDonald and Schrattenholzer

7

IOWA STATE UNIVERSITY
Center for Sustainable Environmental Technologies

Impact of Learning on Brazilian Sugarcane to Ethanol Costs

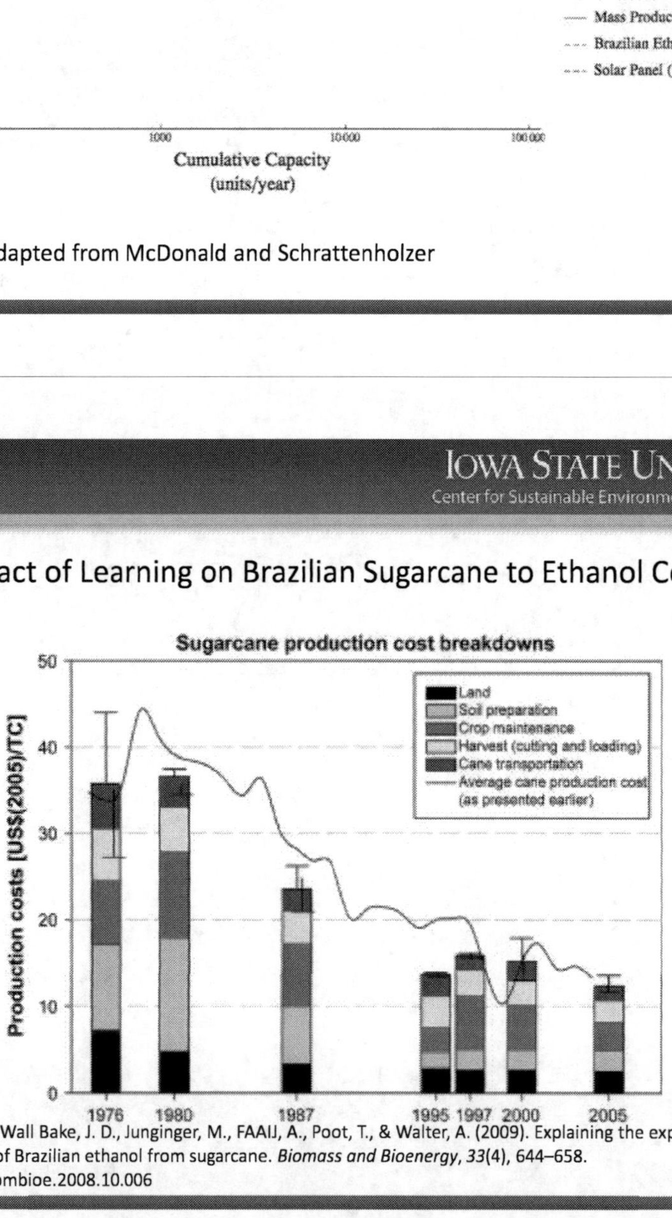

Source: Van Den Wall Bake, J. D., Junginger, M., FAAIJ, A., Poot, T., & Walter, A. (2009). Explaining the experience curve: Cost reductions of Brazilian ethanol from sugarcane. *Biomass and Bioenergy*, *33*(4), 644–658. doi:10.1016/j.biombioe.2008.10.006

8

IOWA STATE UNIVERSITY
Center for Sustainable Environmental Technologies

Impact of Learning on US Corn Ethanol Capital Costs

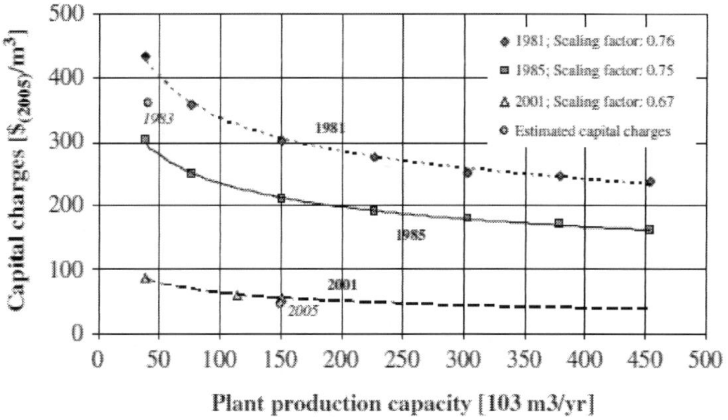

Fig. 7. Scaling factors for capital costs in different years. Capital costs have been reduced by upscaling, but even more by learning and other non-scale-related development. (*Data sources*: LeBlanc and Prato, 1982; Gavett et al., 1986; Whims, 2002).

Source: Hettinga, W. G., Junginger, H. M., Dekker, S. C., & Hoogwijk, M. (2009). Understanding the reductions in US corn ethanol production costs: An experience curve approach. *Energy Policy*.

IOWA STATE UNIVERSITY
Center for Sustainable Environmental Technologies

Scaling Relations for Biorefineries

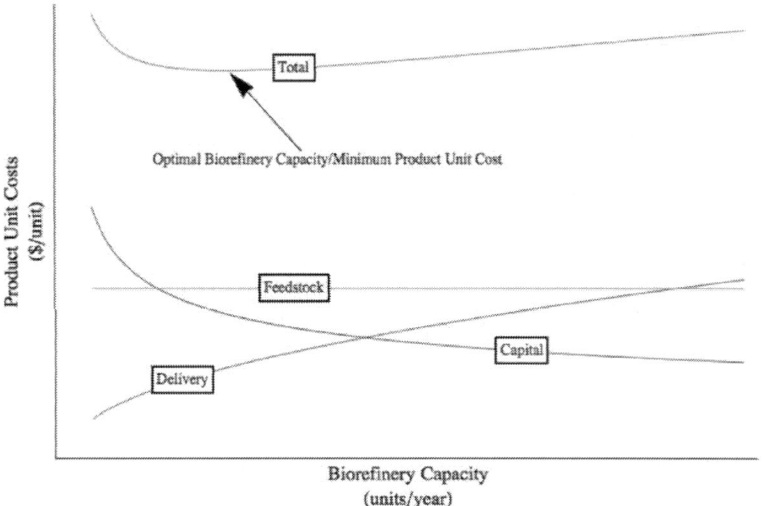

Source: Daugaard T., Mutti L., Wright M., Brown R. C., Componation P. "Learning Rates and Their Impacts on the Optimal Capacities of Biorefineries" (In Press 2014) *Biofuels, Bioproducts & Biorefining*.

10

IOWA STATE UNIVERSITY
Center for Sustainable Environmental Technologies

Scaling and Learning Curve Relations

Economies of Scale:

$$C_T(M) = C_{Po}\left(\frac{M}{M_o}\right)^n + C_{Do}\left(\frac{M}{M_o}\right)^m + C_{Fo}\left(\frac{M}{M_o}\right)$$

Learning Curve:

$$C_t(M_t) = C_o {M_t}^{\log_2(1-LR)}$$

C (Capital Cost):
T (Total unit/facility); P (Plant/Investment); D (Delivery); F (Feedstock); O (Original)
M (Capacity):
t (Total industry/facility); O (Original)
n: facility capital scale factor; m: feedstock delivery scale factor;
LR: Learning Rate

11

IOWA STATE UNIVERSITY
Center for Sustainable Environmental Technologies

Scaling Curve with Learning Effects

$$\underbrace{\frac{C_{po}}{M}\left(\frac{M}{M_o}\right)^n}_{\text{Scaling}}\ \underbrace{\left(\frac{M_t+B_o}{M_o+B_o}\right)^{\log_2(1-LR_p)}}_{\text{Learning}}$$

Scaling Learning

B_O: Prior Knowledge Factor (How many biorefineries has the industry built?)

12

IOWA STATE UNIVERSITY
Center for Sustainable Environmental Technologies

Case study: Gasification and Fast Pyrolysis

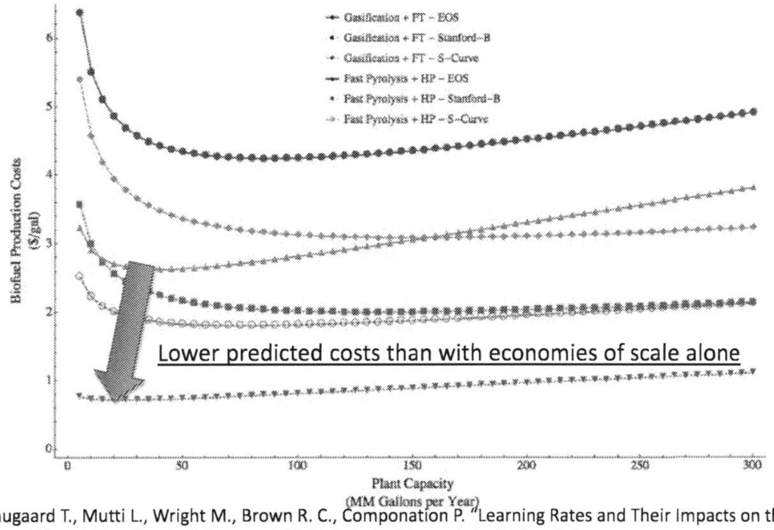

Source: Daugaard T., Mutti L., Wright M., Brown R. C., Componation P. "Learning Rates and Their Impacts on the Optimal Capacities of Biorefineries" (In Press 2014) *Biofuels, Bioproducts & Biorefining.*

13

IOWA STATE UNIVERSITY
Center for Sustainable Environmental Technologies

Research Findings

- Scaling curves with learning effects predict lower biofuel costs than economies of scale alone.
- Fast pyrolysis and grain ethanol platforms have smaller optimal capacities than other technologies.
- Smaller platforms have higher learning rates and a faster approach to lower costs.

14

IOWA STATE UNIVERSITY
Center for Sustainable Environmental Technologies

Smaller facilities foster growth by:

Lowering cost of entry

Increasing learning rates

Lowering overall biofuel costs

How can we benefit

the most from learning rates?

15

IOWA STATE UNIVERSITY
Center for Sustainable Environmental Technologies

Research Approach

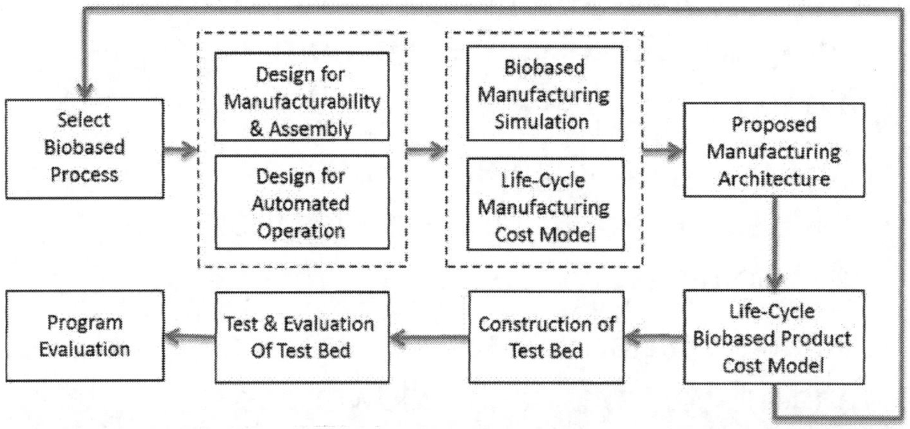

16

Small and Distributed vs. Large and Centralized

17

IOWA STATE UNIVERSITY
Center for Sustainable Environmental Technologies

Distributed Processing Applied to Pyrolysis

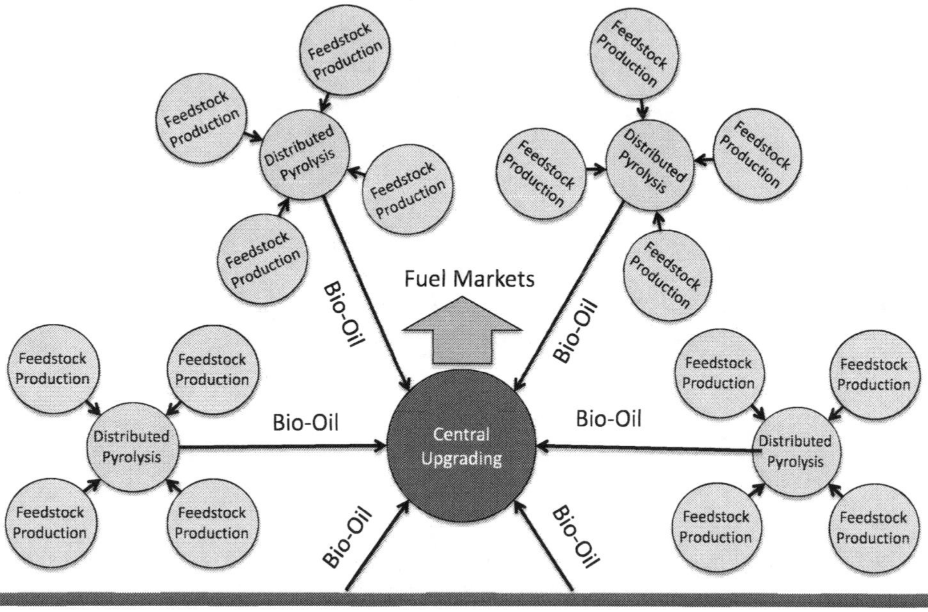

Acknowledgments

Biocentury Research Farm

IOWA EPSCoR

BEI
IOWA STATE UNIVERSITY

Small-Scale Biorefineries: Towards the Biofuel and Bioenergy Sustainability at Local Basis

Souza Aguiar, R., UFSC
Sousa Baudel, P. H., UFSC
Jose, R. A., AMERICA BIOMASS TECHNOLOGIES
de Abreu, C. A. M., UFPE
Baudel, H. M., UFPE
Originally presented on: 7/24/2014 8:48:00 - 9:06:00

Small to medium- scale biomass biorefineries have been identified as a promising route to the creation of local and regional agroindustrial clusters in Brazil. By delivering multiple products from starchy and lignocellulosic biomasses, a biorefinery constituted by integrated plants and processes can make feasible the economic exploitation of a myriad of low-value agricultural and industrial residues. In principle, different biomass components can be converted into sugars and other carbon-rich products, which in turn can be transformed into high-valued chemical products and high-volume biofuels, while generating electricity and process heat for self-consumption. In this scenario, the high-value products enhance profitability, the high-volume fuels contribute to support energy needs, and the power production reduces costs while avoiding greenhouse gas emissions. Hence, the biorefinery concept envisages the maximal value derived from the biomass feedstock at minimal impact to the environment. This paper describes a biorefinery constituted by integrated plants that produce ethanol, active carbon, food-grade carbon dioxide and protein extract (for cattle feed). Agricultural residues such as rice shells and straw as well as non-food energy-rice were used as starchy and lignocellulosic feedstocks. The mentioned biorefinery concept has been built on two different biomass-to-products platforms. The "sugar platform" is based on chemical and biochemical conversion processes, particularly the fermentation of sugars extracted from cellulosics, while the "carbon platform" is based on thermochemical routes with emphasis on the carbonisation of the cellulignin fractions.

PAN-AMERICAN BIOFUELS & BIOENERGY SUSTAINABILITY

UNIVERSITY OF CAMPINAS
School of Chemical Engineering

UNICAMP

Alcoholic Fermentation using Immobilized Cells

FACULDADE DE ENGENHARIA QUÍMICA

July 22 th to 25 th, 2014

Instituto de Química

Introduction

Biomass & Bioenergy

Ethanol

- *Less polluting*
- *Got from renewable source*
- *Environmentally friendly*

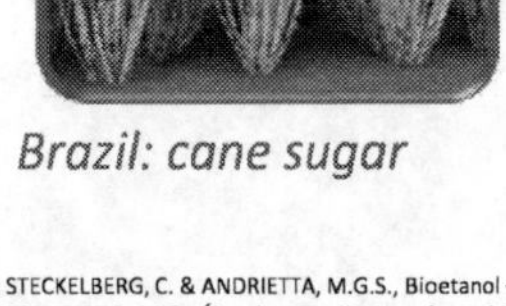
Brazil: cane sugar

USA: corn

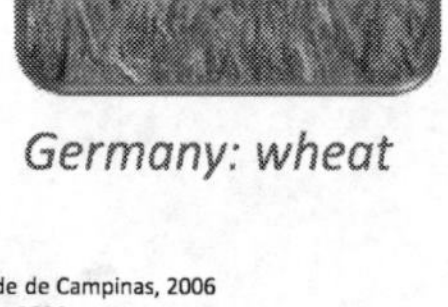
Germany: wheat

ANDRIETTA, S.R.; STECKELBERG, C. & ANDRIETTA, M.G.S., Bioetanol – Brasil, 30 anos na vanguarda. *Multiciência*, Universidade de Campinas, 2006
FURTADO, T.A. & SCANDIFFIO, M.G., Álcool no Brasil – Uma longa história. *Scientific American Brazil*, n. 53, p.66-71, Outubro, 2006

Introduction

R. Wendhausen, A. Fregonesi, P. J. S. Moran, I. Joekes, J. A. R. Rodrigues, E. Tonella, and K. Althoff. Continuous fermantation of sugar cane syrup using immobilized yeast cells. *J. of Biosc. and Bioeng.*, 91(1):48–52, 2001.

Glucose (%)	Theoretical yield based on CO_2 measured (%)	
	Immobilized cells [a]	Free cells [b]
10	87.0	90.0
20	92.0	88.6
30	96.0	77.8
40	72.4	54.6
50	57.8	31.6

[a] – 0,60 g of yeast in o,6 g of Chrysotile
[b] – 0,60 g of yeast

Cell immobilization methods

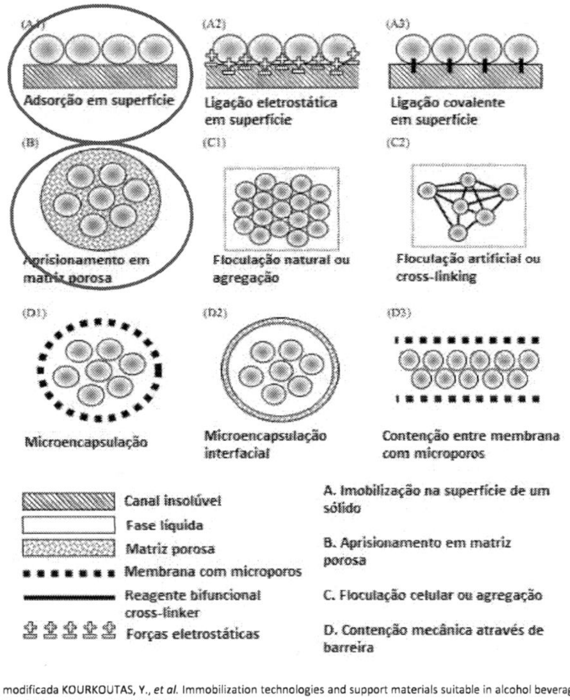

FONTE: Figura modificada KOURKOUTAS, Y., *et al.* Immobilization technologies and support materials suitable in alcohol beverages production: a review. *Food Microbiology*. V.21, p.377-397, 2004.

Methods

Sodium alginate

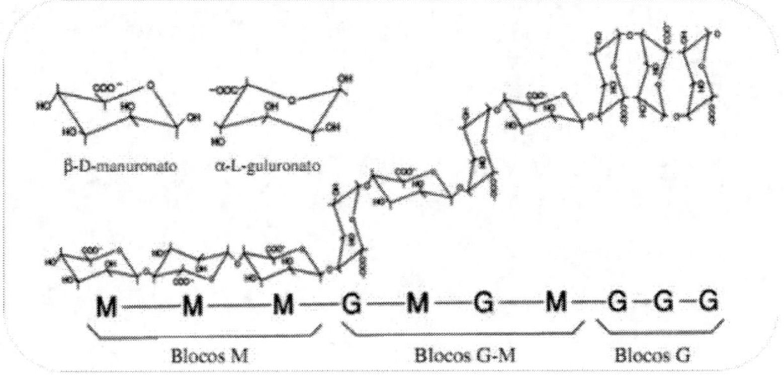

RODRIGUES, A.P., Preparação e caracterização de membranas de quitosana e alginato para aplicação na terapia de lesões. 2008. Tese (Doutorado em Engenharia Química) – Universidade Estadual de Campinas, Campinas, SP.

Methods

Calcium alginate preparation

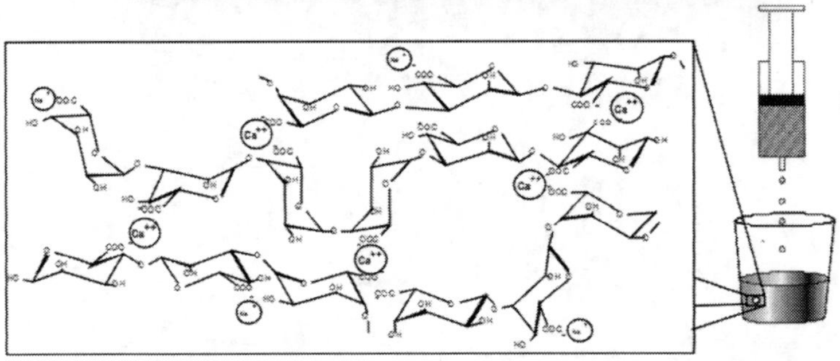

COVIZZI, L.G., et al., Semina: Ciências Exatas e Tecnológicas, Londrina, v.28, n.2, p.143-160, jul./dez., 2007.

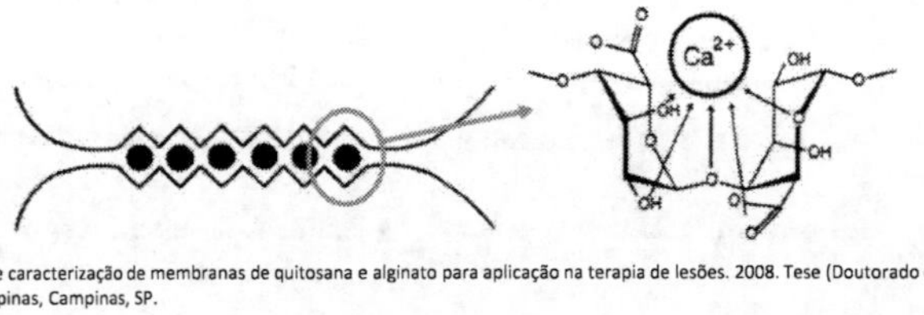

RODRIGUES, A.P., Preparação e caracterização de membranas de quitosana e alginato para aplicação na terapia de lesões. 2008. Tese (Doutorado em Engenharia Química) – Universidade Estadual de Campinas, Campinas, SP.

Methods

Chitosan

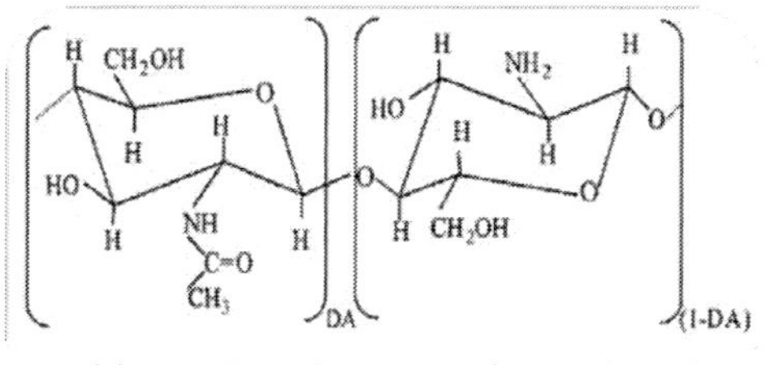

Modificada de DASH, M.; CHIELLINI, F.; OTTENBRITE, R.M.; CHIELLINI, E., Chitosan—A versatile semi-synthetic polymer in biomedical applications, *Progress in Polymer Science*, v. 36, p. 981–1014, 2011.

Methods

Calcium alginate and Chitosan interaction

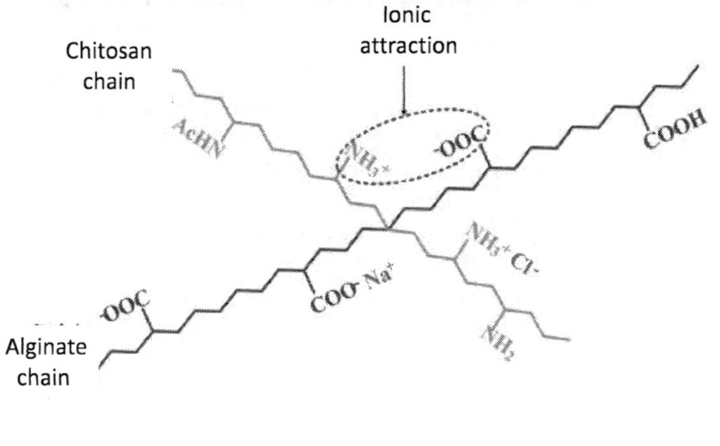

RODRIGUES, A.P., Preparação e caracterização de membranas de quitosana e alginato para aplicação na terapia de lesões. 2008. Tese (Doutorado em Engenharia Química) – Universidade Estadual de Campinas, Campinas, SP.

Experimental

Alcoholic fermentation using immobilized cells

Calcium alginate beads

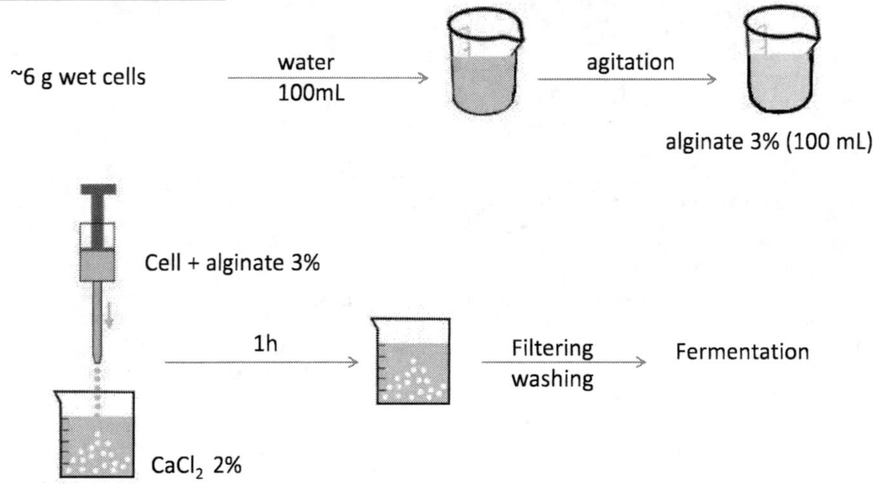

Experimental

Alcoholic fermentation using immobilized cells

chitosan-covered calcium alginate beads

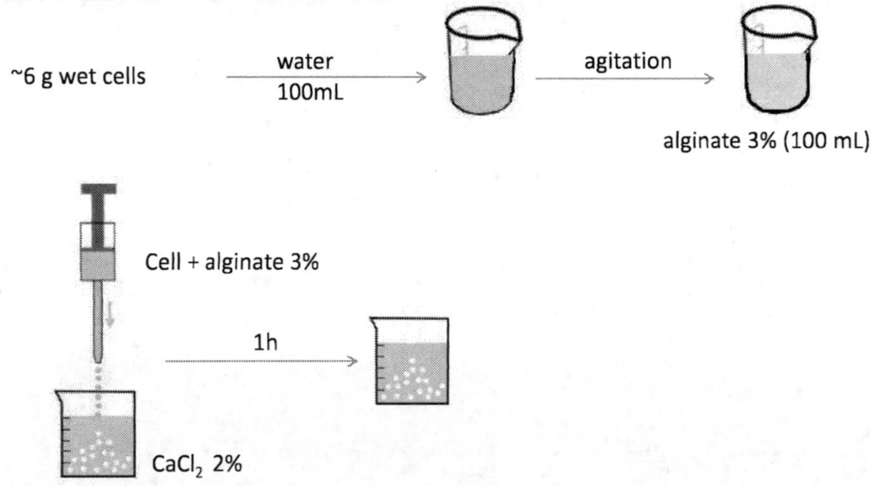

Experimental

Alcoholic fermentation using immobilized cells

chitosan-covered calcium alginate beads

After 1h in $CaCl_2$ solution:

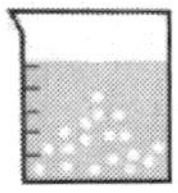

Filtering
washing

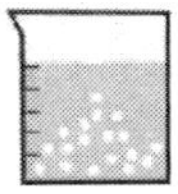

Chitosan solution 0,25% in acetic acid 5% (30 min).

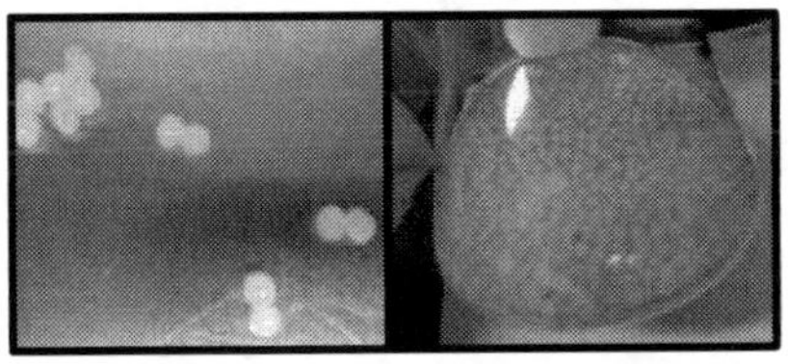

Filtering
washing

Fermentation

Free cells fermentation

Results and discussion

Glucose consumption and ethanol production

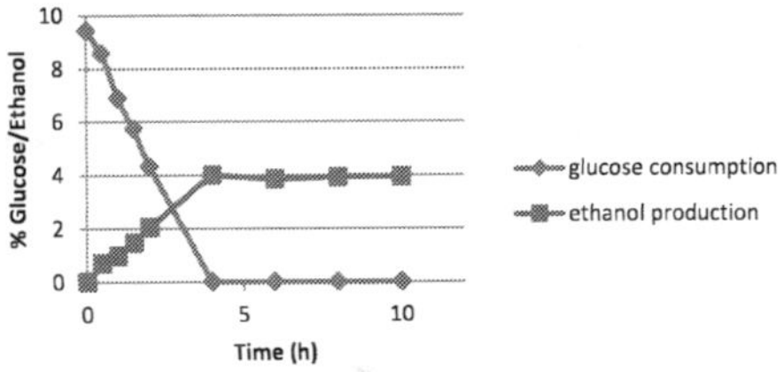

Carbon source: Glucose

Final ethanol concentration: 40 g/L
yield: 78%

Sucrose consumption and ethanol production

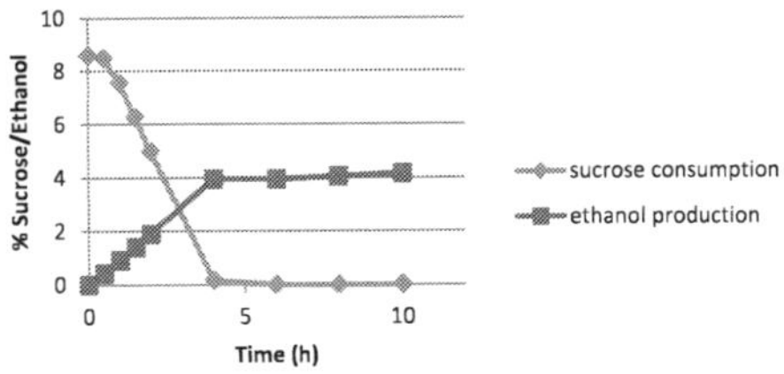

Carbon source: Sucrose

Final ethanol concentration: 40 g/L
yield: 74,3%

Immobilized cells fermentation

Results and discussion

Glucose consumption - immobilized cells in calcium alginate beads

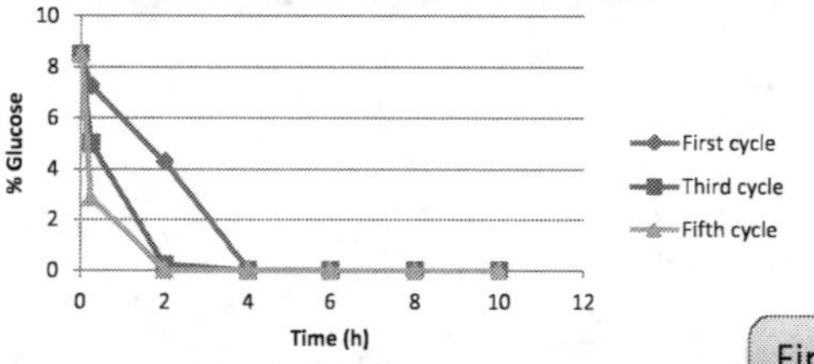

Carbon source
Glucose

Final ethanol concentration: 33 g/L
yield: 64,6%

Ethanol production - immobilized cells in calcium alginate beads

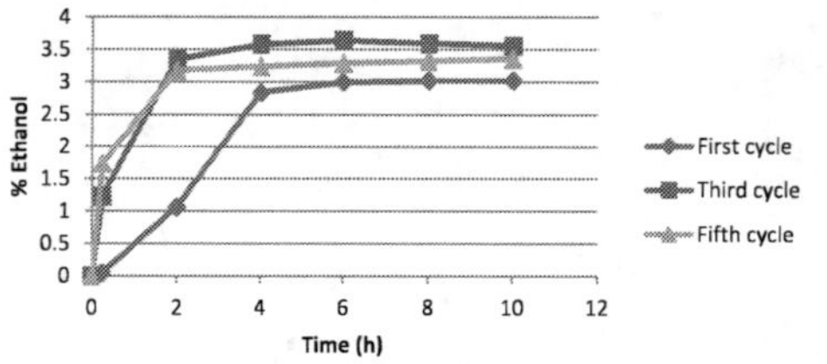

Immobilized cells fermentation

Results and discussion

Glucose consumption - immobilized cells in chitosan covered calcium alginate beads

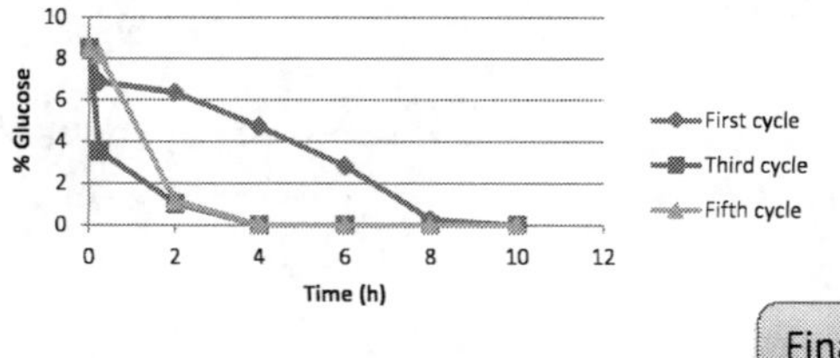

Carbon source
Glucose

Final ethanol concentration: 31 g/L
yield: 60,7%

Ethanol production - immobilized cells in chitosan covered calcium alginate beads

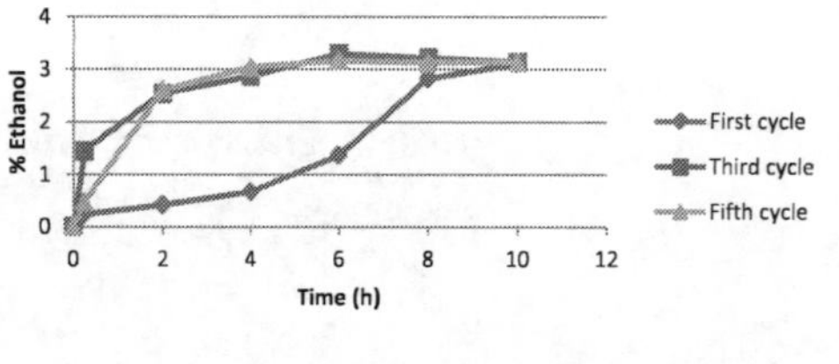

Immobilized cells fermentation

Results and discussion

Sucrose consumption - immobilized cells in calcium alginate beads

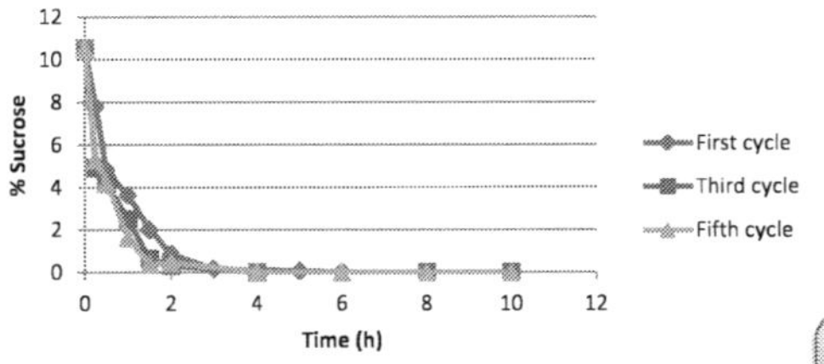

Carbon source
Sucrose

Final ethanol concentration: 33 g/L
yield: 61,3%

Ethanol production - immobilized cells in calcium alginate beads

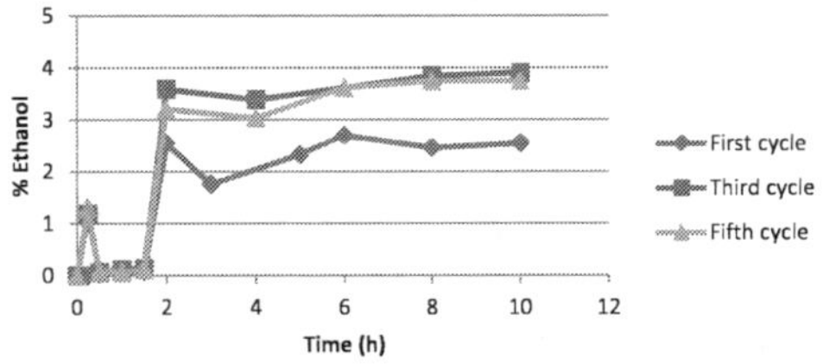

Immobilized cells fermentation

Results and discussion

Sucrose consumption - immobilized cells in chitosan covered calcium alginate beads

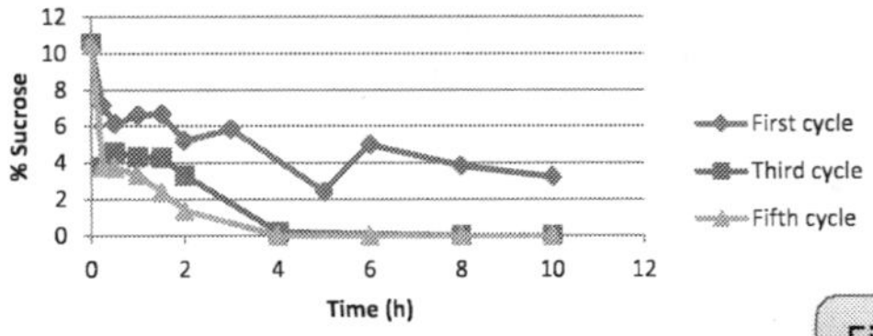

Carbon source
Sucrose

Final ethanol concentration: 32 g/L
yield: 59,5%

Ethanol production - immobilized cells in chitosan covered calcium alginate beads

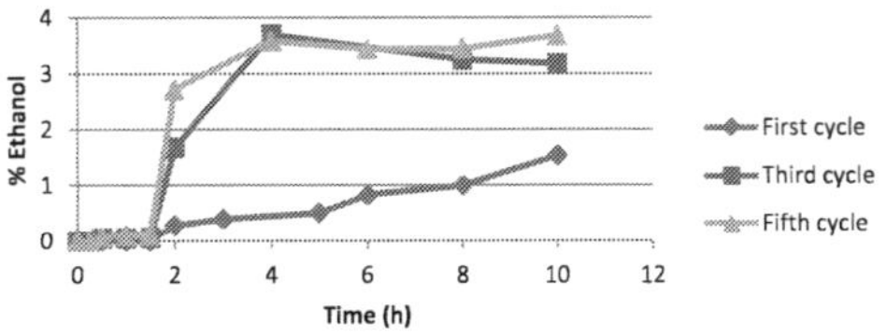

Results and discussion

	Free cells		Immobilized cells CAB		Immobilized cells CCCAB	
	glucose	sucrose	glucose	sucrose	glucose	sucrose
Ethanol production (g/L)	40	40	33	33	31	32
Yield (%)	78	74,3	64,6	61,3	60,7	59,5

Legend:
CAB: Calcium alginate beads
CCCAB: Chitosan covered calcium alginate beads

Conclusions

1. Immobilized cells in calcium alginate and in chitosan-covered calcium alginate beads:

a. Substrate consumption and product production were faster after the first cycle: it suggests **adaptation**

b. After free substrate consumption there was ethanol production: it suggests **difusion** process

c. Using **glucose** as carbon source, maximum ethanol production was reached in 4 and 6 h to CAB and CCCAB, respectively

d. Using **sucrose** as carbon source, maximum ethanol production was reached in 2 and 4 h to CAB and CCCAB, respectively

e. Calcium alginate and chitosan covered calcium alginate beads allowed beads reuse during 8 cycles of approximately 10 hours of fermentation each

Conclusions

f. Ethanol production at the end of each cycle were somewhat similar

g. Chitosan covered calcium alginate beads were less brittle and broke in a smaller amount than calcium alginate beads

h. Chitosan exerted a physical barrier, so the consumption of substrate and product production was slower

Thanks

UNICAMP

FACULDADE DE ENGENHARIA QUÍMICA

IQ Instituto de Química

PAN-AMERICAN BIOFUELS & BIOENERGY SUSTAINABILITY

Prof. J. Roberto Nunhez

Associate Prof.
Gustavo P. Valença

MSc Juliana Canto Duarte

Prof. Paulo J. S. Moran

Prof. J. Augusto R. Rodrigues

Kansas State University | Biological and Agricultural Engineering

Incorporation of sweet sorghum Juice in the current dry-grind ethanol process for improved ethanol yields, energy saving, and water efficiency

RCN Conference on Pan American Biofuels & Bioenergy Sustainability
Recife, Brazil.

Center for Sustainable Energy, Kansas State University

Nana Baah Appiah-Nkansah
Kealin Saul
William Rooney
Donghai Wang

Introduction

Ethanol production in the US

Production, availability, and use of fuel ethanol is expected to increase.

- ❑ Energy Independence and Security Act of 2007

❑Varieties of Ethanol blended fuel exists in the US

❑Varieties of Ethanol blended fuel exists in the US

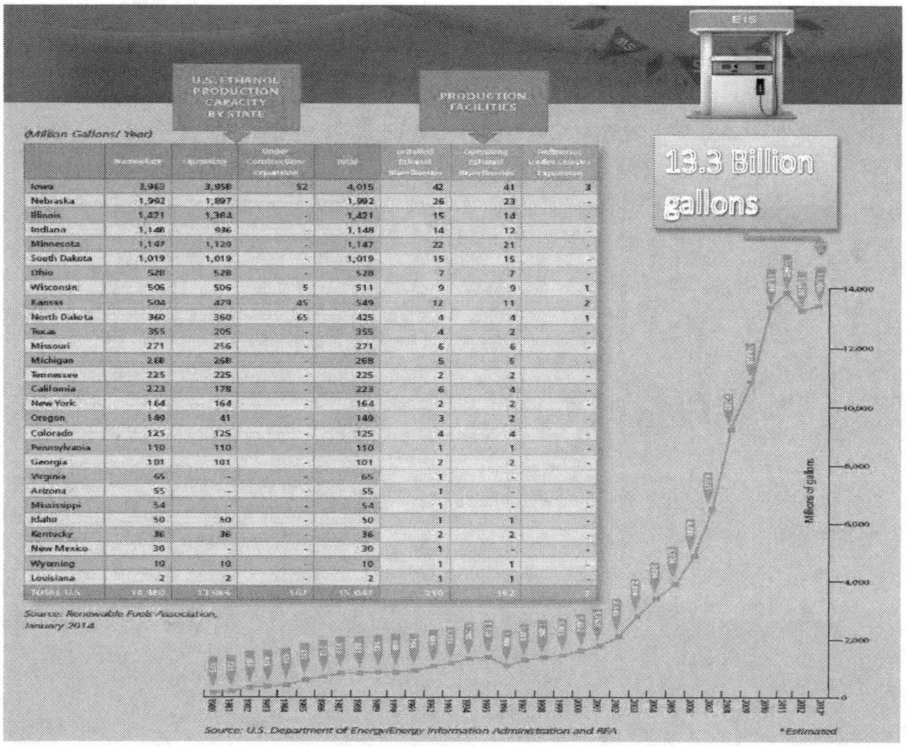

(Million Gallons/ Year)

	[illegible]	[illegible]	[illegible]	[illegible]	[illegible]	[illegible]	[illegible]
Iowa	3,963	3,958	52	4,015	42	41	3
Nebraska	1,992	1,897	-	1,992	26	23	-
Illinois	1,421	1,384	-	1,421	15	14	-
Indiana	1,148	936	-	1,148	14	12	-
Minnesota	1,147	1,129	-	1,147	22	21	-
South Dakota	1,019	1,019	-	1,019	15	15	-
Ohio	528	528	-	528	7	7	-
Wisconsin	506	506	5	511	9	9	1
Kansas	504	479	45	549	12	11	2
North Dakota	360	360	65	425	4	4	1
Texas	355	205	-	355	4	2	-
Missouri	271	256	-	271	6	6	-
Michigan	268	268	-	268	5	5	-
Tennessee	225	225	-	225	2	2	-
California	223	178	-	223	6	4	-
New York	164	164	-	164	2	2	-
Oregon	149	41	-	149	3	2	-
Colorado	125	125	-	125	4	4	-
Pennsylvania	110	110	-	110	1	1	-
Georgia	101	101	-	101	2	2	-
Virginia	65	-	-	65	1	-	-
Arizona	55	-	-	55	1	-	-
Mississippi	54	-	-	54	1	-	-
Idaho	50	50	-	50	1	1	-
Kentucky	36	36	-	36	2	2	-
New Mexico	30	-	-	30	1	-	-
Wyoming	10	10	-	10	1	1	-
Louisiana	2	2	-	2	1	1	-
TOTAL U.S.	[illegible]	[illegible]	[illegible]	[illegible]	[illegible]	[illegible]	7

Source: Renewable Fuels Association, January 2014

Sweet Sorghum

- Stem juice is rich in fermentable sugar - 53-85% sucrose, 9-33% glucose, and 6-21% fructose).
- Short period of growth period (140 days)
- Draught resistant.
- Higher biomass yield (45-80 t/ha).
- Environmentally friendly;
  - Requires less fertilizer, water usage for cultivation.

Objective

To develop a new technology for the current ethanol industry using sweet sorghum juice for ethanol production -

- to improve ethanol yield;
- safe water;
- energy conservation.

Conventional Dry – Grind Ethanol Production Process from Grain Sorghum

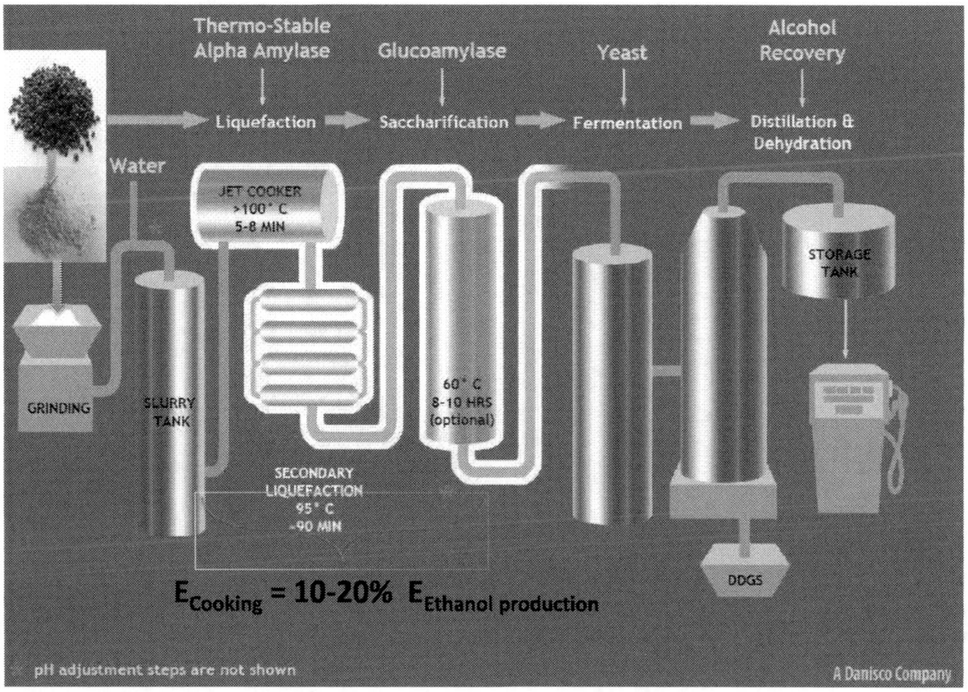

Methodology

Three phases

1. **Varying sorghum grain flour loading**

Three phases

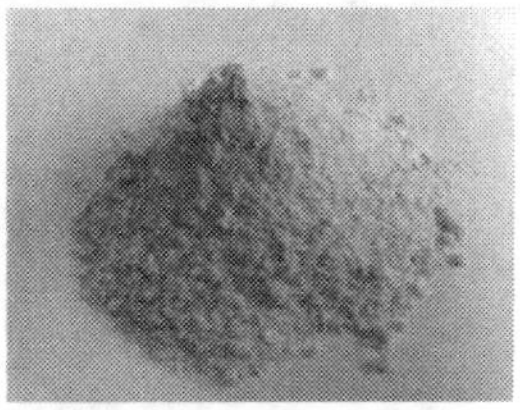

1. **Varying sorghum grain flour loading**

2. **Varying enzymatic hydrolyzing time**

Three phases

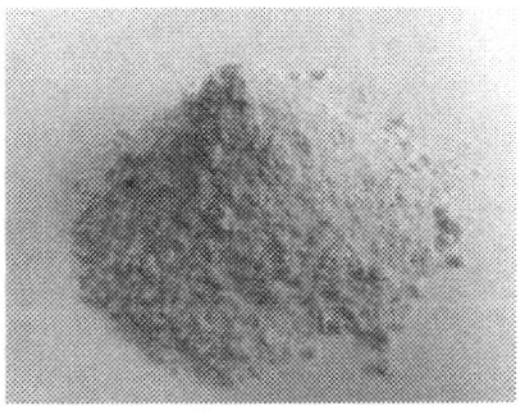

1. **Varying sorghum grain flour loading**

2. **Varying enzymatic hydrolyzing time**

3. **Raw starch hydrolysis** (by GSHE)

(An <u>energy conserving</u> alternative process)

Ethanol Fermentation of sweet sorghum with varying sorghum grain flour loading

Grain Sorghum flour **0 g, 6 g, 9 g, 12 g, 15 g**

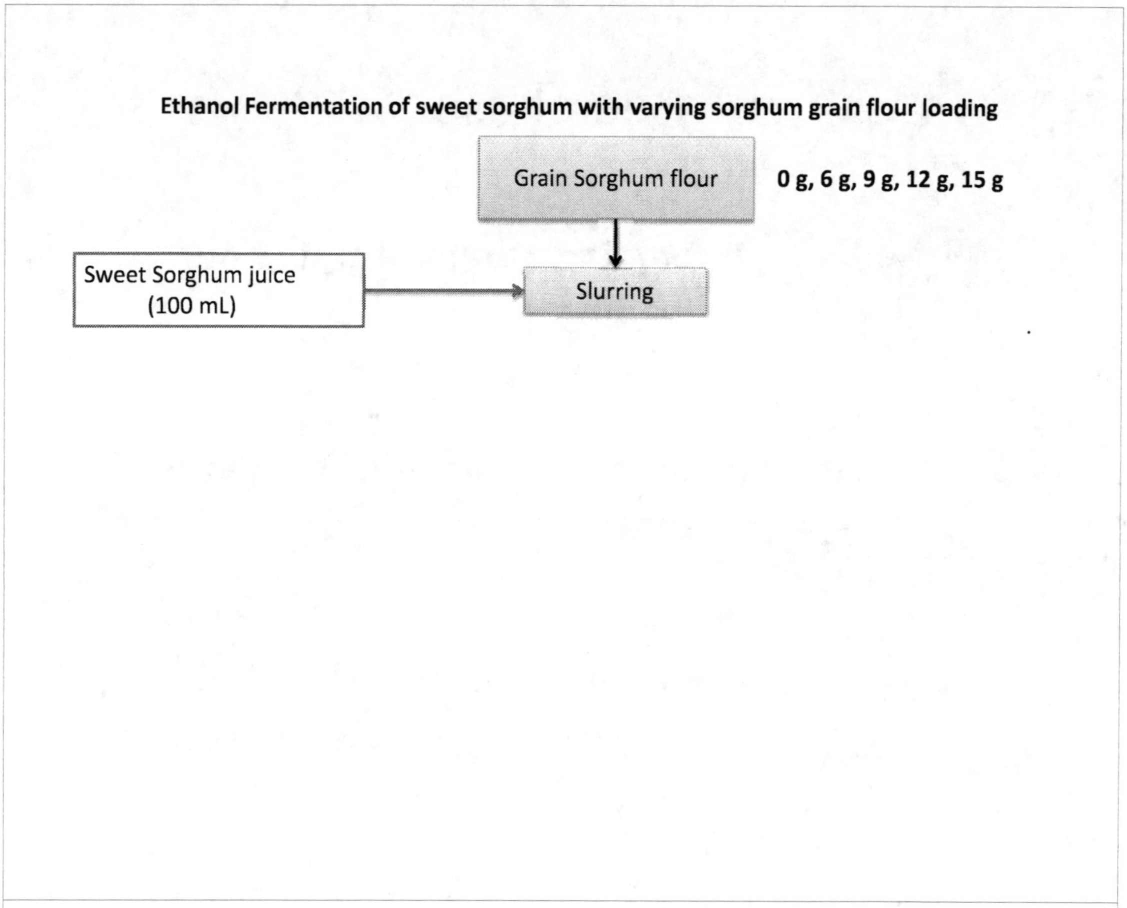
Ethanol Fermentation of sweet sorghum with varying sorghum grain flour loading
Grain Sorghum flour
0 g, 6 g, 9 g, 12 g, 15 g
Sweet Sorghum juice (100 mL)
Slurring

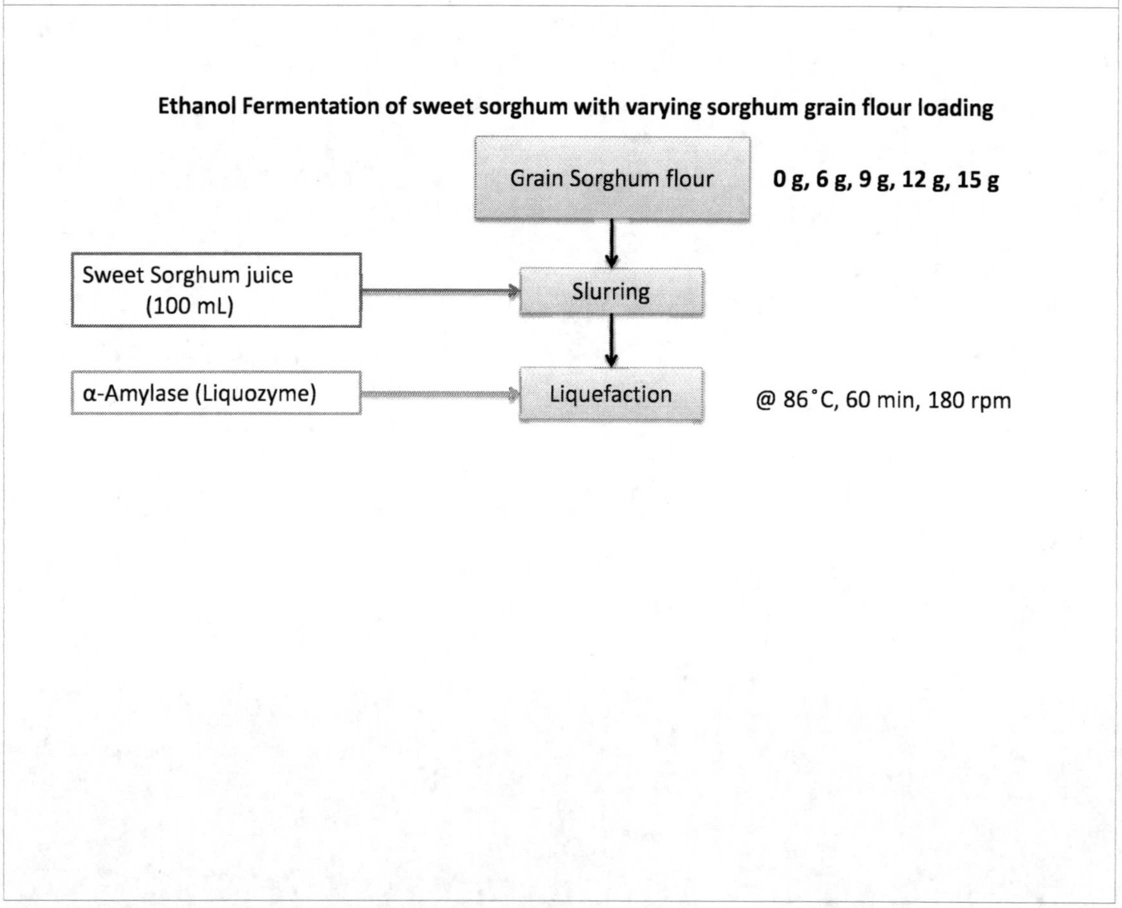
Ethanol Fermentation of sweet sorghum with varying sorghum grain flour loading
Grain Sorghum flour
0 g, 6 g, 9 g, 12 g, 15 g
Sweet Sorghum juice (100 mL)
Slurring
α-Amylase (Liquozyme)
Liquefaction
@ 86˚C, 60 min, 180 rpm

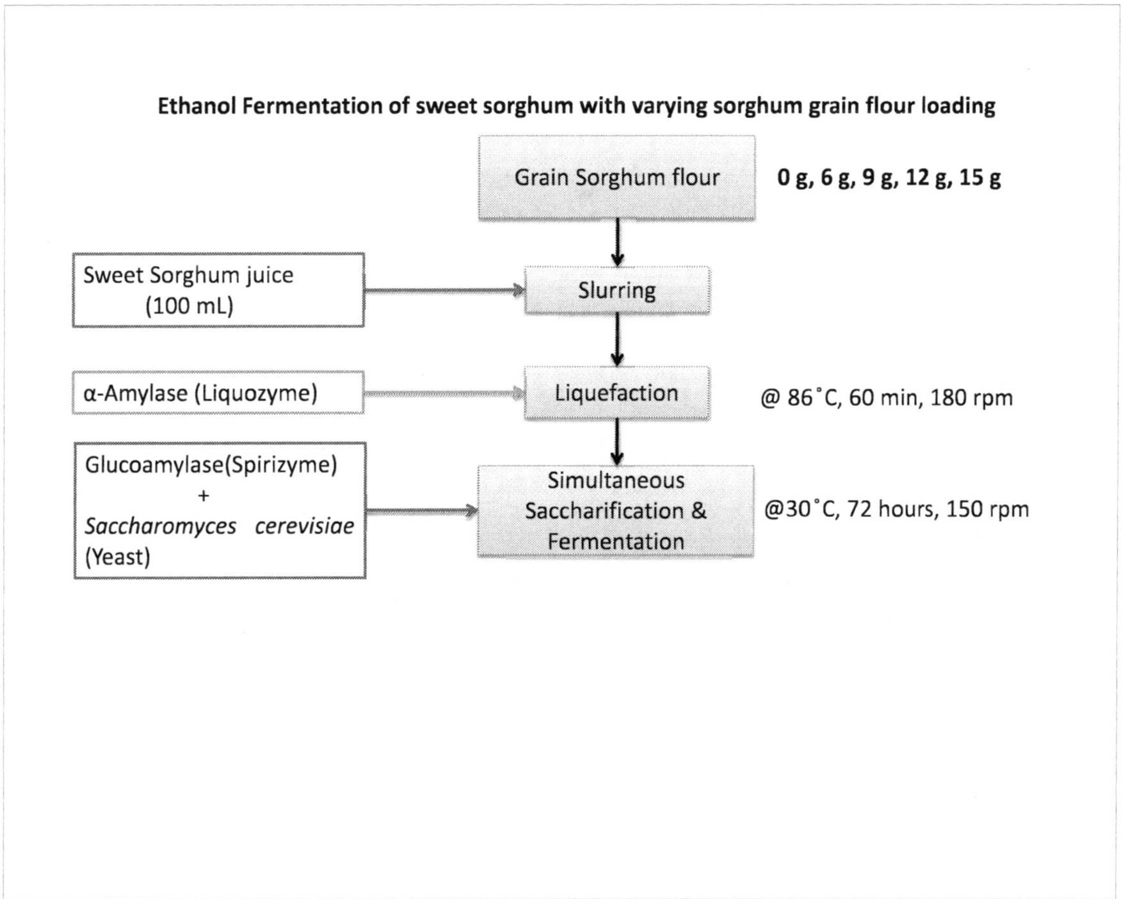
Ethanol Fermentation of sweet sorghum with varying sorghum grain flour loading
Grain Sorghum flour
0 g, 6 g, 9 g, 12 g, 15 g
Sweet Sorghum juice (100 mL)
Slurring
α-Amylase (Liquozyme)
Liquefaction
@ 86˚C, 60 min, 180 rpm
Glucoamylase(Spirizyme) + Saccharomyces cerevisiae (Yeast)
Simultaneous Saccharification & Fermentation
@30˚C, 72 hours, 150 rpm

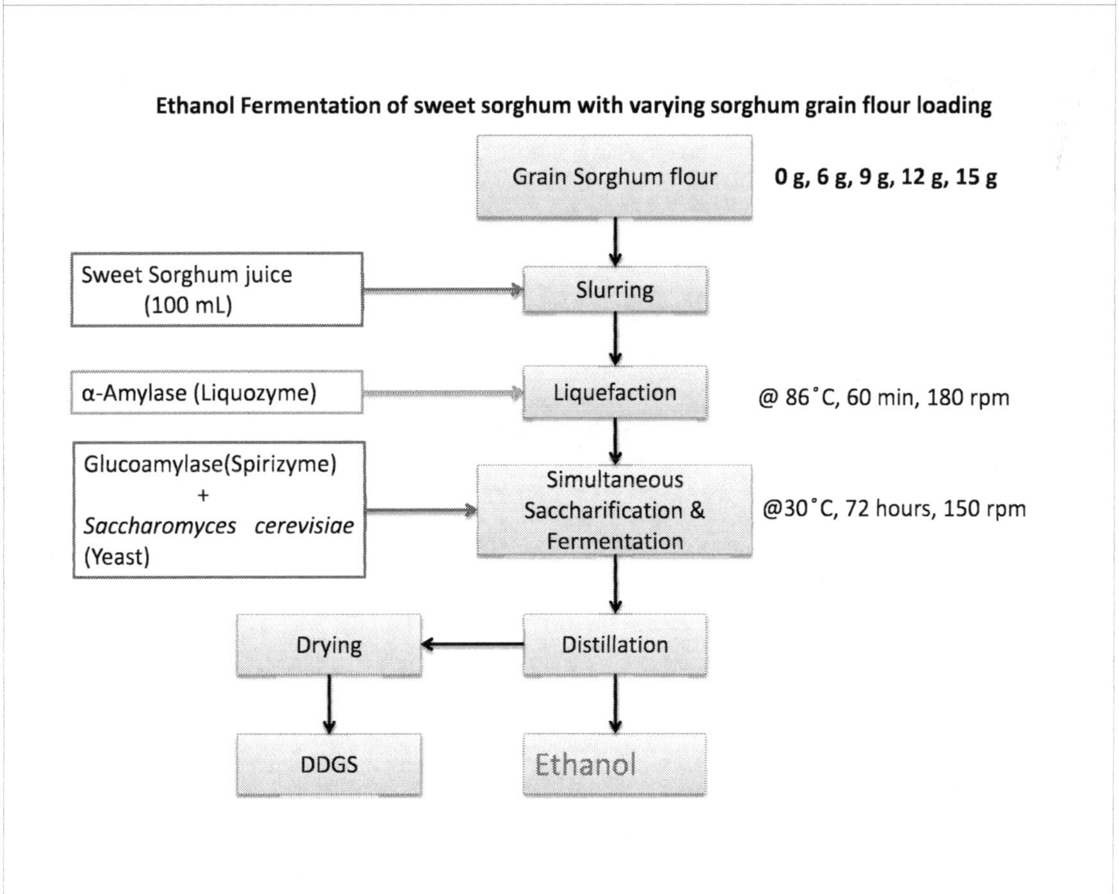
Ethanol Fermentation of sweet sorghum with varying sorghum grain flour loading
Grain Sorghum flour
0 g, 6 g, 9 g, 12 g, 15 g
Sweet Sorghum juice (100 mL)
Slurring
α-Amylase (Liquozyme)
Liquefaction
@ 86˚C, 60 min, 180 rpm
Glucoamylase(Spirizyme) + Saccharomyces cerevisiae (Yeast)
Simultaneous Saccharification & Fermentation
@30˚C, 72 hours, 150 rpm
Drying
Distillation
DDGS
Ethanol

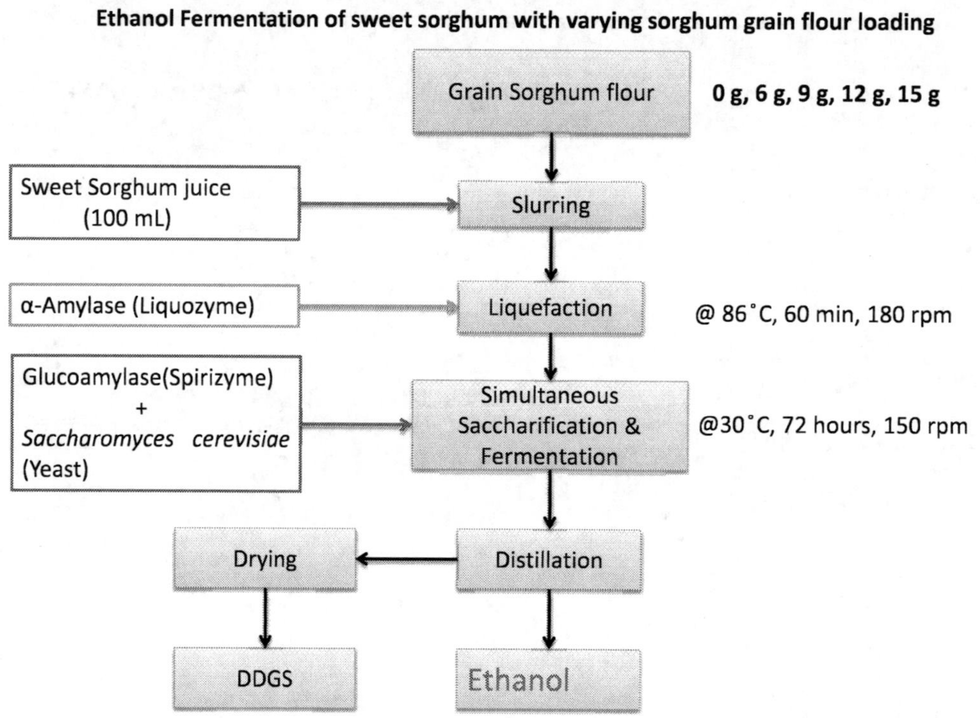
Ethanol Fermentation of sweet sorghum with varying sorghum grain flour loading
Grain Sorghum flour
0 g, 6 g, 9 g, 12 g, 15 g
Sweet Sorghum juice
(100 mL)
Slurring
α-Amylase (Liquozyme)
Liquefaction
@ 86˚C, 60 min, 180 rpm
Glucoamylase(Spirizyme)
+
Saccharomyces cerevisiae
(Yeast)
Simultaneous
Saccharification &
Fermentation
@30˚C, 72 hours, 150 rpm
Drying
Distillation
DDGS
Ethanol

Results

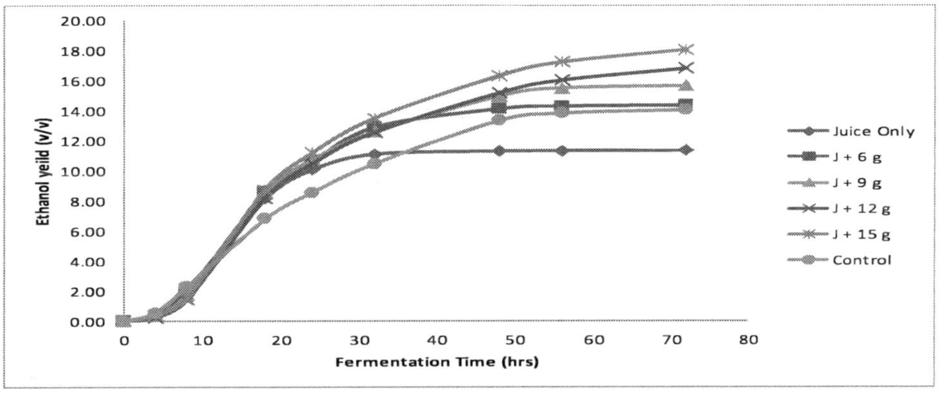

	Juice sugar content (%)	Flour starch content (%)	Theoretical ethanol yield (V/V)	Actual Ethanol yield (V/V)	Ethanol fermentation efficiency (%)
Juice only	18.89	0	12.12	11.29[a]	93.15[b]
Juice + 6 g flour	18.89	71.57	15.21	14.36[b]	94.41[a]
Juice + 9 g flour	18.89	71.57	16.75	15.67[c]	93.55[b]
Juice+ 12 g four	18.89	71.57	18.29	16.81[d]	91.91[c]
Juice + 15 g flour	18.89	71.57	19.95	18.05[e] 28.5 %	90.48[d]
Control- 30 g flour (db)	0	71.70	15.48	14.05[b]	90.75[d]

Means in the same column followed by different superscript letters indicate significant differences ($P \leqslant 0.05$).

Ethanol Fermentation of sweet sorghum with varying hydrolyzing time

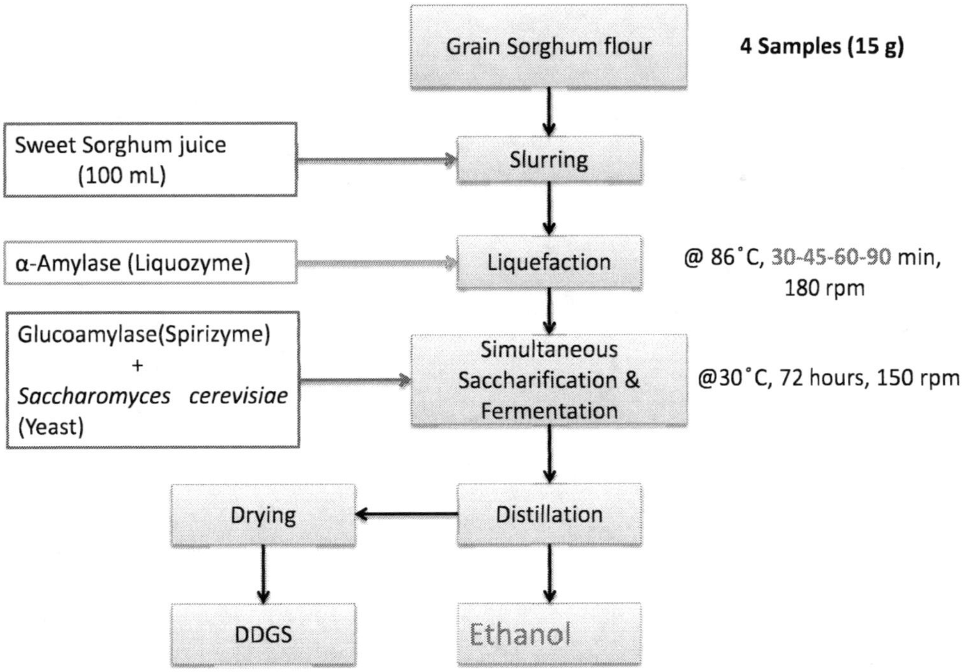

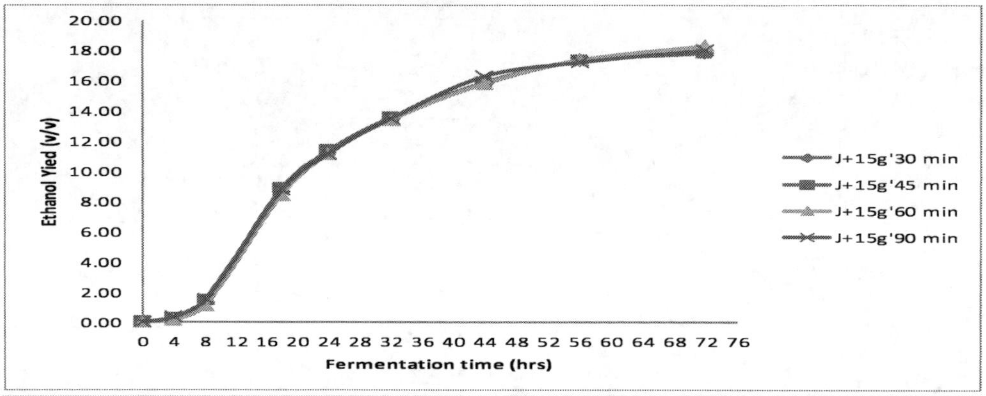

Hydrolysis time (min)	Juice sugar content (%)	Flour starch content (%)	Theoretical ethanol yield (V/V)	Actual Ethanol yield (V/V)	Ethanol fermentation efficiency (%)
30	18.89	71.57	19.95	17.84[a]	89.42[c]
45	18.89	71.57	19.95	17.85[a]	89.47[c]
60	18.89	71.57	19.95	18.33[a]	91.88[a]
90	18.89	71.57	19.95	18.05[a]	90.48[b]

Means in the same column followed by different superscript letters indicate significant differences ($P \leq 0.05$).

Low temperature hydrolysis

Low Energy Ethanol Production Process

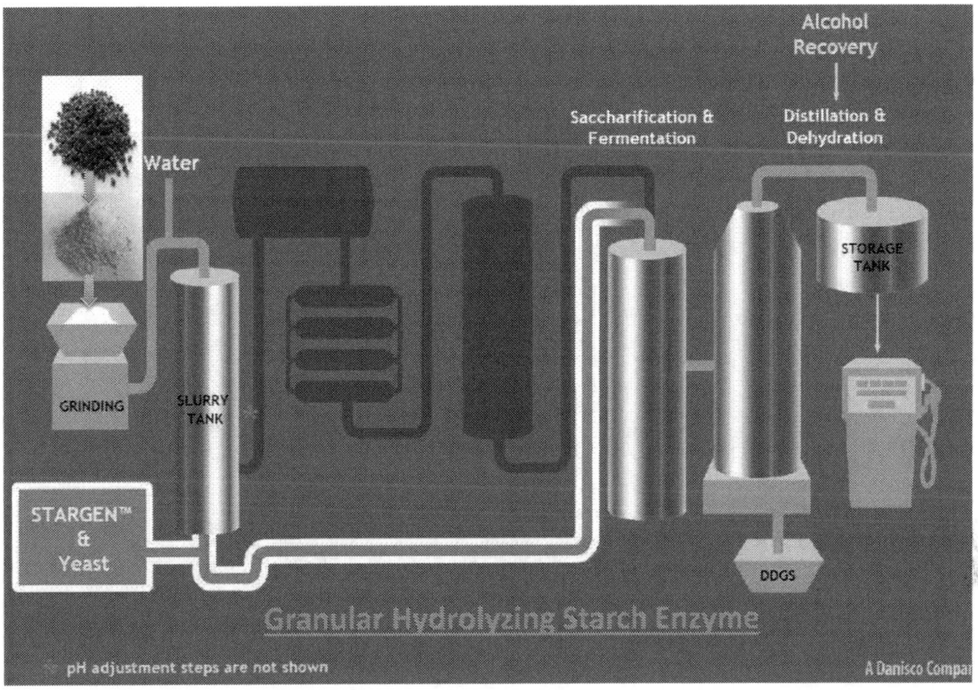

Ethanol Fermentation by Granular Starch Hydrolyzing Enzyme (No Cooking)

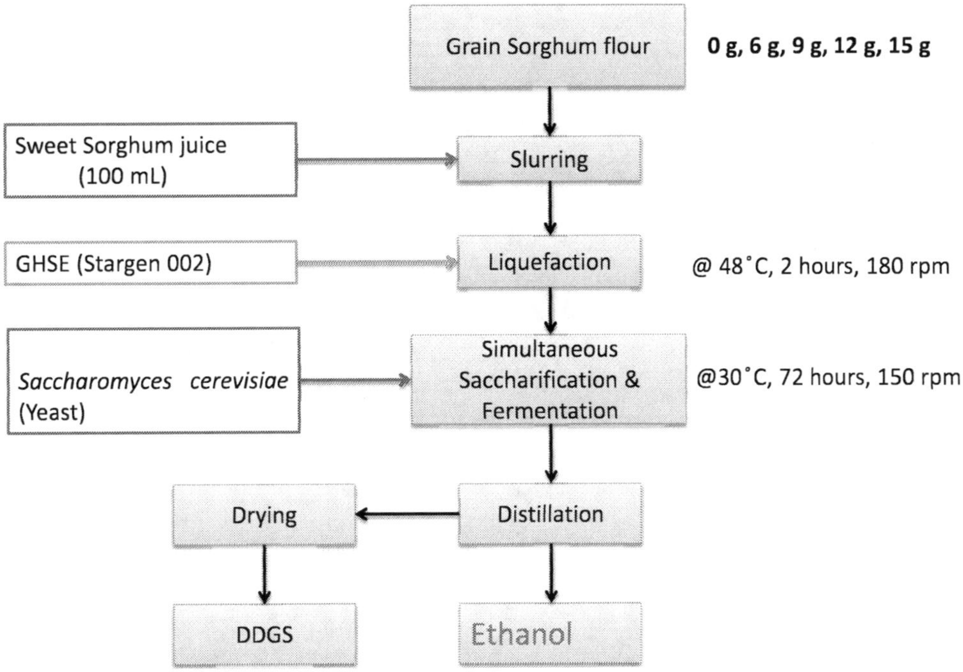

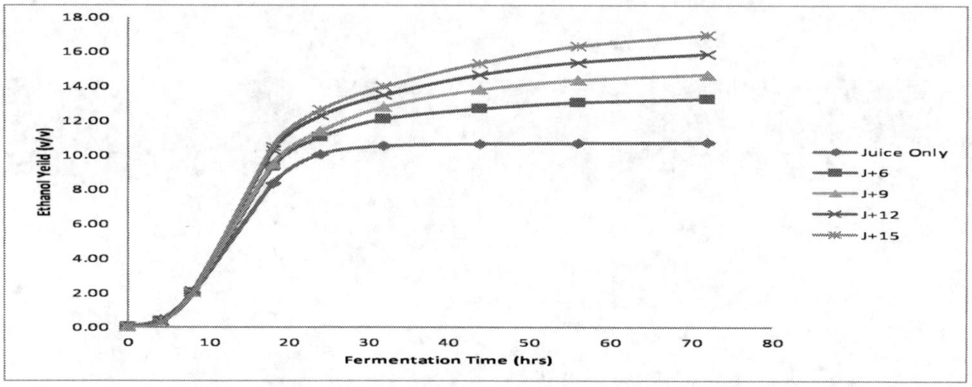

	Juice sugar content (%)	Flour starch content (%)	Theoretical ethanol yield (V/V)	Actual Ethanol yield (V/V)	Ethanol fermentation efficiency (%)
Juice only	17.5	0	11.33	10.73[a]	94.65[a]
Juice + 6 g flour	17.5	71.57	14.42	13.24[b]	91.82[b]
Juice + 9 g flour	17.5	71.57	15.96	14.67[c]	91.92[b]
Juice + 12 g flour	17.5	71.57	17.51	15.87[d]	90.63[c]
Juice + 15 g flour	17.5	71.57	19.05	16.70[e]	87.66[d]

Means in the same column followed by different superscript letters indicate significant differences ($P \leq 0.05$).

Conclusion

- Ethanol yield : 28% higher than conventional ethanol method.
- Enzymatic hydrolysis time reduced by 30 minutes.
- Strong potential for ethanol production from by granular starch hydrolyzing enzyme.

Acknowledgment

This material is based upon work supported by National Science Foundation Grant: From Crops to Commuting: Integrating the Social, Technological, and Agricultural Aspects of Renewable and Sustainable Biorefining (I-STAR); NSF Award No.: DGE-0903701.

Thank You!

(Progress toward) The Continuous-Flow Solid Acid Catalyst Hydrothermal Biorefinery

Michael Timko and Geoffrey Tompsett

Department of Chemical Engineering
Worcester Polytechnic Institute

RCN Conference on Pan American Biofuels & Bioenergy Sustainability
July 24

1

Catalysis and Reaction Engineering Research at WPI
Green Chemistry & Engineering Meets Sustainable Energy

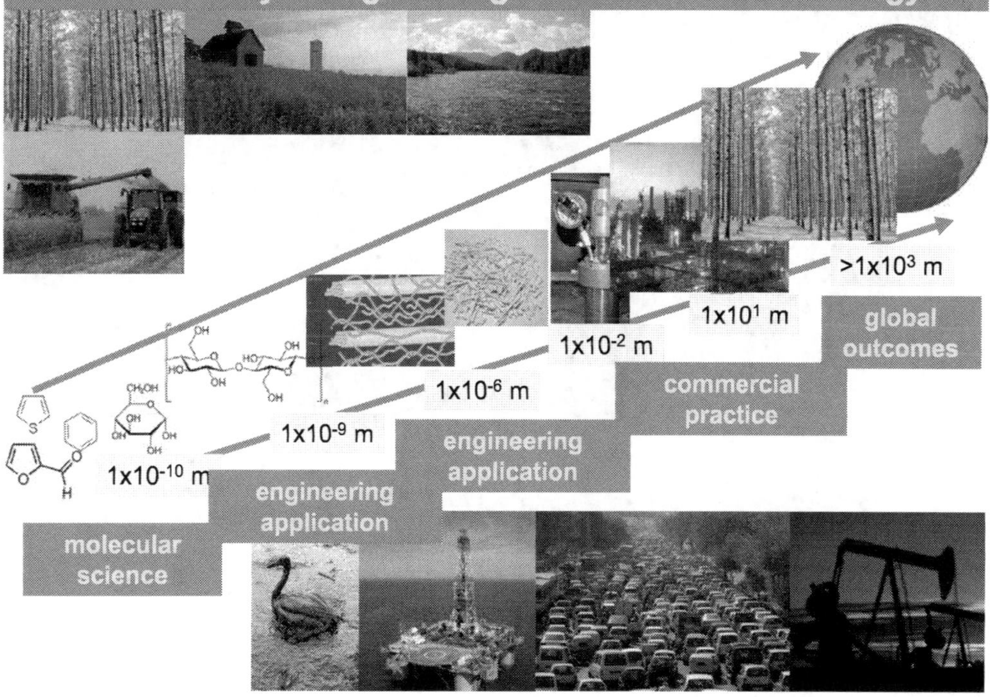

Bioenergy Challenges

- Capital costs of large-scale cellulosic biorefineries are massive (billions of $)
  - Low capital technologies are required
  - Technologies that leverage existing infrastructure are desirable
- Operating costs remain high (about double that for petroleum fuels)
  - Efficient technologies are required
  - Must be feedstock flexible and appropriate for waste biomass
  - Must make use of all biomass components
  - Multiple product streams ($) are required
- Making biofuel production sustainable has not been achieved
  - Land/water use rates
  - Waste production
- Transportation costs and energies remain high
  - Biomass is a distributed resource with low energy density grown in remote, rural areas
  - Small-scale, deployable biomass conversion may reduce transportation costs and energies

High Temperature Water Processes for Valorization of Renewable Resources

- Rapid reaction rates (small reactors)
- No feedstock drying
- Feedstock flexibility (including residues)
- Low oxygen content liquid product with high stability
- Flexibility – pre-treatment, liquefaction, gasification

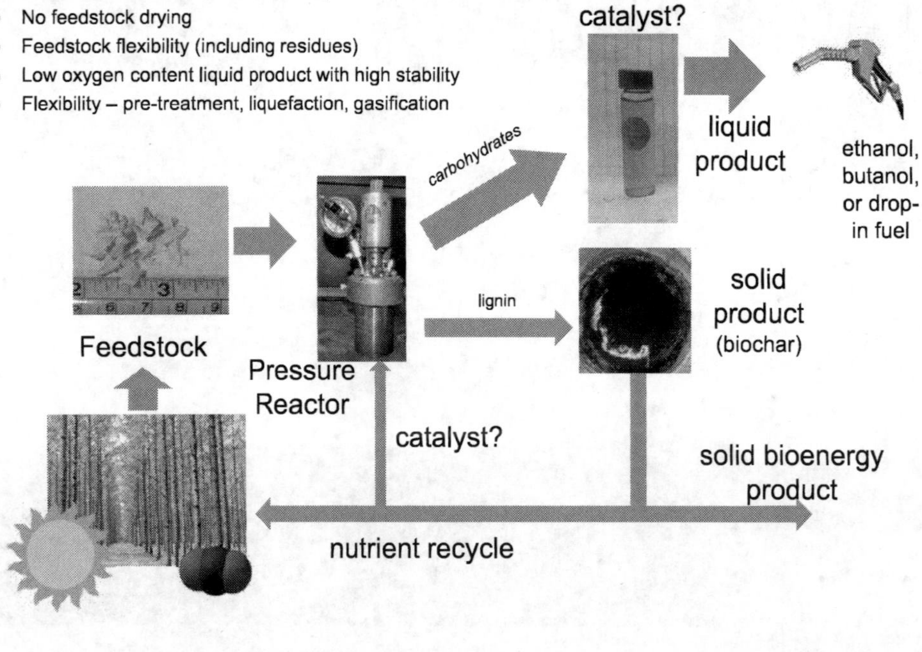

Deployable Biomass Conversion

Advantages:
- Reduced capital costs
- Incremental costs of adding capacity
- Reduced transportation costs and energy consumption due to decreased liquid volumes to be shipped (compared to large solids volumes)
- Appropriate for remote rural areas

Challenges:
- Compact technologies (rapid reactions)
- Product stability and carbon utilization
- Solids handling & feedstock flexibility

Mobile biofuels production platform

Distributed Pre-treatment Concept: Reduced Operations and Capital Costs

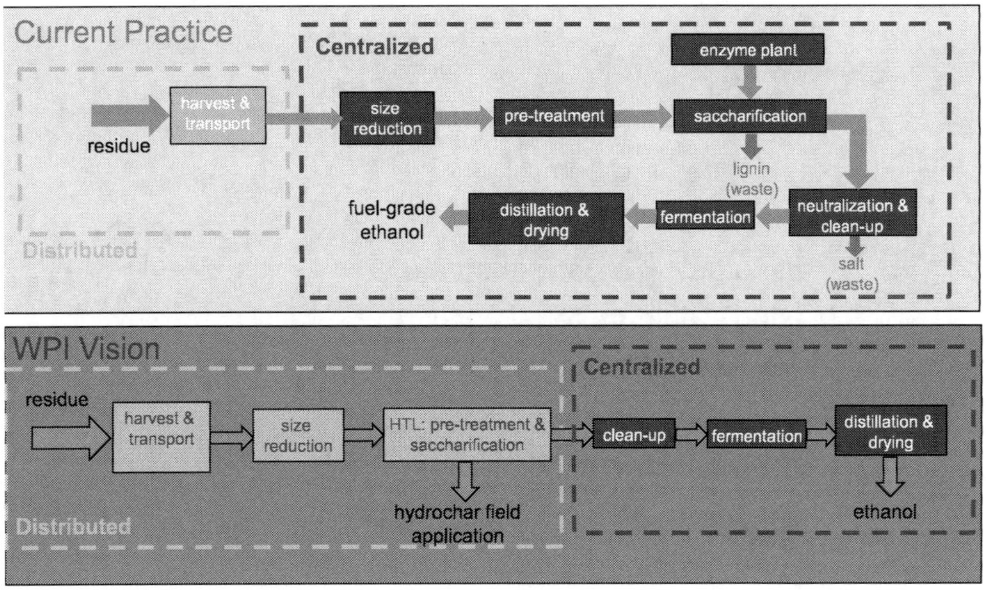

6

Why Hydrothermal?
Techno-Economic Considerations

- ➢ Previous economic analysis suggests that pyrolysis ($2/gge) is more economical than either biochemical ($5/gge) or hydrothermal liquefaction (HTL) ($4/gge)

  Data from Wright et al., 2010 and Zhu et al. 2014
- ➢ Pyrolysis requires drying step and is limited by bio-oil product quality and stability
- ➢ HTL offers flexibility
  - ❑ Liquefaction to produce a bio-oil with superior properties compared to pyrolysis
  - ❑ Pre-treatment to maximize sugar yield for APP or fermentation

Primary Technical Challenge: Controlling chemistry during HTL

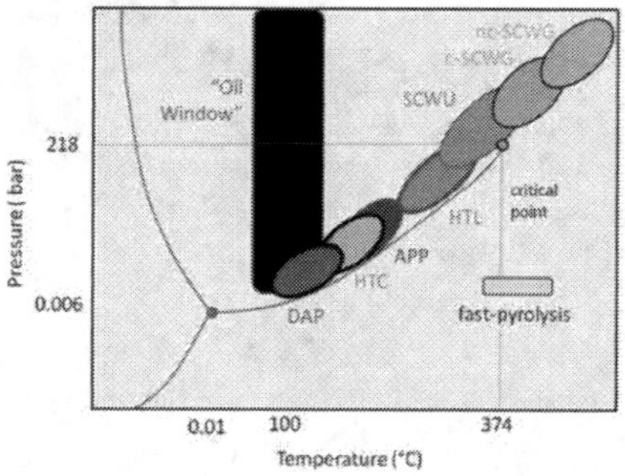

Timko et al. *J SCF*, submitted

Techniques to Control Reactivity

1) Reaction Engineering

   Heat rate

   Residence time control

2) Catalyst Engineering

   Making active catalysts stable

   Making stable materials active

Hydrothermal Batch Reactor

Reactor
500 mL 316-ss Parr reactor
rated to 2000 psi at 350 °C
experiments performed at 300 °C
reaction time = 10 min after heat up
agitation rate = 500 rpm

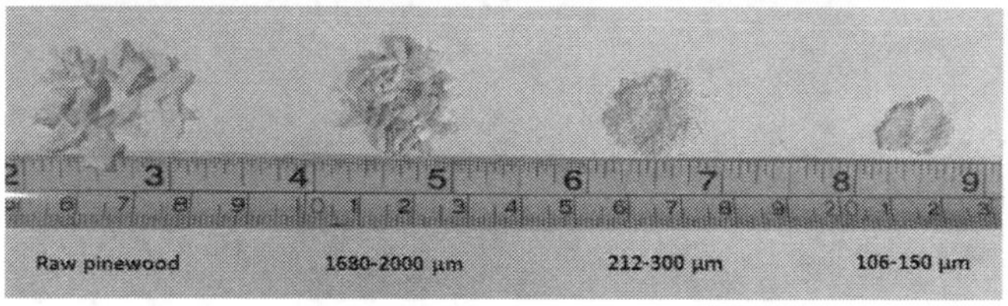

Feed
Size selected pine sawdust
10 wt% slurry

9

Organic Carbon Yield and Heat Rate

- Initially performed to investigate effect of particle size
- Instead, heating rate played an important role
- Qualitatively similar to HTL of grasses (Zhang et al. *J. Applied Anal. Pyrolysis,* 2009)
- Experiments in progress to investigate heating rate effects in batch reactor
- Points to the importance of rapid heat rates in continuous flow reactors

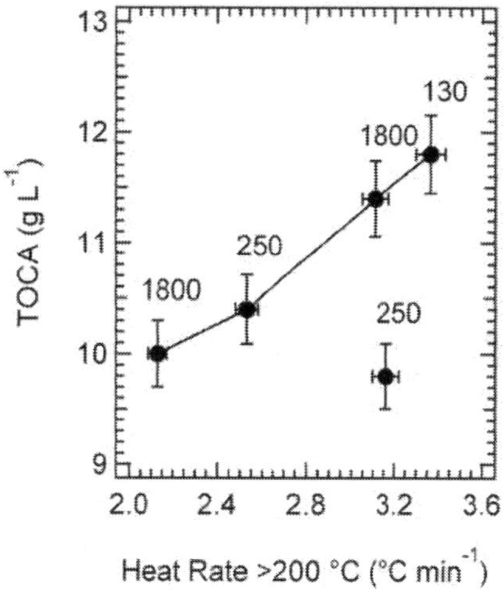

10

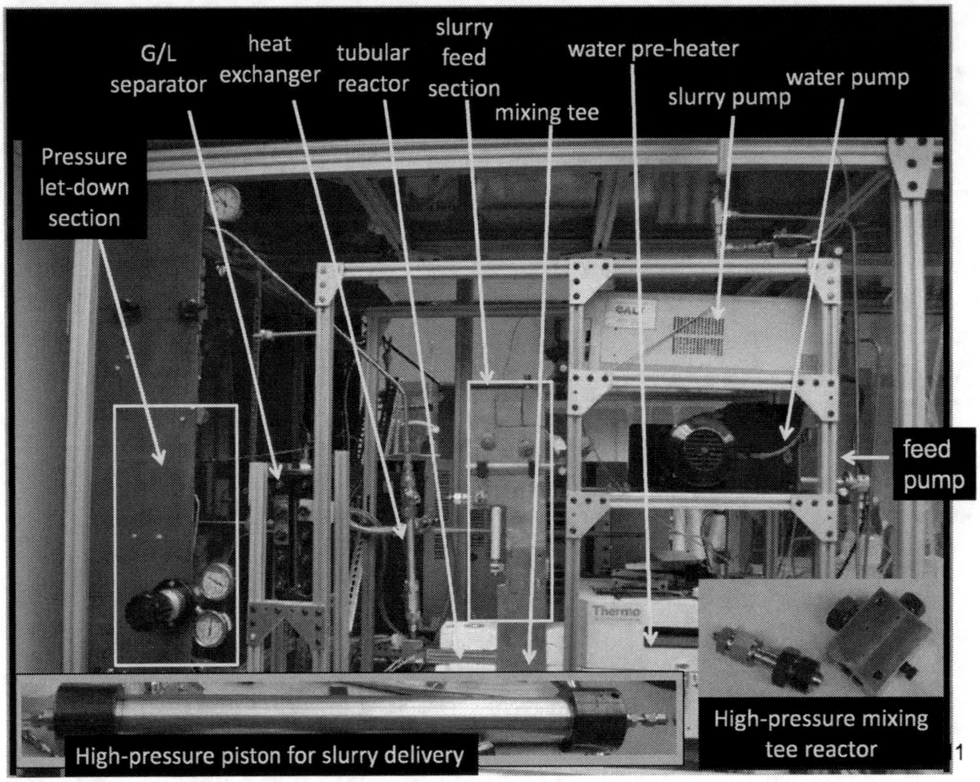

Heat Transfer Rate in Continuous Flow

- In collaboration with MIT in a separate study, the WPI team has been studying rapid heating in a mixing tee reactor
- Biomass fed cold from the top port
- Water superheated (>350 C) to achieve rapid heat rate

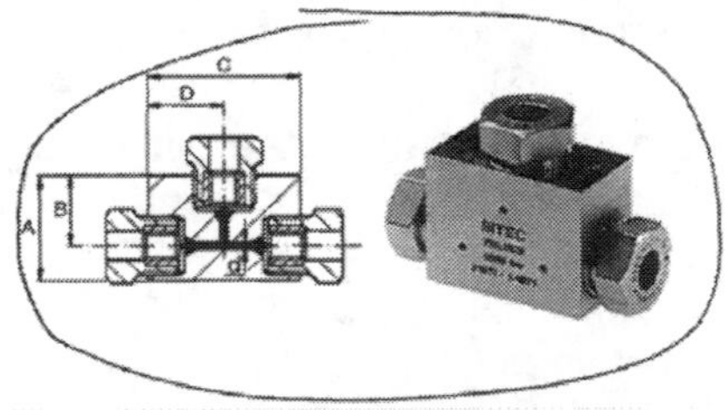

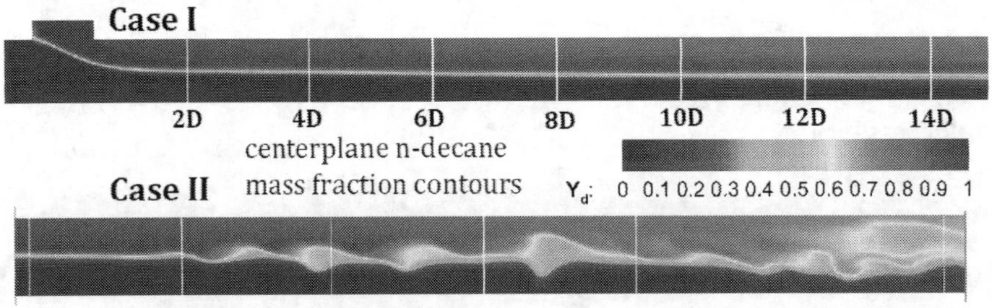

(Rhagavan et al. *J. Supercritical Fluids*, 2014)

Temperature TOS Performance

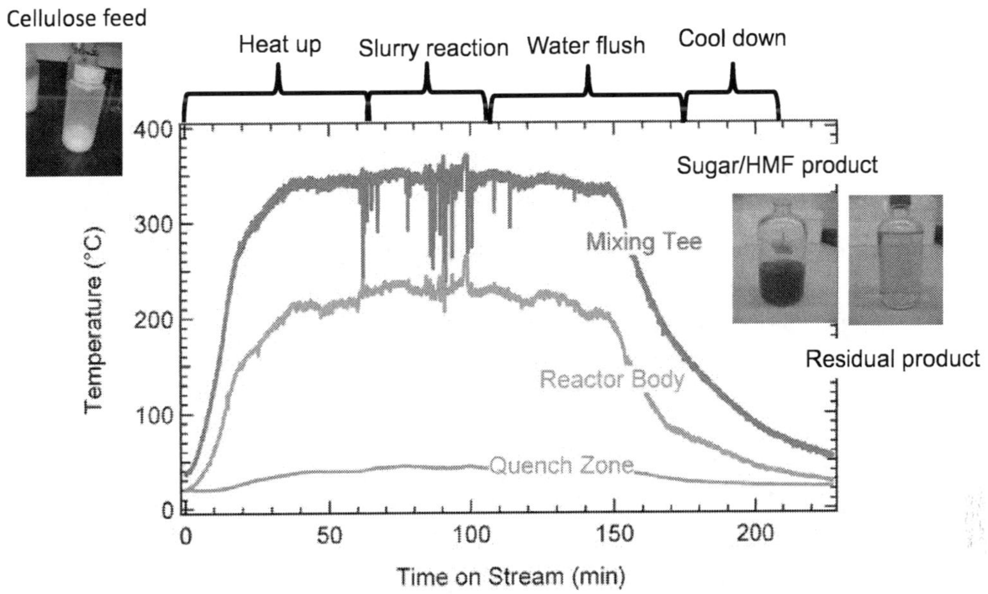

13

Controlling Reactivity - Catalysis

- Catalyst allows temperature to be decreased, improving selectivity
- Catalyst must be – active, selective, and stable

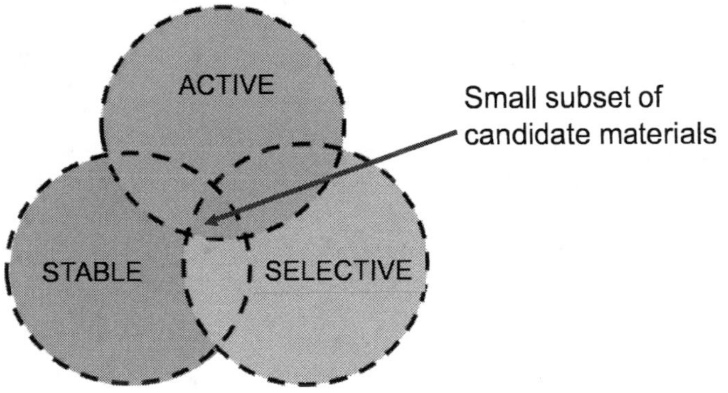

- Additional requirements placed on the material for COST & SUSTAINABILITY

We address stability and activity 2 ways

- Hydrothermal char as a renewable catalyst material
  - The material itself is **stable**
  - **Need – impart activity**
- Organic-modified zeolites as a hydrothermally stable catalyst with well-defined pore structures
  - The material is **active and selective**
  - **Need – impart stability**

Char Materials as Catalysts

✓ *Renewable source (sugars, cellulose, hydrolyzates, biomass)*
✓ *Inherently stable in high-temperature water*
✓ *Can be functionalized (-OH, -SO_3, N)*
x *Low intrinsic catalytic activity*
x *Require harsh conditions to become catalytic (concentrated acid treatment)*

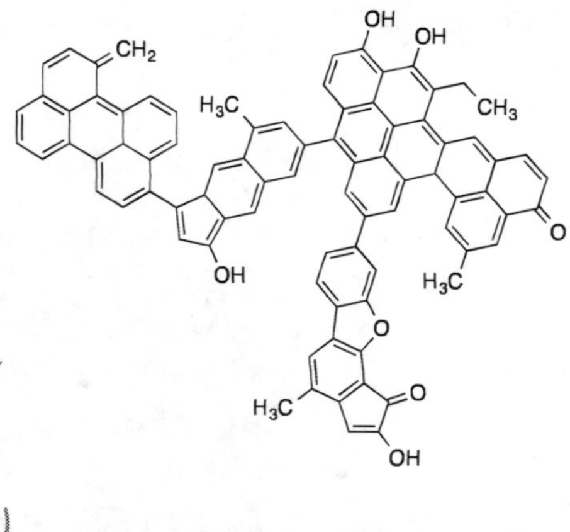

Here, I will focus specifically on hydrothermal chars formed by carbonaceous feedstocks in a water-rich environment

16

Ball-milling to activate hydrochars for catalyst applications

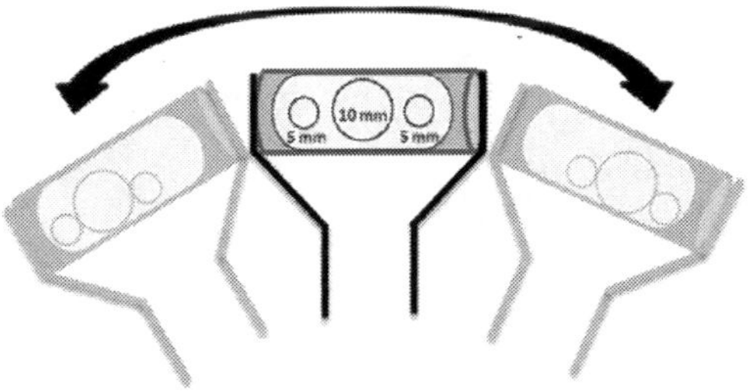

Ball-Milling

- ✓ *Energy imparted by ball impacts*
- ✓ *Solvent-free*
- ✓ *No need for downstream processing*
- x *Mechanism not understood*

We performed fundamental studies to improve our understanding of chemical processes that occur during ball milling

17

Raman Spectroscopy to Analyze Chars

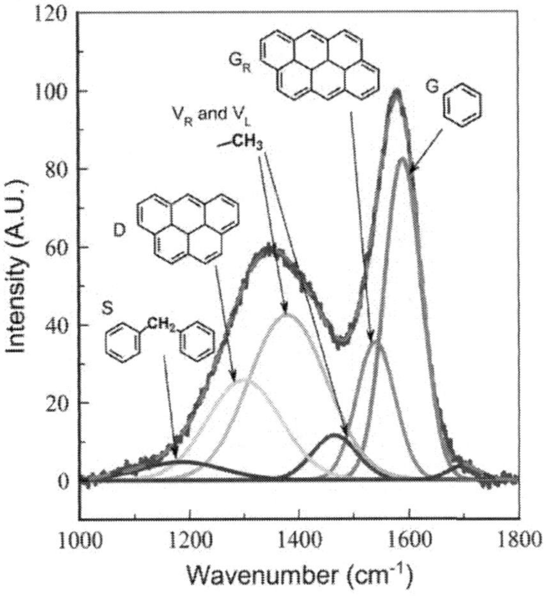

Assignments based on Li et al. (Fuel, 2006)

Vibrational spectra provide significant information on carbon-carbon connectivity

Subject char to ball milling and track structural changes to develop modification protocol

Tompsett et al., Carbon, submitted

Raman Shows Substantial Structural Changes during Ball Milling

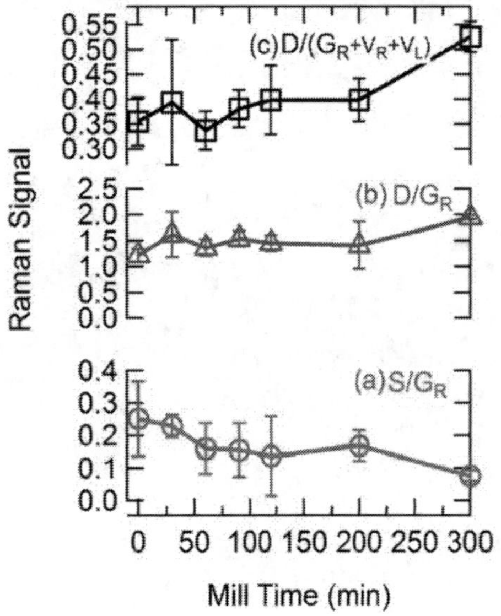

$D/(G_R+V_R+V_L)$ ↑ implies that alkyl chains disappear

D/G_R ↑ implies that small aromatic clusters form larger ones

S/G_R ↓ implies that bridgehead carbons disappear

Tompsett et al., Carbon, submitted 19

Char Reactions and Radical Intermediates

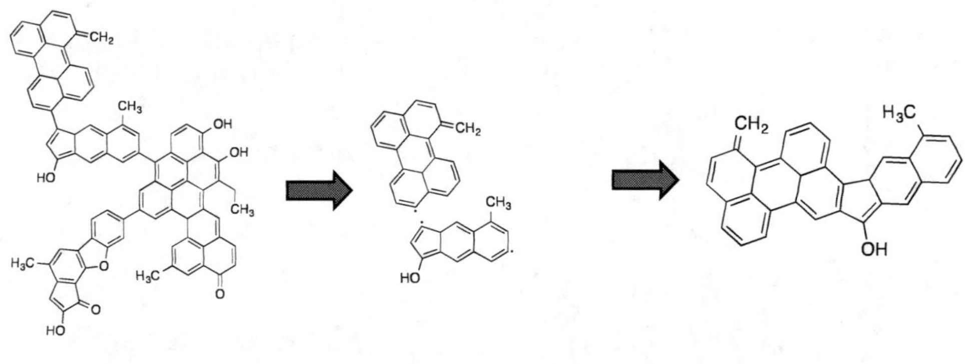

This mechanism explains loss of bridgehead carbons, disappearance of small aromatic clusters, and formation of larger aromatic clusters

Suggests formation of radical intermediates – ESR!

20

ESR Measures Radical Content

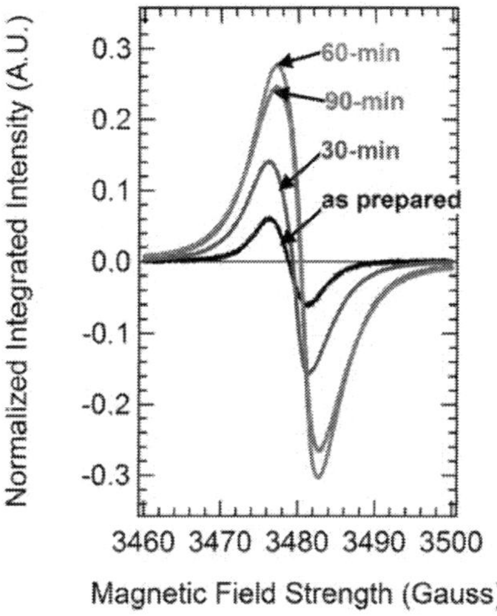

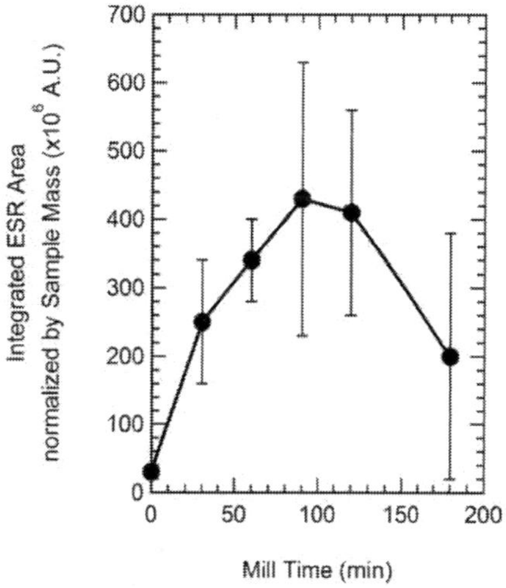

As predicted, radical signal rapidly increases and then decreases with larger milling times

21

Conceptual Kinetic Model

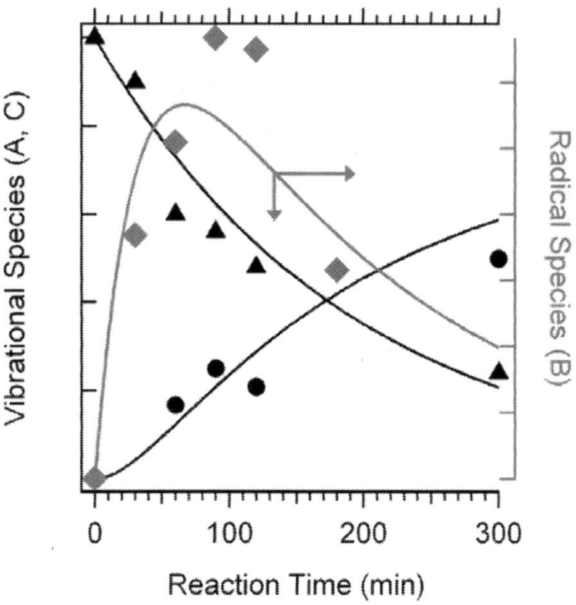

$A \xrightarrow{k_1} B$ Radical formation

$B \xrightarrow{k_2} C$ Radical recombination

$B \xrightarrow{k_3} D$ Radical reaction with air

All rate constants on the order of 1×10^{-2} and 1×10^{-3} min^{-1}

Provides basis for reaction design

22

Organic-Modified Zeolites

Microporous material (<10 Å) with alumina substitution, zeolites are strong Brønsted acids

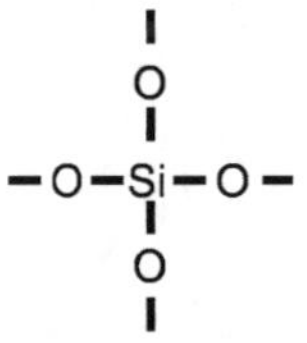

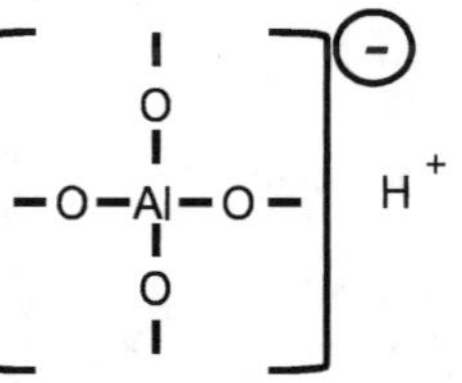

The more alumina, the more active – but the less water stable!

Solution proposed by Resasco et al. – kinetic stability using organic modifiers

Water cannot enter the pores – zeolite remains stable

But, what about reactant access?

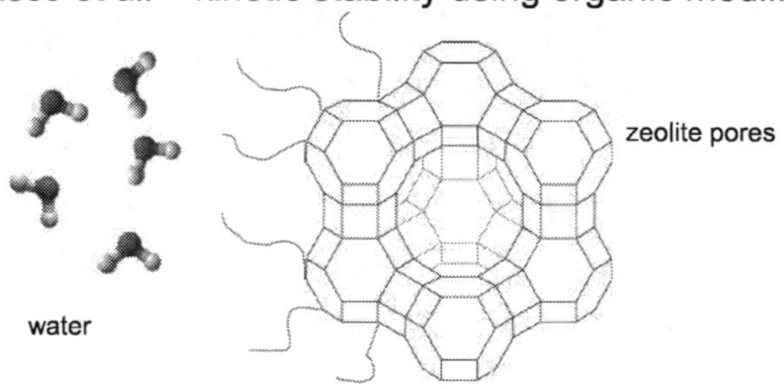

23

Organic-modified ZSM-5

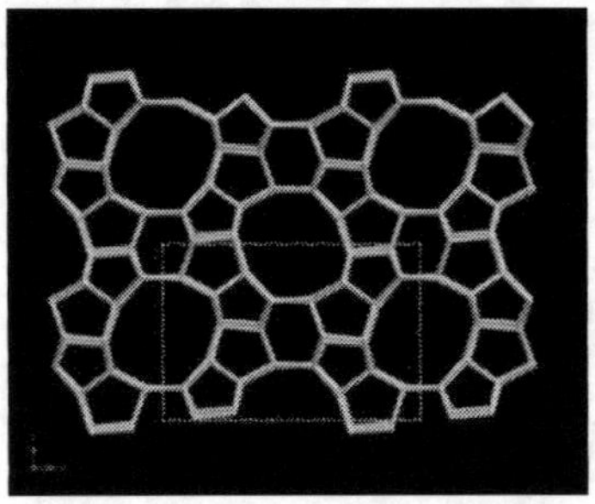

Common industrial zeolite with 4.46 Å pores and 6.36 Å cavities

Stable with high Al_2O_3 substitution (~40:1 SiO_2/Al_2O_3 ratio is commercial)

Use silanation chemistry to coat with hydrophobic molecules

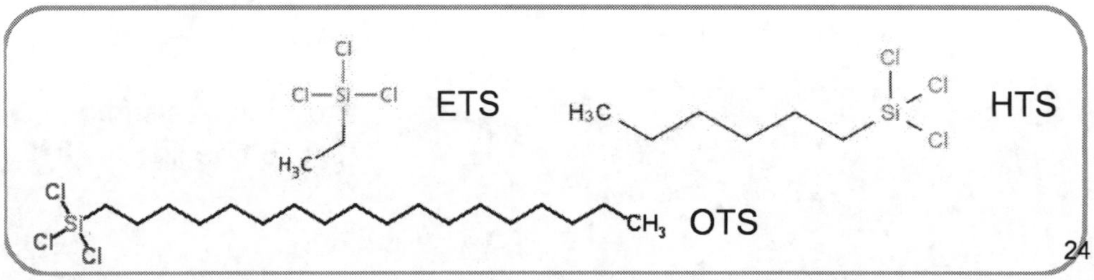

24

Contact Angle – confirms hydrophobic surface

	Uncoated	ETS modified	HTS modified	OTS modified
Modified Using Calcine Zeolite	*[Unable to capture photo because droplet absorbed into pellet instantaneously]*	*[Unable to capture photo because droplet absorbed into pellet instantaneously]*	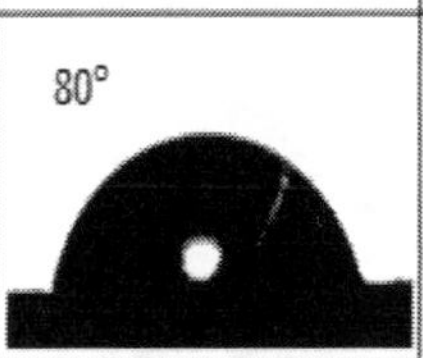	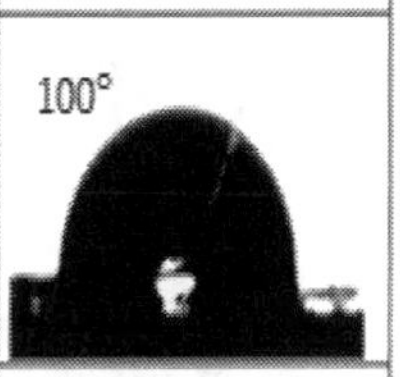
Modified Using Uncalcine Zeolite			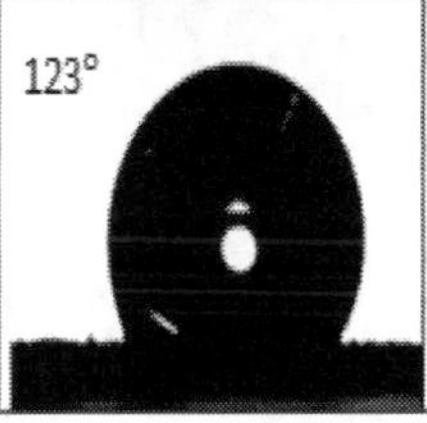	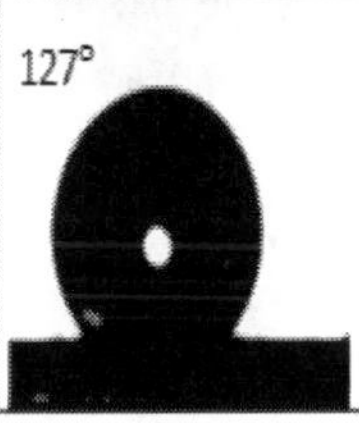

25

Phase behavior – confirms hydrophobic solution behavior

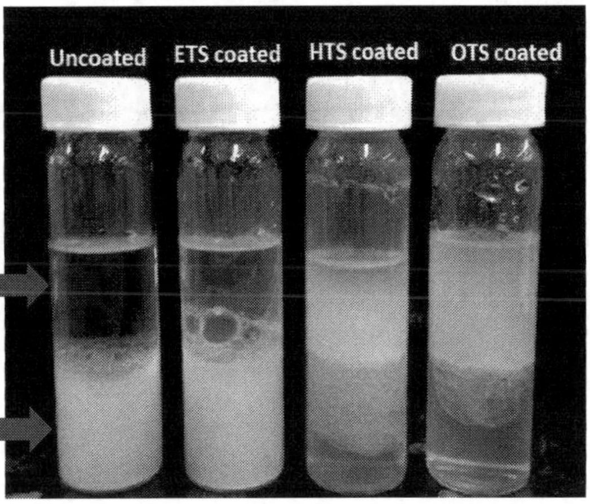

Toluene

Water

hydrophilic ⟹ **hydrophobic**

26

FTIR Confirms Presence of Alkyl Groups

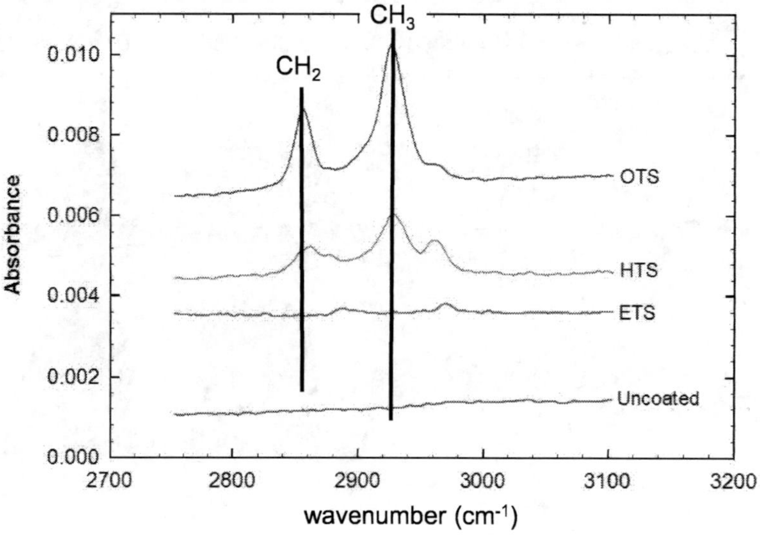

27

TGA Confirms Thermal Stability

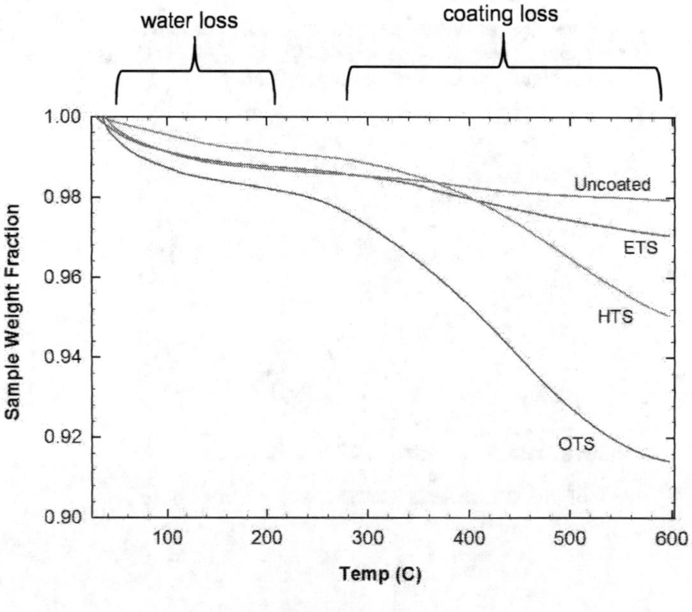

28

Sorption Experiments

Hexanol

μ = 1.55 D S = 5.9 g L^{-1}

Cyclohexanol

μ = 1.46 D S = 0.36 g L^{-1}

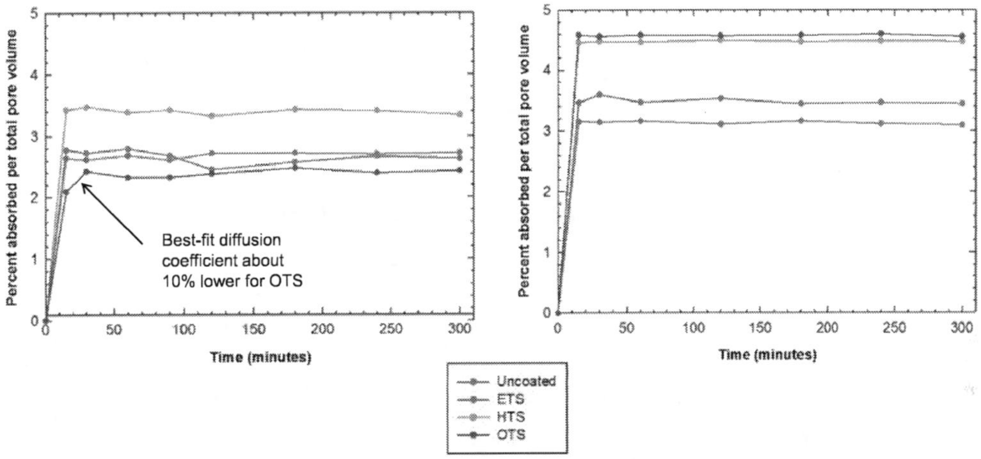

29

Equilibrium Uptake

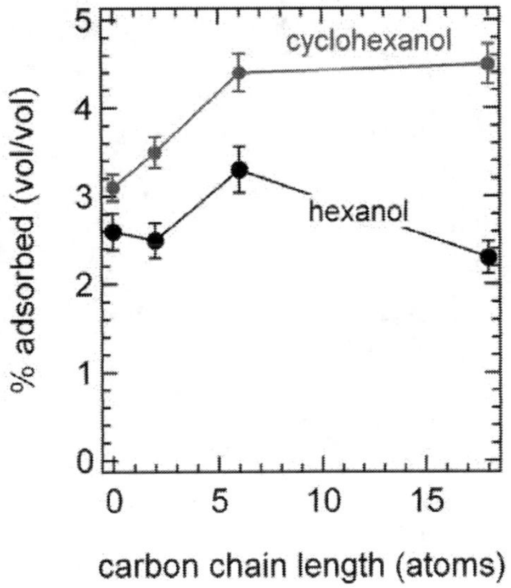

Cyclohexanol uptake increases with coating thickness

Hexanol shows a more modest effect – need to investigate C_6 coating in more detail

Possible explanation, uptake into zeolite pores and into organic coating

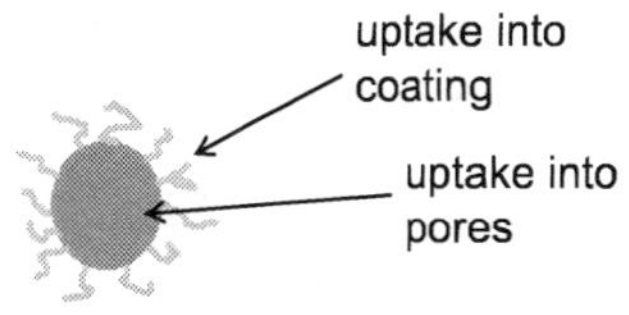

Relative contributions depend on adsorbate sizes and polarities

30

High Temperature Water Processes for Valorization of Renewable Resources

- Rapid reaction rates (small reactors)
- No feedstock drying
- Feedstock flexibility (including residues)
- Low oxygen content liquid product with high stability
- Flexibility – pre-treatment, liquefaction, gasification

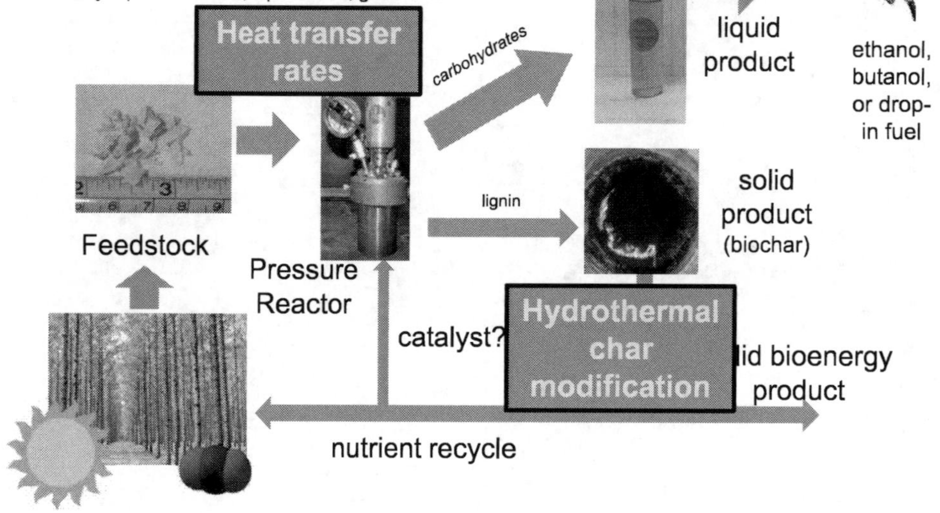

Acknowledgments

Biodiesel Production from Used Cooking Oil using Calcined Sodium Silicate Catalyst

M.O. Daramola, D. Nkazi, K. Mtshali

School of Chemical and Metallurgical Engineering, Faculty of Engineering and the Built Environment, University of the Witwatersrand, Wits 2050, Johannesburg, South Africa

RCN Conference on Pan American Biofuels and Bioenergy Sustainability, Brazil, July 22-25, 2014

Motivation

Sources: Guo et al, 2012; Aransiola et al., 2012; Sivozhelezova et al, 2009; Papayannakos, 2013; Li et al, 2014; Cheah et al, 2009

- Conversion of waste to useful product such as Biodiesel
- Homogeneous catalysts give very high conversion for biodiesel production but the consumption and recovery of the catalyst constitute a challenge
- Enzymatic conversion is promising but huge cost of enzyme and separation of product pose a challenge
- The use of heterogeneous catalyst has eliminated equipment corrosion and waste water discharge, and appear to be the most promising and sustainable processes
- Need to develop heterogeneous base catalyst for the production of biodiesel with higher tolerance for FFA's content in the feedstock

2

Research Questions

- What could be the conversion yield of biodiesel obtained from Used (waste) Cooking (UCO) Oil over sodium silicate catalyst?
- What is the effect of reaction time and temperature on the conversion?

3

Objectives

- Synthesize and characterize sodium silicate catalyst
- Evaluate the FFA content of the UCO sample before the production of biodiesel
- Use the prepared catalyst for biodiesel production from UCO
- Investigate the effect of reaction temperature and reaction time on the conversion

4

Biodiesel from used cooking oil

Source: Green Cape, 2013

New SA regulation to kick in on the 1st of October 2015

Fuel producers will be required to blend a minimum of 5 percent biodiesel with diesel fuel

Challenges:

- SA UCO market: poor supplier
- Transportation effect of UCO over long distances on the system's sustainability.

5

Renewable energy of South Africa

Source: Green Cape, 2013

Status of Biofuel in October 2013

Company Name	Crop / Feedstock	Capacity (million litres/yr)	Location	Licence status
BIOETHANOL				
Mabele Fuels	Sorghum	158	Bothaville, FS	Issued
Ubuhle Renewable Energy	Sugarcane	50	Jozini, KwaZulu Natal	Issued
E10 Petroleum Africa cc	Sugarcane and other crops	4.2	Germiston, Gauteng	Granted
ARENGO 316 (Pty) Ltd	Sorghum and sugar beet	180 (in two phases of 90 each)	Cradock, Eastern Cape	Granted
TOTAL BIOETHANOL CAPACITY		**392.2**		
BIODIESEL				
Rainbow Nation Renewable Fuels	Soya Bean	288	Port Elizabeth, Eastern Cape	Issued
Exol Oil Refinery	Waste Vegetable Oil	12	Krugersdorp, Gauteng	Granted
Phyto Energy	Canola	>500	Port Elizabeth, Eastern Cape	Applying for licence
Basfour 3528 (Pty) Ltd	Waste Vegetable Oil	50	Berlin, Eastern Cape	Granted
TOTAL BIODIESEL CAPACITY		**850**		

6

Sodium Silicate Catalyst

Sources: Dai et al, 2010; Guo et al, 2012

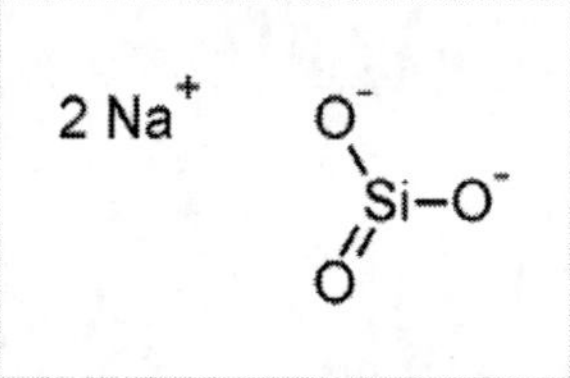

- The use of heterogeneous catalyst showed higher tolerance of FFA
- Sodium silicate as transesterification catalyst has higher catalytic activity after calcination
- Hydrolysis reaction suppresses the formation of soap and decrease water content

7

Biodiesel production: Transesterification

Source: *Schuchardt et al, 1998*

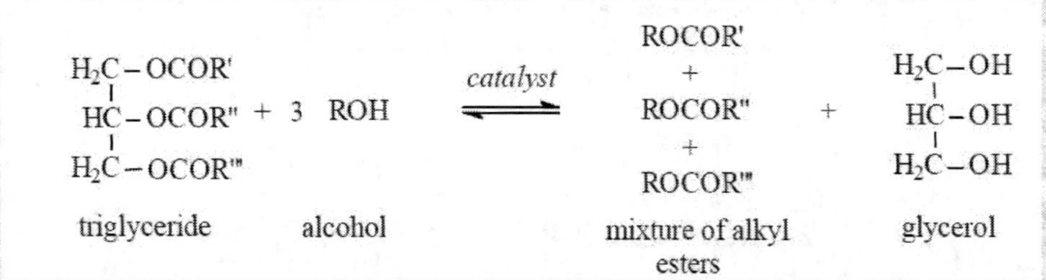

Reaction for transesterification

8

Experimental Procedure

1. Catalyst preparation

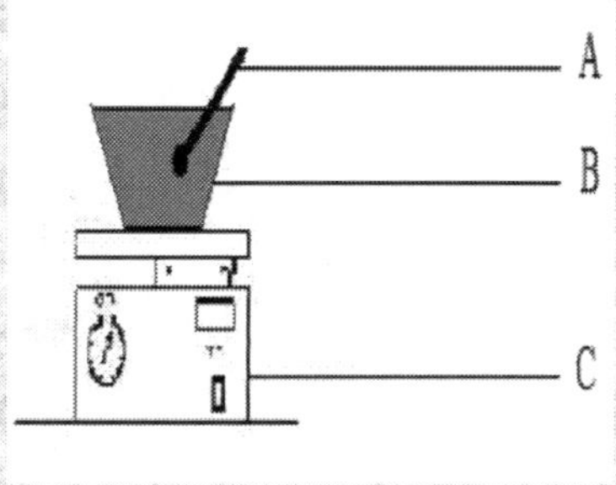

NaOH + H_2O SiO_2

Mixing at constant temperature → Calcination $200/350\ ^0C$

Catalyst Preparation Experimental Set-up

A: spatula used to ensure the uniform mixing of reagents, B:steel bowl used to perform the reaction and C: magnetic heater and stirrer, used to obtain the required reaction temperature.

Experimental Procedure

- Biodiesel production over the catalyst

Na_2SiO_3 (2.51 g) + Methanol (214.3 ml)

WCO (35.7 ml)

Reaction at varied Temperature (25-70°C) and time (0-180 minutes) Under continuous stirring

The equipment used in the transesterification setup is a Liebig condenser that is connected to a running tap.

Catalyst Characterization

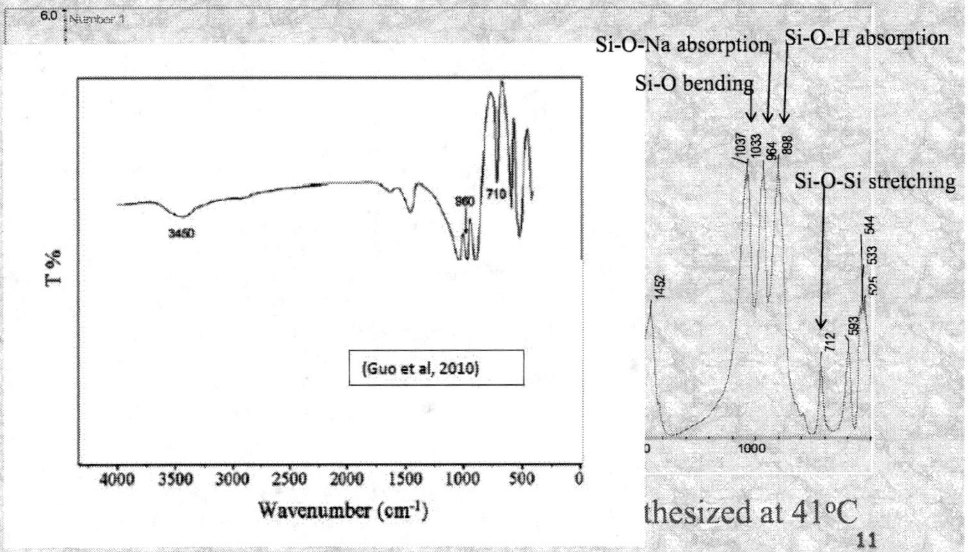

thesized at 41°C

11

UCO Characterization

- Physico-chemical characteristics of the UCO

Moisture content (%)	High heating value (MJ/kg)	Melting point (°C)	Specific gravity
<0.05	40	30	0.87

12

FFA Content of the UCO

The number of titrations conducted along with the volume of titrant used for the solution to reach end-point. (Titration method used by Ding et al (2012))

Titration	Volume used (ml)
1	1.7
2	1.95
3	1.9
4	1.8
Average titration volume (ml)	1.8375 ↔ FFA % = 1.41 % > 1 %
%volume of acid (%)	0.115
Average density of acid used (kg/m³)	900

13

Effect of Time & Temperature

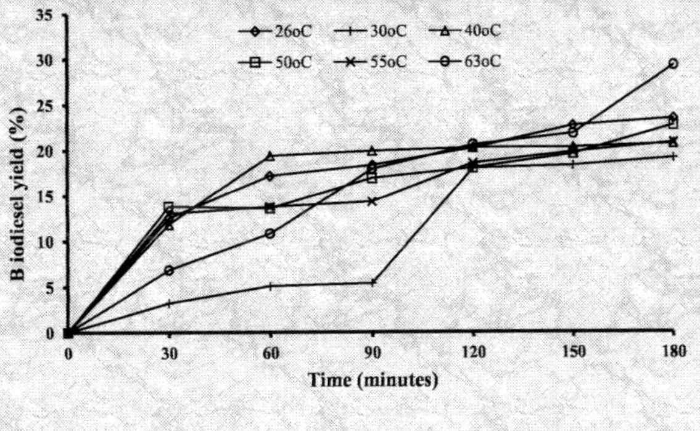

Biodiesel yield as function of reaction temperature and reaction time

14

Effect of Time & Temperature

❑Yield of fatty acid methyl esters (FAME) during the reaction increased with time and about 30% FAME was obtained after 180 minutes and at reaction temperature of 63°C.

❑The increase in FAME yield at increasing reaction time is consistent with results from previous research efforts on the use of a homogeneous catalyst (NaOH) for the conversion of Jatropha Curcas oil seeds and palm oil to biodiesel (Aransiola et al, 2012 & 2013; Alamu et al, 2008) and on the conversion of WCO to biodiesel over RHC-SO_3H and Ambertyst-15 catalysts (Li et al, 2014).

❑Increase in the yield of FAME at increasing reaction time, and reaction temperature could be attributed to the kinetics of the reaction.

15

Results Compared with Literature

Catalyst	Nature	Reaction T (°C)	Reaction Time (h)	Methanol/oil ratio	BD yield (%)	Ref.
Ambertyst-15	acid	110	3	20:1	~25	Li et al, 2014
RHC-SO_3H	acid	110	3	20:1	42	Li et al, 2014
Na_2SiO_3	base	63	3	6:1	30	This study

16

Results Compared with Literature

- Based on the same reaction time scale, the yield of FAME obtained in this study was about 5% higher that the FAME yield obtained for Ambertyst-15 catalyst. However, the performance of our catalyst in terms of the yield of FAME is lower than that of RHC-SO_3H catalyst by about 12%.
- The lower reaction temperature and lower methanol-to-oil ratio at which transesterification was conducted in this study could explain the lower performance of our catalyst when compared to RHC-SO3H.
- Since reaction rate increases with increasing temperature, it is expected that increase in reaction temperature beyond 63°C in this study might enhance the yield of FAME beyond 30% while keeping the time and the methanol-to-oil ratio the same.

17

Conclusions & Recommendations

- Solid sodium silicate catalyst was synthesized and evaluated for biodiesel synthesis from UCO.
- Results show that the calcined Na_2SiO_3 catalyst is able to convert UCO to biodiesel at mild temperature. The results also compare well with literature
- More in-depth studies on the characterization, activity, and the kinetics of the catalyst in transforming UCO to biodiesel to improve the yield of FAME are on-going in our lab.
- At the same time, improvement of the synthesis protocol of the catalyst via optimization study is essential to improving the activity of the catalyst.
- Evaluation of performance stability and optimization of the transesterification operating conditions are essential to achieving optimized process.

18

Acknowledgments

Wits FEBE Top-slice Grant

&

UNIVERSITY OF THE WITWATERSRAND
JOHANNESBURG

19

Thank You for Your Attention

20

Food Security and Biofuels: Can Policy Flexibility Reduce Food Price Crisis for the Poor?

Part 1: Global Markets and Brazil Sugarcane Production Case

Manoel Regis L. V. Leal[1], Keith L. Kline[2], Daniel Capitani[3]
[1]CTBE – Brazilian Bioethanol Science and Technology Laboratory
[2]ORNL – Oak Ridge National Laboratory, Climate Change Science Institute
[3]UNICAMP – University of Campinas

RCN Pan American Biofuel and Bioenergy Sustainability Conference

Recife, PE – July 23-25, 2014

Contents

- Food vs Fuel context
- Points of view
- Bioenergy demand
- Sugar and ethanol markets
- Public policies
- Final comments

World Hunger

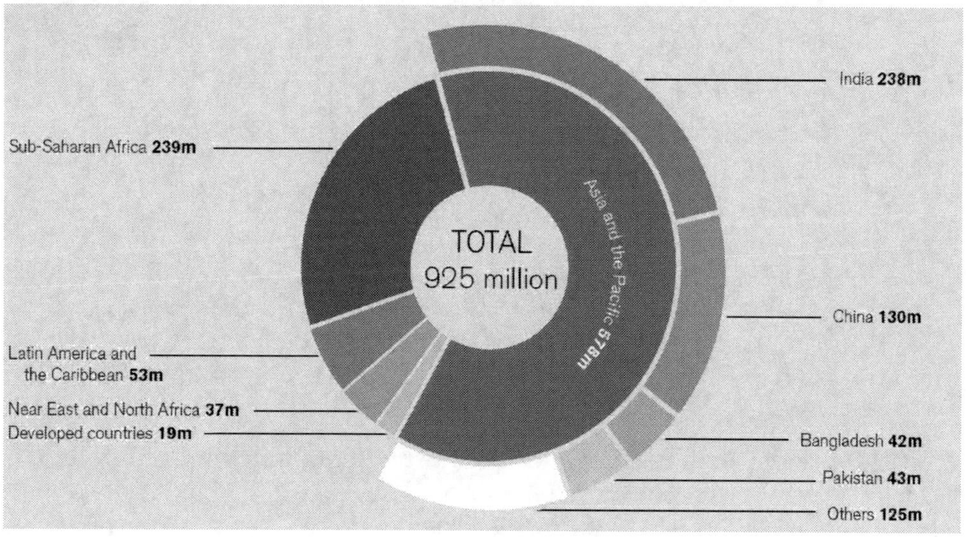

Food Security Definition

Food security exists when all people, at all times, have physical and economic access to sufficient safe and nutritious food that meets their dietary needs and food preferences for an active and healthy life (1996 World Food Summit)

Four main dimensions

- Availability
- Access
- Utilization
- Stability

Food vs Fuel

- Current media coverage has been negative.
- Media and popular discourse reinforce ideas that:
  - Food is being used for biofuels
  - Biofuels contribute to global hunger and food insecurity
- Negative opinions undermine public and political support, and have direct effects on policies and investment
  - Example: EU RED revised to cap biofuels produced from conventional (food) feedstocks
  - Popular pressure to reduce or eliminate the USA RFS ethanol requirements

The Food vs Fuel Discussion

CTBE Brazilian Bioethanol Science and Technology Laboratory

Common Proposed "Solutions"

CNPEM

1. **Simply forget biofuels**
   - Focus on solar, wind and conservation
2. **Only use "non-food" crops** for biofuels
   - Several NGOs and governments testing this idea
3. **Change policies so they adjust biofuel production** in response to food crises
   - Recommendation from the November 2013, 40th Meeting of the United Nations Committee on Global Food Security

CTBE Brazilian Bioethanol Science and Technology Laboratory

Common Proposed "Solutions"

CNPEM

1. **Simply forget biofuels**
   - Focus on solar, wind and conservation

- Solar and wind can play important roles
- But they cannot, in near to medium term, alleviate needs for liquid transport fuels
- Conservation should always be first priority
- However, all research suggests need for dense liquid fuels for transport...

Do we need biofuels?

Brazilian Bioethanol Science and Technology Laboratory

CNPEM

After efficiency, biofuels are most important option available to cut emissions from transport sector

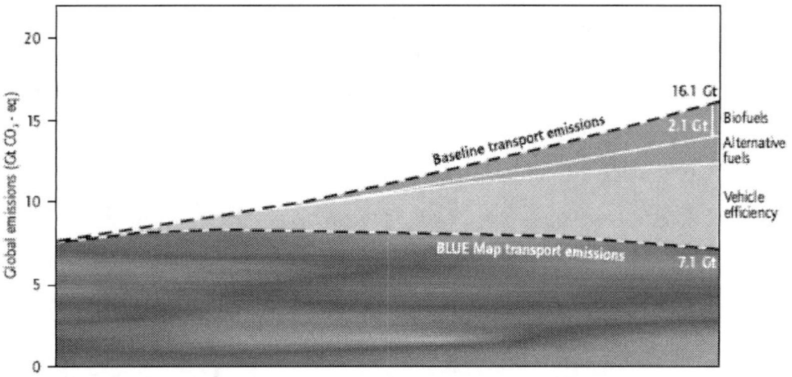

Note: Modal shifts (not included) could contribute an additional 1.8 Gt CO_2-eq. of emission reductions.
Source: IEA, 2010c.

Do we need biofuels?

Brazilian Bioethanol Science and Technology Laboratory

CNPEM

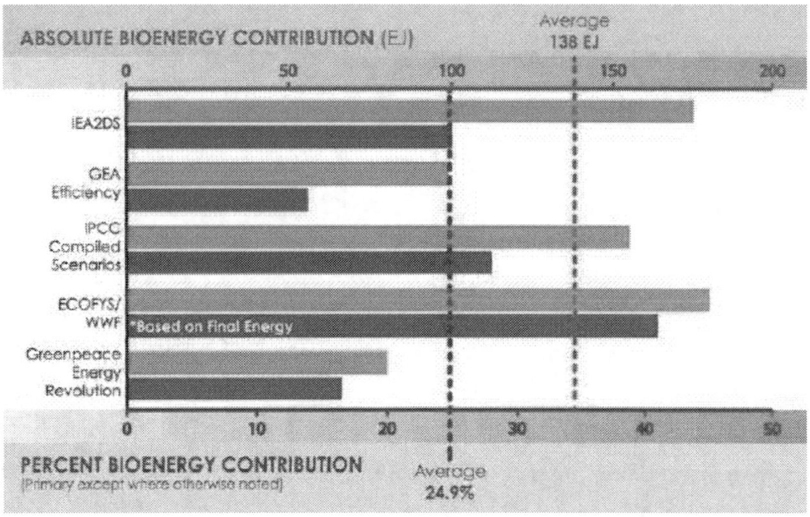

Source: B. Dale et al., 2014

So, several important institutions consider that biofuels will be needed

Common Proposed "Solutions"

CTBE Brazilian Bioethanol Science and Technology Laboratory — CNPEM

2. Only use "non-food" crops for biofuels

- Several NGOs and governments testing this idea

- Evidence to date…
  - Non-food crops lack improved seeds and markets
    - First RSB-certified "Sustainable" biofuel project in Latin America (Yucatan Jatropha) was bankrupt before it produced first liter of fuel
    - Costs and risks are high (algae and other "next-gen" feedstocks)
  - **Fact:** You cannot eat "non-food crops"
    - Therefore, using land, labor and water to produce inedible crops could have opposite effect of that desired

Common Proposed "Solutions"

CTBE Brazilian Bioethanol Science and Technology Laboratory — CNPEM

3. Change policies so they adjust biofuel production in response to food crises

- Recommendation from the November 2013, 40th Meeting of the United Nations Committee on Global Food Security
- Food "crises" typically occur suddenly, exemplified by spikes in local staples in affected countries
- Food "crisis" have other causes than competition for feedstocks

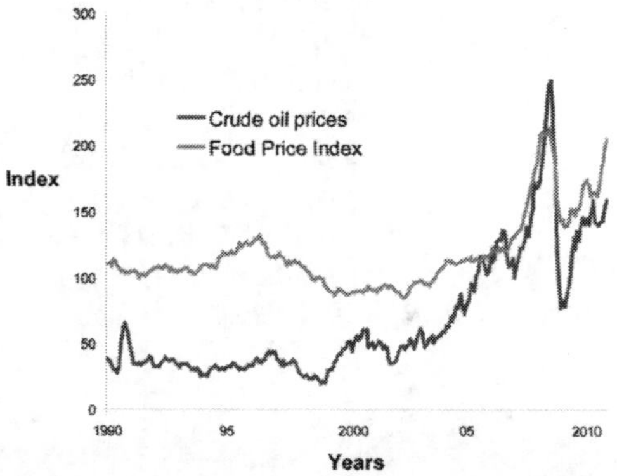

Brazilian Bioethanol Science and Technology Laboratory

Common Proposed "Solutions"

CNPEM

3. Change policies so they adjust biofuel production in response to food crises

- Major ***assumptions that*** -
  - Current conventional feedstock production systems can respond quickly
  - Suspension of biofuel policy/mandate will "free-up" food feedstocks
  - The biofuel response in US or Brazil will help avert or mitigate food crises in distant, affected nations

Our project explores these assumptions...

Brazilian Bioethanol Science and Technology Laboratory

CNPEM

Our Project

We review recent research and data relevant to the world's two largest biofuel production systems:

- Brazil sugarcane ethanol (this talk)
- US maize (corn) ethanol (next talk)

These two systems represent nearly 90% of the global ethanol production. We will attempt to answer the following questions:

- How does the feedstock use for biofuel interact with food security and global food prices?
- How do government policies affect the natural equilibrium of food prices and prices volatility?

Finally, we will provide recommendations for moving forward constructively.

Flexibilities in Brazilian Sugar/Ethanol Sector (1)

- **Production side**: most mills have the flexibility to direct the amount of sugar in cane for sugar or ethanol (around 40-60% range); it is possible for most distilleries to adjust the ratio hydrous/anhydrous.
- **Demand side**: >60% of Brazil light duty vehicles are Flex-Fuel (FFV's) that can be operated using:
  - 100% hydrous ethanol
  - Gasoline-C (or gasohol) with 18-25% ethanol blend
  - or any mixture of these two fuels, e.g., from 18% blend to 100% ethanol.

New LDV sale

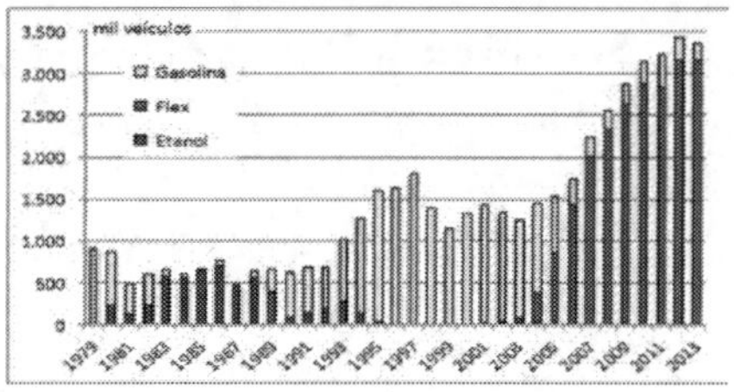

Brazilian Sugarcane Sector Profile

Mill type	Number of mills	Milled cane (Mtc/y)	% milled cane
Sugar/ethanol	257	378.4	67.4
Ethanol distillery	127	161.2	28.7
Sugar mill	18	21.4	3.8
Total	402	561.0	100.0

Source: CONAB, 2012

Notes:

1. More than two thirds of the cane is milled in sugar/ethanol units; this number tends to grow
2. These mixed mills have, on average, a flexibility to produce 40% to 60% of either product
3. The international model of sugar only has almost disappeared in Brazil

Brazilian Bioethanol Science and Technology Laboratory

CNPEM

Flexibilities in Brazilian Sugar/Ethanol Sector

- **Feedstock production flexibility:** Although mills have some flexibility, annual cane production is relatively fixed, being a product of the areas previously planted and weather.
  - After fields are prepared for cane, it takes normally two seasons before first harvest begins
  - Harvests then continue for 5 or more years
  - So increases in planted area require foresight and large investments
- **Ability to adjust processing in response to markets:**
  - Decisions to produce sugar or ethanol are based on market analysis, risk and long-term contracts
  - The contracts are usually set before cane processing season, (or safra) May-Dec, begins.
  - This limits ability for mills to make changes in sugar-ethanol mix on short notice.

Brazilian Bioethanol Science and Technology Laboratory

CNPEM

2007-2014 Sugar and Ethanol Production

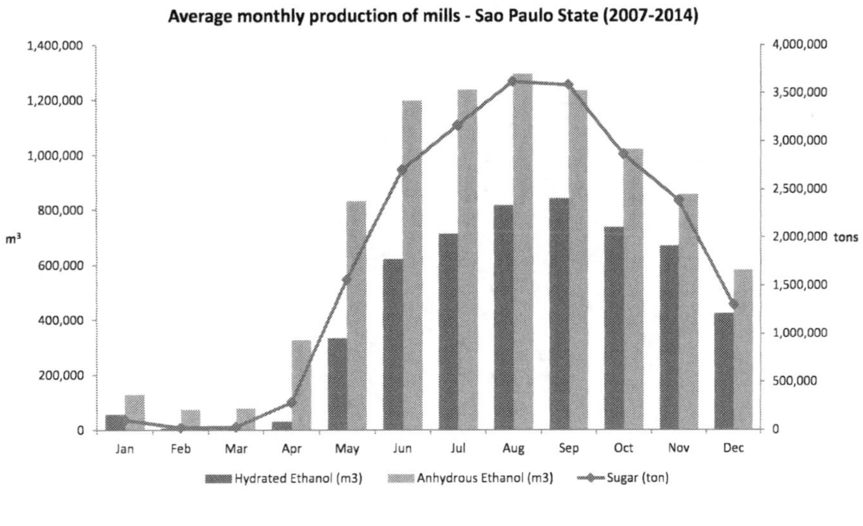

Source: Capitani et al., 2014

Impacts of Government Policies

- On ethanol
  - 1931: mandate 5% blend on imported gasoline
  - 1975: Proalcool Program – subsidies, parity prices of ethanol with sugar and gasoline, support to increasing blend ratio and neat (100%) ethanol vehicles
  - 1986: oil shock – decreasing subsidies and stagnation of ethanol production
  - 2000: end of government control in the sector
  - 2003: FFV supported by government policy
  - 2005 to present: government subsidizes gasoline
- On sugar
  - 1933: created IAA that controlled all production and prices (including ethanol)
  - 1990: closed IAA and initiate the sector deregulation
- Conclusions:
  - Sugar-ethanol sector grew when industry was allowed to respond to markets and government price controls were removed
  - Government support policies (e.g. favoring FFVs) are important, especially in the initial phase of the program

Transport Fuel Consumption in Brazil

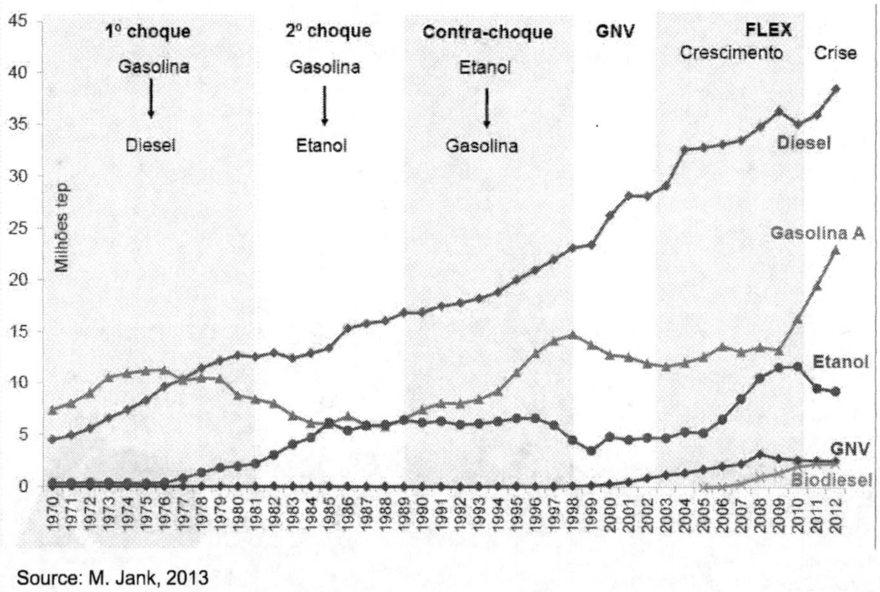

Source: M. Jank, 2013

Gasoline and Ethanol Prices Evolution Under Gasoline Price Control

CTBE Brazilian Bioethanol Science and Technology Laboratory — CNPEM

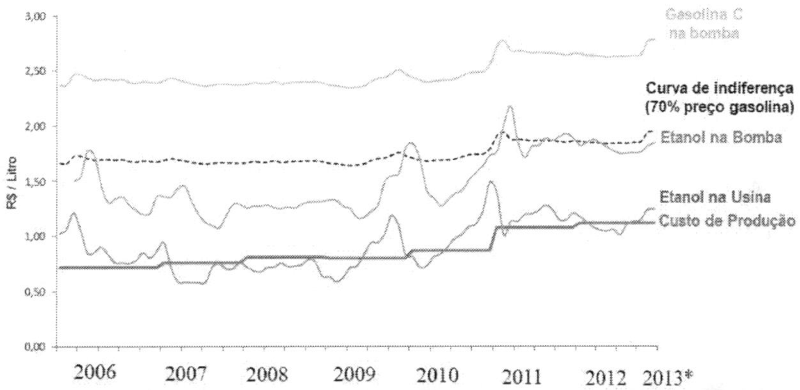

- Desde 2006, a relação de preços com a gasolina se tornou um limite virtual no preço do etanol.
- Ausência de transparência na política de precificação da gasolina afasta investimentos para expandir a produção de etanol.

Source: M. Jank ,2013 (ANP, CPEA, Agroconsult)

CTBE Brazilian Bioethanol Science and Technology Laboratory — CNPEM

Sugar and Ethanol Margins and Mix (2007-2014)

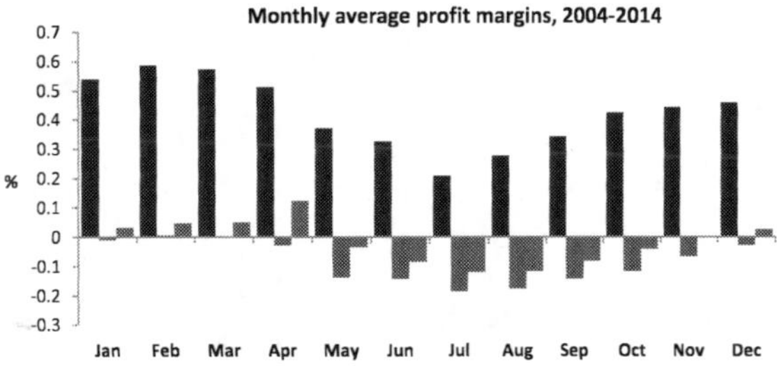

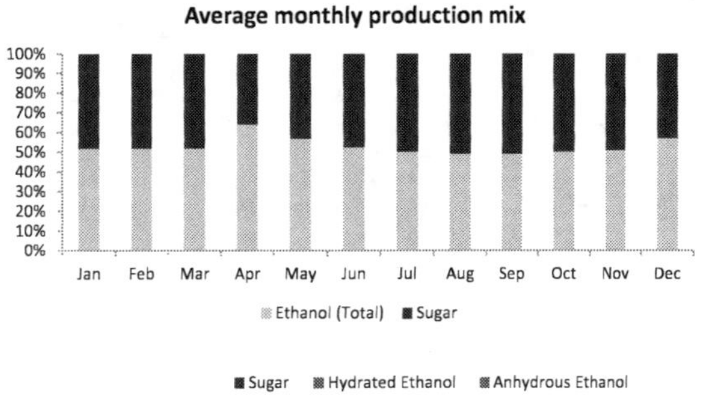

Source: Capitani, 2014

CTBE Brazilian Bioethanol Science and Technology Laboratory

CNPEM

World Sugar Prices and Hydrous Ethanol Parity

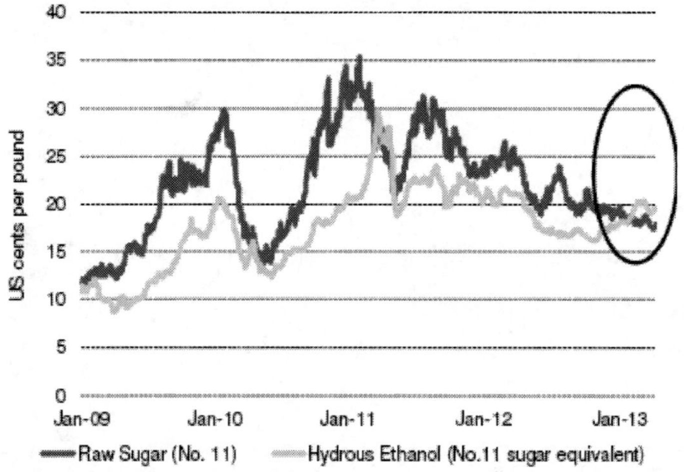

Source: LMC International, 2013

CTBE Brazilian Bioethanol Science and Technology Laboratory

CNPEM

International Sugar Supply and Demand

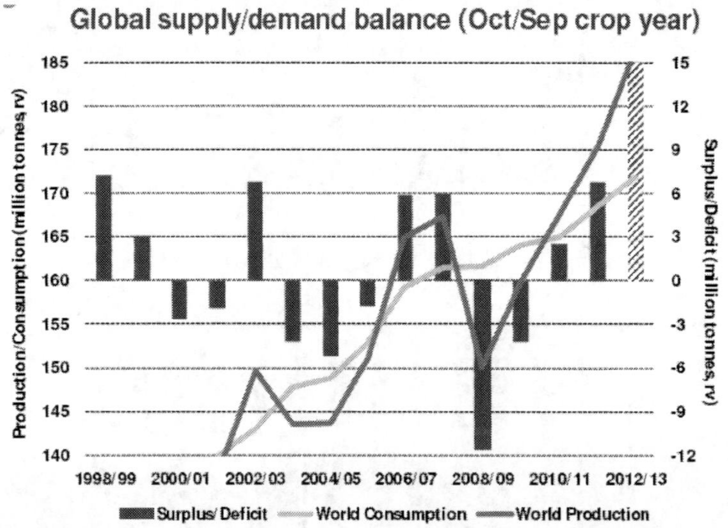

Source: LMC International, 2013

How Do Sugar Producers Respond to Prices?

It depends on

- The alternative crops
- Ratoon cycle

There are significant differences among producing countries with respect to these two cases

Ratoon Cycle in Different Countries

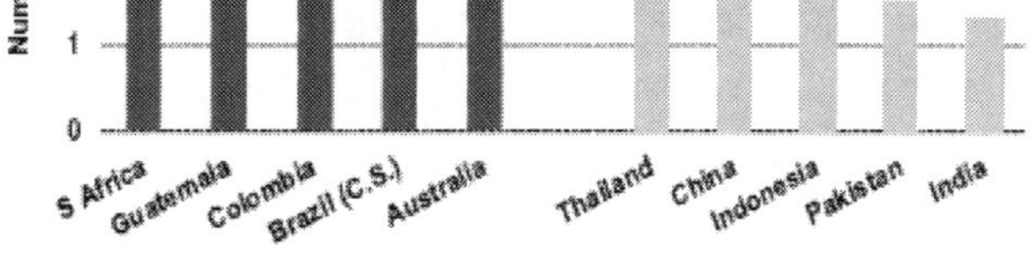

Source: FMC International, 2013

Sugar/Ethanol Production in C/S Brazil

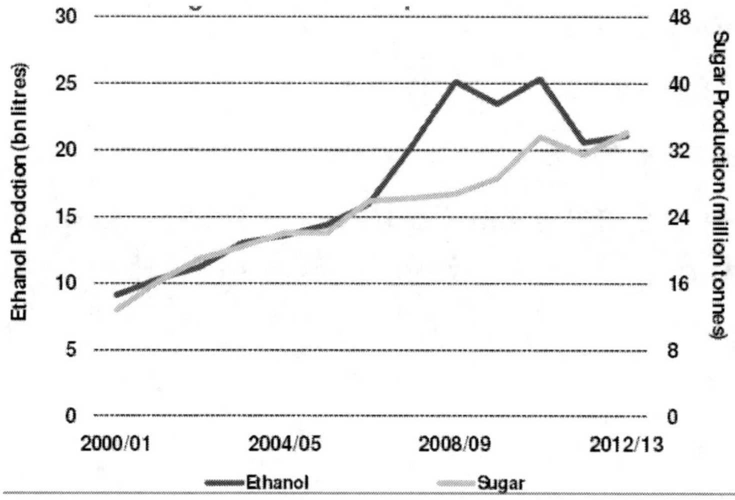

Source: FMC International, 2013

Final Comments (1)

- Government policies dominated the sugarcane sector in Brazil from 1933 to 1991 (for sugar) and until 2000 (for ethanol): established production quotas, prices and export
- Today, government can influence markets through control of mandated blending rate for ethanol in gasoline
- Government is having significant impact via control (subsidies) of the price of gasoline (1st time)
- Government can also adjust taxes (not done very often)
- Ethanol prices have a cap represented by the parity price with gasoline (70% of gasoline price) and helps to bring down the sugar parity price in the international market
- Government can set tariffs, duties and taxes or create other incentives to produce, export or retain, sugarcane products, but has not intervened through these mechanisms lately

CTBE Brazilian Bioethanol Science and Technology Laboratory

CNPEM

Final Comments (2)

- Sugar sector suffered overproduction and lacked competitiveness over most of 20th century despite government intervention
- Pro-Alcohol Program had significant impacts on investment, technology and improvements (cane breeding, management, processing efficiency, etc.)
- Government and industry contributed to infrastructure improvements that were critical for socio-economic development (ports, railroads, communications, etc.)
- Sugarcane feedstock represents 2/3 of total final product costs (ethanol or sugar) so improvements in sugarcane genetics and management has been very influential in competitiveness
- Sugar and ethanol in Brazil are partners not competitors for the same feedstock

CTBE Brazilian Bioethanol Science and Technology Laboratory

CNPEM

Common Proposed "Solutions"

3. **Change policies so they adjust biofuel production** in response to food crises? This assumes quick adjustment can free food for distant lands, but*:*

   – Brazil has limited capacity to adjust processing at mills each season

   – The ethanol-sugar mix is already driven by market prices so policy does not need to change

   – Adjustments are unlikely to affect food availability in distant nations suffering food crisis

Brazilian Bioethanol Science and Technology Laboratory

CNPEM

Thank you!

regis.leal@bioetanol.org.br

CTBE Brazilian Bioethanol Science and Technology Laboratory | UNICAMP | CNPEM | Oak Ridge National Laboratory — Managed by UT-Battelle for the Department of Energy

Food Security and Biofuels: Can Policy Flexibility Mitigate Food Price Crises for the Poor?

Presented by Manoel Regis L. V. Leal[1]

Keith L. Kline[2], M. Regis Leal[1] and Daniel Capitani[3]

[1]CTBE – Brazilian Bioethanol Science and Technology Laboratory
[2]ORNL – Oak Ridge National Laboratory, Climate Change Science Institute
[3]UNICAMP – University of Campinas

2nd Brazilian BioEnergy Science and Technology Conference
BBEST 2104

Campos do Jordão, SP – Brazil, October 20-24, 2014

CTBE Brazilian Bioethanol Science and Technology Laboratory | CNPEM | Oak Ridge National Laboratory — Managed by UT-Battelle for the Department of Energy

Overview

We consider current food security issues, review recent research with a focus on drivers, and consider the world's two largest biofuel production systems: Brazil sugarcane and US maize (corn) ethanol. These two systems represent nearly 90% of global ethanol production. We attempt to answer the following questions:

- How does feedstock use for biofuels interact with food security and global food prices?
- Can biofuel policy changes in these nations reduce the likelihood of food price crises for the poor?

We conclude with observations and recommendations for moving forward constructively.

Food Security is ***only recently*** being defined and measured in more consistent manners

- Food security: when all people at all times have physical and economic access to sufficient, safe and nutritious food…for an active and healthy life (FAO 1996 World Food Summit). Four main dimensions:
  - ***Availability***
  - ***Access***
  - ***Utilization***
  - ***Stability***
- Food security is basic human right
- Key causes of hunger include poverty, lack of social safety net (weak governance), disasters (weather, political), and lack of local production capacity (FAO SOFI reports 2009-14).

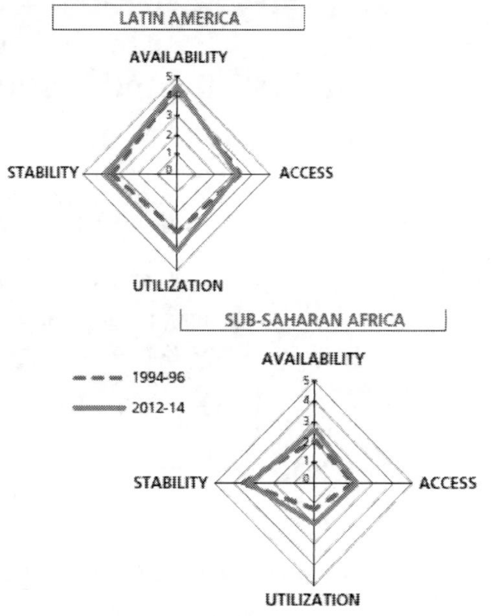

Source: FAO State of Food Insecurity in World (SOFI) 2014

World hunger is decreasing. Most under-nourished people (92%) are in Asia and Africa

The changing distribution of hunger in the world: numbers and shares of undernourished people by region, 1990–92 and 2012–14

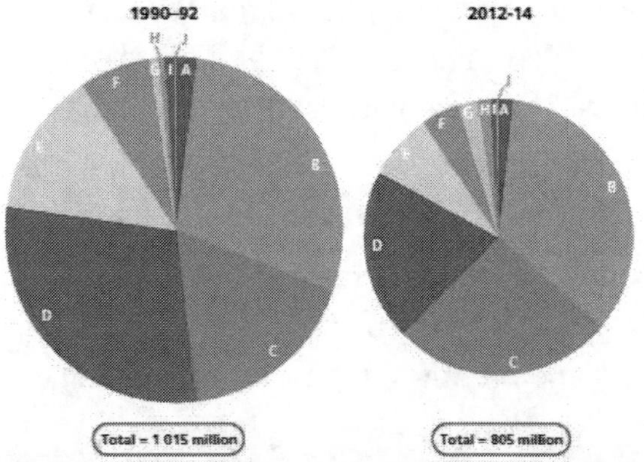

	Number (millions)		Regional share (%)	
	1990–92	2012–14	1990–92	2012–14
Developed regions	20	15	2.0	1.8
Southern Asia	292	276	28.8	34.3
Sub-Saharan Africa	176	214	17.3	26.6
Eastern Asia	295	161	29.1	20.0
South-Eastern Asia	138	64	13.6	7.9
Latin America and the Caribbean	69	37	6.8	4.6
Western Asia	8	19	0.8	2.3
Northern Africa	6	13	0.6	1.6
Caucasus and Central Asia	10	6	0.9	0.7
Oceania	1	1	0.1	0.2
Total	1015	805	100	100

Note: The areas of the pie charts are proportional to the total number of undernourished in each period. Data for 2012–14 refer to provisional estimates. All figures are rounded.
Source: FAO.

Source: FAO State of Food Insecurity in the World (SOFI) 2014

Experiences with food insecurity indicate that:

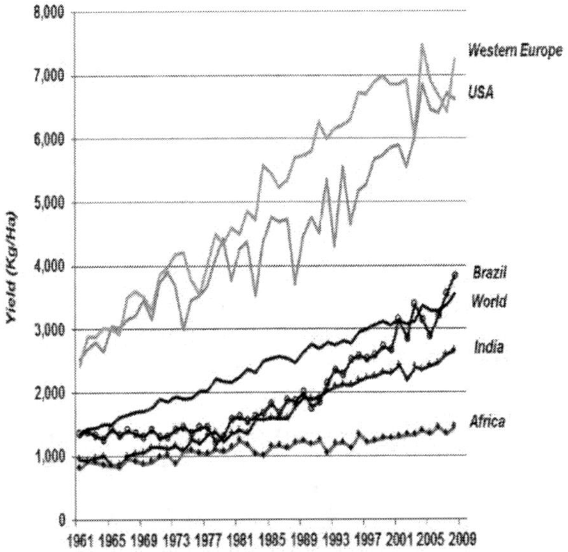

1. At global scale, there is plenty of land and food production; distribution and cost are key issues for poor.
2. Local investment in agriculture and effective social safety nets for the poor are essential to improve food security.
3. Additional factors contributing to food insecurity:
   - 40+ years of food aid, over-production
   - Lack of market incentives
   - Lack of access to financial resources
   - Price volatility
   - Yield gaps

Sources: World Bank (Chart); and FAO Annual reports on State of Food Insecurity (SOFI) in the World

Experiences with food insecurity indicate that:

Key research question: How do biofuel policies actually interact with the conditions and processes that determine food insecurity?

1. At global scale, there is plenty of land and food production; distribution is an issue for the poor.
2. Local investment in agriculture and effective social safety nets for the poor are essential to improve food security.
3. Additional factors contributing to food insecurity:
   - 40+ years of food aid, over-production
   - Lack of market incentives
   - Lack of access to financial resources
   - Price volatility
   - Yield gaps

Sources: FAO Annual reports on State of Food Insecurity (SOFI) in the World

Land Availability – huge areas suited for rain-fed agriculture, especially Africa and Latin America

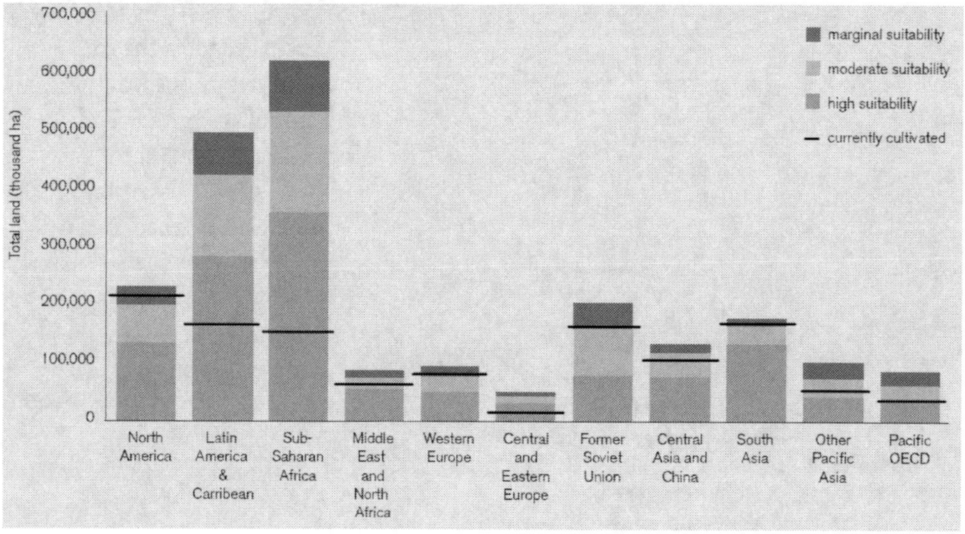

FAO 2007

Low Agricultural Productivity correlates with more severe Food Insecurity

FIGURE 6
Share of countries in each malnutrition category, by level of agricultural productivity

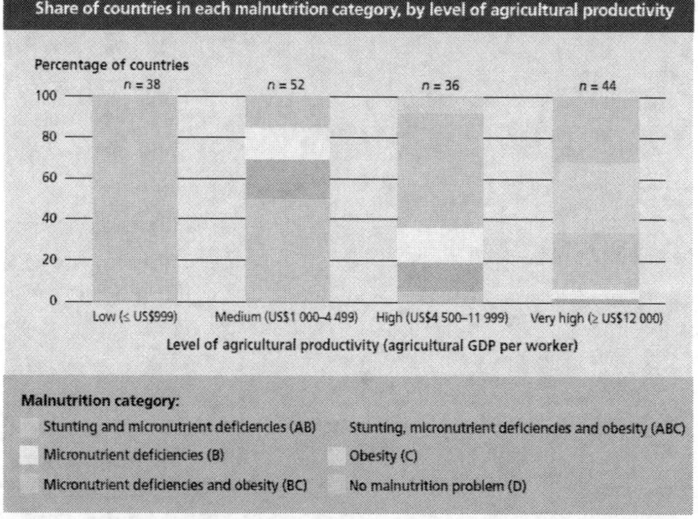

Notes: n is the number of countries characterized by each level of agricultural productivity. Agricultural productivity is derived by dividing agricultural GDP (in 2010 measured in current US dollars) by the population economically active in agriculture. Malnutrition categories are those illustrated in Figure 4.
Sources: Authors' calculations using agricultural GDP data from the United Nations (2012) and data on agricultural workers from FAO, 2013. Sources used to determine malnutrition categories are those used for Figure 4.

Source: FAO (2013 State of Food and Agriculture; SOFA)

Food security remains a problem for 800 million people. How can we reduce food crises?

Some proposed "solutions" have targeted biofuels:

1. **Simply forget biofuels**
   - Focus on solar, wind, conservation
2. **Only use "non-food" crops** for biofuels
   - Several NGOs and governments testing this idea
3. **Change policies so biofuel production adjusts** in response to food crises
   - Recommendation from the November 2013, 40th Meeting of the United Nations Committee on Global Food Security

Let's consider each…

Commonly proposed "Solutions" to mitigate food price crises

1. **Simply forget biofuels to focus on solar, wind and conservation**
   - Solar and wind can play important roles
     - But they cannot, in near to medium term, alleviate needs for liquid transport fuels
     - Aviation and long-haul shipping require dense liquid fuels
   - Conservation should always be first priority – but alone it is insufficient to provide energy security and meet societal development goals

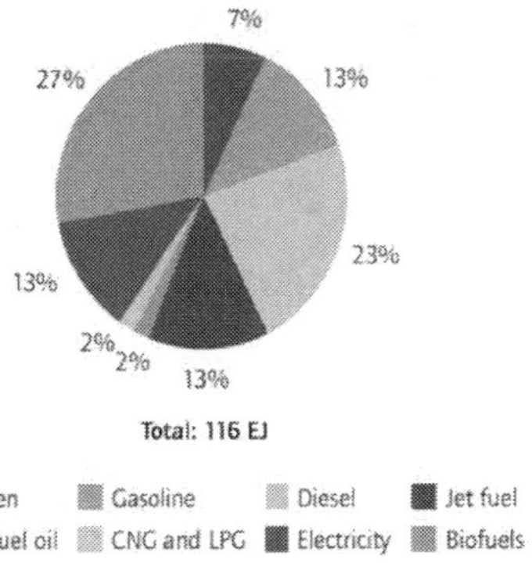

Chart: IEA 2011 Roadmap: Biofuels role in projected 2050 transport fuels

Studies that prioritize other renewables still rely on bioenergy for large, essential role to help society address climate concerns

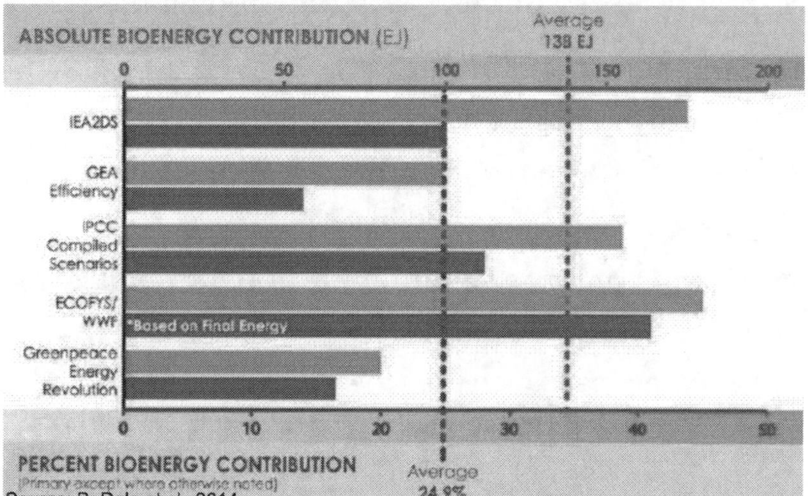

Source: B. Dale et al., 2014

To address climate change, we need bioenergy and biofuels

Commonly proposed "Solutions" to mitigate food price crises

2. Only use "non-food" crops for biofuels

- Several NGOs and governments testing this idea

- Evidence to date...
  - Non-food crops lack investments to improve seeds, develop infrastructure and markets
  - Evaluations (e.g., NL Agency 2010 *Jatropha Assessments*) of multiple efforts to develop biofuels using non-food crops found consistent short-comings and failures
  - Costs and risks are high for new crops such as algae and other "next-generation" feedstocks

National Public Radio
www.npr.org: "How A Biofuel Dream Called Jatropha Came Crashing Down"
(August 21, 2012)

Commonly proposed "Solutions" to mitigate food price crises

2. Only use "non-food" crops for biofuels

Fact: You cannot eat "non-food crops"

- Therefore, using land, labor and water to produce inedible crops could have the ***opposite effect of that desired***
- Biofuel production with traditional crops builds a ***"production cushion" that can reduce local price volatility*** and allows markets to respond in times of crisis
- Non-food prices, products with limited markets, may be ***MORE volatile*** than food, creating added risk for producers of non-food products…

Non-food commodity prices rose more than those for food in the 2008 crisis

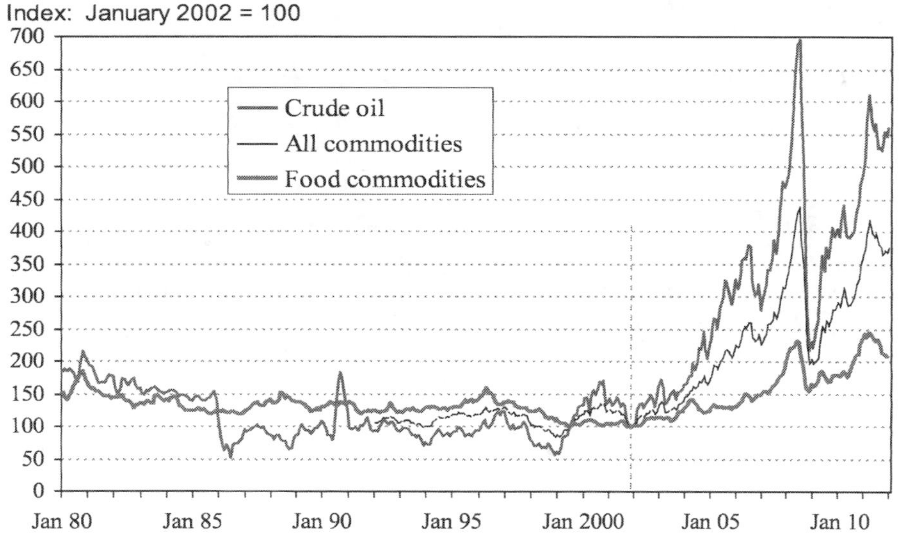

Source: Ron Trostle, ERS; based on International Monetary Fund: International Financial Statistics, Jan 2012p

14

Commonly proposed "Solutions" to mitigate food price crises

3. Change policies so that biofuel production adjusts in response to food crises

- This concept ***assumes*** that
  - Current policies do not allow such adjustment
  - Current conventional feedstock production systems can respond quickly to a food price crisis
  - Government interventions in markets (to change policies in "times of crisis") will not have detrimental side-effects on future production
  - Suspension of biofuel policies will "free-up" the food that's needed and/or reduce prices of required food stuffs
  - A biofuel policy response in US or Brazil that cuts biofuel production will help avert or mitigate food crises for poor populations in distant, affected nations

Available data raise doubts about each of these assumptions

Sugar/Ethanol Production Mix

Brazil adjusts sugar-to-ethanol output ratio to respond to markets, but it takes time

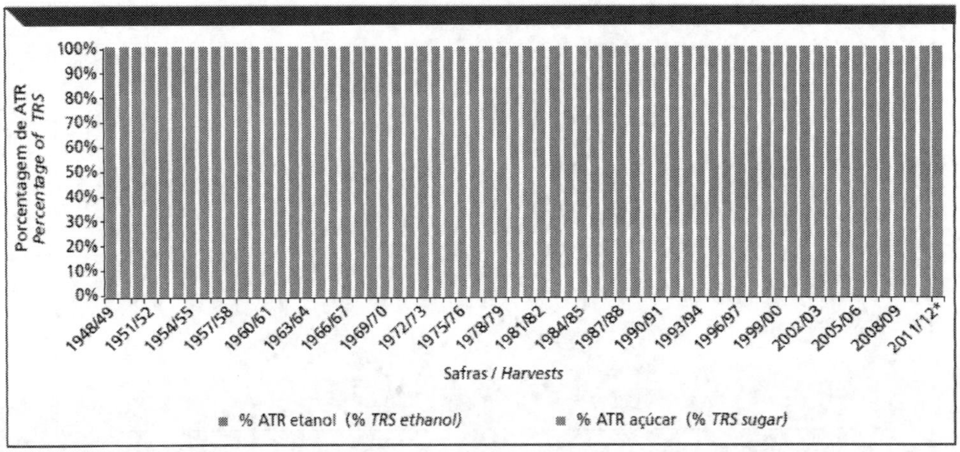

Source: MAPA, 2013

World Sugar Prices and Hydrous Ethanol Parity

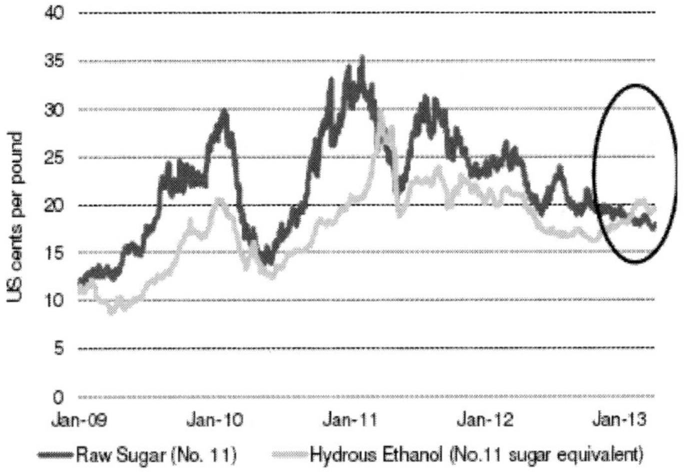

Sugar and ethanol prices respond to common external drivers. Sugar and hydrous ethanol parity prices tend to converge to a common value.

Source: LMC International, 2013

Is Ethanol Affecting Food Commodity Prices in Brazil? If not, how can ethanol impact food prices in other nations?

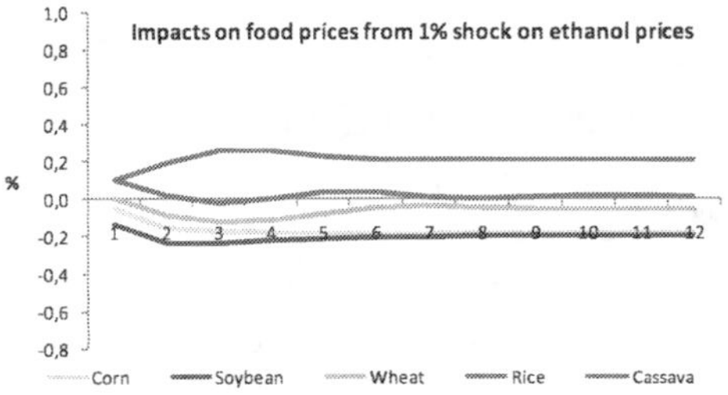

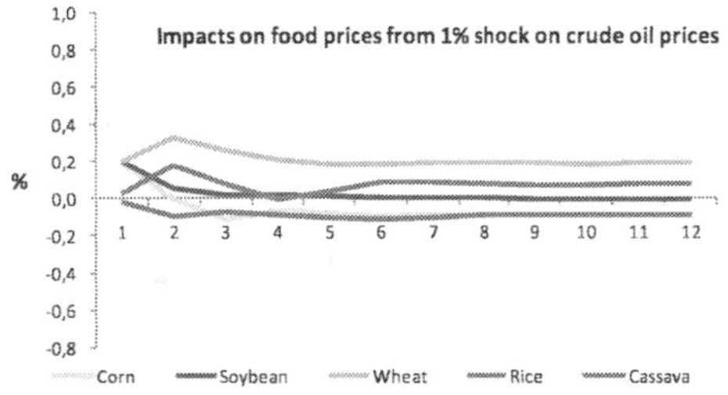

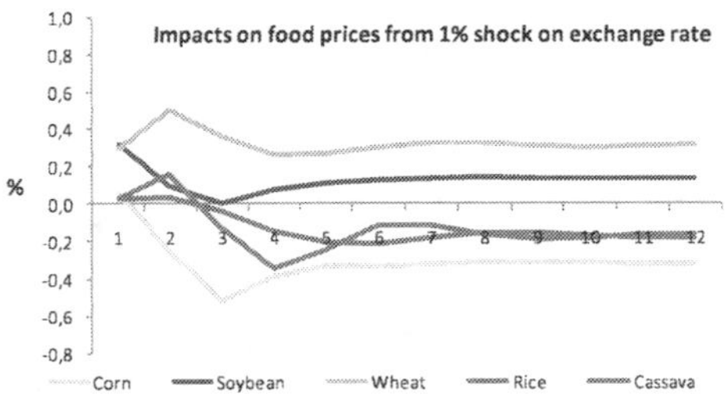

Preliminary results using time series *"Auto-regressive Vector with error correction model"* point out weak impacts from positive shocks on ethanol/sugarcane prices:

- ❑ There are no significant variations on domestic food commodities prices.
- ❑ Oil and exchange rate exhibit similar or greater impacts on food commodities prices

Source: Capitani et al., 2014

Flexibilities in Brazilian Sugar/Ethanol Sector

- **Feedstock production flexibility:** Although mills have some flexibility, annual cane production is relatively fixed, being a product of the areas previously planted and weather.
  - After fields are prepared for cane planting, it takes 18 months before first harvest begins and harvests continue seasonally for 5 or more years
  - Increases in planted area require foresight, large investments and about two years advance planning
- **Ability to adjust processing in response to markets:**
  - Decisions to produce sugar or ethanol are based on market analysis, risk and established long-term contracts.
  - Contracts are usually set before cane processing season begins.
  - Contracts limit ability for mills to make large changes in sugar-ethanol mix on short notice, making it difficult to respond quickly to sudden price crises.
- **USA** has similar limitations in terms of large, quick responses in corn production, but adjusts annually to market signals.
- In both Brazil and USA: **larger supplies** create market "**shock-absorbers**" and opportunities exist for a **second corn crop** if prices rise (as demonstrated by the historic US drought in 2012).

Source: Authors' analysis

Food Price Index of Selected African Countries

Figure 5: Evolution of global food price index and food price indices of selected ESA countri

Of note: Malawi, the least-impacted nation, is one of few in Africa with long-established biofuel programs

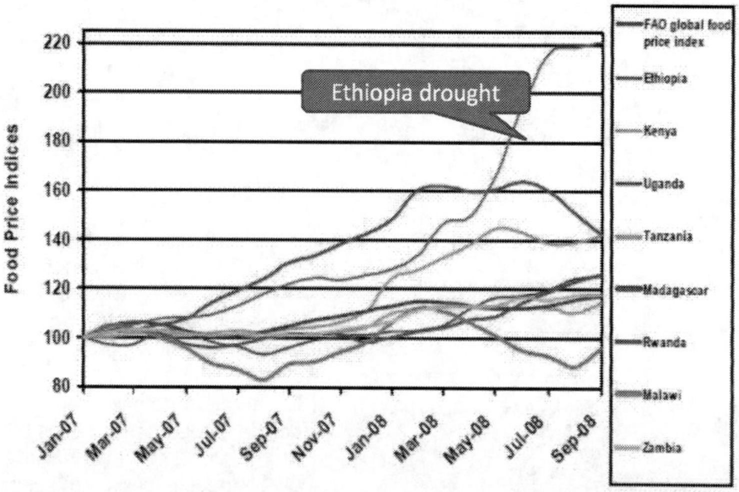

Source: Reproduced by kind permission of CGIAR. From 'Responding to Food Price Crisis in Eastern and Southern Africa: Policy Options for National Regional Action' (Karugia, et. al. 2009)

Source of chart: UK Defra 2010 – from CGIAR 2009

Commonly proposed "Solutions" to mitigate food price crises

3. Change policies so they adjust biofuel production in response to food crises

This was proposed in the November 2013, 40th Meeting of the United Nations Committee on Global Food Security. But…

- Food "crises" typically occur suddenly, exemplified by spikes in local staples in affected countries
- Food "crisis" have other causes – including ***significant role of crude oil prices****
- Shifts in corn and sugar use by US and Brazil are unlikely to have rapid influence on local food stuff prices in distant developing nations

*** If biofuel production helps moderate spikes in oil price, more biofuel production rather than less, may be best option to mitigate future food price crises.**

Grain commodity price volatility is far different from "food price" volatility (US Consumer Food Price Index, CPI) 1976-2012

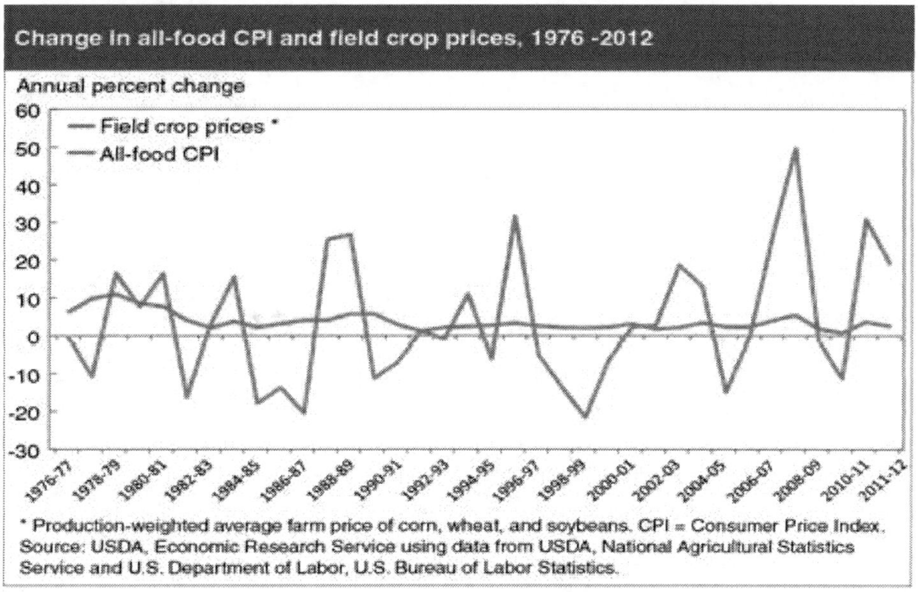

Chart: http://www.ers.usda.gov/data-products/chart-gallery/detail.aspx?chartId=41700&ref=collection&embed=True&widgetId=39734

Primary factors affecting global grain commodity prices[1] (2010–2012) – Weather events and import/export policies

Index: January 2002 = 100

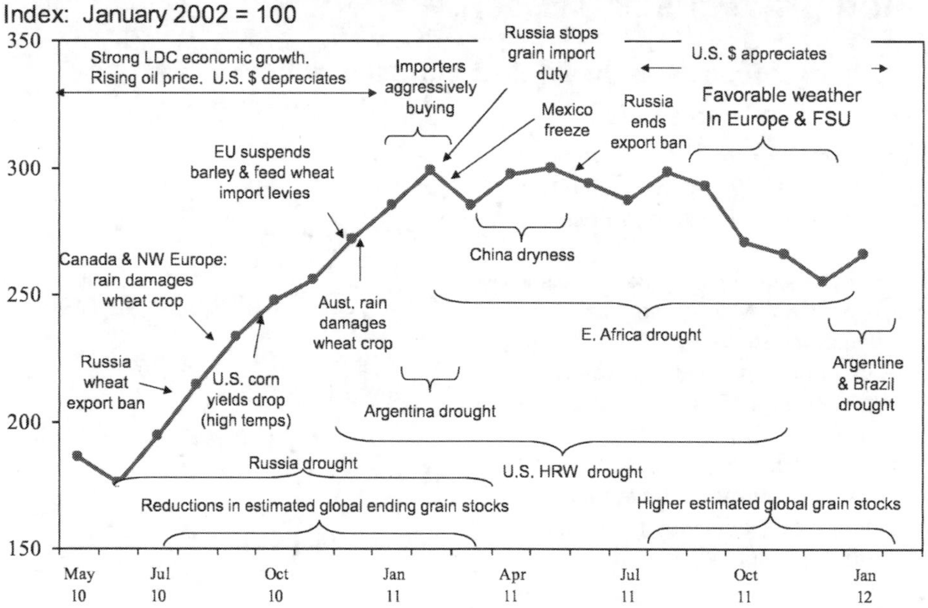

[1]4-crop monthly price index: Wheat, rice, corn, & soybean prices; based on IMF price and trade share data.

Source: Ron Trostle, ERS

CTBE Brazilian Bioethanol Science and Technology Laboratory

CNPEM

World Commodity Prices

Figure 1: Index of world commodity prices, January 2000 – April 2009

Note major influence of oil price (black) and largest peak: rice (red) not biofuel crop – but see next slide

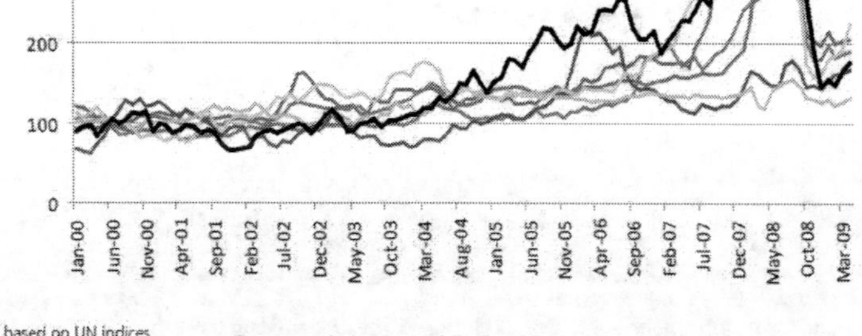

Source: Defra, based on UN indices

Source of chart: UK Defra 2010 – based on UN

Common policies in response to food price crisis: Export restrictions drive prices even higher for rice, basic food staple in poor nations

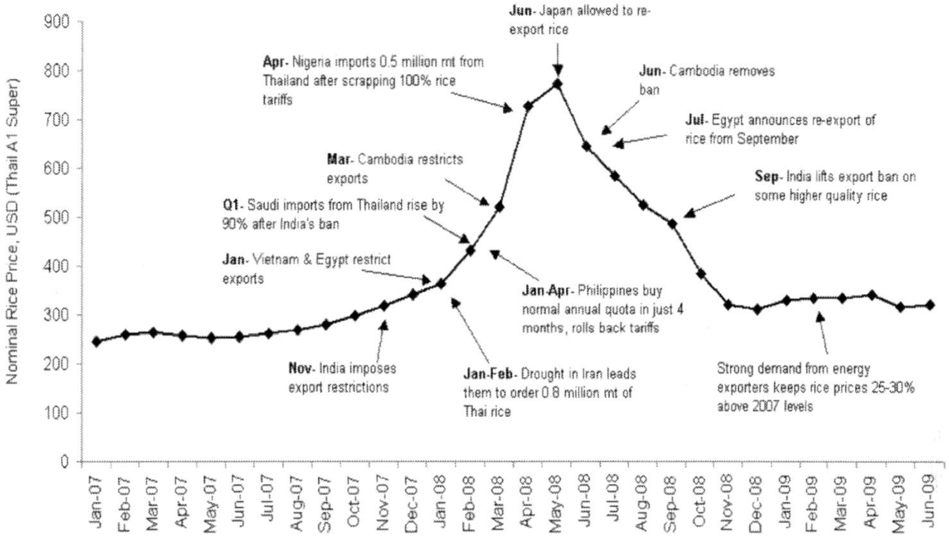

Source of chart: UK Defra 2010, based on IFPRI 2009.

Common Policy Actions in Response to Food Price Spikes

Figure 7: Policy actions to address high food prices (sample of 77 countries by type of action)

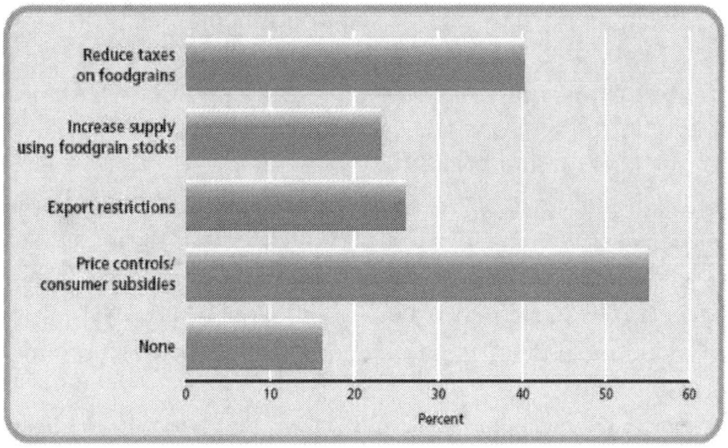

Source: FAO (2009)

These policies exacerbate food insecurity; they address symptoms, not causes.

Source of chart: UK Defra 2010 – based on FAO 2009

Real Price Indices: long-term trend reflects periods of volatility but declining price over time

Figure 2: Real price indices January 1960 – December 2008

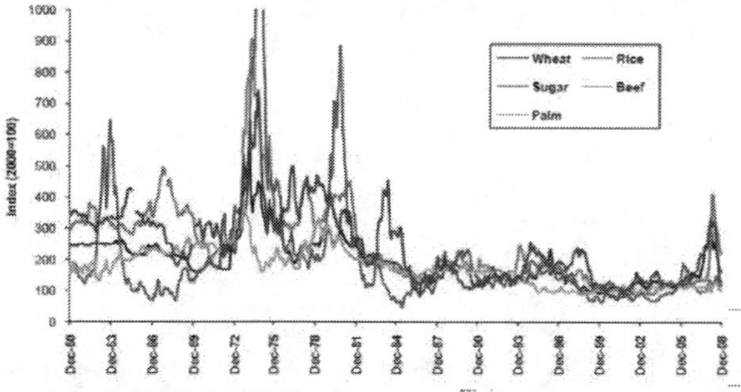

Source of chart: UK Defra 2010

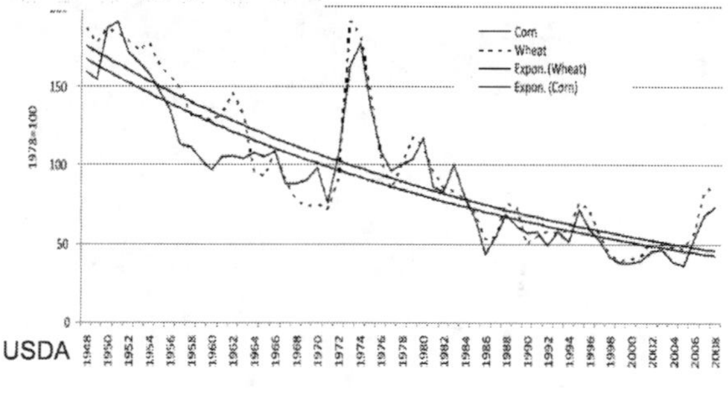

Source: USDA

Sugar, Maize and Ethanol Markets in Brazil and USA

- "Futures markets" and speculation play important roles in price volatility, influenced by oil price volatility, economic growth swings, weather extremes, and unexpected policy changes.
- Current ethanol and feedstock markets are interconnected in ways that appear to dampen maize, sugar, and ethanol price volatility (note Brazil←ethanol→USA trade).
- The rate of growth in sugar and corn feed production (including exports) was not reduced by growth in ethanol production.
- Assuring demand for ethanol can support a "price floor" for producers (the "shock absorber" effect goes both ways)
- A policy opportunity often applied in Brazil is to adjust ethanol blend rate in gasoline.
- Policies that create incentives to maintain high levels of production and flexibility (multiple markets; ability to substitute), help dampen price volatility in linked markets.

Source: Authors' analysis

Index of 4 Dimensions of Food Security

(The Economist Intelligence Unit, 2014)

- Analysis and ranking of 109 nations
- Increasing food security correlates well with democracy
- and less with obesity; more complex
- Food security correlates strongest with HDI

Source: The Economist Intelligence Unit Food Security Index 2014

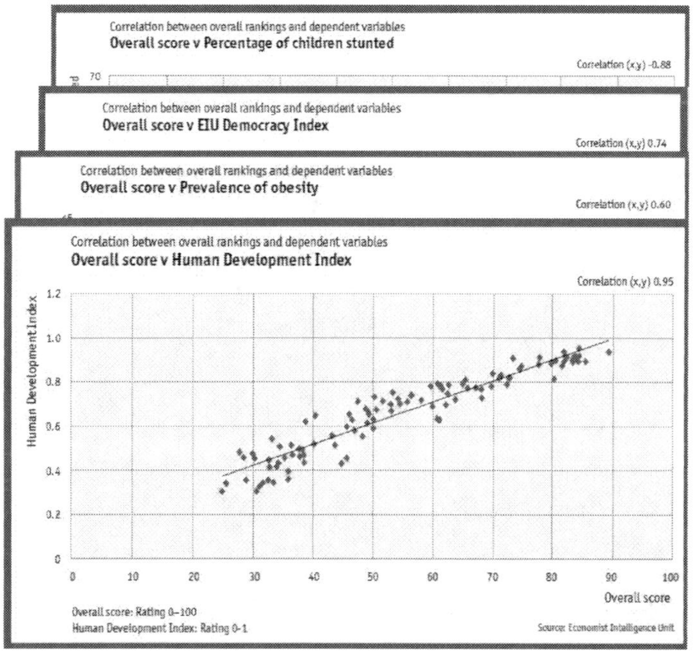

Food security has improved in some nations, surpassing MDG goals to cut malnutrition by 50%

- Malawi, an extremely poor land-locked African nation and
- Brazil

Both met the MDG goal for cutting hunger by at least 50% - ***while producing biofuels***

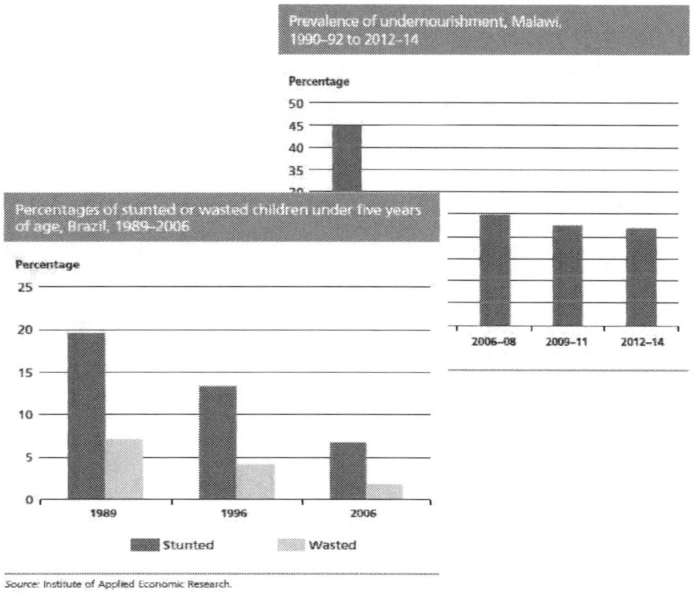

Source: FAO State of Food Insecurity in the World 2014

Wrapping up – Food Security, Biofuels and the Sustainable Development – what is needed to address food security concerns?

- Food security: Four main dimensions: ***Availability, Access, Utilization, Stability***
  - Causes of hunger include poverty and lack of social safety net (weak governance)
- Principles for sustainable development
  - Include stakeholders in process – define priorities
  - Sustainable employment
    - Reliable, sufficient household income provides food security
    - Improved rural livelihoods
  - Rural economies with more resilient production systems
    - Threats include climate change, extreme weather
    - Solutions include adaptive land management, land-use efficiency
    - Incentives to increase land-use efficiency and productivity
    - Systems that can minimize negative effects of extreme weather and disturbance
    - Fire and pests affect over 500 million Ha of biomass per year

Wrapping up: Food Security, Biofuels and the Sustainable Development needed to address food security concerns

- Food security: Four main dimensions: ***Availability, Access, Utilization, Stability***
  - Causes of hunger include poverty and lack of social safety net (weak governance)
- Principles for sustainable development
  - Include stakeholders in process – define priorities
  - Sustainable employment
    - Reliable, sufficient household income provides food security
    - Improved rural livelihoods
  - Rural economies with more resilient production systems
    - Threats include climate change, extreme weather
    - Solutions include adaptive land management, land-use efficiency
    - Incentives to increase land-use efficiency and productivity
    - Systems that can minimize negative effects of extreme weather and disturbance
    - Fire and pests affect over 500 million Ha of biomass per year

Biofuel policies, developed with care, can interact with these conditions and processes to improve food security.

Concluding remarks, food and biofuels (1)

- Research on potential effects often begins with assumption that land is the limiting factor. This assumption is misleading policy makers.
- Oil price spikes, commodity market speculation, US$ devaluation, export restrictions, social conflict, and weather, all had greater impact on food security for poor than biofuels (e.g. WB 2010, CBO 2009, Defra 2010).
  - Institutional/governance capacity is essential to improve food security
  - Needed: Incentives for investment in improved land management for agriculture and other services including waste reduction

Source: Kline presentation to "Pathways to Climate Solutions: Assessing Energy Technology and Policy Innovation" Workshop organized by the Aspen Global Change Institute; 24-28 February, 2014. Aspen CO.

Concluding remarks, food and biofuels (2)

- Food –vs- Biofuels has become emotionally charged, political issue
  - Media and popular discourse are predominantly negative, assuming:
    - Food is being used for biofuels
    - Biofuels contribute to global hunger and food insecurity
  - Lots of modeling of concerns , e.g. *"What might go wrong if…?"*
  - Little modeling of opportunities, e.g., *"What could be improved if…?"*
  - Very little analysis of decade of recent experience, e.g., *"What actually occurred?"*
- Negative opinions undermine public and political support and have direct effects on policies and investment
  - Example: EU RED revised to cap biofuels produced from conventional (food) feedstocks
  - Pressures to reduce or eliminate the USA ethanol requirements
- With more data, recent studies find diminishing role for biofuels in the 2007-2008 commodity price increase and, more important, the potential to mitigate food price volatility (e.g., lack of global food security crisis despite historically severe drought in USA 2012).

Source: Authors' analysis

Concluding remarks, food and biofuels (3)

- Research and analysis (FAO SOFI reports) indicate that food price crises are largely caused by sudden events affecting local access to food by poor in places where historic prices/incentives undermined local production.
- Food price crises are not caused by a lack of total food supply or lack of productive land at global scales.
- Commodity price changes for maize (US) and sugar (Brazil) represent tiny fractions of total food basket price, especially in less developed nations susceptible to food price crises.
- The potential for food crops for bioenergy to act as a buffer stock when unforeseen crises arise merits more study. This appears to occur in Brazil sugar and US corn where ethanol production is a partner rather than competitor for feedstock.
- Fossil fuel price is a major driver of global economy and food prices.
- There is no consensus about relative importance or direction of effects of US and Brazil biofuel policies on global food price indices.
- Biofuels 'done right' (Kline et al. 2009) can be a driver for agricultural sector investments that improve food security: technology, energy services, jobs, ports, railroads, communications, and more.

Source: Authors' analysis

CTBE Brazilian Bioethanol Science and Technology Laboratory — Oak Ridge National Laboratory, managed by UT-Battelle for the Department of Energy — CNPEM

Acknowledgements

The authors wish to thank the São Paulo Research Foundation – FAPESP (Process no. 2012/00282-3), Oak Ridge National Laboratory (ORNL), the US National Science Foundation (NSF) Project for Environmental and Social Sustainability Assessment of Bioenergy in Pan America, and the U.S. Department of Energy for the support to this research.

Keith Kline's research is supported by the U.S. Department of Energy (DOE) Bio-Energy Technologies Office and performed at Oak Ridge National Laboratory (ORNL). Oak Ridge National Laboratory is managed by the UT-Battelle, LLC, for DOE under contract DE-AC05-00OR22725.

The views in this presentation are those of the author(s) who are responsible for any errors or omissions.

Brazilian Bioethanol Science and Technology Laboratory

OAK RIDGE NATIONAL LABORATORY

MANAGED BY UT-BATTELLE FOR THE DEPARTMENT OF ENERGY

CNPEM

Thank you!

regis.leal@bioetanol.org.br

klinekl@ornl.gov

Bibliography and references

- Arezki et al. 2014. Understanding international commodity price fluctuations. Journal of International Money and Finance 42 (2014) 1–8
- Babcock, B. A. (2011 June). The impact of US biofuel policies on agricultural price levels and volatility. . International Centre for Trade and Sustainable Development (35)
- Charles, C. (2012 April). Should we be concerned about competition between food and fuel? International Institute for Sustainable Development.
- Defra 2010 The 2007/08 Agricultural Price Spikes: Causes and Policy Implications. HM Government, United Kingdom.
- de Gorter, H., & Just, D. R. (2010). The social costs and benefits of biofuels: The intersection of environmental, energy and agricultural policy. Applied Economic Perspectives and Policy.
- Economic Research Service, Amber Waves, 10(2 (June)), 2012.
- Durham, C., Davies, G., & Bhattacharyya, T. (2012, June). Can biofuels policy work for food security? Department for Environment Food and Rural Affairs.
- FAO, IFAD, IMF, OECD, UNCTAD, WFP, the World Bank, the WTO, IFPRI, and the UN HLTF. (2011, June 2). Price volatility in food and agricultural markets: Policy responses. Policy Report for the G-20 .
- Fischer, G., van Velthuizen, H. & Nachtergaele, F. 2011. GAEZ v3.0 – Global Agro-ecological Zones Model documentation, (mimeo). IIASA, Luxemburg.
- King R (Oxfam), Kelbert A (IDS), Chisholm N (University College Cork), Hossain N (IDS). 2014. Help Yourself - Food Rights and Responsibilities: Year 2 findings from Life in a Time of Food Price Volatility. Joint Agency Research Report. www.ids.ac.uk and www.oxfam.org
- Langeveld, H., Dixon, J., and van Keulen, H. 2014. Biofuel cropping systems: carbon, land and food. Routledge Earthscan, Abingdon, UK. ISBN13: 978-0-415-53953-1
- Locke, A., Wiggins, S., Henley, G., & Keats, S. (2013 April). Diverting grain from animal feed and biofuels. London : Overseas Development Institute.
- McPhail, Lihong Lu, & Du, X. (2012). Ethanol Strengthens the Link Between Agriculture and Energy Markets.
- Peterka, A. (2012 31-July). Livestock groups urge EPA to waive ethanol mandate. Governors' Biofuels Coalition.
- Schafer, E. (2012 8-March). Proposed RFS changes spark food vs. fuel debate. Feed & Grain.
- Strassburg BBN, Latwiec AE, et al., 2014. When enough should be enough. Improving the use of current agricultural lands could spare natural habitats in Brazil. Glob.Env.Change 28 84-97.
- Tyner, W. E., Taheripour, F., & Hurt, C. (2012 16-August). Potential impacts of a partial waiver of the ethanol blending rules. Farm Foundation and Purdue University.
- Wright, B. (2011, February). Biofuels and food security: Time to consider safety valves? . IPC Policy Focus, International Food and Agricultural Trade Policy Council.

OAK RIDGE National Laboratory

For more in formation:

- Alexandratos, N. and J. Bruinsma. 2012. *World agriculture towards 2030/2050: the 2012 revision*. ESA Working paper No. 12-03. Rome, FAO
- CBES. 2009. *Land-use change and bioenergy*. Report from the 2009 workshop, ORNL/CBES-001, U.S. Department of Energy, Office of Energy Efficiency and Renewable Energy and Oak Ridge National Laboratory. http://www.ornl.gov/sci/besd/cbes.shtml .
- Dale VH, KL Kline, LL Wright, RD Perlack, M Downing, RL Graham. 2011. Interactions among bioenergy feedstock choices, landscape dynamics and land use. *Ecological Applications* 21(4):1039-1054.
- Dale, VH, RA Efroymson, KL Kline, MH Langholtz, PN Leiby, GA Oladosu, MR Davis, ME Downing, MR Hilliard. 2013. Indicators for assessing socioeconomic sustainability of bioenergy systems: A short list of practical measures. Ecological Indicators 26: 87-102.
- Dornburg et al. 2010. Bioenergy revisited: Key factors in global potentials of bioenergy. Energy Environ. Sci., 2010,3, 258-267..
- Efroymson, R. A., V. H. Dale, K. L. Kline, A. C. McBride, J. M. Bielicki, R. L. Smith, E. S. Parish, P. E. Schweizer, D. M. Shaw. 2012. Environmental indicators of biofuel sustainability: What about context? Environmental Management DOI 10.1007/s00267-012-9907-5
- FAO and IIASA. 2007. Mapping biophysical factors that influence agricultural production and rural vulnerability. Food and Agriculture Organization and International Institute for Applied Systems Analysis, Rome, Italy.
- Giglio L., J. T. Randerson, G. R. van derWerf, P. S. Kasibhatla, G. J. Collatz, D. C. Morton, and R. S. DeFries. Assessing variability and long-term trends in burned area by merging multiple satellite fire products. Biogeosciences, 7, 1171–1186, 2010.
- IPCC 2012 Special Report on Renewables and Climate Change Mitigation.
- Kline KL, Dale VH, Lee R, Leiby P. 2009. In Defense of Biofuels, Done Right. Issues in Science and Technology 25(3): 75-84. http://www.issues.org/25.3/kline.html
- Langholtz M, Eaton L and Turhollow A. (in press). 2013 Feedstock Supply and Price Projections and Sensitivity Analysis. (BioFPR 2014).
- McBride A, VH Dale, L Baskaran, M Downing, L Eaton, RA Efroymson, C Garten, KL Kline, H Jager, P Mulholland, E Parish, P Schweizer, and J Storey. 2011. Indicators to support environmental sustainability of bioenergy systems. *Ecological Indicators* 11(5) 1277-1289.
- Oladosu D, KL Kline, P Leiby, R Martinez, M Davis, M Downing, L Eaton. 2012. Global economic effects of the US biofuel policy and the potential contribution from advanced biofuels. Biofuels 3(6):703-723. http://www.future-science.com/doi/pdfplus/10.4155/bfs.12.6
- Parish ES, M Hilliard, LM Baskaran, VH Dale, NA Griffiths, PJ Mulholland, A Sorokine, NA Thomas, ME Downing, R Middleton. 2012. Multimetric spatial optimization of switchgrass plantings across a watershed. *Biofuels, Bioprod. Bioref.* 6(1):58-72.
- USDOE 2011. U.S. Billion-Ton Update: Biomass Supply for a Bioenergy and Bioproducts Industry. ORNL. http://www1.eere.energy.gov/bioenergy/pdfs/billion_ton_update.pdf
- USDoe State of Technology updates: http://www1.eere.energy.gov/bioenergy/key_publications.html

Questioning the Social Dimensions of Sustainability: the Weakening of Traditional Communities and the Biofuel Industry in Brazil

ANA LUIZA DE CAMPOS PAULA – KANSAS STATE UNIVERSITY

VANESSA ZUCHETTO – UNIVERSITY OF ALBERTA

Traditional Populations

Convention 169 of ILO from 1989 protects the rights of indigenous and tribal people in independent countries. Article 14:

"The rights of ownership and possession of the peoples concerned over the lands which they traditionally occupy shall be recognized. In addition, measures shall be taken in appropriate cases to safeguard the right of the peoples concerned to use lands not exclusively occupied by them, but to which they have traditionally had access for their subsistence and traditional activities."

Traditional Communities in Brazil

- Represented by indigenous peoples, river and delta dwellers, rubber tappers, grazers, fisher-folk, peasants, and the quilombolas (descendants of runaway slave communities of African origins).
- Have been part of a historical process of expropriation and exploitation.
- Complex class/race relationship dictates their lack of power.

What is **Quilombola**

- Descendents of Afro- Brazilians who lived in **quilombos** (communities of run away slaves).
- Called *parenques, gafitunas, marron, etc.* in other languages.
- *Conception/identity problem: "Quilombos are African-descendant communities. Regardless of their specific history or whether they began as maroons, these communities share an ethnic identity and a collective view of land ownership"* Quilombola Jo in Bowen 2010.

Debates Over the Concept of Quilombo

- Palmares: The biggest: between 15 to 30 thousand people - from 1630 until 1695. Zumbi: the lider

Long-Term Titling process

INCRA- National Land Reform Institute	Ministry of Culture- Instituto Palmares
• delimit, demarcate and title quilombo territory	• identify, recognize

By 2013, only 139 quilombola communities had their land titles, while 1264 processes where open

Old vs. New

- Currently, around 2 million people distributed in 2,500 communities
- Land vs. territory
- Rural vs. urban
- Race vs. poverty

N° de comunidades quilombolas por estado

- Maranhão 642
- Bahia 396
- Pará 294
- Minas Gerais 135
- Pernambuco91
- Rio Grande do Sul 90
- Piaui 78
- São Paulo 70
- Rio Grande do Norte 64
- Mato Grosso 59
- Ceará 54

Socio-Economic Position

Brazilian population by color or race (%)

Color or race	1940	1950	1960	1980	1991	2000	2010
White	63.5	61.7	61	54.2	52	53.74	47.7
Brown	21.2	26.5	29.5	38.8	42	38.45	43.1
Black	14.6	11	8.7	5.9	5.0	6.21	7.6
Native	X	X	X	X	0.3	0.3	0.6
Yellow	0.6	0.6	0.7	0.7	0.4	0.5	1

Source: IBGE (2011)

Population with high school degree (%)

Color/ race by gender	1960		1980		1991		2000	
White	2.6	2.3	8.43	9.53	12.65	14.28	17.86	20.26
Black	.11	.12	1.84	2.03	5.06	5.87	9.01	11.11
Yellow	1.43	2.22	20.44	19.4	24.62	24.6	27.24	26.62
Brown	.35	.29	2.79	3.33	5.7	7.31	9.39	12.09
Native	—	—	—	—	2.5	2.9	7.77	8.54
Total	1.7	1.51	5.95	6.86	9.39	11.06	13.94	16.05

Source: Beltrao and Teixeira (2005)

p.s.: women in red men in blue

Population with college degree (%)

Color/ race by gender	1960		1980		1991		2000	
White	1.38	.22	3.84	2.52	6.03	5.25	6.57	6.6
Black	.03	0.1	0.27	.23	.93	.96	1.24	1.58
Yellow	.55	.19	10.55	6.61	18.89	14.79	21.11	18.33
Brown	.11	.02	.54	.43	1.15	1.16	1.29	1.55
Native	—	—	—	—	.78	.73	1.37	1.4
Total	0.87	.14	2.41	1.64	3.72	3.42	4.19	4.46

Source: Beltrao and Teixeira (2005)

p.s.: women in red men in blue

Conflict of Interests

Indigenous Lands: the Guaranis

- The Guarani Indians in Brazil are divided into three groups: Mbyá, Kaiowá and Ñandeva. The Kaiowá and the Ñandeva live in the state of Mato Grosso do Sul, on the border with Paraguay.
- The Guarani live in extended family groups and each has its own land called tekohá which refers to the whole space occupied by natural resources: land, rivers, forests and gardens which are integral to sustaining their way of life.

- Before the arrival of the Europeans in the sixteenth century, the Guarani occupied a vast region in the south-center and south-east of South America. There were an estimated 1,500,000 Indians in a territory of about 350,000 square kilometers. The Guarani population in Brazil now numbers approximately 43,000. Waves of violent invasions have converted their homeland into a vast network of cattle ranches, soya farms, and sugar cane plantations for Brazil's biofuels market (Survival 2010).

- Today many Guarani live in chronically overcrowded reserves, for example Dourados Reserve where 12,000 Guarani are living on 3,000 hectares of land. Here they are no longer self-sufficient because they have very little land on which to hunt, fish and grow crops. Lack of opportunity, migrant labour outside the community and cramped conditions have led to social tensions, high rates of internal violence, alcoholism and disease.
- Some Guarani communities have no land at all, and live camped by roadsides in appalling conditions with no access to clean water and food. For example, the Guarani of Laranjeira Nanderu who were evicted from their land in September 2009 and whose village was set on fire by unidentified people, and the Guarani of Apyka'y whose roadside camp was attacked and torched in the same month (http://intercontinentalcry.org/second-guarani-village-burned-in-mato-grosso-do-sul/).
- The process of expulsion of the Guarani has forced them to take up temporary employment in the region's large farms and alcohol refineries, thus separating them from the extended family and the traditional form of social organisation. Profoundly affected by their huge loss of land, the Guarani of Mato Grosso do Sul suffer a wave of suicide unequalled in South America. They also suffer from high rates of unfair imprisonment, exploitation in the work place, malnutrition, violence, homicide and assassination.

- ***"I always remember one old man said, 'The whites – they're going to finish us off. They're going to finish off our houses, finish our fish, even our crops. And once all our forest is gone, we as a people will be finished. It's all going to change and our land will become very small. And you know, that man, all those years ago, calculated absolutely right'"*** (Paulito, a Guarani shaman, summed up his people's situation in an interview with Survival in 1998).

Other peasants and fisherman

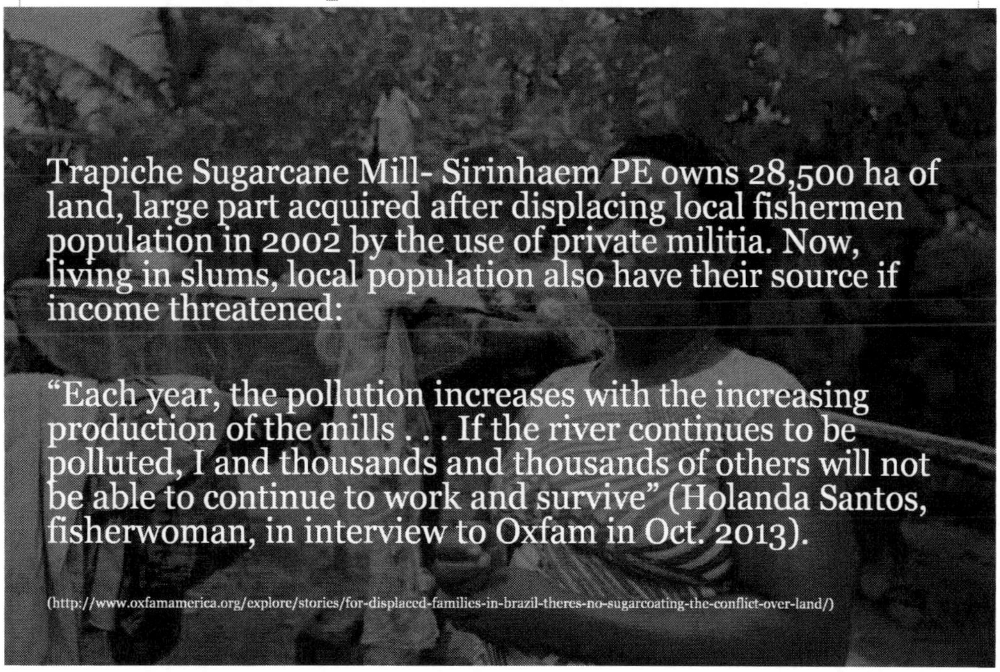

Total land area Brazil	851.487.659 ha
Used for agriculture in 2006	330.000.000 ha of which 24% owned by family farmers and 76% owned by corporations
Agricultural stablishments in 2006	5.175.489 units of which 85% are family and 15% are corporate
Annual gross product value 2006	71.5 billion dollars (38% produced by family farmers and 62 % by agribusiness)
Employment	17.074.025 people (74% by family farms and 26% by agribusiness)

Fernandes et al. 2013.

Land Grabbing

- Land grabbing in Brazil was facilitated in the 1990s by drastic changes made to land regulation pertaining to foreigners. In 1995, Brazil's National Congress approved an amendment that effectively removed barriers to the amount of land foreign corporations were able to purchase in in the country, by allowing national companies controlled by international capital to continue operating as Brazilian entities, regardless of their respective degree of foreign ownership and capital (Oliveira 2010).
- Three years later, in 1998, another major alteration was made, by way of an Informed Opinion (GQ181) issued by the Office of Solicitor General (AGU) at the direct request of President Cardoso, which saw the federal government relinquishing 'any form of effective control on land purchase by foreign companies in Brazil' (Pretto 2009: 7).

- Expansion of soybean has created space for the involvement of transnational agro-industrial giants, such as ADM, Bunge, Cargill, Dreyfus, Monsanto, Syngenta and Dupont. These corporations, among many others, have reaped considerable benefits from Brazil's soybean boon, gaining control over national agricultural land and processing facilities through an intricate and ongoing process of corporate mergers and acquisitions. Between 1995 and 2005, the share of international capital in Brazil's agro-industrial grain sector increased dramatically, jumping from 16 per cent to 57 per cent (Sauer and Leite 2012).
- There has been a flurry of mergers and acquisitions among national and international firms in the energy, agriculture, biotechnology and chemical sectors, such that foreign entities today control about 22 per cent of Brazilian sugarcane and ethanol companies (Geiver and Jessen 2010).
- As a result of the purchase of land by foreigners in Brazil, land prices have increased significantly across the country.

Sugarcane and Soybeans in Brazil 2010

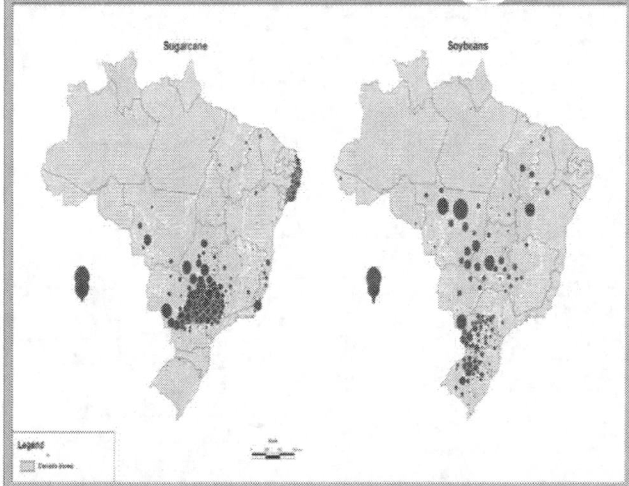

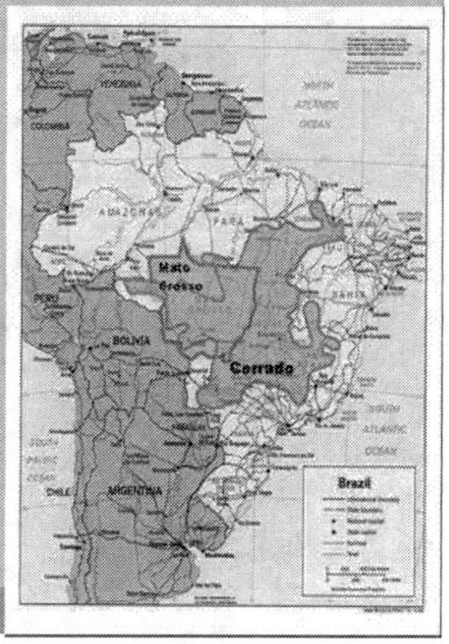

Source: Clements and Fernandes 2013.

- Thank you!

THE SUGARCANE INDUSTRY AND THE GLOBAL ECONOMIC CRISIS

A report by the Network for Social Justice and Human Rights

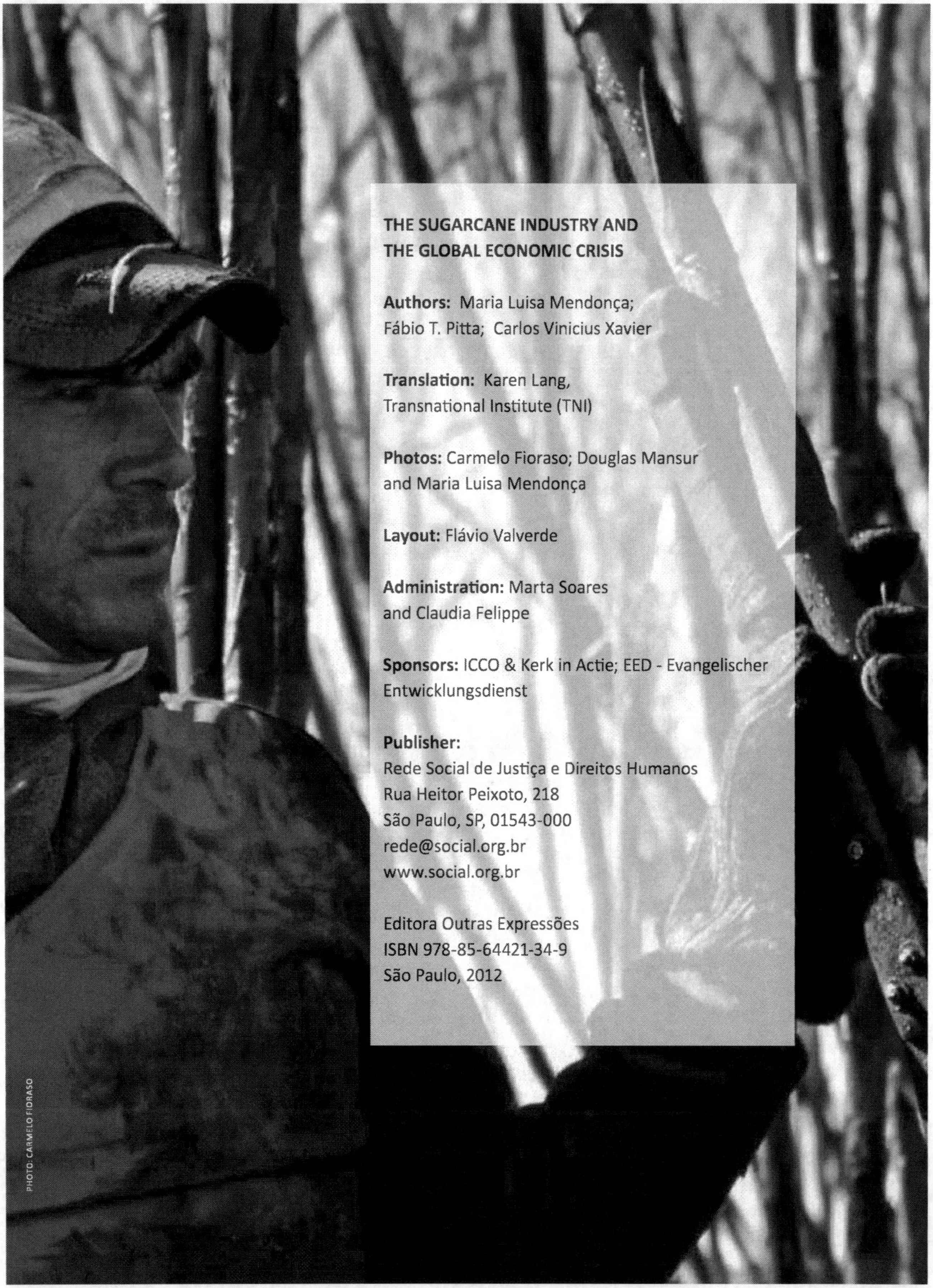

THE SUGARCANE INDUSTRY AND THE GLOBAL ECONOMIC CRISIS

Authors: Maria Luisa Mendonça; Fábio T. Pitta; Carlos Vinicius Xavier

Translation: Karen Lang, Transnational Institute (TNI)

Photos: Carmelo Fioraso; Douglas Mansur and Maria Luisa Mendonça

Layout: Flávio Valverde

Administration: Marta Soares and Claudia Felippe

Sponsors: ICCO & Kerk in Actie; EED - Evangelischer Entwicklungsdienst

Publisher:
Rede Social de Justiça e Direitos Humanos
Rua Heitor Peixoto, 218
São Paulo, SP, 01543-000
rede@social.org.br
www.social.org.br

Editora Outras Expressões
ISBN 978-85-64421-34-9
São Paulo, 2012

THE SUGARCANE INDUSTRY AND THE GLOBAL ECONOMIC CRISIS

Recent trends in the Sugarcane Industry

This article presents information on the latest trends in ethanol production in Brazil and their relation to the global economic crisis. We highlight the role of financial capital, its linkage to the territorial expansion of agribusiness and the impacts of this expansion on labour relations and disputes over the land of indigenous peoples and peasant farmers.

In rural Brazil, we have observed that the expansion of monocropping for the production of agrofuels, namely sugarcane ethanol, continues. Ethanol made from sugarcane is said to be Brazil's main source of agro-energy, considering the volume produced, the total area used for sugarcane production and the amount invested in the expansion of the sugar-energy industrial park.

Data from the National Supply Company (Companhia Nacional de Abastecimento or CONAB in Portuguese) allow us to observe recent trends in the amount of sugarcane processed in order to produce sugar and ethanol:

Table 1

Production of the Sugar-Energy Sector - Brazil

Harvest-Year	Production area (ha)	Sugarcane produced (tons)	Sugar produced (tons)	Ethanol produced (one thousand liters)
2004/05	5 625 300	415 694 500	26 621 221	15 416 668
2005/06	5 840 300	431 413 400	26 713 539	16 997 433
2006/07	6 163 200	474 800 400	30 223 600	17 471 138
2007/08	6 963 600	495 723 279	31 279 800	22 526 824
2008/09	7 057 800	571 434 300	31 049 206	27 512 962
2009/10	7 409 600	604 513 600	34 636 900	25 866 061
2010/11	8 056 000	623 905 100	38 675 500	27 699 554
2011/12	8 368 400	571 471 000	36 882 600	22 857 589

Source: CONAB – Companhia Nacional de Abastecimento.
The data has been taken from CONAB's harvest monitoring documents (always from the third recording of each harvest year). Org. XAVIER, C. V.

PHOTO: CARMELO FIORASO

Described as a process to open *new* frontiers, the expansion of sugarcane production has been concentrated, in terms of production volume, in the Centre-South region, principally in the States of Minas Gerais, Goias and Mato Grosso do Sul, Parana and Sao Paulo[1].

Current estimates indicate that sugarcane monocropping occupies an area of approximately 8.4 million hectares in Brazil. Of this total, 86.1% of cultivated land is concentrated in the Centre-South region. The State of Sao Paulo is the largest producer, with 52.2% of its total area (the equivalent of 4.37 million hectares) dedicated to sugarcane production, followed by Minas Gerais with 8.87% (742 650 hectares), Goias with 8.1% (678 420 hectares), Parana with 7.3% (611 440 hectares), Mato Grosso do Sul with 5.70% (480 860 hectares), Alagoas with 5.45% (463 650 hectares) and Pernambuco with 3.89% (326 110 hectares)[2].

We observed rapid growth in the sector's production levels, mainly between the 2004-05 and 2010-11 harvest years. During this period, there was a 50.08% increase in the amount of sugarcane processed. Also in this period, the production of sugar and ethanol increased 45.2% and 79.6% respectively. However, there was a change in this trend during the 2011-12 harvest: while the volume of production began to decline, the territorial expansion of sugarcane monocropping continued.

The global financial crisis brought significant changes to the sugarcane industry in terms of the patterns of expansion that had been traced over previous years. Taking advantage of the increase in the value of the Brazilian Real (R$), many sugarcane plants took out cheap loans in US dollars and used the funds to speculate in foreign exchange derivatives. When the dollar's value rose again in relation to the Real, many factories went bankrupt and the sector accumulated more than R$4 billion in losses. Companies stopped investing in, for example, the renewal of sugarcane plantations, crop handling and fertilization, which are needed to maintain high production levels. This is why in January 2012, the Brazilian government released R$4 billion in funds to be used specifically for plantation renewal. In addition to the decline in productivity levels, we have observed the monopolistic internationalisation of the sector, an increase in the need for subsidized credit, the expropriation of small producers' and indigenous peoples' land and the substitution of food crops with sugarcane.

1 While these are the main regions of sugarcane production, it is also important to note the recent advance of monocropping in the Northern part of the country. Even though the area used to produce sugarcane represents a small portion of the total area in terms of percentages (close to 0.41%), the North is currently the region with the highest growth rate: the amount of land occupied by this monoculture increased nearly 77.2% between the 2010/11 and 2011/12 harvests. This jump is namely due to the establishment of a new processing unit in the State of Tocantins.

2 CONAB - Companhia Nacional de Abastecimento. Acompanhamento de safra brasileira: cana-de-açúcar, terceiro levantamento, dezembro/2011 - Companhia Nacional de Abastecimento. Brasília, Conab 2011.

The role of the State

PHOTO: MARIA LUISA MENDONÇA

BNDES gave credit to corporations that cultivated sugarcane on Indigenous land

The recent expansion of the sugarcane industry in Brazil was made viable mainly through funding from the National Bank for Economic and Social Development (BNDES). In 2010 alone, the BNDES provided R$7.4 billion to the sector. During the eight years of the Lula administration, ethanol plant owners received a total of R$28.2 billion in loans[3]. These disbursements represent a much more intense growth in investments than in any other economic sector in the country.

With the current crisis in the sector and productivity in decline, the Dilma government recently announced that it could increase the mix of ethanol in gasoline from 20% to 25% to boost demand in the domestic market[4].

State support for agribusiness comes in the form of frequent rollovers of billions of

[3] *Na Era Lula, usineiros receberam mais de R$ 28 bi do BNDES*. Disponível em: http://www.ihu.unisinos.br/noticias/40982-na-era-lula-usineiros-receberam-mais-de-r-28-bi-do-bndes.

[4] *Folha de São Paulo*, Governo propõe zerar tributos do etanol. 10/7/2012.

PHOTO: CARMELO FIORASO

Reals in debt, fiscal incentives, loans with subsidized interest rates, measures to ensure market security, as well as the negotiation of international trade agreements and the provision of incentives for agrofuels production, as had been done in the past. This policy gained in importance in 2004, when soybean prices fell sharply on the international market and large producers were rescued by the National Biofuels Program. A similar thing occurred with the sugarcane plants: the injection of new public funds in ethanol production "resuscitated" many plants that had been stagnant or bankrupt since the Institute of Sugar and Alcohol's closure in 1990.

These measures have been justified by the idea that agrofuels are a "clean and renewable" source of energy and the possibility of a significant increase in the global demand for ethanol. According to the companies' estimates, in order for predictions on expansion to become reality over the next ten years, investments of approximately R$43.8 billion are required. Of this amount, R$24.5 billion is needed for land acquisition and R$19.2 billion to establish plantations and build operational infrastructure[5].

[5] MOREIRA, C. *Limite de venda de terras a estrangeiro barrou entrada de US$ 15 bi no País*. Agência Estado. São Paulo, 18/4/2011. Disponível em: <http://economia.estadao.com.br/noticias/neg%C3%B3cios,limite-de-venda-de-terras-a-estrangeiro-barrou-entrada-de-us-15-bi-no-pais,63309,0.htm>. Acesso em: 1º/6/2012.

The role of financial capital and multinational corporations in the sugarcane agribusiness

PHOTO: MARIA LUISA MENDONÇA

The current scenario reinforces the tendency of corporations to seek to raise funds on financial markets, in addition to obtaining ongoing State support, in order to guarantee the expansion of their sugarcane production. The Cosan Group, which is currently the largest company in the sector and whose shares have been traded on the Bovespa stock market since the mid-2000s, serves as an example of this. The influx of foreign capital, mainly from investment funds, led to an exponential increase in Cosan's assets, which it used to set up subsidiary firms, such as Cosan Açúcar e Álcool, Rumo Logística, Cosan Combustíveis e Lubrificantes (CCL) and Radar Propriedades Agrícolas.

In addition to attracting investors from the financial system, the sugar-energy sector also engaged in a series of mergers with oil and commodity trading corporations. The entrance of the Royal Dutch Shell oil company in the sector, through its establishment of a joint venture with the Cosan Group, is part of this process. This association resulted in the creation of the Raizen Corporation, which has quickly become one of the five largest economic groups in the country, with an estimated market value of US$20 billion. The corporation is responsible for producing 2.2 billion litres of ethanol and 4 million tons of sugar annually. Since its creation, one of its stated goals is to increase its ethanol production to 5 billion litres by 2014. To reach this goal, it is seeking to increase its control over "new" sugarcane producing regions, such as the north-western part of the State of Sao Paulo and the States of Goias and Mato Grosso do Sul[6].

6 RAIZEN, 2011. Available at: <http://www.raizen.com.br/pdfs/apresentacao.pdf>.

British Petroleum (BP) is another oil company that has established itself in the sugarcane industry. In 2008, through its agrofuels affiliate, BP Biofuels, the corporation acquired 50% of Tropical BioEnergy, a joint venture between LDC Bioenergia and the Maeda Group. The plant operated by Tropical BioEnergy is located in the Edeia municipality in the south of Goias. The unit is not currently operating at full capacity, but it is able to produce up to 435 million litres of ethanol per harvest (BRITO, 2011).

Petrobras is also one of the main corporations operating in the sugar-energy sector. Through the Nova Fronteira Bioenergia firm, which is the result of a merger between Grupo Sao Martinho and Petrobras Biofuels (PBio), the company plans to invest around R$520.7 million to enlarge the Boa Vista plant located in Quirinopolis, Goias. The goal is to create the largest sugarcane ethanol plant in the world by increasing the unit's processing capacity from the 3 million tons it currently processes to 8 million by the 2014-15 harvest year. During the most recent harvest period, the plant was operating at idle capacity, processing only around 2.4 million tons, which is 600 thousand tons below its current capacity level.

One of the agribusiness trading companies active in the sugar-energy sector is Cargill. In 2006, Cargill acquired 64% of the Companhia Energética do Vale do Sapucaí (Cevasa). In mid 2011, the corporation announced its joint venture with the Usina São João, for two plants in Goias: one already in operation and the other, in the final stages of being built. ADM (Archer Daniels Midland) has also established its presence in southern Goias and Minas Gerais, namely by establishing joint ventures with existing plants. This strategy was also adopted by the Sojitz Corporation, which acquired 33% of ETH in 2007, together with the Odebrecht Group.

Bunge also established its participation in the sugarcane industry by acquiring existing companies. First, in 2007, it acquired Usina Santa Juliana, located in Minas Gerais. Then, in 2008, it set up a company to commercialise sugar with the Tate & Lile group, thereby becoming one of the biggest exporters of the commodity in the country. Over the next few years, the corporation continued to pursue its acquisition strategy and by 2011, it controlled eight plants already in operation and one in the process of being finalized. This structure has the

PHOTO: MARIA LUISA MENDONÇA

Ethanol plant in Mato Grosso do Sul pollutes water sources

PHOTO: MARIA LUISA MENDONÇA

Entrance of the town of Caarapó where Raízen (Cosan-Shell) cultivated sugarcane on Indigenous land

capacity to grind 20 million tons of sugarcane per harvest.

Investment funds have also become involved in the sector both by fully acquiring plants and by establishing associations or mergers. The creation of Infinity Bio-energy is one example of this kind of full acquisition; it was set up by the Kidd & Company, Stark and Och Zitt Management Funds from the U.S. and the Merrill Lynch bank. In 2006 and 2007, the company finalised a deal to purchase eight plants and announced the construction of five more. The total cost of these acquisitions amounted to approximately R$1 billion. This venture is linked to US$1.5 billion obtained in May 2006 on the London-based AIM stock market, which is meant to provide resources for companies in the early stages of development[7].

With regards to the strategy of forming associations and partnerships, one case that draws our attention is CEB (Clean Energy Brazil), which raised R$400 million on the London stock exchange in 2006. In 2007, the company acquired 49% of Usaciga Açúcar, Álcool e Energia Elétrica located in the State of Parana. In 2009, it formed a joint venture with Unialco S.A., acquiring 33% of shares in two plants in Mato Grosso do Sul.

The main strategy adopted by foreign corporations that have succeeded in entering the sugarcane industry is to strengthen their presence in regions identified as *new* areas of production. It has been noted that corporations such as Raizen (Shell – Cosan), Nova Fronteira (Petrobras – Sao Martinho), BP Biofuels (Tropical BioEnergy and CNAA), Cargill, Bunge, ADM, ETH (Sojitz Corporation and Odebrecht), Infinity Bio-energy, among others, have concentrated their investments in these regions.[8]

The growing participation of transnational corporations in the sector is causing an increase in the concentration of capital. Their presence is apparently related to their search for investments that they believe are "healthy", due to a supposed 'stability' in Brazil's land market. The move towards greater capital concentration is, however, the result of the global economic crisis itself, as we will see shortly.

7 STEFANO, F. O comprador de usinas. Revista Exame, 19 de outubro de 2007.

8 Information on commodity trading firms and investment funds were obtained from the following reference: PINTO, M. J. A. Investimentos diretos estrangeiros no setor sucroenergético. 2011. Dissertação de Mestrado – Faculdade de Administração, Economia e Contabilidade de Ribeirão Preto, Universidade de São Paulo, Ribeirão Preto, 2011.

The expansion of agribusiness as determined by the world economic crisis

PHOTO: MARIA LUISA MENDONÇA

To develop a critical analysis of the sugarcane sector's expansionary process, it is essential to situate the current economic crisis in its historical context. The current moment has been marked by the way financial capital creates the illusion that money has the ability to move about autonomously. In its "autonomous" form, money is negotiated on the financial markets as a merchandise whose price is determined by interest rates. The term "financial industry" arose as a by-product of this "miraculous" system, in which money is believed to have the ability to generate more money, without going through the productive process. The financial markets offer a wide variety of "services" to generate more money for their clients.

The sugarcane industry reaffirms that in order to ensure its expansion, it has a constant need to attract more and more resources from the financial system, always with the goal of covering its previous loans. In other words, it aims to guarantee its expansion by accumulating increasingly high volumes of debt. Therefore, the source of this expansion can be defined as fictitious capital – that is, when companies take out new credit to pay off their bad loans.

In this context, then, what we see is a growing dependency of real capital on credit. The extremely high initial costs of establishing competitive production processes demand that sugarcane corporations take out increasingly high amounts of subsidies. As a result, the sugarcane industry expands territorially by using enormous amounts of credit for mergers, construction and modernization of processing units, plantations or to obtain control over land by purchasing it or signing partnership or leasing contracts.

It is clear that financial market actors' involvement in the sector is motivated by their quest for returns on their investments. Even though this capital originated in this process of fictionalised reproduction – in which the volume of capital that has been accumulated is generated by the illusion that money can reproduce itself on its own, free from its abstract substance, labour – the huge amounts of loan money to be incorporated into the expansion of sugarcane production lead investors to believe that the numbers will effectively rise. Furthermore, the incorporation of these investments into broadening the assets of the major corporations in the sector ends up forging a seemingly healthy reproduction process. Thus, investors are counting on promises of capital gains, when there is, in fact, a crisis scenario. This is made evident by the constant need for credit in order to sustain the productive process.

In this context of fictitious reproduction of capital, the companies' assets and profits are inflated and detached from real values. The apparent "health" of many corporations, then, masks the hidden portion of their debt. This strategy characterises the transition from industrial capitalism to the current period in which speculative capital prevails.

The history of ethanol production in Brazil as the genesis of the financialization of capital

The Pro-Alcohol program (1975-1990) represented one of the Brazilian State's main policies for the industrialisation of rural areas, as it provided subsidized credit to the sugar-energy sector. This industrialisation process is related to the country's foreign debt and the defaults[9] of private capital in the sector. The accumulation of foreign debt in the late 1980s led to the "collapse" of the Institute of Sugar and Alcohol (Instituto do Açúcar e do Álcool or IAA), the bankruptcies of various sugarcane plants and the decrease in the production of ethanol-based fuel and vehicles run on ethanol. Other consequences of the debt crisis in the 1980s were hyperinflation and the Brazilian State's moratorium in 1986.

The modernisation process in Brazil, promoted by the military dictatorship, was rooted in foreign debt. Starting in the 1950s, during the Juscelino Kubitscheck administration (1956-1960), there was an increase in the flow of international financing to the country. The Brazilian State's recourse to foreign loans responded to idle international capital's need to appreciate. In 1973, with the so-called "First Oil Crisis", various periphery countries saw their foreign deficits skyrocket, which moved their governments to seek alternatives in order to pay their debt and maintain their global liquidity.

In Brazil, the government coordinated its attempts to replace the energy matrix with the automobile industry, which led to the creation of vehicles run on ethanol. The government set up subsidies for the production, purchasing and consumption of ethanol. The Pro-Alcohol

[9] TRIBUNAL DE CONTAS DA UNIÃO (TCU). *Proálcool: Relatório de Auditoria Operacional*. Brasília: TCU, 1990.

PHOTO: DOUGLAS MANSUR

program promoted the industrialisation of agriculture by introducing the mechanisation of sugarcane harvesting and seeding, which reduced the need for labour. Thus, the entire production chain received subsidized credit. The industrialisation of agriculture in general, and not only the sugar-energy sector, got its greatest boost from the military dictatorship's economic policies in the 1970s.

It is interesting to examine Ricci's[10] description of the changes occurring throughout the Pro-Alcohol program's lifespan:

> "The first activities to be mechanised on sugarcane plantations were soil preparation and seeding. For both, the impacts of mechanisation were the reduction in the time it takes to carry out the activity and the decline in the use of workers. Sugarcane harvesting is made up of three independent phases: cutting, carrying and transporting the cane to the plant. The mechanisation of sugarcane harvesting has been very slow, reaching the transportation of sugarcane first... [...]. Then, the work to carry sugarcane was mechanised. Up until the 1950s, the workers cut and bundled sugarcane, and tied the load with the leaves [...]. The men carried the cut and bundled sugarcane on their backs to the little trucks or carts pulled by animals [...]. In the next stage, at the end of the 1960s, manual carriers were replaced by trucks, on which sugarcane was piled and carried to truck trailers [...]".

The industrialisation of agriculture in Brazil was driven by industrial and financial capital that was seeking to generate profits through the appropriation of land income. In order to maintain productivity at levels compatible with the international market, Brazilian agribusiness took on enormous debt, which eventually generated the sector's economic crisis in the 1980s. The production of sugar and ethanol,

[10] RICCI, Ruda *et al. Mercado de trabalho do setor sucroalcooleiro no Brasil*. Brasília, Ipea, 1994; pg. 104-107.

for both exportation and domestic consumption, assumed the role of "rolling over" Brazil's external debt. In the late 1980s, after a period of modernisation, companies were unable to repay their debts. As for the State, it had to eliminate its policy of subsidising the industry, which caused a series of plants to go bankrupt[11].

The 1990s were marked by stagnation. People stopped buying cars run on ethanol and the price of sugar remained low due to fluctuations in the international commodities market and slow growth rates in urban areas. The sector only started receiving a significant amount of incentives again in 2004.

The "new" expansion of the sugar-energy sector and the conversion of the Brazilian State's external debt into internal debt

According to data from CONAB (see Table 1 above), between 2004 and 2009, the sugar-energy sector expanded strongly in absolute terms, something that had only happened during the Pro-Alcohol program. From the 2004-2005 harvest to that of 2008-2009, the total area planted in sugarcane grew from 5 625 300 hectares to 7 057 800 hectares. Also during that time, sugarcane production increased from 415 694 500 tons to 571 434 300 tons; sugar production from 26 621 221 tons to 31 049 206 tons; and that of ethanol from 15 416 668 litres to 27 521 962 litres. However, this rate of expansion lasted only five years, between the 2004-2005 and 2008-2009 harvests, when the crisis' impacts begin to affect the sector. From 2010 on, only the size of the cultivated area continued to grow, whereas productivity dropped significantly. This process is related to the international economic crisis, which led to a fall in commodity prices, the valorisation of the dollar and the lack of liquidity in the financial market, which plays a key role in the sector's expansion.

The most recent period of growth in ethanol production in Brazil began when new opportunities to access financial capital were created. During the 1990s, core countries that were previously creditors of Third World debt developed securitisation mechanisms, which created new possibilities for financing. New speculative financial products, known as "derivatives", were created. These possibilities expanded the financial sector's capacity to create money and its leverage. This generated huge amounts of international liquidity, as well as an exponential increase in debts, yet this time in the form of countries' internal debts.

a) The securitisation of debt

The securitisation process allowed investors to purchase bonds tied to countries' debts, which can be paid in different ways, like, for instance, through real estate, car or university loans. Brazil renegotiated its external debt with U.S. banks in 1994, during the Itamar Franco government, through an agreement called the

[11] TRIBUNAL DE CONTAS DA UNIÃO (TCU). *Proálcool: Relatório de Auditoria Operacional.* Brasília: TCU, 1990.

PHOTO: MARIA LUISA MENDONÇA

"Brady Plan": "The agreement also resulted in the "securitisation" of credits against Brazil – that is, in their transformation into bonds or bonuses that are more easily negotiated on secondary markets[12]."

Securitisation makes it possible for creditor banks to remove outstanding debts from the assets recorded on their balance sheets. This mechanism is essential for guaranteeing the circulation of financial capital, as it increases lending capacities. It also stimulates consumption of the State and among the general population, which generates apparent economic growth. Such a procedure expanded the banks' lending capacities far beyond previous legal limits, as creditors holding securitised bonds are no longer included in the financial reports of the banks selling the securitisation packages.

In Brazil, the Brady Plan was assessed in the following way:

> "Some advocates of the Brady Plan and its implementation in the Brazilian case (...) claim that the fundamental advantage of this type of agreement resides is something less tangible, which is a gain in "credibility" that would result in, among other things, the broadening of the debtor country's access to international financial markets[13].

The financial market is characterised by the predominance of "idle" capital seeking possible returns. The Pro-Alcohol program made the rolling over of Brazil's external debt possible during the 1980s. The Brady Plan allowed Brazil to broaden the rolling over of its external debt and to transform it into "internal" debt, which was no longer calculated in dollars, but rather as National Treasury bonds whose value was established in Reals (Brazilian currency). The Brazilian State's capacity to rollover its debts is related to the availability of funds it can use to stimulate the economy. As such, the National Bank for Economic and Social Development (BNDES) plays a central role in encouraging new loans.

Economist Jose Roberto Afonso explains how the mechanisms used to rollover Brazil's internal debt worked between 2003 and 2010:

> "By paying the highest interest rate in the world, the country attracts an increasing amount of foreign capital, accumulates more and more foreign reserves, and raises the Real's value higher and higher. The earnings and confidence of foreign and national investors increase. The current arrangement combines a rapid increase of credit from public sources – the state -controlled banks' slice is worth 19.8% of GDP, while the National Treasury is a creditor of 14.5% of the GDP of these public banks. In practice, the Treasury became the largest bank in the country. However, contrary to a regular bank, it is not subject to prudential regulation. To put itself in this position, it increased between August 2008 and June 2011 the amount of extraordinary credit it conceded to official banks (basically the BNDES) by 6.5 points of GDP, while gross debt increased 5.8 points in the same period, jumping to 65.3% of GDP. Just like any other bank, there is a limit to the Treasury's ability to continue going deeper into debt in order to continue handing out loans. [...] For this system to work, the government accumulates, directly or indirectly, an increasing amount of debt. And for what? Mainly to give more credit to public banks, which use 95% or more of these funds to give loans to the private sector"[14].

[12] JÚNIOR, Paulo Nogueira Batista; & Rangel, Armênio de Souza. "O Brasil no Plano Brady". Available at: http://revistas.fee.tche.br/index.php/indicadores/article/viewFile/872/1151; consulted on July 5, 2012; p. 46.

[13] Idem, p. 45.

[14] AFONSO, José Roberto. *Tesouro se Transformou no maior banco do país*. "O Estado de São Paulo", Caderno Econômico, 29 de agosto de 2011. Available at: http://m.estadao.com.br/noticias/impresso,tesouro-se-transformou-no-maior-banco-do-pais,765399.htm; consulted on July 5, 2012.

The burst of the bubble on the Nasdaq technology index in 2001 made investors search for new options. At that time, internal debt bonds of periphery countries looked like a good deal. With the United States keeping its interest rates so low, investors have been attracted to Brazil's internal debt bonds, where interest rates are much higher. This mechanism, known as *carry trade*, encourages private agents to take out loans in dollars and use the funds to buy State bonds of countries that pay higher interest rates. This mechanism was used not only in Brazil, but also in other Latin American countries and the BRICS, a group of countries made up of Brazil, Russia, India, China and South Africa.

In Brazil, many sugar and ethanol plants acquired credit on foreign markets, in dollars, in order to take advantage of the low interest rates in effect on the U.S. financial market[15]. With ample liquidity, or a good borrowing capacity, the Brazilian government began to offer, via subsidized credit, generous loans to the private sector, namely the agribusiness industry. The offering of incentives for sugarcane, sugar and ethanol production, which began in 2003, was part of this process.

Brazil's public debt (both internal and external) neared the three trillion-real mark in 2011. In 1994, the internal debt was close to zero. Today, it represents the greatest portion of the country's debt, as the external debt is only approximately R$350 billion. The public debt amounts to 75% of the national GDP, which was near R$4 trillion in 2011. Brazil spends more than R$700 billion – that is, more than 45% of the Federal government's budget (R$1.571 trillion in 2011) –a year on servicing its debt.

"Brazil's internal debt, which amounted to R$892.4 billion when Lula assumed office in 2003, reached R$1.4 trillion in 2009 and, according to the limits established by the government itself, it could reach R$1.73 trillion at the end of 2010, almost double the amount. A growth of 94% in eight years of government", explains the economist Waldir Serafim[16]. While the debt doubled in size during the Lula government, the GDP grew approximately 32%, which allows us to highlight the interrelation between them, as well as the need to rollover the national debt in order to keep the country solvent. The difference between the interest rate the State pays on government bonds (called the Selic Rate, or Taxa Selic in Portuguese) and the amount it charges as interest on loans to the private sector (including agribusiness) represents yet another way the government provides subsidised credit through public banks such as the BNDES, which is the main basis for the expansion of the sugar-energy sector.

Various studies confirm the existence of subsidized credit for the sector. The economist Pedro Ramos[17] affirms that the obligation to add ethanol to gasoline can be

[15] "Perdas com derivativos nas usinas atingem até R$ 4 bilhões". *Jornal Valor Econômico*, 9 de outubro de 2009. Available at: http://www.jornalcana.com.br/noticia/Jornal-Cana/31304+Perdas-com-derivativos-nas-usinas-atingem-ate-R$-4-bi; consulted on July 1, 2012.

[16] SERAFIN, Waldir. *Dívida Interna: perigo à vista.* "SóNotícias", February 11, 2010. Available at: http://www.sonoticias.com.br/opiniao/2/100677/divida-interna-perigo-a-vista; site consulted on July 5, 2012.

[17] RAMOS, Pedro. *Financiamentos subsidiados e dívidas de usineiros no Brasil: uma história secular e ... atual.* Available at: http://www.google.com.br/search?q=Financiamentos+subsidiados+e+d%C3%ADvidas+de+usineiros+no+Brasil&rls=com.microsoft:pt-br&ie=UTF-8&oe=UTF-8&startIndex=&startPage=1&redir_esc=&ei=bkr3T-e2HOKQ6wHA3YzsBg, 2011; consulted on June 6, 2012.

considered a subsidy. The geographer Antonio Thomaz Jr.[18] shines light on the successive pardoning of the sugarcane industry's debts as proof of the "benefits" it enjoys. In August 2008, the government decreed that the Long Term Interest Rate (6.25% per year) would substitute the Selic rate (13% per year at the time) in the renegotiation of the agricultural sector's R$75 billion debt, which dates back to the 1980s and 1990s. This amount is close to the R$65 billion allocated to the 2008-2009 Agriculture Plan (Plano Agricola) for investments in the agribusiness industry, with fixed rates of 6.75% per year, which represented an increase of 12% from the previous year.

The BNDES' investments in the sugar-energy sector during the Lula government are massive when compared to other sectors. The total amount was R$28.2 billion. In 2010 alone, R$7.4 billion was disbursed to finance a range of activities, from the planting of sugarcane (R$953 million) to the production of sugar and alcohol (R$5.6 billion) and the co-generation of energy (R$665 million). The amount of these loans is greater than those provided to other sectors of the economy that year, such as the paper, cellulose and extractive industries together (R$3.1 billion), the mechanical industry (R$5.3 billion), the metalworking industry (R$4.9 billion) and the clothing and textiles industry (R$2.1 billion). In 2008, due to the financial crisis, even the working capital of companies in financial difficulty was financed. These are exorbitant amounts when compared to what the sector received in subsidized credit (with negative real interest rates) during the entire Pro-Alcohol program: more than 7 billion dollars between 1975 and 1990[19].

b) The derivatives market

> "With the end of the Bretton Woods agreements and the increased volatility of interest and exchange rates, financial derivatives were created and propagated, initially with the goal of covering risks. However, the use of these mechanisms did not end up being restricted to this purpose, as they became the preferred instrument for speculation, due to the possibility of high capital gains. (...) a derivative is a financial contract whose value is derived from the value of an underlying asset, commodity price, index, rate or event. (...) One characteristic that is particular to the ensemble of derivatives is that one negotiates now what the value of an asset will be in the future."[20]

This quote makes it clear that derivatives are related to what is known as the "futures markets". Agricultural commodities are negotiated on this market and their physical sale on the "export markets". Indexes, currencies, interest rates and commodity prices are negotiated on future markets, but their actual delivery is not.

Derivatives markets arose during the 1970s and 1980s in a midst of strong fluctuations in interest rates and, as a result, exchange rates. International traders began to buy and sell currencies and commodities

[18] THOMAZ JR., Antonio. *Por trás dos canaviais:* os nós da cana. São Paulo, Annablume/Fapesp, 2002.

[19] XAVIER, C .V; PITTA, F. T.; MENDONÇA, L. M. *Monopólio da produção de etanol no Brasil: a fusão Cosan-Shell. Rede Social de Justiça e Direitos Humanos*. São Paulo, 2011. Available in Portuguese at: http://www.social.org.br/revistacosanshel.pdf. And in n English: http://www.social.org.br/ethanol_monopoly_brazil.pdf.

[20] Fahri, Maryse. *Operações com derivativos financeiros das corporações de economias emergentes no ciclo recente*. Trabalho apresentado no II encontro internacional da Associação Keynesiana Brasileira, setembro de 2009. Available at: http://www.ppge.ufrgs.br/akb/encontros/2009/02.pdf; consulted on July 5, 2012; p. 2.

in advance, which guaranteed them a favourable price, regardless of whether there were any abrupt changes in the markets or not. At the same time, these variations enabled investments in the derivatives markets to generate large capital gains. This is a fundamental point.

Speculation on both agricultural commodity prices and currencies cause strong impacts on and oscillations in prices, which reflect movement on the futures markets. Nowadays, the prices of certain kinds of merchandise or goods may be the simple expression of speculators' future bets, as they raise or lower prices according to what they foresee as the best option for their financial gain. What is more, these variations affect exchange and interest rates, as they stimulate investments in these derivatives markets. This, in turn, contributes even further to price volatility.

Capital gains earned by betting on financial markets began to integrate the profits of companies that used to invest only in the production of goods. The so-called "hedge" operations are speculative in nature, as the futures exchange is the speculative market *par excellence*. This kind of financial structure has gained even greater importance today, as it allows its participants to enhance their wins and losses. There is an interconnection between "productive" companies and speculative capital, as many play both roles. For example, various commodity-trading firms take on debt in apparently favourable conditions and apply the resources in commodity speculation, so as to "close" their books with the earnings they obtain from negotiating future prices.

Another speculative activity in which a growing number of companies engage in order to enhance their assets is entering in the stock market. Various plants in Brazil, such as Cosan, Açúcar Guarani, Usina Costa Pinto and Usina São Martinho, used the Initial Public Offer (IPO) mechanism to enter the stock exchange. The stock market began to expand in the 1970s and allowed financial capital to seek gains based on the flow of money, which generates more money, without participating in the productive process or the exploitation of labour.

Companies opt for this kind of procedure with the intention of increasing their capital and other assets, such as machinery, land, subsidiaries, etc. As a result, share prices end up being a fundamental part of companies' market value and a parameter for obtaining credit. The sugar-energy sector used loans in U.S. dollars from private banks during its recent expansion. Access to credit and new financial instruments augmented the sector's capacity to borrow and invest, opening the way for the technological transformation that widened the gap between investments in fixed capital and the workforce that was to be exploited. This move made companies even more incapable of generating profit through the exploitation of labour.

The main consequence of this process was the mechanisation of sugarcane harvesting. In the State of Sao Paulo, the country's biggest producer, 75% of sugarcane is harvested mechanically. In other states where sugarcane monocropping is expanding, like Mato Grosso do Sul and Goias, 50% has been mechanised[21]. As a result, the number of cases of over-

[21] To see this data, consult: BACCARIN, José Giacomo. "Boletim – Ocupação formal no setor sucroalcooleiro em São Paulo". Jaboticabal, UNESP, 2009-2011, números 1-21. Disponível em <fcav.unesp.br/baccarin>. Consulted on 10/08/2011.

exploitation of workers, including tractor and combine drivers, is on the rise.

The sector's crisis is cyclical, like a spiral that spins deeper and deeper. A company's capacity to acquire new debt in order to continue investing increasingly high amounts is directly tied to its size – that is, the value of its assets. The need to buy land and machines, combined with the tendency to speculate on financial markets to obtain higher-than-average profits, are defining the sector's expansion. As they modernise the production process, they replace workers with machines, which makes accumulation even more difficult. This, in turn, contributes to further indebtedness, expansion and financial speculation, which are determining factors in the over-exploitation of labour.

The impacts of the 2008 crisis in the sugar-energy sector

PHOTO: CARMELO FIORASO

PHOTO: CARMELO FIORASO

As seen above, the production of sugar and ethanol depends on regular debt rollovers and financing. The 2008 global economic crisis was characterised by defaults on real estate loans, principally in the United States, which limited companies' ability to obtain credit. For many economists, this process meant the end of financial market liquidity.

Brazil's sugarcane industry was hit hard by the crisis. The sector had gone through several crises in its history, all with different characteristics. In the second half of the 1980s, the Pro-Alcohol program went under. According to the report of the Federal Auditor General[22], which investigated the sugarcane plants' debts, it was the end of subsidised credit from the government that led many companies into bankruptcies and mergers. Subsidies had to be cut due to the State's inability to rollover the external debt. Initially, in the 1980s, this incapacity was compensated by a policy of hyperinflation. Later, in 1986, the government declared a moratorium and it became more difficult to access credit.

The sugarcane industry's most recent growth cycle lasted from 2003 up to the 2008 financial crisis. During this period, companies relied mainly on loans from the BNDES, but also heavily on private credit in U.S. dollars. The sector also used financial mechanisms, like exchange rate derivatives, to generate profit. The 2008-2009 crisis combined the impossibility of accessing resources for debt rollovers with losses from investments in exchange rate derivatives. Many plants went under, which intensified the process of acquisitions by and mergers with transnational corporations.

The crisis was reflected in the fall of sugarcane production in the 2011-2012 harvest year. With the decline in ethanol production, in 2011, Brazil had to import corn ethanol from the United States. This was important symbolically, since Brazil presents on the international scene the idea that it would become the "Saudi Arabia

[22] TRIBUNAL DE CONTAS DA UNIÃO (TCU). *Proálcool: Relatório de Auditoria Operacional*. Brasília: TCU, 1990.

of ethanol". The main argument ethanol plants used to justify the decline in productivity was their inability to invest in plantation renewal, fertilization and chemicals. One way of compensating the reduction in investments was through the territorial expansion of sugarcane monocropping. Despite the crisis, the BNDES continued to finance plants and, in early 2012, it made R$ 4 billion available to them through a line of subsidised credit to be used specifically for plantation renewal[23].

The subsidised credit provided by the BNDES was not, however, enough to sustain productivity levels, which became evident with the global financial crisis. Taking out loans in U.S. dollars and speculating with exchange rate derivatives had become common practice for companies in the sector. The economic context established prior to the crisis provided these companies with speculative gains, at a time when idle international capital was looking for ways to appreciate on the financial market. Furthermore, the United States' low interest rates encouraged *carry trade* ´that is, generating gains by earning dollars and applying them in investments that take advantage of Brazil´s high interest rates. This caused the dollar to depreciate vis-à-vis the Real, which stimulated investments in exchange rate derivatives, as investors bet on the value of Brazil´s currency remaining high.

The gains earned from exchange rate derivatives increased the flow of dollars to Brazil, which contributed to the strengthening of the Real vis-à-vis the U.S. dollar and generating more earnings with these derivatives. Specula-

PHOTO: DOUGLAS MANSUR

[23] "BNDES Prorenova apoia com R$ 4 bi renovação e implantação de canaviais". BNDES, 11 de janeiro de 2012. Available at: http://www.bndes.gov.br/SiteBNDES/bndes/bndes_pt/Institucional/Sala_de_Imprensa/Noticias/2012/energia/20120111_prorenova.html; consulted on July 12, 2012..

tion in commodities allowed the agribusiness industry to experience strong sales on the international market, including sugar.

Between the 2004-2005 and 2008-2009 harvest years, the sugar-energy sector experienced growth in absolute terms, which meant companies had gone even deeper into debt. Despite the lack of transparency in the level of indebtedness in the sector, bankruptcies and mergers involving many sugarcane plants revealed that their debts had reached unmanageable proportions[24]. Various companies had taken advantage of the low interest rates to take on debt in dollars and applied this money in exchange rate derivatives. Thus, they made money when the dollar devaluated in relation to the Real. Having obtained these gains, the companies invested in their productive processes, paid the interest on their debts and were able to take on new credit.

With the crisis in 2008, these factories suffered major financial losses. Due to the insecurity generated on the global market by defaults on sub-prime mortgages, many foreign investors withdrew their speculative capital from the economies of periphery countries and bought U.S. treasury bonds, which were considered more secure. This move led the dollar to appreciate abruptly vis-à-vis the Real, generating sudden fluctuations in the exchange rate. The companies´ debts in dollars jumped significantly and those who had invested funds in exchange rate derivatives incurred enormous losses. In the case of the sugarcane agro-industry corporations, this loss exceeded R$4 billion[25]. The following quote describes this process:

> "The deepening of the international financial crisis in the second half of 2008 (...) exposed productive companies' leveraged positions, by intensifying the volatility of asset prices. The crisis (...) caused the dollar to appreciate in relation to other currencies. However, this valorisation was much more accentuated in emerging economies. Directly affected by another sudden halt in foreign capital flows, these economies' asset prices and exchange rates became important targets for companies seeking secure investments. In this context, the enormous financial losses incurred by major emerging economies companies through their holdings on the foreign exchange derivatives markets were revealed. They had conducted operations involving large sums in these markets, betting on the belief that their national currency would not depreciate against the dollar. This decision was apparently made in the first half of 2008, a time when the dollar devalued greatly in relation to all other currencies, which contributed to a strong increase in commodity prices. However, the deepening of the crisis caused commodity prices to fall significantly and generated a new tendency for the dollar to appreciate internationally. It was then that the companies' losses brought on by their speculative bets surfaced"[26].

In Brazil, several plants went bankrupt, which intensified the process of mergers and acquisitions already in place. Both existing plants that had incorporated other units to boost their production, like Santelisa Vale and

[24] For more information, see: "Perdas com derivativos nas usinas atingem até R$ 4 bilhões". Jornal Valor Econômico, 9 de outubro de 2009. Available at: http://www.jornalcana.com.br/noticia/Jornal-Cana/31304+Perdas-com-derivativos-nas-usinas-atingem-ate-R$-4-bi; consulted on July 1, 2012..

[25] *Idem, ibidem.*

[26] Fahri, Maryse. *Operações com derivativos financeiros das corporações de economias emergentes no ciclo recente.* Trabalho apresentado no II encontro internacional da Associação Keynesiana Brasileira, setembro de 2009. Available at http://www.ppge.ufrgs.br/akb/encontros/2009/02.pdf; consulted on July 5, 2012; p. 6.

PHOTO: MARIA LUISA MENDONÇA

Moema, and new plants, like those known as the "greenfields projects", declared bankruptcy.

The case of the Nova America group, owner of the União brand, is quite significant. The company was not able to shoulder its debts and was bought by the Cosan group in 2009. Shortly after, Cosan negotiated its merger with Shell, the Anglo-Dutch corporation, to form the Raizen Corporation. The Nova America de Caarapó unit, in Mato Grosso do Sul, was incorporated into the Cosan Group and is now part of the process to expand sugarcane plantations into the Centre-West region of Brazil.

Nova America was denounced for planting sugarcane on indigenous land officially recognized by the Federal government[27], with funding from the BNDES, as reported in an article published on March 23rd, 2011:

> "According to the Federal Public Prosecutor's Ministry, Cosan/Shell leased the Santa Claudina estate, which runs onto the land of the Guyraroca indigenous people, in Caarapo. The area has already been the object of the National Indian Foundation's (FUNAI) identification and demarcation studies. The company also exploited gravel from the Taquara indigenous people's land, without the community's authorisation, which it used for the roads the plant's trucks circulate on"[28].

In spite of these denouncements, bankruptcies and defaults, the BNDES continues to fund the sugarcane industry. The companies argue there is not enough credit available and pressure the State to provide more subsidies, as they are incapable of reproducing without their link to the financial system.

[27] For more details on the process to officially recognise indigenous lands and on the homologation by large landowners in Mato Grosso do Sul, see : Homero, M. & Guimarães, V. "Multicultural, mas esquizofrênico; a mão que afaga é a mesma que apedreja: o Estado e o estímulo ao desenvolvimento e seus impactos sobre as terras indígenas em Mato Grosso do Sul". Em: *As Violências contra os povos indígenas em Mato Grosso do Sul, e as resistências do Bem-viver por uma terra sem males*. Relatório do Conselho Indigenista Missionário (CIMI), MS, 2011.

[28] "BNDES financia usinas que usam cana de área indígena". Available at: http://uniaocampocidadeefloresta.wordpress.com/2011/03/23/bndes-financia-usinas-que-usam-cana-de-areas-indigenas; consulted on July 8, 2012.

Land ownership and the crisis of the sugarcane agribusiness

PHOTO: DOUGLAS MANSUR

To maintain its expansion, sugarcane agribusiness requires enormous amounts of credit and state subsidies in order to cover costs ranging from the implantation of processing units to plantation development and the implementation of projects to restructure the sector. However, as discussed above, this process has caused the sector's debt to grow exponentially, which, in turn, encourages its territorial expansion through constant incorporation of *new* areas for monocropping.

In other words, this expansionary tendency – historically seen as being intrinsic to the sugarcane industry – has gained speed in the midst of growing debt accumulation and appears as a supposed solution for the sector's earning crisis. However, it is this expansionary context that has brought on the crisis. Agribusiness seeks to expand with the goal of remunerating capital that has been invested, although in recent years, this capital has been increasingly immobilised in the form of mechanisation and supplies, which tends to reduce profits and results in companies' de-capitalisation. It is in this scenario that the sugarcane industry enters into a spiral of crisis-expansion.

The ramifications of this process were made evident by the recent slowdown in the production growth, as can be observed in the table below:

Table 2

Sugar-energy Sector - Brazil: Comparison of Area, Productivity and Total Sugarcane Produced

Harvest-Year	Production area (ha)	Productivity (Kg/ha)	Sugarcane produced (tons)
2004/05	5 625 300	73 897	415 694 500
2005/06	5 840 300	73 868	431 413 400
2006/07	6 163 200	77 038	474 800 400
2007/08	6 963 600	78 969	495 723 279
2008/09	7 057 800	80 965	571 434 300
2009/10	7 409 600	81 585	604 513 600
2010/11	8 056 000	77 446	623 905 100
2011/12	8 368 400	68 289	571 471 000

Source: CONAB – Companhia Nacional de Abastecimento.
Data from CONAB's harvest monitoring documents (this data are always taken from the third recording of each harvest-year). Org. XAVIER, C. V

Companies from the sugarcane sector blame the decline in productivity on factors from the current conjuncture, mainly those linked to climatic conditions, such as the excess of rain in the 2009-10 harvest or the prolonged drought in 2010-11. While it is true that these environmental factors influence yields, this explanation alone is insufficient. The economic context that directly influences this process must be analysed in order to understand the expansion-crisis cycle[29].

In this context, gaining control over land – by purchasing, leasing or setting up partnerships – becomes increasingly fundamental for the corporations. They use this strategy in an attempt to raise the value of their assets by speculating on the perspectives of the land values increasing.

The case of Radar Propriedades Agricolas, a subsidiary of the Cosan Group whose role is to intermediate property deals, provides an example of this process. It was created in 2008 with Cosan S.A. owning approximately 18.9% of its share capital and the remaining 81.1% in the hands of other investors (such as the U.S.-based Teachers Insurance and Annuity Association of America – TIAA pension fund). While Cosan initially contributed only US$35 million to this

[29] The National Supply Company (CONAB) lists the following factors that contribute to the recent decline in productivity: "the failure to renew plantations at the right moment, the decrease in the quantity of products applied due to the decapitalisation of smaller producers and the mechanization of harvesting, which makes it necessary to renew sugarcane plantations more often. It is worth highlighting that the productivity of sugarcane plantations is lower after the second or third cut. Today, it is easy to find fields where production has gone on uninterrupted for over ten years; here, productivity levels are only 30% of those of sugarcane that has been cut twice" (CONAB, 2011, p. 10). CONAB - Companhia Nacional de Abastecimento. Acompanhamento de safra brasileira: cana-de-açúcar, terceiro levantamento, dezembro/2011 - Companhia Nacional de Abastecimento. Brasília, Conab 2011.

PHOTO: DOUGLAS MANSUR

venture and other investors provided US$150 million, it retained administrative control over the firm, as it holds the majority of votes in the Board of Directors[30].

According to the corporation, "the objective is to buy property that has strong potential to increase in value and lease it to large rural producers. After its value has risen significantly, the property is put up for sale. The initial focus is on land geared towards the production of sugarcane, soybean, corn and cotton. In the first years of its operation, the company has already invested US$ 400 million. Today, it manages nearly 106 000 hectares of land"[31].

Other corporations in the sector employ the same strategy in order to draw financial investors into the land acquisition process, as was the case of Infinity Bio-Energy, Açúcar Guarani S/A, Adecoagro S/A, among others[32]. With the emergence of the earnings crisis, these companies are seeking to attract new volumes of loan money, which will also be invested in businesses involved in land ownership, as part of a process to inflate the value of their assets.

However, if the goal of obtaining large areas of land is to raise the value of various companies' assets, this mechanism proves to be, in reality, yet another symptom of the crisis in the productive process, as the sugar-energy sector's growing indebtedness extends to the entire production chain, affecting producers, suppliers and outsourcing firms, as well as land owners, partners and lease-holders.

30 COSAN. Comunicado ao mercado. Available at: http://www.acionista.com.br/home/cosan/280808_constituicao_subsidiaria.pdf.

31 RADAR, 2012. Available at: http://cosan.com.br/cosan2009/web/conteudo_pti.asp?idioma=0&conta=45&tipo=35726.

32 For more details on land purchasing deals, see: HERNANDES, T. Estudo sobre processos, causas e efeitos da concentração e estrangeirização das terras no Brasil – estrutura de mercado. Projeto de Cooperação Técnica "Apoio às políticas e à participação social no desenvolvimento rural sustentável" (PCT IICA/MDA – NEAD), Brasília, NEAD, 2009.

The over-exploitation of labour in the harvesting of sugarcane

PHOTO: CARMELO FIORASO

During the 1950s and 1960s, agriculture in Brazil was mainly based on the exploitation of the work of settlers, inhabitants, dwellers and squatters. Few of them received a salary, as they worked for landowners in exchange for access to land. With the industrialisation of agriculture, however, the majority of these workers were expelled from the farms and began to be hired back as salaried employees[33]. The industrialisation of agriculture led to the mechanisation of various stages of sugarcane production. The sugarcane labourer, known as ´bóia-fria´ in Brazil and who was a predominate figure on sugarcane plantations, was overexploited due to the high level of unemployment in rural areas. Having no other option for survival, these workers were forced to subject themselves to degrading working conditions.

With the most recent expansion of the sugarcane agro-industry in the 20th century, the conditions of sugarcane workers have not improved, despite investments and the "modernisation" of the sector. This is precisely where the contradiction in the modernisation process lies: in the work of sugarcane labourers. The mechanisation of the harvest is now predominant in the sector and has promoted the dismissal of workers, whose numbers have decreased in absolute terms. Unemployment intensified pressure on workers who began to compete for access to jobs.

On average, one combine replaces 120 manual workers[35]. According to the Formal Employment in the Sugar-Alcohol Sector in Sao Paulo newsletter[36], "it can be calculated that the average number of people employed (...) fell 11 186 between 2007 and 2008; 9 788 between 2008 and 2009; 3 658 between 2009 and 2010 and 14 468 between 2010 and 2011. The major decline in the latest period is striking and is probably associated to technological changes that have been taking place[37].

33 PITTA, Fábio T. *Modernização retardatária e agroindústria sucroalcooleira paulista: o Proálcool como reprodução fictícia do capital em crise*. 2011. Dissertação de Mestrado (Mestrado em Geografia Humana) – Departamento de Geografia, Faculdade de Filosofia Letras e Ciências Humana, Universidade de São Paulo, São Paulo, 2011. See especially chapter 2 of the thesis: "O fetiche da história, o espaço como abstração, o fetiche da mercadoria".

34 For a description of work in the sugarcane harvesting process, please see our report: XAVIER, C .V; PITTA, F. T.; MENDONÇA, L. M. *Monopólio da produção de etanol no Brasil: a fusão Cosan-Shell*. Rede Social de Justiça e Direitos Humanos. São Paulo, 2011.

35 RAMOS, Pedro. "O trabalho na lavoura canavieira paulista: evolução recente, situação atual e perspectivas". *Emprego e trabalho na Agricultura Brasileira*. Coordenadores Antonio Márcio Buainain e Claudio S. Dedecca, Série Desenvolvimento Rural Sustentável, v. 9, p. 304-325. Brasília, IICA, 2008.

36 BACCARIN, José Giaccomo. Boletim – Ocupação formal no setor sucroalcooleiro em São Paulo. Jaboticabal, UNESP, 2009-2011, números 1-21. Available at <fcav.unesp.br/baccarin>. Consulted on: 10/08/2011.

37 Idem, p. 2 e 3.

Sao Paulo – evolution of seasonal workers' salary, pay rate, physical and monetary performance in sugarcane cutting and minimum wage – 1969 – 2005 (monetary values expressed in the July 1994-value of the R$)

Year	Average daily wage of a seasonal worker (1)	Payment in R$/ ton of cane harvested (2)	Average yield in tons cut/day (3)	Daily wage (2) x (3))	Daily minimum wage in Sao Paulo (4)
1969	3,86 (March only)	2,73	2,99	8,16	4,94
1970	4,36	2,02	3,05	6,16	4,92
1972	5,11 (March only)	2,50	3,00	7,50	4,98
1973	5,90	2,51	3,30	8,28	5,02
1977	7,59	2,57	3,77	9,69	5,33
1980	6,60	2,29	3,97	9,09	6,03
1982	6,23	2,17	4,50	9,77	5,68
1985	5,72	1,92	5,00	9,60	5,51
1988	3,70	1,25	5,00	6,25	3,86
1990	3,95	0.96	6,10	5,86	2,30
1992	3,12	0,84	6,30	5,29	3,11
1994	5,67 (November only)	0,83	7,00	5,81	2,22 (= in US$)
1996	6,36	1,05	7,00	7,35	2,52
1998	6,27	1,06	7,00	7,42	2,63
2000	5,40	0,88	8,00	7,04	2,47
2002	5,13	0,88	8,00	7,04	2,67
2004	4,54 (c. v.: R$ 15.42)	0,86 (c.v.:R$ 2.93)	8,00	6,88	2,48 (c.v.:R$ 8.44)
2005	4,83 (c. V.: R$ 17.47)	0,86 (c.v.:R$ 3.11)	8,00	6,88	2,64 (c.v.:R$ 9.56)

Source: IEA, Informações Estatísticas e Anuários Estatísticos, several years. (c. v. = current value); organization of data by Pedro Ramos.[38] (1) Average of both sets of data (April and November), with the exceptions as indicated. (2) As minimum wage and other monetary values, corrected based on FGV's General Price Index – Internal Availability (IGP-DI da Conjuntura Econômica/FGV).

The impact on workers can be noted both in the decrease in the average wage and the staggering increase in the amount of sugarcane they cut. Both stem from the need to try to maintain jobs at a time when the number of positions is being reduced.

Average of both sets of data (April and November), with the exceptions as indicated. (2) As minimum wage and other monetary values, corrected based on FGV´s General Price Index ´ Internal Availability (IGP-DI da Conjuntura Econômica/FGV).

Even though the data available is not very recent, it is possible to demonstrate that there is a tendency for the price paid per ton of sugarcane cut to decrease while, at the same

[38] RAMOS, Pedro. *O Uso de mão-de-obra na lavoura canavieira: da legislação (agrária) do Estado Novo ao trabalho super-explorado na atualidade.* Anais II Seminário de História do Açúcar: Trabalho População e Cotidiano. Itu, SP, 11-15 novembro de 2007. São Paulo, Editora do Museu Paulista da USP, 2007.

time, the average number of tons cut daily increases. The reduction in the price in Reals/ton is often in accordance with the agreements between plants and trade unions, which shows that the pressure of this tendency is growing. It is currently estimated that a worker cuts an average of 12 tons of sugarcane per day, but reports from workers indicate that this number could be higher.

With regards to labour negotiations, the plants seek to comply with the terms of collective agreements related to working conditions, as required by the Public Ministry. For example, they provide individual protection equipment (IPE) and avoid delays in the payment of wages. However, over-exploitation occurs on a structural level, as the system of paying according to productivity leads workers to intensify their efforts. Due to the growing mechanisation of the sector, only the most productive workers are able to keep their jobs. The economist Pedro Ramos, from UNICAMP, explains that:

> "It has clearly become explicit that workers on sugarcane plantations have to intensify their efforts to increase their productivity, yet they have not succeeded in preventing current real daily wages from being lower than those received in the second half of the 1970s and the early 1980s. It is these efforts that ensure that a cutter's daily pay remains higher than both the daily minimum wage and the average wage of temporary workers used by plantations in Sao Paulo. As reported by the media, this effort may be the main cause of recent deaths in sugarcane plantations in Sao Paulo. Although few employers (plant owners, suppliers and especially intermediaries) are willing to confirm this, there is no hiring of workers who do not reach a certain minimum daily yield and there are researchers who state that the real cutting average is currently situated at around 15 tons"[39].

PHOTO: CARMELO FIORASO

PHOTO: DOUGLAS MANSUR

According to denunciations of the Migrant Pastoral Service in Guariba[40], "birola" – which is when a person experiences generalized cramps throughout the body and then dies – brought on by overexertion on the job has caused at least 21 deaths in sugarcane plantations in Sao Paulo between the 2004 and 2007 harvests. The deaths occurred precisely during the process of modernising the sector, with the support of government subsidies. Thus, even when the plants appear to be respecting labour agreements, over-exploitation does, in fact, persist.

Mechanisation has also led to the over-exploitation of machinery drivers and other workers employed in the industrial process. It is estimated that one combine substitutes 120 manual cutters, and between 6 and 10 people

[39] RAMOS, Pedro. *O Uso de mão-de-obra na lavoura canavieira: da legislação (agrária) do Estado Novo ao trabalho super-explorado na atualidade*. Anais II Seminário de História do Açúcar: Trabalho População e Cotidiano. Itu, SP, 11-15 novembro de 2007. São Paulo, Editora do Museu Paulista da USP, 2007, p. 16.

[40] To accompany denunciations of over-exploitation of workers and cases of deaths on sugarcane plantations, see the Guariba Migrant Pastoral Service's website: <www.pastoraldomigrante.org.br>.

are employed in the mechanised cutting of cane. Competition to get these jobs is intense and the working conditions are degrading. A recent example was the over-exploitation of combine drivers in land leased by Raizen (Cosan-Shell). In this case, the Public Labour Ministry observed problems with outsourcing, low wages, failure to comply with safety norms and excessively long workdays, as recorded in the inspectors' report:

> "The investigation found that Raizen had committed fraud, with the clear objective of reducing the costs of the production process. At least 10 workers hired by the Marca de Ibaté outsourcing firm had had an employment relationship with Raizen. Their contract was terminated with the corporation, which was Cosan at that time, on July 28th, 2011 and they were rehired by the outsourcing firm the following day, on July 29th, 2011 to carry out the same tasks. The comparison of pay slips allowed [the inspectors] to confirm that outsourcing increases precariousness, as the salaries of the workers registered with the outsourcing firm correspond, in average, to 63% of the salary paid by Raizen. Due to outsourcing, workers who had previously worked [for Raizen] stopped receiving life insurance and production bonuses, which were given to the plant's machine operators. Furthermore, outsourced workers were submitted to precarious working conditions and had lower salaries. (...) The strongest evidence of increased precariousness due to outsourcing is reflected in the tractor drivers' working conditions. The report affirms that they do not have access to washrooms, a place to have meals, shelter from bad weather (like sun and rain), drinking/fresh water and first aid supplies, which are essential in case of accidents. The inspectors also identified excessively long working days for outsourced workers. There are reports of workers that go for more than ten days without a break. The legislation obliges companies to provide at least 24 consecutive hours of rest per week"[41].

Another case of the violation of the rights of workers employed in mechanised sugarcane harvesting was found on the Santa Laura estate in Goias. There, the existence of slave laboured was discovered, as the follow report highlights:

> "For the first time in the history of Brazil, workers who were subjected to a regime analogous to slavery in the mechanised harvesting process have been liberated. In total, 39 workers were rescued. They operated sugarcane-cutting machines on an estate in the city of Goiatuba, Goias. Their work schedule alternated 24-hour workdays (27, when one includes the 3 hours of travel to and from plantations) with a break of 21 consecutive hours. At least two accidents caused by fatigue at the wheel were registered. The accidents involved two drivers that had been operating machines for more than 20 hours straight. No delays in the payment of wages were recorded, but extra hours and weekly breaks were not paid. The rescue operation took place on the Santa Laura estate, which belonged to the Associação dos Fornecedores de Cana da Usina Bom Sucesso. The estate owner will have to pay the workers a total of nearly R$1 million in severance pay, which does not include social security benefits". [42]

[41] "Raízen (antiga Cosan) é Processada pelo MPT". Jornal de Araraquara – SP, 21 de abril de 2012. Available at: http://www.brasilagro.com.br/index.php?noticias/detalhes/3/43091; consulted on July 6, 2012.

[42] "Primeiro resgate de trabalhadores escravizados em colheita mecanizada ocorre no país". RádioAgência NP, 22 de dezembro de 2011. Available at: http://www.radioagencianp.com.br/10474-primeiro-resgate-de-trabalhadores-escravizados-em-colheita-mecanizada-ocorre-no-pais; consulted on July 9, 2012.

These reports demonstrate that slave labour and degrading working conditions persist, even with the mechanisation of the harvesting process. In regions where mechanisation is more predominant, working conditions are worse, as cutters need to reach increasingly high productivity targets in order to keep their jobs. The companies use mechanisation as blackmail to prevent workers from demanding better salaries and working conditions. As workers are paid according to what they produce, they are forced to cut more and more to try to reach a target that continues to grow with mechanisation.

What is more, cutters have no control or means to monitor the process by which the number of metres cut is converted into tons. This facilitates the appropriation of the unpaid work by the plants and "gatos" or "cats", as the intermediaries who recruit workers are known. Thus, workers continue to have to face outsourcing, the lowering of their wages and robbery at the time when the sugarcane they have cut is weighed. Therefore, it is not the "backwardness" of the sector that leads to the failure to respect labour rights, but rather the modernisation process imposed by the crisis of this model of production.

Impacts of the expansion of agribusiness in Mato Grosso do Sul

PHOTO: MARIA LUISA MENDONÇA

Indigenous community in Dourados, Mato Grosso do Sul state

The State of Mato Grosso do Sul (MS) is known for its biodiversity and abundant water sources, as it is located in a region that is home to the Pantanal and part of the Guarani Aquifer. However, it is also renown for its history in the violent colonisation of indigenous people. Agribusiness corporations are currently expanding their economic power in the region, which now looks like a big ranch with its immense sugarcane, soybean and, more recently genetically-modified corn plantations, not to mention extensive livestock raising. This agricultural model combines extensive monocropping with the intensification of the use of machines and chemicals, and is based on the monopoly of a handful of national and foreign corporations. Therefore, the concentration of land ownership goes together with the monopolisation of food production and trade.

This land ownership structure began to be established between 1915 and 1928 with the confinement of the Guarani, Terena and Kadiweu people to so-called "reserves". The reserves fulfil the purpose of exploiting indigenous labour, as the areas are too small to meet their subsistence needs. In the Dourados Reserve, for example, 12 000 people live on 3 500 hectares of land. In such a situation, the indigenous population is forced to subject itself to the worst of working conditions.

As they are forced to live in these conditions, the indigenous people of Mato Grosso do Sul are subject to violence, racism and total negligence on behalf of the State. The Indigenous Missionary Council's (CIMI)[43] report on the genocide against these people presents shocking figures. From 2003 to 2010, 250 indigenous people were assassinated in MS, more than the 202 indigenous people killed in the rest of the country. Data on suicide are also alarming: 176 cases were registered between 2003 and 2010, which represents 83% of all suicides among indigenous people in Brazil. Over the last eight years, more than 4 000 indigenous children suffered malnutrition in MS. Furthermore, in the past five years, it was reported that 1 787 people from the indigenous communities in the State died from lack of access to adequate medical attention. From 2003 to 2005, more than 70 conflicts arising from land ownership disputes between indigenous people and landowners were recorded. Currently, three indigenous leaders and four entire communities are suffering death threats and are under the protection of the SDH/PR's Protection of Human Rights Defenders Program.

The spread of agribusiness in the region is making this situation worse. Incentives from the National Bank for Economic and Social Development (BNDES) for the sugar-energy sector include support for sugarcane plantations established on indigenous land claimed by the Guarani-Kaiowa people and officially recognized by the Federal government itself, which is prohibited. The Raizen corporation, created by the merger between Cosan and Shell, recently signed a Term for Change in Conduct agreement with the Public Prosecutor's Ministry, for having used sugarcane planted by the Nova America estate on indigenous land in Caarapo, south of Dourados. The territorial expansion of monocropping occurs mainly in areas with access to infrastructure and in regions with vast water basins, like the Cerrado.

Currently, plants in Mato Grosso do Sul are only employing indigenous labourers. With

[43] Rangel, Vera. "As Violências em números, gráficos e mapa". Em: *As Violências contra os povos indígenas em Mato Grosso do Sul, e as resistências do Bem-viver por uma terra sem males.* Relatório do Conselho Indigenista Missionário (CIMI), MS, 2011.

PHOTO: CARMELO FIORASO

the advances in mechanisation, it is estimated that manual cutting has declined 40%. Of the 10 000 indigenous workers employed in the harvesting of sugarcane until 2006, there is now only approximately 6 000. Denunciations of irregularities in hiring practices, housing and food arrangements, safety and transportation of migrant workers coming from the North East of Brazil has led the plants to opt for using local indigenous labour.

Workers are recruited by so-called "headhunters" or "chiefs" who assume the role of the "cats" or intermediaries, play a leadership role in the villages and receive a percentage of the cutters' salaries. Headhunters also have the task of demanding higher production levels from workers, as, with the growing unemployment, only the most productive workers keep their jobs.

Indigenous workers in Mato Grosso do Sul have recently affiliated to the Central Única dos Trabalhadores (CUT) trade union central. One of the indigenous union leaders, Evanildo da Silva, explains that the union's struggle combines the objective of "[providing] stronger protection [and ending] workers' submission to and dependency on the private sector and the State, in order to preserve the indigenous identity". The union defends the immediate demarcation of their land and compliance with Convention 169 of the International Labour Organisation (ILO) on the protection of the rights of traditional indigenous and rural Afro-Brazilian communities.

Evanildo points out that work on sugarcane plantations generates illnesses and physical exhaustion: "Workers knock themselves out cutting sugarcane and suffer from the impacts of the use of chemicals on their health and the environment, which are long-term". He remembers that indigenous people have historically been exploited and given the hardest work. They even built the state's infrastructure, such as the railroads. "It is common to find indigenous youth working on the estates with fake documents. When I was 17, I worked cutting sugarcane and I can still feel the effects of that today. We were housed in plastic tents. I didn't have a contract and we had to pay for all of the equipment and food", he states.

The union leader explains that prejudices against indigenous peoples vary according to what is convenient at the time, switching between the colonial image of the "lazy Indian" and the idea currently permeating corporate discourse that they are the most "apt" for heavy labour. One employee from the Agrisul/CBAA plant in Sidrolandia, who did not want to identify herself, explained that the company currently hires "100% indigenous labourers because they aren't ambitious". Since 2009, with the expansion of mechanisation, the demand for greater productivity from workers has caused frequent accidents resulting in cuts, fractures and dislocated limbs. Unemployment increases job insecurity and leads workers to accept temporary three-month contracts that free companies from the obligation of providing severance pay when harvest is over.

In the Dourados region, where the number of conflicts over land is the highest in the state, the indigenous people who have reoccupied their traditional land and are demanding its demarcation also have to subject themselves to working at the plants. As their villages are surrounded by monocultures that use a large quantity of chemicals, food production is hindered by soil contamination and the high number of plagues proliferating in the environmental imbalance caused by neighbouring plantations. This enables companies to guarantee that a workforce is available for the plantations, even when they offer degrading working conditions.

PHOTO: MARIA LUISA MENDONÇA

Reginaldo, who now works at the local health clinic, tells about how the companies discount the costs of clothing, food, water and equipment like boots and thermoses from workers' pay. Many under-aged youth falsify their pieces of identity and start working when they are adolescents. Occupational diseases such as back, shoulder, arms and hand problems are common, as are lung diseases and tuberculosis caused by exposure to pollution in the plantations and the unsanitary conditions of lodging at the factories.

Other indigenous people from the village tell similar stories. João[44] says that he started cutting sugarcane when he was 16. He left the village to stay at the lodge and only had a day off once every 45 days. The workday went from 5:30 am to 4:30 pm and workers were paid 10 cents per metre cut. With all of the deductions for transportation, food, medicine and equipment, the average monthly salary did not reach R$400 and workers had no right to pension, unemployment insurance or health care. Delays in the payment of wages are common and generally, workers are only paid when they go on strike. In the regions where burning sugarcane has been prohibited, cutting "raw" cane exposes workers to contact with snakes, scorpions and other insects.

The vulnerability of indigenous people, who are subject to discrimination and repression, makes it easier for companies to circumvent problems linked to the registering of workers and the verification of working conditions. In the municipality of Dourados alone, 1 400 complaints against plants concerning labour rights were filed with the Public Prosecutor's Ministry. Public attorneys have recently filed suits demanding that the companies pay workers for the time it takes them to get to and from the plantations. Despite their legal obligation to do so, the companies do not pay for these hours, which in one of these cases alone amounted to R$350 million. Some plants went so far as to propose "funeral support" in exchange for this right. Another type of irregularity that is common is deforestation, which the corporations hide by burying the trees to avoid fines.

To get around problems with debt and to avoid defaults, many plants adopt another legal name and registration number by merging with large economic groups. Agrisul/CBAA, for example, declared bankruptcy and was able to resume its activities through a special process that allowed it to legally recuperate the company. This status authorises the companies to rollover debts and avoid paying the fines they receive for failing to comply with labour or environmental rights. Delays in paying employees are frequent, as are the strikes cutters organise to force corporations to pay their salaries. The plants' strategy serves, in practice, as a way of curtailing workers' freedom, given the absence of other forms of subsistence. Monopoly over the best quality land also prevents the development of other modes of production.

In areas of peasant farm production, the lack of basic infrastructure, such as housing, water, transportation, energy, credit and technical support, often creates dependency on and submission to work on sugarcane plantations. The non-indigenous rural population is generally hired to operate the mechanical cutters, tractors and combines. These machine-operators generally have long and gruelling workdays. The lack of appropriate support

[44] The names of the interviewees have been changed in order to preserve their integrity and safety.

Indigenous children receive food at school in Mato Grosso do Sul state and, consequently, the difficulties in making a living with an alternative agricultural model leads these farmers to lease their land for planting sugarcane for corporations, which often leaves them in debt.

Mauro lives in one of the oldest agrarian reform settlements of the region and had always been able to sustain his production. In 2007, plant representatives convince him to grow sugarcane for them, promising that he would get a good financial return on the deal. He sold part of his cattle to invest in planting sugarcane and to cover all the costs of supplies and harvesting. But the plant never paid Mauro, and he was left with a loss of R$56 000, not to mention the work he had to do to recuperate the quality of the soil on his land. He still faces serious problems due to the fires from neighbouring sugarcane plantations, which often reach environmental protection areas.

In one of the newer agrarian reform settlements, Pedro reports that he leased land for sugarcane production in the past, but he did not receive payment either. "We only suffered losses. We spent money on the crop, fuel and labour and we didn't receive anything in return. We ended up with R$14 000 in debt and since the plants do not sign contracts, they are able to avoid going to court". He also says that he used to produce manioc, beans, corn and wheat, but due to the lack of incentives he had to lease his land. Faced with no other options, Pedro also worked as a tractor driver at the plant: "They don't pay for lunch hours and we have to eat inside the tractor while driving. They don't accept it when the machines are stopped and we can't go to the bathroom. We only get a day off every seven days and the contract lasts six months, during harvest. It's like slavery".

Peasants also look for work as machinists in the sugarcane grinding process. Joilson reports that it is common for him to be paid late, and it has been two months since he last received his salary. "We go to the plant out of necessity, because the working conditions there are terrible. Salaries are low. We aren't paid extra for night shifts or unsanitary conditions. They don't pay employment insurance and now, we are without pay. That's why we went on strike and paralysed the plant". Another problem is the lack of safety equipment. For example, the brakes on many tractors do not work and accidents are frequent.

The agrarian reform settlers' need to look for work comes from the lack of infrastructure required to make a living on the land. After eight years of organizing land occupations to demand agrarian reform, the area was expropriated in 2005, yet the farmers still do not have access to credit for production. Romilson plants manioc, corn, okra, pumpkin, fruit and vegetables and produces milk and eggs, yet it is difficult for him to maintain production above sustenance levels, as the biggest portion of money from sales goes to the middlemen. To reverse this situation, peasants need to receive support for production, such as credit, guaranteed market access and prices, and access to infrastructure, transportation, water and energy.

However, Brazil's agriculture policy in support for agribusiness promotes the continuing grabbing of natural resources. Indigenous and peasant communities are surrounded by monocropping that continues to expand, causing deforestation, intoxication from farm chemicals, land and water contamination. This production model does not prioritize food production, thereby generating extreme situations of repression, hunger and poverty. One can see, then, that the "product" of agribusiness is not sugarcane, soybean, sugar or ethanol, but rather an enormous financial, social and environmental debt.

PHOTO: MARIA LUISA MENDONÇA

Conclusion

As we have noted in this study, the Brazilian economy's apparent "growth" has been determined by the global economic crisis, at a time when financial capital is predominant. The Brazilian government's efforts to turn ethanol into a commodity, so it can be negotiated on the futures markets, could deepen the impact of land speculation in Brazil, in a way that is similar to the crisis generated by the bubble on the United States and European real estate markets. We have observed this trend in Brazil's land market, which is accompanied by speculation on agriculture and raw material commodities. The recent deindustrialisation of the Brazilian economy combines the speculative movement of idle capital on the financial market with the tendency of this capital to "take refuge" in strategic resources like land, water, oil and mines.

Our study shows the tendency of large corporations to acquire land through leasing in order to expand monocropping, which is resulting in the replacement of food production with sugarcane and the expansion of the agricultural frontier into protected areas. This process is linked to a certain type of flow of financial capital, which generates speculative bubbles and causes sharp increases in land and food prices. The discourse in favour of the "modernisation" of ethanol production serves to mask the rise in labour exploitation.

The so-called "competitiveness" of Brazilian ethanol on foreign markets is based on the exploitation of labour, as the plants attempt to compensate their high level of debt and defaults. The model that has been historically adopted in the sector, based on the remuneration of sugarcane cutters according to their yields and not the number of hours worked, has generated degrading working conditions, which are structural. In other words, the over-exploitation of workers is not a specific or isolated incident; it is systematic. To improve their public image, mainly to ensure their access to foreign markets, corporations are seeking to further the mechanisation process.

The sugar-energy industry justifies the territorial expansion of monocropping by alluding to the possibility of growth in the global demand for ethanol. But with the international economic crisis, predictions on this demand have changed. Another argument for the expansion is the potential increase in the domestic consumption of ethanol, which has been boosted by the automobile industry's use of *flex fuel* technology. These arguments, however, have several limitations, as they ignore the structural crisis in the sector.

The search for competitiveness on the world market has led the sugarcane industry, and other agribusiness corporations in Brazil, to continue assuming an increasingly high amount of debt in order to maintain productivity at an acceptable level. As in the past, the recent expansion process depended on State subsidies. The provision of this support can be interpreted as a continuation of the policies from the 1970´s Pro-Alcohol period, which is contradictory to the common idea that agribusiness is continuously improving its "efficiency". The transformation of the current agricultural model is thus necessary in order to overcome the simultaneous crises in the economic, social and environmental spheres.

Looking for Sustainability in Biofuel and Bioenergy Industries: Some Challenges and Opportunities
Cunha, M., TBA
Originally presented on: 7/24/2014 8:30:00 - 8:48:00

Marcelo Pereira da Cunha – Institute of Economics – Unicamp – Brazil

marcelocunha@eco.unicamp.br

Looking for sustainability in biofuel and bioenergy industries: some challenges and opportunities

The adoption of biofuels and bioenergy have been recognized as a possible sustainable alternative for replacing fossil fuels, mainly from the beginning of this century. Several countries have set policies to implement and to enlarge the use of those sustainable options. Bioethanol and biodiesel are the most important biofuels produced and used in North America and South America – United States, Brazil and Argentina are among the main countries in this field. Cosidering that there are many differences in the regulation, development, agricultural conditions and the markets in the American countries, the opportunities and challenges faced in the biofuels and bioenergy industries are regionally dependent. The objective of this presentation is to display some opportunities and challenges in this industry considering the three pillars of sustainability. In economic terms, energy policy and regulation can be an incentive to replace part of fossil fuels in domestic or international markets; on the other hand, changes on them (and sometimes the near absence of them) can create barriers to enlarge the bioenergy adoption. In this aspect, the current situation of ethanol in Brazil competing in unfavorable conditions with gasoline is an example. Still in the economic dimension, the limits of how industries of the sector can explore the production of co-products need to be careful analyzed. Considering the social pillar, special attention has to be placed on the business model adopted by a bioenergy company because there are different implications on sustainability depending on it. For example, the possible positive impacts in jobs and income generation, mainly in the agricultural phase, will be influenced by the technological level adopted in agriculture but also on the verticalization (or not) in this key phase of the production chain. A particular analysis is provided in a recent sugarcane ethanol expansion area in Brazil. Another one is the example of how the biodiesel industry has been operating with feedstock supplied by family farmers. Taking into account the environmental issues, the industry has been capable to improve, in many cases, the GHG emissions by through its production process. One of the possibilities is to use biodiesel in the agricultural phase replacing mineral diesel oil in an integrated sugarcane ethanol and biodiesel production. On biodiesel sector, the use of beef tallow (second most important feedstock in Brazil) is another other example of improving environmental benefits, as well as economics. Finally, other important issue is regard to the methods applied over biofuel and bioenergy sustainability assessment, which may (and shall) contribute to evaluate in the best possible way the benefits (or not) of this diversified sector, mainly because biofuels and bioenergy are not the same around the world.

Achieving Sustainable Biodiesel Production in Brazil: Challenges and Opportunities

Minelli, J.,

Originally presented on: 7/24/2014 8:48:00 - 9:06:00

Abstract – Julio Minelli

Director of Aprobio – Brazilian Biodiesel Producers Association

julio.minelli@aprobio.com.br

Achieving sustainable biodiesel production in Brazil: challenges and opportunities

Governments are searching for renewable energy alternatives to diminish the dependence on fossil fuel, with different commitment levels. They are fostering the use of renewable energy, but this transition has many challenges. Biofuels appears as an alternative energy resource but has, by definition, a limited flow. Moreover, it is important to have positive outcomes in Social, Economic and Ecological aspects. The impacts to the society may be managed by the right regulatory framework and assessed by indicators as employment, prices, ecological benefits, etc.

The Biodiesel production in Brazil generates employment, midland economic growth, social inclusion of more than 80 thousand families, costs comparable to imported diesel. The main raw materials used today are soybean oil and tallow. The use of biodiesel also reduces GHG emissions and pollutants, as SOx, CO, particulate matter and hydrocarbons.

The food & fuel debate is a gut issue, normally addressed as an incompatible matter. APROBIO would see it as synergistic solution as most oilseed crops produce feed at the same time. In this sense, increasing oilseed crops increases the feed production, important nutrition element for animal protein. The limiting factor for food, feed and bioenergy production may be the land availability. The policies and efforts should seek the best resource allocation and efficiency.

Challenges and opportunities:

To increase the public awareness that bioenergy has benefits in the mid long term, to obtain a consensus on sustainability evaluation schemes, to increase the oil availability with competitive costs and endorsement to run diesel motors on B100 or at least B20.

The average income in the developing countries is rising and so is the demand for animal protein, therefore more feed demand. Excess of vegetal oil are expected to happen, and this surplus may be directed to biofuels with higher mandatory blending.

Green House Savings Produced By Argentine Biodiesel Use

Hilbert, J. A. Sr., Instituto Nacional de Tecnología Agropecuaria
Galbusera, S., Sec Medio Ambiente

Originally presented on: 7/24/2014 9:06:00 - 9:24:00

One answer to the environmental concern regarding Greenhouse gases emitions control has been the promotion of different biofuels as mandatory targets in many countries. Argentina as one of the world main producer of vegetable oils found a great opportunity to use a byproduct of its important seed processing industry to develop efficient biofuels for domestic and international markets.

The Argentinean soybean industry is extremely strong: more than 70,000 producers for 18 million of cultivated hectares and 30 million tons of products (2008-2009). The sector provides 13 billion dollars and represents an important source of foreign exchange for the State. The export taxes were 5% for biofuel and 35% for grain in 2007: this difference prompted soy exporters to invest in biodiesel plants for export. In this way, the first phase of biofuel development started. Economies of scale and the efficiency of the soy chain are exploited to make the Argentinean biodiesel a competitive product. Despite the increase in the biodiesel tax to 20% (2008), the strong international demand encouraged new investments. The production is concentrated around Rosario (province of Santa Fe), on the Paraná River, in the soybean oil exportation complex. Thus, the province of Santa Fe has 80% of the national production, against 8% for Buenos Aires and 7% for Santiago del Estero this gives additional advantages regarding GHG emitions since distances from the production area to the transforming complex and export ports don't exceed 300 km.

Large national companies (the oil manufacturers, General Deheza, Vicentin, and Eurnekian Citrusvil[1]) and transnational corporations (Dreyfus, Glencore and Bunge) build efficient industrial plants followed by medium private investors with smaller facilities in different parts of the country mainly concentrated on domestic market. The capacity 2013 is over 4 million tons per year.. Such volumes can be competitive in the global market. However, biodiesel exportation is fragile due to international and local changes in rules. The internal market is more recent and starts with an obligation to add 5%.

Since the 1970s, (bio) technological changes improve the production and productivity of the soybean chain. Now, biodiesel is diversifying its by-products, customers and locations. It also awakes the interest and creativity of engineers, chemists, and researchers since it offers an opportunity to put old and new industrial plants in operation.

Soybean production is characterized for the employment in more than 82 % of no tilling together with other modern technologies as presition agriculture. This gives an important advantage in GHG emitions savings and energy balance.

The industrial plants responsible for the principal market share of biodiesel are characterized by its high scale and efficiency. Most are located beside the processing complex and ports, which gives enormous advantages from the energy and emitions results. Raw material is coming from a radius no larger than 300 km, which also helps to increase efficiency.

In the last years, new bio refineries were developed in order to get higher value products of the biodiesel process as glycerin and sub products. This enlarges the benefits of the chain and increases the countries income.

The last eight years were characterized by huge investments in biodiesel industry using soybean oil as the main feedstock. The first investments were produced by the new emerging European market and more recently for a growing domestic one, triggered by increasing mandatory blend now reaching 10 % in the country.

In order to give an answer to increasing concerns regarding the green house savings produced by soybean oil biodiesel several international and national studies were performed giving a realistic evaluation of a very efficient transformation chain in the country.

Several studies have been completed in order to clearly calculate the GHG emissions of the Argentine biodiesel sector according to generally accepted international methodologies. Since the country has different agro-ecosystems and distances from the ports, different analyses were completed for different regions of the country. The key element to assess the Greenhouse Emission Saving (GES) was the soybean supply area. (Panichelli, 2012) With respect to GHG emissions (Kg CO_2 eq/km), emissions reduction averaged 76% (0.0447 Kg to 0.0464 CO_2 eq/km). The scenario that showed more GHG emissions was South East of Bs.As. (0.0447 Kg CO_2 eq/km). Comparatively and percentage with conventional diesel, its reductions of GHG emissions were of 75.5 %. The scenario that showed less GHG emissions was placed in South of Córdoba (0.0464 Kg CO_2 eq/km). Comparatively and percentage with conventional diesel, its reductions of GHG emissions are of 76.5%.

A deep analysis was performed of the studies looking at criteria, methodology employed, allocation factors, coproducts evaluation and data origin in order to obtain a realistic value of greenhouse savings that could represent soybean oil biodiesel produced under the particular Argentinean conditions.

A comparative analysis showed that the crop yields and the oversea transport are the main sources of variability between studies and years. The industry values are very constant and no significative difference was detected between the differen literature sources.

As methodological framework the ACM0017 (production of biodiesel to be used as a fuel) was used. Specific calculations were performed in order to obtain the total ammount of conventional diesel fuel replaced in each market.

Based on export and domestic market commercialization a national greenhouse savings calculation was performed for the last four years (2010-2013). For each case the total emitions were taken into account savings ranged 4 million tons per year.

---

[1] This group privileges the installation of a biodiesel plant near its oil factory (at Frías, Province of Santiago del Estero), becoming the only mega-plant located far from Rosario.

Bioethanol Dehydration Assessment with Environmental Issues

Colombo, M., Universidad Nacional de Tucuman
Mele, F. D., Universidad Nacional de Tucuman
Hernández, M. R., Universidad Nacional de Tucuman
Kahwaji Janho, M., Cleveland State University
Gatica, J. E., Cleveland State University
Originally presented on: 7/24/2014 9:24:00 - 9:42:00

ABSTRACT

Environmental effects and health hazards posed by fossil-fuel based technologies complemented by changes in the global economy have further demanded the need for developing "cleaner" and more efficient technologies that rely on renewable or synthetic resources. An alternative, commonly referred to as bio-fuels, has significantly matured and today's economy recognizes the significance of being able to produce ethanol from renewable resources such as biomass. Moreover, the potential of ethanol to be further converted to hydrogen makes it a very attractive alternative to replace or complement fossil fuels as sources of energy.

Argentina has recently enacted legislation to promote the use of bio-fuels to ameliorate its ever increasing annual gasoline demand, currently bordering 4 billion liters. This new legislation (*Law 26.093 of Biofuels Promotion*) would result in a demand of over 200 million liters of dehydrated alcohol. This quota would be equivalent to produce ethanol out of all the sugar that is currently exported. With fifteen (15) active Sugar Cane processing plants, a mill capacity of 15 million tons/year, and thirteen (13) distilleries with 1.5 million liters/year of hydrated alcohol, Tucumán (located at North West of Argentina) could meet more than 50% of the ethanol demands for Argentina.

Process simulation allows a systematic analysis that leads to an advantageous combination of all the values of process variables. In particular, since they demand significant energy influx, an important subsystem for analyzing is the separation of alcohol-water mixtures. Though many techniques for ethanol dehydration are known; adsorption, distillation, hybrid processes, and pervaporation, are the most common technologies in practice. Two alternatives for ethanol dehydration technologies are considered in this work. The first is based on the combination of distillation and azeotropic distillation, while the second relies on hybrid distillation and pervaporation processes, so it was necessary to develop a simulation module for calculating the membranes performance, typically not included in commercial process simulators. Both alternatives are simulated and their optimal design and operating parameters are identified by means of rigorous simulation using Aspen Plus®.

To achieve sustainable designs process integration is a critical task. For this, we consider the dehydration step as an integral part of the overall scheme of alcohol / water separation, taking into account potential energy integration scenarios considering different concentrations of the products obtained in the different units, to analyze its impact on global energy consumption and other environmental indicators. Particular attention is given to energetic integration by performing a pinch-analysis to each of the fundamental stages to find minimum energy consumptions.

This paper uses an ad-hoc pervaporation module to assess the production of fuel-grade bio-ethanol in distillery installations complementing a sugar mill. A baseline condition is defined as a case-study to illustrate the effect of structural and parametric changes on membrane separation units.

The user-defined module is integrated in commercial process simulators to study the environmental and energetic impact of different process configurations. Different schemes meeting equivalent performance standards in terms of purity and recovery are examined in this paper. Particular attention is given to energetic demands as comparison metrics.

This paper formulates an approach that accounts for environmental issues in explicit form. The approach makes use of Life Cycle Assessment (LCA) as described by the ISO 14000 series. Unlike most common approaches that consider environmental impact by focusing on reducing effluents, this methodology also considers the impact associated with all the involved processes in the FPD. The software used to evaluate the environmental impact is SimaPro®.

Process alternatives that meets same specifications of purity and recovery in exit streams, are examined in this paper. Of particular relevance is the assessment of costs and environmental impact of these processes. A baseline case is defined in our case-study, and structural and parametric changes are made. Different scenarios with respect to economic and environmental issues taking into account the specifications of the final product are generated. The impact of different operating policies for the production of anhydrous alcohol on the environment is examined in detail.

The membrane area, the number of plates in the distillation units, the heat exchange area, and the pressure manipulators are considered as structural variables. Among the most relevant operational variables considered one can mention reflux ratio, pressure ratio, as well as compositions and flow rates of streams.

One could consider the addition of environmental aspects within energy systems optimization as a promising contribution to the energetic optimization and LCA.

Keywords: Bio-fuels, Ethanol Dehydration, Life-cycle Assessment, Process Simulation

Opportunities and Challenges Considering the Socioeconomic Dimension for Brazilian Ethanol Industries in the Recent Past

Marques, A., Amarques & Associados

Originally presented on: 7/24/2014 9:42:00 - 10:00:00

Opportunities and challenges considering the socioeconomic dimension for Brazilian ethanol industries in the recent past.

The sugarcane sector in its evolution have a history of ups and downs that was influenced by several factors, sometimes causing structural changes, sometimes superficial in industry dynamics . However, the transformations experienced in the last 10 years have had a significant impact on the sector. In this latest expansion process, new elements and obstacles were put to groups of entrepreneurs and the new ways to overcome those obstacles have created a new profile for the sector with significant impacts to the communities where the projects are implemented.

Among the variables to be considered on an expansion greenfield project there are elements in the agricultural , industrial and commercial steps that can be more or less aligned with the best practices of sustainability into force . The set of criteria for adoption of these elements often depends on technological factors, endafoclimáticos, legal or even structural elements of the chosen region (topography, availability of manpower).

This paper presents a description of these agricultural, industrial and commercial stages of a sugarcane company with the various technological and procedural options possible. The following focuses attention on one of the fundamental aspects of the business that has impact on the whole processing chain and in the communities where the company operates. This aspect concerns the way how the access to sugar cane has being done

In fact, the format of access to sugarcane (and so to say, land) has changed significantly in the recent expansion cycle (2000-2012). That comes to changing the role of land ownership. Before, the cultivation areas and expansion were the only states in the Northeast and São Paulo, now the crop expands to the Midwest and South. Before, planting on their own land used to be the preference and dominance of the sector who has justified by the dependence of the flow of raw materials, now, other forms of access to cane become the preference of some groups and land ownership becomes a subsidiary or complementary strategy in portfolio of raw material to be processed.

When it analyses the expansion to the state of Goiás where this author worked for 4 years, the differences leap to the eye. A simple implementation of a performance model from São Paulo to the new territory had to undergo several changes. The most striking and impactful to society differences concern the impact on agrarian structure of the region since this element brings to the company and external community consequences of long-term duration.

Then, when it analyses the expansion to the state of Goiás where this author worked for 4 years, the differences leap to the eye. A simple implementation of a performance model from São Paulo to the new territory had to undergo several changes. The most striking and impactful to society differences concern the impact on agrarian structure of the region since this element brings to the company and the external community the consequences of long-term duration.

This vertical integration strategy, when used for the expansion of the sector in the Midwest, this brings about the concentration of income in the hands of a few producers mill owners. Economic activity in the region decreases due to the increased use of larger suppliers capable of endereças the demands of large-scale owner / mill owner.

On the other hand, the strategy of horizontal coordination, where land acquisition gives rise to new contractual arrangements through the lease and formalizing partnerships production, causes a greater extent, the landowner and agricultural entrepreneur, continue ahead of production, and thus using more intensively local resources as manpower, suppliers of implements, products or services. The economic activity of such regions tends to be more dynamic and diverse.

But the question is what leading the adoption of vertical or horizontal strategy by sucroalcoleiras companies in the process of expanding into the region in southern Goiás? Which specificities found on the sector´s expansion to the Midwest have caused or intensified such behavior?

The hypothesis with which it works is that from the mid-2000s changed the profile of new entrants who favour new contractual arrangements in order to guarantee the flow of sugarcane production, at the expense of the vertical integration model adopted until that time by traditional entrants.

The reasons for this change is based on 2 elements: the change of profile of the owner-investor and prior presence of an established culture of using these institutes of lease and partnership that allows the new business owner to access the areas of planting without necessarily having to acquire the land

But why the new business owner profile do not accept or not prefer land acquisition anymore? Research shows that this type of investor now has a profile of predominantly large business group , who pay attention on the scale production and at generating business cash flow instead of traditional family business which has in the real estate valuation of the land one of the attractive for the business.

The consequences of these differences for the community that receive the projects are various and belongs to social and economic dimension. To illustrate the two types of situation, the research presents two case studies of mill that expanded to the territory of Goias after 2000s - The Vale do Verdão - family management mill with access to vertical cane (sugarcane own strategy on own land acquired from the owners of the region) and the Tropical Bioenergy, a BP group business that uses horizontalized strategy, ie, does not have its own land for planting sugarcane and its supply is 100 % coming from the partnership , independent leasing and cane supply contracts .

The methodology used for the research was based on analysis of primary data collected from the companies presented, representation of business institutions. In addition, secondary data helped explain some impacts initially planned in the surrounding communities. Such analyzes and conclusions were validated with two field visits where interviews with community members and businesses surveyed supported the conclusions presented here.

Diesel displacement in the sugarcane ethanol life-cycle: a comparative analysis of different integrated systems

Simone P. Souza [1,2]; Joaquim E. A. Seabra [2]

[1] Brazilian Bioethanol Science and Technology Laboratory
[2] Faculty of Mechanical Engineering, University of Campinas

CTBE Centro de Ciência e Tecnologia do Bioetanol

UNICAMP

The sugarcane sector in Brazil

- High diesel consumption

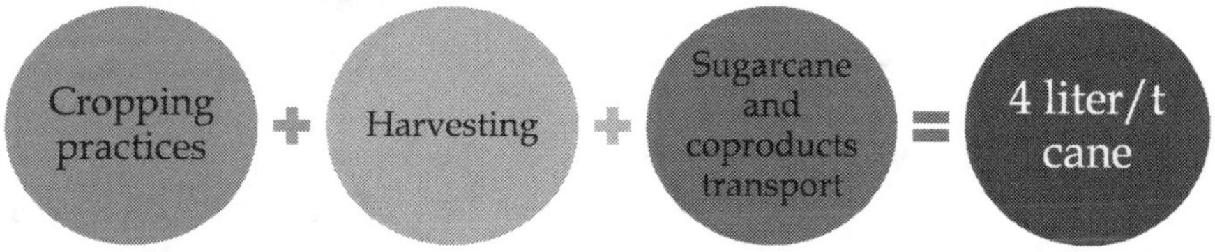

<u>How much did the sector consume?</u>

→ 2.7 billion liters of diesel in 2014 season → 4% of the national consumption (EPE, 2013).
→ Upward trend due to the mechanical harvesting.

<u>Brazil</u>:
→ 3rd /4th largest producer of biodiesel

Goal

- This study aims to explore three integrated production systems applied to the sugarcane sector in Brazil focused on diesel displacement. We assessed the energy balance and the GHG emissions.
- The integrated systems are:
  - Soybean-sugarcane
  - Algae-sugarcane
  - Palm-sugarcane

Methodology

- Life Cycle Assessment → Well-to-gate analysis
  - GHG emission
  - Fossil Energy Use
- Uncertainty: Monte-Carlo simulation → Crystal Ball®
- Data
  - Sugarcane and soybean sectors
    - Sugarcane mills, biodiesel industries, sugarcane suppliers in the States of São Paulo, Mato Grosso and Goiás.
  - Algae
    - Literature and SAT Company (NDA)
  - Palm sector
    - Literature and Agropalma

SYSTEM 1: Soybean-Sugarcane Integrated Production

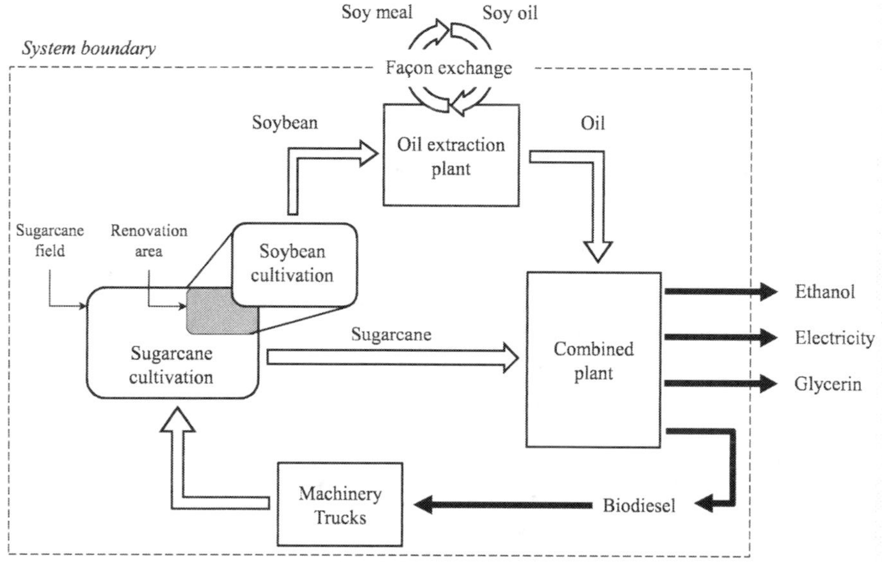

SYSTEM 1: important remarks

- Sugarcane (Conab, 2014/15)
  - Area: 9,000,000 ha
  - Production: 672,000,000 tons
- Soybean (Conab, 2014/15)
  - Area: 30,000,000 ha
  - Production: 86,000,000 tons
  - 80% of the biodiesel source

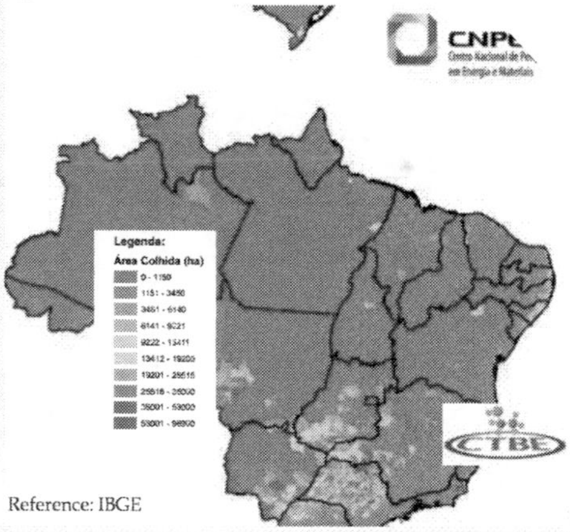

Reference: IBGE

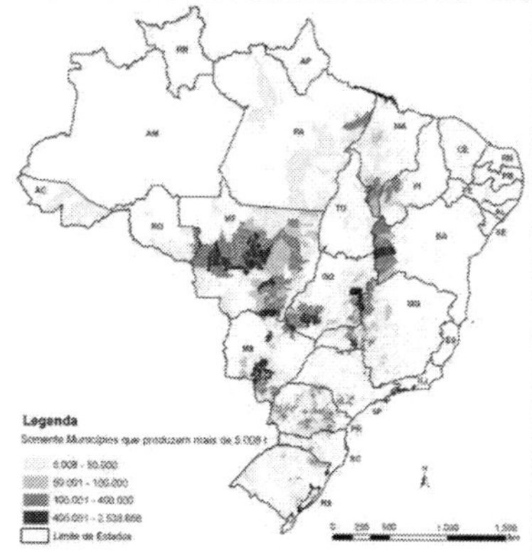

Reference: CONAB, 2014.

SYSTEM 2: Algae-Sugarcane Integrated Production

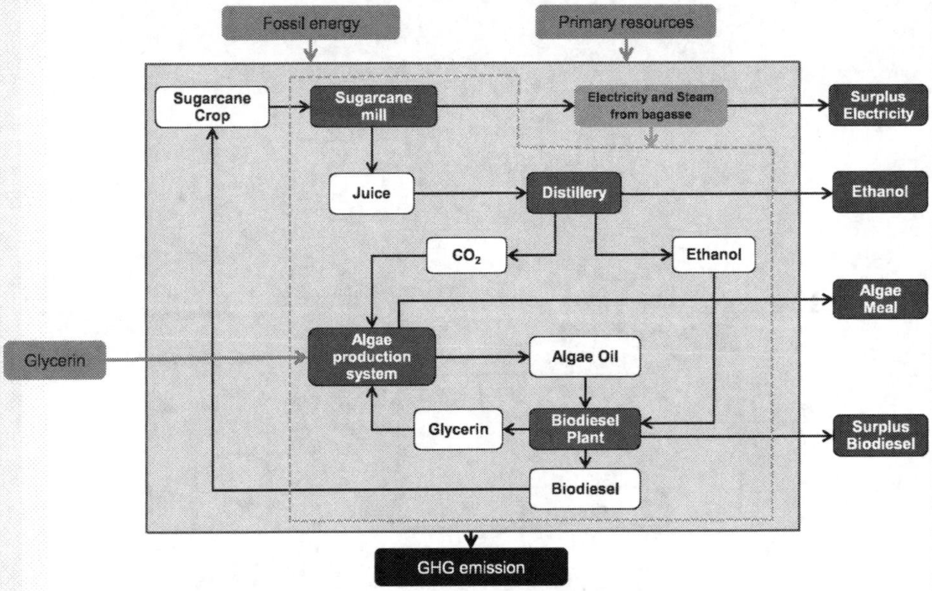

* **Note**: dark grey boxes: processes; white boxes: products consumed inside the system; green boxes: output products; orange boundary: energy flow, which supplies all the process; and black arrows: mass flow.

SYSTEM 2: important remarks

- CO_2 available from fermentation (pure) and bagasse burning.
- Biodiesel production is limited by the <u>CO_2 availability</u>.
- Due to a favorable climate, Brazil has the potential to become a major algae producer (Franz et al., 2012).

SYSTEM 3: Palm-Sugarcane Integrated Production

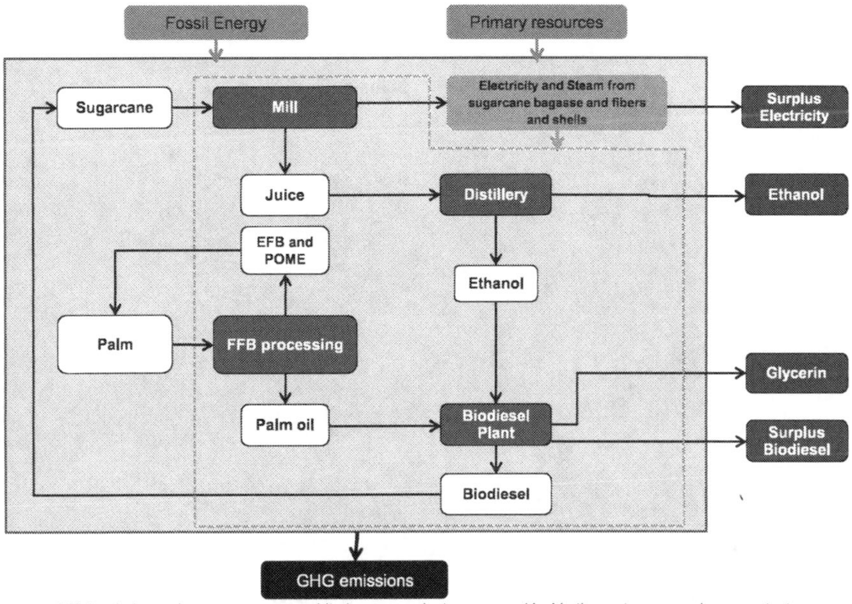

* **Note**: dark grey boxes: processes; white boxes: products consumed inside the system; green boxes: output products; orange boundary: energy flow, which supplies all the process; and black arrows: mass flow.

SYSTEM 3: important remarks

- Palm → main feedstock used for oil production worldwide (food and chemical industries).
- Over ten years, the palm oil has remained as the lowest price in the international market (MPOB, 2012).
- Embrapa experiment: Promising results of palm under irrigated conditions in Cerrado region.
  - Irrigation: POME and vinasse

RESULTS…

Energy Balance (EROI)

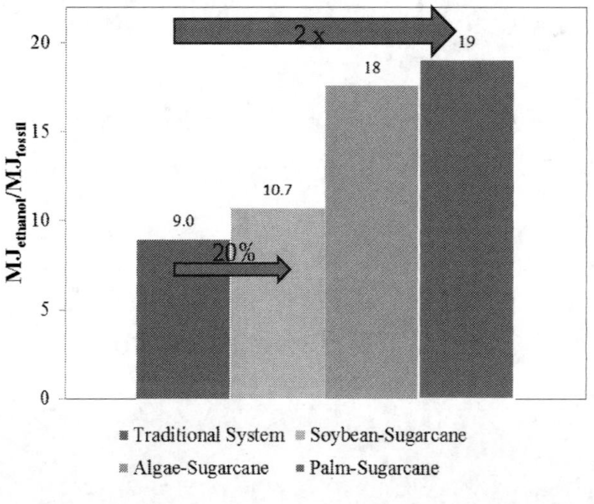

Energy Flow

REFERENCE: 1 TON OF SUGARCANE = 18.8 kg ALGAE (DW)

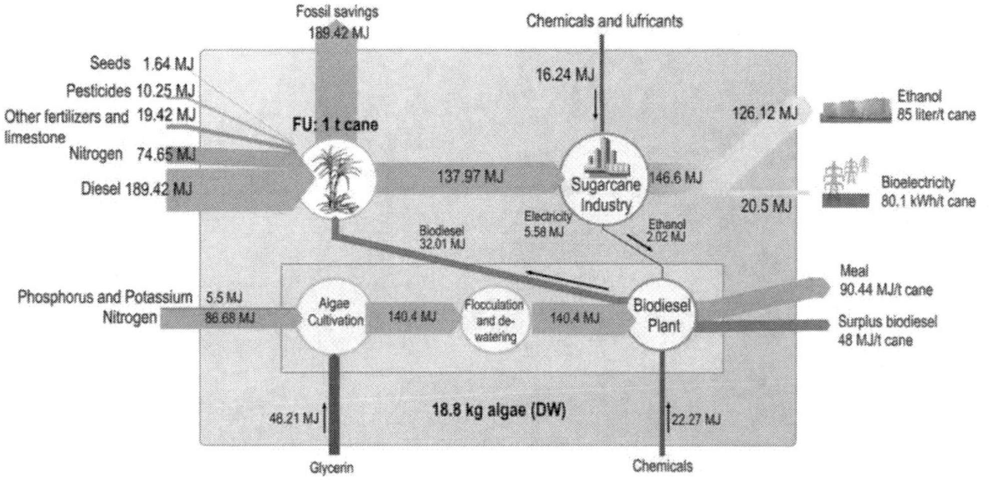

Fossil Energy Use

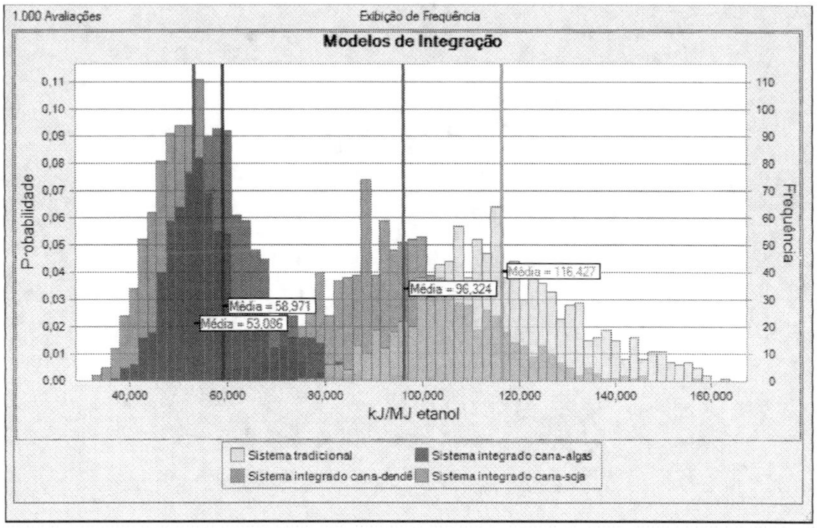

GHG emissions

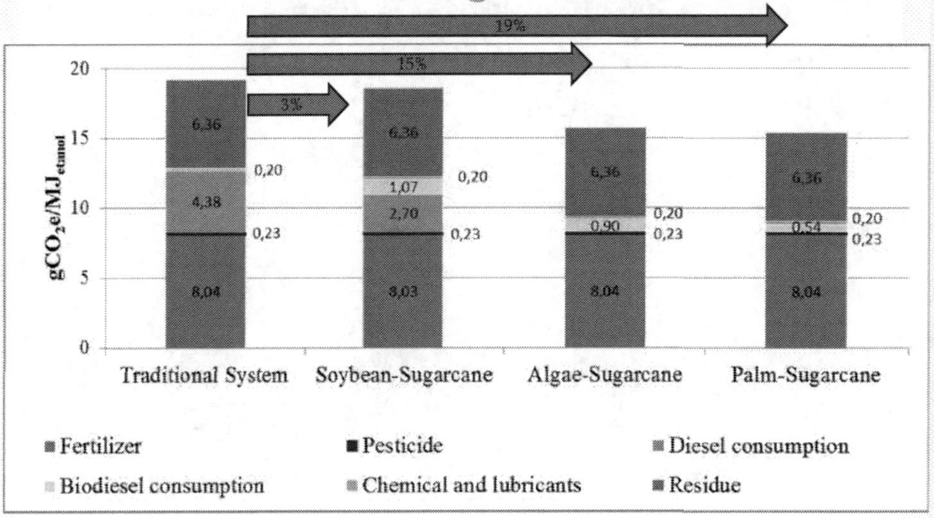

GHG emissions

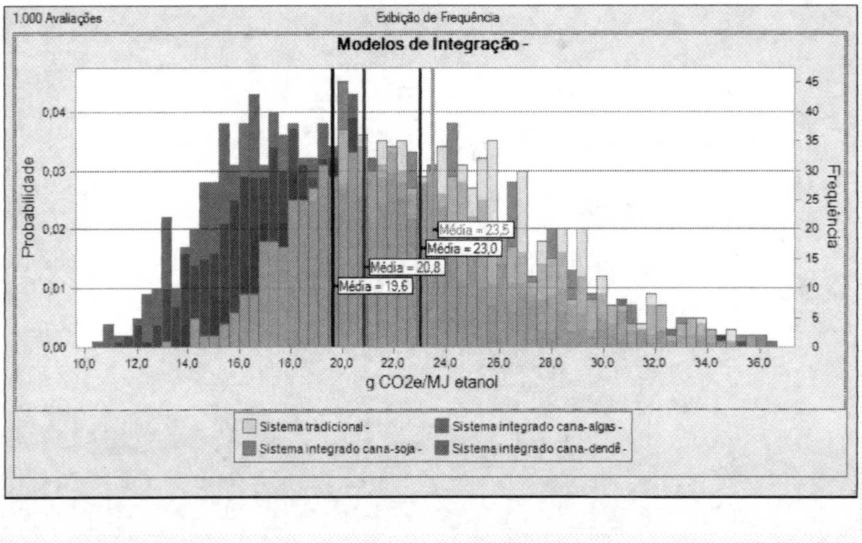

Final Remarks

- Algae-Cane and Palm-Cane integration are the best options in terms of environmental aspects.

HOWEVER...

- Algae technologies are still under development
- There are many uncertainties regarding palm cultivation under irrigated conditions
- Soybean is a well-established culture and in short-term it seems to be the best option to be integrated to the sugarcane sector.
- Other opportunity costs can discourage the diesel displacement. Ex: biodiesel auction, high oil price market, etc.

Takeaway message

- The use of ethyl route is an important aspect of the proposed integrated system
  - 80% of the biodiesel factories use methanol → US$ 250 million was spent in methanol importation in 2012
- Local consumption of biodiesel
  - Avoid logistic problems
  - Avoid long distance transportation of biodiesel
- Diesel imports
  - Brazil imported 8.1 billion liters of diesel, equivalent to 6.6 billion dollars
- Diesel consumption will increase 65% until 2030 [MME, 2007] → 8 million ha of soybean (B5)

Acknowledgment

SEE ALGAE Technology

AGROPALMA

UNICAMP

USP

CTBE
Centro de Ciência e Tecnologia do Bioetanol

BERKELEY LAB

CNPq
Conselho Nacional de Desenvolvimento Científico e Tecnológico

FAPESP

THANK YOU FOR YOUR ATTENTION!!!

Contact:
Simone P Souza
sp.souza@yahoo.com.br
sps@fem.unicamp.br

Direct and Indirect Impacts of Agricultural Land-Use Due to Increased Production of Grain-Based Ethanol in Kansas, USA: A County-Level Analysis

Bergtold, J., Kansas State University
Lauer, B.,
Originally presented on: 7/24/2014 10:18:00 - 10:36:00

Applied economists have dedicated much literature to the design and specification of acreage allocation models for land use decisions. These models can play an important role in understanding how acreages might shift in the event of new policy or in a changing agricultural landscape, especially with the increasing demand for biofuel production. Furthermore, the increase in ethanol production over the past two decades has impacted markets for corn and grain sorghum in Kansas, USA, especially with the prospect of grain sorghum being classified as an advanced biofuel feedstock. The increased production of the ethanol within Kansas has had an impact on local markets and the intensity of agricultural production in these areas. The presence of an ethanol plant can impact the prices of crops in the local market; the outlets for marketing grain by farmers and elevators; and the intensity of crop production in the vicinity of the plant. The direct changes in land-use production from a stronger ethanol industry in the state will likely lead to indirect changes in the use of agricultural land, as well.

The purpose of this paper is to examine the direct and indirect impact on agricultural land-use from increased ethanol production in Kansas.

The study utilizes acreage allocation models designed in the agricultural economics literature and builds on these previous acreage allocation studies by correcting for spatial autocorrelation; spatial proximity to ethanol refineries using capacity as a function of distance to a refinery as an explanatory factor in the model; and by incorporating changes in agricultural land-use in neighboring counties to capture potential indirect land-use changes. Bioenergy is an important agricultural topic to the state of Kansas, and understanding how the presence of ethanol plants effects producers' acreage decisions is critical as the bioenergy industry moves forward and new policies are developed. This study uses a 1996-2009 Kansas county level dataset for analysis. Variables that will be examined include shifts in acreage allocated to crops in neighboring counties, expected input prices, production costs, livestock populations, government programs, site and soil characteristics, weather, and ethanol plant characteristics such as proximity to a county and size of the plant. Models will be estimated for dryland corn, wheat, sorghum and soybean. The methods in this paper will follow methods proposed by Wu and Brorsen (1995) which extended the HEAR model proposed by Kmenta (1996) to a set of seemingly unrelated equations. These methods will be expanded to take account of spatial dependence.

Results indicate that own- and cross-price effects of crops and the capacity/proximity to an ethanol refinery have had significant impacts on the crop acreage allocated to the production of corn and grain sorghum. In addition, using the spatial lags of shifts in crop production in neighboring counties allows for the derviation of spillover marginal effects of these on within-county acreage allocation to help capture indirect effects on crop land allocation decisions. Results from the paper will help to assess the agricultural land-use impacts from the expansion of the ethanol markets in the U.S. and the impacts of U.S. bioenergy policies.

Activists Follow as Markets and States Transnationalize Northeast Brazilian Sugarcane, 1954-1989

Rodrigo G. Pinto, Ph.D.

Assistant Professor of International Relations
James Madison College of Public Affairs
Michigan State University

Outline: Activists Follow as Markets and States Transnationalize... Sugarcane...

- Introduction
- Theory on a defensive activist tactic
- Evidence on a 1954-89 campaign
- Cross-check of theory and evidence
  - Impact assessment: process-tracing causality
- Further research on a 1954-89 campaign
  - Theory & evidence on another tactic—radiation
- Conclusion

Introduction

- Historical legacies and patterns

Centro Josué de Castro

Introduction

- Historical legacies and patterns
- Mismatch—theory and evidence
  - State *and market* aspirin*s* impact organizing?
    - Cuban preview
    - Brazilian preview

Theory on a defensive activist tactic

- Defensive transnationalism or internalization

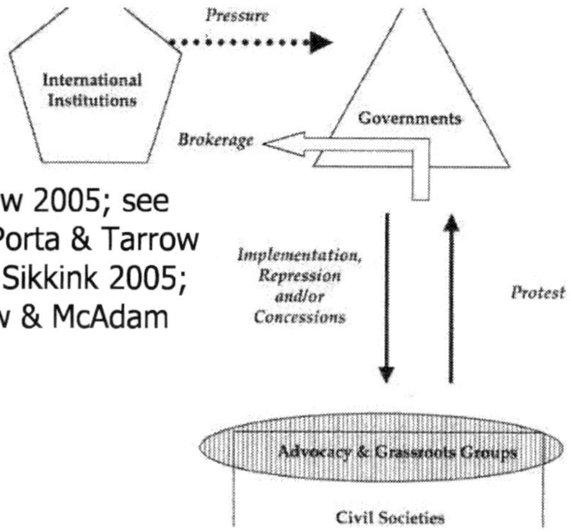

  - (Tarrow 2005; see della Porta & Tarrow 2005; Sikkink 2005; Tarrow & McAdam 2005)

Evidence on a 1954-89 campaign

- Campaign for the peasantry and rural labor on the sugarcane fields of Northeast Brazil, 1954-89
  - Rise in global sugar prices
  - UN FAO aids peasant forum
  - Cuban revolution and 'bay of pigs' intensify activism
  - Regimes de- then re-mobilize
  - Oil price, debt and ethanol rise

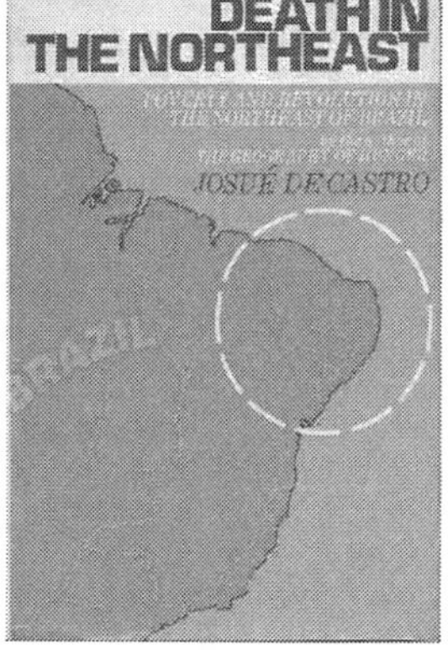

Cross-check of theory and evidence: Process-tracing market and state impact

- 1954-74
  - Pressure and implement—through sugar exports
  - Protest—through the Peasant Leagues, unions
  - Broker and concede—sugar quotas, premium
  - Repress and implement—thru de-democratization
- 1973-89
  - Pressure—through petroleum imports plus debt
  - Implement—through pro-ethanol program
  - Concede and protest—via cane-worker strikes

Cross-check of theory and evidence: Process-tracing market and state impact

- 1954-62: Pressure and implement

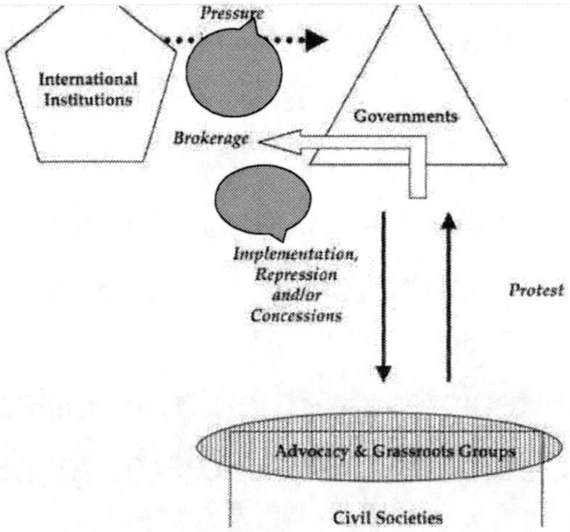

Cross-check of theory and evidence: Process-tracing market and state impact

- 1954-64: Protest

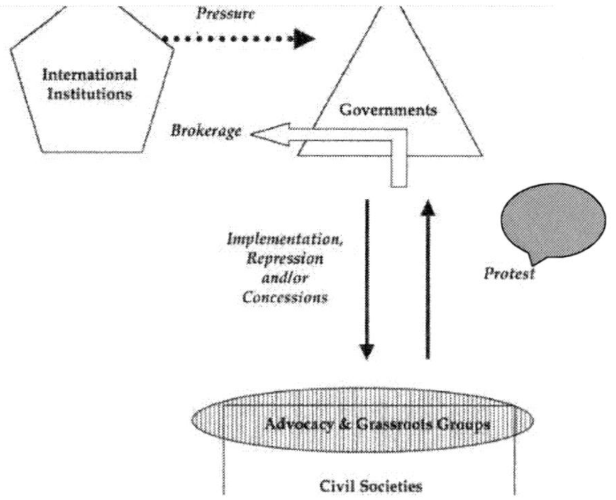

Cross-check of theory and evidence: Process-tracing market and state impact

- 1960-64: Broker and concede

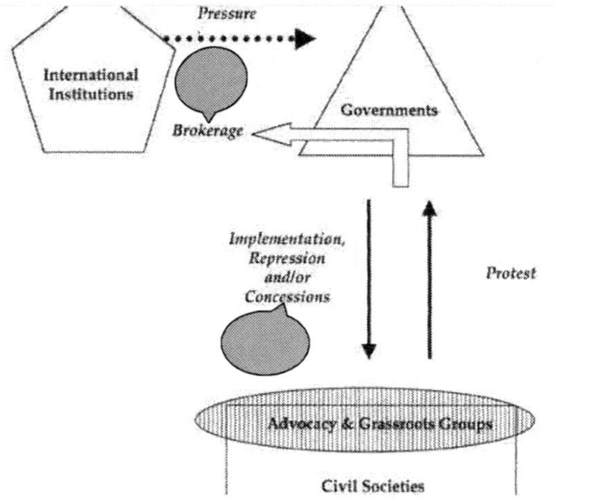

Cross-check of theory and evidence: Process-tracing market and state impact

- 1964-74: Repress and implement

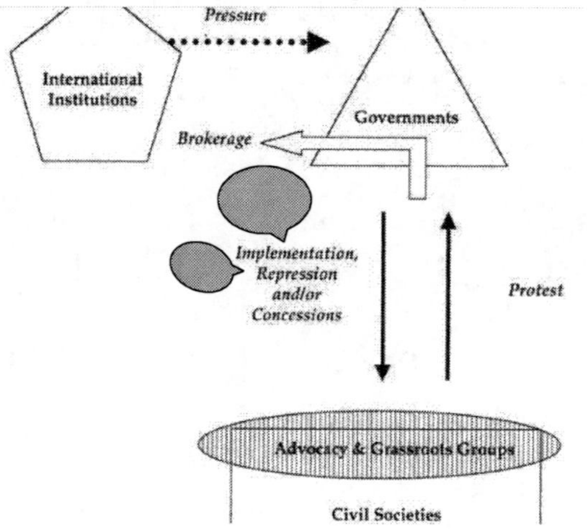

Cross-check of theory and evidence: Process-tracing market and state impact

- 1973-89: Pressure

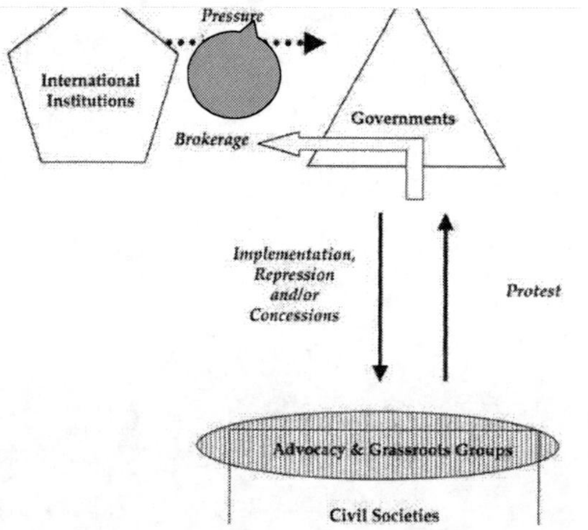

Cross-check of theory and evidence: Process-tracing market and state impact

- 1974-89: Implement

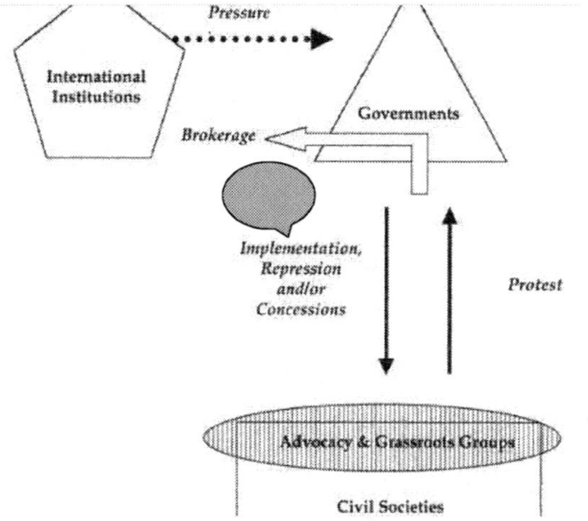

Cross-check of theory and evidence: Process-tracing market and state impact

- 1978-85: Concede and protest

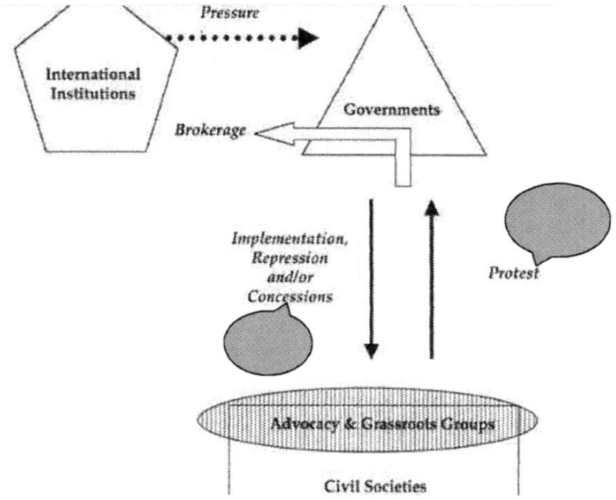

Further research on a 1954-89 campaign

- Theory on another tactic—radiation

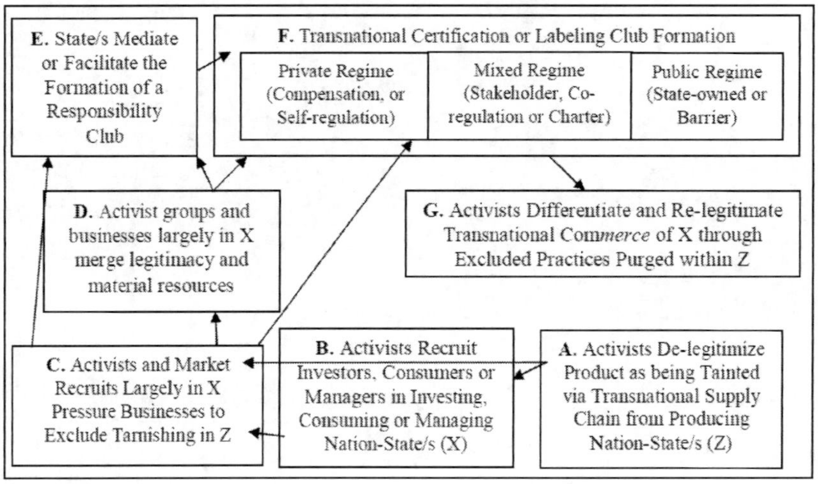

Further research on a 1954-89 campaign

- Cross-check of theory and evidence
  - Impact assessment
    - Process-tracing causality

Further research on a 1954-89 campaign

- Cross-check theory & evidence: mechanism A

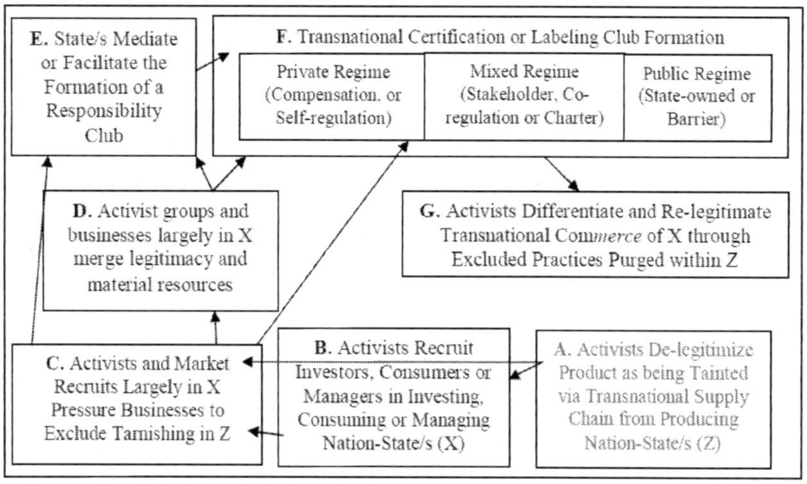

Further research on a 1954-89 campaign

- Cross-check theory & evidence: mechanism B

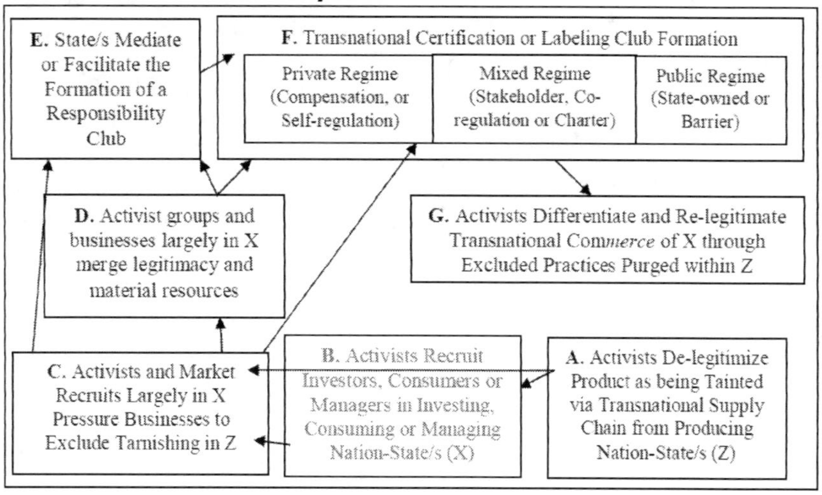

Further research on a 1954-89 campaign

- Cross-check theory & evidence: mechanism C

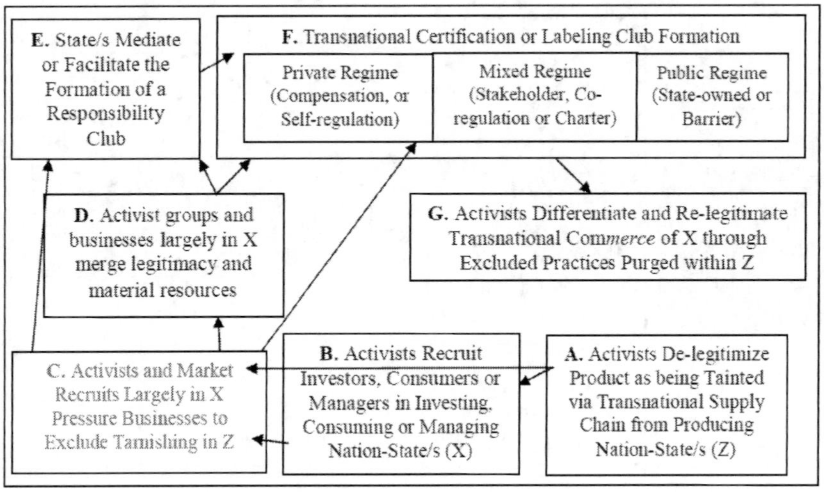

Further research on a 1954-89 campaign

- Cross-check theory & evidence: mechanism D

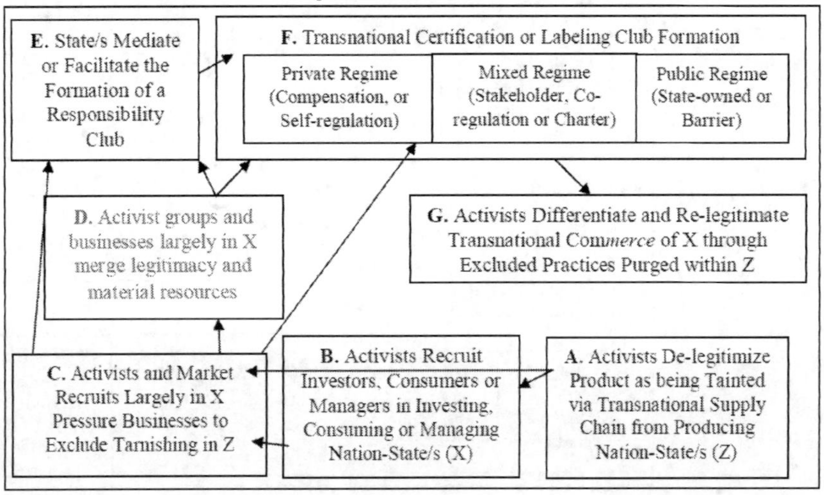

Further research on a 1954-89 campaign

- Cross-check theory & evidence: mechanism E

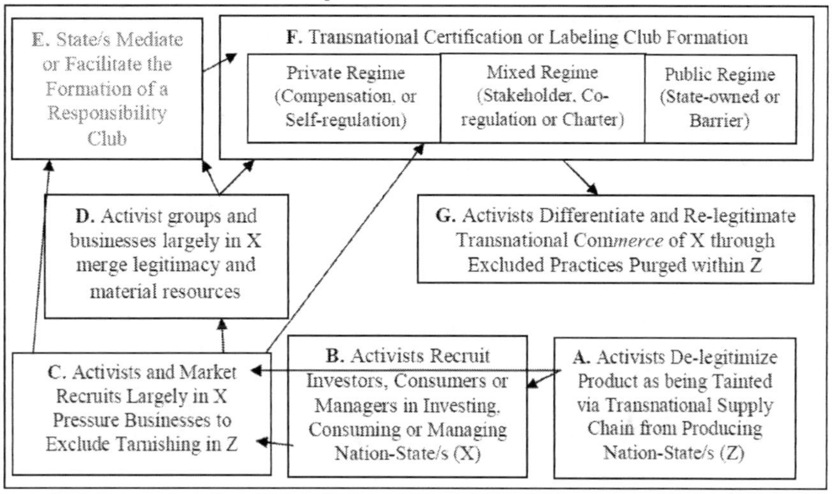

Further research on a 1954-89 campaign

- Cross-check theory & evidence: mechanism F

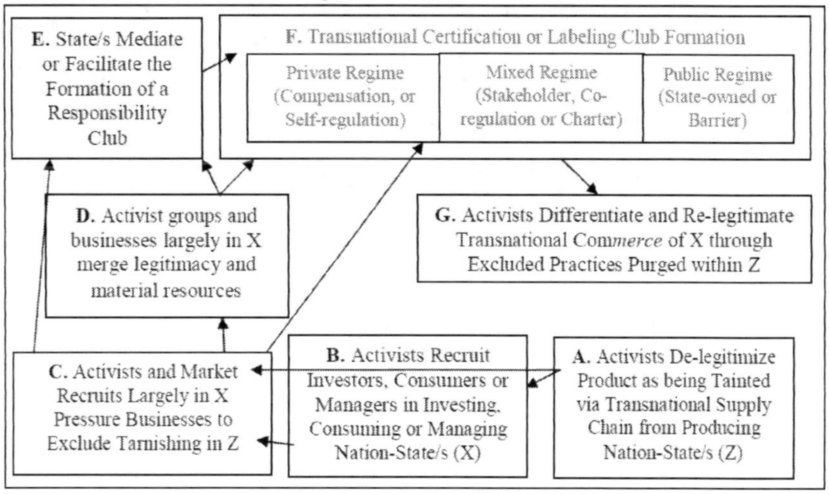

Further research on a 1954-89 campaign

- Cross-check theory & evidence: mechanism G

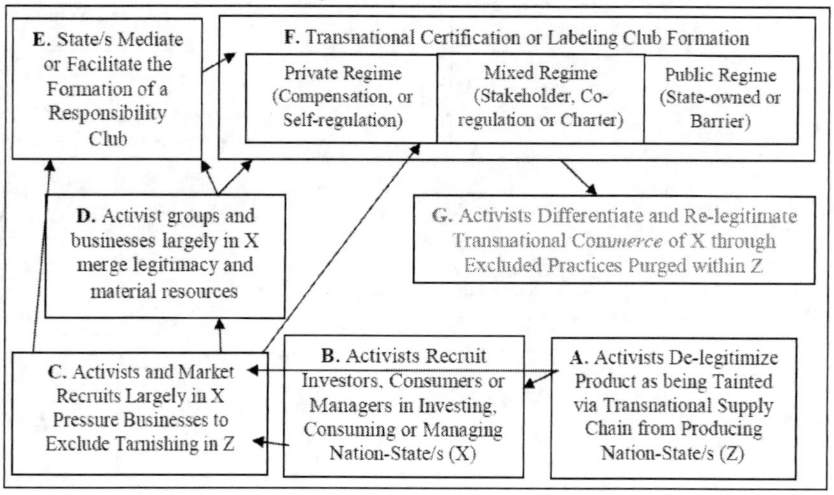

Conclusion: Activists Follow as Markets and States Transnationalize… Sugarcane…

- Conclusion
  - (Bioenergy) markets and states positively impact transnational (bioenergy) activism
    - The more markets globalize and the more states interact, the more activism occurs in transnational (bioenergy) relations

Outline: Activists Follow as Markets and States Transnationalize... Sugarcane...

- Introduction
- Theory on a defensive activist tactic
- Evidence on a 1954-89 campaign
- Cross-check of theory and evidence
  - Impact assessment: process-tracing causality
- Further research on a 1954-89 campaign
  - Theory & evidence on another tactic—radiation
- Conclusion

Activists Follow as Markets and States Transnationalize Northeast Brazilian Sugarcane, 1954-1989

Thank you,

Rodrigo G. Pinto
pintor[at]msu.edu

Sustainability of ethanol production in Mexico

Carlos A. García, Fabio Manzini

RCN Conference on Pan American Biofuels & Bioenergy Sustainability

Recife, Brazil. July 24, 2014.

Gasoline and diesel imports in Mexico

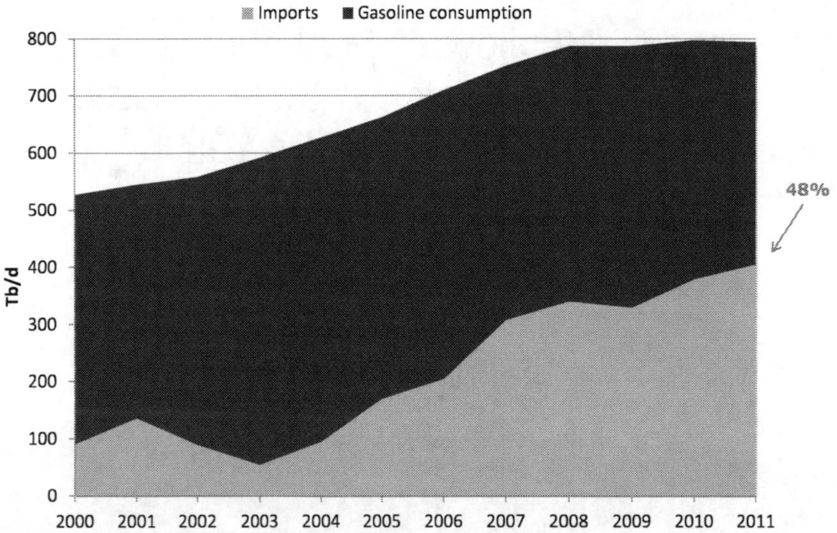

Source: Energy Information System. SENER.

2

Gasoline and diesel imports in Mexico

Potential for first-generation ethanol production in Mexico.

Energy crops	Mha	Yield t/ha	Production ML	Production PJ	% Gasoline consumption (2011)
Molasses (C molasses)	1.664 **a		391[a]	8	0.6%
Sugarcane	2.9*	40-120*	14449*	338*	22.5%
Corn	4.5*	0.8-2.5*	2028*	44*	2.9%
Sorghum	2.6*	2.2-4.2*	3900*	84*	5.6%
Total			14523	460	31.6%

Source: García et al. 2014

3

Ethanol Modalities

Crop	*Ethanol source /Technology*	*Water source*	*Fertilization kg/(ha*Y)*	*Yield (t/ha)*	*Energy source for etOH*	*Products / co-products*
SUGARCANE	**C Molasses**	**Rainwater**	**126:41:150**	**70**	**Bagasse**	**Sugar/Anhydrous Ethanol**
SUGARCANE	**Juice**	**Rainwater**	**126:41:150**	**70**	**Bagasse**	**Anhydrous Ethanol/Electricity**
SORGHUM MEDIUM	**Dry mill**	**Rainwater**	**0**	**2.5**	**Natural Gas**	**Anhydrous Ethanol/DDGS**
SORGHUM HIGH	**Dry mill**	**Irrigation**	**189:70:00**	**5.4**	**Natural Gas**	**Anhydrous Ethanol/DDGS**
CORN LOW	**Dry mill**	**Rainwater**	**0**	**1**	**Natural Gas**	**Anhydrous Ethanol/DDGS**
CORN MEDIUM	**Dry mill**	**Irrigation**	**210:00:00**	**5**	**Natural Gas**	**Anhydrous Ethanol/DDGS**
CORN HIGH	**Dry mill**	**Irrigation**	**263:161:00**	**9**	**Natural Gas**	**Anhydrous Ethanol/DDGS**

4

Indicators

Energy Indicator (Ie)
(EROEI) Ethanol E/ Fossil E (GJ)

Environmental Indicator (Ia)
kg CO_2e/GJ

Water use Indicator (Iua)
m_3/GJ

Economic Indicator (Iec)
Production cost USD/L

Social Indicator (Is)
Employments/TJ

Land use indicator (Ius)
ha/GJ

5

Methodology (1)

- For air and energy indicators (Ia, Ie), EU Dir-RES was applied. Frontiers of the system

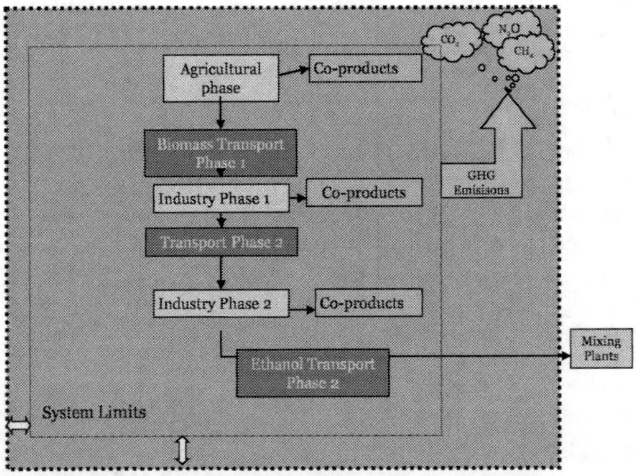

- Emission Allocations by energy content.

6

Methodology (2)

- Water Use (Iua): irrigation water (blue water) calculation using CROPWAT software and its weather database CLIMWAT. Water allocated by energy content. (m3/GJ)

- Land Use (Ius):(crop yield X factory yield X ethanol LHV)$^{-1}$ (ha/GJ)

7

Methodology (3)

- Economic Indicator (Iec): Net Present Value calculation. Input biomass production costs were considered, in the industrial phase: investment, O&M costs, fuel costs, electricity costs. Benefits from co-product sales. (USD/L)

- Social Indicator (Is): direct hours of employment were calculated for biomass production, transport, and industrial transformation. 1 job = 1800 hours/year. (Jobs/TJ)

8

Ia Life cycle assessment of greenhouse gas emissions

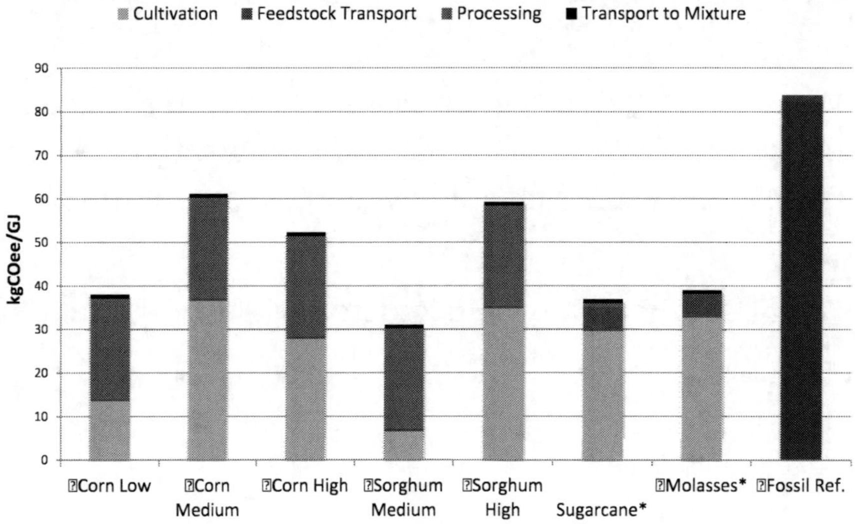

Source: García et al. 2014

9

Ie (EROEI)

	Molasses	*Sugarcane*	*Corn Low*	*Corn Medium*	*Corn High*	*Sorghum Medium*	*Sorghum High*
GJ ethanol/ GJ fossil	**4.4**	**4.8**	**2.7**	**2.1**	**2.2**	**2.9**	**2.1**

10

Iua (Water consumption)

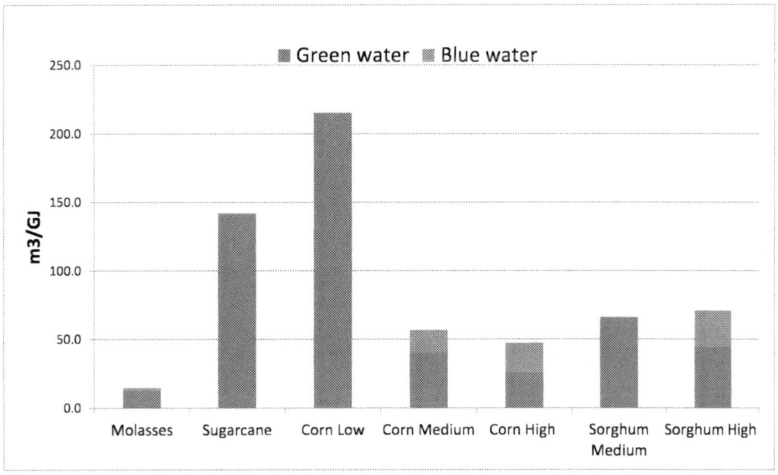

11

Ius (Land Use)

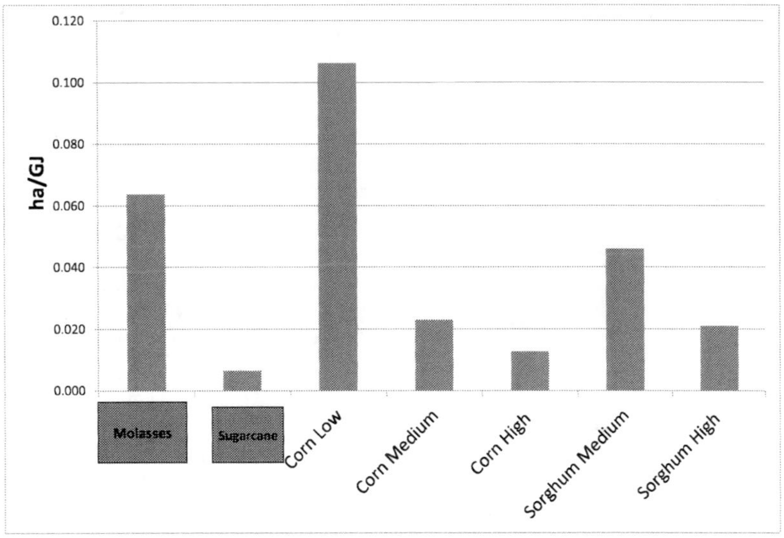

12

Iec (Production Costs)

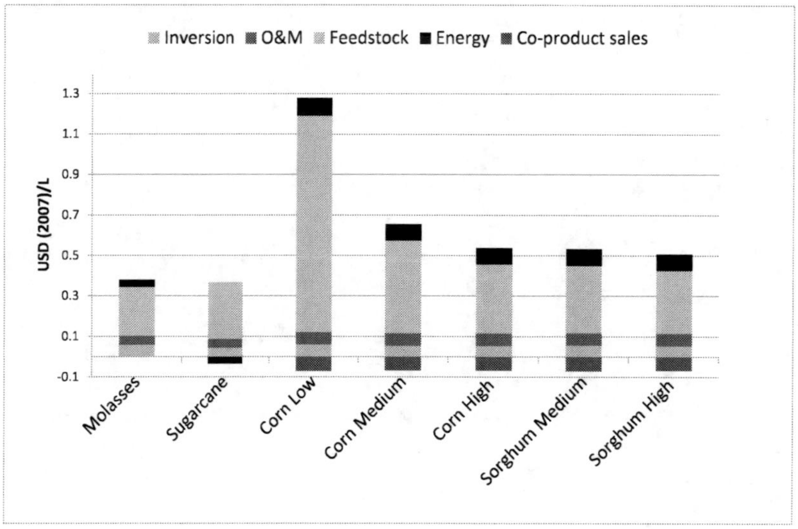

13

Is (Jobs Generated)

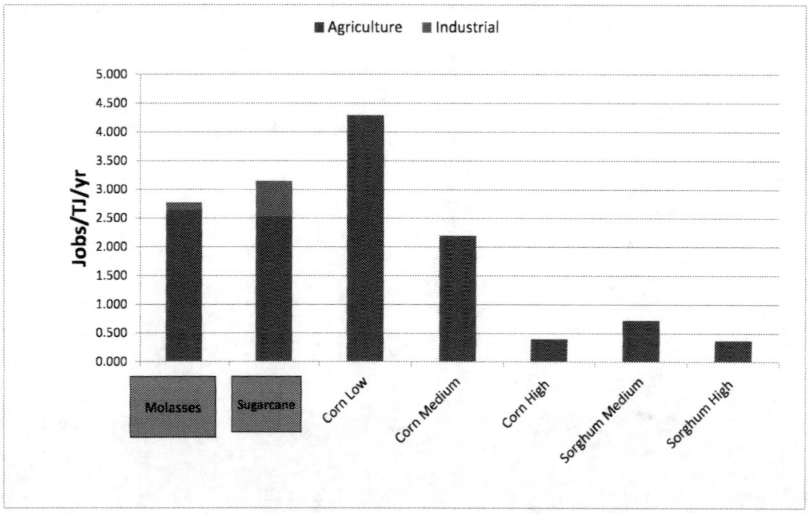

14

Indicators

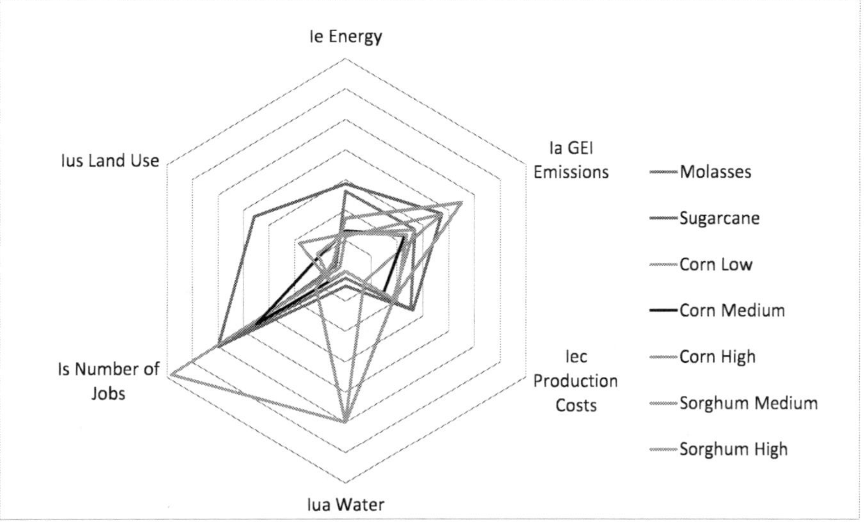

The greater the area covered by each of the options the more sustainable it is.

15

Sustainability Index (Results)

$$I_{SUS} = \sum_{c} w_c x_c$$

	Ia	Iec	Is	Ie	Iua	Ius
Weighting factor *w*	23%	22%	17%	15%	12%	11%

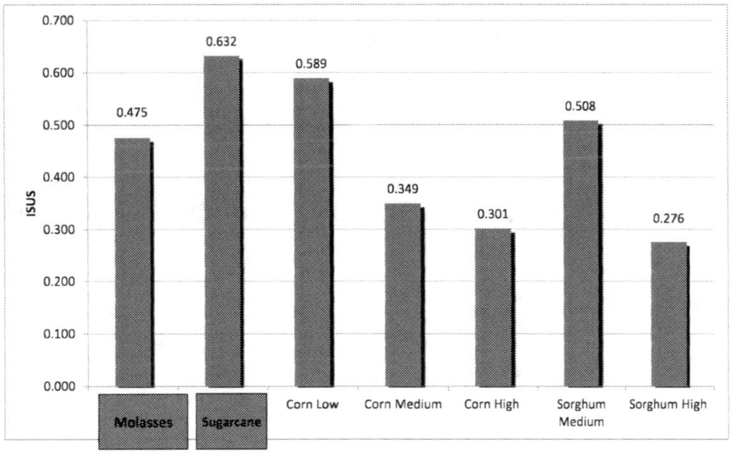

16

Conclusions

- Mexico needs to decrease its gasoline and diesel consumption and imports.
- Mexico has limited land resources, but there is an important potential for biofuel feedstocks, especially sugarcane for ethanol production.
- Ethanol from sugarcane comes out as the best (most sustainable) option for Mexico.

17

Conclusions

- To improve ethanol sustainability it is necesary to:

- Improve agricultural practices: increase fertilization efficiency, decrease diesel use in field vehicles and machines, increase crop yields, avoid or minimize irrigation.
- Avoid fossil fuels for energy generation in industrial production.
- Promote co-products.

18

Thank you!

carloscieco@gmail.com

Industrial processes of ethanol production using sugarcane

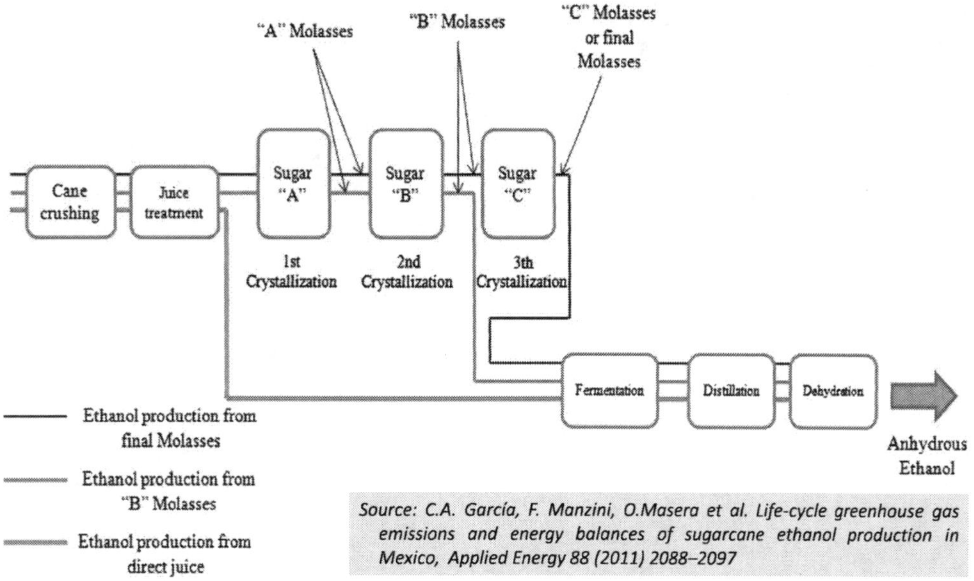

Source: C.A. García, F. Manzini, O.Masera et al. Life-cycle greenhouse gas emissions and energy balances of sugarcane ethanol production in Mexico, Applied Energy 88 (2011) 2088–2097

Ia (emissions with direct LUC)

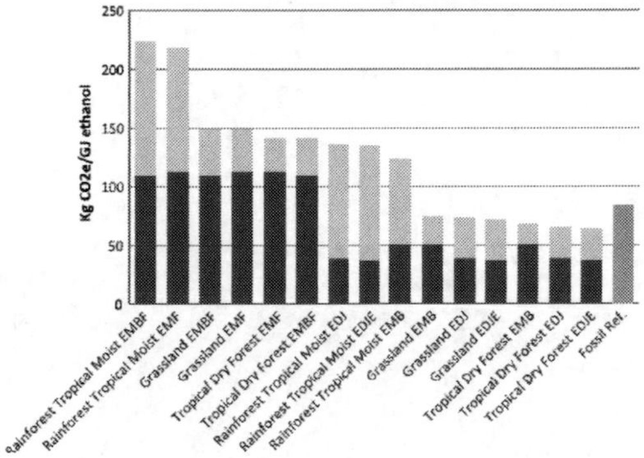

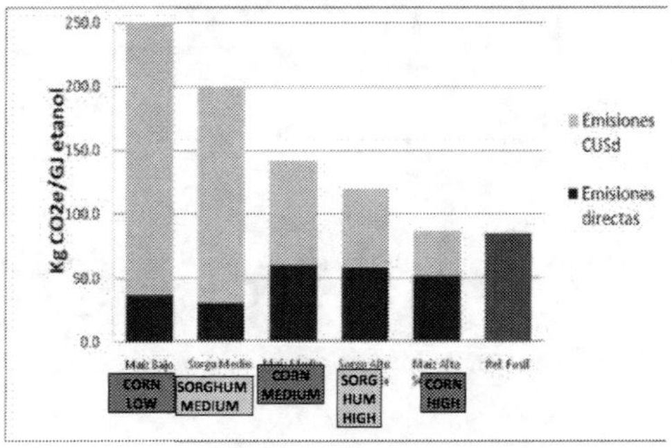

EMF: Ethanol C Molasses Fueloil
EMBF: Ethanol B Molasses Fueloil
EMB: Ethanol C Molasses Bagasse
EDJ: Ethanol Direct Juice
EDJE: Ethanol Direct Juice +Electricity

RCN CONFERENCE ON PAN AMERICAN BIOFUELS AND BIOENERGY SUSTAINABILITY

BIOFUEL IN BRAZIL: SOYBEAN AND PALM OIL IMPACTS ON SOCIOECOLOGICAL SYSTEMS

MSc. candidate Fernanda Martinelli
Prof. Dr. Rodrigo Medeiros
MSc. candidate Tatiana Martins

Presentation Overview

- Biodiesel in Brazil
- Strategies and challenges for agro energy production
- Soybean and palm oil production impacts
- Conclusions

Biodiesel production in Brazil

- Soybean: 70% of biodiesel production
- Large scale biodiesel production
- Soybean expansion
  - Started in 1970
  - Drive by global market forces
  - Government actions: subsides and investments in transportation infrastructure
- World second biggest producer
- Economic benefits vs. Environmental impacts
  - Cerrado and Amazon rainforest replacements

Energy and develpoment

- Political measures to promote rural, social and environmental development
- Palm oil as a major contributor
- Production area doubled during 2001-2009
  - 46 to 109 thousand ha
- Projected scenarios predict major expansion in the next few years

Strategies to avoid the same impacts of soybean expansion

- Palm Oil Agroecological Zoning
  - Only to degraded lands without environmental legal restrictions
  - Removal of native vegetation is forbidden
- Social Fuel Seal
  - Partnerships with small farmers by contractual agreements
  - Tax benefits and priority in biodiesel auctions
- Important measures but targets might not be achieved
- Land use changes challenges and unanticipated effects
- Challenges
  - Remoteness for supervision of palm production poles
  - Displacement of farmers who do not accept palm crop
  - Environmental consequences for communities regardless indirect and illegal direct deforestation

Objectives

- Comparison of soybean and palm oil production impacts
  - Biodiversity
  - Ecosystem services
  - Local community
- Evaluation
  - Land use change
  - Agrochemicals use
  - Soil, Water use and GHG emissions
  - Job creation and inclusive growth
- Indicators assessed in a literature-based inventory of the sustainability of major biofuel production systems, based on current certification and production practices (RSB, RSPO, RTRS) in major production areas.

Land conversion: direct and indirect LUC

- Native forest replacement
  - Lack of accurate data on land cover change
  - Incomplete understanding of its complex causes
- Timeline
  - Soybean - 70's
  - Palm oil - 90's
- But our intention is to verify what type of plantation can be **more productive** and **use less land extension** starting from now
- The scenario of **soybean-based biodiesel** production to meet future global biodiesel demand would likely result in the **highest amount of habitat loss compared with alternative scenarios of oil palm-based biodiesel production**

Land conversion: direct and indirect LUC

- Yield productivity for biofuel
  - Palm oil ranks the highest compared to soy
  - Soybean occupies 220 times more land to produce only 57 times more oil than palm.
- On the other hand
  - Infrastructure for production and transportation of soybean is well consolidated
  - Palm oil has increasing faster than for soybean which may drive indirect land use change.
- Indirect land use change
  - Food crops farmers
  - Cattle raising

Agrochemicals use

- The runoff from large use of agrochemicals
- Impacts aquatic biodiversity, rivers and “igarapés” water quality and the health of local communities that lives and cultivates food nearby
- In all agrochemical indicators, palm oil crops present the least amount of input needed and also the least amount of runoff to soil and water
- Less amounts of agrochemicals to the crops reflects in less harm on Amazon and Cerrado ecosystems and on human health

Soil, Water and GHG Emissions

- Soil erosion is a serious problem due its irreversibility and the fact that threats the agriculture productivity by losing plant nutrients and organic matter
- Based on a case study comparing soybean in Brazil and palm oil in Malaysia, palm crops promote more loss of soil than soybean.
- Volume of surface, ground water and rainwater consumed.
- Results from Thailand:
  - Water footprint is similar
- Fossil energy balance
  - Palm oil crops would emit less carbon than soybean
  - 9 times the amount of energy required for its production

Job creation and inclusive growth

- Job creation
- Rural development by family farming inclusion
- Job generation
  - Soybean = 2 workers per 100 ha
  - Palm oil = 1 worker per 10 ha
- Palm oil and income security
  - Has high yield potential and lower production cost than the other major oilseeds
  - Is a permanent crop and requires less labor-intensive which permits the development of other agricultural activities

Conclusions

- Government incentives and increasing energy demands make the expansion of biofuels production in Brazil a virtual certainty
- Although many questions remain, the results of our study suggest that an increase in soybean-based biodiesel production could have greater negative social and environmental effects than those of palm oil crops in Brazil
- From **17 indicators** evaluated, **palm oil was better on 12** regarding biodiversity, ecosystem services and local community impacts
- Soybean production, however, is still the most profitable oil crop and the only one available for large-scale biodiesel production
- Although we used the most recent available data to build this framework, new technologies and best practices might change the indicators results for each case and improve biofuel production
- Crop management practices aimed at increasing yields and/or minimizing ecological costs will further influence the role biofuels play on ecosystem and biodiversity, and partnerships with local smallholders might bring several community benefits.

Indicators	Soybean		Oil Palm		Source / Reference
LUC and ILUC					
Habitat loss (million ha)	36.4–114.2		0.4–5.4		Koh, 2007
Yield productivity for biofuel (t/ha)	0.36		3.68		Basiron and Foong-Kheong, 2013; Basiron and Kook Weng, 2003; Costa e Santos (2008)
Land use in Brazil (thousand ha)	24000		109		IBGE 2011
Average yield (thousand kg/ha)	3.1		12		IBGE, 2011
Area Harvested Growth Rate (2000-2012)	Cumulative	Geometric	Cumulative	Geometric	FAO, 2013
	0.83	0.05	1.42	0.08	
Agrochemicals use (kg)					
Nitrogen input	315		47		FAO, 2013
Phosphorus input	77		8		FAO, 2013
Pesticide/herbicide input	29		2		FAO, 2013
Nitrogen output	32		5		FAO, 2013
Phosphorus output	23		2		FAO, 2013
Pesticide/herbicide output	23		0.4		FAO, 2013
Soil, Water and GHG Emissions					
Fossil energy balance	1 - 4		9		FAO, 2008
Water footprint (m^3/T)	1,875		965 - 2,353		Kaenchan and Gheewala, 2013
Soil loss (t/ha/year)	8		14		Mattsson et al, 2009
Carbon emissions (kg CO_2/t biodiesel produced)	1,387		1711		Zutphen et al, 2011
Job Creation					
Work force (worker/ha)	2/100		1/10		FAO, 2013 Reporter Brasil, 2008

THANK YOU!

Tatiana Martins Souza
tsouza@conservation.org

Fernanda Martinelli - fmartinelli@conservation.org

Process Integration Case Study for the Second Generation Ethanol Production

PanAm RCN Bioenergy Sustainability Conference
July 22-25, 2014
Paulo Brito - Granbio
Sophia Rouzinou, Vesa Pylkkanen – American Process Inc

Objectives

- Increase the ethanol production by integrating 1st and 2nd generation plants with energy saving, without any additional biomass

Methodology

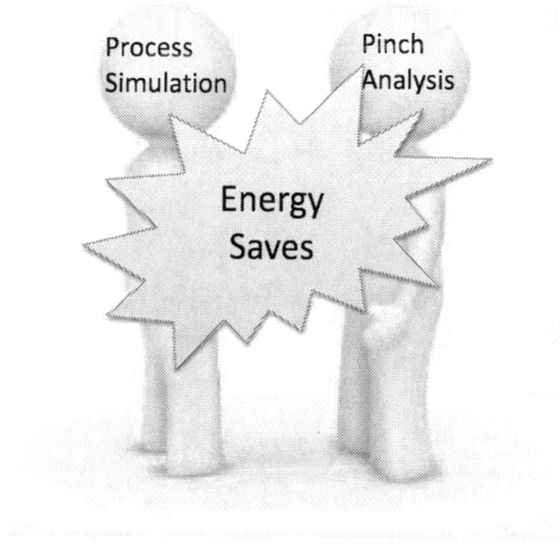

- Integration of 1st and 2nd Generation plants
- Increase ethanol production
- Improve process efficiency

Applications of Process Integration

Pinch Analysis is integral part of Process Integration, when

- Integrating heat needs with waste heat
- Generating steam/heat "appropriately",
- Reusing / recovering raw materials,
- Integrating many processes / Co-production / Technology synergies,
- Upgrading byproducts

Principle of Pinch Analysis

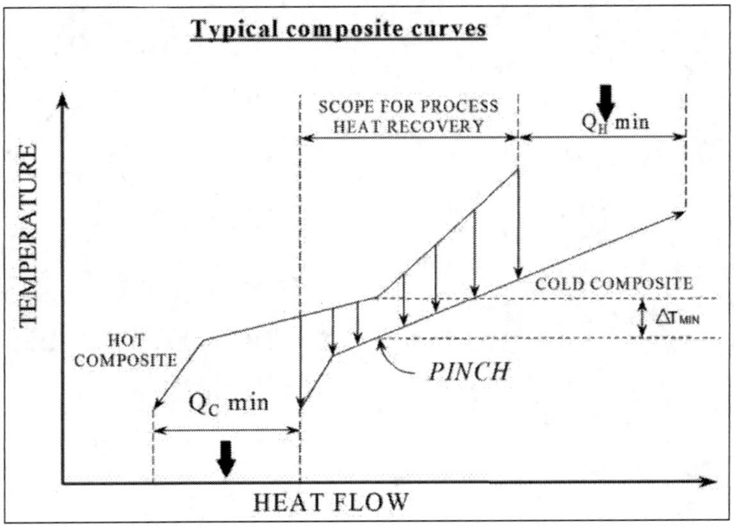

Pinch Analysis Example

First Generation Sugar Cane Ethanol Plant
3 million tons crush during 180-day harvest

Sugarcane Ethanol Plant

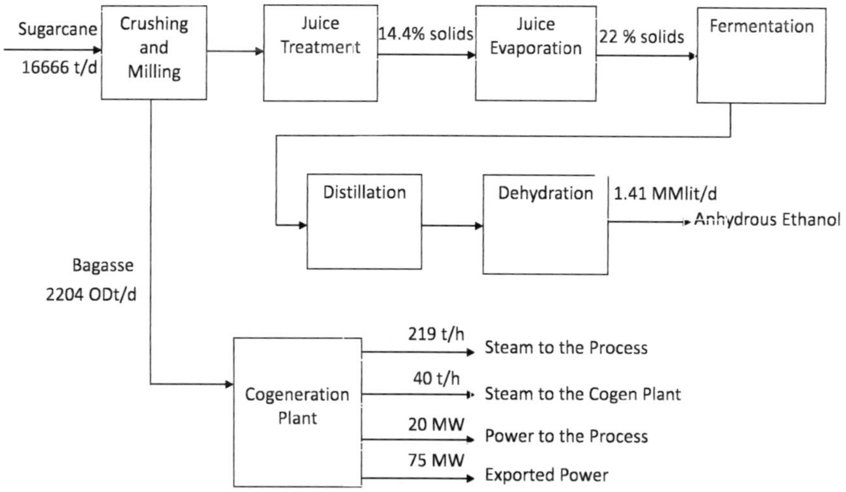

Sugarcane Plant Main Heat Duties

Description	Heat demand, GJ/h	Steam consumption, GJ/h
Juice preheating before liming	87	24
Juice heating after liming	89	76
Imbibition and wash water heating	38	0
Juice evaporation	141	141
Wine preheating (to distillation)	79	0
Distillation (stripper column)	161	161
Distillation (rectifier column)	79	79
Molecular Sieves heating	7	3
Boiler Feed Water heating	41	76
Totals	**723**	**560**

Sugarcane Plant Composite Curves

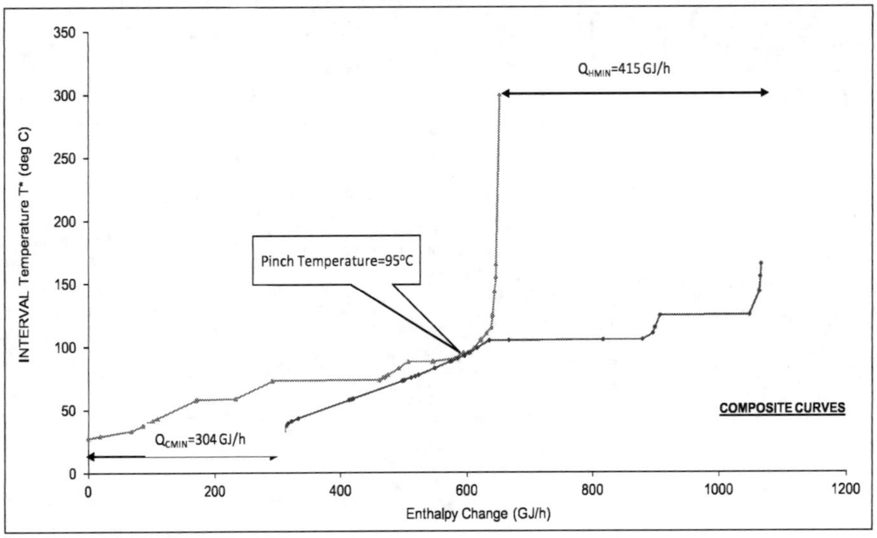

Sugarcane Plant Pinch Results

• Pinch Temperature	95 °C
• Minimum ΔT selected	10 °C
• Minimum heating utility(QHMIN)	415 GJ/h
• Actual process steam consumption	554 GJ/h
• Maximum scope for heat recovery	**139 GJ/h**

Second Generation Ethanol Plant

Cellulosic Ethanol and Acetic Acid
from Bagasse Feedstock

Second Generation Ethanol Plant

- American Process has developed hemicellulose extraction process termed "Green Power +™"
  - Liquid Hot Water digestion of Bagasse
  - Hemicelluloses are acid hydrolyzed to Pentose sugars and Acetic Acid
  - Acetic Acid is separated by evaporation & purified
  - The sugars are fermented to beer
  - The beer is distilled to produce ethanol
  - Unreacted solids material are concentrated and burned in the boiler.

Green Power+™ 2nd Gen. Plant

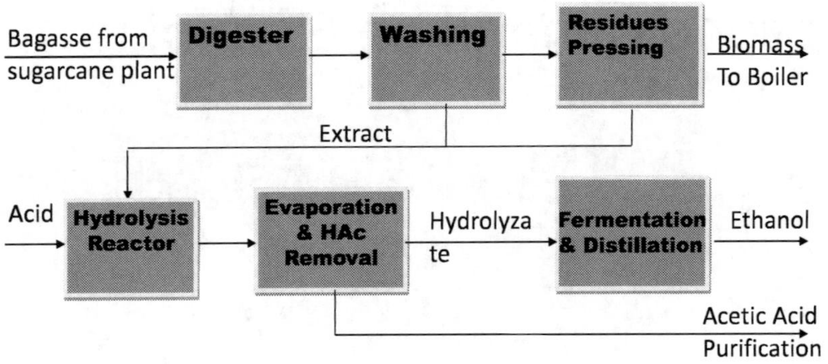

GP+ Plant Main Heat Duties

Description	Heat Demand, GJ/h	Power demand, MW
Biomass presteaming	18	
GP+ Digester heating	167	7.4
Biomass wash water heating	36	
Hydrolysis reactor heating	37	
Sugar solution evaporation (MVR)	33	5.8
Beer preheating	26	
Distillation (beer column)	40	
Distillation (rectifier column)	40	
Acetic Acid plant (MVR)	38	3.6
Totals	**436**	**16.8**

The sugar solution evaporation and concentration of acetic acid is performed in Mechanical Vapor Compression (MVR) Evaporator

GP+ Plant Pinch Results

- Pinch Temperature 101 °C
- Minimum ΔT selected 10 °C

- Minimum heating utility(QHMIN) 103 GJ/h

Combined 1st & 2nd Generation Plants

	Host sugarcane ethanol plant	Host ethanol plant combined with GP+
Sugarcane crush, t/d	16666	16666
Anhydrous ethanol production, MMlit/d	1.41	1.65
Acetic Acid production, t/d	-	123
Bagasse generated from process, ODt/d	2203	2203
Steam generated from Bagasse, GJ/h	1183	791
Steam generated from Bagasse, t/h	414	292
Specific steam generation from bagasse, t/t	2.16	2.61
Steam consumed in the plant, t/h	259	259+47 (at least) =306
Steam consumed in the process, t/h	219	219+47 (at least)=266
Power consumption, MW	20	37

The steam consumption is more than available from leftover bagasse solids!

Pinch Analysis for the Integrated plant

First Generation Sugarcane Ethanol Plant with Green Power+ Bagasse Extraction to Cellulosic Ethanol and Acetic Acid

PanAm RCN Bioenergy Sustainability Conference

Composite Curves for Integrated Plant

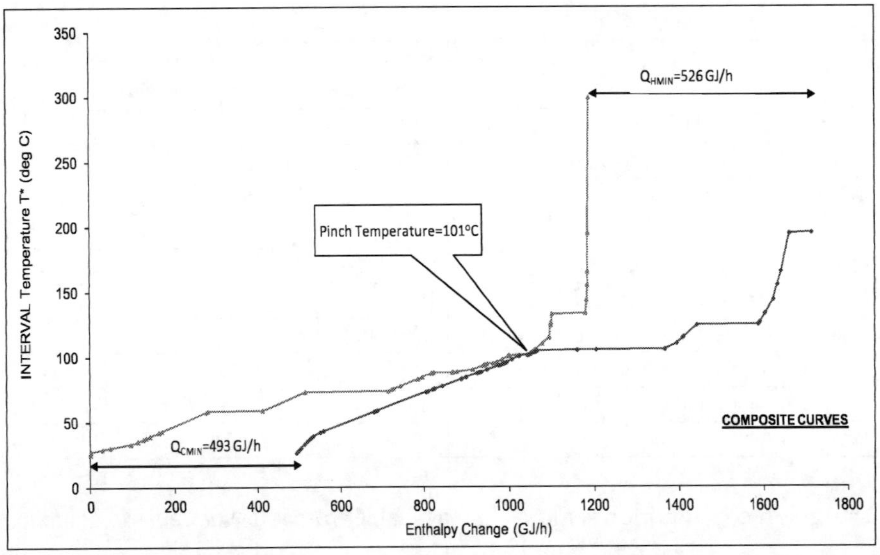

PanAm RCN Bioenergy Sustainability Conference

Integrated Plant Pinch Results

- Pinch Temperature 101 °C
- Minimum ΔT selected 10 °C

- Minimum heating utility(QHMIN) 526 GJ/h

Grant Composite Curve

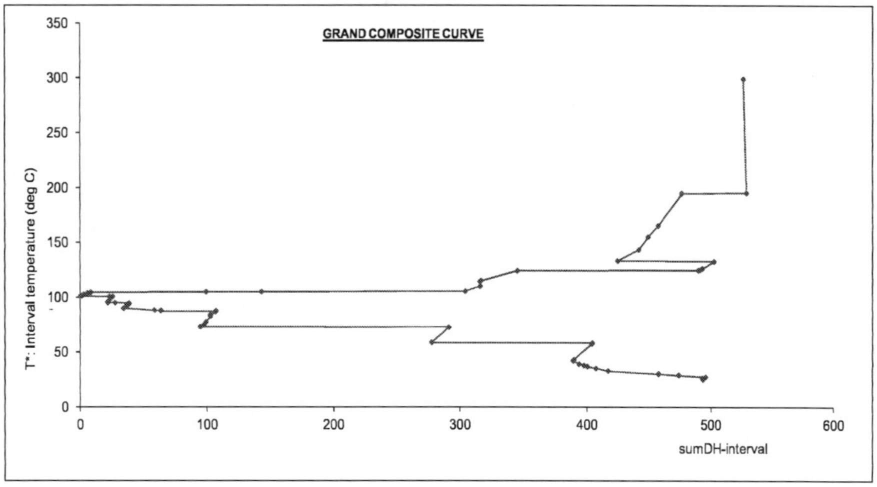

Comparison of Performance before and after Integration of 2nd Gen. Ethanol

	Host sugarcane ethanol plant	Integrated plants with 2nd generation ethanol
Sugarcane crush, t/d	16656	16656
Anhydrous ethanol production, MMlit/d	1.41	1.65
Acetic Acid production, t/d	-	123
Bagasse generated from process, ODt/d	2203	2203
Steam generated from Bagasse, GJ/h	1183	791
Steam generated from Bagasse, t/h	414	292
Specific steam generation from bagasse, t/t	2.16	2.61
Steam consumed in the plant, t/h	259	244 (306)
Steam consumed in the process, t/h	219	233 (266)
Steam consumed in the cogeneration plant, t/h	40	11
Process steam consumption per ton of cane, kg/t	316	336
Steam to the turbine condenser, t/h	158	51
Total Power generation, MW	95	54
Power consumption, MW	20	29 (37)
Net exported power, MW	75	25 (17)

PanAm RCN Bioenergy Sustainability Conference

Simulation Model in apiMAX™

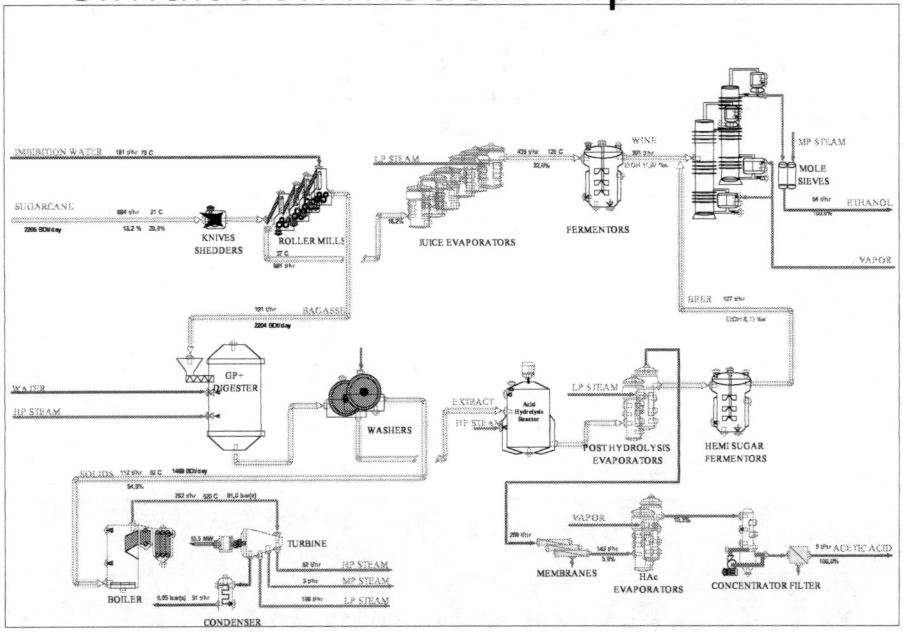

Results Summary

- Pinch Analysis was applied to stand-alone sugarcane ethanol mill to find **25% steam energy savings**
- The integration resulted in **17% increase in ethanol** and by-product acetic acid production **without any additional biomass**

Conclusions

- Combination of two optimized plants can be improved by process integration
- Reducing steam consumption allows leveraging of existing boiler and turbines
- Common distillation system can be used for 17% incremental ethanol production
- Additional cellulosic ethanol can be made from bagasse without bringing additional biomass to the site
- Techno-economic evaluation is necessary to calculate trade-off with power sales and the investment payback

RCN CONFERENCE
PAN AMERICAN BIOFUELS & BIOENERGY SUSTAINABILITY

A PRODUÇÃO DE ELETRICIDADE ATRAVÉS DE MICROALGAS: Uma atividade em conjunto com as usinas de etanol do país

M.Sc. Laryssa Ramos de Holanda – Doutoranda PPGEP - UFPE

Docteur. Francisco de Sousa Ramos – Professor Adjunto - UFPE

RECIFE, JULHO/2014

1

Introdução | Crescimento Econômico e Demanda por Eletricidade | Microalgas como Fonte de Energia | Processo Produtivo | Metodologia | Resultados | Conclusão

INTRODUÇÃO

Problemática

- Crescimento econômico ⟺ consumo de energia elétrica.

- Crescimento do PIB > Crescimento da OIE;

- Matriz energética concentrada na fonte hidráulica;

- Os últimos apagões e o apagão de 2001.

2

INTRODUÇÃO

Problemática

- Entre tantas fontes de energia surge também a possibilidade de produzir energia através da combustão da biomassa de microalgas.

3

INTRODUÇÃO

Justificativa

- A adoção dessa nova fonte de energia soluciona problemas como:
  - Dependência energética;
  - Aumento da demanda de energia;
  - Efeito estufa, mudanças climáticas e degradação do meio ambiente.

4

AS MICROALGAS COMO FONTE DE ENERGIA

- Algas:
  - Macroalgas – visíveis a olho nú;
  - Microalgas:
    - Microrganismos unicelulares algais que têm o corpo formado por um talo;
    - Possuem clorofilas e/ou outros pigmentos fotossintéticos;
    - São a forma mais primitiva das plantas unicelulares;
    - Podem ser encontrados em meio marinho, água doce e no solo.

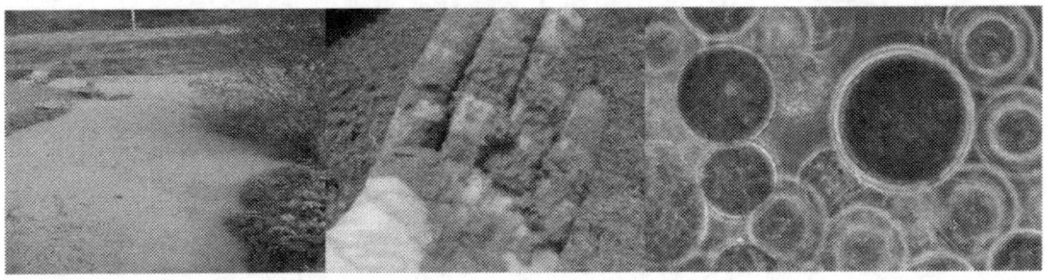

Introdução Crescimento Econômico e Demanda por Eletricidade Microalgas como Fonte de Energia Processo Produtivo Metodologia Resultados Conclusão

AS MICROALGAS COMO FONTE DE ENERGIA

- Alimentam-se com dióxido do carbono, e com utilização da luz solar transformam o carbono em açúcares e posteriormente em gorduras. No processo liberam oxigênio, igual a todas as outras plantas, mas de forma mais eficiente.

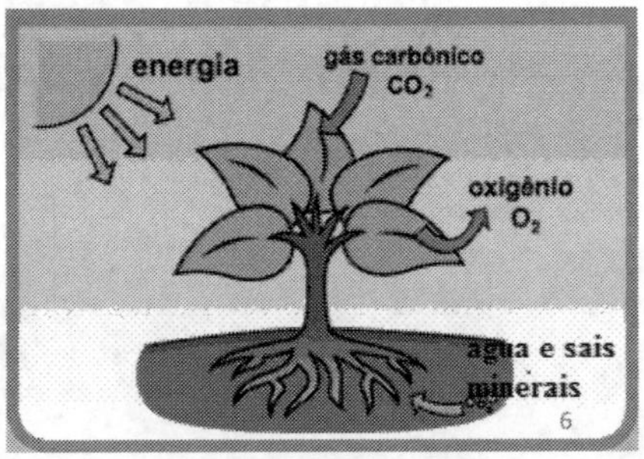

6

Introdução Crescimento Econômico e Demanda por Eletricidade **Microalgas como Fonte de Energia** Processo Produtivo Metodologia Resultados Conclusão

AS MICROALGAS COMO FONTE DE ENERGIA

- Possibilidade de aplicação comercial em indústrias:
  - Alimentar,
  - Química,
  - Cosmética, e
  - Farmacêutica.

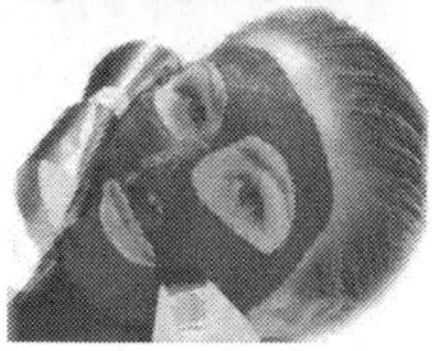

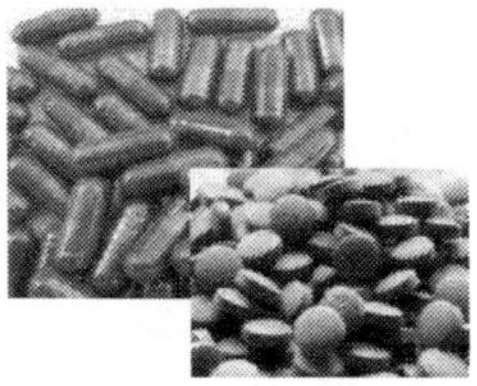

Introdução Crescimento Econômico e Demanda por Eletricidade **Microalgas como Fonte de Energia** Processo Produtivo Metodologia Resultados Conclusão

AS MICROALGAS COMO FONTE DE ENERGIA

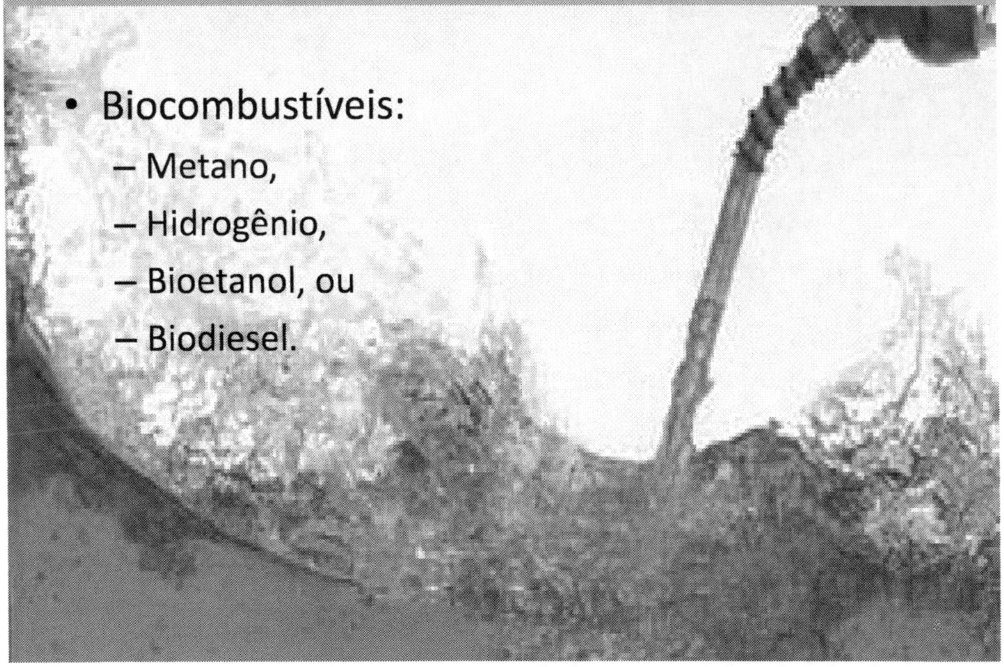

- Biocombustíveis:
  - Metano,
  - Hidrogênio,
  - Bioetanol, ou
  - Biodiesel.

AS MICROALGAS COMO FONTE DE ENERGIA

- Energia elétrica
  - Queima da biomassa de microalgas;
    - 1 ton de biomassa queimada ⟹ 8,12 MWh são gerados;
  - Energia de "emissão zero".

9

AS MICROALGAS COMO FONTE DE ENERGIA

- Cultivo:
  - Não necessita de terras aráveis ou férteis;
  - O solo pode ser desértico, com baixo valor econômico e alta irradiação solar;
  - Pode ser em meio salino, doce, ou em águas residuais de processos industriais ou urbanos;
  - Produção contínua, ou seja, não segue regime de safra e a colheita é diária.

10

Introdução Crescimento Econômico e Demanda por Eletricidade **Microalgas como Fonte de Energia** Processo Produtivo Metodologia Resultados Conclusão

AS MICROALGAS COMO FONTE DE ENERGIA

- Cultivo,

Necessitam de:

- CO_2 :
  - 1 ton de biomassa algal consome 1,7 ton de CO_2;
- Luz;
- Temperatura entre 25°C a 35°C;
- Nutrientes:
  - Ferro, potássio, fósforo e nitrogênio.

11

Introdução Crescimento Econômico e Demanda por Eletricidade **Microalgas como Fonte de Energia** Processo Produtivo Metodologia Resultados Conclusão

AS MICROALGAS COMO FONTE DE ENERGIA

- Tecnologias de cultivo:
  - A céu aberto;

  - Em fotobiorreatores.

12

AS MICROALGAS COMO FONTE DE ENERGIA

- Em fotobiorreatores:
  - Matriz de tubos transparentes que são montados e alinhados em postos;
  - Tubos feitos de plástico, PET, acrílico ou vidro;
  - Proporciona um volume maior da produção;
  - Ausência de contaminação;
  - Ambiente propício em termos de luminosidade e temperatura;
  - Consomem pouca água;
  - Ocupam pouco espaço.

13

AS MICROALGAS COMO FONTE DE ENERGIA

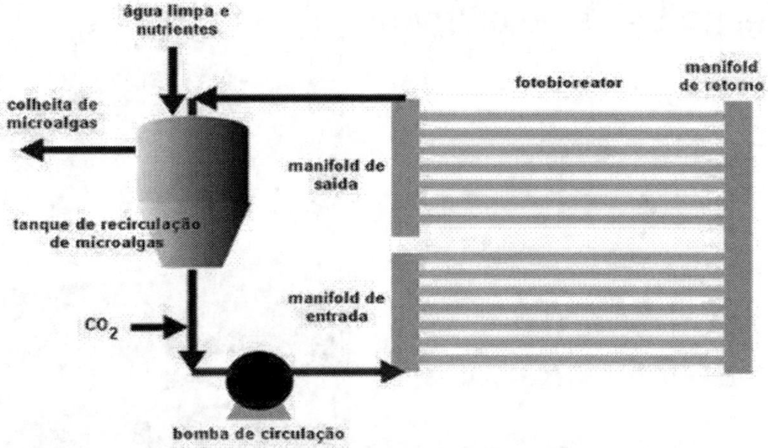

Figura 3.3: Esquema dos fotobiorreatores

14

PROCESSO PRODUTIVO PROPOSTO

- Com o intuito de maximizar a produção dessa fonte de energia os pesquisadores Jan Mulder (UEPB) e Paulo Oliveira (UFPE) desenvolveram um projeto do processo produtivo de modo a se realizar em conjunto com a produção de etanol de uma usina de cana-de-açúcar que usa o bagaço da cana para a geração de energia elétrica como subproduto.

PROCESSO PRODUTIVO PROPOSTO

- Os custos de produção de microalgas para a geração de bioeletricidade são reduzidos:
  - Todo CO_2 necessário é absorvido do processo de fermentação do etanol;
    - 1.000 litros de álcool ⟹ 800 kg de CO_2 ⟹ 470 kg de biomassa algal ⟹ 3,82 MWh.
  - A água utilizada pode ser a água residual do sistema de lavagem da cana.

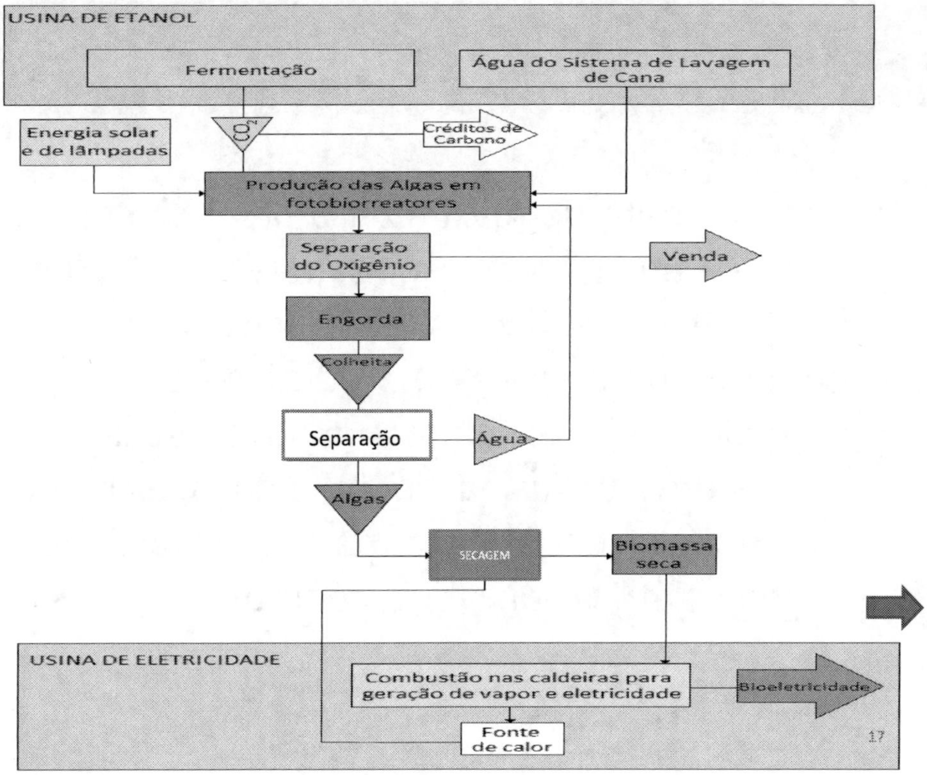

Introdução Crescimento Econômico e Demanda por Eletricidade Microalgas como Fonte de Energia **Processo Produtivo** Metodologia Resultados Conclusão

PROCESSO PRODUTIVO PROPOSTO

18

Introdução Crescimento Econômico e Demanda por Eletricidade Microalgas como Fonte de Energia **Processo Produtivo** Metodologia Resultados Conclusão

PROCESSO PRODUTIVO PROPOSTO

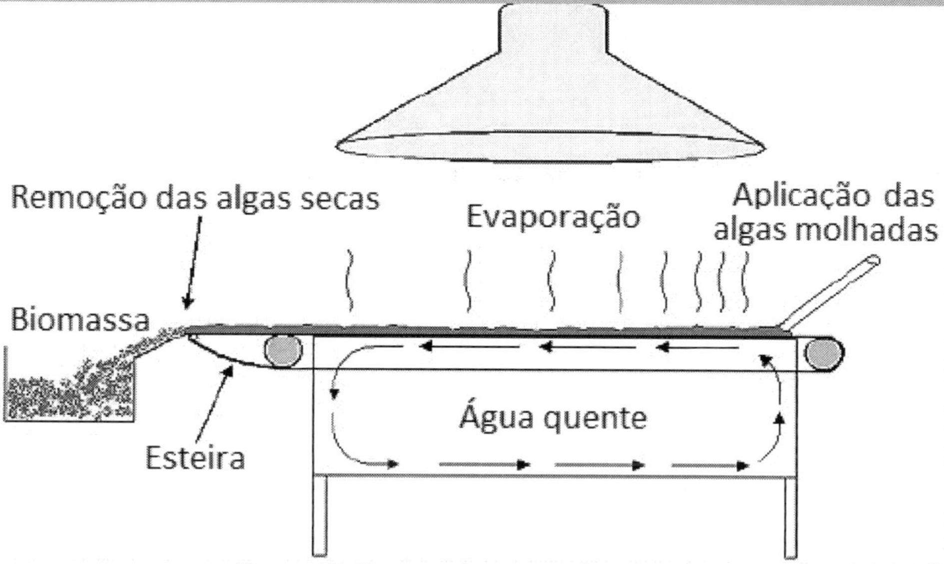

Figura 4.2: Processo de secagem das microalgas

19

Introdução Crescimento Econômico e Demanda por Eletricidade Microalgas como Fonte de Energia **Processo Produtivo** Metodologia Resultados Conclusão

PROCESSO PRODUTIVO PROPOSTO

- A produção em conjunto maximiza a produção de microalgas:
  - Reduzindo custos;
  - Gerando receitas:
    - Créditos de carbono;
    - Venda de O_2;
    - Venda de eletricidade.

20

PROCESSO PRODUTIVO PROPOSTO

- A produção em conjunto aumenta a produtividade da indústria de etanol:
  - Produzir durante a ociosidade dos equipamentos;
  - Utilização dos resíduos;
  - Reduz o índice de emissão da usina, melhorando a percepção de sustentabilidade da produção de etanol.

21

PROCESSO PRODUTIVO PROPOSTO

Estimação do Investimento Necessário

10 MWh de bioeletricidade

→ 30 T de microalgas/dia

→ 51 T de CO_2

→ 64.000 L de etanol/dia

22

PROCESSO PRODUTIVO PROPOSTO

Estimação do Investimento Necessário

Tabela 4.2: Estimativa dos investimentos necessários para a produção de microalgas a preços correntes de 2010.

Processo	Investimento (R$)
Fotobiorreatores	1.080.180,00
Iluminação	100.000,00
Separação	200.000,00
Secagem	200.000,00
Mão-de-obra	1.300.000,00
Total	2.880.180,00

PROCESSO PRODUTIVO PROPOSTO

Estimação do Investimento Necessário

- 50 fotobiorreatores;
  - Cada fotobiorreator:
    - 3 reatores;
    - Cada reator:
      - 30 tubos de altura;
      - 15 m de comprimento;
      - Serpentina.

Introdução Crescimento Econômico e Demanda por Eletricidade Microalgas como Fonte de Energia **Processo Produtivo** Metodologia Resultados Conclusão

PROCESSO PRODUTIVO PROPOSTO

Estimação do Investimento Necessário

- Investimentos em fontes alternativas de energia:

Eólica ⇨ 8MWh ⇨ R$ 150 milhões

Solar ⇨ 10MWh ⇨ R$ 115 milhões

Microalgas ⇨ 10MWh ⇨ R$ 2.880.180,00

25

Introdução Crescimento Econômico e Demanda por Eletricidade Microalgas como Fonte de Energia **Processo Produtivo** Metodologia Resultados Conclusão

PROCESSO PRODUTIVO PROPOSTO

Estimação dos Custos de Produção

Tabela 4.3: Estimativa dos custos de operação do processo de produção das microalgas a preços correntes de 2010

Descrição	Custo/Mês (R$)	Custo/ano (R$)
CO_2	0	0
Água	0	0
Nutrientes	12.000,00	80.000,00
Iluminação dos Fotobiorreatores	0	0
Energia do Processo	6.000,00	40.000,00
Energia da Secagem	0	0
Mão-de-obra	6.000,00	40.000,00
Total	24.000,00	160.000,00

26

PROCESSO PRODUTIVO PROPOSTO

Estimação das Receitas

Tabela 4.4: Receitas oriundas do processo produtivo de microalgas

Produção	Preço da unidade	Quantidade/ Mês	Receita/ Mês (R$)	Quantidade/ Ano	Receita/ Ano (R$)
Bioeletricidade	156/ MWh	7.200 MWh	1.123.200	48.000 MWh	7.488.000
Créditos de Carbono	31,5/ tonelada	1.530 toneladas	48.195	10.200 toneladas	321.300
Total	-		1.171.395		7.809.300

METODOLOGIA

- Critérios da análise de investimento:
  - Valor Presente Líquido,
  - Taxa Interna de Retorno e, a
  - Razão Benefício Custo.
- Todos os critérios levam em consideração a **taxa mínima de atratividade**, que é a taxa de mercado financeiro, a qual remunera o capital aplicado no fluxo de caixa.

RESULTADOS

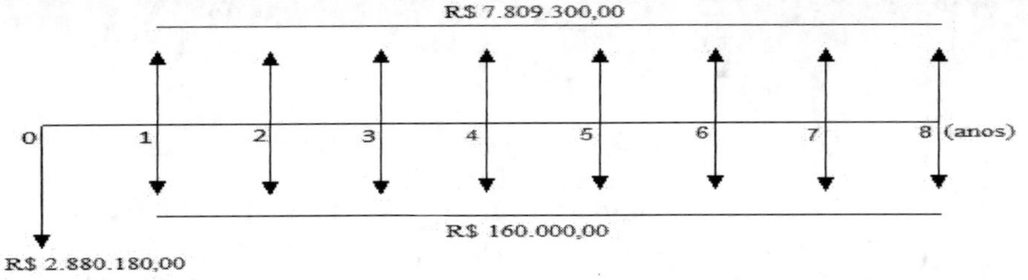

- i_M = 12% a.a.

$$VPL\ (i_M) = -I + \sum_{t=0}^{n} \frac{R_t - C_t}{(1+i_M)^t} \qquad VPL\ (12\%) = -2.880.180 + \sum_{t=1}^{8} \frac{7.809.300 - 160.000}{(1+0{,}12)^t}$$

$$VPL\ (12\%) = -2.880.180 + \sum_{t=1}^{8} \frac{7.649.300}{(1{,}12)^t}$$

$$VPL\ (12\%) = R\$\ 35.118.786{,}87$$

VPL > 0 = Projeto do processo produtivo de microalgas é viável.

RESULTADOS

- TIR;
- i_M = 12% a.a.

$$VPL\ (i) = -I + \sum_{t=0}^{n} \frac{R_t - C_t}{(1+i_M)^t} = 0$$

$$0 = -2.880.180 + \sum_{t=1}^{8} \frac{7.649.300}{(1+i)^t}$$

$$i = 265{,}57\%\ a.a$$

$i > i_M$ = viável

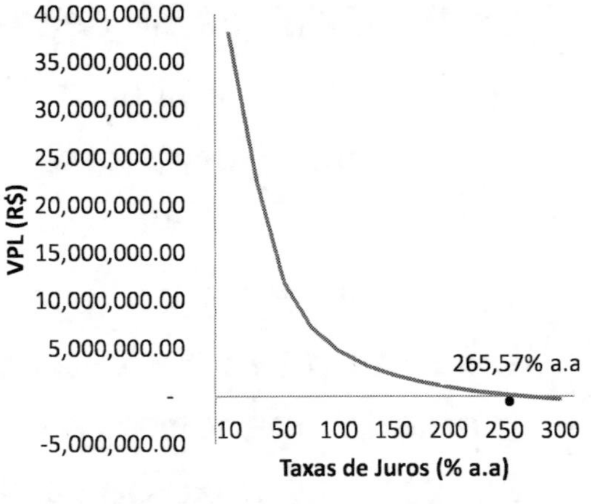

Gráfico 6.1: Valor presente líquido do projeto correspondente às taxas de juros

CONCLUSÃO

- O processo produtivo de microalgas em conjunto com a produção de uma usina de etanol:
  - Maximiza a produção de etanol, pela oportunidade de obter ganhos além dos meses de safra, reduzindo a ociosidade do maquinário;
  - Maximiza a produção de microalgas, reduzindo custos de investimento e produção. Além disso, pôde-se verificar a oportunidade de agregar valor a produção de microalgas através dos créditos de carbono e da produção de oxigênio.

CONCLUSÃO

- Através dos critérios de análise de investimento obteve-se o mesmo resultado, a afirmação que o projeto do processo produtivo de microalgas é rentável, ou seja**, apresenta viabilidade econômica**.

- O retorno do investimento se dá logo no primeiro ano de produção.

Obrigada!

Socioeconomic Aspects of Sugarcane Producing Municipalities in Brazil

Gerber Machado, P. Sr.,
Originally presented on: 7/24/2014 14:00:00 - 14:20:00

SOCIOECONOMIC ASPECTS OF SUGARCANE PRODUCING MUNICIPALITIES IN BRAZIL

Pedro Gerber Machado, M.Sc *
* UNICAMP – Cidade Universitária "Zeferino Vaz", Campinas, Brazil

Introduction

Government policies to support the production of ethanol from sugar cane , large availability of land and a favorable climate , ie tropical, with good rainfall and temperatures , have made Brazil a global leader in this field . In 2009, Brazil accounted for 33 % of global production of ethanol and was also the largest exporter (HERRERAS, 2013). The importance of the sector to the economy has taken large proportions, especially in the last four decades. In 2008 participation in the Brazilian GDP reached 2 %, generating more than 1,280,000 formal jobs (UNICA, 2011). The number of plants in 2010 was 413, with 103 producing only ethanol, 11 producing sugar and 297 mixed (MAPA, 2010).

The scale of production of sugarcane and ethanol, as well as the growth of consumption, which is a consequence of the introduction and success of flex vehicles in the Brazilian fleet, justifies the analysis not only of environmental aspects, but also social and economic issues, including its contribution to the development of the country and the regions in which it operates

This article aims to analyze the socio-economic aspects associated with the production of sugarcane in two major producing states: São Paulo, the largest producer in the Center-South region, and in the country, and Alagoas, the largest producer in the Northeast. To do so, regularly compiled municipal indicators were used and disclosed by official Brazilian government agencies. The cluster method was used to separate in a non-biased way two groups: with higher and lower socioeconomic indicators.

Methodology

Data collection for this study was performed prioritizing official databases from the Brazilian government and state institutions, with public access. Priority was given to electronic databases of IPEADATA, the Institute for Applied Economic Research (IPEA). Three categories of data are available: Macroeconomic, regional and social. Given the nature of research, the social category was prioritized, and six indicators and two indices for the years 1980, 1991, 2000 and 2010 were selected.

Despite the effort in creating methodologies for sustainability assessment, whether in public, private or community environments, the choice of indicators depends mainly on the availability of data, especially when it is a temporally and spatially extensive study. Following Oliveira (2011), the choice of indicators for this study was intended to encompass several subdivisions of social, economic and also taking into account the existence of data in years and municipalities. Frame 1 shows the selected indicators and their corresponding category.

Frame 1
Selected socioeconomic indicators for this study

Category/Indicator	Definition
Education	
Illiterate - people aged 15 and over - (%)	Percentage of persons aged 15 and more years of age who cannot read and write a simple note. Called "Illiterate".
Health	
Life expectancy at birth - (Years)	Years of life expectancy of a person born in the year of reference, assuming that mortality rates estimated in previous years, by age remained constant in subsequent years. Called "Life Expectancy".
Child mortality (per thousand live births)	Number of people per thousand live born in the year of reference which should not complete a year of life. Called "Child mortality".
equity	
Theil index L	Measures the degree of inequality in the distribution of household income among individuals. It is the logarithm of the ratio between the arithmetic and geometric

	averages of individual incomes, being zero when income inequality between individuals does not exist, and tending to infinity when inequality tends to the maximum. Called "Theil L".
Wealth	
Poor people - (%)	Percentage of people with household income below R $ 75.50, equivalent to half of the minimum wage in August 2000. Called "poor people".
Infrastructure	
Percentage of households with electricity (%)	Percentage of households with electricity. Called "Electricity".
Percentage of households connected with general sewer system (%)	Percentage of households with toilets connected to the sewage system. Called "Sewer system".
Development	
Municipal Human Development Index (HDI-M)	It is obtained by simple arithmetic average of three sub-indices relating to the dimensions Longevity (HDI-Longevity), Education (HDI-Education) and Income (HDI-Income). Varies from zero to one. Called "HDI".

After the selection of indicators, a cluster analysis is performed within the sugarcane producing municipalities, in order to divide them in two groups: One with better indicators, hence better quality of life, and one with worst.

In order to verify the differences between these two groups, as economic activity goes, the mean of five indicators is calculated and then compared using a Mann-Whitney's test: Participation of sugarcane in local GDP, Participation of Industry in local GDP, Participation of Services in local GDP, Participation of agriculture in Local GDP and Percentage of rural population.

In cluster analysis the aim is to find patterns in data to group the observations of the sample in great clusters, in which these observations are similar, but the groups are as different as possible.

There are three methods to divide observation into groups: joining or tree clustering, two-way joining and K - means clustering. Only the latter was used in this work and therefore will be treated here. In the technique of cluster K –means, first K centroids are chosen, usually the average of the sample, where K is the number of centroids specified by the researcher or user. The sample points are then assigned to groups whose centroids are closest, and a new centroid is calculated. The centroids are calculated until no individual is relocated to a different group (TAN et al., 2006).

Results

The results for the average comparison between the groups with highest and lowest quality of life (represented by the unbiased groups formed after the clustering) show two particularly different scenarios for the state of São Paulo and Alagoas.

The state of São Paulo has a systematic evolution throughout the years, with the best group with lower averages of sugarcane participation in the GDP. Also, the group with higher industry participation in its GDP is the best group, except for 1991. As it is for sugarcane, the group with higher agriculture participation in the GDP is the worst group.

Table 1 - Comparison of averages between groups for the state of São Paulo 1970-2010 (decimal)

Variables	1980		1991		2000		2010	
	Best group	Worst group	Best group	Worst group	Best group	Worst group	Best group	Worst group
Sugarcane participation	0.16±0.16	0.24±0.22*	0.04±0.04	0.03±0.07	0.05±0.05	0.10±0.06*	0.05±0.07	0.08±0.09*
Industry participation	0.43±0.18*	0.32±0.15	0.24±0.24	0.22±0.15	0.29±0.13*	0.24±0.13	0.24±0.11*	0.20±0.12
Services participation	0.34±0.10	0.32±0.12	0.47±0.47	0.48±0.10	0.61±0.10	0.59±0.10	0.54±0.11	0.53±0.11
Agriculture	0.22±0.16	0.34±0.17*	0.27±0.27	0.28±0.19	0.08±0.07	0.16±0.08*	0.09±0.10	0.19±0.12*

participation								
Percentage of rural population	0.23±0.12	0.36±0.15*	0.13±0.13	0.25±0.12*	0.09±0.06	0.14±0.08*	0.11±0.11	0.20±0.14*

*Highest value, when statistically different (at 5%).

The state of Alagoas, on the other hand, has a different and interesting evolution. Until 2000, it follows the same patterns as São Paulo, but in 2010 the group with best development has higher sugarcane participation in its GDP, in contrast with agriculture participation (the worst group has higher average).

Table 2 - Comparison of averages between groups for the state of Alagoas 1970-2010 (decimal)

Variável	1980	1991	2000	2010				
Best group	Worst group	Best group	Worst group	Best group	Worst group	Best group	Worst group	
Sugarcane participation	0.27±0.29	0.54±0.19*	0.14±0.06	0.24±0.11*	0.15±0.07	0.18±0.07	0.06±0.06*	0.05±0.07
Industry participation	0.41±0.10*	0.26±0.06	0.41±0.16	0.38±0.06	0.21±0.11	0.16±0.16	0.15±0.10*	0.12±0.09
Services participation	0.41±0.10*	0.26±0.06	0.31±0.08	0.23±0.16	0.51±0.11	0.48±0.08	0.66±0.12	0.68±0.09*
Agriculture participation	0.18±0.20	0.47±0.12*	0.28±0.12	0.38±0.13*	0.28±.11	0.35±0.14	0.14±0.06	0.15±0.07*
Percentage of rural population	0.34±0.28	0.64±0.12*	0.45±0.19	0.51±0.15	0.38±0.19	0.38±0.14	0.28±0.19	0.55±0.20*

*Highest value, when statistically different (at 5%).

Conclusion

The states of São Paulo and Alagoas are two important sugarcane producers in Brazil. The importance of this culture for the local development in both states is historically known. The numbers, on the other hand, show that most of the activities with higher value added (as the industry sector) have better impacts on the local quality of life, than those of the primary sector.

The year 2010 in the state of Alagoas shows a different scenario. In this particular year, the sugarcane participation indicates a positive impact in the quality of life of sugarcane production. Possibly due to the higher importance of the culture in the state, and also to the value it represents for local economy, and the growth in sugarcane prices over the last decade.

References

HERRERAS, Sara et al. Analysis of socio-economic impacts of sustainable sugarcane–ethanol production by means of inter-regional Input–Output analysis: Demonstrated for Northeast Brazil. Renewable And Sustainable Energy Reviews, Utrecht, v. 1, n. 28, p.290-316, ago. 2013.

UNICA. Relatório de Sustentabilidade 2010. São Paulo, 2011. 128 p.

MAPA. Relação das unidades produtoras cadastradas no Departamento da Cana-de-açucar e Agronergia. Brasilia, 2010.

IPEADATA. Dados sociais e regionais do Brasil. Disponível em: <http://www.ipeadata.gov.br>. Acesso em: 10 jan. 2012.

OLIVEIRA, Janaína Garcia. Estudo dos impactos socioeconômicos regionais do setor sucroalcooleiro. 2011. 211 f. Tese (Doutorado) - Unicamp, Campinas, 2011.

TAN, Pang-ning; STEINBACH, Michael; KUMAR, Vipin. Cluster Analysis: Basic Concepts and Algorithms. Em: TAN, Pang-ning; STEINBACH, Michael; KUMAR, Vipin. Introduction to Data Mining. Michigan State University: Michigan State University, 2006. p. 487-588.

Analysis of the Viability of Ethanol Production in Brazil: Economical, Social and Environmental Implications

Isabel Alvarez Murillo – MCGILL UNIVERSITY

RCN Conference on Pan American Biofuels and Bioenergy Sustainability

22-25 July 2014, Recife, Brazil

Presentation[1] outline:

- Background
  - Biofuel industry
  - Ethanol
  - Brazil's production
- Repercussions
  - Economical
  - Environmental
  - Social
- Discussion

[1] *Based on* Alvarez Murillo, I. 2013. Analysis of the viability of ethanol production in Brazil: Economical, social & environmental implications. Energy and Environment Research 3 (2): 166-175. doi:10.5539/eer.v3n2p166

Background

Overview of Global Energy Production

- Global energy consumption grew by 2.5% in 2011
- Brazil leader in biofuel exports in 2007
- Global biofuel production only grew by 0.7% in 2011

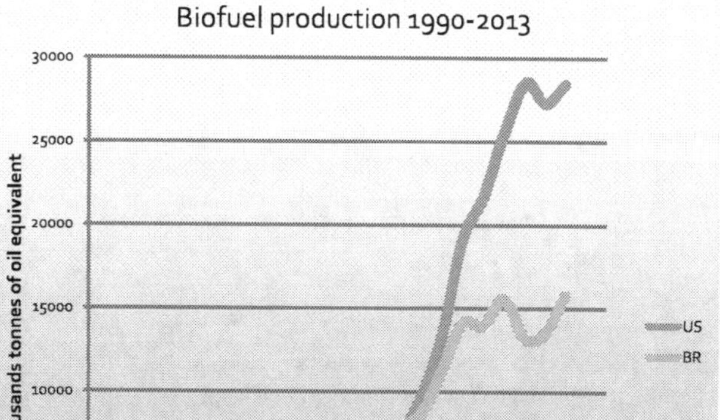

British Petroleum (2014) & British Petroleum (2012) in Alvarez Murillo(2013)

First-generation Biofuels

- Corn:
  - Average production rate 3.3 thousand liters per hectare[1]
  - Net energy yield of only 20%
  - Advantages: fairly easy conversion from starch to ethanol
- Sugarcane
  - Average production 6.1 thousand liters per hectare[1]
  - Advantages: more easily conversion only fermentation
  - Carbon dioxide emissions can be lowered by 90% compared to gasoline

[1] biofuel.org.uk

Second-generation biofuels

- Feedstock not suitable for human consumption grown in marginal lands
- Conversion: thermo-chemical and biochemical
- Feedstock: grasses, jatropha, waste vegetable oil and municipal solid waste.
- Advantages:
  - Low fertilizer needs
  - Direct use as biomass
  - Net energy yield to energy input ratio 5.4
- Disadvantages:
  - Lengthy growth
  - High water demand

Third-generation Biofuels

- Average production yield 84.2 housand-liters per hectare[1]
- Advantages:
  - Algae can be genetically manipulated to directly produce: biodiesel, butanol, gasoline and diesel fuel.
  - Capable of growing anywhere where temperatures are high

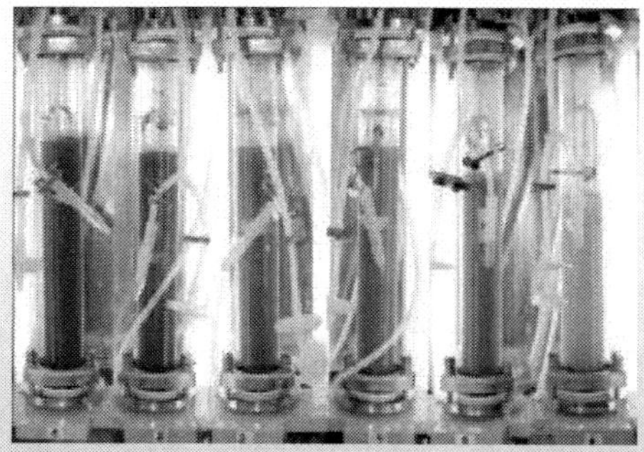

- Disadvantages:
  - Require significant amounts of water and nutrients to grow
- Cultivated in: open ponds, closed-loop systems, photo-bioreactors
- Advancement in plant biology has improved crop yields

[1] biofuel.org.uk

Fourth-generation Biofuels

- Carbon-negative
- Bioconversion process: fermentation, gasification and fast-pyrolisis
- Carbon dioxide captured using pre-combustion, oxyfuel or post-combustion processes
- Accumulated in depleted oil and gas fields or saline aquifers
- New developments

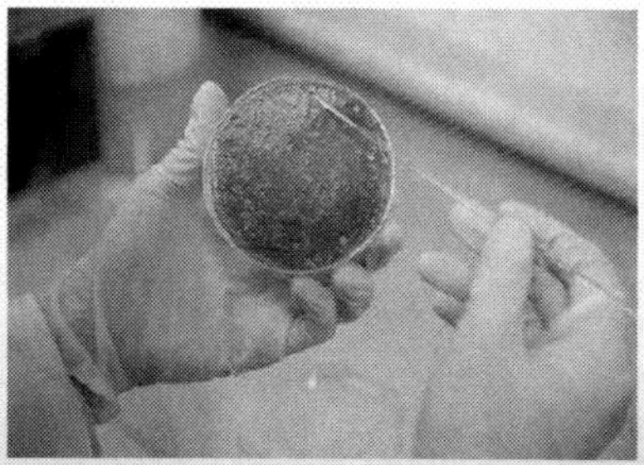

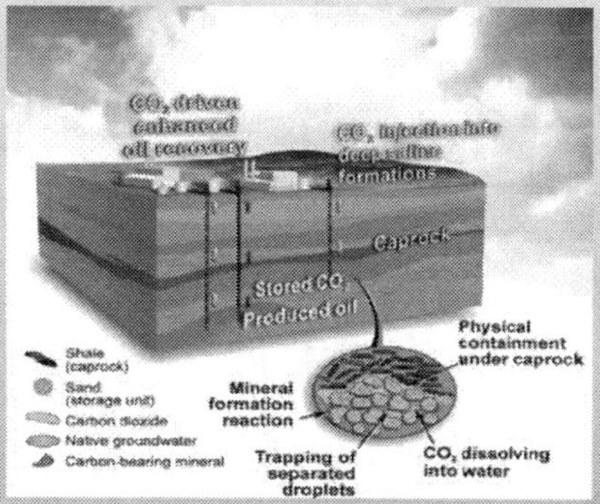

[1] biofuel.org.uk

Brazilian ethanol production

- 1930: Bio-ethanol production started
- 1973: Arab oil embargo
- 1974: Brazilian National Alcohol Program
- 2007: International interest in global warming
- Advantages:
  - Reduction in greenhouse gas emissions
  - Lesser dependence of fossil fuels
  - Job creation
  - Income and foreign investment opportunities

Table 1: Top 25 Countries, Ethanol Production Capacity[1]

Rank	Country	Million Liters	Million Gallons
1	United States	51,415.97	13,584.14
2	Brazil	26,887.52	7,103.70
3	China	2,699.48	713.20
4	France	1,821.03	481.12
5	Canada	1,494.50	394.85
6	India	1,420.92	375.41
7	Poland	1,079.00	285.07
8	Germany	916.97	242.26
9	Thailand	868.50	229.46
10	Jamaica	832.70	220.00
11	T & T	757.00	200.00
12	Indonesia	683.38	180.55
13	Spain	546.00	144.25
14	Austria	485.00	128.14
15	Belgium	485.00	128.14
16	Netherlands	480.00	126.82
17	United Kingdom	470.00	124.17
18	US Virgin Islands	387.50	102.38
19	Colombia	352.00	93.00
20	Vietnam	318.11	84.04
21	Australia	292.70	77.33
22	Czech Republic	280.00	73.98
23	El Salvador	247.10	65.28
24	Paraguay	237.25	62.68
25	Argentina	237.20	62.67
	Total	95,694.83	25,282.65

[1] Global Biofuels Center (2012) *in* Alvarez Murillo (2013)

Brazilian ethanol production (2011)

- Ethanol is the main biofuel used in the transportation sector
- Brazil is the 2nd major ethanol producer and greatest exporter
- Sugar and ethanol industry represent 2.3% of the Gross Domestic Product
- Generates 4.5 million jobs
- Sugarcane occupies 8 million hectares
- Sugarcane production of 600 million tons per year
- Largest sugar cane producer worldwide

Basso et al. (2011) http://cdn.intechopen.com/pdfs-wm/20058.pdf

Repercussions

Economical Impacts

- In 2012, Brazil's ethanol output showed its greatest decline since 1965 (-15.3%)
- By 2015, more biofuel production is expected to use cellulose as an input, rather than food crops.
- Global ethanol production expected to reach 180 billion liters per year by 2021.
- Three major leaders: United States, Brazil and European Union
- Possibility of their future replacement by advanced biofuels

British Petroleum (2012), Sawyer (2008), FAO (2012) *in* Alvarez Murillo (2013)

Economical Impacts

- The policies developed in the United States by the Environmental Protection Agency (EPA) will affect the ethanol market on a global scale
- Three alternative implementation options
  - Lowering the total and advanced mandates by the shortfall in the cellulosic mandate
  - **Maintaining both the advanced and total mandates**
  - Maintaining the total mandate and lower the advanced mandate by the shortfall in cellulosic production

Economical Impacts

- Baseline projections: cellulosic ethanol production gradual increase 2011-2021 to reach 16 billion liters per year

1. Expansion of 9% in sugarcane acreage compared to baseline
2. Brazilian ethanol demand will decrease significantly
3. Increase imports to meet domestic demand expected to reach 18 billion liters per year

- Brazil's dependency on foreign policies will create a source of economic vulnerability given the United States' policies significant influence on Brazil's ethanol and sugar markets.

FAO (2012) *in* Alvarez Murillo (2013)

Environmental Impacts

Greenhouse Gas Emissions

- Comparative energy analysis of biofuels vs. fossil fuels
- Fertilization of sugarcane production can lead to the soil-mediated release of nitrous oxide (NO_2) and methane (CH_4)
- Deforestation arising from land-use shifts towards agriculture contributes to 75% of Brazil's carbon dioxide (CO_2) emissions[1]
- Brazil contributes to 3% of the global greenhouse gas emissions (4th)[1]
- Production of biofuels are not totally independent from fossil fuels

[1] Ross (2012) *in* Alvarez Murillo (2013)

Environmental Impacts

Greenhouse Gas Emissions

- Expansion of sugarcane into cattle farms has created pressure towards further deforestation
- No restrictions on purchasing land to raise cattle

- Exacerbates deforestation and increases methane emissions from cattle

Sawyer (2008) *in* Alvarez Murillo (2013)

Environmental Impacts

Water Requirements and Water Pollution

- Most sugarcane grown under rainfed conditions
- It takes 2000 L of water to produce 1 L of biofuel from sugarcane
- In the northeast of Brazil agriculture is supported through irrigation

- Fertilizers and pesticides used in sugarcane plantations contain nitrogen and phosphorus

Hoogeveen *et al.* (2009), Ross (2012) *in* Alvarez Murillo (2013)

Environmental Impacts

Climate Change and Biodiversity

- Cerrado region is the world's most diverse savannah
- Currently 2-3 times as much annual deforestation compared to 2005-2006
- Clearing the Cerrado can have severe repercussions on global biodiversity

- Shifting rainfall patterns and increasing evaporation rates will increase the risk of desertification

Ross (2012), Sawyer (2008) *in* Alvarez Murillo (2013)

Social Impacts

Land Tenure and Workers

- Expulsion of small farmers from their lands by large-scale ethanol producers (1970 -1980)
- Concentration of income to producers and processors
- Displacement of family workers

- Employees work conditions undermines health
- 80% of sugarcane is cut manually by approximately 1 million seasonal workers

Ross (2012), Sawyer (2008) *in* Alvarez Murillo (2013)

Social Impacts

Food Security

- Increasing demand for biofuels resulted in an increasing demand for agricultural products
- Dramatic increase in food prices 2005-2008 had a devastating effect on the urban or landless poor
- In 1970, Pernanbuco State led the country in sugarcane production but ranked among the poorest regions in the world
- Increase vulnerability of the poorest sectors since a small increase in food prices can dramatically change food availability to households

Hoogeveen *et al.* (2009), Ross (2012) *in* Alvarez Murillo (2013)

Social Impacts

Indigenous Populations: Guarani-Kaiowa

- Mato Grosso do Sul is currently seeing rapid expansion of soybean and sugarcane cultivation
- Threatened by the expansion of farms and landowners' violence
- Landowners have resorted to armed attacks on native encampments

Frayssinet (2012) *in* Alvarez Murillo (2013)

Social Impacts

Indigenous Populations: Guarani-Kaiowa

- Sep 29th 2012: The Federal Court of Navarai dispatched an order to expel communities from the riverside
- Petition to the Brazilian Government to enact their collective death

- Lack access to food, proper settlements, medical services or protection from farmer's gunman
- These lands have not yet been demarcated by the federal government and the legal vacuum has fueled conflict

Frayssinet (2012), Void Mirror (2012) *in* Alvarez Murillo (2013)

Discussion

Discussion

- Biofuels cannot be seen as the 'miracle solution' to the world's dependency on fossil fuels
- The implications of biofuels in Brazil are enormous
- Brazil is considered one of the most economically unequal countries in the world
- The economic benefits go to the wealthy countries that can purchase ethanol and the wealthy landowners, producers and processors that are part of this market
- The viability of ethanol's market in Brazil needs to be reconsidered

UNDP(2013), Haslam *et al.* (2012) *in* Alvarez Murillo (2013)

Thank you

UFRRJ
UNIVERSIDADE FEDERAL RURAL DO RIO DE JANEIRO

CONSERVAÇÃO INTERNACIONAL Brasil

RCN CONFERENCE ON PAN AMERICAN BIOFUELS AND BIOENERGY SUSTAINABILITY

SOCIOENVIRONMENTAL ASPECTS OF THE CULTIVATION OF PALM FOR BIODIESEL PRODUCTION IN THE CENTER OF ENDEMISM IN BELÉM – PA, BRAZIL

PhD. candidate Gustavo Simas
Prof. Dr. Rodrigo Medeiros
Prof. Dr. Bruno Coutinho
MSc. candidate Tatiana Martins

Presentation Overview

- Biodiesel in Brazil
- Palm oil production in Brazil
- Belém Endemism Center production
- Socioenvironmental aspects

Brazil and Bioenergy

- Since 1950
- National Program of Biodiesel Use and Production (PNPB)
- 2005 – 2011: production jumps from 736 m^3 to 2,6 millions m^3
- During this period the acreage of soybeans increased by 29.4% bringing the total to 24 million hectares, producing over 75 million tons of grain

Evolution of current production, compulsory demand and nominal capacity

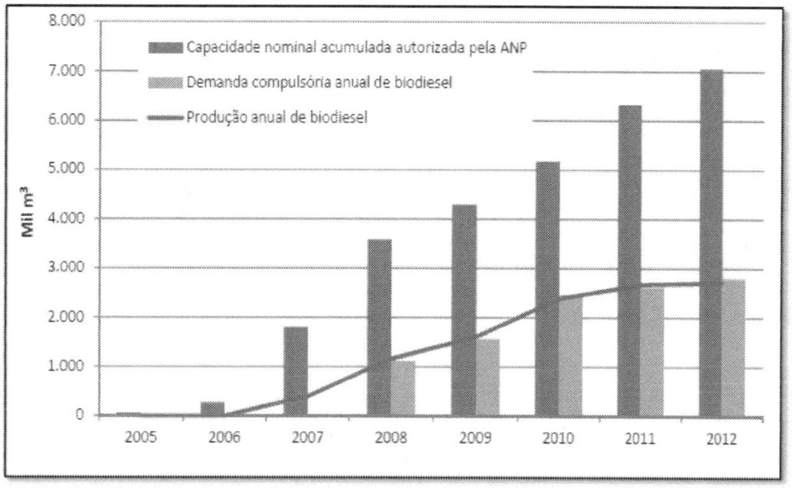

Source: ***ANP, 2013***

Palm oil for biodiesel production

- High productivity
- High potential to promote social inclusion
- Suitable for the soil and weather conditions of Brazil

Palm oil in the World

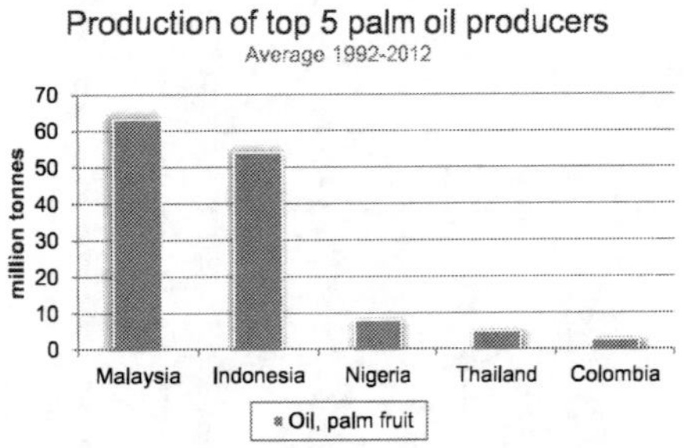

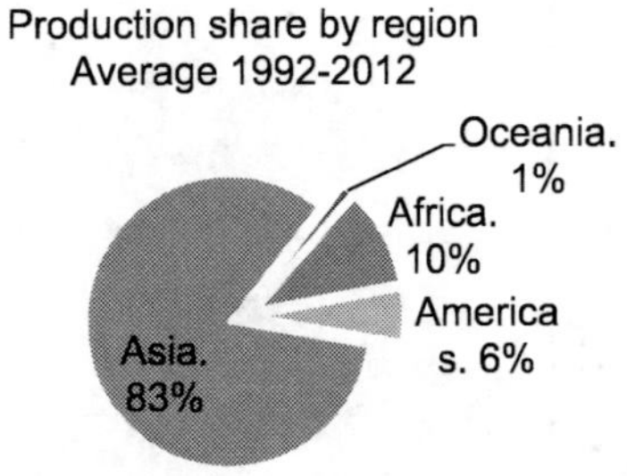

Source: FAO 2012 - http://faostat.fao.org

National Program of Biodiesel Use and Production & Social Fuel Seal

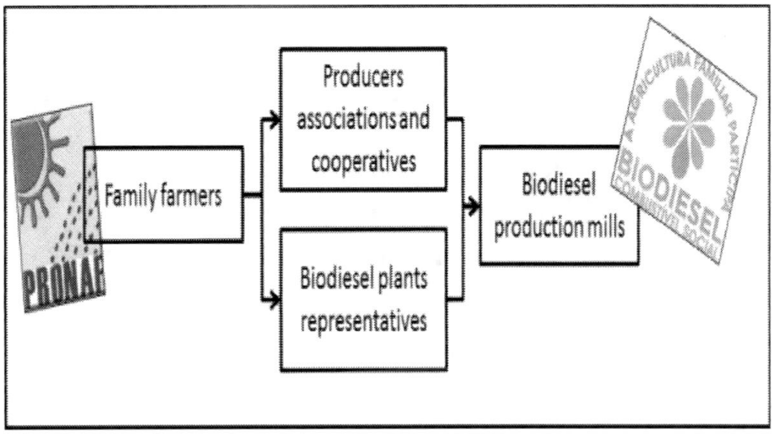

Source: Modified from César & Batalha 2013

National Program of Sustainable Palm Oil Production

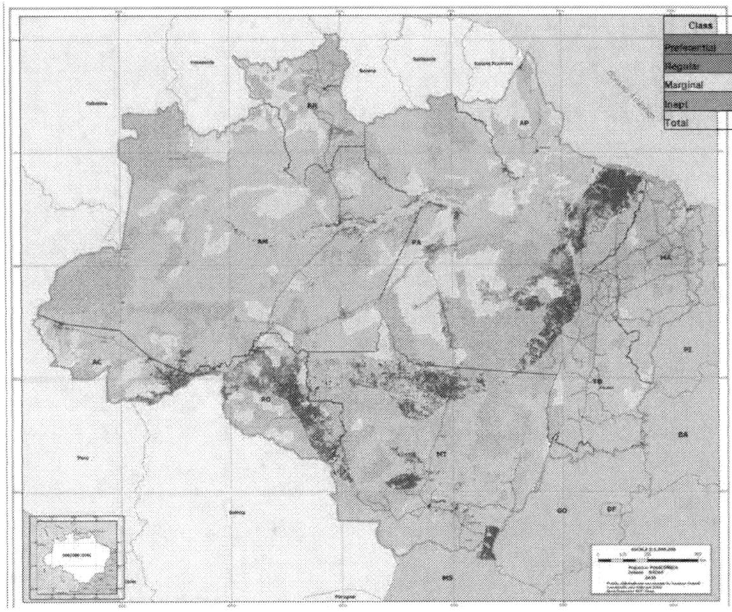

Class	hectare	Km2
Preferential	7.296.279	72.963
Regular	21.637.101	216.371
Marginal	2.299.816	22.998
Inept	39.173.381	391.734
Total	70.406.577	704.066

Hydrography
Protected areas
Indigenous area
Forest area

Source: Embrapa – 2010
www.cnps.embrapa.br/zoneamento_dende/

Palm oil expansion: suitable areas

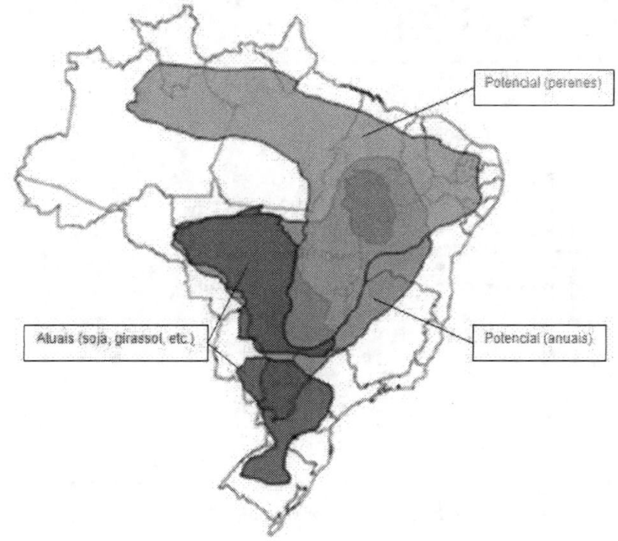

- 704.066 km² suitable for palm oil expansion
- 14% of Legal Amazon
- 250.513 km² in Pará state

Source: PNA - 2011

Palm oil production in Brazil

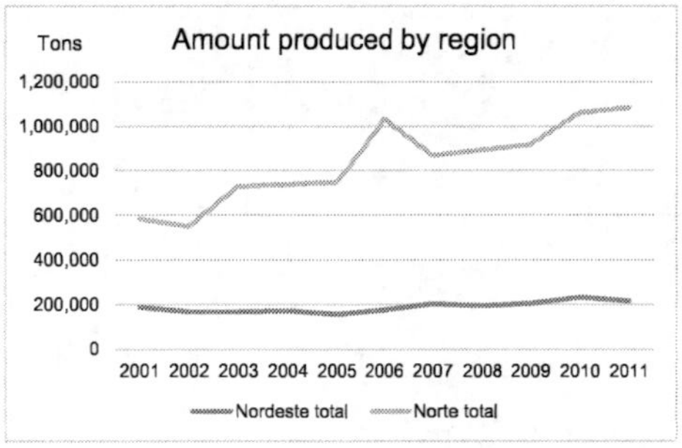

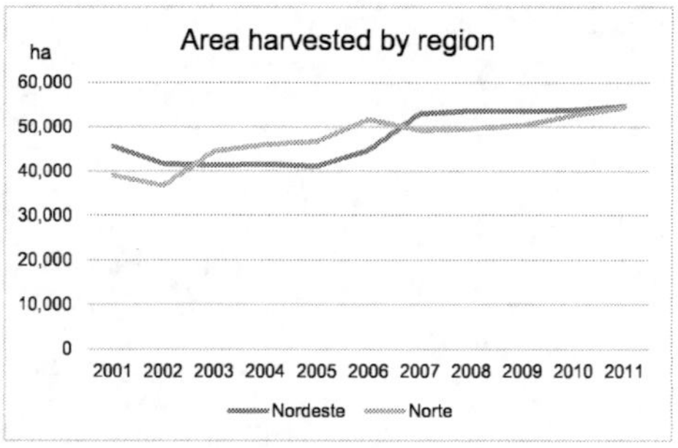

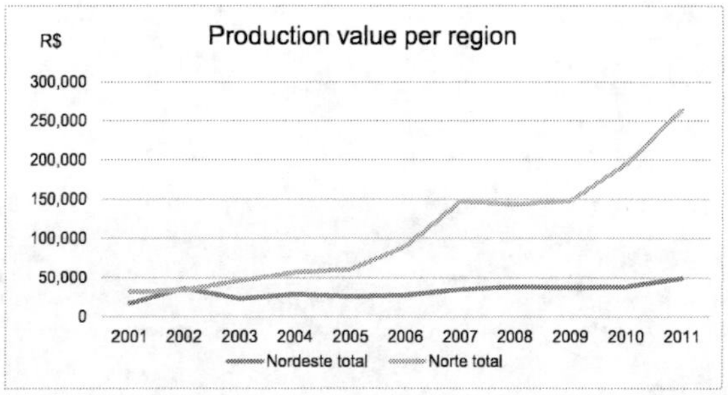

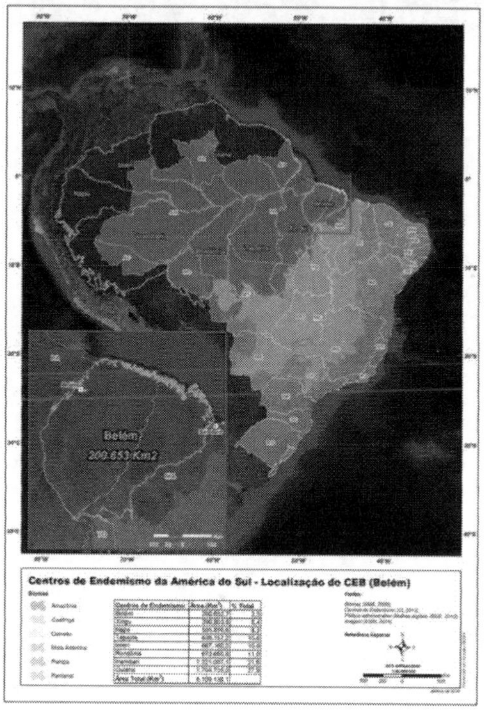

- Identify and characterize the expansion areas of the palm in Belém Endemism Center
- Micro region of Tomé Açu: Acará; Concórdia do Pará; Moju; Tailândia; e Tomé-Açú

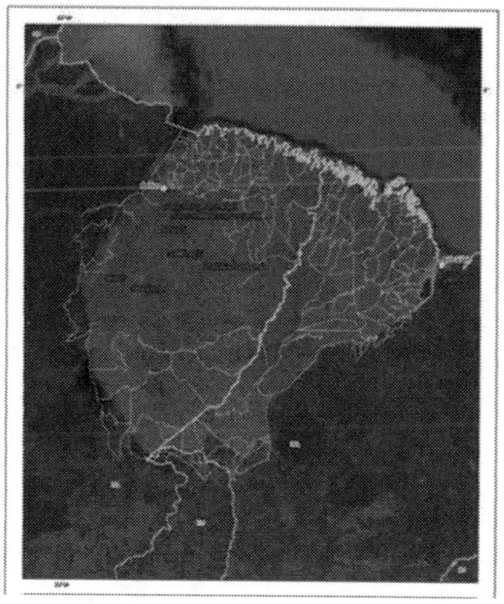

Belém Endemism Center

- Principal units of production of biodiesel from palm oil
- 140 thousand hectares, reaching almost 470 thousand hectares after the consolidation of expansion projects by 2020.
- 2010 – 2012: over 581 families (in 5,810 ha) added to the partnership programs of Banco da Amazônia
- It is estimate palm oil expansion enabled the inclusion of 2,191 new households (in 21,110 ha) in the last three years.

Belém Endemism Center Production

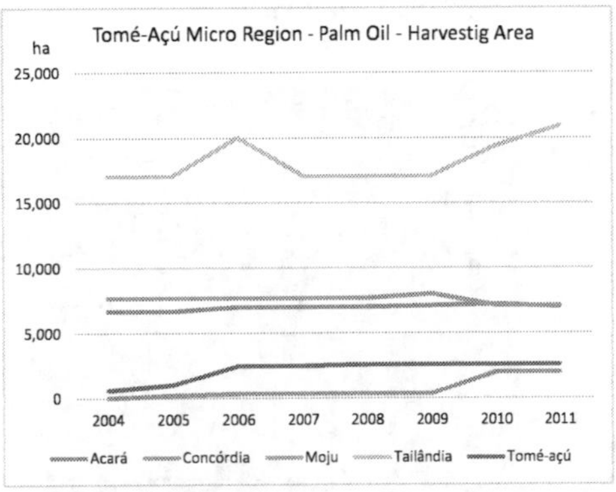

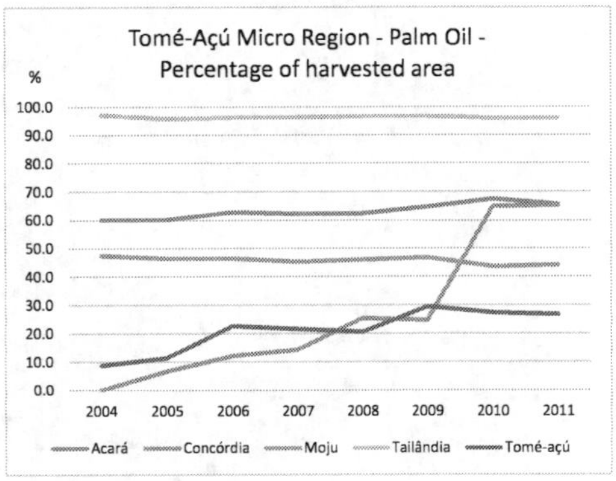

Source: Modified from IBGE, 2014

Field analysis

- 2 groups evaluated
  - interest group - producers of palm oil
  - control group - no palm producers

- Stratified samples
  - Not Productive palm (DNP) under 2 years of cultivation;
  - Palm Oil Production (DP) over 4 years of cultivation;
  - No producers Dende (ND);
  - No palm producers belonging to Quilombo community (CHI).

Preliminary findings

- 4 municipalities
  - Moju, Acará, Concórdia do Pará and Tomé-Açu
- 23 families interviewed
  - 15 with palm oil
  - 8 with other crops
- On average each family occupies **10 hectares** and can get an average of **R$ 2,000 monthly** between the fifth and eighteenth year of life of the palm when it reaches the height of its productivity, totaling **R$ 24,000 per year**
- Other crops: cassava, black pepper, urucum and others
- Companies partnerships: technical assistance, seedlings, fertilizers and transportation

Preliminary findings (cont.)

Activity	DNP (%)	DP (%)	ND (%)	QUI (%)	Average (%)
Palm oil	28,6	57,1	0,0	0,0	**27,3**
Cassava	42,9	28,6	20,0	66,7	**36,4**
Other crops	14,3	14,3	20,0	0,0	**13,6**
Retirement	0,0	0,0	60,0	0,0	**13,6**
Pension	14,3	0,0	0,0	0,0	**4,5**
Employment	0,0	0,0	0,0	33,3	**4,5**
	100	**100**	**100**	**100**	100

Main economic activity of the family

Motivation	DNP (%)	DP (%)	Average (%)
Long term income	50	50	**50**
Incentive of the community	25	20	**22**
Incentive from the company	25	30	**28**
	100	**100**	**100**

Main reason for growing palm oil

Preliminary findings – Small farmers perception

- Time dedicated to other cultures
  - Palm oil is very time consuming
  - Is not a generalized view
  - Depends on how many family members participate in the production
- Employment
  - Number of persons employed increased due to the arrival of biodiesel companies
- Land prices increased
- Community structure
  - Motorcycles and brick houses
  - Small farmers became employees
- Social organization
  - Intend to form a cooperative to discuss contracts with companies
- Food prices
  - Cassava prices increased
  - Seen as a good thing for those who continue producing and selling because increased income

Preliminary findings – Small farmers perception

- Impact on quality of life
  - Varies with the number of people that the farmer sustains
- Production cost
  - Varies greatly depending on how many member of the family work in the production and how many are hired
- Ecosystems Services
  - Clean air
  - Climate
  - Water
- Use of agrochemicals
  - Impact on water quality
- Positive effects on the environment
  - Improved the climate because covered areas previously deforested
  - Decreases the number of burns because it is a long cycle crop
- Reasons for not producing palm oil
  - Labor intensive
  - Lack of information

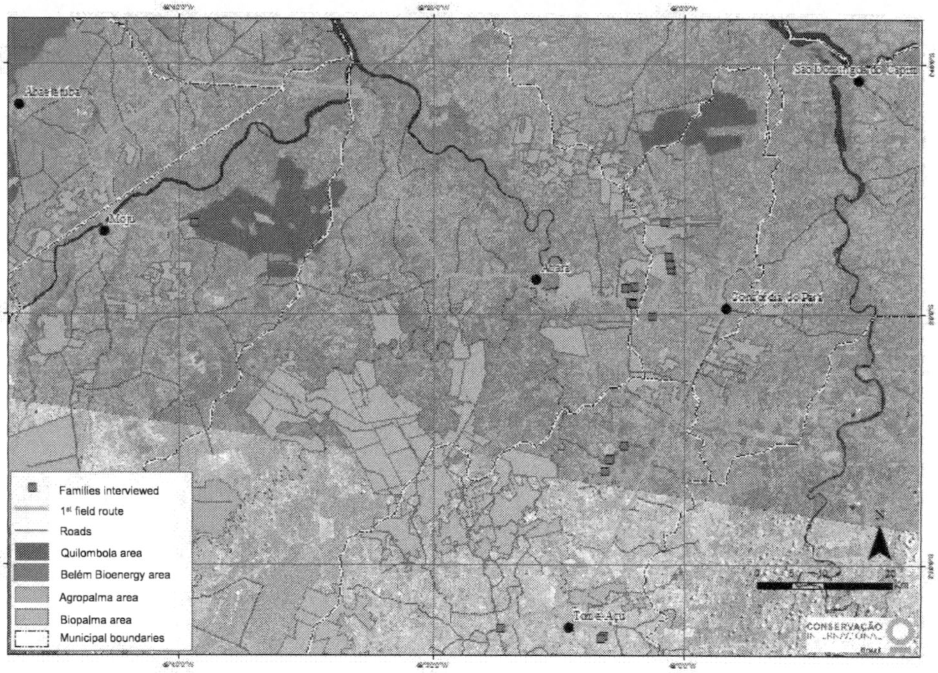

THANK YOU!

Tatiana Martins Souza
tsouza@conservation.org

Gustamo Simas - gustavosimas@gmail.com

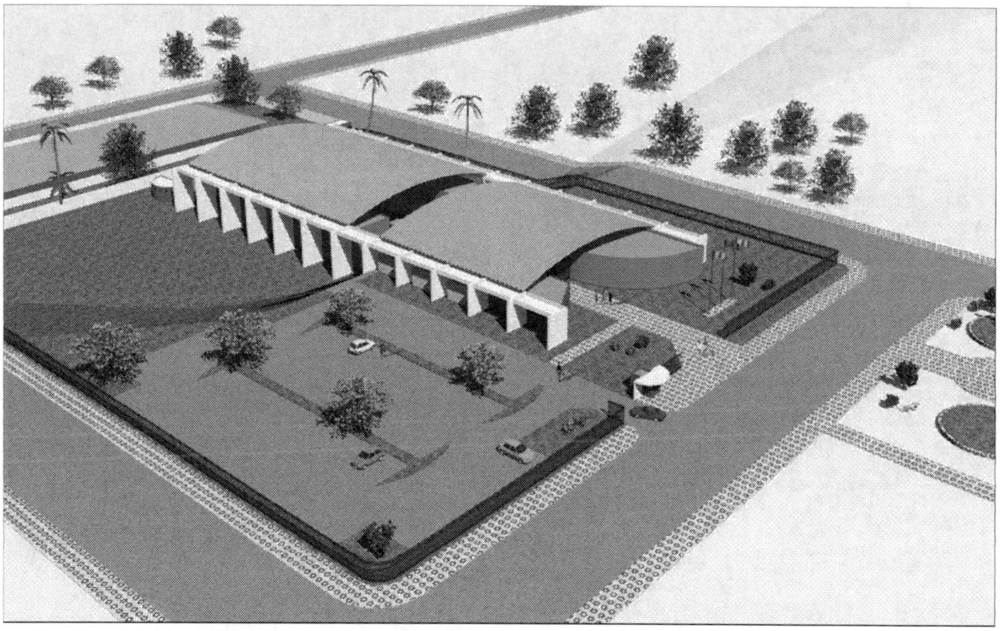

Centro Tecnológico do Araripe – CTA
Rua Antônio Alexandre Alves, 112
Bairro: Vila Santa Isabel, Araripina - PE

Secretaria de Ciência e Tecnologia PERNAMBUCO GOVERNO DO ESTADO

ITEP INSTITUTO DE TECNOLOGIA DE PERNAMBUCO

MODELO DE GESTÃO

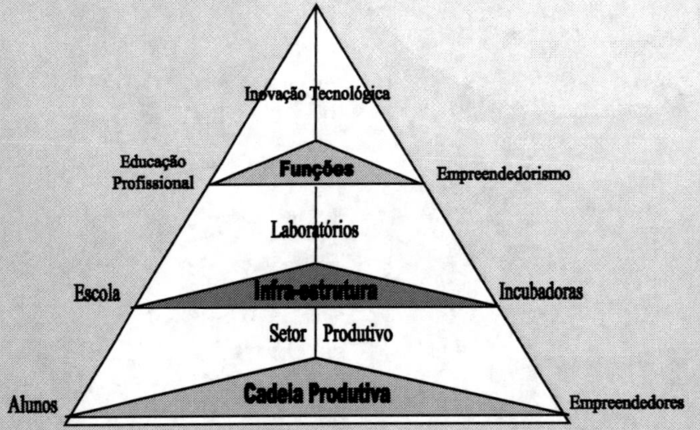

www.itep.br

Secretaria de Ciência e Tecnologia PERNAMBUCO GOVERNO DO ESTADO

ITEP INSTITUTO DE TECNOLOGIA DE PERNAMBUCO

EQUIPE

- ✓ **Gestora:** Cibele Rejane Lucena Lopes
- ✓ **Equipe Técnica:**
- ▪ 04 Profissionais
- ✓ **Equipe Administrativa**
- ▪ 02 Profissionais

INFRAESTRUTURA

Laboratórios: 11
Salas de Aulas: 05
Outros: Auditório, Sala para Reuniões e Planta de Calcinação

www.itep.br

Secretaria de Ciência e Tecnologia
PERNAMBUCO
GOVERNO DO ESTADO

ITEP
INSTITUTO DE TECNOLOGIA DE PERNAMBUCO

Educação Profissional

www.itep.br

Secretaria de Ciência e Tecnologia
PERNAMBUCO
GOVERNO DO ESTADO

ITEP
INSTITUTO DE TECNOLOGIA DE PERNAMBUCO

Cursos de Formação Inicial ou Continuada – FIC

- Auxiliar Técnico em Análises Químicas.
- Minerais Industriais e Desmontes de Rochas com foco no Polo Gesseiro.
- Aperfeiçoamento em Processos de Formalização de Empresas
- Eletricista Industrial
- Instalador Predial de Baixa Tensão
- Controlador Lógico Programável (CLP) e Programação Ladder

www.itep.br

Secretaria de Ciência e Tecnologia PERNAMBUCO

ITEP INSTITUTO DE TECNOLOGIA DE PERNAMBUCO

Figura 1 - Aulas práticas dos cursos FIC

www.itep.br

Secretaria de Ciência e Tecnologia PERNAMBUCO GOVERNO DO ESTADO

ITEP INSTITUTO DE TECNOLOGIA DE PERNAMBUCO

Cursos Técnicos:

Modalidade Presencial:

✓ Técnico em Química

✓ Técnico em Eletroeletrônica

Modalidade de EAD:

✓ Informática, Biblioteconomia, Recursos Humanos, Administração, Logística e Segurança do Trabalho, Secretaria Escolar e Multimeios Didáticos.

www.itep.br

ITEP
INSTITUTO DE TECNOLOGIA DE PERNAMBUCO

Secretaria de Ciência e Tecnologia
PERNAMBUCO
GOVERNO DO ESTADO

Figura 2 - Aulas práticas de Química

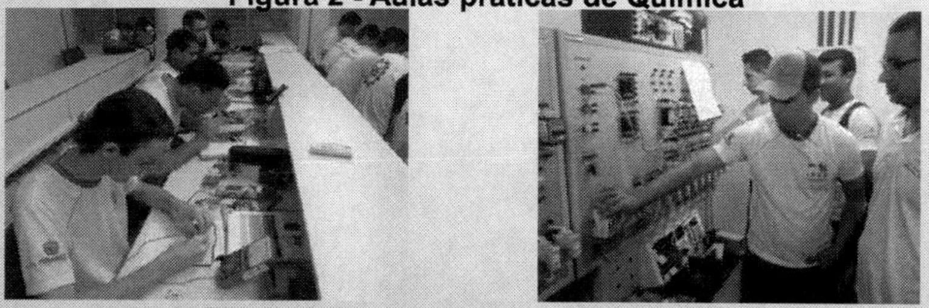

Figura 3 - Aulas práticas de Eletroeletrônica

www.itep.br

ITEP
INSTITUTO DE TECNOLOGIA DE PERNAMBUCO

Secretaria de Ciência e Tecnologia
PERNAMBUCO
GOVERNO DO ESTADO

Investimento Laboratorial e Prestação de Serviços

www.itep.br

Secretaria de Ciência e Tecnologia
PERNAMBUCO
GOVERNO DO ESTADO

ITEP
INSTITUTO DE TECNOLOGIA DE PERNAMBUCO

- Contratação de Técnicos;
- Adequação da infraestrutura laboratorial;
- Calibração e Recuperação de Equipamentos;
- Aquisição de Reagentes e Insumos;
- Prestação de serviços ao polo gesseiro:
  - o Ensaios Laboratoriais
  - o Melhoria da Qualidade de Produto
  - o Consultoria Empresarial

www.itep.br

Secretaria de Ciência e Tecnologia
PERNAMBUCO
GOVERNO DO ESTADO

ITEP
INSTITUTO DE TECNOLOGIA DE PERNAMBUCO

Realização de ensaios em termo-derivados de gipsita, no Laboratório de Controle da Qualidade e Desenvolvimento de Novos Produtos (LCQ), em atendimento do Polo Gesseiro.

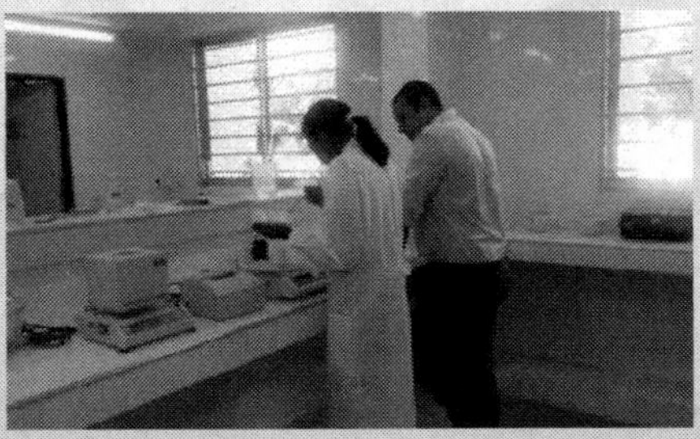

www.itep.br

Secretaria de Ciência e Tecnologia
PERNAMBUCO
GOVERNO DO ESTADO

ITEP
INSTITUTO DE TECNOLOGIA DE PERNAMBUCO

Planta de Calcinação

www.itep.br

Secretaria de Ciência e Tecnologia
PERNAMBUCO
GOVERNO DO ESTADO

ITEP
INSTITUTO DE TECNOLOGIA DE PERNAMBUCO

- Projeto de reforma no prédio da Planta de Calcinação;
- Adequação as NR's;
- Manutenção das máquinas e equipamentos existentes;
- Aquisição de Insumos;
- Previsão de prestação de serviços:
  - o Desenvolvimento de Tecnologia;
  - o Aulas práticas;
  - o Mudança na Matriz Energética.

www.itep.br

Secretaria de Ciência e Tecnologia
PERNAMBUCO
GOVERNO DO ESTADO

ITEP
INSTITUTO DE TECNOLOGIA DE PERNAMBUCO

Figura 5: Planta de Calcinação.

www.itep.br

Secretaria de Ciência e Tecnologia
PERNAMBUCO
GOVERNO DO ESTADO

ITEP
INSTITUTO DE TECNOLOGIA DE PERNAMBUCO

Ações Prioritárias

www.itep.br

ITEP
INSTITUTO DE TECNOLOGIA DE PERNAMBUCO

Secretaria de Ciência e Tecnologia
PERNAMBUCO
GOVERNO DO ESTADO

Apoio ao PROAPL

Figura 6: Reunião do PROAPL do Gesso

www.itep.br

ITEP
INSTITUTO DE TECNOLOGIA DE PERNAMBUCO

Secretaria de Ciência e Tecnologia
PERNAMBUCO
GOVERNO DO ESTADO

Atualização das Normas Técnicas

www.itep.br

Secretaria de Ciência e Tecnologia
PERNAMBUCO
GOVERNO DO ESTADO

ITEP
INSTITUTO DE TECNOLOGIA DE PERNAMBUCO

Parceria ITEP/OS e ABNT

- ITEP/OS está liderando o processo de Atualização das Normas Técnicas do setor gesseiro.
- Implantação da ISO/IEC 17.025 no **LCQ**.
- Aproximação do ITEP com os empresários do Araripe

www.itep.br

Secretaria de Ciência e Tecnologia
PERNAMBUCO
GOVERNO DO ESTADO

ITEP
INSTITUTO DE TECNOLOGIA DE PERNAMBUCO

Apoio ao Projeto Selo Branco

Figura 7: Visita às empresas.

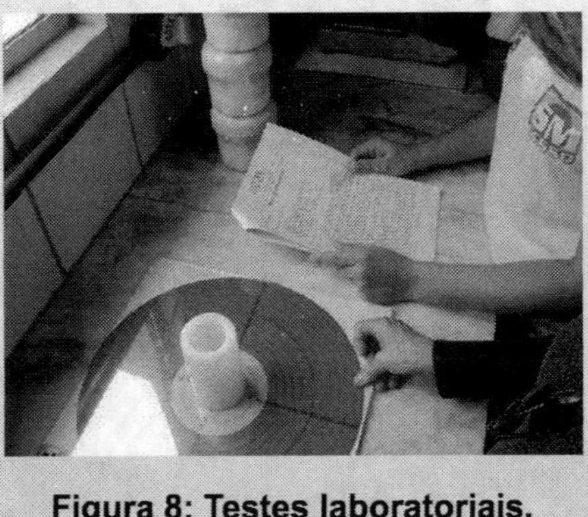

Figura 8: Testes laboratoriais.

www.itep.br

Secretaria de Ciência e Tecnologia

PERNAMBUCO

GOVERNO DO ESTADO

ITEP

INSTITUTO DE TECNOLOGIA DE PERNAMBUCO

Centro Tecnológico do Araripe

Direção: (87)3873-8377

Secretaria Escolar/Fax: (87)3873-8376

Secretaria Administrativa: (87)3873-8378

E-mail: ctararipe.itep@gmail.com

www.itep.br

RCN CONFERENCE ON PAN AMERICAN BIOFUELS AND BIOENERGY SUSTAINABILITY

SINDAÇÚCAR

PANORAMA DO SETOR DE BIOCOMBUSTÍVEIS E BIOENERGIA NO BRASIL

Marcelo Cavalcanti Guerra

Recife, 22 de julho de 2014.

RCN CONFERENCE ON PAN AMERICAN BIOFUELS AND BIOENERGY SUSTAINABILITY

SINDAÇÚCAR

O SETOR SUCROENERGÉTICO HOJE

- ✓ **ESTRUTURA PRODUTIVA**: 385 Unidades Produtoras
- ✓ **PRODUTORES DE CANA – DE –AÇÚCAR**: 65.000
- ✓ **EMPREGOS DIRETOS**: 1 milhão
- ✓ **PIB setorial:** US$ 48 bilhões
- ✓ **Exportações:** US$ 15 bilhões
- ✓ **% Matriz Energética:** 18% (2ª fonte)

1º PRODUTOR MUNDIAL DE AÇÚCAR

- ✓ 25% Produção mundial
- ✓ 50% Exportações mundiais

2º PRODUTOR MUNDIAL DE ETANOL

- ✓ 20% Produção mundial
- ✓ 20% Exportações mundiais

RCN CONFERENCE ON PAN AMERICAN BIOFUELS AND BIOENERGY SUSTAINABILITY

SINDAÇÚCAR®

O SETOR SUCROENERGÉTICO HOJE – REGIÃO NORDESTE

- ✓ **ESTRUTURA PRODUTIVA**: 77 Unidades Produtoras
- ✓ **PRODUTORES DE CANA – DE –AÇÚCAR**: 25.000
- ✓ **EMPREGOS DIRETOS**: Cerca de 330.000
- ✓ **PRODUÇÃO DE CANA**: 63 Milhões Tons. (Safra 2010/2011)
- ✓ **PRDOUÇÃO DE AÇÚCAR**: 4,6 Milhões Tons. (Safra 2010/2011)
- ✓ **PRODUÇÃO DE ETANOL**: 1,9 Milhões M^3 (Safra 2010/2011)

RCN CONFERENCE ON PAN AMERICAN BIOFUELS AND BIOENERGY SUSTAINABILITY

SINDAÇÚCAR®

Matriz Energética Brasileira (2012)

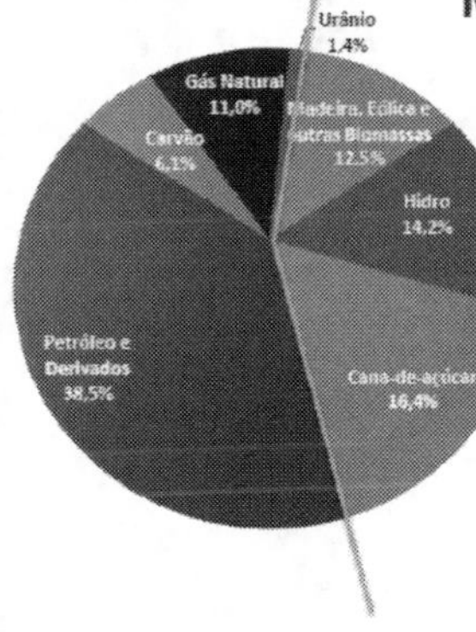

RENOVÁVEIS / MATRIZ ENERGÉTICA
BRASIL: 43%
MUNDO: 13%
OCDE: 7.6%

Energias renováveis na matriz energética

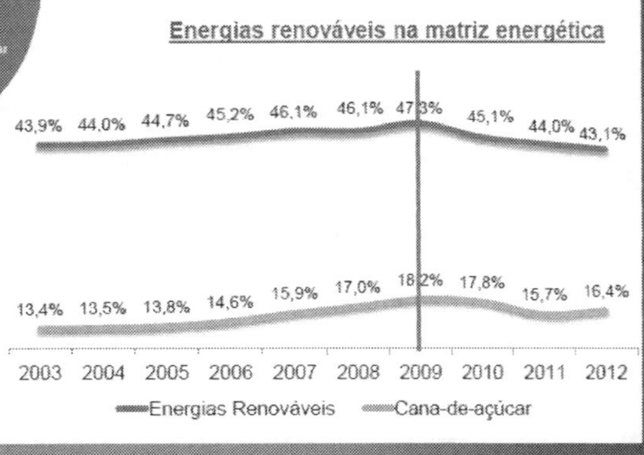

O Brasil tem a matriz energética mais limpa do planeta, mas a cana-de-açúcar perdeu relevância nos últimos 3 anos!!

RCN CONFERENCE ON PAN AMERICAN BIOFUELS AND BIOENERGY SUSTAINABILITY

SINDAÇÚCAR®

PRODUÇÃO / CONSUMO ETANOL BRASIL
ÚLTIMOS 5 ANOS

ANO	ANIDRO (M3)		HIDRATADO (M3)	
	PRODUÇÃO	CONSUMO	PRODUÇÃO	CONSUMO
2013	9.693.072,00	10.341.315,00	13.762.534,00	11.754.963,00
2012	8.623.622,00	9.924.428,00	14.077.260,00	9.850.180,00
2011	8.025.559,00	7.098.251,00	19.574.260,00	10.899.221,00
2010	7.065.250,00	5.968.733,00	18.625.675,00	15.074.300,00
2009	9.630.574,00	5.081.818,00	18.058.226,00	16.470.948,00

Fonte: ANP

RCN CONFERENCE ON PAN AMERICAN BIOFUELS AND BIOENERGY SUSTAINABILITY

SINDAÇÚCAR®

Cenário Base
Consumo de Etanol Hidratado e Anidro + Exportações

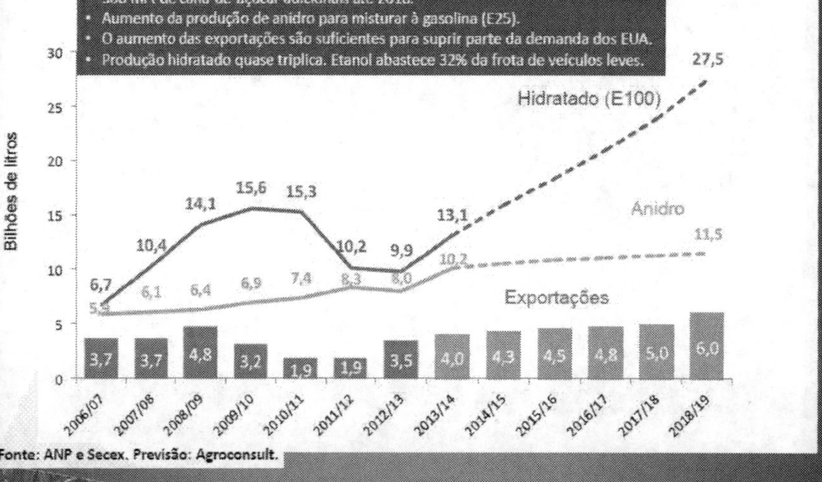

Fonte: ANP e Secex. Previsão: Agroconsult.

RCN CONFERENCE ON PAN AMERICAN BIOFUELS AND BIOENERGY SUSTAINABILITY

SINDAÇÚCAR®

VEÍCULOS E MOTOCICLETAS FLEX FUEL

GM
PEUGEOT
Ford
FIAT
CITROËN
KIA MOTORS
JAC MOTORS
HYUNDAI
RENAULT
MITSUBISHI
YAMAHA
TOYOTA
NISSAN
CHERY
HONDA

Hoje 15 fabricantes oferecem 212 modelos *flex fuel* e 2 empresas vendem 5 modelos de motocicletas *flex*

RCN CONFERENCE ON PAN AMERICAN BIOFUELS AND BIOENERGY SUSTAINABILITY

SINDAÇÚCAR®

Veículos Flex Fuel (FF) no Brasil

	Fabricante	% modelos & versões FF*
1	VW	81,8
2	Fiat	93,8
3	Renault	100,0
4	Peugeot	84,2
5	Citröen	73,7
6	GM	76,2
7	Ford	74,1
8	Toyota	75,0
9	Honda	66,7
10	Nissan	100.0
11	Mitsubishi	25,0
12	Hyundai	14,3
13	Kia	23,0
14	JAC	16,6
15	Chery	16,6

173 modelos & versões FF disponíveis

FF: > 90% vendas de veículos leves Otto em 2012

Uso do Etanol:

- Emissão reduzida de CO_2
- Menor emissão de SOx e MP
- Menor toxidez dos COV

Legenda
CO2 – Dióxido de Carbono;
Sox – Óxido de Enxofre;
MP – Material Particulado
COV – Compostos Orgânicos Voláteis

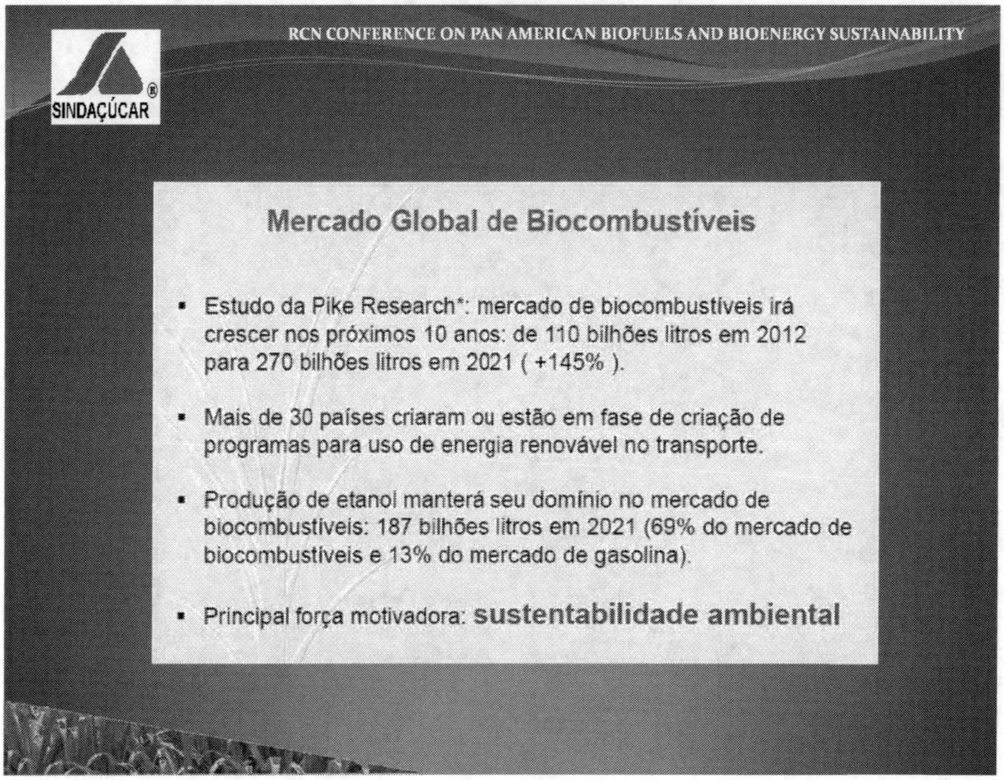
RCN CONFERENCE ON PAN AMERICAN BIOFUELS AND BIOENERGY SUSTAINABILITY
SINDAÇÚCAR®
Mercado Global de Biocombustíveis
Estudo da Pike Research*: mercado de biocombustíveis irá crescer nos próximos 10 anos: de 110 bilhões litros em 2012 para 270 bilhões litros em 2021 (+145%).
Mais de 30 países criaram ou estão em fase de criação de programas para uso de energia renovável no transporte.
Produção de etanol manterá seu domínio no mercado de biocombustíveis: 187 bilhões litros em 2021 (69% do mercado de biocombustíveis e 13% do mercado de gasolina).
Principal força motivadora: sustentabilidade ambiental

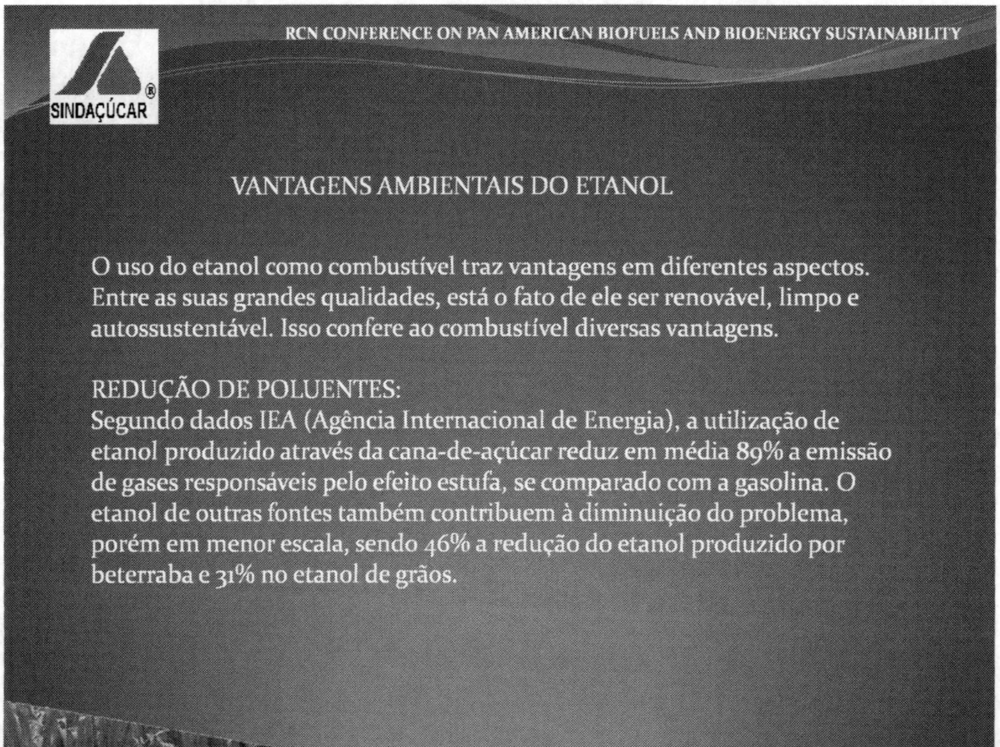
RCN CONFERENCE ON PAN AMERICAN BIOFUELS AND BIOENERGY SUSTAINABILITY
SINDAÇÚCAR®
VANTAGENS AMBIENTAIS DO ETANOL
O uso do etanol como combustível traz vantagens em diferentes aspectos. Entre as suas grandes qualidades, está o fato de ele ser renovável, limpo e autossustentável. Isso confere ao combustível diversas vantagens.
REDUÇÃO DE POLUENTES:
Segundo dados IEA (Agência Internacional de Energia), a utilização de etanol produzido através da cana-de-açúcar reduz em média 89% a emissão de gases responsáveis pelo efeito estufa, se comparado com a gasolina. O etanol de outras fontes também contribuem à diminuição do problema, porém em menor escala, sendo 46% a redução do etanol produzido por beterraba e 31% no etanol de grãos.

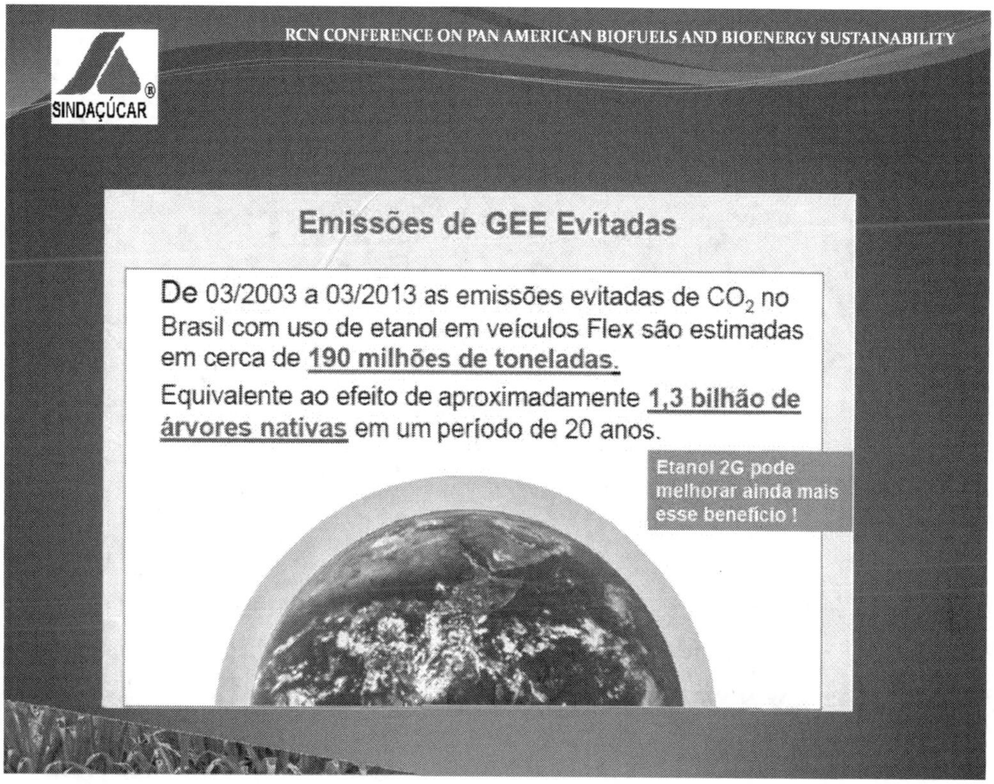
RCN CONFERENCE ON PAN AMERICAN BIOFUELS AND BIOENERGY SUSTAINABILITY
SINDAÇÚCAR
Emissões de GEE Evitadas
De 03/2003 a 03/2013 as emissões evitadas de CO_2 no Brasil com uso de etanol em veículos Flex são estimadas em cerca de 190 milhões de toneladas.
Equivalente ao efeito de aproximadamente 1,3 bilhão de árvores nativas em um período de 20 anos.
Etanol 2G pode melhorar ainda mais esse benefício !

RCN CONFERENCE ON PAN AMERICAN BIOFUELS AND BIOENERGY SUSTAINABILITY
SINDAÇÚCAR
Aspectos da Bioeletricidade da Cana de Açúcar: Situação em 2012
• Autossuficiência na safra: todas as mais de 385 usinas do setor sucroenergético
• Venda de excedentes de energia elétrica: em torno de 160 usinas
2.865 MW médios
1.381 MW médios
Venda de excedentes
1 Uruguai + 1 Jamaica (3% do Brasil)
Cogeração: Bagaço e palha na Caldeira
1.484 MW médios
Autossuficiência
Mais da ½ de Portugal ou 6% do Brasil
1 Bolívia + 1 Paraguai (3% do Brasil)
Fonte: MME e UNICA (2013) e The World Factbook (2012).

RCN CONFERENCE ON PAN AMERICAN BIOFUELS AND BIOENERGY SUSTAINABILITY

SINDAÇÚCAR®

Bioeletricidade: considerações

Qual será a participação da bioeletricidade na matriz energética?

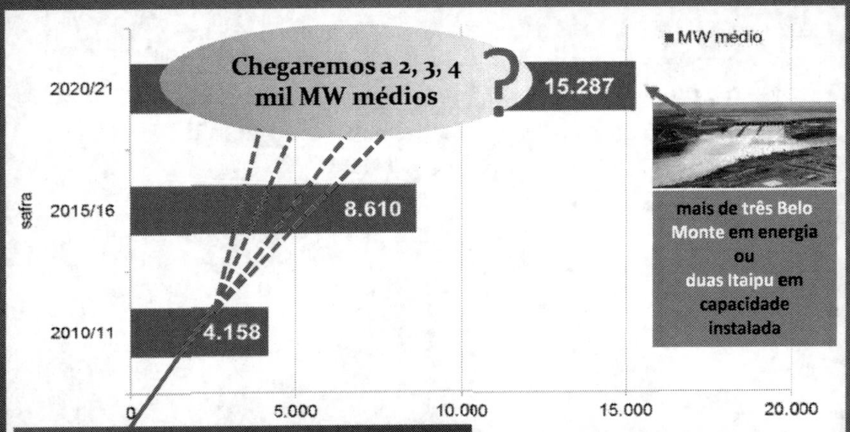

Realizado 2012: só 1.400 MW médios

Política setorial de longo prazo
Estímulo para investir em bioeletricidade, etanol e açúcar

RCN CONFERENCE ON PAN AMERICAN BIOFUELS AND BIOENERGY SUSTAINABILITY

SINDAÇÚCAR®

Mandatos de mistura de etanol na gasolina

(E5 a E25)												
País	Mistura Etanol	Uso	País	Mistura Etanol	Uso	País	Mistura Etanol	Uso	Estado	Mistura Etanol	Estado	Mistura Etanol
Países com misturas obrigatórias ou disponível para uso opcional						União Europeia			Estados Unidos* (Somente os estados com mistura obrigatória)			
Austrália	E10	Opcional	Índia	E5	Obrigatória	Áustria	E10	Opcional	Florida	E10	Minnesota	E10
Brasil	E20 a E25	Obrigatória	Jamaica	E10	Obrigatória(5)	Dinamarca	E5	Opcional	Hawaii	E10	Missouri	E10
Canadá	E5/E10	Opcional	Nova Zelândia	E10	Opcional	Finlândia	E5	Opcional	Iowa	E10	Montana	E10
China	E10	Nove províncias	Paraguai	E12	Obrigatória	Suécia	E5	Obrigatória	Kansas	E10	Oregon	E10
Colômbia	E10	Obrigatória	Tailândia	E10/E20	Obrigatória				Louisiana	E10	Washington	E10
Costa Rica	E7	Obrigatória							*Ainda que somente é obrigatória em 10 estados, as misturas de baixo teor de etanol estão disponíveis em outros estados opcionalmente ou misturado para oxigenar			

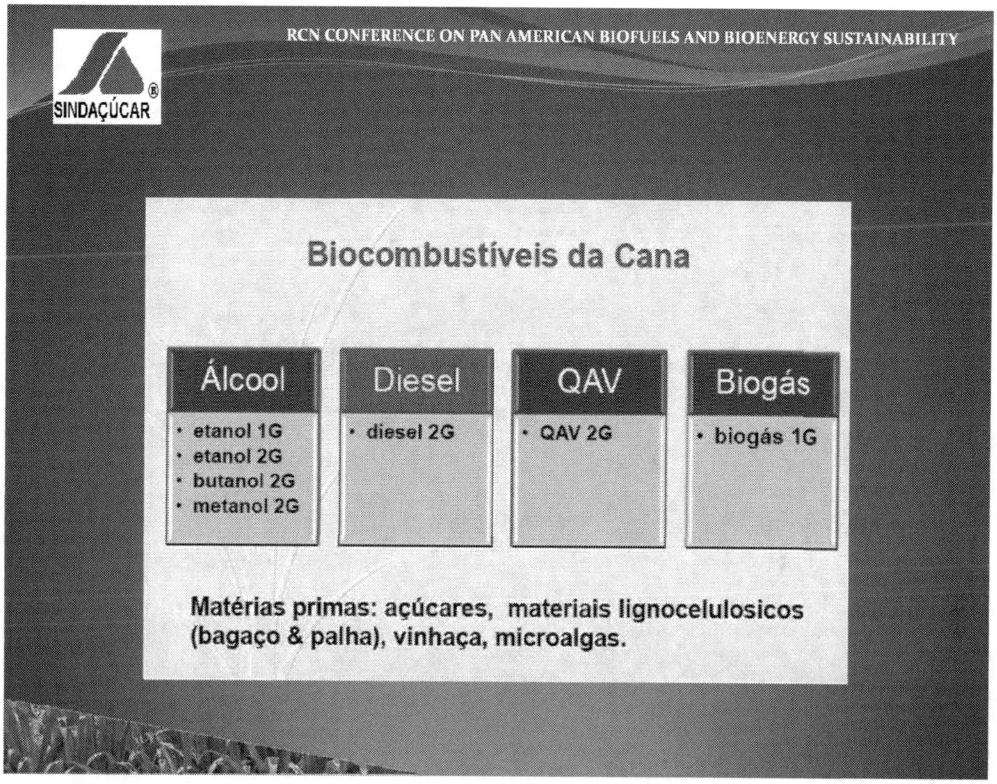
SINDAÇÚCAR
RCN CONFERENCE ON PAN AMERICAN BIOFUELS AND BIOENERGY SUSTAINABILITY
Biocombustíveis da Cana
Álcool
• etanol 1G
• etanol 2G
• butanol 2G
• metanol 2G
Diesel
• diesel 2G
QAV
• QAV 2G
Biogás
• biogás 1G
Matérias primas: açúcares, materiais lignocelulosicos (bagaço & palha), vinhaça, microalgas.

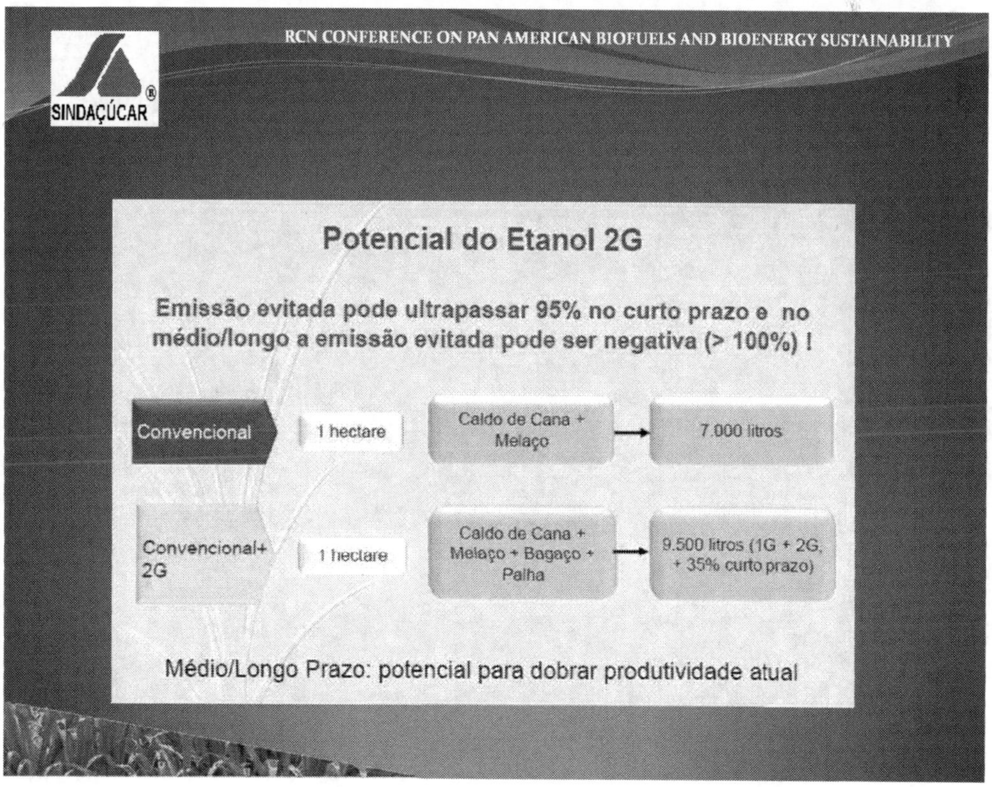
SINDAÇÚCAR
RCN CONFERENCE ON PAN AMERICAN BIOFUELS AND BIOENERGY SUSTAINABILITY
Potencial do Etanol 2G
Emissão evitada pode ultrapassar 95% no curto prazo e no médio/longo a emissão evitada pode ser negativa (> 100%) !
Convencional
1 hectare
Caldo de Cana + Melaço
7.000 litros
Convencional+ 2G
1 hectare
Caldo de Cana + Melaço + Bagaço + Palha
9.500 litros (1G + 2G, + 35% curto prazo)
Médio/Longo Prazo: potencial para dobrar produtividade atual

RCN CONFERENCE ON PAN AMERICAN BIOFUELS AND BIOENERGY SUSTAINABILITY

SINDAÇÚCAR®

OPORTUNIDADE: RIN (RENEWABLE IDENTIFICATION NUNBER

A Renewable Identification Number (ou RIN) é um número de série atribuído a um lote de biocombustível com a finalidade de acompanhar sua produção, uso e comercialização, conforme exigido pela Agência de Proteção Ambiental dos Estados Unidos 's Renewable Fuel Standard implementado de acordo com a Política Energética Act de 2005.

Sob a Lei de Política Energética 2005, a EPA está autorizado a estabelecer quotas anuais ditar qual a percentagem do montante total de combustíveis consumidos em os EUA devem ser representados por biocombustível misturado em combustíveis fósseis . As empresas que refinam, importar ou misturar os combustíveis fósseis são obrigados a cumprir determinadas quotas RFS indivíduo com base no volume de combustível que introduzem no mercado.

Ao cumprir estes requisitos, os projetos da EPA que a indústria irá satisfazer coletivamente a quota nacional em geral eles definido. Para garantir o cumprimento, as partes sujeitas a obrigação sejam periodicamente obrigados a demonstrar que conheceram a sua quota RFS através da apresentação de uma certa quantidade de RINs com a EPA. Como cada uma destas RINs representam uma quantidade de biocombustíveis que foi misturado com combustíveis fósseis, os RINs submetidos à EPA por partes obrigadas são uma representação quantitativa da quantidade de biocombustíveis que foi misturado com os combustíveis fósseis utilizados na América.

RCN CONFERENCE ON PAN AMERICAN BIOFUELS AND BIOENERGY SUSTAINABILITY

SINDAÇÚCAR®

POLÍTICAS PÚBLICAS NECESSÁRIA PARA O SETOR NO CURTO PRAZO:

- TRIBUTAÇÃO DIFERENCIADA QUE ESTIMULE O CONSUMO DO ETANOL COMBUSTÍVEL;
- AUMENTO DA MISTURA DE ETANOL NA GASOLINA
- ESTUDO SOBRE A EFICIÊNCIA DE MOTORES USADOS NOS CARROS FLEXS

RCN CONFERENCE ON PAN AMERICAN BIOFUELS AND BIOENERGY SUSTAINABILITY

SINDAÇÚCAR®

OBRIGADO

Marcelo Cavalcanti Guerra

mguerra@sindacucar.com.br

Bioenergy and Biofuels: Presentation

Sicsú, A., UFPE

Originally presented on: 7/24/2014 13:40:00 - 14:00:00

The proposal of this specific table of discussion on Government and Industry interface for Bioenergy and Biofuels on the view that the action of the state in regulating and supporting the sector in Brazil has always been strategic and essential to the definition of a pattern of sectorial development. In this direction, as political issues for the feasibility of public and private investment; impacts in relation to the environment; labor relations; government regulation for the production and trading of electricity; promotion of research; agricultural zoning are the key points. The table was structured with the following proposal: a brief opening presentation. Eric will make a presentation of the important international experience in the commercial exploitation of energy forests and debate on ethanol and biodiesel. Suani also doing a presentation on biofuel sustainability in various aspects. As the topic of debate is vast it seems that will focus on the possibilities of introduction of raw materials and innovations that increase productivity and reduce the differences in agricultural productivity in certain regions. The need for public policies to support the development of supply chains is another important point. It seems very unlikely that the operation of the market alone would reach these guidelines. Geraldo Eugenio will put in context the perspective of the Ethanol Industry in Brazil, including the outputs of the current crisis and Marcelo Guerra, with a practical view of those who follow the daily lives of the sector, will present the view of the productive sector about the current situation and Brazilian prospect. Romulo Meneses will report on recent studies on the potential for energy recovery from biomass sources in the Northeast region with emphasis on biodiesel. He will present the reality of this specific sector and its prospects, including the technological stand point. In this debate four strands emerge as guiding practices. Agronomic Technology and agro-climatic conditions defined for the type of crop or raw material being used; agro industrial and industrial technology established and developed for production in small, medium and large scale; logistics and infrastructure for the production and circulation of goods in the economic production chain; the minimum scale of production, complementarity in production processes and simultaneity of production shares. All this items will be discussed for the development of an institutional arrangement that is being built between the State and Industry as effective partnerships.

Envergent
TECHNOLOGIES
A Honeywell Company

RTP™ Rapid Thermal Processing:

MJCleveland
Vice President and General Manager Latin America
Recife, Brazil
July 24, 2014

Envergent Technologies LLC – UOP / Ensyn Joint Venture

- **Formed in October 2008**
- **Delivers Rapid Thermal Processing (RTP™) technology for energy generation**

UOP A Honeywell Company	ENSYN
• Independent oil, gas and renewables technology provider	• Over 20 years of commercial fast pyrolysis operating experience
• Commercialized first heterogeneous catalyst and synthetic zeolite.	• Developers of innovative RTP™ fast pyrolysis process
• Modular process unit supplier	• Seven commercial RTP units designed and operated
• Global strength of Honeywell & UOP sales and service teams	

Second Generation Renewable Energy Company – Global Reach

RTP – Second Generation Residues to Energy

Envergent Technologies
A Honeywell Company

RTP Green Fuel

- Transportable fuel
- Energy densification relative to biomass
- Maximum liquid yield 65 – 75 wt%

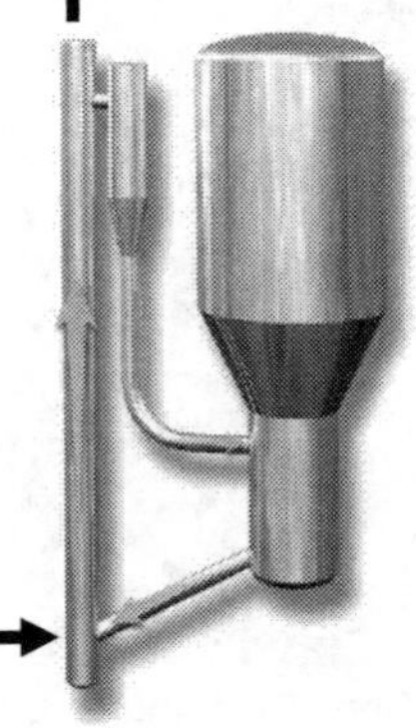

Fuel Oil Substitution

Electricity Production

Upgrade to Transport Fuels (Gasoline, Jet & Diesel)

Decouples biomass conversion from energy generation

Envergent Technologies
A Honeywell Company

Renewable Feedstock Sources

Envergent Technologies
A Honeywell Company

- Forest Industry
  - ✓Wood chips, sawdust and bark
  - ✓Forest residues
- Agricultural
  - ✓Residues – corn stover, expended fruit bunches from palm (EFB), bagasse,
  - ✓Purpose-grown energy crops – miscanthus and elephant grass

Second Generation Feedstock Widely Available

History and Commercial Experience

Envergent Technologies
A Honeywell Company

- **Commercialized in the 1980's**
- **7 units designed and operated in the US and Canada**
- **Continuous process with >90% availability**

New Projects Under Development:

Location	*Application*	*Size (TPD)*
Northern Europe	District Heating	Up to 3 X 400
South America	Industrial Process Heat	400
Malaysia	Industrial Process Heat	150
North America	Building Heat	400
North America	Refinery Co-Processing	100
North America	Refinery Co-Processing	400

RTP Green Fuel Energy Applications

Envergent Technologies
A Honeywell Company

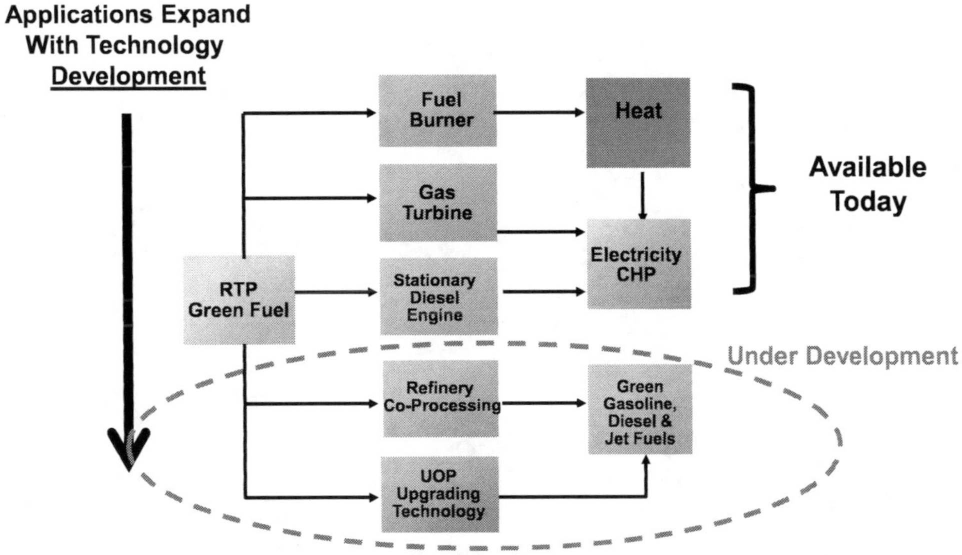

Increasing Applications Grow Market for RTP Green Fuel

ENV 5233-09

Stand Alone Upgrading of RTP Green Fuel To Transportation Fuel

Envergent
TECHNOLOGIES
A Honeywell Company

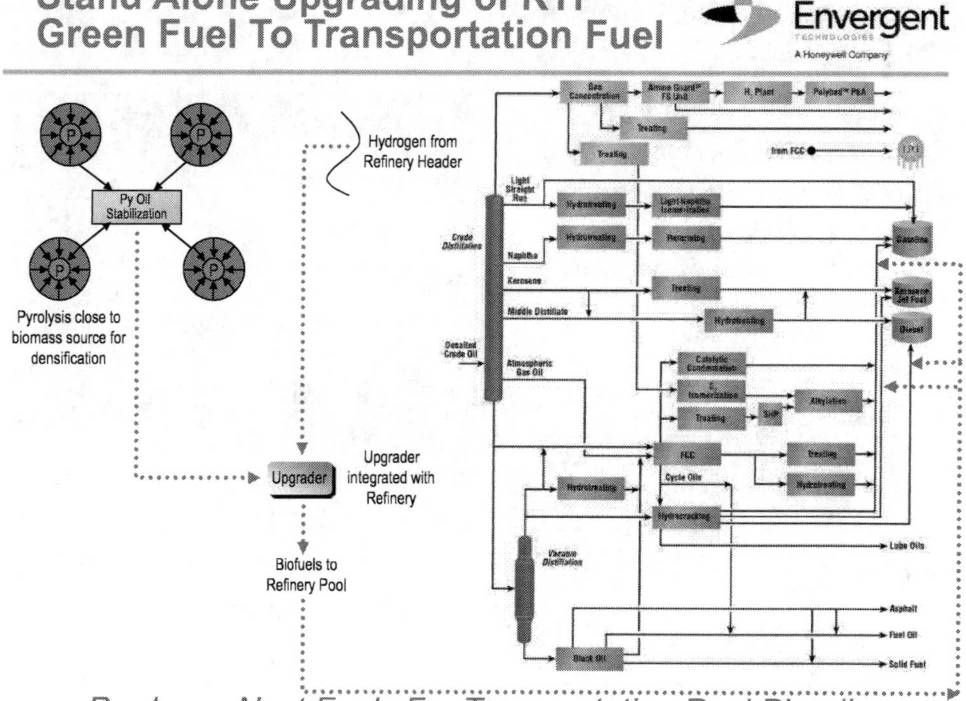

Produces Neat Fuels For Transportation Pool Blending

Upgrading RTP Green Fuel To Transportation Fuels

Envergent
TECHNOLOGIES
A Honeywell Company

Objectives:

- Remove oxygen as water and CO_2 by hydrogen & catalyst
- pH neutral fuel with viscosity equivalent to refined fuels
- Produce high octane gasoline, or diesel/jet components

Two Stage Hydrodeoxygenation

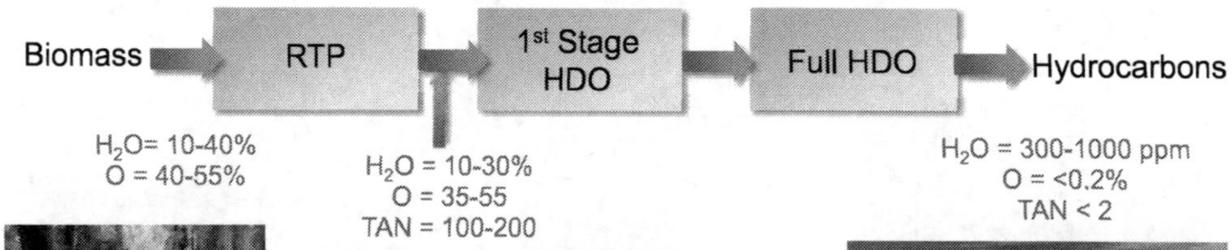

Upgraded Pyrolysis Oil Products

Envergent Technologies
A Honeywell Company

RTP Green Fuel to Fuels Feed/Product Analysis			
	RTP Green Fuel	Upgraded Fuel	Gasoline Requirements
Water, %	~25	0.03	<0.1
O, %	51	<0.1	<2.0
TAN, meq/g	91	<0.1	<0.1

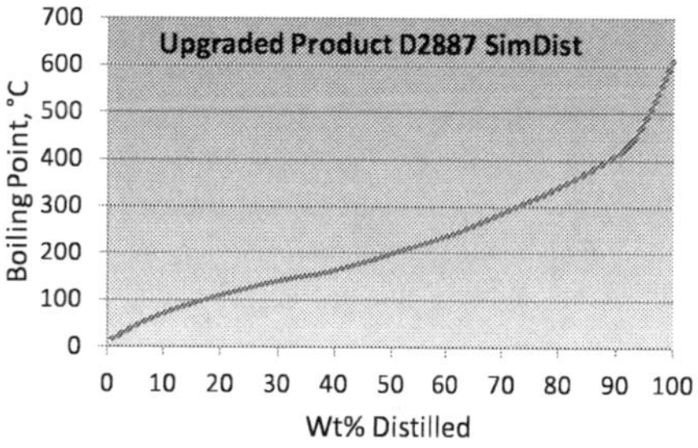

RTP Green Fuel Transportation Fuel Yield[1]	
	Overall Yield, % of pyrolysis oil feed
Mass	41
Volume	60[2]

1. Demonstrated yield at multiple equipment scales.
2. Equals > 90 gallons per dry MT for woody biomass.

- ~50% of material in gasoline boiling range 40-200°C
- RON of gasoline ~85-90
- ~40% of material in diesel/jet boiling range

Renewable Gasoline GHG Emissions

Envergent Technologies
A Honeywell Company

PRELIMINARY MODEL RESULTS (WOODY FEEDSTOCK)

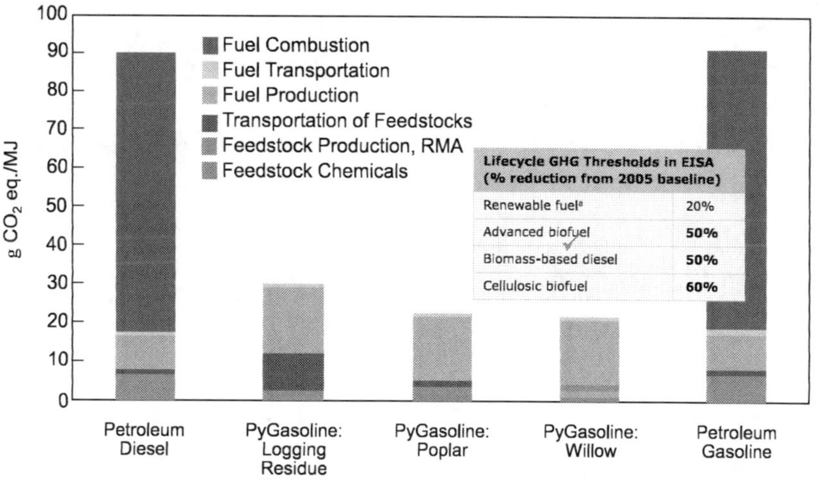

Lifecycle GHG Thresholds in EISA (% reduction from 2005 baseline)	
Renewable fuel[a]	20%
Advanced biofuel	**50%**
Biomass-based diesel	**50%**
Cellulosic biofuel	**60%**

Upgrading RTP Green Fuel Makes Cellulosic Biofuels

ENV 5233-19

Co-Processing Of RTP Green Fuel To Produce Transportation Fuel

Envergent Technologies
A Honeywell Company

Pyrolysis close to biomass source for densification

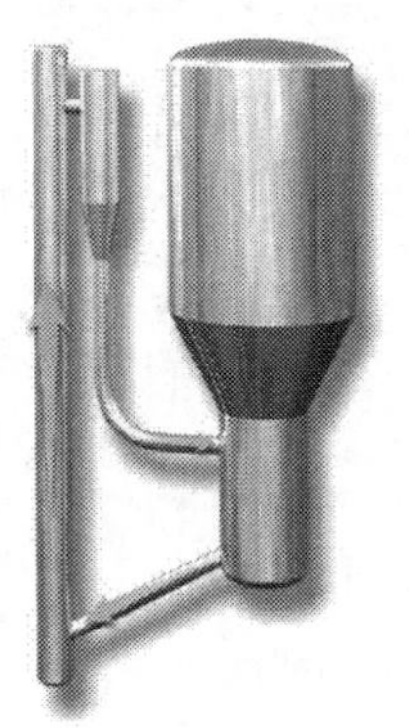

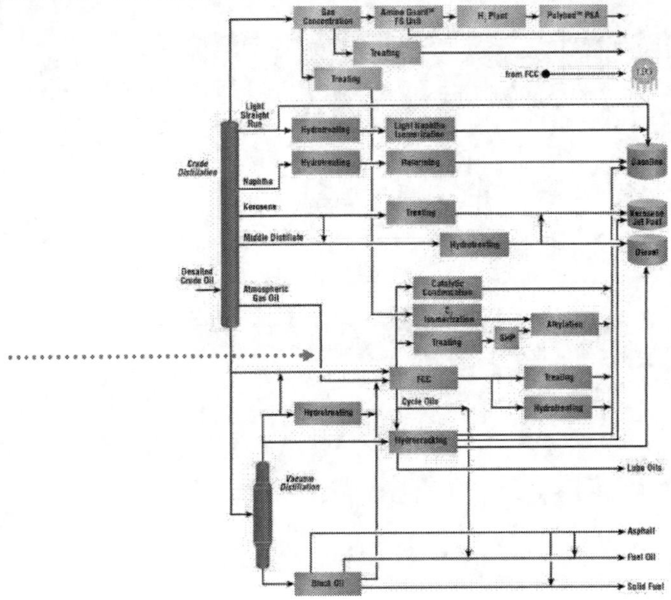

Commercial Application is Distributed Model

Co-Processing Of RTP Green Fuel To Produce Transportation Fuel

Envergent Technologies
A Honeywell Company

- In a co-processing application RTP green fuel is processed in a refinery
  FCC unit along with the traditional vacuum gas Oil (VGO) feed
- Feed blends containing up to 5% RTP green fuel are being considered
  by refiners
- Produces the same gasoline and diesel products with a renewable
  component
- Qualifies as a cellulosic biofuel under US RFS2
- Qualifies for double counting under European RED
- Low capex renewable fuel solution for the refiner
- High value added application for the RTP owner

RTP Summary

Envergent
TECHNOLOGIES
A Honeywell Company

- Commercially proven fast pyrolysis process for converting biomass into a renewable fuel
  - Seven units and 20 years of commercial experience
- Decouples biomass conversion from energy generation
  - RTP located near biomass – energy generation located near the consumer
  - Enables on-demand renewable energy production
- High yield of liquid product
  - Produces a transportable and storable liquid fuel for heat and power generation
  - Can be co-processed in refinery
  - Can be upgraded to transportation fuels at high yields on biomass

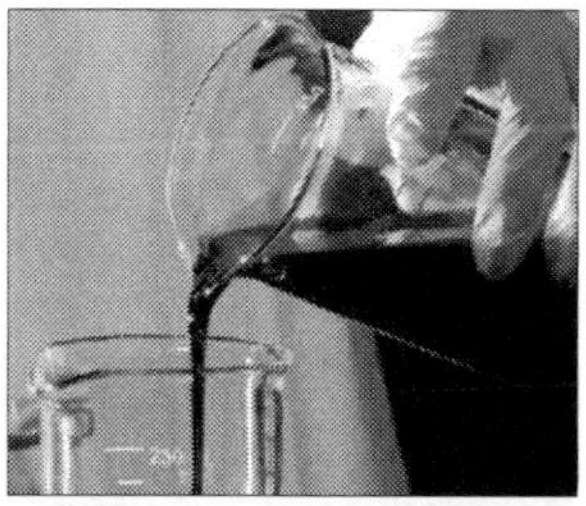

ENV 5233-04

Barriers to Solution Adoption

UOP
A Honeywell Company

- ***Is there a need for a solution***
- Practicality of the solution
- Government regulations / support – both in support and blocking
- Capital availability
- Robustness of the solution
- Financial returns
- Long term feedstock availability/cost
- Risk tolerance
- First mover advantage (and disadvantage)

Investors Look for Stability and Long Term Returns

Envergent
TECHNOLOGIES
A Honeywell Company

Thank You

RCN Conference on
Pan American Biofuels
& Bioenergy Sustainability
Golden Tulip Recife Palace,
Recife, Brazil July 22-25, 2014

Session: Industry-Government Perspectives on Biofuel and Bioenergy Sustainability

The sustainability of biofuels in Brazil

Suani T. Coelho, J. Goldemberg

Recife, July 24, 2014

General overview

Concept of sustainability: evolved considerably in the 20th century (from landscape/biodiversity conservation in IC´s).

Population growth and higher access to goods in former undeveloped countries:
- pressures on mineral resources,
- scarcity
- environmental problems (urban air pollution and global warming).

Sustainability concerns :
- Use of resources in a more rational fashion,
- Increased efficiency
- Shifting the resources used, and
- Use of renewable resources, particularly in the case of energy. (replacement of fossil fuels.
- Liquid biofuels, in which Brazil in the United States are the main producers.

However, liquid biofuels sustainability: still seen as a controversial issue considering environmental and social aspects.

Consensus (?) today that "bioenergy is not bad or good" but it depends on how it is produced.

- **CENBIO – the Brazilian Reference Center on Biomass**

- Created in 1996

- Studies on Biofuels and Bioenergy

↓

Bioenergy sustainability

http://cenbio.iee.usp.br

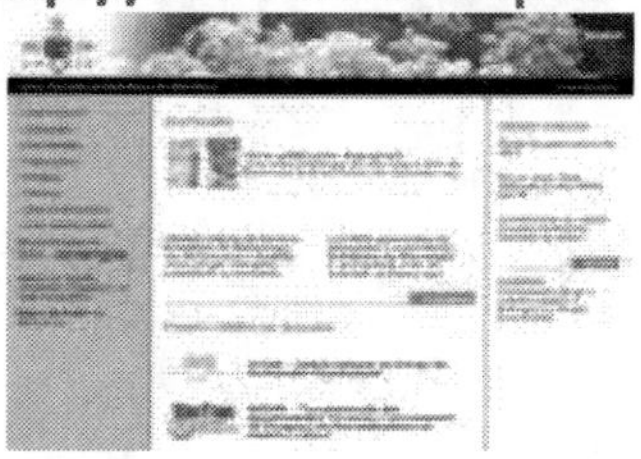

BIOACV Project – Life Cycle Assessment of Biodiesel from Soybean and Animal Fat through Methylic and Ethylic Routes (2013)

Funding Agency:

CNPq

Conselho Nacional de Desenvolvimento Científico e Tecnológico

- Environmental sustainability of biodiesel in Brazil. Érica Castanheira, Renata Grisoli, Fausto Freire, Vanessa Pecora, Suani Coelho. **Energy Policy**, v. 65,2014,p.680-691
- **Energy for Sustainability Conference 2013** Sustainable Cities: Designing for People and the Planet Coimbra, 8 - 10 September 2013
- Emissões de Gases de Efeito de Estufa no Ciclo de Vida do Biodiesel de Soja Produzido no Brasil. R.,Grisoli, A. Nogueira, É. G. Castanheira, F. Freire , G. A. Silva, S. Coelho. **III Congresso Brasileiro em Gestão do Ciclo de Vida de Produtos e Serviços.** 03 a 06 de setembro de 2012. Maringá – PR - Brasil

COMPARISON OF MUNICIPAL SOLID WASTE TECHNOLOGIES THROUGH LCA METHODOLOGY AS A TOOL FOR ADEQUATE POLICIES

Suani T. Coelho, Gil Anderi da Silva, Cristiane L. Cortez, Vanessa Pecora, Manuel Moreno, Alex Nogueira, Carlos Alberto da Silva, J.Goldemberg

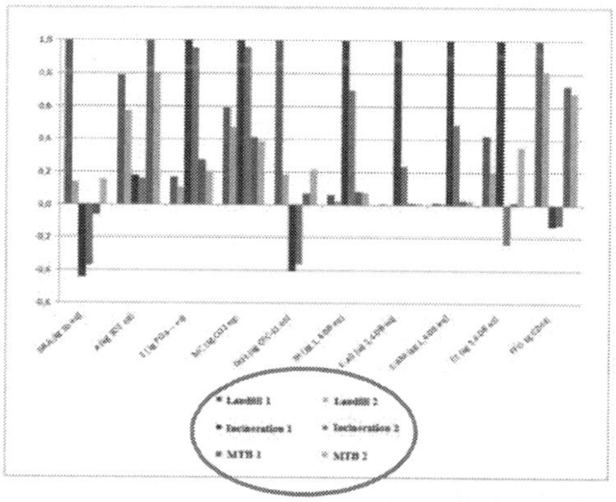

R&D Project CENBIO/IEE/USP/EMAE nº 0393-00611, 2011 – 2013, developed under the ANEEL (Brazilian Regulatory Agency of Electric Energy) regulation

Funded by EMAE

GBEP – The Global Bioenergy Partnership

The Global Bioenergy Partnership (co-chaired by Brazil and Italy) - FAO (Food and Agriculture Organization) - UNEP (United Nations Environmental Program)

- 24 bioenergy sustainability indicators (environmental, social and economic), accepted by all its member countries.
- Several countries already implementing this methodology (Europe, Africa and Latin America) for different types of bioenergy.

Brazil - a new study is starting to be developed for sugarcane ethanol mills in São Paulo State by the University of São Paulo, funded by the Government of Italy/Forum of the Americas with the support of the Brazilian Federal Government and the Secretariat for Environment of São Paulo.

Several certifications schemes already in place, but:

- GBEP indicators methodology seems to be a consensus among the governments members of GBEP and
- Good experience for comparing bioenergy among the countries.

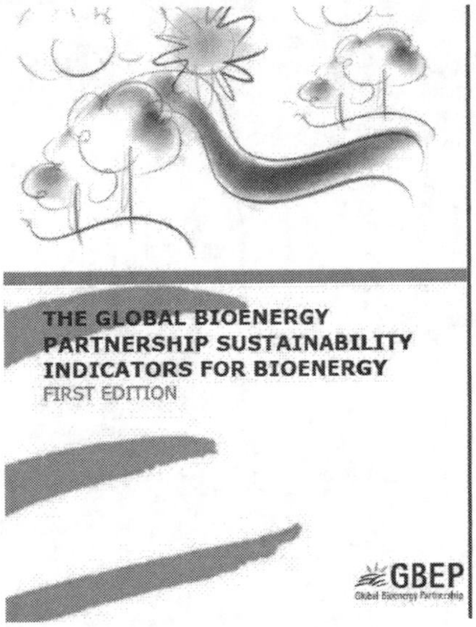

CENBIO

PILLARS GBEP's work on sustainability indicators was developed under the following three pillars, noting interlinkages between them:		
Environmental	**Social**	**Economic**
THEMES GBEP considers the following themes relevant, and these guided the development of indicators under these pillars:		
Greenhouse gas emissions, Productive capacity of the land and ecosystems, Air quality, Water availability, use efficiency and quality, Biological diversity, Land-use change, including indirect effects.	Price and supply of a national food basket, Access to land, water and other natural resources, Labour conditions, Rural and social development, Access to energy, Human health and safety.	Resource availability and use efficiencies in bioenergy production, conversion, distribution and end use, Economic development, Economic viability and competitiveness of bioenergy, Access to technology and technological capabilities, Energy security/Diversification of sources and supply, Energy security/Infrastructure and logistics for distribution and use.
INDICATORS		
1. Lifecycle GHG emissions	9. Allocation and tenure of land for new bioenergy production	17. Productivity
2. Soil quality	10. Price and supply of a national food basket	18. Net energy balance
3. Harvest levels of wood resources	11. Change in income	19. Gross value added
4. Emissions of non-GHG air pollutants, including air toxics	12. Jobs in the bioenergy sector	20. Change in consumption of fossil fuels and traditional use of biomass
5. Water use and efficiency	13. Change in unpaid time spent by women and children collecting biomass	21. Training and requalification of the workforce
6. Water quality	14. Bioenergy used to expand access to modern energy services	22. Energy diversity
7. Biological diversity in the landscape	15. Change in mortality and burden of disease attributable to indoor smoke	23. Infrastructure and logistics for distribution of bioenergy
8. Land use and land-use change related to bioenergy feedstock production	16. Incidence of occupational injury, illness and fatalities	24. Capacity and flexibility of use of bioenergy

CENBIO

- Countries already evaluating the indicators for biofuels:
  - Germany,
  - The Netherlands,
  - Colombia,
  - Indonesia,
  - Ghana and
  - Jamaica.
- **Brazil/São Paulo – starting 2014**

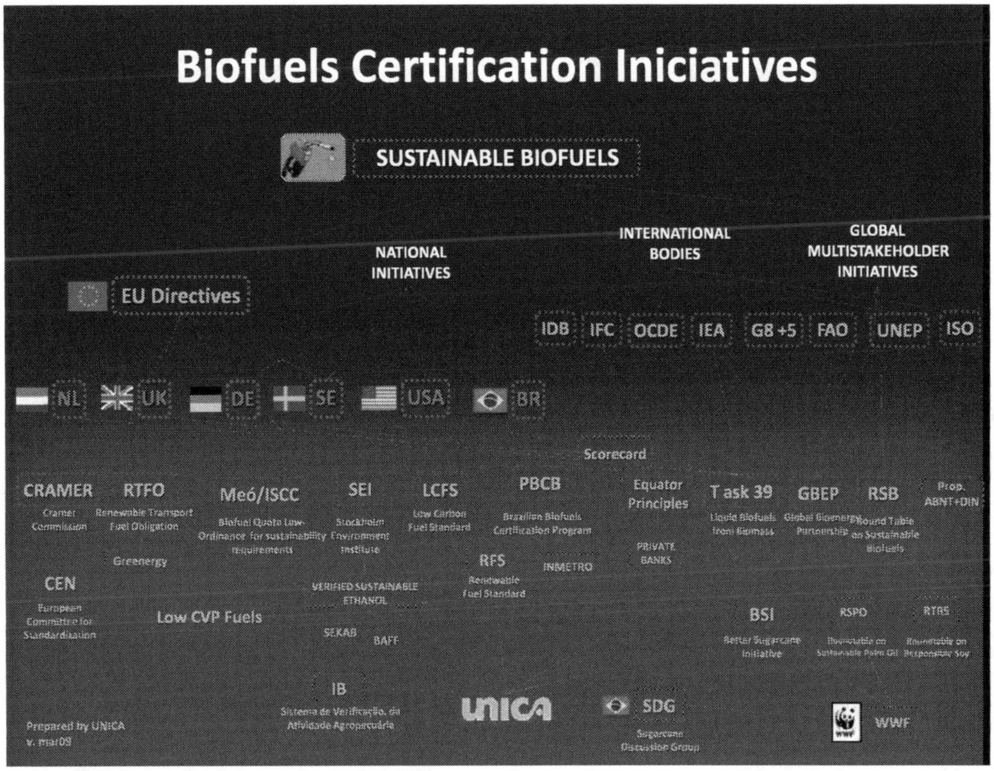

Bonsucro certification in Brazil

- Feb 2013 – 28 mills
- Feb 2014 – 36 mills

CENBIO

te	Production	Organization	Group Name	City	St
014	Sug. & Ethan	Unidade Serra	Raízen Energia S/A	Ibate	SI
2013	Ethanol	U. Morro Vermelho	Odebrecht Agroindustrial	Mineiros	G
2013	Sugarcane	Industrial Severínia	Guarani	Severínia	SI
2013	Sugarcane	Unid. Ind.Cruz Alta	Guarani	Olimpia	SI
2013	Sug. & Ethan	Unidade Junqueira	Raízen Energia S/A	Igarapava	SI
2013	Sugarcane	Santa Cruz S/A	Santa Cruz S/A	Américo Brás.	SI
2013	Sug. & Ethan	Unid. Dois Córregos	Raízen Energia S/A	Dois Córregos	SI
013	Sug. & Ethan	Unidade Junqueira	Alto Alegre	Colorado	PI
2013	Sug. & Ethan	Usina São Luiz S/A	Copersucar	Ourinhos	SI
2013	Sug. & Ethan	Unidade Univalem	Raízen Energia S/A	Valparaiso	SI
013	Sug. & Ethan	Usina Monte Alegre	Adecoagro	Monte Belo	M
013	Ethanol	Usina Rio Claro	Odebrecht Agroindustrial	Caçu	G
2012	Sug. & Ethan	Gasa	Raízen Energia S/A	Andradina	SI
2012	Sug. & Ethan	Tropical BioEnergia	BP Biocombustíveis	Edeia	G
2012	Sug. & Ethan	Unidade Bonfim	Raízen Energia S/A	Guariba	SI
2012	Sug. & Ethan	Usina Iracema	Grupo São Martinho	Iracemapolis	SI
2012	Sug. & Ethan	Unidade Sta Elisa	LDC SEV Bioenergia S/A	Sertãozinho	SI
2012	Sug. & Ethan	Usina Alta Mogiana	Usina Alta Mogiana S.A	São Joaquim	SI
012	Sugarcane	Usina Guariroba	Bunge	Pont. Gestal	SI
012	Sugarcane	Adecoagro Ivinhema	Adecoagro	Angélica	M
012	Sug. & Ethan	Destilaria Alcídia	Odebrecht Agroindustrial	T. Sampaio	SI
2012	Sugarcane	Usina Itapagipe	Bunge	Itapagipe	M
2012	Sug. & Ethan	USJ S.A.	USJ	Araras	SI
2012	Ethanol	Unidade Jatai	Raízen Energia S/A	Jatai	G
2011	Sugarcane	Usina Bom Retiro	Raízen Energia S/A	Capivari	SI
2011	Sugarcane	Usina Costa Pinto	Raízen Energia S/A	Piracicaba	SI
2011	Sugarcane	Conquista do Pontal	Odebrecht Agroindustrial	Paranapanema	SI
2011	Sug. & Ethan	Usina Moema.	Bunge	Orindiuva	SI
2011	Sug. & Ethan	Usina Frutal	Bunge	Frutal	M
2011	Sug. & Ethan	Usina Quatá	Copersucar (Zilor)	Quatá	SI
2011	Sug. & Ethan	Usina São Manoel	Copersucar	São Manoel	SI
2011	Sug. & Ethan	Usina Santa Adélia	Copersucar	Jaboticabal	SI
2011	Sug. & Ethan	Barra Gde Lençois	Copersucar (Zilor)	Lençois Paul.	SI
2011	Sug. & Ethan	Zillo Lorenzeti	Copersucar (Zilor)	Macatuba	SI
011	Sug. & Ethan	Equipav Mill	Renuka do Brasil S.A.	Promissão	SI
011	Sug. & Ethan	Usina Maracai	Raízen Energia S/A	Maracai	SI

Application of sustainability indicators of the Global Bioenergy Partnership (GBEP) in ethanol mills in the State of São Paulo, Brazil

(2014-2016 – CENBIO/IEE/USP – Forum of the Americas)

- Funding agency: Forum of the Americas (Government of Italy);
- Support: UNICA/SP, Secretariat for Environment of Sao Paulo, Brazilian Federal Government
- Coordinator: Suani T. Coelho (CENBIO/USP)
- Collaborators: J. Goldemberg (USP), C. Cerri, C.A. Cerri, Marcia Azanha (ESALQ/USP), among others
- Partners from Industries: importante support from industries
  - Odebrecht Agro Industrial
  - Grupo Raizen

Main Challenge

- Allocation of impacts;
- Production of both sugar and ethanol in some mills;
- Allocation on energy basis, mass basis, market basis

Three GBEP Sustainability Pillars

1. Environmental Sustainability
2. Social Sustainability
3. Economic Sustainability

Not applicable →

PILLARS

GBEP's work on sustainability indicators was developed under the following three pillars, noting interlinkages between them:

Environmental	Social	Economic
THEMES GBEP considers the following themes relevant, and these guided the development of indicators under these pillars:		
Greenhouse gas emissions, Productive capacity of the land and ecosystems, Air quality, Water availability, use efficiency and quality, Biological diversity, Land-use change, including indirect effects.	Price and supply of a national food basket, Access to land, water and other natural resources, Labour conditions, Rural and social development, Access to energy, Human health and safety.	Resource availability and use efficiencies in bioenergy production, conversion, distribution and end use, Economic development, Economic viability and competitiveness of bioenergy, Access to technology and technological capabilities, Energy security/Diversification of sources and supply, Energy security/Infrastructure and logistics for distribution and use.
INDICATORS		
1. Lifecycle GHG emissions	9. Allocation and tenure of land for new bioenergy production	17. Productivity
2. Soil quality	10. Price and supply of a national food basket	18. Net energy balance
3. Harvest levels of wood resources	11. Change in income	19. Gross value added
4. Emissions of non-GHG air pollutants, including air toxics	12. Jobs in the bioenergy sector	20. Change in consumption of fossil fuels and traditional use of biomass
5. Water use and efficiency	13. Change in unpaid time spent by women and children collecting biomass	21. Training and requalification of the workforce
6. Water quality	14. Bioenergy used to expand access to modern energy services	22. Energy diversity
7. Biological diversity in the landscape	15. Change in mortality and burden of disease attributable to indoor smoke	23. Infrastructure and logistics for distribution of bioenergy
8. Land use and land-use change related to bioenergy feedstock production	16. Incidence of occupational injury, illness and fatalities	24. Capacity and flexibility of use of bioenergy

Reduction of GHG emissions

- Ethanol from sugarcane reduces GHG emissions by 90%, to replace gasoline

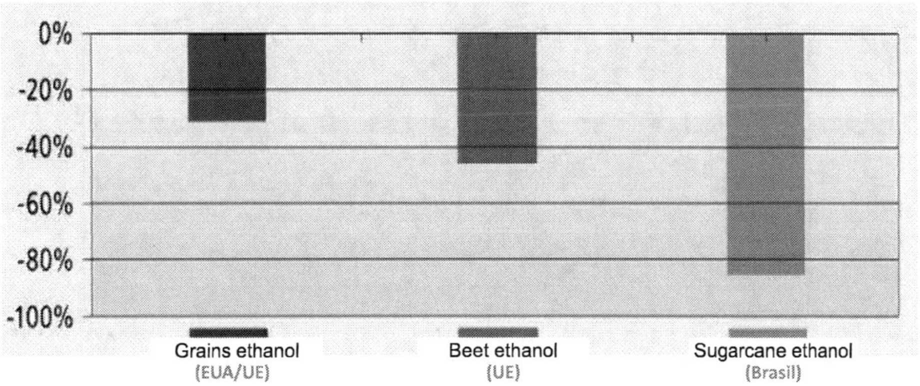

Source: IEA – International Energy Agency (2004). Elaboration: UNICA.

Water use in sugarcane ethanol

- Irrigation: most of the sugarcane produced in Brazil does not need irrigation.
- Industrial processes:
  - reduction on water consumption:
    1997: 5 m^3 /t sugarcane
    2004: 1.83 m^3/t sugarcane (average in São Paulo).
  - water recycling.
  - high efficiency in water treatment: 98%
  - environmental agency requirements : less than 1 m^3/t sugarcane
  - sugarcane washing process = 5m^3/tc (replaced by dry cleaning).

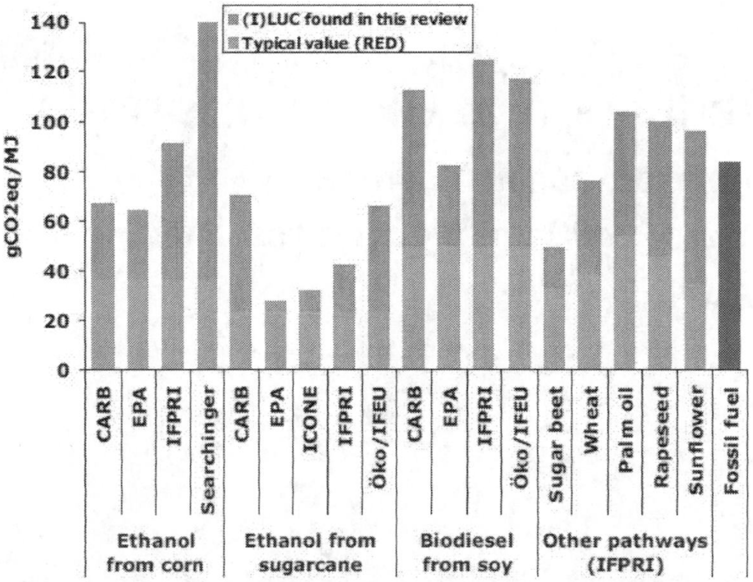

Figure 2 - 2 Graphical representation of the emissions caused by (I)LUC, direct and indirect land use change, for different biofuel pathways and different studies. For reference, typical non-land-use change emissions for the different pathways and a fossil reference from the EU Renewable Energy Directive (RED) have been added.

Location of the sugarcane production in Brazil

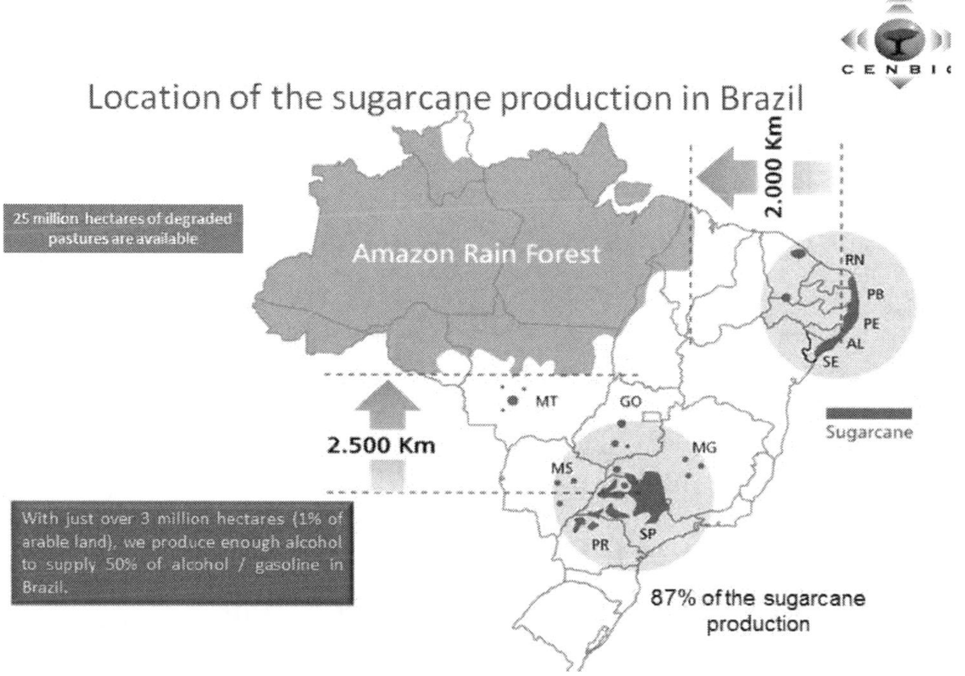

Land use in Brazil (20017-2011)

Million hectares (2007-2011)		TOTAL %	ARABLE LAND %
BRAZIL	851		
TOTAL ARABLE LAND	354.8		
1. CULTIVATED AREAS	76.7	9.0%	21.6%
SOY	20.6	2.4%	5.8%
CORN	14.0	1.6%	3.9%
SUGARCANE	7.8-9.6	0.9%-1.26%	2.2%-3.08%
SUGARCANE FOR ETHANOL	3.4-4.8	0.4%-0.56%	1.0%-1.4%
ORANGE	0.9	0.1%	0.3%
2. CATTLE	172.3	20.2%	48.6%
3. AVAILABLE LAND (ARABLE LAND – CULTIVATED LAND – CATTLE AREA)	105.8	12.4%	29.8%

Evolution of the pasture area in the State of Sao Paulo

	2001	2002	2003	2004	2005	2006	2007	2008
Cattle (million heads)	13,15	13,46	13,76	13,77	14,07	13,75	12,20	11,95
Pastures (million hectares)	10,29	10,10	10,11	10,12	10,01	9,71	9,12	7,64
Density (heads of cattle/ha)	1,28	1,33	1,36	1,36	1,41	1,42	1,34	1,56

Trend for more intensive use

Source: Secretaria de Estado de Agricultura. Elaboration: Cenbio

Evolution of the pasture area in the State of Sao Paulo (Cont)

	2009	2010	2011	2012	2013
Cattle (million heads)	11,00	11,37	11,15	11,08	9,84
Pastures (million hectares)	7,77	7,86	7,43	7,41	7,14
Density (heads of cattle/ha)	1,42	1,45	1,50	1,50	1,38

Trend for more intensive use

Source: Sec. Estado Agric. Instituto de Economia Agrícola. Elaboration: Cenbio

Pastures in Brazil

Current situation		
Area	**Heats of cattle**	**Density (Heats of cattle/ha)**
172 million hectares	169 million	0,98
Prospects		
108 million hectares	169 million heats of cattle	1,56 (SP - 2008)

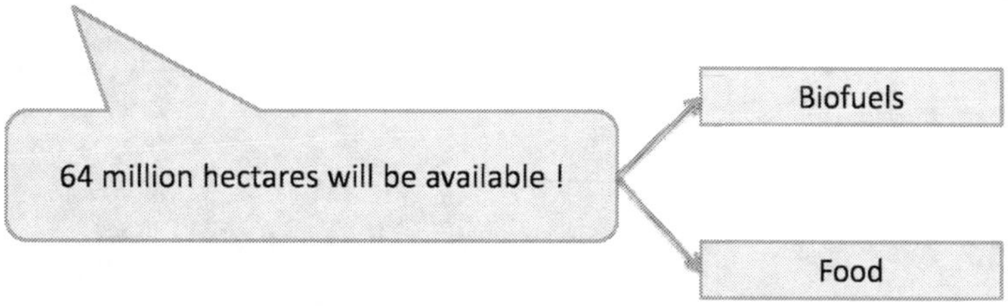

Agro-ecological Zoning of Sugarcane

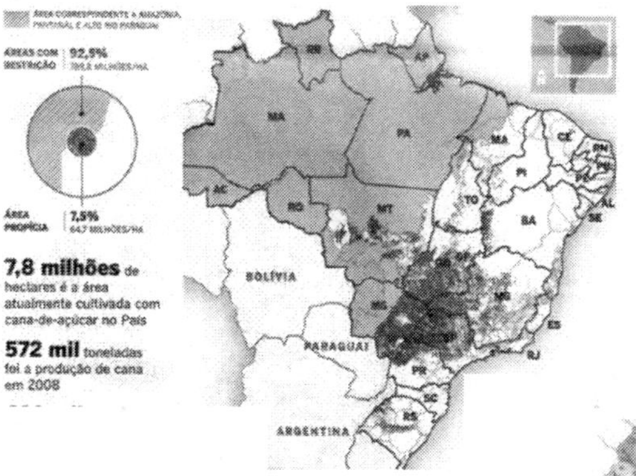

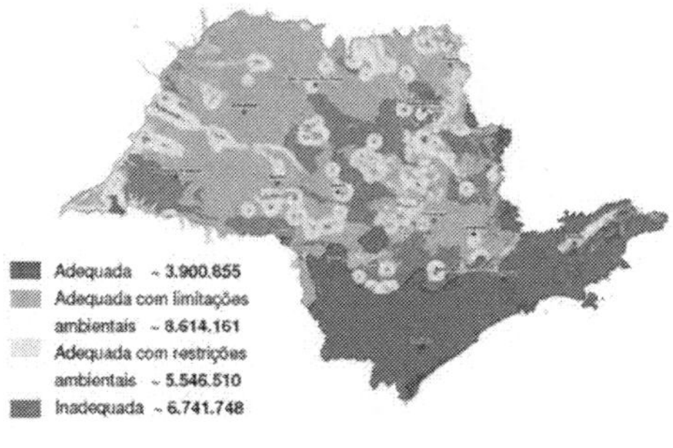

Three Sustainability Issues

1. Environmental Sustainability
2. **Social Sustainability**
3. Economic Sustainability

PILLARS

GBEP's work on sustainability indicators was developed under the following three pillars, noting interlinkages between them:

Environmental	Social	Economic
THEMES GBEP considers the following themes relevant, and these guided the development of indicators under these pillars:		
Greenhouse gas emissions, Productive capacity of the land and ecosystems, Air quality, Water availability, use efficiency and quality, Biological diversity, Land-use change, including indirect effects.	Price and supply of a national food basket, Access to land, water and other natural resources, Labour conditions, Rural and social development, Access to energy, Human health and safety.	Resource availability and use efficiencies in bioenergy production, conversion, distribution and end use, Economic development, Economic viability and competitiveness of bioenergy, Access to technology and technological capabilities, Energy security/Diversification of sources and supply, Energy security/Infrastructure and logistics for distribution and use.
INDICATORS		
1. Lifecycle GHG emissions	9. Allocation and tenure of land for new bioenergy production	17. Productivity
2. Soil quality	10. Price and supply of a national food basket	18. Net energy balance
3. Harvest levels of wood resources	11. Change in income	19. Gross value added
4. Emissions of non-GHG air pollutants, including air toxics	12. Jobs in the bioenergy sector	20. Change in consumption of fossil fuels and traditional use of biomass
5. [illegible]	13. Change in unpaid time spent by women and children collecting biomass	21. Training and requalification of the workforce
6. Water quality	14. Bioenergy used to expand access to modern energy services	22. Energy diversity
7. Biological diversity in the [illegible]	15. Change in mortality and burden of disease attributable to indoor smoke	23. Infrastructure and logistics for distribution of bioenergy
8. Land use and land-use change related to bioenergy feedstock production	16. Incidence of occupational injury, illness and fatalities	24. Capacity and flexibility of use of bioenergy

N.A.

Percentage of informal jobs in agricultural sector. Brazil, 2006

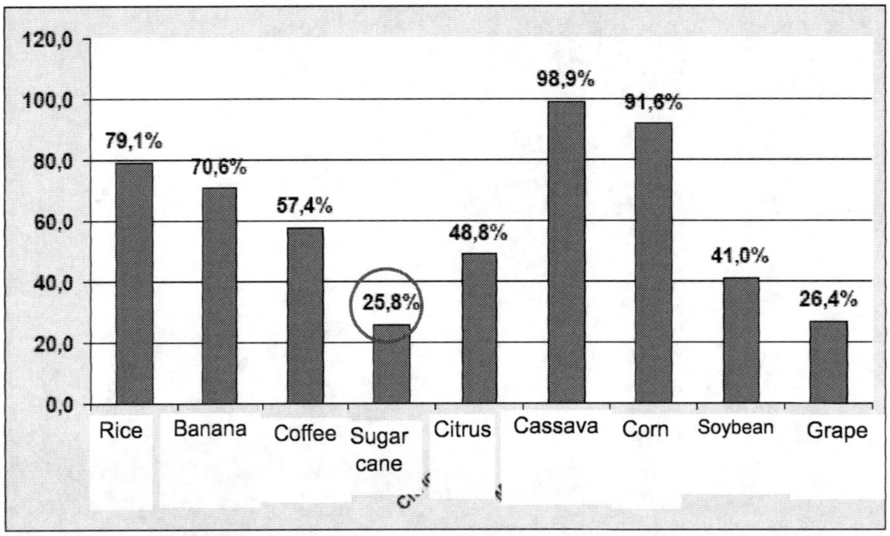

Evolution of the average income of employees in several crops. Brazil and São Paulo, 2002-2006 (BRL - August 2007)

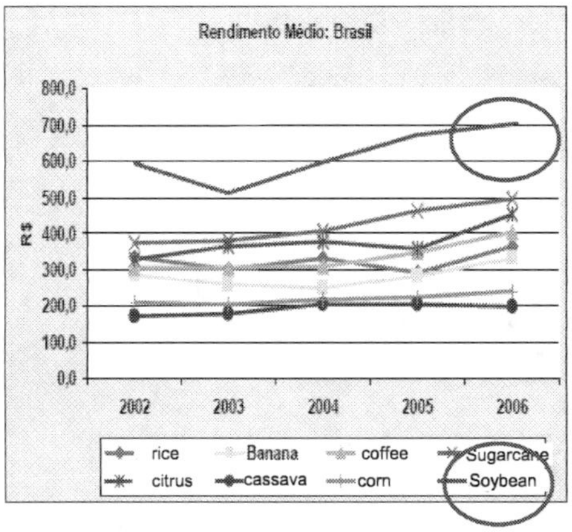

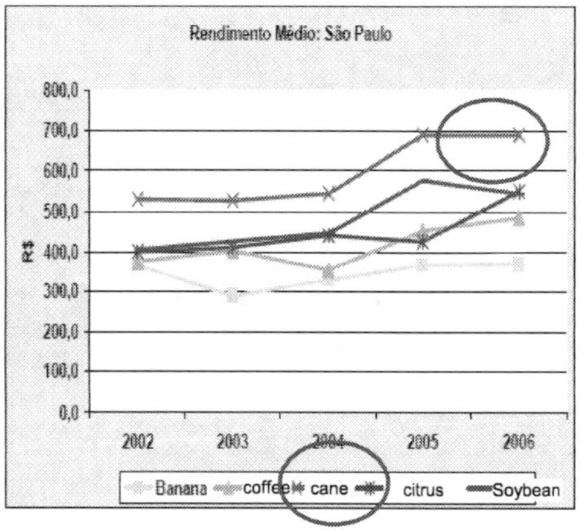

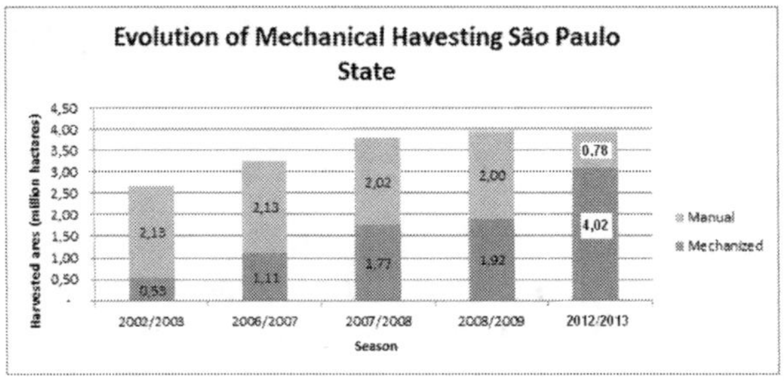

Fig. 3.12 Mechanical harvesting of green cane. (Photo courtesy of Agricef Soluções Tecnológicas Para Agricultura Ltda, Brazil; reprinted with permission)

Mechanical harvesting in São Paulo:

May 2013: 72.6% of the areas that could be mechanized
were harvested without burning, corresponding to 3.38 million hectares, against
34.2% in 2006

Source: Coelho, S. T., Guardabassi, P. "Ethanol". In: B. D. Solomon, R. Bailis (eds.), *Sustainable Development of Biofuels in Latin America and the Caribbean,* DOI 10.1007/978-1-4614-9275-7_3 © Springer Science+Business Media NewYork

Evolution of jobs for manual harvesting of sugarcane in São Paulo

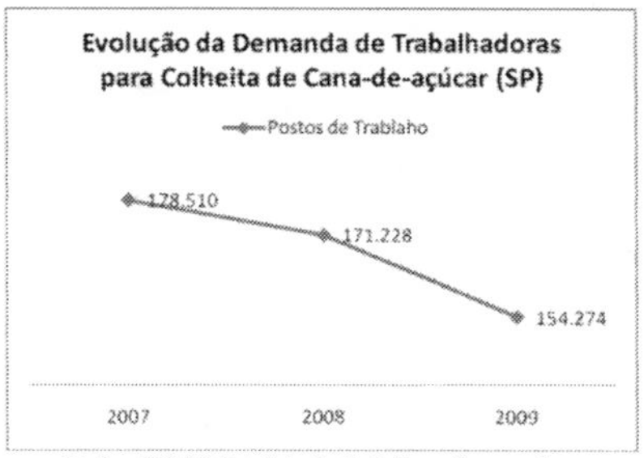

Dados: TEM (Ministério do Trabalho e Emprego)

Social impacts from mechanized harvesting

- Lost of jobs in compliance with the elimination of burning.
- Since 2007 Unica associates **retrain more than 5,000 people**.
- New requalification program:
  - Unica, Feraesp and companies in the production chain,
  - support of the Interamerican Development Bank (IADB) - a program of training and retraining of workers
  - workers and members of the communities in six regions of São Paulo.
  - capacity building: drivers, tractors operators and harvesters, other areas (mechanic, electrician and welder); programs for other sectors such as forestry, horticulture, handicrafts, construction, computing, sewing, catering and tourism.
- ***Renovação*** **project** : 4,350 workers qualified in 2012/2013 season (http://www.unica.com.br)

Three GBEP Sustainability Pillars

1. Environmental Sustainability
2. Social Sustainability
3. **Economic Sustainability**

PILLARS GBEP's work on sustainability indicators was developed under the following three pillars, noting interlinkages between them:		
Environmental	Social	Economic
THEMES GBEP considers the following themes relevant, and these guided the development of indicators under these pillars:		
Greenhouse gas emissions, Productive capacity of the land and ecosystems, Air quality, Water availability, use efficiency and quality, Biological diversity, Land-use change, including indirect effects.	Price and supply of a national food basket, Access to land, water and other natural resources, Labour conditions, Rural and social development, Access to energy, Human health and safety.	Resource availability and use efficiencies in bioenergy production, conversion, distribution and end use, Economic development, Economic viability and competitiveness of bioenergy, Access to technology and technological capabilities, Energy security/Diversification of sources and supply, Energy security/Infrastructure and logistics for distribution and use.
INDICATORS		
1. Lifecycle GHG emissions	9. Allocation and tenure of land for new bioenergy production	17. Productivity
2. Soil quality	10. Price and supply of a national food basket	18. Net energy balance
3. Harvest levels of wood resources	11. Change in income	19. Gross value added
4. Emissions of non-GHG air pollutants, including air toxics	12. Jobs in the bioenergy sector	20. Change in consumption of fossil fuels and traditional use of biomass
5. Water use and efficiency	13. Change in unpaid time spent by women and children collecting biomass	21. Training and requalification of the workforce
6. Water quality	14. Bioenergy used to expand access to modern energy services	22. Energy diversity
7. Biological diversity in the landscape	15. Change in mortality and burden of disease attributable to indoor smoke	23. Infrastructure and logistics for distribution of bioenergy
8. Land use and land-use change related to bioenergy feedstock production	16. Incidence of occupational injury, illness and fatalities	24. Capacity and flexibility of use of bioenergy

Ethanol productivity (liters per hectare)

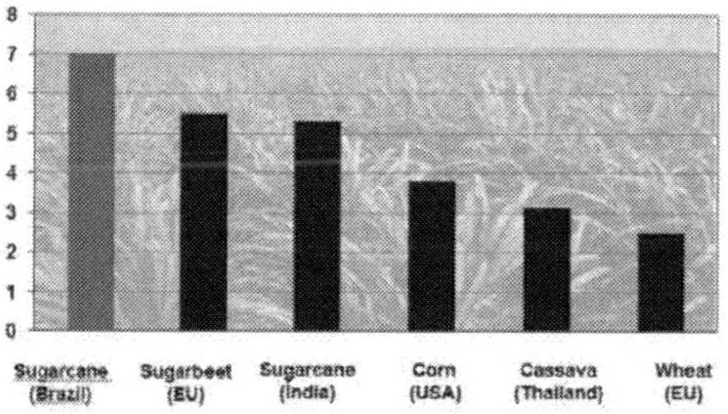

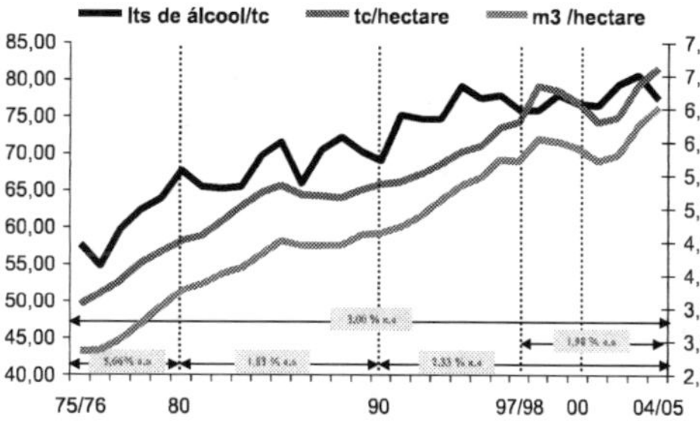

Source: Rodrigues, Unicamp .2005

Sugarcane ethanol in Brazil
Growth rate 3.8% per year

Current situation of ethanol in Brazil due to gasoline prices´ control

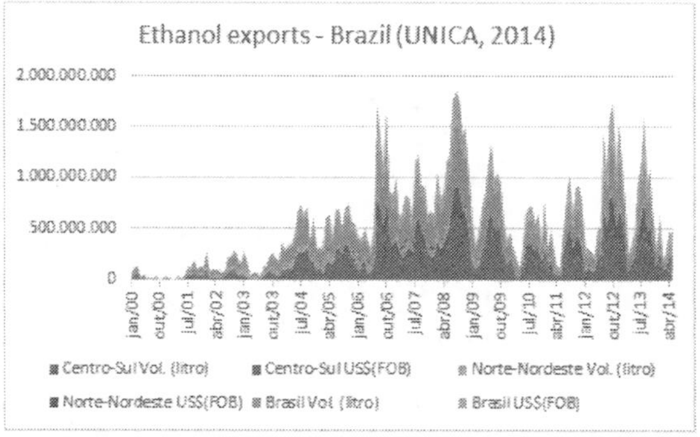

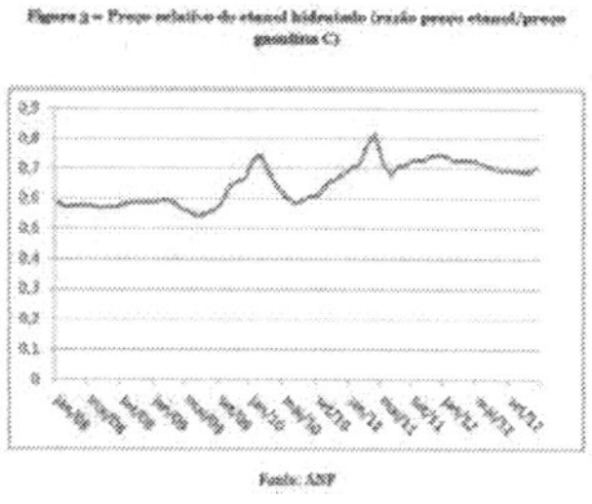

Increase on ethanol blend to gasoline 25% in 2014

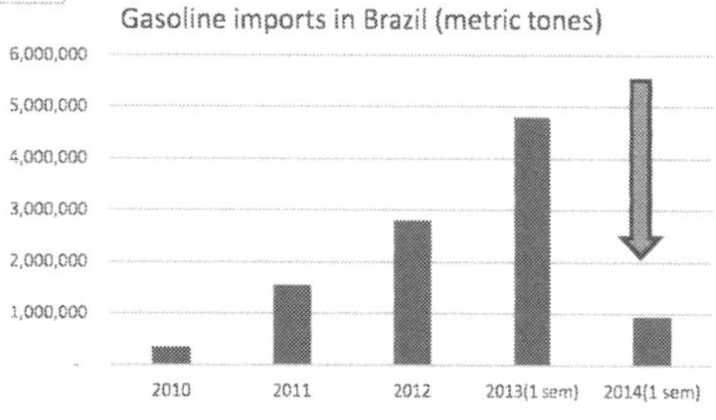

INSTITUTO DE ENERGIA E AMBIENTE
UNIVERSIDADE DE SÃO PAULO

IEE USP

USP 80 1934 2014

CENBIO

Obrigada / Thank you

RCN Conference on
Pan American Biofuels
& Bioenergy Sustainability
Golden Tulip Recife Palace,
Recife, Brazil July 22-25, 2014

PAN-AMERICAN
BIOFUELS &
BIOENERGY
SUSTAINABILITY

Prof Suani T. Coelho, PhD
Cenbio/IEE/ USP
suani@iee.usp.br
Http://cenbio.iee.usp

"Bioethanol Industry Perspectives in Brazil"

RCN Conference on Pan American Biofuels & Bioenergy Sustainability

Geraldo Eugênio – DTC Itep
geugenio1@terra.com.br
Recife, July 24th 2014

Issues to be addressed

- Recent efforts by the Brazilian private and government sectors
- Technology advancements
- R&D investments
- Market: price and exemptions

Recent efforts on bioenegy and biofuel iniciatives

- 450 years of sugarcane production
- 40 years as an feedstock for energy – ethanol
  - The ethanol production
  - The flex fuel vehicle
  - The cogenaration

Technology advancements

- Sugarcane breeding – 1% increase yield/yr
- Agriculture mechanization
- Industry efficiency for sugar, ethanol and cogeneration
- Flex-fuel vehicles

R&D investments

- Sugarcane breeding programs – RIDESA, IAC, CTC
- Mechanization – CTBE, Univesities
- Industry efficiency for sugar, ethanol and cogeneration (fermentation) – Private industry
- Flex-fuel vehicles – Auto industry

R&D demands

- 2nd and 3rd generation biofuels and bioproducts – Feedstock production and new options, conversion, environmental and social and economic issues
- Research and Innovation networks – RCN, BiofuelNet Canada
- Science without Borders – 2nd phase

Market issues

- Land price
- Ethanol cost production
- Sugar demand and price
- High demand for investments
- Tax and exemptions
- Debt ceiling for the bioethanol sector
- Public perception

World sugar market indicators

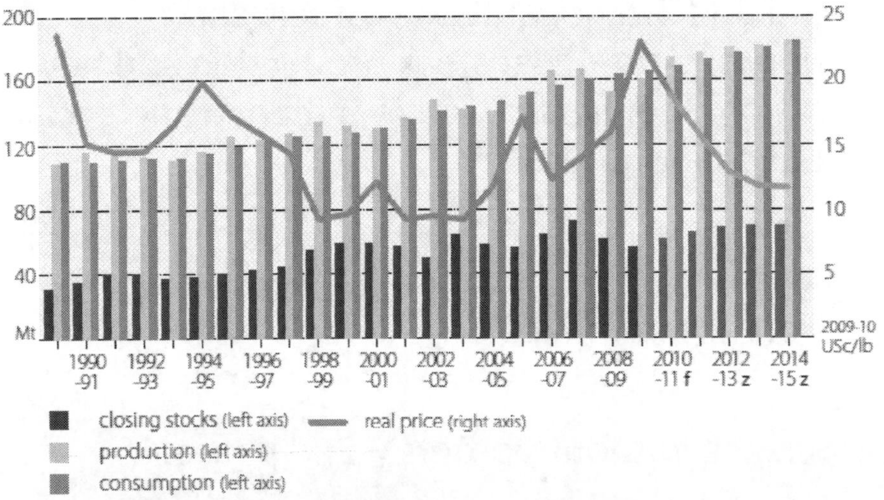

www.thebioenergysite.com

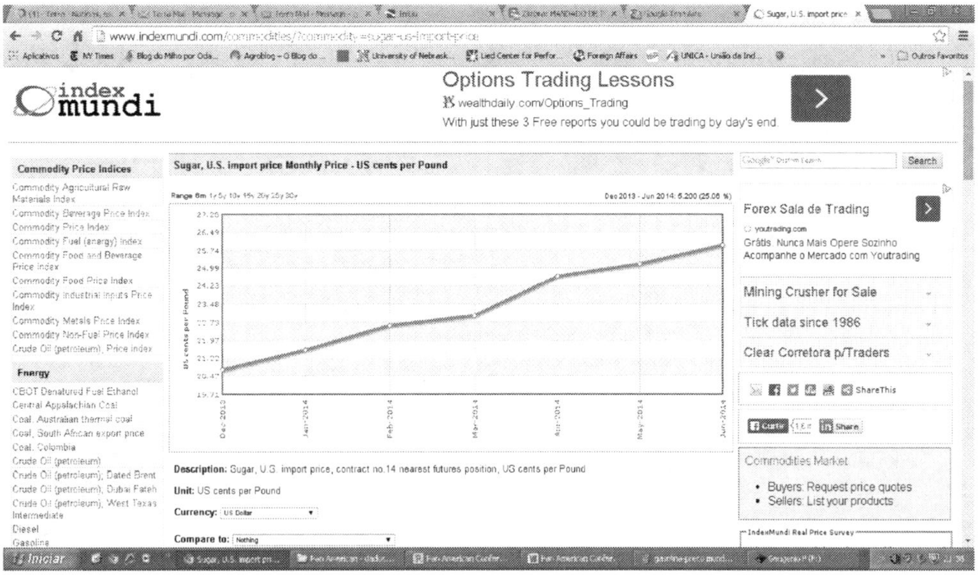
www.indexmundi.com
index mundi
Options Trading Lessons
wealthdaily.com/Options_Trading
With just these 3 Free reports you could be trading by day's end.
Commodity Price Indices
Sugar, U.S. import price Monthly Price - US cents per Pound
Energy
Forex Sala de Trading
Grátis. Nunca Mais Opere Sozinho
Acompanhe o Mercado com Youtrading
Mining Crusher for Sale
Tick data since 1986
Clear Corretora p/Traders
Commodities Market
Buyers: Request price quotes
Sellers: List your products
Description: Sugar, U.S. import price, contract no.14 nearest futures position, US cents per Pound
Unit: US cents per Pound
Currency:
Compare to:
Iniciar

INS EXEMPLOS DE PREÇO DA GASOLINA COMUM NOS POSTOS (por litro)
Oslo (Noruega) R$ 4,49
Roma (Itália) R$ 4,10
lres (Reino Unido) R$ 3,91
Paris (França) R$ 3,89
Tóquio (Japão) R$ 3,17
São Paulo (Brasil) R$ 2,90
Na lista de 35 países, o Brasil ocupa a 13ª posição
Santiago (Chile) R$ 2,71
Sidnei (Austrália) R$ 2,70
rgo (África do Sul) R$ 2,32
Toronto (Canadá) R$ 2,13
Pequim (China) R$ 2,08
Aires (Argentina) R$ 2,07
(Estados Unidos) R$ 1,71
Moscou (Rússia) R$ 1,56
México (México) R$ 1,29
Cairo (Egito) R$ 0,55
O CUSTO DA GASOLINA NO BRASIL
Em cada litro de gasolina consum
o motorista brasileiro paga em m
R$ 1,13 em impostos
Custo total do litro:
R$ 2,90
R$ 0,99 Margem da Petrobras (34% do total)
R$ 0,81 ICMS (28% do total)
R$ 0,32 Cide, PIS/P e Cofins (11% do total)
R$ 0,52 Distribui e revenc (18% do
R$ 0 Etanc anidr (9% d
* Os dados do estudo foram convertidos de dólares amer

Gasoline and diesel retail prices in the USA and Brasil

USA – gasoline and diesel - retail ( ), 5/15/14 ($/gallon)			
Regular gasoline	U.S. average	3.65	100
Diesel	U.S. average	3.94	100

Brasil – gasoline and diesel - retail ( ), 5/15/14 ($/gallon)			
Regular gasoline	Recife	5.13	140
Premium gasoline	Recife	5.35	
Diesel	Recife	4.66	118

Diferencial de preços dos combustíveis no Brasil

Utilização de etanol pela frota flex

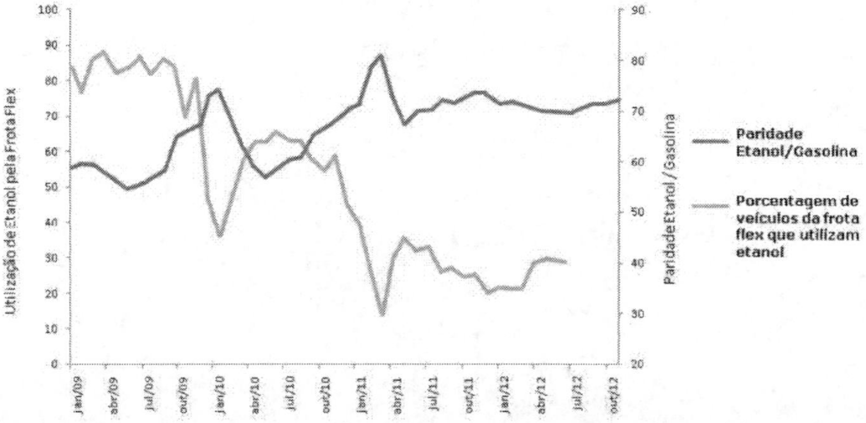

rudaricci.blogspot.com

Origem	Motor	IPI Anterior	IPI Reduzido
Nacional	Até 1.000 cilindradas	7%	0%
	Até 2.000 cilindradas flex	11%	5,5%
	Até 2.000 cilindradas gasolina	13%	6,5%
	Até 2.000 cilindradas utilitários	4%	1%
Importado	Até 1.000 cilindradas	37%	30%
	Até 2.000 cilindradas flex	41%	35,5%
	Até 2.000 cilindradas gasolina	43%	36,5%
	Até 2.000 cilindradas utilitários	34%	31%

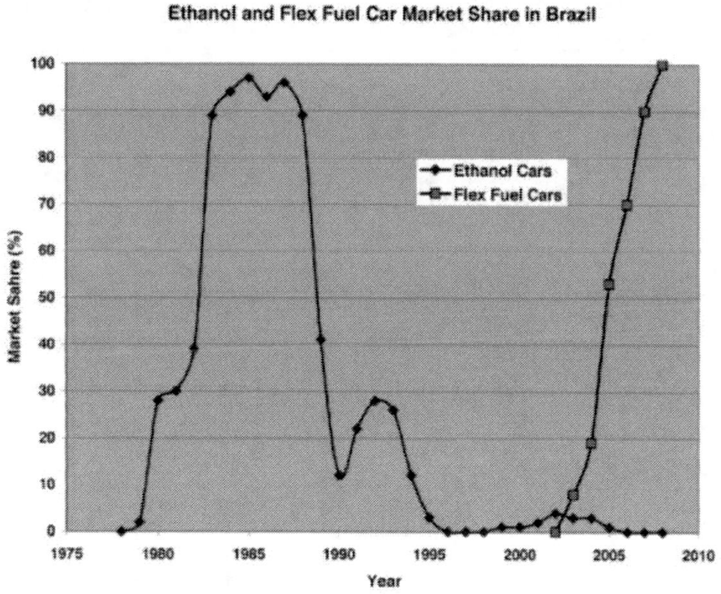

Brazilian sugar cane production and allocation a

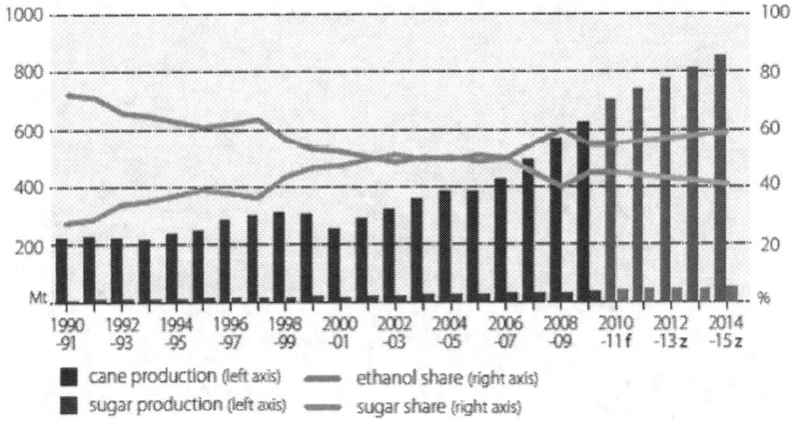

a April to March years.

www.thebioenergysite.com

Sugarcane, sugar and ethanol production UNICA sugar mills

Produtos		Safras 2012/2013	Safras 2013/2014	Variação (%)
Cana-de-açúcar [1]		349.583,3	406.262,8	↑ 16,21%
Açúcar [1]		21.798,6	22.921,0	↑ 5,15%
Etanol anidro [2]		5.357,3	7.364,5	↑ 37,47%
Etanol hidratado [2]		8.333,9	9.876,2	↑ 18,51%
Etanol total [2]		13.691,2	17.240,8	↑ 25,93%
ATR [1]		46.217,0	53.491,1	↑ 15,74%
ATR/ tonelada de cana [3]		132,21	131,67	↓ -0,41%
Mix (%)	açúcar	49,50%	44,97%	↓
	etanol	50,50%	55,03%	↑
Litros etanol/ tonelada de cana		39,16	42,44	↑ 8,36%
Kg açúcar/ tonelada de cana		62,36	56,42	↓ -9,52%

Fonte: UNICA. Nota: [1] - mil toneladas; [2] - milhões de litros; [3] - kg de ATR/ tonelada de cana; dados sujeitos a pequenos ajustes.

Perspectives

- Bioethanol will continue to be an strategic fuel option for Brasil
- There will be a balance between the local and the international ethanol markets as well as the sugar demand and prices
- There will be a high demand for the bagass – Energy vs 2nd generation ethanol
- Industry reliability and public confidance
- Land use, deforestation and water efficient use are issues to be tackled in the next few years

RCN Conference on Pan American Biofuels & Bioenergy Sustainability

Bioenergy in Northeastern Brazil: Opportunities and Challenges

Rômulo S. C. Menezes and Claudemiro de Lima Júnior

UNIVERSIDADE FEDERAL DE PERNAMBUCO

Acknowledgements

- Organizing Committee of RCN Conference
- Professors Abraham Sicsú and Marcia Alcoforado
- Fundação de Amparo à Ciência e Tecnologia do Estado de Pernambuco (FACEPE)

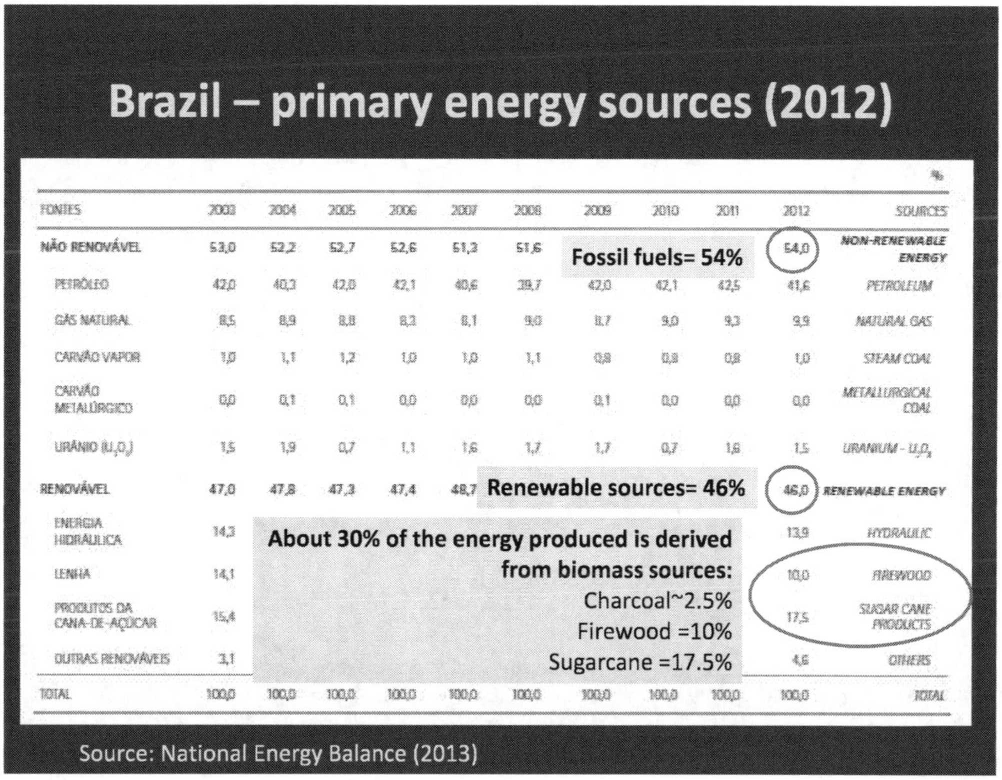

Brazil – primary energy sources (2012)

%

FONTES	2003	2004	2005	2006	2007	2008	2009	2010	2011	2012	SOURCES
NÃO RENOVÁVEL	53,0	52,2	52,7	52,6	51,3	51,6	[illegible]	[illegible]	[illegible]	54,0	***NON-RENEWABLE ENERGY***
PETRÓLEO	42,0	40,3	42,0	42,1	40,6	39,7	42,0	42,1	42,5	41,6	*PETROLEUM*
GÁS NATURAL	8,5	8,9	8,8	8,3	8,1	9,0	8,7	9,0	9,3	9,9	*NATURAL GAS*
CARVÃO VAPOR	1,0	1,1	1,2	1,0	1,0	1,1	0,8	0,8	0,8	1,0	*STEAM COAL*
CARVÃO METALÚRGICO	0,0	0,1	0,1	0,0	0,0	0,0	0,1	0,0	0,0	0,0	*METALLURGICAL COAL*
URÂNIO (U_3O_8)	1,5	1,9	0,7	1,1	1,6	1,7	1,7	0,7	1,6	1,5	*URANIUM - U_3O_8*
RENOVÁVEL	47,0	47,8	47,3	47,4	48,7	[illegible]	[illegible]	[illegible]	[illegible]	46,0	***RENEWABLE ENERGY***
ENERGIA HIDRÁULICA	14,3	[illegible]	[illegible]	[illegible]	[illegible]	[illegible]	[illegible]	[illegible]	[illegible]	13,9	*HYDRAULIC*
LENHA	14,1	[illegible]	[illegible]	[illegible]	[illegible]	[illegible]	[illegible]	[illegible]	[illegible]	10,0	*FIREWOOD*
PRODUTOS DA CANA-DE-AÇÚCAR	15,4	[illegible]	[illegible]	[illegible]	[illegible]	[illegible]	[illegible]	[illegible]	[illegible]	17,5	*SUGAR CANE PRODUCTS*
OUTRAS RENOVÁVEIS	3,1	[illegible]	[illegible]	[illegible]	[illegible]	[illegible]	[illegible]	[illegible]	[illegible]	4,6	*OTHERS*
TOTAL	100,0	100,0	100,0	100,0	100,0	100,0	100,0	100,0	100,0	100,0	*TOTAL*

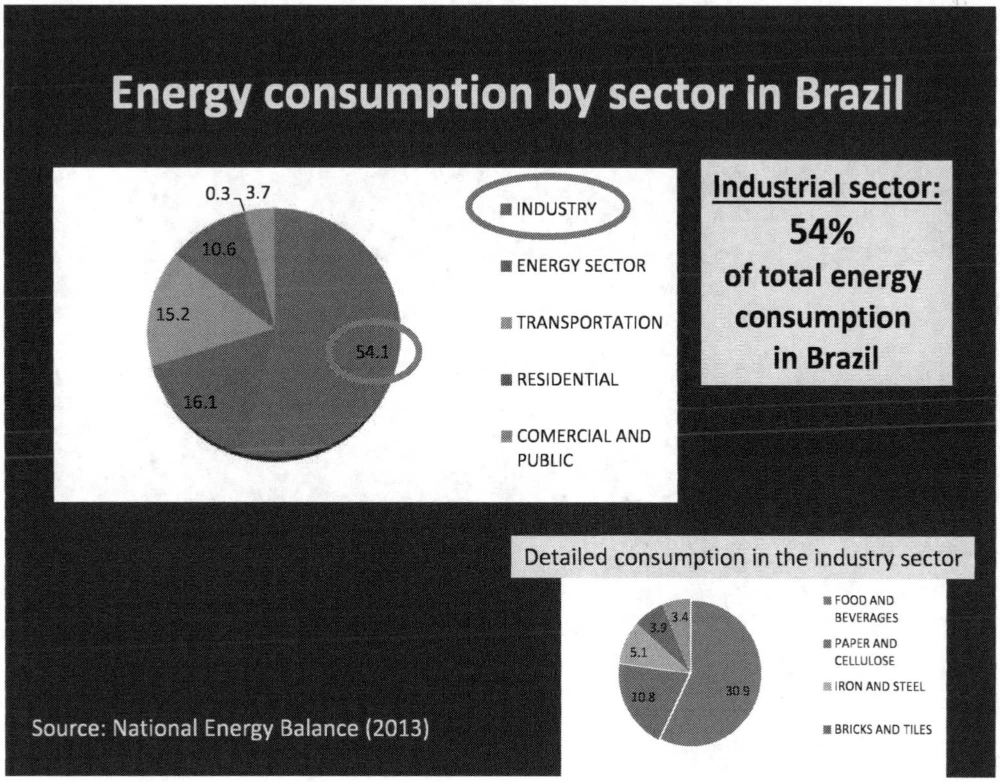

Biomass contribution for energy supply to the industrial sector in Brazil

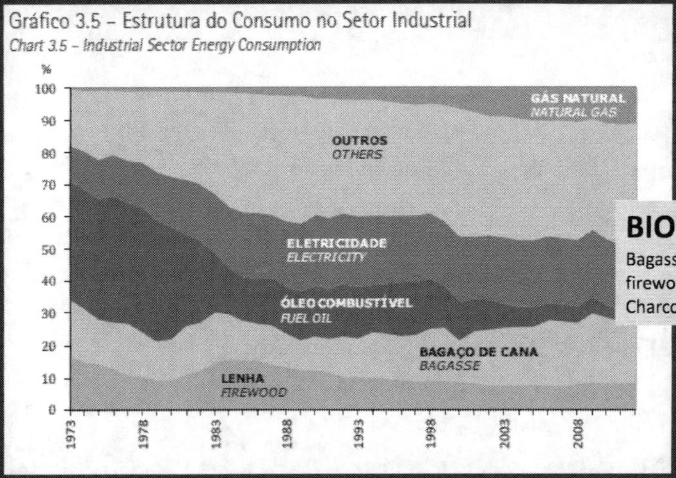

BIOMASS = 33,1%

Bagasse=20,1
firewood=8,4
Charcoal=4,6

Source: National Energy Balance (2013)

Background – NE region of Brazil

NE region of Brazil

Total area: 1.561.177 km²
Equivalent to the areas of Germany, Holand, Belgium, France, Spain, and Portugal combined.

Population ~ 55 million people

- **Bioenergy in NE Brazil**
  - Biomass is an important energy source in Brazil;
  - The NE region of Brazil is very large but relatively little studied in this respect.

NE Region of Brazil: High environmental variability

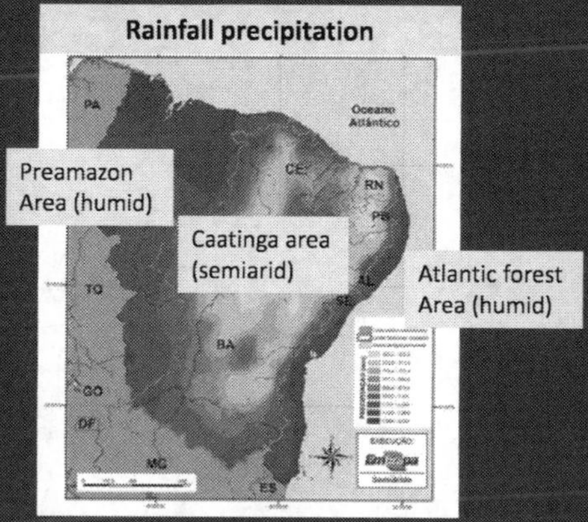

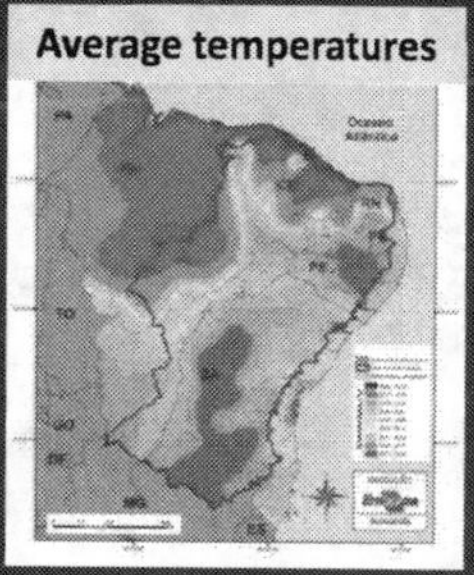

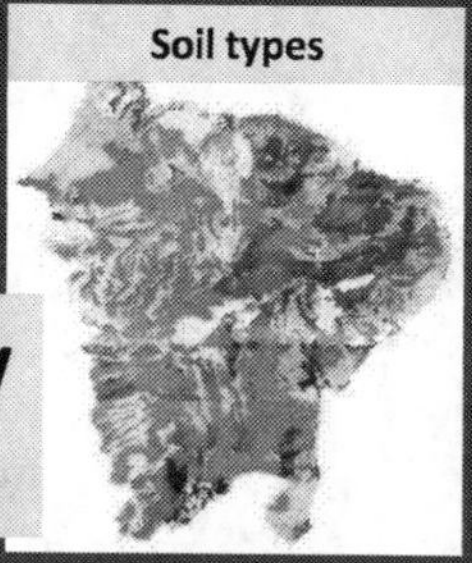

Biomass sources and availability vary according to each area

Main points for this talk

- What are the <u>LESS KNOWN OPPORTUNITIES</u> about biomass sources for energy recovery in NE Brazil?
- What are the main <u>RESEARCH AND POLICY GAPS</u> that need to be approached?

Bioenergy Atlas of Brazil

_ Elaborated by the National Biomass Center (Cenbio)_

- Published first version in 2009 and a second and expanded version in 2012;
- For the NE region considered only: sugarcane, pig manure, forest residues, crop residues (rice, coconut, peanut) and palm oil.

Are there other relevant biomass sources that could be used for energy recovery in NE Brazil?

- How much is produced from each biomass source?
- Where this biomass is produced?
- How much energy may be recovered from these sources?
- What are the main routes to process these biomass sources?
- **What are the opportunities and challenges for the industrial sector and government institutions to allow the use of these biomass sources?**

"Bioenergy Atlas of NE Brazil"

Elaborated by our research group

"BIOENERGY ATLAS OF NORTHEASTERN BRAZIL"

Sources of data:

- National Bureau of Statistics (IBGE)
- Ministry of Agriculture
- Literature review

Biomass sources evaluated

Sugarcane (bagasse, ethanol and vinasse)
Firewood (from caatinga and from planted forests)
Manure (cattle, pigs, goats, sheep, poultry)
Municipal Solid Waste (Organic fraction)
Crop residues (rice husk, coconut)
Oil crops (soybean, castor bean, sunflower, peanuts)
Perenial species that produce oil (palm oil, babaçu)

Estimates of potential energy recovery from biomass sources

- <u>Briefly, for each source, we considered:</u>
  - The amount produced annualy per municipality;
  - The calorific value;
  - Conversion routes (direct combustion, fermentation, anaerobic biodigestion, pyrolisis, etc);
  - Conversion efficiencies to produce electricity.

Annual production of biomass sources in Northeastern Brazil (million tons year^{-1})

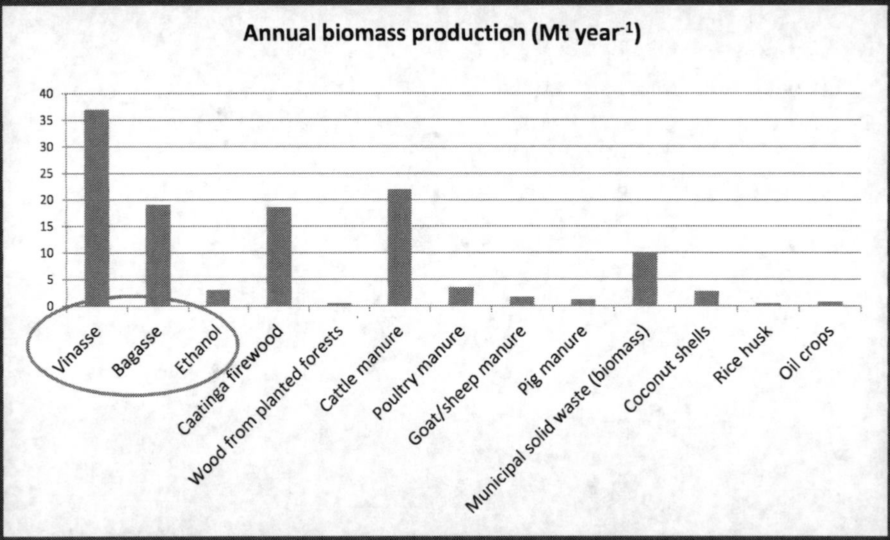

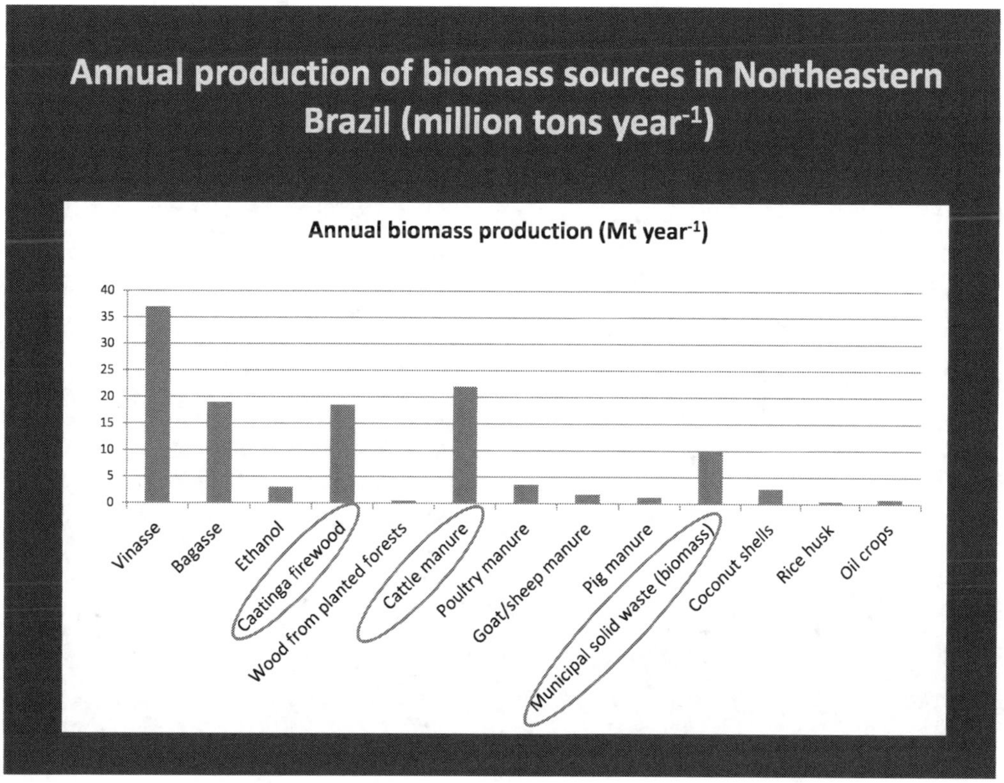
Annual production of biomass sources in Northeastern Brazil (million tons year-1)
Annual biomass production (Mt year-1)
40
35
30
25
20
15
10
5
0
Vinasse
Bagasse
Ethanol
Caatinga firewood
Wood from planted forests
Cattle manure
Poultry manure
Goat/sheep manure
Pig manure
Municipal solid waste (biomass)
Coconut shells
Rice husk
Oil crops

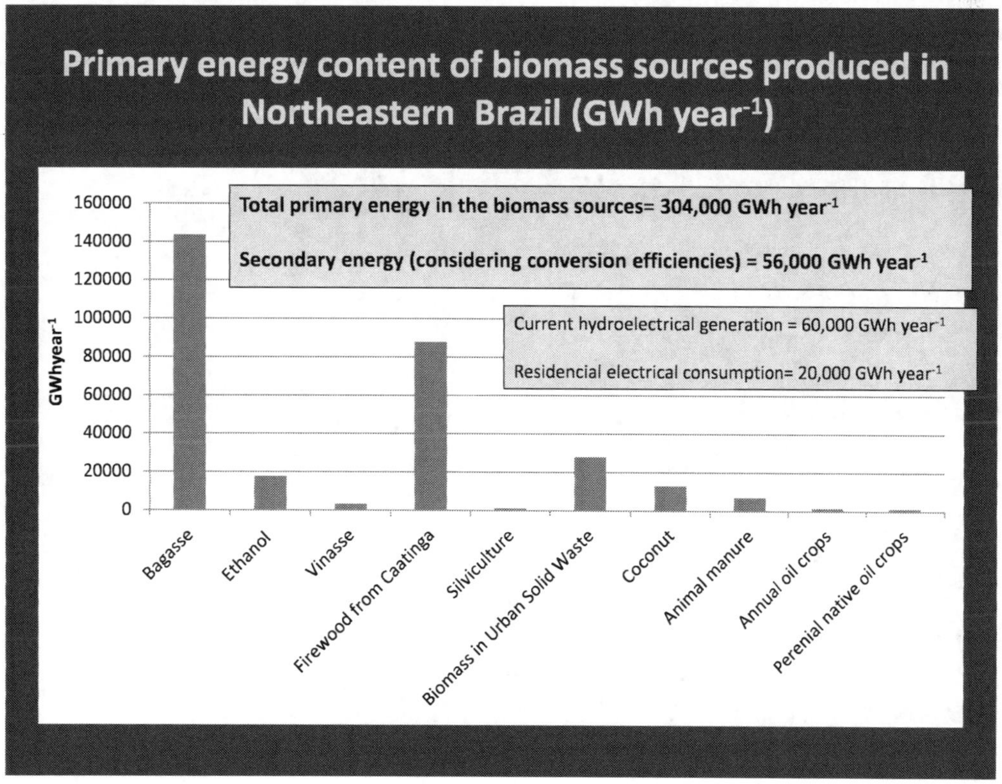
Primary energy content of biomass sources produced in Northeastern Brazil (GWh year-1)
Total primary energy in the biomass sources= 304,000 GWh year-1
Secondary energy (considering conversion efficiencies) = 56,000 GWh year-1
Current hydroelectrical generation = 60,000 GWh year-1
Residencial electrical consumption= 20,000 GWh year-1
GWhyear-1
160000
140000
120000
100000
80000
60000
40000
20000
0
Bagasse
Ethanol
Vinasse
Firewood from Caatinga
Silviculture
Biomass in Urban Solid Waste
Coconut
Animal manure
Annual oil crops
Perenial native oil crops

Spatial distribution of biomass sources

Distribution of sugarcane biomass production in NE Brazil (t year^{-1})

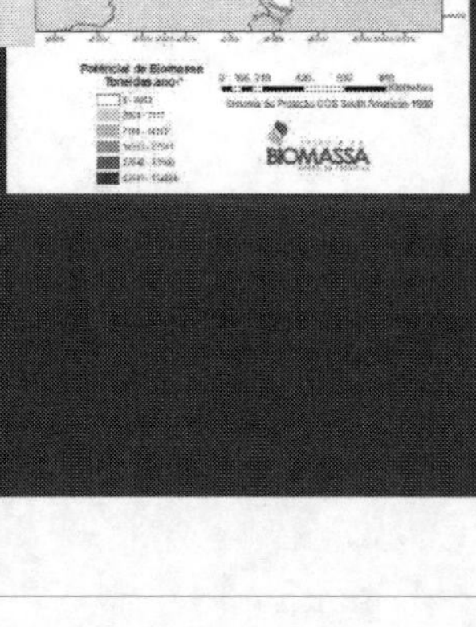

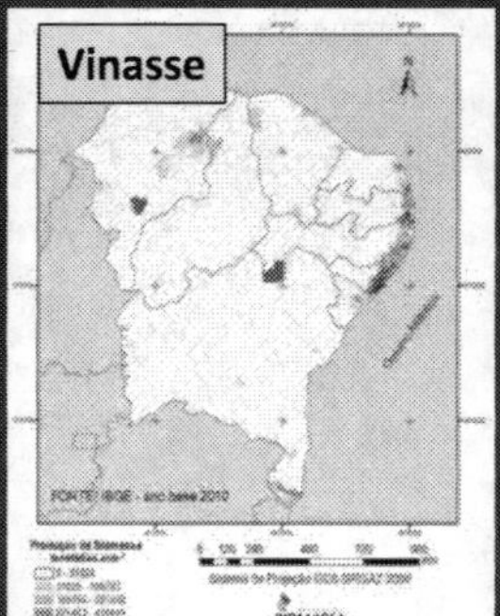

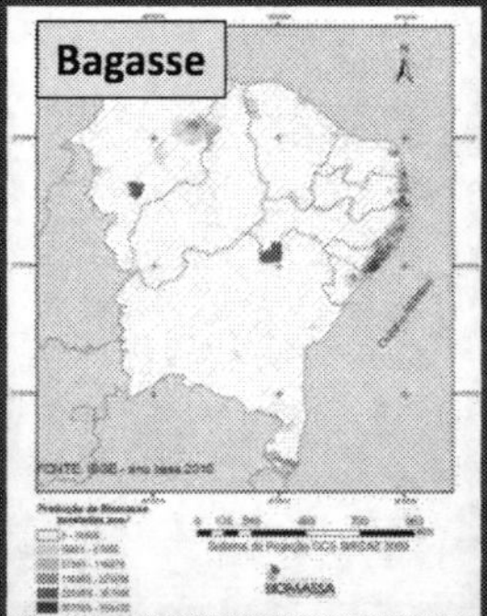

Sugarcane biomass was already discussed in a previous talk.

Primary energy content of biomass sources produced in Northeastern Brazil (GWh year^{-1})

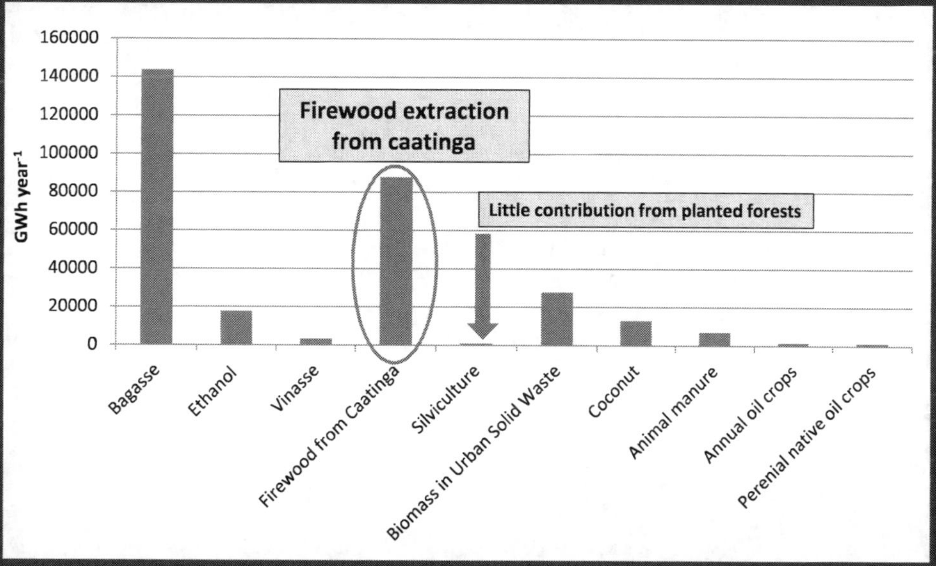

Distribution of the potential for extraction of firewood from caatinga in NE Brazil (t year^{-1})

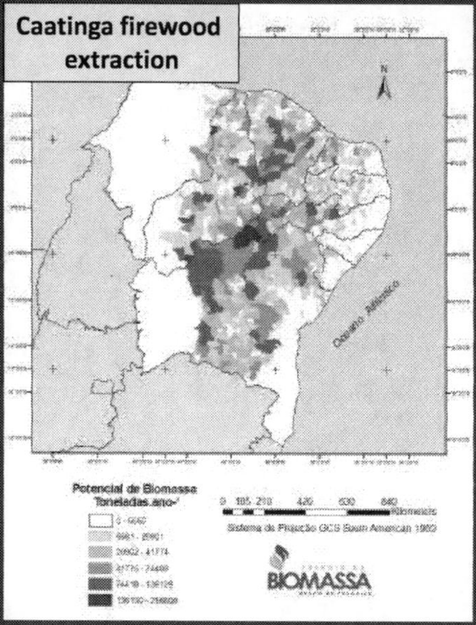

- Largely demanded for industrial, commercial and residential uses;
- Most extraction is illegal and not sustainable;
- Need for incentives for sustainabe production;

Sustainability of firewood extraction systems

- Single recomendation for all environmental situations in the region is not adequate;
- Tools for decision support systems are necessary;
- Ex: Simulation Modelling

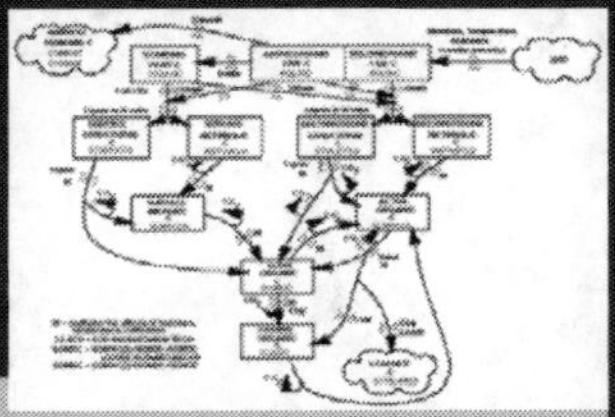

Primary energy content of biomass sources produced in Northeastern Brazil (GWh year^{-1})

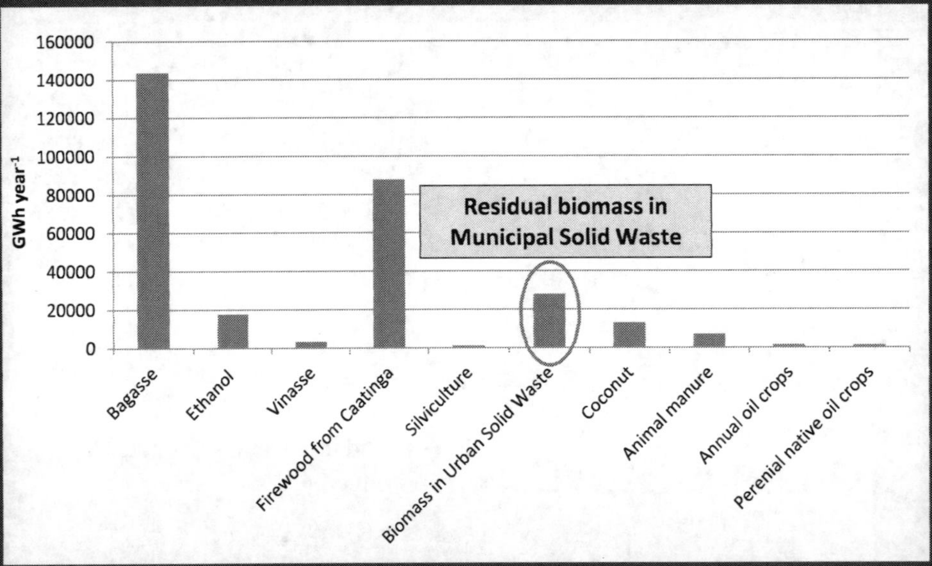

Distribution of the production of municipal solid waste in NE Brazil (t year^{-1})

Municipal solid waste (organic fraction)

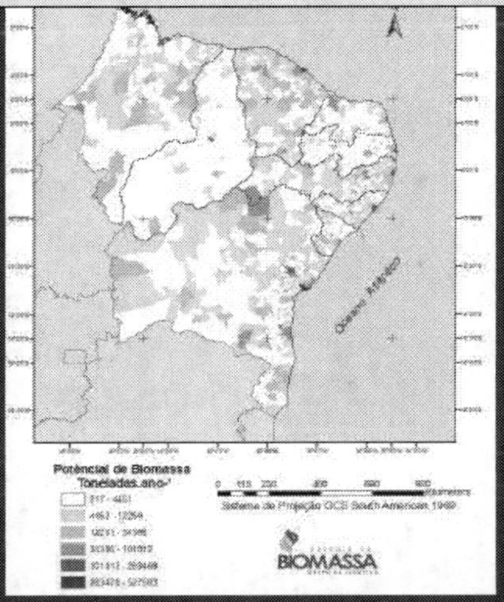

The Law of Municipal Solid Waste may bring changes to this sector;

Opportunities for the industrial sector: energy recovery from residual biomass may be more viable than disposal in landfills;

Incentives for effective segregation in the source are crucial.

Primary energy content of biomass sources produced in Northeastern Brazil (GWh year^{-1})

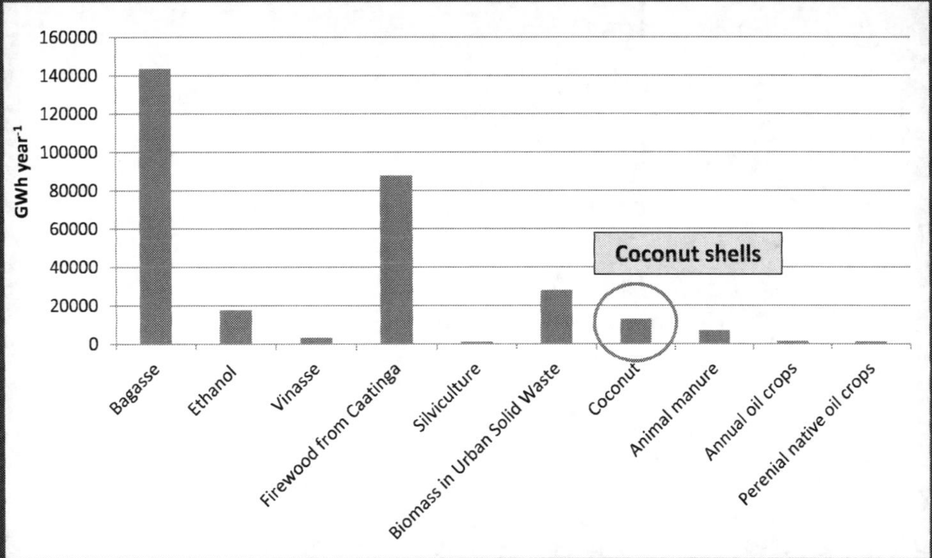

Coconut

Coconut shells

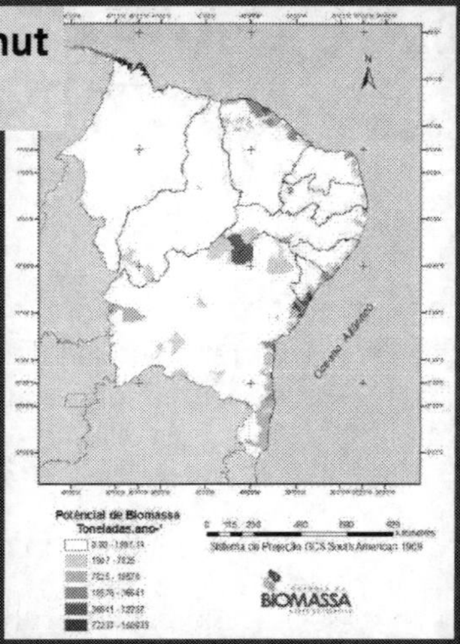

Coconut farms in the coastal area will be affected by urbanization, labor cost, tourism and industrialization

Coconut

Coconut shells

Irrigated areas in the São Francisco Valley

Coconut water processing industries

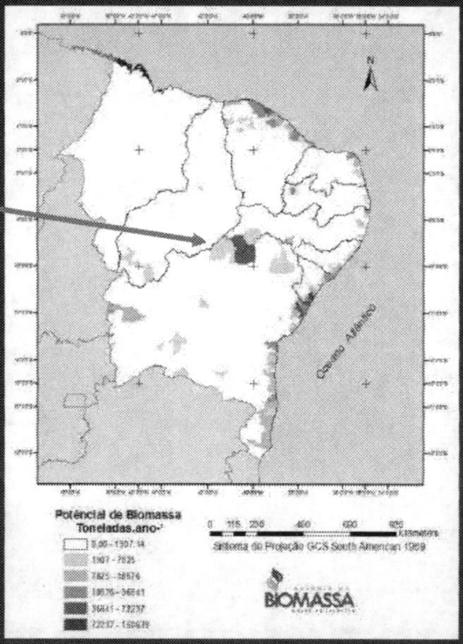

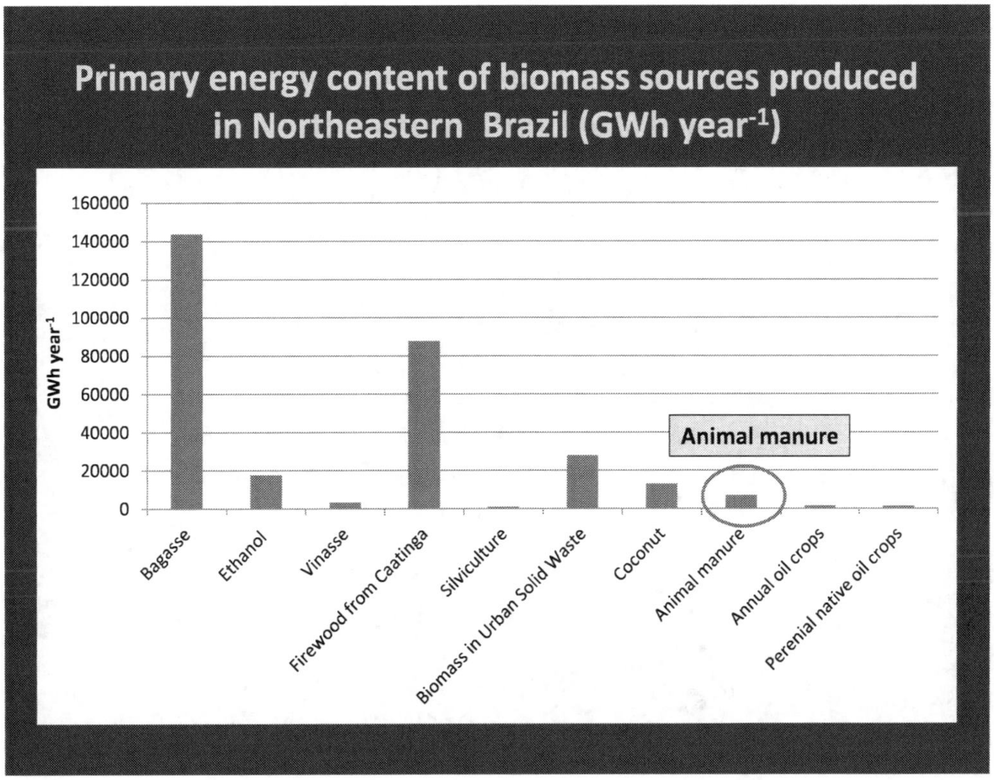
Primary energy content of biomass sources produced in Northeastern Brazil (GWh year^{-1})
160000
140000
120000
100000
80000
60000
40000
20000
0
GWh year^{-1}
Animal manure
Bagasse
Ethanol
Vinasse
Firewood from Caatinga
Silviculture
Biomass in Urban Solid Waste
Coconut
Animal manure
Annual oil crops
Perenial native oil crops

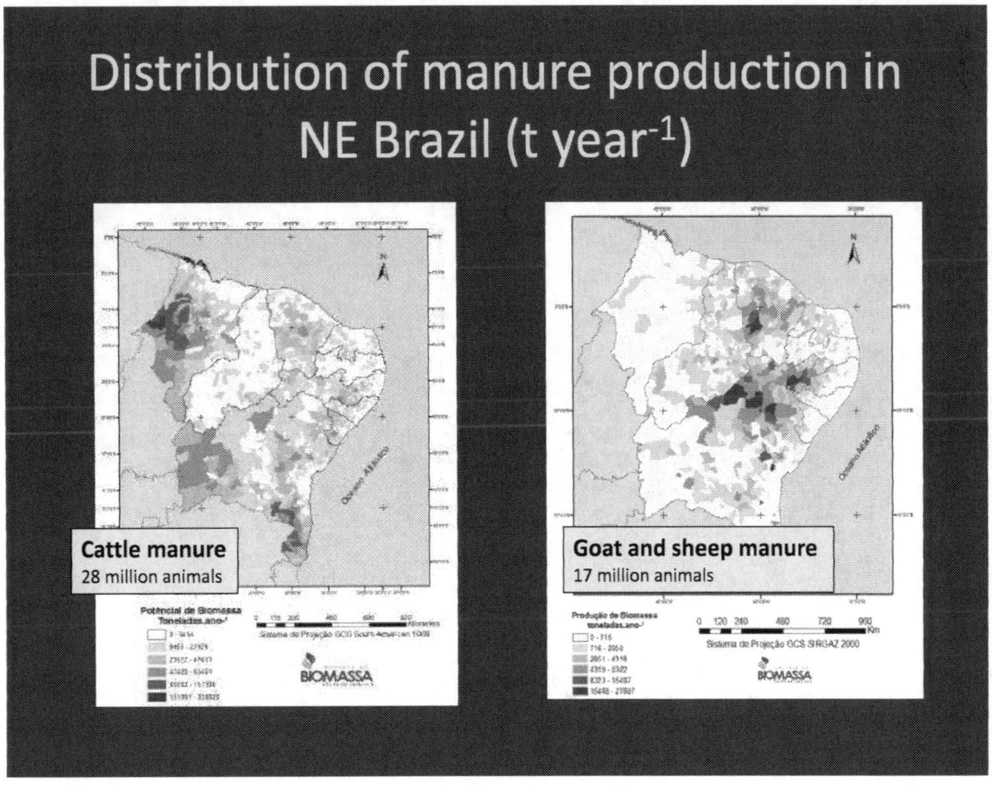
Distribution of manure production in NE Brazil (t year^{-1})
Cattle manure
28 million animals
Goat and sheep manure
17 million animals
BIOMASSA
BIOMASSA

Anaerobic biodigestion

- **Biodigestion is not relevant in NE Brazil, despite the large amount of biomass sources that could be processed through this route.**

- **There is need for capacity building in all sectors (academic, technical and the productive sector);**
- **Little research is done, few systems are operating;**
- **Has good potential for widespread microgeneration systems.**

The electricity that could be generated through digestion of vinasse, municipal solid waste, and manures could supply nearly 20,000 GWh $year^{-1}$.

Primary energy content of biomass sources produced in Northeastern Brazil (GWh $year^{-1}$)

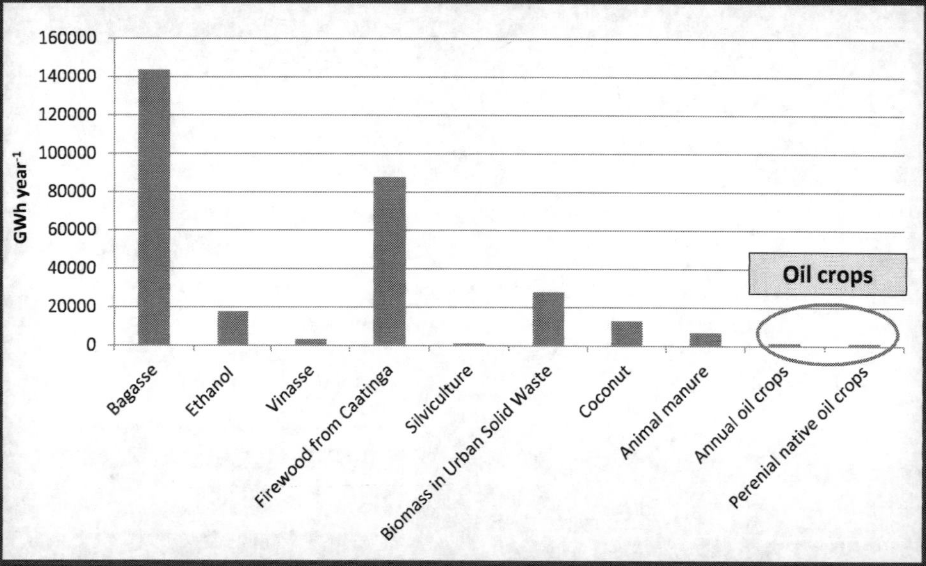

Pre-Amazon region

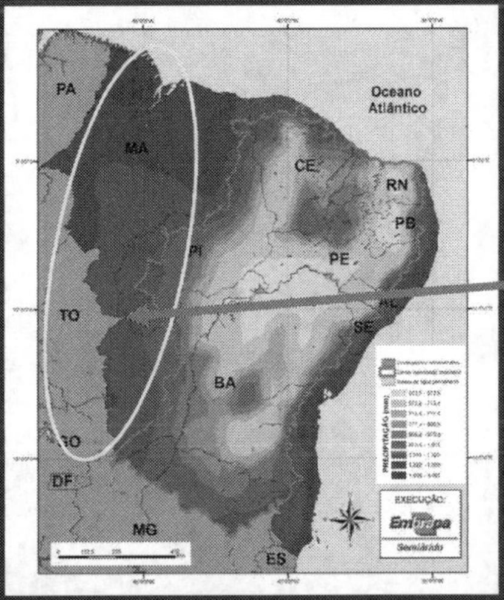

High-input crop systems in the Cerrado and Preamazon areas

Soybeans and cotton

Native species with potential for oil production

Native oil species

- **Babaçu:**
  - Occupies nearly 5 million hectares in NE Brazil;
  - Energy potential underutilized

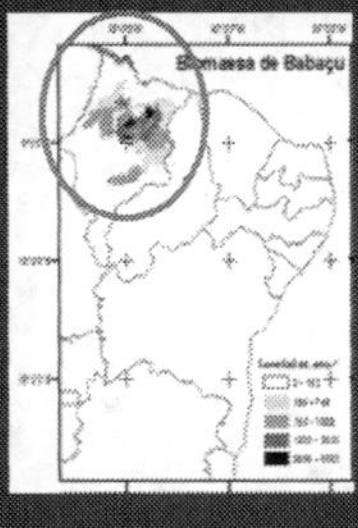

Fonte Marcos Alexandre Teixeira, Departamento de Energia - FEM- UNICAMP, mteixeira@fem.unicamp.br. - www.fem.unicamp.br/~mteixeira

Concluding remarks

- Sugarcane biomass and firewood from caatinga are important energy sources in NE Brazil. Both sources face nowadays severe sustainability issues;
- Other biomass sources are produced in significant amounts and should be considered for energy purposes, such as: animal manure, vinasse and residual biomass in municipal solid waste;
- Research priorities:
  - Sustainability of firewood extraction systems;
  - Overcome constraints for anaerobic digestion systems;
- Public policy priorities:
  - Incentives to sustainable production and trade of firewood;
  - Reduction of constraints for anaerobic biodigestion systems;
  - Estimulate residual biomass segregation for energy use in urban areas.

Thank you

Biomass Harvesting and Site Productivity: Is Policy ahead of Science?

Eric D. Vance, National Council for Air and Stream Improvement (NCASI)
W. Michael Aust, Virginia Tech
Robert E. Froese, Michigan Technological University
Robert B. Harrison, University of Washington
Larry A. Morris, University of Georgia
Brian D. Strahm, Virginia Tech

RCN Conference on Pan American Biofuels and Bioenergy
July 22-25, 2014
Recife, Brazil

Why Harvest Residues?

- Renewable energy
  - Electrical, thermal (pellets)
  - Biofuels
- Particle board, composite materials
- Forest health (fire, insects, disease)
- Forest operations (site preparation)

Environmental Questions

- Removal of biomass traditionally left on site
- Shorter rotations, more frequent site entries
- Site productivity
- Water
- Wildlife

Common U.S. Biomass Harvesting Guideline Provisions

- Retain 15-30% or "as much as possible"
- Restrictions on shallow, coarse-textured, nutrient-poor soils
- Minimize area in roads, landings, skid trails

Forest Guild Biomass Retention and Harvesting Guidelines

(February, 2012)

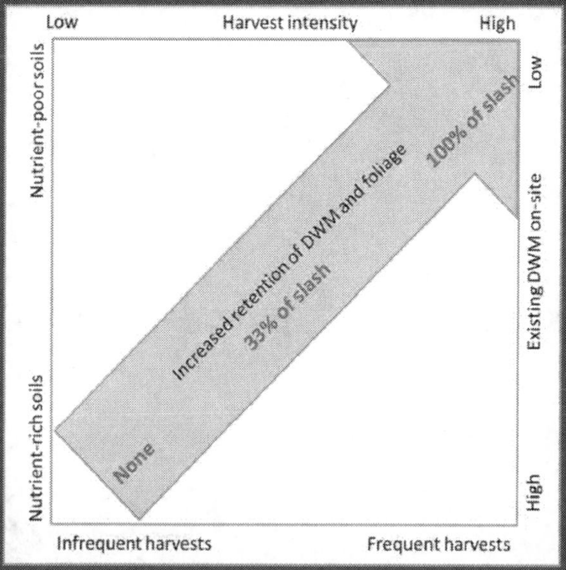

Forest Guild Biomass Retention and Harvesting Guidelines

- "Where science remains inconclusive, we relied on field observation and professional experience"

Frames of Reference

- “Natural” is optimal, more retention is better
- Conventional harvests retain residues
- Biomass harvesting removes all residues
- No beneficial practices
- Changes in State = Changes in Function

Perceived Benefits of Traditional Residue Management

25 to 80% of Non-Bole Residues Left after Operational Biomass Harvesting Due to Technical Limitations, Incidental Breakage

(Nurmi 2007, Ralevic et al. 2010, Klockow et al. 2013)

Sites in Minnesota following biomass harvesting
(The Forestry Forum)

Ecological Effects of Harvest Residue Management in Southern Pine Plantations

(USDA-AFRI, NC State Univ., Univ. of Georgia, Weyerhaeuser, Plum Creek, Georgia Pacific, NCASI)

- 8 sites in NC and GA
- Residues retained
- Residues removed
- 15% or 30% retained
- Clumped or dispersed

Harvest Residue Management in Southern Pine Plantations

- Small mammals, reptiles, amphibians, arthropods
- Plant community composition
- Soil properties, variability
- Technical input on BHGs
- Logger/manager survey

Residues Retained After Operational Biomass Harvesting in North Carolina, USA

S. Fritts, C. Moorman, D. Hazel, B. Jackson
NC State Univ., Univ. of Georgia

NO BHG, 15CLUS, 15DISP, 30CLUS, 30DISP, BOHarv

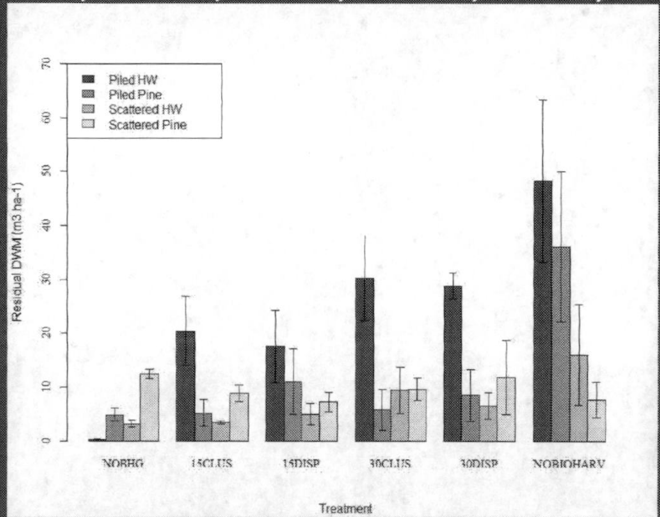

New Technologies will Increase Harvesting Efficiency

Analysis of 53 Temperate and Boreal Forest Studies

(Thiffault et al. 2011)

- Biomass harvesting can reduce soil nutrient pools, particularly in the forest floor
- "No clear impact of whole-tree harvest on soil C"
- "No consistent, unequivocal and universal effects of biomass harvesting on soil productivity"

Forest Service Long-Term Soil Productivity Project

(Ponder et al. 2012)

- 10 years, 45 installations across North America
- Whole-tree harvest: "little consistent effect on any response variable, including tree growth"
- Vegetation control consistently increased tree biomass
- Soil nutrient demands may increase

Ecological Effects of Logging Residue Harvest in Great Lakes Aspen Stands

(Michigan Tech, Plum Creek, NCASI)

("Residues removed" treatment; Michael Premer, Michigan Tech)

- 40 year chronosequence of harvested stands
- With or without residue removal

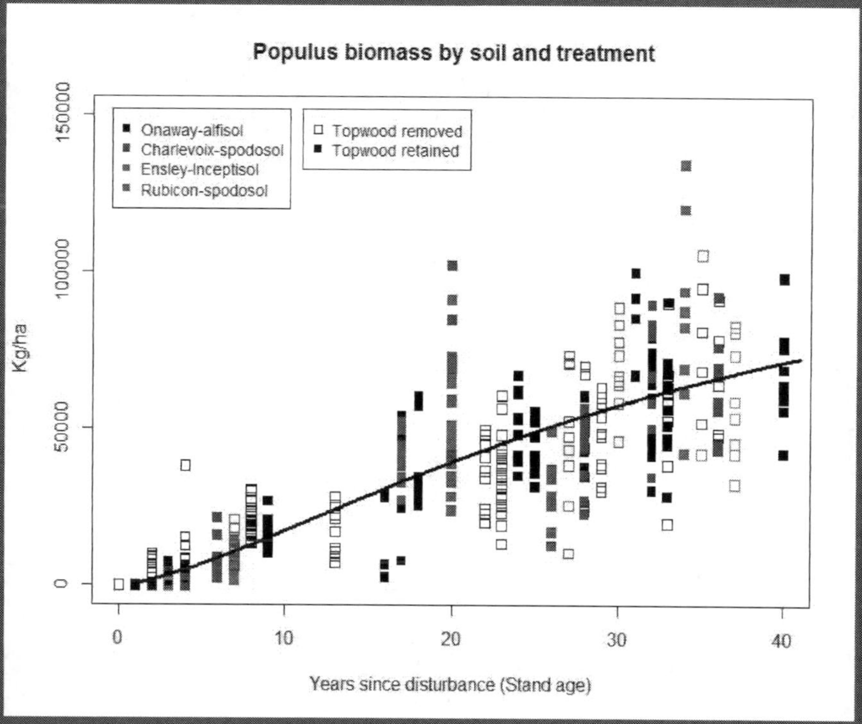

Fall River Long-Term Soil Productivity Project

(Univ. of Washington, Weyerhaeuser, USFS, NCASI)

- Conventional bole-only removal

- Total stem (bole-only to a 5cm top)
- Total-tree removal
- Total-tree + legacy-wood removal

Fall River: Effects of Harvesting, Compaction and Vegetation Control on Age 10 Tree Volume

(Holub et al. 2013)

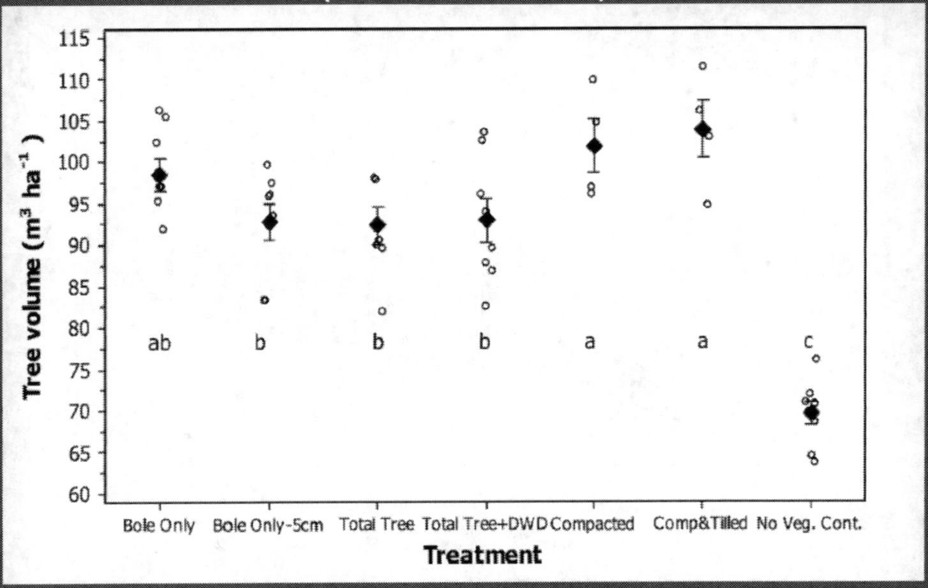

Can Nutrient Deficiencies be Prevented or Corrected?

- Evidence suggests yes
  (e.g., Helmisaari et al. 2011, Jacobson et al. 2000, Smith et al. 2000)
- Lack of evidence for uncorrectable declines
- Many studies do not include fertilization
- Fertilization not an option on some lands
  - Management restrictions
  - Economics

Soil Erosion May Impact Soil Productivity and Water Quality

- Decades of Research: BMPs for water quality effective with 90% implementation
  - (Ice et al. 2011, Schilling et al. 2009)
- Biomass Harvesting Can Exacerbate Disturbance
  - (Curran and Howes 2011)
- Extra precautions
  - Minimize bare soil
  - Retain residues where needed
  - Monitor

Moving Forward in the Face of Uncertainty

- Default to operational expertise, best management practices
- Monitor Function rather than State
- Assess economic and bureaucratic deterrents
- Identify sensitive sites
- Mitigate, modify, or suspend practices if problems occur

Discussion

Sustainability in Biofuel and Bioenergy Sectors: The Role of Public Policy

Hayri Önal

University of Illinois at Urbana-Champaign

PAN-AMERICAN BIOFUELS & BIOENERGY SUSTAINABILITY

July 22-25, 2014, Recife, Brazil

1

Recent Situation in Three Major Biofuel Markets

- **EU**: The Renewable Energy Directive requires that by 2020 the share of biofuels to be >10% of the conventional fuel consumption. Biofuels must reduce GHG eemissions by at least 35% relative to the replaced fossil fuel. In 2013, the EU parliament proposed a cap that restricts bifouels from food crops to be <7% of the mandate.
- **US**: RFS mandates blending **136** blt of renewable biofuels with oil-based fuels by 2022. Up to **56** blt can be corn-based ethanol, **>80** blt must be <u>advanced</u> biofuels, of which **61** blt must be <u>cellulosic</u> biofuels.

RCN, July 22-25, Recife, Brazil 2

RFS Biofuel Blending Mandates (BGY, 2008-2022)

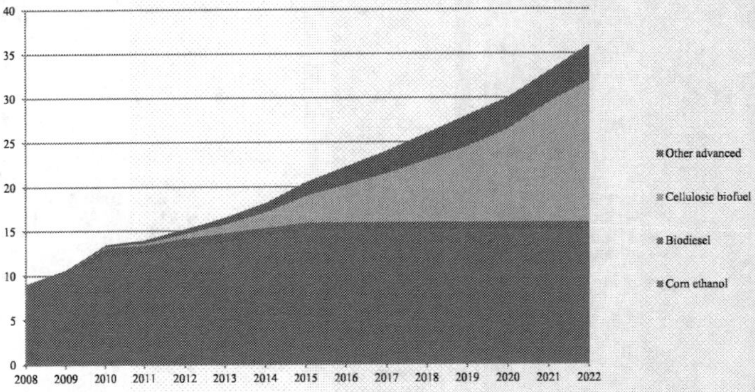

Where We Are Now

- Corn ethanol production tripled from **14.7** blt in 2005 to about **53.2** blt in 2011. Current production capacity is about **55.6** blt ≈ maximum capacity envisioned by RFS.
- Cellulosic biofuel mandates could not be met since the inception of the RFS;
- At present, there is no biorefinery producing cellulosic biofuel at commercial scale, some small scale test production has occurred in the past few years, first commercial production is scheduled for 2014 (?).

Progress in US Ethanol Plant Construction and Production Capacity

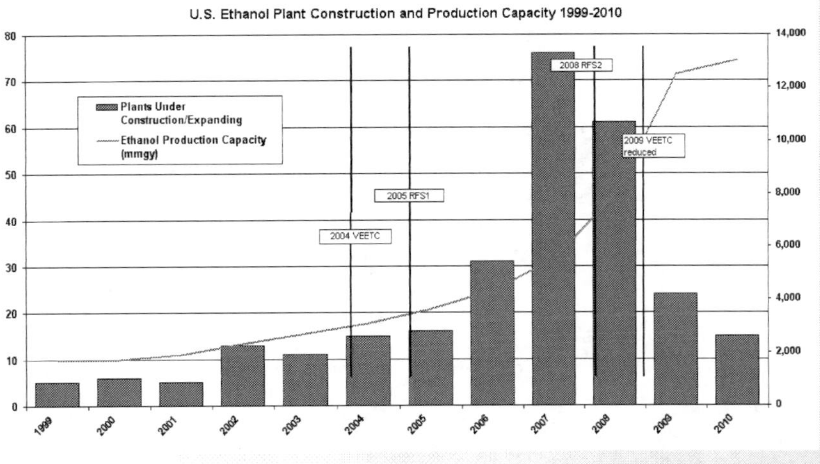

RCN, July 22-25, Recife, Brazil 5

US Corn Production and Use for Ethanol

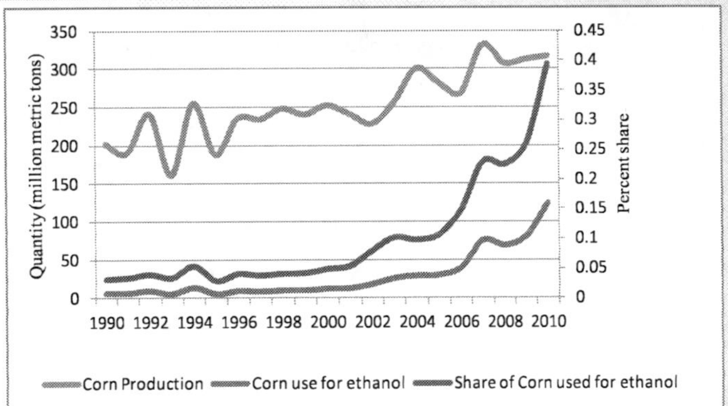

Figure 1: Corn production in US and corn use for fuel ethanol, 1990-2010.
Source: USDA

RCN, July 22-25, Recife, Brazil 6

Crop Price Trends and Volatility in US

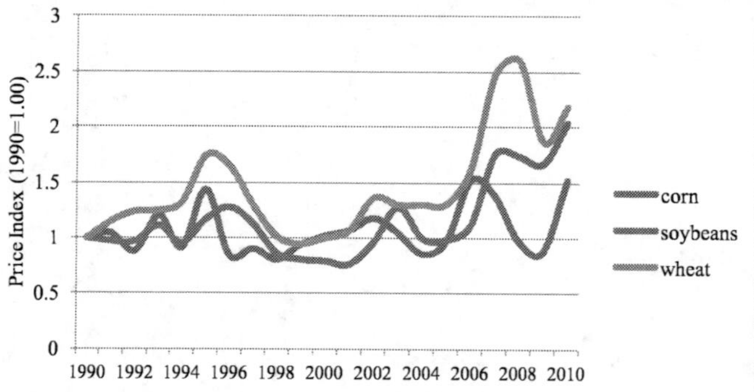

Figure 2: Crop price trends and volatility in the US, 1990-2010. Source: USDA

Trends in Farm Real Estate Value

Average U.S. farm real estate value, nominal and real (inflation adjusted), 1969-2011

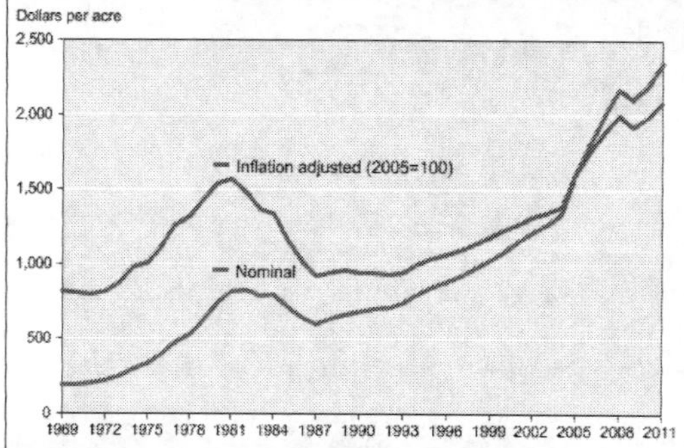

Note: The GDP chain-type price index is used to convert NASS current-dollar statistics to 2005=100 equivalents (Bureau of Economic Analysis, Dept. of Commerce).
Source: USDA-NASS.

Implementation of the RFS

Each year EPA revises the mandated volumes of biofuels. Since 2008, the annual cellulosic biofuel blending targets have been waived or adjusted drastically.

Year	Target (mg)	Revised (mg)
2010	100	6.0
2011	250	6.6
2012	500	8.65
2013	1,000	6.0 (down from 14)
2014	1,750	17 (?)

RCN, July 22-25, Recife, Brazil 9

Cellulosic Biofuel Plants Under Construction & Planned Production Capacity (mg/year)

Facility	Fuel	Year	Cap
Abengoa, KS	Ethanol	2013	25
Fiberight, IA	Ethanol	2013	6
Ineos Bio, FL	Ethanol	2013	8
DuPont, IA	Ethanol	2014	30
Fulcrum, NV	Biodiesel	2013	10
Mascoma, MI	Ethanol	2014	20
Kior ,MS	Eth/Diesel	2013	11
POET, IA	Ethanol	2013	25
BP, FL	Ethanol		(36, cancelled)

RCN, July 22-25, Recife, Brazil 10

Responses to Policy Changes

Corn prices during 2006 through the 2012 were supported by a rapidly growing domestic ethanol industry. Relatively poor corn yields in 2010, 2011, and 2012 kept corn supplies very tight.

Recently, EPA proposed to reduce the 2014 blending mandates for both the advanced and corn-based components (41 and 16%, respectively).

Commodity Price Changes in US, 2005-2013

Ratio of farm commodity prices relative to corn (Pi/Pc)

	2005	Avg. of 2006-12	2013
soybeans	2.83	2.28	2.92
wheat	1.71	1.34	1.54
soymeal	2.44	2.06	3.02

Effects of Biofuel Policies on Fuel Consumption, Fuel Mix and GHG Emissions

Scenarios	BAU	Carbon tax	RFS	LCFS
Cumulative Fuel Consumption, 2007-2030 (Billion gallons)				
US gas consumption	3155.1	3056.0	2898.6	3024.3
		(-3.1)[1]	*(-8.1)*	*(-4.1)*
US gasoline imports	2031.2	1949.6	1821.5	1923.3
		(-4.0)	*(-10.3)*	*(-5.3)*
US petro diesel consumption	989.2	954.7	969.3	980.1
		(-3.5)	*(-2.0)*	*(-0.9)*
ROW gas consumption	4421.0	4467.5	4532.9	4477.8
		(1.1)	*(2.5)*	*(1.3)*
First generation biofuels[2]	115.7	117.5	384.7	82.4
Cellulosic Ethanol			160.4	253.9
BTL diesel[3]				5.8
Reduction in Cumulative GHG Emissions relative to Baseline, 2007-2030 (Billion tons)				
US GHG emissions		-2.4	-2.4	-2.4
		(-4.2)	*(-4.2)*	*(-4.2)*
Global GHG emissions		-1.8	-0.2	-1.4
		(-3.1)	*(-0.4)*	*(-2.6)*

Welfare Effects of Biofuel Policies

Scenarios	Carbon Tax	RFS	LCFS
US fuel consumers (a)	-1377.9	42.7	89.6
US fuel producers (b)	-164.9	-203.5	-87.3
US agricultural consumers (c)	-12.9	-124.0	3.7
US agricultural producers (d)	19.6	296.6	10.9
US government (e)	1669.8	35.0	16.4
Reduction in US externality costs (f)[2]	75.1	63.4	58.5
US economic surplus (Δ=a+b+c+d+e)	133.7	46.7	33.4
US social welfare (Δ+f)	208.9	110.1	91.9
	(0.8)	(0.4)	(0.3)

Biofuel Policies in Brazil

- Blending rate: in the past decade varied within 18- 25% range. Due to supply shortages, in 2011 the rate was reduced from 25% to 20%, in May 2013 it was raised back to 25%.
- Tax rates : >100% for gasoline, <40% for ethanol until 2011. In 2011, gasoline tax was 80%, in 2012 it was reduced to 53% and then totally eliminated. Ethanol tax is also eliminated in May 2013.
- Price controls: PETROBRAS regulates gasoline & diesel prices. In the past decade gasoline price was almost fixed at a level 19% lower than the world price. In 2013, to reduce losses from subsidized sales the price was raised by 11%.

Ethanol Market

- Ethanol price is not regulated and fluctuates depending on the fuel blending ratio, tax rate and sugarcane/sugar prices.
- Despite the tax incentives provided to ethanol consumers, low gasoline prices in recent years weakened the competitiveness of ethanol and reduced the share of ethanol in the total fuel consumption from 56% in 2009 to 39% by volume in 2012. Currently it is about 50%.

Trends in Fuel and Sugar Prices

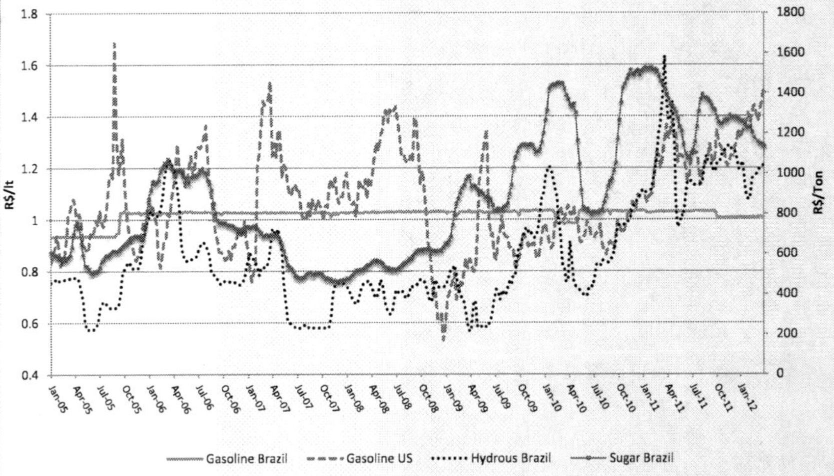

RCN, July 22-25, Recife, Brazil 17

Ethanol Blending Ratio in Brazil

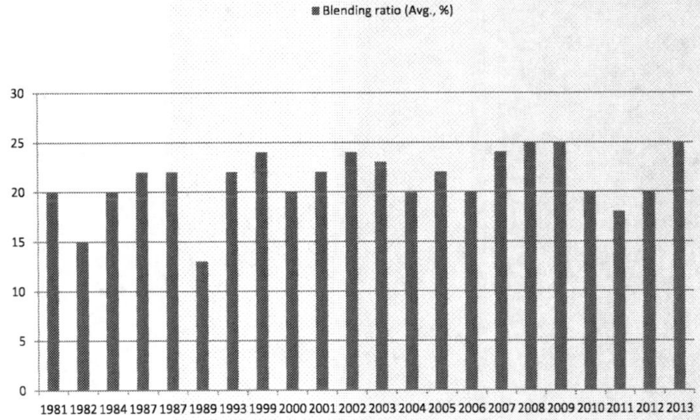

RCN, July 22-25, Recife, Brazil 18

Simulation Results Under Different Market Conditions and Policy Scenarios - 2022

- Benchmark Case: Strong ethanol and strong sugar demand in global markets, average sugarcane yield, 25% blending rate, and base year fuel tax rates
- Four market conditions are considered by coupling two demand and two supply conditions
  - Strong vs Weak demand for ethanol and sugar
  - Historical average sugarcane yield vs low yield
- Each case is run assuming blending mandates ranging over 15-25% and tax rate reductions ranging over 0-100%

RCN, July 22-25, Recife, Brazil 19

Percent Changes wrt Benchmark	Strong Ethanol Demand + Strong Sugar Demand					Weak Ethanol Demand+ Average Sugar Demand				
Sugarcane Yield	Average		Low			Average		Low		
Blending Rate	15%	15%	25%	25%	15%	15%	15%	25%	25%	15%
Tax Rate	bt	nt	bt	nt	bt	bt	nt	bt	nt	bt
Vehicle Km Traveled	-0.4	5.0	-0.3	5.1	-0.6	0.3	5.1	0.2	5.4	-0.2
Eth Consumption Total	22.2	-77.2	-4.0	-62.0	-2.9	29.5	-77.2	6.0	-61.9	24.8
Gas Consumption Total	-22.9	88.9	3.5	72.8	2.2	-28.9	89.1	-5.7	73.4	99.0
E100 Consumption by FFV	44.8	-100.0	-6.1	-100.0	8.6	55.0	-100.0	9.4	-100.0	48.5
Gas Consumption by FFV	-39.9	130.2	5.4	110.5	-1.2	-49.1	130.5	-8.7	111.3	-43.3
E100 Consumer Price	6.6	6.6	2.3	11.0	4.6	1.3	-4.0	-1.4	-3.0	4.8
Gasohol Consumer Price	7.5	-32.3	0.5	-35.3	8.6	6.8	-33.0	-0.3	-37.2	7.4
Ethanol Exports	-4.3	10.1	-29.1	-13.9	-29.2	-96.6	-67.7	-100.0	-95.0	-100.0
Sugarcane Area	3.8	-15.9	-6.6	-15.5	-6.4	-40.0	-55.0	-42.9	-60.2	-37.7
GHG Emissions	-4.8	18.9	-0.8	14.5	-1.1	-12.0	14.4	-7.1	10.1	-11.2
Ethanol Producers' Surplus	20.0	-81.5	-4.2	-73.3	-6.9	29.6	-76.1	11.1	-58.8	27.4
Government Tax Revenue	-1.9	-68.2	0.6	-69.7	4.4	-2.3	-68.2	-1.0	-69.6	-2.1
Social Surplus	-0.7	1.0	-0.7	0.8	-1.2	-5.2	-3.1	-4.5	-3.5	-5.1

RCN, July 22-25, Recife, Brazil 20

Highlights from Model Results

- Fuel taxes seem to be the most important policy instrument in Brazil's biofuel economy. Removing the taxes makes gasohol a more appealing fuel than ethanol. The total consumption of ethanol goes down, driving goes up, GHG emission goes up, tax revenue goes down, yet social surplus goes up wrt the corresponding scenario with higher taxes.

Highlights from Model Results (continued)

- Impacts of a reduced blending mandate on the fuels sector and social welfare vary across the scenarios
- Impact on VKT demand insignificant, but
- Composition of the fuel mix changes significantly, more ethanol is consumed by FFVs switching from gasohol to ethanol because gasohol becomes more expensive since it has a more of an expensive fuel (gasoline) in its mix.

An Unintuitive Result

- A smaller blending rate increases the GHG emissions from transportation fuels since this increases the amount of gasoline in the fuel consumption by CVs. BUT, a simultaneous effect is the possibility of switching from gasohol to E100 by FFVs. The net effect is ambiguous.
- The simulation results show that under the same market conditions the adverse effect of a more gasoline-intensive mix on GHG emissions is dominated by the emission reduction effect of increased ethanol consumption resulting from FFVs' switching to E100. Thus, reducing the blending mandate w/o tax rate change would decrease the GHG emissions!

RCN, July 22-25, Recife, Brazil 23

Concluding Remarks

- Lowering the blending rate temporarily to cope with a short supply of ethanol may be a sound policy in the short run, but if continued in the long run it is not a good policy
- Lowering the tax rate on gasoline may reduce some of the adverse effect on consumers, but it is harmful for the environment.
- Sound policies: invest in agriculture (e.g. livestock intensification) and invest in infrastructure development (e.g. ethanol pipelines, roads, rail, irrigation).

RCN, July 22-25, Recife, Brazil 24

Concluding Remarks (Continued)

- Under a lower blending rate CVs will be driven less, while FFVs and EDVs will not be affected much. Reducing the blending rate alone reduces consumers' surplus, increases producers' surplus.
- Fuel taxes are strong and effective policy instruments. Lowering the tax rates reduces competitiveness of ethanol against gasoline blends. Low fuel taxes are beneficial for consumers, but lower producers' surplus.
- A reduced tax policy may have dramatic environmental impacts by increasing GHG emissions from transportation fuels.

Current State of The Biofuels Policy World

THANK YOU!

RCN, July 22-25, Recife, Brazil 27

Biofuels and Bioenergy Sustainability Assessment Methods

Schuster, D., Institute for Sustainability
Alexander, L.,
Originally presented on: 7/24/2014 13:18:00 - 13:36:00

This presentation from AIChE's Institute for Sustainability will explore biofuels and bioenergy sustainability assessment methods, focusing on supply chain and value chain, and safety, economic, social, and environmental concerns. Various popular methods will be compared on the basis of cost- and time-effectiveness, ease of use, flexibility, clarity of results, and other strengths and weaknesses. Also discussed will be the needs for and benefits of such assessments for bioenergy producers and users. The presentation will identify areas where more research is needed and what challenges exist for implementing assessment methods across the supply chain. Applicability will be compared for different types of energy producers, and specific recommendations will be made based on size, feedstock, location, process, products, and resource availability. The ways in which these factors are likely to affect attitudes, operations, impacts from a sustainability standpoint will be explored.

This paper will come from educational materials offered as part of the AIChE Credential for Sustainability Professionals.

RCN CONFERENCE ON PAN AMERICAN BIOFUELS & BIOENERGY SUSTAINABILITY

Biogas Generation from Sanitary Landfill Leachate

Universidade Federal de Pernambuco
Departamento de Engenharia Civil

Maurício Alves da Motta Sobrinho

RECIFE/2014

LIMA - UFPE
Laboratório Interdisciplinar de Meio Ambiente

PAN-AMERICAN BIOFUELS & BIOENERGY SUSTAINABILITY

GRS UFPE
Grupo de Resíduos Sólidos

Introduction

- Energetic Sustainability;
  - Employment of alternative energy sources;
  - Renewable energy sources in Brazil: 45%;
- Energy resulting from treatment of effluents;
  - Collaboration between research institutuions and industries

cetrel

Introduction

- Anaerobic degradation process:
  - Methane generation: 50 a 70%;
- Problem: treat the leachate;
  - Complexity, toxicity, recalcitrance;

Introduction

- BMP Testing - Biochemical Methane Potencial
  - Initial residue composition;
  - Biodegradability conditions;
  - Pressure Build up;
  - F:M Ratio (Organic Load/microorganism).

Aim

- Evaluate the influence of anaerobic degradation in solid contaminants and liquids in the production of biogas through a 2^3 factorial design in BMP trials.

Materials and Methods

- Leachate COD
  - Method: Colorimetric - SMEWW 5220 C (Closed reflux);
    - Threaded Test tubes (reflux);
    - Reagent: Potassium dichromate- $K_2Cr_2O_7$ (1,5mL);
      Silver sulphate+ sulphuric acid(3,5 mL);
    - Thermoreaktor CR 2200;
    - Spectrophotometer: Thermo Scientific Genesis 10 - S - UV - Vis 600nm;

Materials and Methods

- Volatile Suspended Solids: (VSS) of the sludge
- ABNT /NBR 10664/Apr 1989
  - 10 g of sludge in vacuum filter NALGENE maximum pressure of 1 bar;
  - Vacuum pump: Quimis 03558;
  - Fiberglass microfilter GF-1 47mm Macherey-Nagel transfered to a porcelaine capsule and transfered to an oven Brasdonto – Modelo 3 at 105°C;
  - Calcinate at 550°C/1 h in oven Muffle Furnace Quimis;
  - Weighed on analytic balance Bioprecisa FA-2104N;

Materials and Methods

- 2^3 Full Factorial Design

	Lower level	Center	Upper level
Sludge volume(mL)	10	30	50
Headspace (mL)	50	100	150
pH	6	7,5	8,5

- Control
  - 50 mL of sludge without leachate ;

Materials and Methods

- pH adjust (Figure 1);
  - Bench pH-metro : Quimis Q400AS;
  - NaOH 40%;
  - HCL PA;

Figure 1- pH Adjustment .

Materials e Methods

- Anaerobic Sludge: Brasil Kirin Brewery - Guabiraba PE (Figure 2);
- Leachate: Leachate treatment station at the landfill Muribeca(Figure 3);
- BMP trial in triplicate (Figure 4);

Figure 2 - Anaerobic Sludge

Figure 3 - leachate

Figure 4 -triplicate trials

Materials e Methods

- BMP trials (Figure 5)
  - Glass vials(250 mL);
  - Nylon screwed lid with rubber ring;
  - Manometer ASTA NBR 14105-1 graduation in bar;
  - Gas collection valves and for pressure readings;
  - Aluminum foil covering;

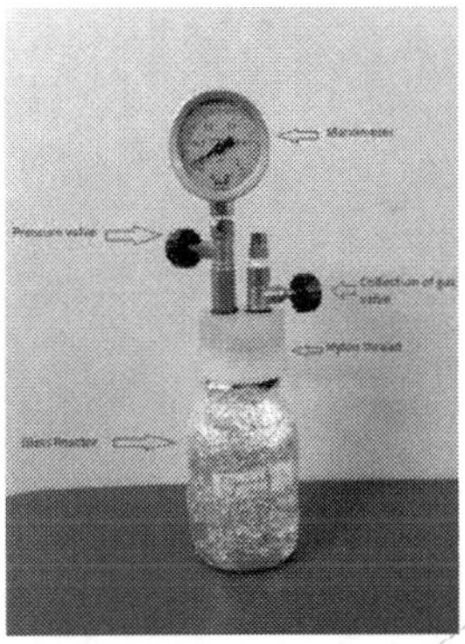

Figure 5 - BMP trials

Materials e Methods

- **BMP trials**
  - Inoculate: Sludge + Leachate (according too experimental planning);
  - Recirculation of nitrogen during 2 minutes(Figure 6) ;
  - Temperature: 35°C - Oven Fanem - 320 - SE;
  - Daily readings;

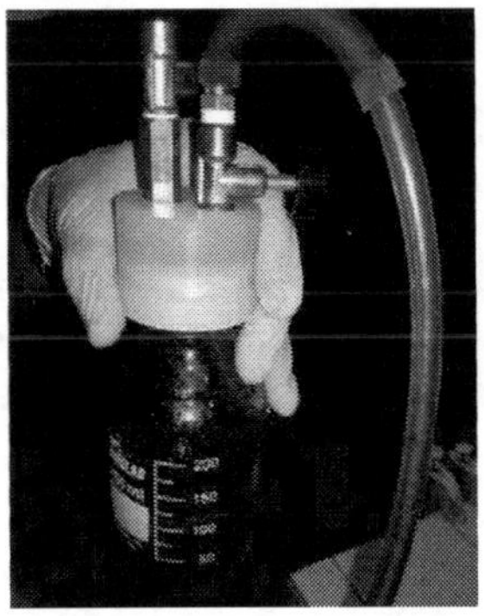

Figure 6 - Recirculation of Nitrogen

Materials e Methods

- Gas analysis
  - Methane (CH_4);
  - Carbon Dioxide(CO_2);
  - Chromatograph: CG Master/Gas;
  - Triplicate

Results

- COD leachate: 3.066 mgO_2/L;
- VSS Sludge: 152,27 g/L;
- Initial pH of the mixture: 8,3;
- Control : Production rate: 6,42mL/d;
  Total volume produced: : 154 mL;

- Best Factorial planning:
  - 50 mL anaerobic sludge + 50 mL leachate;
  - pH: 7,5;

Results

- Biogas Production
  - Production rate: 11,54 mL/day;
  - Total volume produced: 277 mL;
- Biogas Composition
  - 67,07% $CH_{4;}$
  - 32,91% CO_2;

Conclusion

- There was no Inhibition as the total gas production volume from the sludge mixture with leachate was greater than the control (pure sludge). The best experimental condiction is the one in which the volumes of sludge and leachate are 50 mL with pH of 7,5 and headspace of 150mL. This shows that the process can be influenced by pH nearer to neutrality.
- The production of biogas with approximately 70% of the total volume of gas produced, shows the viability of adopting de producion by this method.

Our thanks to

GPTA
Grupo de Processos e Tecnologias Ambientais
DEQ - UFPE

UFPE

GRS UFPE

cetrel

CNPq
Conselho Nacional de Desenvolvimento Científico e Tecnológico

CAPES

FACEPE
Fundação de Amparo à Ciência e Tecnologia do Estado de Pernambuco

FINEP
FINANCIADORA DE ESTUDOS E PROJETOS
MINISTÉRIO DA CIÊNCIA E TECNOLOGIA

Thank you!

RCN CONFERENCE ON PAN AMERICAN BIOFUELS & BIOENERGY SUSTAINABILITY

Biogas Generation from Sanitary Landfill Leachate

Universidade Federal de Pernambuco
Departamento de Engenharia Civil

Maurício Alves da Motta Sobrinho

RECIFE/2014

UNIVERSIDADE FEDERAL DE PERNAMBUCO
CENTRO DE TECNOLOGIA E GEOCIÊNCIAS
DEPARTAMENTO DE ENGENHARIA CIVIL
PROGRAMA DE PÓS-GRADUAÇÃO EM ENGENHARIA CIVIL
ÁREA DE TECNOLOGIA AMBIENTAL

UFPE

PAN-AMERICAN BIOFUELS & BIOENERGY SUSTAINABILITY

M. V. C. Paiva*, S. M. S Barbosa[1], E. A. Pastich*

The potential use of waste stabilization ponds for biofuel production by the microalgae biomass

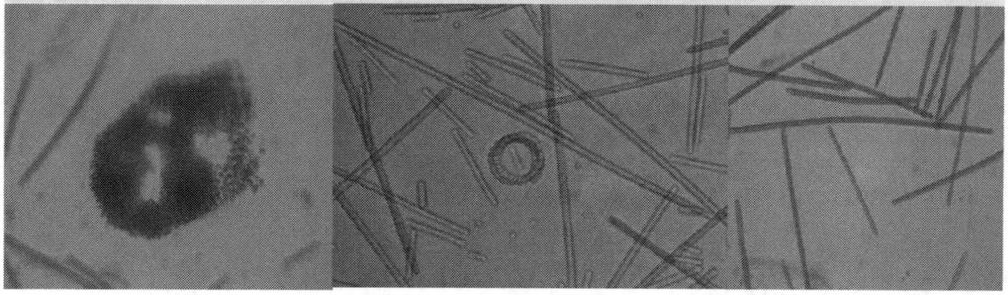

Lagoas de Estabilização

PAN-AMERICAN BIOFUELS & BIOENERGY SUSTAINABILITY

INTRODUÇÃO

Lagoas de Estabilização

INTRODUÇÃO

- Crescimento elevado organismos fitoplanctônicos, podendo afetar a eficiência do sistema do tratamento, causando desequilíbrio no corpo receptor

- Floração de cianobactérias- Riscos a saúde e ao meio ambiente

Lagoas de estabilização

INTRODUÇÃO

Pós-tratamento de lagoas de estabilização para remoção da biomassa de algas:

- Ozonização
- Flotação por ar dissolvido
- Biofiltros aerados
- Microfiltração

Filtros de pedra

Qual a solução para biomassa de algas produzidas em lagoas de estabilização?

PAN-AMERICAN BIOFUELS & BIOENERGY SUSTAINABILITY

Potencial para o uso de microalgas como fonte de energia renovável, principalmente para a produção de biocombustíveis

INTRODUÇÃO

Biocombustíveis

INTRODUÇÃO

Biocombustíveis de 1° geração

Biocombustíveis de 3° geração

O rendimento médio de produção de biodiesel
a partir d microalgas pode ser de 10 a 20 vezes maior o que o
rendimento o tido a partir de sementes oleaginosas e/ou ó os vegetais
(Gouveia, 2009)

Impacto na dispo
alimen

Alta demanda por ág
nutrien

Dependente da
climáti

dade de biomassa e
culturas de plantas
ginosas

dade de área para
doce e nutrientes

climas variáveis

Produção de biodiesel

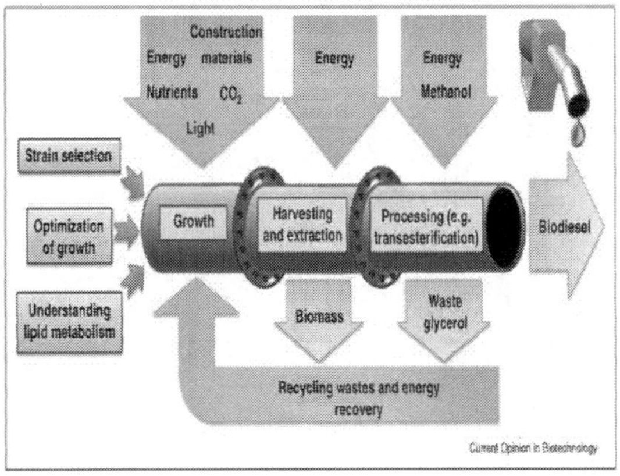

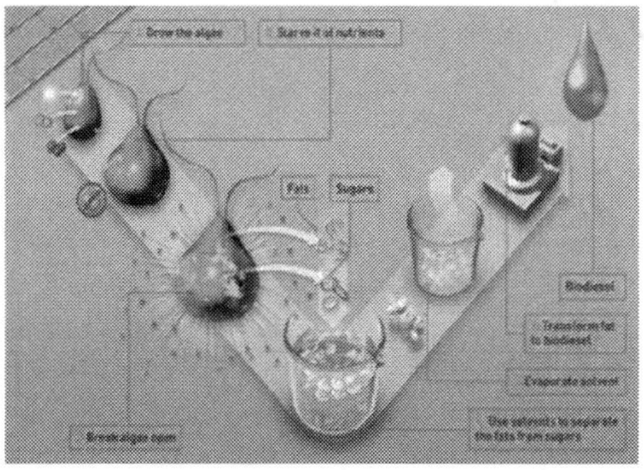

Utilização de algas para produção de Biocombustíveis

INTRODUÇÃO

Remoção de CO_2 emitidos pela indústrias através da biofixação

Redução da emissão de gases do efeito estufa

Remoção de NH_4^+, NO_3^- , PO_4^{-3} através da utilização de águas residuárias para o crescimento de algas.

Produzir alimentação de gado, fertilizante, queimado para geração de energia elétrica e calor

Utilização de algas para produção de Biocombustíveis

Espécies de microalgas	Conteúdo de óleo (% da matéria seca)	Produtividade de óleo (mg L^{-1} d^{-1})
Chlorella protothecoides	23-55	1214
Chlorella vulgaris	5-50	-
Chlorella sorokiniana	19-22	45
Nannochloropsis oculata	22-30	84-142
Dunaliella salina	14-20	116
Neochloris oleoabundans	35-65	90-134
Phaeodactylum tricornutum	18-57	45
Spirulina maxima	4-9	-
Chlorococcum sp.	19	54

Scott *et. al.*, 2010

Fonte de biomassa	Tipo de combustível produzido	Produtividade (bep ha^{-1} ano^{-1})	Eficiência fotossintética (%)
Milho	Etanol	20	0,2
Cana-de-açúcar	Etanol	210-250	2-3
Soja	Biodiesel	13-22	0,1-0,2
Girassol	Biodiesel	8,7-16	0,1-0,2
Microalgas	Biodiesel	390-700	4-7

Schenk *et. al.*, 2008

Produção e armazenamento de lipídeos

INTRODUÇÃO

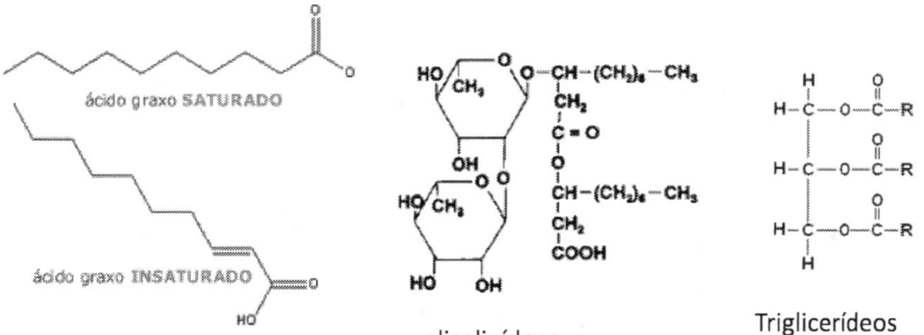

glicolipídeos

Triglicerídeos

Quantidade de lipídios produzidos dependerá das espécies de microalgas e as condições de crescimento (Chisti, 2007; Griffiths e Harrison, 2009;. Hu et al, 2008).

Utilização de águas residuárias

INTRODUÇÃO

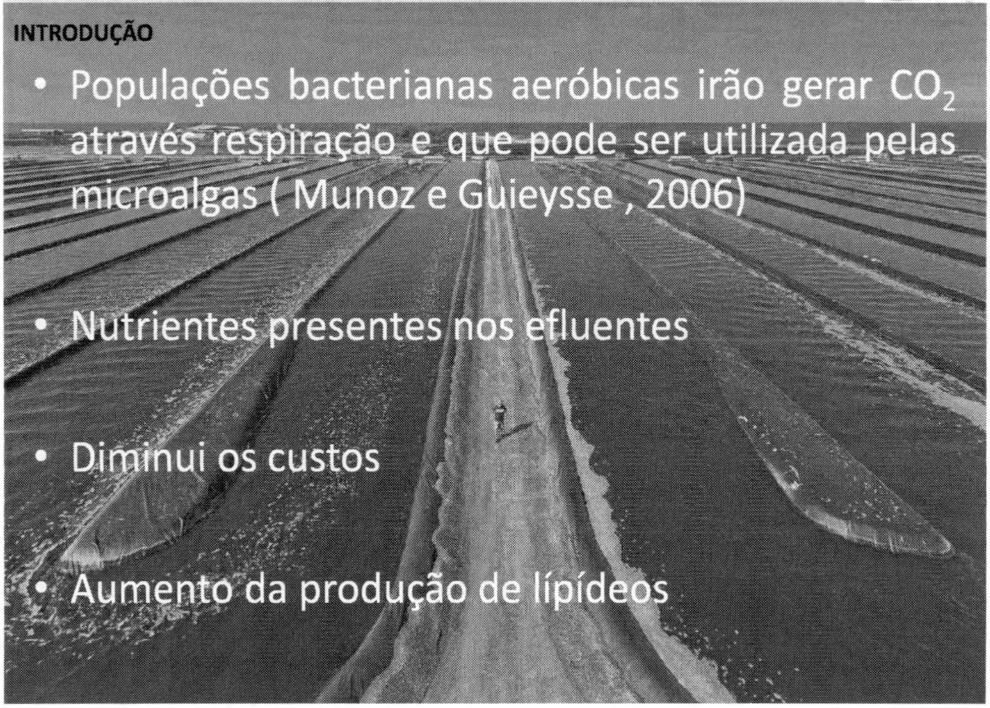

- Populações bacterianas aeróbicas irão gerar CO_2 através respiração e que pode ser utilizada pelas microalgas (Munoz e Guieysse , 2006)
- Nutrientes presentes nos efluentes
- Diminui os custos
- Aumento da produção de lípideos

Cultivo em lagoas ou bioreatores?

INTRODUÇÃO

Lagoas	Bioreatores
Vantagens:	**Vantagens:**
Baixo custo de construção e operação	Minimização da contaminação Maiores densidades de culturas Maior controle das condições físico-químicas
Desvantagens:	**Desvantagens:**
Baixa produção de biomassa Dificuldade de colheita Dificuldade de controlar as condições físico-químicas	Alto custo de construção e operação

Área de Estudo

MATERIAIS E MÉTODOS

Popu. 22.140 hab

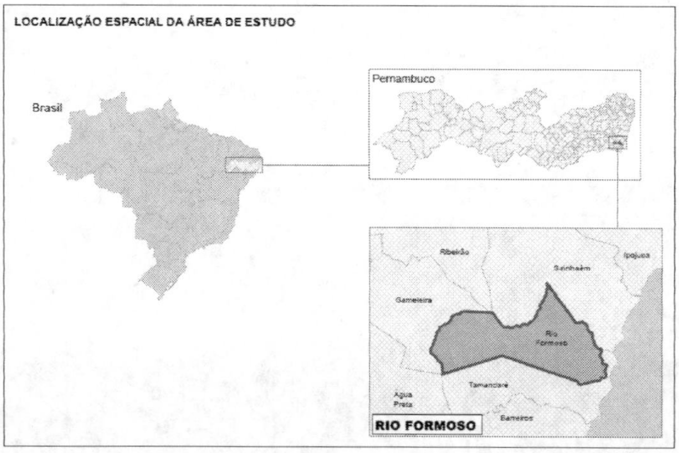

ETE- UASB+ Lagoa de polimento + Filtro de pedra

Área de Estudo

MATERIAIS E MÉTODOS

Dados do Projeto	Unidade	Reator UASB (3 UNIDADES)	Lagoa de Polimento	Filtros de pedra (4 UNIDADES)
Comprimento	(m)	11,6	167	120
Largura	(m)	16	110	120
Profundidade útil (m)	(m)	5,3	1,50	0,55
Volume	(m^3)	984	28.050	7.920
Área	m^2	186	14110	14400
Vazão	m^3/dia	40	40	40
TDH	dia	0,3	8	2

Parâmetros analisados

MATERIAIS E MÉTODOS

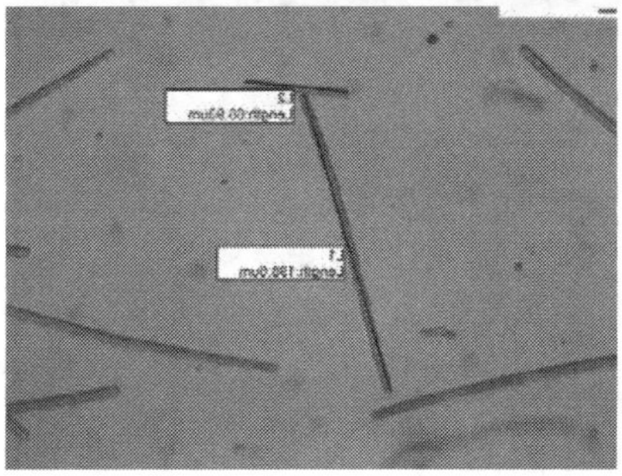

Coletas : 6 meses
Horários 14 e 2h
Superfície e fundo

Medições de campo: temperatura, oxigênio dissolvido, pH, condutividade

Análises laboratoriais: fósforo total, ortofosfato, NTK, nitrogênio amoniacal, nitrito, DQO, DBO

Análise das microalgas: identificação dos táxons , densidade e biomassa

Análise da composição de microalgas

RESULTADOS E DISCUSSÃO

CONTRIBUIÇÃO DAS DIVISÕES PARA A RIQUEZA TOTAL

■ CYANOPHYTA ■ CHLOROPHYTA ■ EUGLENOPHYTA

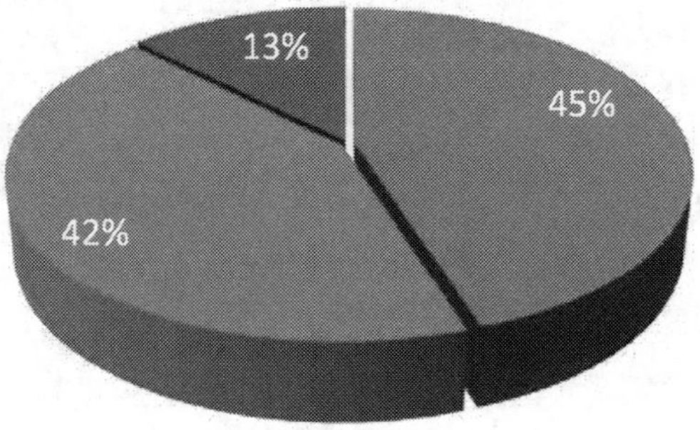

Divisão Cyanophyta

RESULTADOS E DISCUSSÃO

- *Pseudoanabaena, Oscillatoria, Microcystis, Dolichospermun, Anabaenopsis, Coelomoron, Aphazinomenon, Raphidiopsis, Aphanocapsa, Coelospharium, Merismopedia, Choroococcus, Radiocystis, Sphaerocavum, Eucapsis, Arthrospira*

- Não contêm grandes quantidades de lipídeos (cerca de 20%)
- Tem uma produtividade relativamente elevada de biomassa
- Composição da biomassa pode ser manipulado por vários fatores ambientais e operacionais para produzir mais células.
- (Balasubramanian *et al.*, 2010)

Divisão Chlorophyta

- *Scenedesmus, Sphaerocystis, Monoraphidium, Closteriopsis, Tetradesmu, Desmodesmus, Eudorina, Chlamydomonas, Coelastrum, Oocystis, Keratococcus, Radiococcus, Golenkia, Franceia*

Scenedesmus obliquo tem uma taxa de crescimento melhor em águas residuais municipais (Ruiz-Marin et al., 2010).

Remoção quase completa da amônia, nitrato e P total em tratado secundário (Martinez et al., 2000; Ruiz- Marin et al., 2010; Zhang et al., 2008)

Análise da composição de microalgas

RESULTADOS E DISCUSSÃO

Densidade total

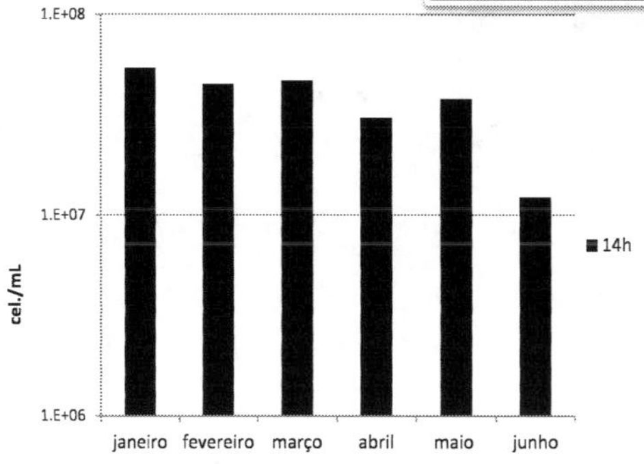

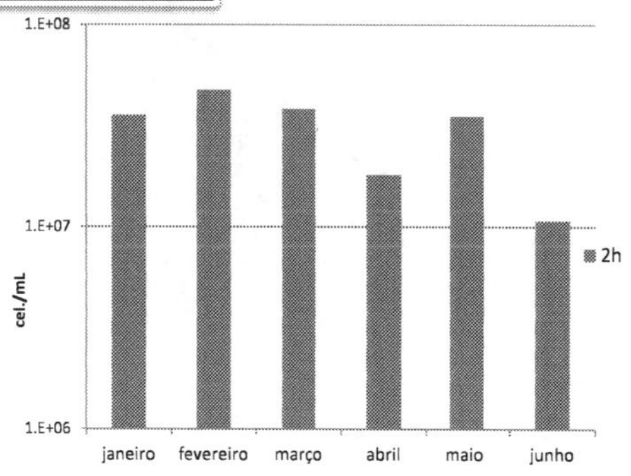

14h

Maior densidade (5.4 x 10^7 cel./mL)
Menor densidade(1.2 x 10^7 cel/mL)

2h

Maior densidade(4.7 x 10^7 cel./mL)
Menor densidade(1.1 x 10^7 cel./mL).

Análise da composição de microalgas

RESULTADOS E DISCUSSÃO

Biovolume

14h

2h

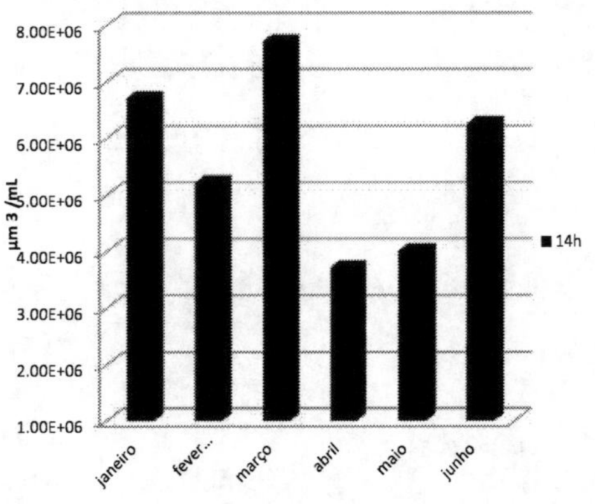

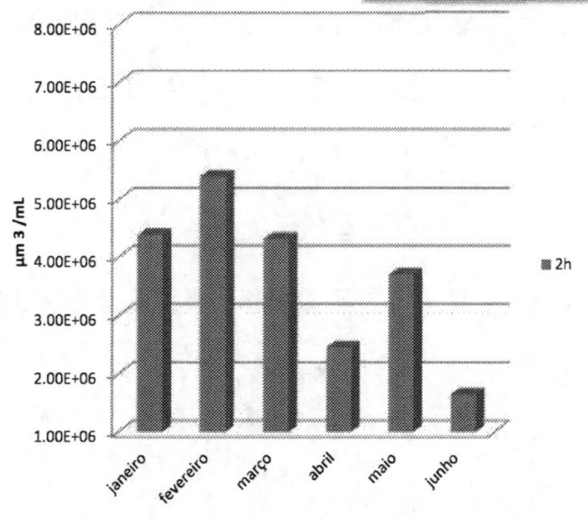

Maior biovolume: 7,7 x 10^6 µm ^{3} /mL
Menor biovolume: 3,7 x 10^6 µm ^{3} /mL

Maior biovolume: 5, 3 x 10^6 µm ^{3} /mL
Menor biovolume: 1,6 x 10^6 µm ^{3} /mL

Remoção de nutrientes

Nitrogênio amoniacal

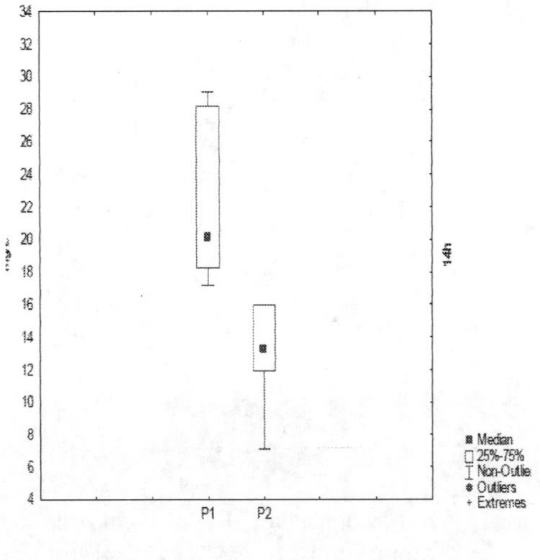

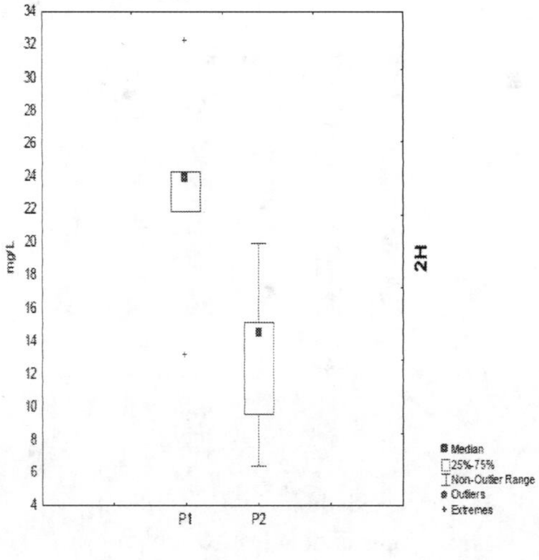

Remoção de nutrientes

NTK

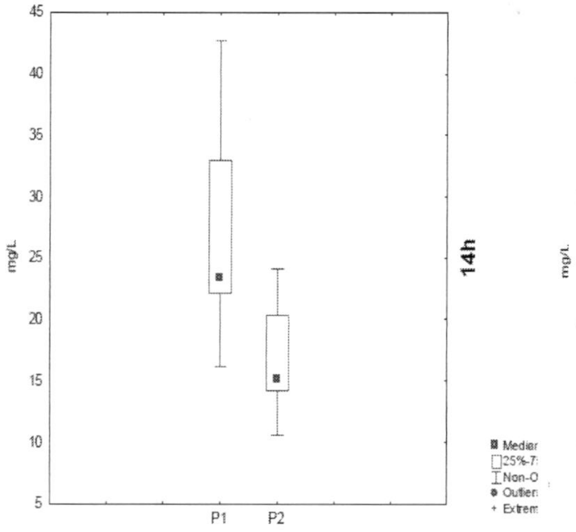

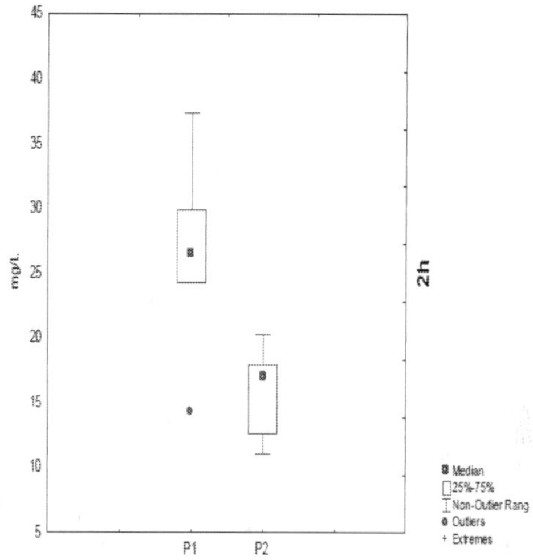

Análise da composição de microalgas

RESULTADOS E DISCUSSÃO

Cianobactérias e algas verdes (Chlorophyta)- fotoautotrófico

Para manter uma certa taxa de crescimento, tem de recorrer a um modo de quimiotrófica para geração de energia.

Metabolismo heterotrófico e mixotrófico

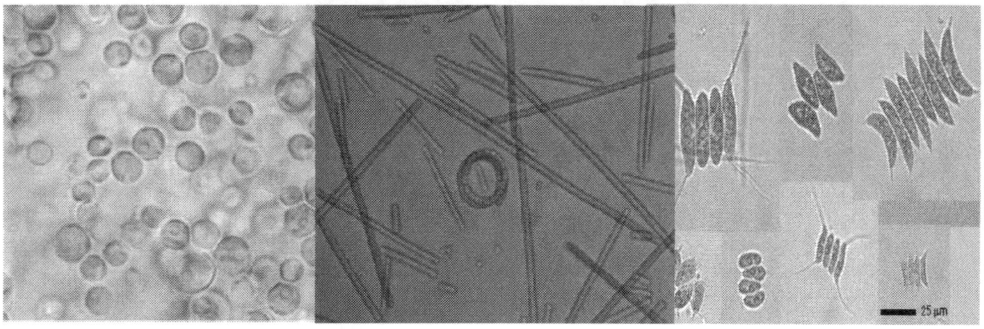

Análise da composição de microalgas

RESULTADOS E DISCUSSÃO

Mais do que 25% da biomassa produzida durante o dia, pode ser perdida durante a noite por causa da respiração (Chisti, 2007).

Não há uma queda significativa nem na densidade de algas, nem para a medição da biomassa, devido às várias formas de metabolismo realizados por espécies de microalgas presentes

Soluções para aumento da produtividade

Iluminação artificial

Introdução de CO_2

Transformação das lagoas de estabilização em lagoas de alta taxa ou raceways ponds

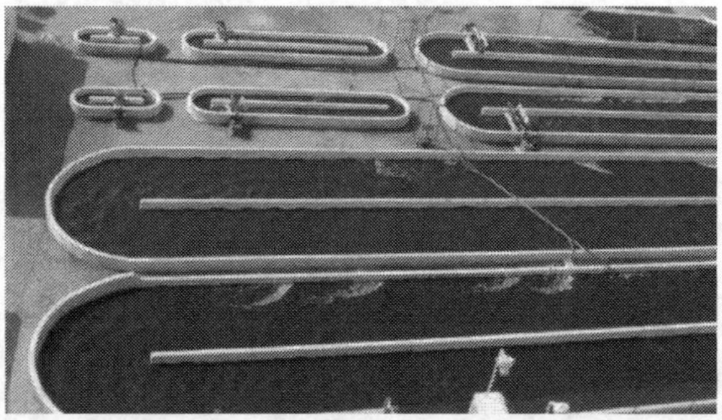

Outros usos: Fertilizante, alimentação animal, queima para geração de energia elétrica

Conclusão

Obstáculos técnicos

- Altos custos de produção
- Dificuldades para produção de elevada biomassa e teor de lipídeos
- Dificuldades para colheita, extração de óleo e produção de biodiesel

Soluções

- Identificação de condições que induzem a elevada acumulação lipídios e biomassa
- Desenvolver tecnologias de baixo custo para colheita eficiente de biomassa e extração de óleo

UFPE

PAN-AMERICAN BIOFUELS & BIOENERGY SUSTAINABILITY

Obrigada!!!

marcellavcpaiva@yahoo.com.br

Brazilian Bioethanol Science and Technology Laboratory

CNPEM

The Sustainability of Brazilian Sugarcane Ethanol: The Contribution of CTBE

Manoel Regis L. V. Leal
CTBE – Brazilian Bioethanol Science and Technology Laboratory

RCN Pan American Biofuel and Bioenergy Sustainability Conference

Recife, PE, July 23-25 2014

Brazilian Bioethanol Science and Technology Laboratory

CNPEM

Contents

- ✓ Sugarcane ethanol sustainability issues
- ✓ CTBE Programs and Impacts on ethanol sustainability
  - Agriculture
  - Industrial
  - Technology Assessment
  - Basic Science
  - Sustainability

GEBEP Sustainability Indicators

INDICATORS		
Environmental	**Social**	**Economic**
Lifecycle GHG emissions	**Allocation and tenure of land for new energy production**	Productivity
Soil quality	**Price and supply of a national food basket**	Net energy balance
Harvest levels of wood resources	Change in income	Gross value added
Emissions of non-GHG air pollutants, including air toxics	Jobs in the bioenergy sector	Change in consumption of fossil fuels and traditional use of biomass
Water use and efficiency	**Change in unpaid time spent by women and children collecting biomass**	**Training and re-qualification of the workforce**
Water quality	**Bioenergy used to expand access to modern energy services**	Energy diversity
Biological diversity in the landscape	**Change in mortality and burden of disease attributable to indoor smoke**	**Infrastructure and logistics for distribution of bioenergy**
Land use and land use change related to bioenergy feedstock production	**Incidence of occupational injury, illness and fatalities**	**Capacity and flexibility of use of bioenergy**

Source: GBEP, 2011

Energy and GHG Emissions Balances

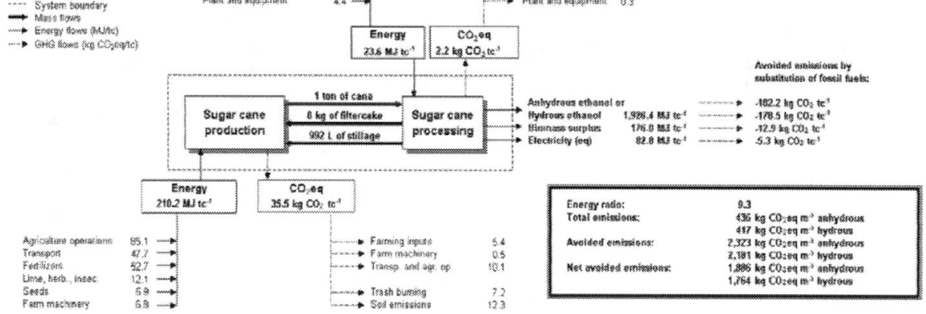

Source: Macedo et al., 2008

CTBE Brazilian Bioethanol Science and Technology Laboratory

CNPEM

CTBE Brazilian Bioethanol Science and Technology Laboratory

A National Laboratory on Bioethanol

CNPEM

Own team dedicated to Research, Development and Innovation	Infrastructure available to research institutions (universities and industries)	Focus on innovation and integration of new technologies towards a more complex Bio-refinery of sugarcane

Numbers:

. Implanting Federal Funding (2008-2010): **US$ 50 million**
. Buildings: **8.722,28 m^2**
. Research team by 2013 (biologists, physicists, chemists and engineers): **110 employees**

Brazilian Bioethanol Science and Technology Laboratory

Programs

CNPEM

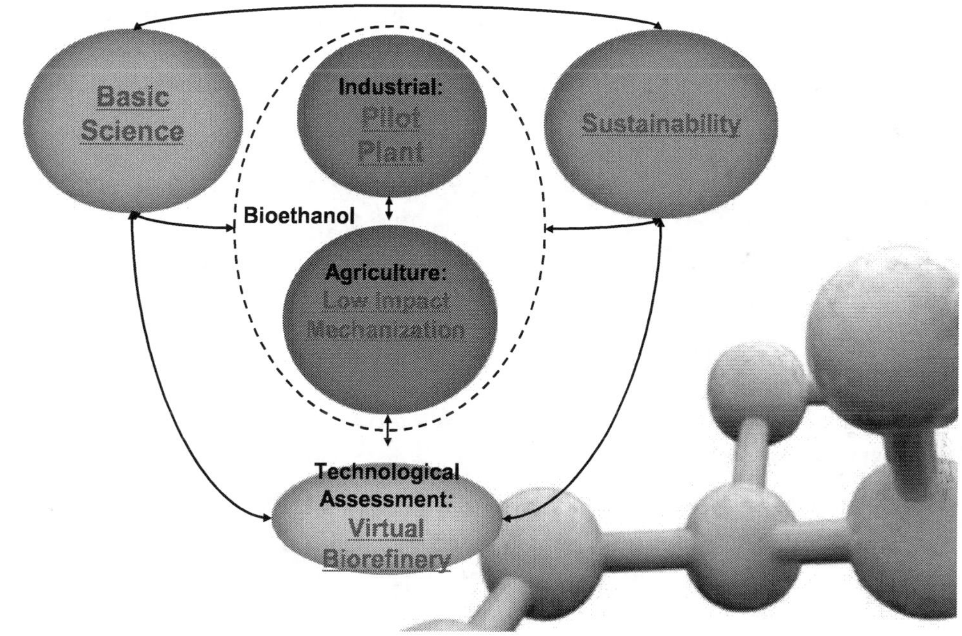

Brazilian Bioethanol Science and Technology Laboratory

Agricultural Program

CNPEM

Low Impact Mechanization to Reduce Soil Compaction

1. No-till farming
Structure for Controlled Traffic farming (ETC)

2. Precision farming
IT for Agricultural Management - Data acquisition, communication and agricultural modeling

3. Trash recovery
Quality and cost of recovered trash; amount to be left on the ground

4. Mechanized Planting
Seed cane quality and distribution in the furrow.

Expected Results: improve soil quality, increase sugarcane yield, reduce fossil fuels and fertilizers use, reduce GHG emissions, reduce harvest losses, reduce sugarcane production costs, recover the straw integrated with sugarcane.

ETC: Low Impact Mechanization System

CTBE Brazilian Bioethanol Science and Technology Laboratory — CNPEM

- Help in the implementation of no-till farming of sugarcane (soil protection and costs and GHG emission reduction)
- Introduce Precision Agriculture
- Develop mechanization to reduce traffic on planted area from 60% to 13%
- Permit sugarcane cultivation with optimized row spacing

- Tests in the field will be conducted with the participation of collaborating mills
- This is a joint project with a Brazilian Industry JACTO with financial support of US$ 9.4 million from BNDES

Low Impact Agricultural Mechanization

CTBE Brazilian Bioethanol Science and Technology Laboratory — CNPEM

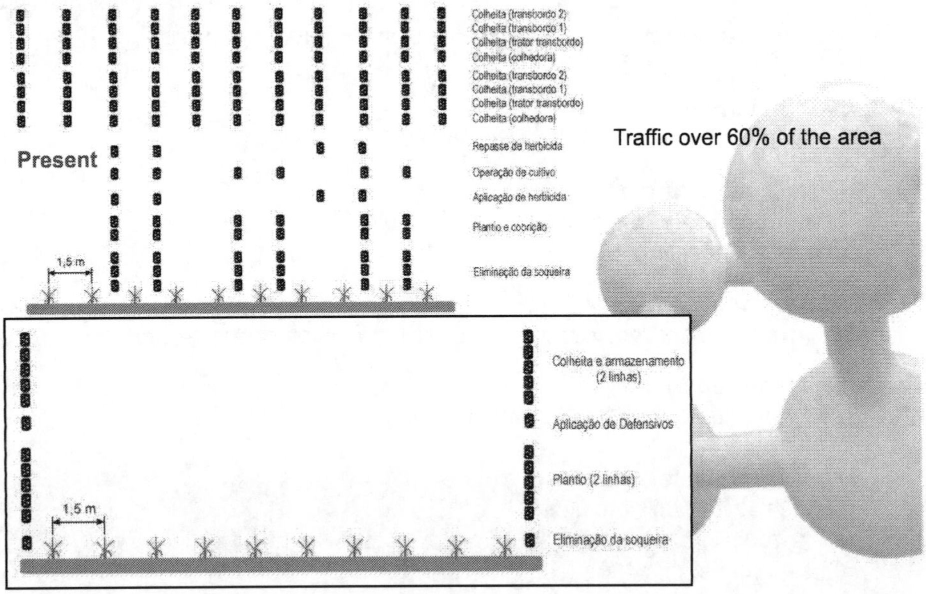

Brazilian Bioethanol Science and Technology Laboratory

CNPEM

Better Conditions to Propagate the Root System

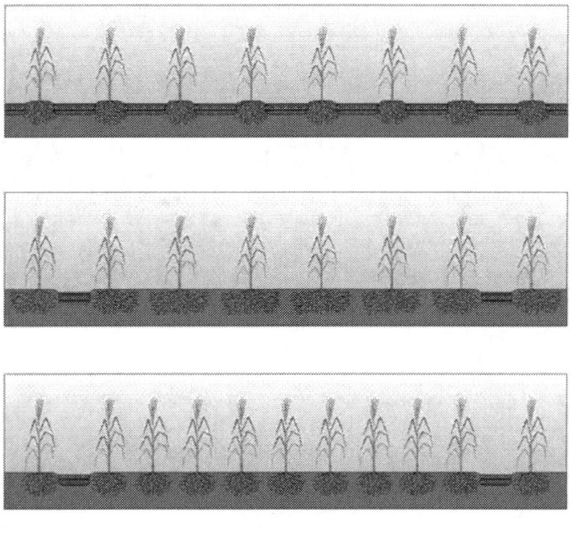

Brazilian Bioethanol Science and Technology Laboratory

CNPEM

Impacts of Row Spacing on Cane Yield

BUREAU OF SUGAR EXPERIMENT STATIONS

QUEENSLAND, AUSTRALIA

Figure 4 Impact of row spacing on cane yield in the plant crop in three cultivars

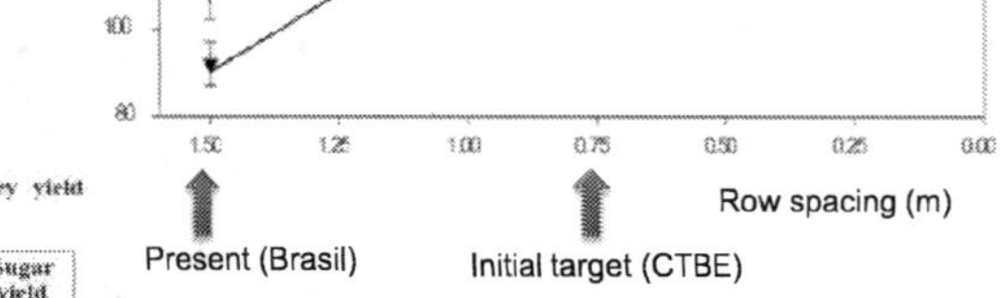

Table 5 Relative impact of inter-row and intra-row spacing on key yield parameters of cultivars Q124, Q151 and Q170[ⓟ]

Row spacing (m)	Within-row space (eyes/m)	Eyes planted (,000/ha)	Stalk number (,000/ha)	Weight /stalk (kg)	Cane yield (t/ha)	CCS	Sugar yield (t/ha)
1.5	2.5	17	70.6	1.25	88	16.6	15.6
1.5	5	33	80.7	1.19	96	16.9	17.0
1.5	10	67	97.7	1.12	112	17.4	20.1
1.0	2.5	25	113.2	1.07	123	16.6	23.3
1.0	5	50	122.3	1.13	138	16.6	24.4
1.0	10	100	131.3	1.04	137	15.8	23.9
0.5	2.5	50	128.4	1.17	150	16.0	24.3
0.5	5	100	146.6	1.08	159	16.2	27.2
0.5	10	200	157.1	1.04	164	16.8	31.1

October 2002

FINAL REPORT - SRDC PROJECT BSS212

INVESTIGATION OF THE LIMITS TO HIGH DENSITY PLANTING

by

J L Collins

CTBE Brazilian Bioethanol Science and Technology Laboratory

CNPEM

Mechanized Planting

Quality of seed cane and seed spacing in the forrow

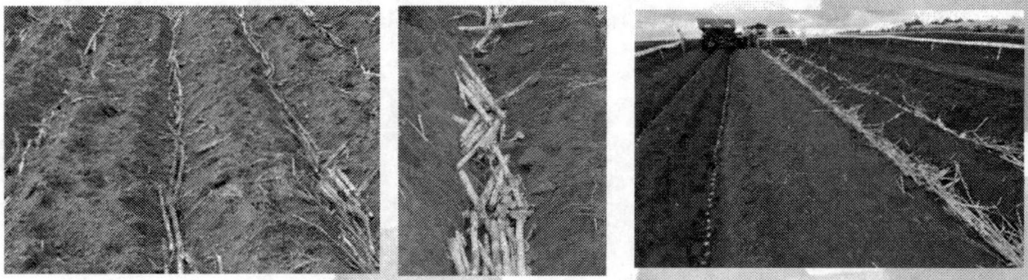

Seed distribution in manual and mechanical planting

CTBE Brazilian Bioethanol Science and Technology Laboratory

CNPEM

Precision Farming

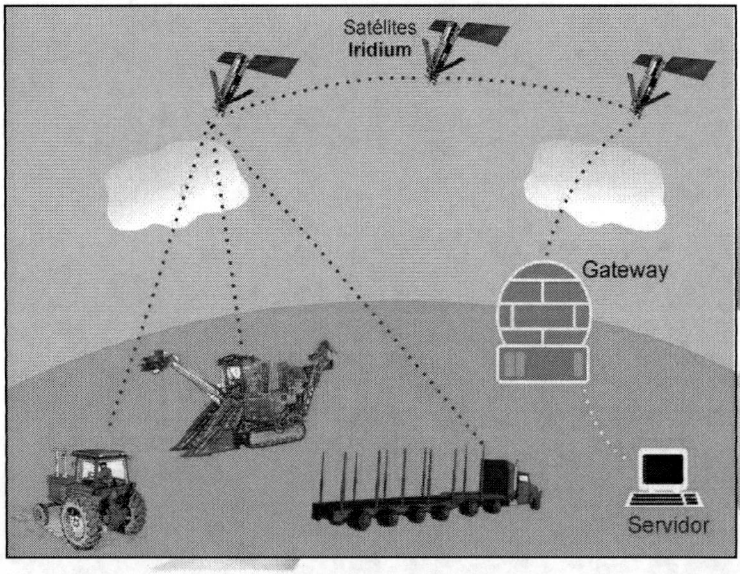

Brazilian Bioethanol Science and Technology Laboratory

Precision Farming

CNPEM

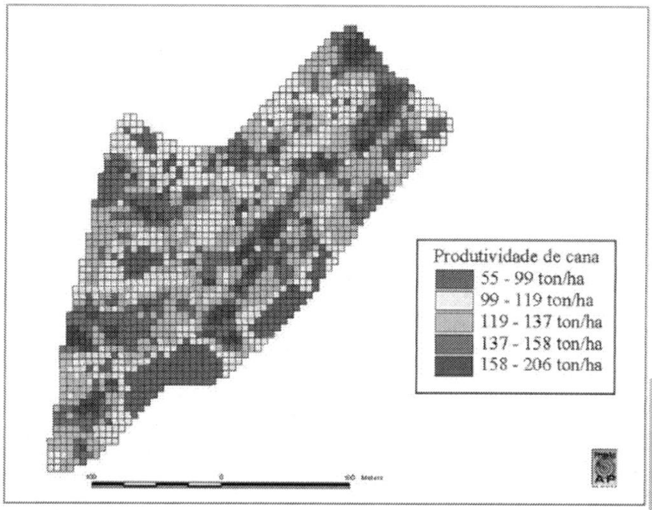

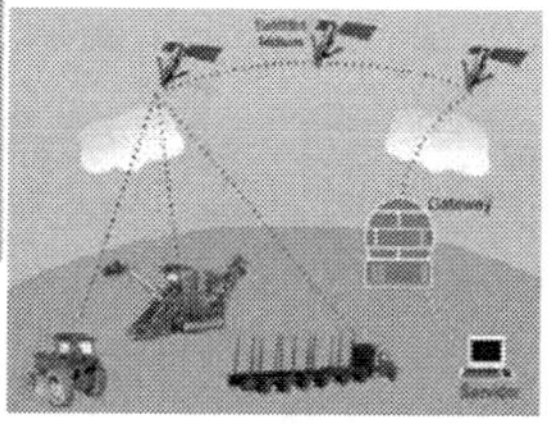

Source: Molin, J.P.

Brazilian Bioethanol Science and Technology Laboratory

Industrial Program

CNPEM

Pilot Plant for Process Development (PPDP)

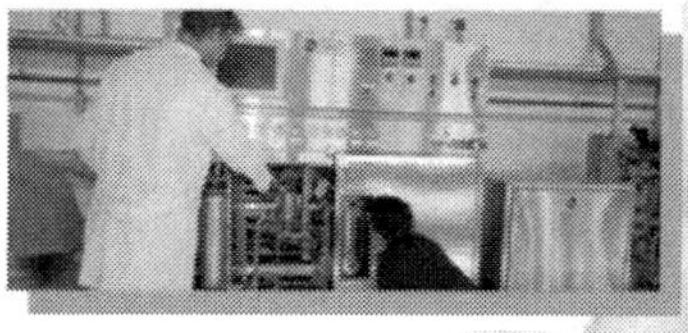

- Development of technologies for cellulosic ethanol (estimated increase of 40% in ethanol production).
- Complex for technological development is open to external groups.
- Offer *"scaling up"* to scientific community and industry
- Deep scientific knowledge to overcome technological challenges pointed out by the productive sector.

CTBE Brazilian Bioethanol Science and Technology Laboratory

Conversion of Bagasse to Ethanol

CNPEM

Critical steps

- Physical and/or Chemical pretreatment of sugar cane bagasse;
- An enzymatic complex (hydrolases) tailored for conversion of cellulosic raw materials into fermentable sugars;
- Development of enzymatic hydrolysis process;
- Microorganisms for fermentation of pentoses to ethanol.

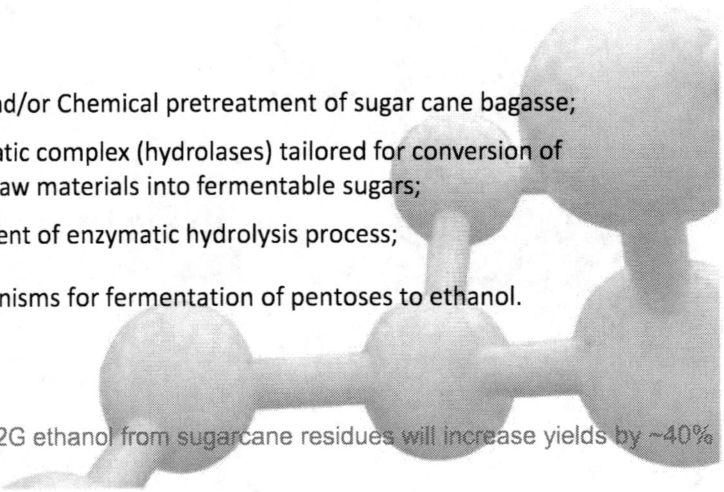

Once commercial the 2G ethanol from sugarcane residues will increase yields by ~40%

CTBE Brazilian Bioethanol Science and Technology Laboratory

Pilot Plant for Process Development PPDP

CNPEM

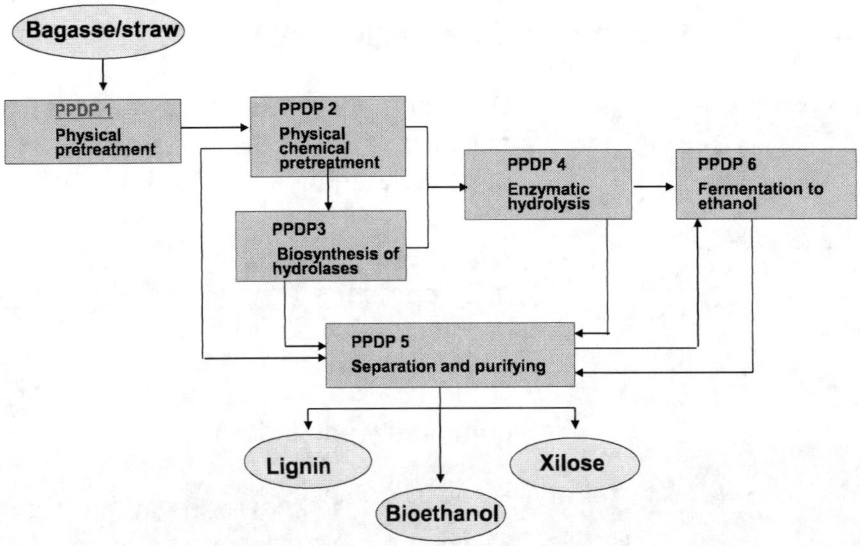

Biorefineries Integrated Evaluation

CTBE Brazilian Bioethanol Science and Technology Laboratory — CNPEM

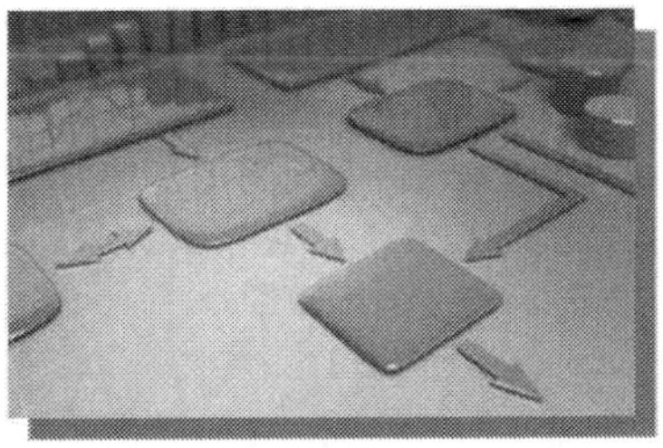

- Assessing impacts of new technologies on the ethanol production cycle
- Optimization of processes and integration of new technologies in the present industry
- Analysis of priorities for investment planning
- Investigating the possibility of using sugarcane biomass as a carbon source

Concept

CTBE Brazilian Bioethanol Science and Technology Laboratory — CNPEM

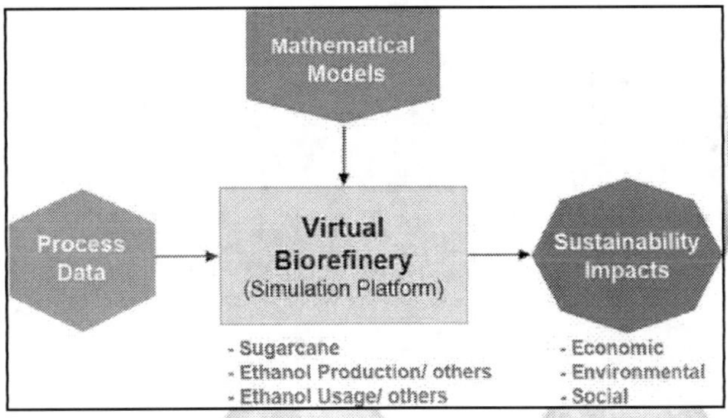

Virtual Biorefinery Flowchart

CTBE Brazilian Bioethanol Science and Technology Laboratory

Basic Science

CNPEM

- Agenda that goes from sugarcane photosynthesis to the deconstruction of cellulose structure into fermentable sugars (controlled recovery of sugars from bagasse and trash cellulose for ethanol production)
- In addition to its own research agenda, <u>supporting</u> other CTBE programs

CTBE Brazilian Bioethanol Science and Technology Laboratory

Sugarcane Physiology - Leaves: How does CO_2 turn into sugars?

CNPEM

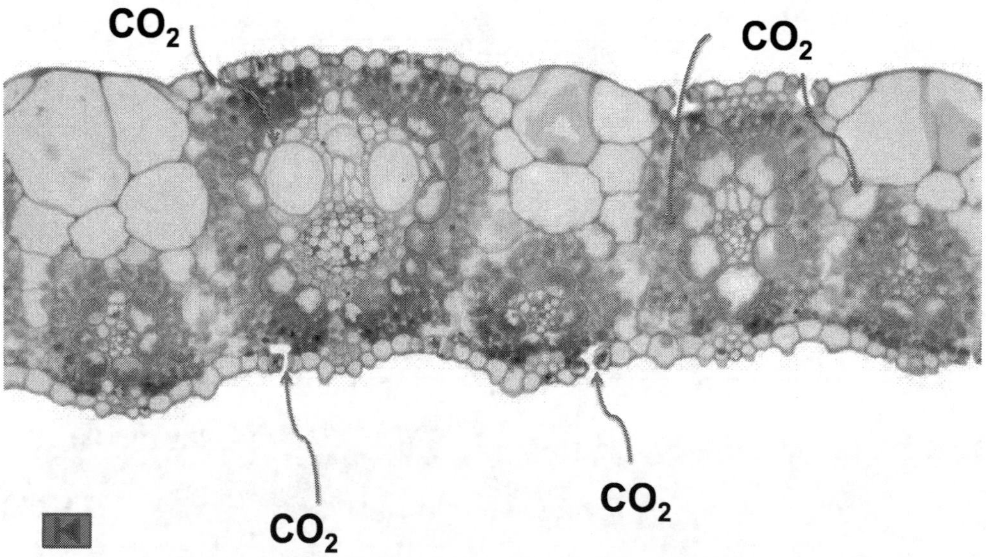

Brazilian Bioethanol Science and Technology Laboratory

CNPEM

Sustainability

- Ethanol sustainability evaluation, considering present and future technologies.
- Generating data for public policies.
- Georefenced data basis

- Focus (obtaining data in the sugarcane cycle for a scientific discussion on):
  - Energy balance and GHG emissions
  - Direct and indirect land use change (LUC and ILUC)
  - Soil carbon stock and soil emissions (CO_2 and N_2O)
  - Socioeconomic impacts
  - Impacts on water resources quality and availability
  - Impacts on biodiversity
  - Integrated sustainability analysis

Brazilian Bioethanol Science and Technology Laboratory

CNPEM

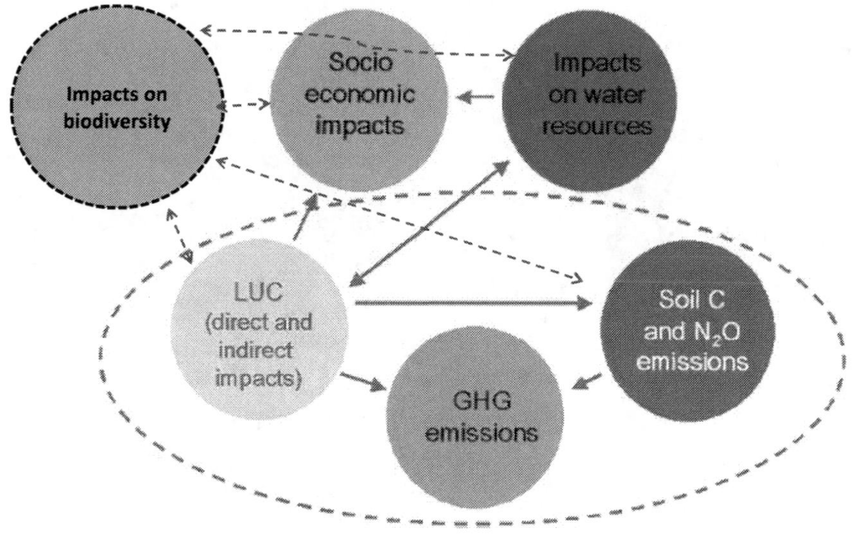

CTBE Brazilian Bioethanol Science and Technology Laboratory

Soil Carbon Stocks

CNPEM

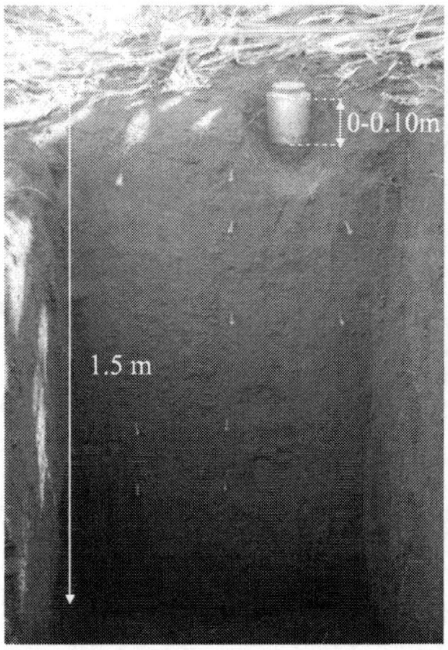

CTBE Brazilian Bioethanol Science and Technology Laboratory

Soil GHG Fluxes

CNPEM

CTBE/UTA/BP

Brazilian Bioethanol Science and Technology Laboratory

CNPEM

Assessing Water Resources in Data Limited Regions and Changing Land Use

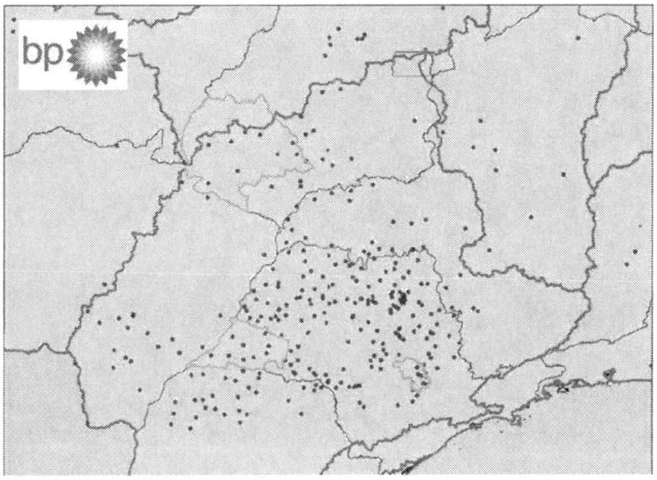

- Satellite images;
- Hydrological models
- Meteorological information;

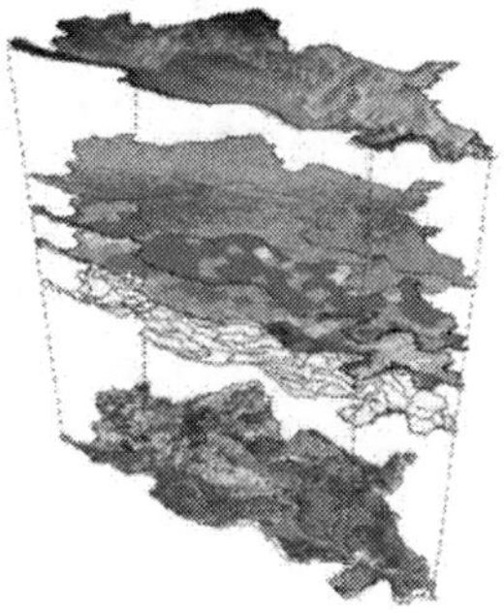

Water Balance Modeling - SWAP

Brazilian Bioethanol Science and Technology Laboratory

CNPEM

Calibrating and validating an agro-hydrological model to simulate rainfed sugarcane growth

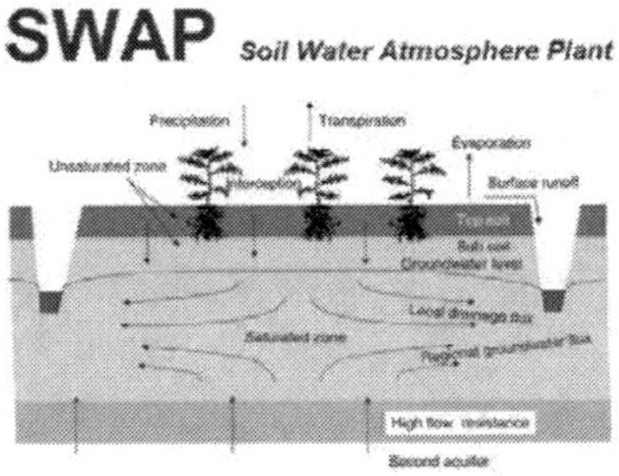

Field scale

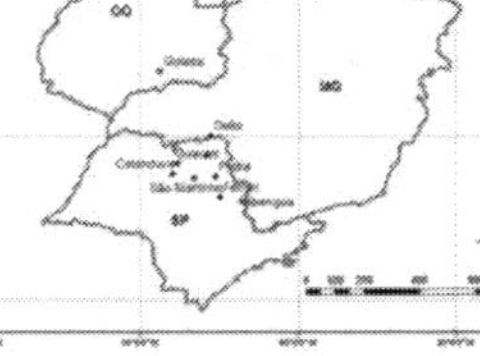

cena

IAC

Goiás (2000)

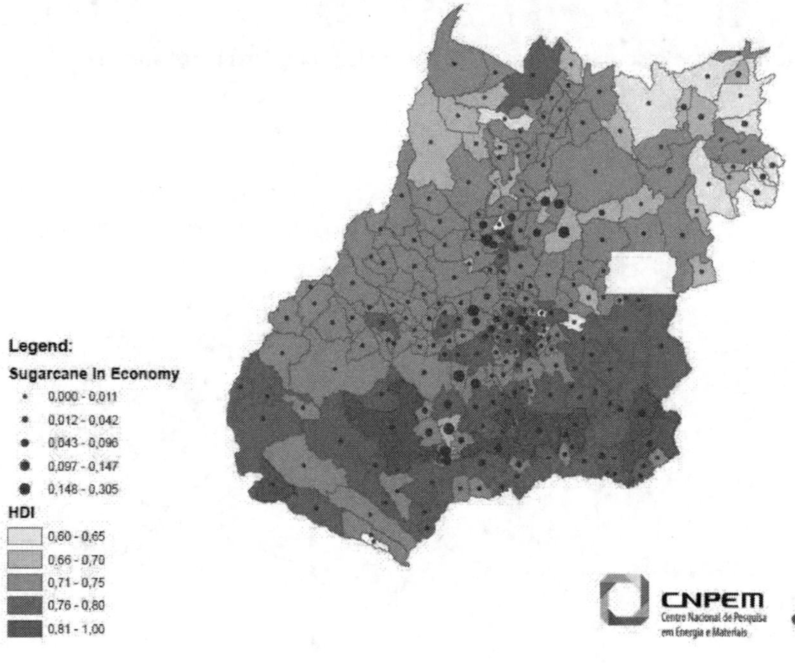

CNPEM Centro Nacional de Pesquisa em Energia e Materiais CTBE

CTBE Brazilian Bioethanol Science and Technology Laboratory CNPEM

Data will be available to the community through a web page, that would allow users: to access a single source of information and consequently to get assertiveness in actions, to frequently monitor the results and to easily handle data and results.

CTBE IBGE INPE Ministério da Agricultura e do Abastecimento NASA INMET ipea Instituto de Pesquisa Econômica DELTA O2 Sustentabilidade Ambiental DATASUS MINISTÉRIO DA SAÚDE

CTBE Brazilian Bioethanol Science and Technology Laboratory

CNPEM

Thank you for your attention!

regis.leal@bioetanol.org.br

Comparison of Acid and Enzymatic Hydrolysis of Jerusalem Artichoke Tubers for Fermentative Butanol Production

Sarchami, T.,

Originally presented on: 7/24/2014 14:30:00 - 14:48:00

Abstract

In this study, a central composite design and response surface methodology were used to study the effect of various hydrolysis variables on both acid and enzymatic hydrolysis of Jerusalem artichoke-derived inulin. It was found that quadratic model was able to predict inulin conversion as a function of all investigated factors in both types of hydrolysis. The models were confirmed by additional experiments and via analysis of variance (ANOVA). Subsequently, numerical optimization was used to maximize the inulin conversion (94.5% in enzymatic hydrolysis and 96% in acid hydrolysis) of Jerusalem artichoke powder within the experimental range. The optimum conditions for enzymatic inulin conversion were a temperature of 48°C, pH of 4.8, substrate concentration of 60 $g.l^{-1}$, and enzyme loading of 10 unit. $g^{-1}{}_{substrate}$ in 24 hours. For acid hydrolysis using Sulphuric acid as catalyst, the optimum conditions were achieved at a temperature of 97°C, pH of 2.0, and in 35 minutes.

The hydrolysates of Jerusalem artichoke were fermented via solventogenic clostridia to acetone- butanol- ethanol (ABE). ABE yields of 0.33 $g.g^{-1}$ and 0.31 $g.g^{-1}$ were obtained using enzymatic and acid hydrolysates of Jerusalem artichoke, respectively. Acid hydrolysis produced HMF as fermentation inhibiting byproduct resulting in a lower yield compared to enzymatic hydrolysis. Therefore, enzymatic hydrolysis of Jerusalem artichoke was found to be a more effective method of hydrolysis for butanol production.

Understanding Concentration and Pressure Changes during Ultrasonication-enhanced Blending of Petrol-Ethanol Blended Fuel

D. Nkazi, M.O. Daramola, S Iyuke

School of Chemical and Metallurgical Engineering, Faculty of Engineering and the Built Environment, University of the Witwatersrand, Wits 2050, Johannesburg, South Africa

RCN Conference on Pan American Biofuels and Bioenergy Sustainability, Brazil, July 22-25, 2014

Motivation

Source: DOESA, 2009

- Increasing demands in alternative clean fuels due to environmental pollution calls for fuels modifications
- Blending of petrol with alcohols, such as ethanol is a promising alternative due to its ability in the reduction of carbon monoxide emissions and thus resulting in less greenhouse effects
- Significantly higher research octane number of ethanol-blended fuel compared to unblended petrol
- Therefore, understanding production of petrol-ethanol blended fuel is very essential

2

Research Problem

Sources: Shad, 2011; Hughes, 2009

- ❑ Hygroscopic nature of ethanol causes phase separation in the ethanol-petrol blends

- ❑ Production of homogeneous mixture at low energy and control of phase separation is a challenge

3

Objectives

- ❑ Investigate effect of the position of ultrasonicator's horn on the concentration and pressure changes during ultrasonication-enhanced blending of petrol-ethanol fuels
- ❑ Understand the concentration and pressure changes during the blending of ethanol-petrol fuels

  - to achieve a perfectly homogenized petrol-ethanol blend
  - for design of blender for the purpose

4

Ultrasonic & Ethanol-Petrol Blends

Source: Suslick et al, 2004

- Generation of intense, ultrasonic waves
- formation, oscillation,
  implosion bubbles in liquids
- induce high temperatures and pressures within imploding bubble
- allows for extraordinary physical + chemical conditions

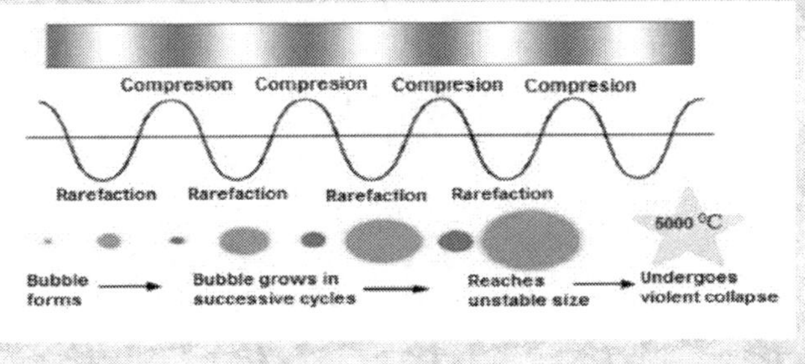

5

Experimental Procedure

Ultrasonication-enhanced blending of ethanol hydrate - gasoline mixture

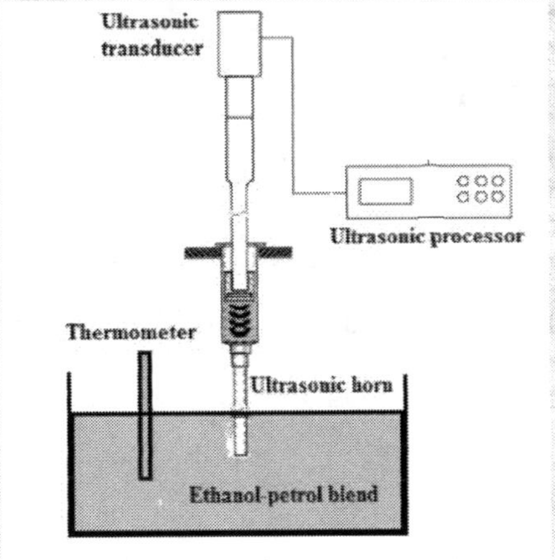

Frequency of horn	24 KHz
Cycle	1
Amplitude	30
Diameter horn	0.7 cm

Experimental set up for ultrasonication-enhanced blending of ethanol hydrate-gasoline mixture

6

Experimental Procedure

Analysis of the blended ethanol-petrol mixture

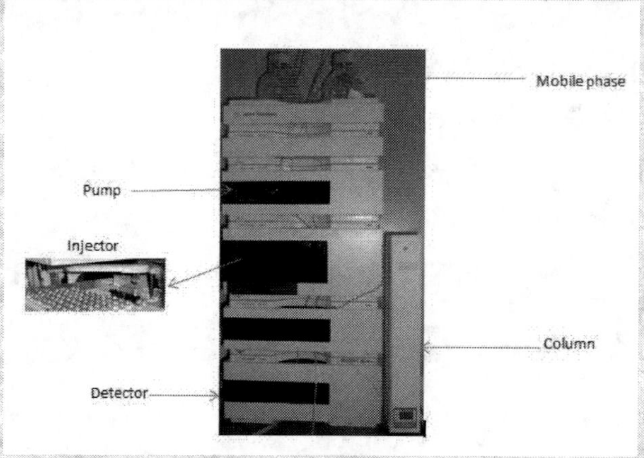

Agilent High-Pressure Liquid Chromatography
(Agilent 1200 Series HPLC System)

7

Results: Concentration Profile

Variation of concentration with the horizontal distance and time

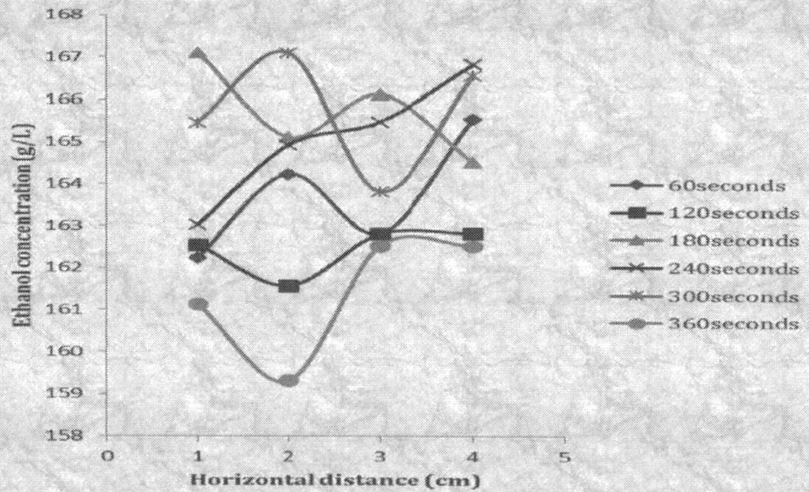

Variation of concentration with distance and time in the horizontal direction for a 30% ethanol mixture

8

Results: Concentration Profile

Variation of concentration with the horizontal distance and time

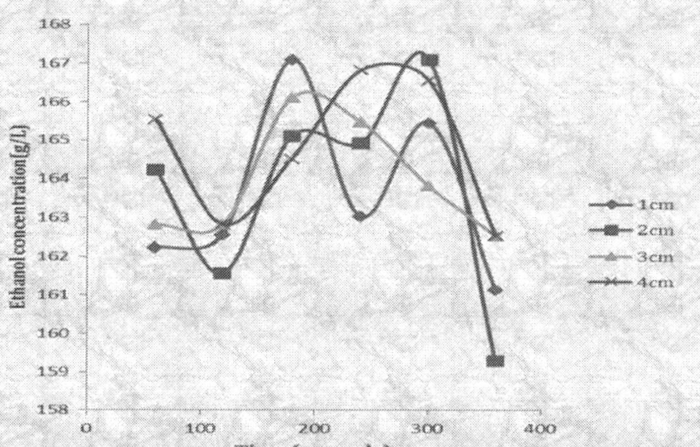

Variation of concentration with distance and time in the horizontal direction for a 30% ethanol mixture 9

Results: Concentration Profile

Variation of concentration with the vertical distance and time

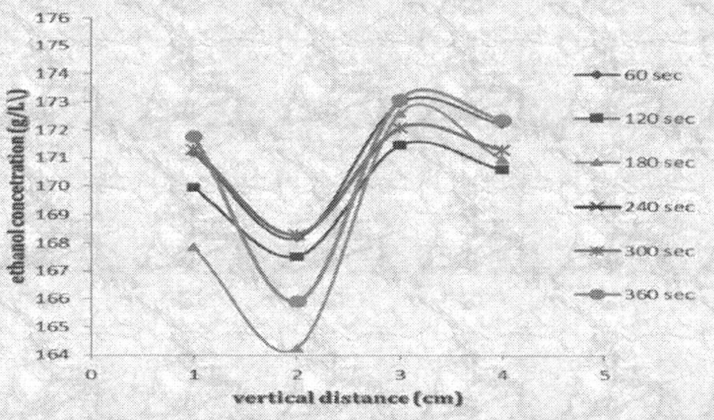

Variation of concentration with distance and time in the vertical direction for a 30% ethanol mixture 10

Results: Concentration Profile

Variation of concentration with the vertical distance and time

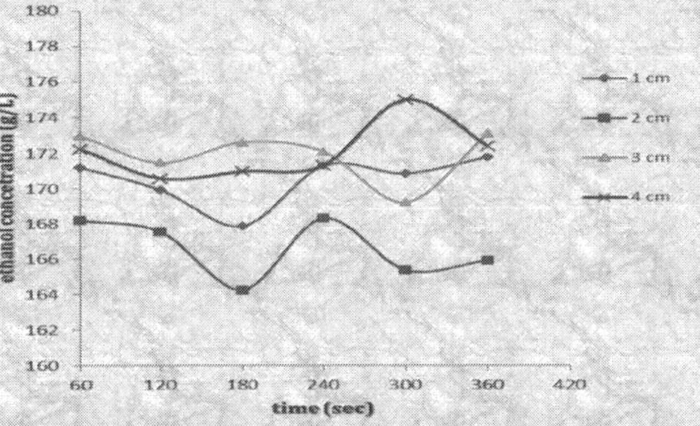

Variation of concentration with distance and time in the vertical direction for a 30% ethanol mixture

11

Results: Concentration Profile

Sources: Hielscher, 2005; Raman, 2006

- Wave behaviour: growth and collapse of bubbles due to the energy absorbed
- Non uniform amplitude pattern due to different size of bubbles and wave superposition (reflection off the reactor walls)
- Concentration gradient and diffusion rates were higher in the vertical direction, making the diffusion in the horizontal direction a rate-limiting step. The dimensions of the blender for an homogenized mixture should have larger height-to-diameter ratio.

12

Results: Pressure Profile

Variation of pressure with the horizontal distance and time

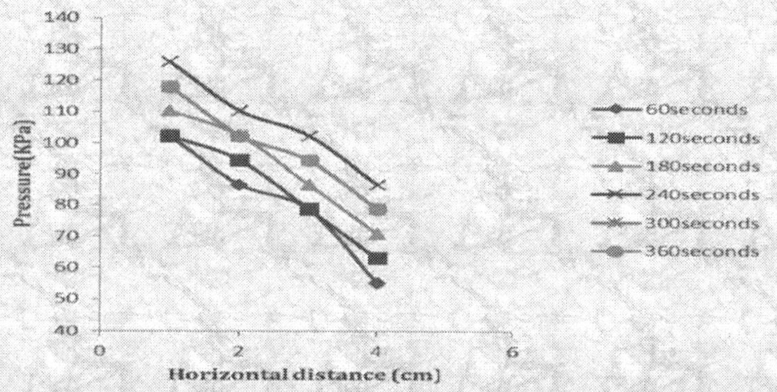

Variation of pressure with distance and time in the horizontal direction for a 30% ethanol mixture

13

Results: Pressure Profile

Variation of pressure with the horizontal distance and time

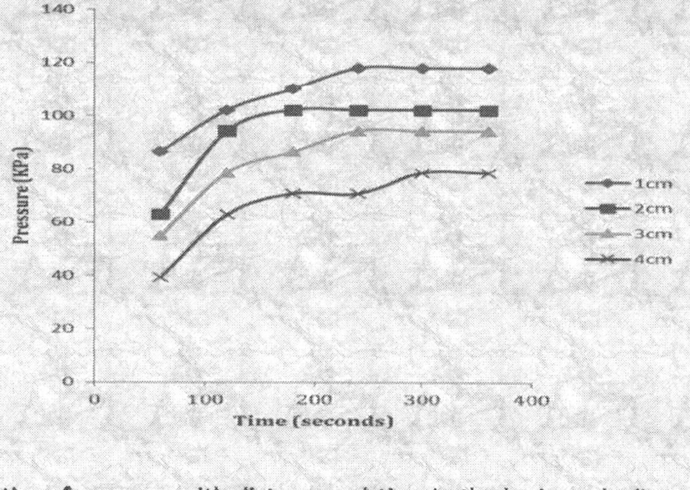

Variation of pressure with distance and time in the horizontal direction for a 30% ethanol mixture

14

Results: Pressure Profile

Variation of pressure with the vertical distance and time

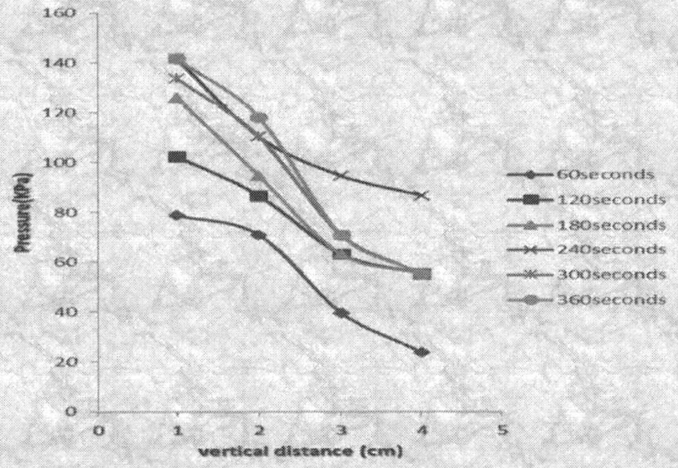

Variation of pressure with distance and time in the vertical direction for a 30% ethanol mixture

15

Results: Pressure Profile

Variation of pressure with the vertical distance and time

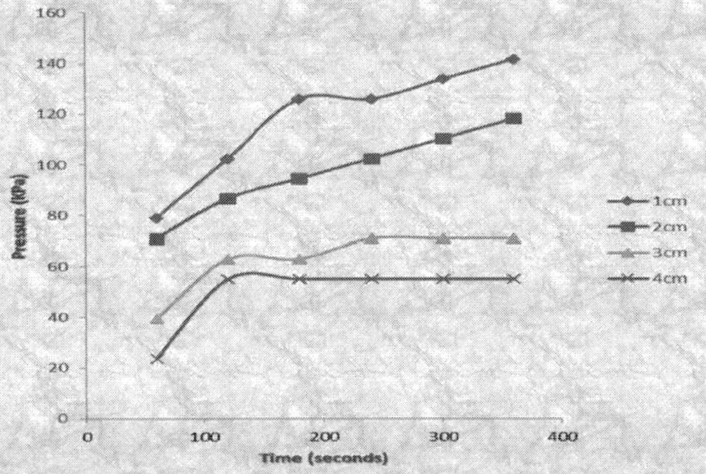

Variation of pressure with distance and time in the vertical direction for a 30% ethanol mixture

16

Results: Pressure Profile

Sources: Goksen et al, 2004; Klima et al, 2006

- This behaviour shows that the intensity of energy is assumed to decrease accordingly with an increasing area into which it is spread ($I \sim 1/A$)
- Ultrasonic energy is consumed within a volume near the horn
- Between 120 and 360 seconds, the pressure gradient became zero (a plateau), indicating a constant distribution of ultrasound energy, hence perfectly homogenization of the fuel blend.

17

Conclusions & Recommendations

- The concentration profile is a wave function for both horizontal and vertical direction. Furthermore, highest concentration gradient was obtained in the vertical direction.
- Ultrasonication has an impact on both the horizontal and vertical diffusion since the concentration profiles are waves confirming the formation of cavitation in the mixture.
- Ultrasonication increases pressures which in turn enhances diffusion rate and thus mixing possibilities of ethanol and petrol.
- The experimental results on the concentration gradient conclude that if the blender geometry was to be proposed, it will have a greater height than the radius since the vertical diffusion seem to be favoured.

18

Conclusions & Recommendations

- Further investigation on the influence of ultrasonicator parameters and mixing mechanism should be done in order to understand the effect of ultrasound on the phase behaviour.
- Further study in the area of optimization might be required to obtain optimum operating variables, such as, the amplitude, the pulse rate, the temperature, the pressure, the blending time and the frequency, that could be related to the force required to break the molecular bonds of ethanol-water mixtures at various concentrations

19

Acknowledgments

Wits FEBE Top-slice Grant

&

20

Thank You for Your Attention

21

BACK-UP SLIDES

22

Phase separation

- Ethanol + petrol blends sensitive to moisture

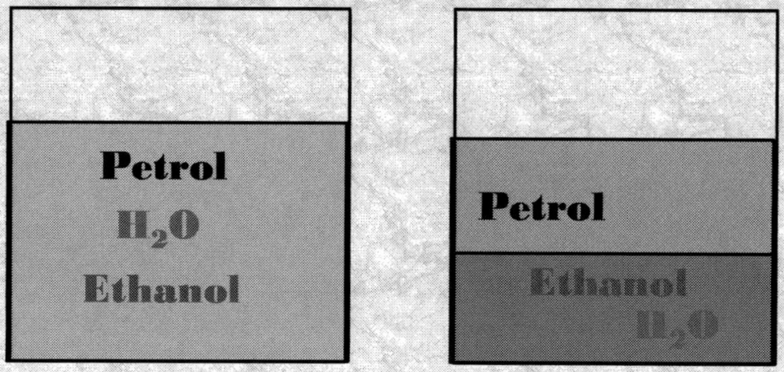

Poverty in the Pirapama Basin: The Role of Management of Hydrological Resources

Viana, J. H. N.,
Moraes, M. M. G. A., UFPE – Pernambuco, Brazil

The sustainable use of water resources has become a very popular theme in scientific literature along with a broader discussion about the sustainable use of natural resources. More specifically, there is an increased interest in how climate change may affect water availability. The social and economic impact of different water allocation strategies has recently been modeled and measured (Bhatia et al, 2006, Moraes et al, 2009, CALZADILLAA et al, 2013). Many studies have shown that the anthropogenic emissions of greenhouse gases may be responsible for changes in weather conditions on the planet (IPCC, 2007 COMOU; ROBINSON & Rahmstorf, 2013). The economic impact resulting from these changes is likely to be strongly influenced by the increased water availability for agricultural production.

Water availability for agriculture is related not only to the natural conditions of the climate, but also to water resources management practices that define supply and demand conditions. Thus, the rational and efficient use of water mainly through demand management – using water allocation policies - becomes urgent. There are many benefits that can be derived from the application of adequate water allocation policies, such as: environmental, for example, an improvement in water quality; or economic, for example higher economic returns from activities that use water, such as agriculture.

The probable diminishing of rainfall rates, resulting from climate change, is potentially disastrous. This is true especially for individuals living in regions already in a socioeconomic context of poverty and whose incomes are closely linked to the availability of natural resources.

This study aims to develop an integrated framework to assess the effects of different management practices of water resources on the economy of the Pirapama Riverbasin, especially the effects on the poorest strata of the population.This will be done by joining the hydro-economic model for this region already developed at the Federal University of Pernambuco (UFPE) with a Computable General Equilibrium Model (CGEM) being formulated for the same hydrographic region.

Starting from the results obtained from the first model, concerning water allocation for agricultural users and their resulting crop production, the effects on all sectors of the regional economy including possible changes in the condition of poor families will be measured . These results are expected to support decision-making about different allocation policies in the region.

The Pirapama River basin was mainly chosen as the area of study because of its economic importance for Pernambuco, since one hand is the most important water source for the metropolitan region of Recife, capital of Pernambuco and supplies water for others major users such as alcohol distilleries and sugar cane irrigation. Thus, there are already many existing conflicts between different uses in the region, such as electricity generation, irrigation, and human consumption.

Water quality is also a major concern in the Pirapama basin since throughout the length of the river most currently monitored water quality standards cannot be met, mostly because of sugarcane production together with fertirrigation, which is a process of irrigation and fertilization that uses water to carry and distribute fertilizers to the sugar cane. On top of that, two recently constructed reservoirs on the basin (the last one completed in 2010) might also contribute to further water quality deterioration through eutrophication processes (MORAES ET AL, 2010).

Furthermore, the region where the Pirapama River basin is located (*Zona da Mata*) is not associated with the climate variability and drought as the semi-arid regions in northeastern Brazil, therefore the basin is less used to extreme drought periods making it particularly vulnerable to future climate changes (ENGLE & LEMOS, 2010). These conditions make good water management practices in the region indispensable

Given the main goal highlighted above we have the following specific objectives:

- To build a Social Accounting Matrix for the Pirapama River basin;
- To develop and calibrate a CGEM for the Pirapama River basin with different types of representative families;
- To include the results from the hydro-economic model in the general equilibrium model;
- To simulate through the integrated framework how different decisions on water management might affect the Regional Economy, especially those areas with poorest population.

Effects of Maize Weevil Infestation on Corn and Utilization of Infested Corn for Ethanol Production

Corrêa, G. A., Kansas State University

EFFECTS OF MAIZE WEEVIL INFESTATION ON CORN AND UTILIZATION OF INFESTED CORN FOR ETHANOL PRODUCTION

Abstract

The maize weevil, *Sitophilus zeamais* (Motschulsky) (Coleoptera: Curculionidae), is a cosmopolitan pest of stored cereal grains, especially maize in tropical and subtropical regions. The focus of our laboratory study was to characterize physical and chemical damage caused by *S. zemais* to corn over a six month period, and to determine ethanol yield from infested corn.

Clean, certified organic whole yellow corn (5 kg) of 11.9% moisture in 20-L round plastic with ventilated lids was infested with 20 unsexed adults of mixed ages of *S. zeamais*. There were a total of 36 buckets, of which 18 were infested and 18 were uninfested (controls). It was stored for periods of 1, 2, 3, 4, 5 and 6 months. The moisture content, test weight, dead and live insects, and proximate analysis were conducted using the AOAC (Association of Official Analytical Chemists) and Ankom methods, were measured monthly. The ethanol production was evaluated using two procedures, the enzyme digestion and fermentation processes, to measure the amount of glucose at initial time of the experiment and evaluated the glucose complete conversion into ethanol and its amount, respectively. The amount of glucose and ethanol was analyzed using the High Performance Liquid Chromatography (HPLC) technique.

The corn moisture content of uninfested corn fluctuated between 10.6 and 11.9% during the six-month study period, whereas that of infested corn increased from 11.9 to 14.8% at the end of the study. The number of live insects increased exponentially in infested corn and reached a maximum of 6,985 by the end of the study. Dockage in 1000 g of uninfested corn was 1.2 to 2.2 g but steadily increased in infested corn from 1.2 g to a maximum of 154 g at the end of six months. The weight of 100 uninfested corn kernels normalized to 11.0% wet moisture basis remained unchanged (35.8 g) but this weight in infested corn decreased gradually to 22.6 g at six months. However, dry matter losses were marginal for both uninfested and infested corn over the study period. The crude protein, crude fat, crude fiber, and ash of uninfested and infested corn over time showed minor variation that are not biologically significant. The starch content of uninfested and infested corn ranged from 58 to 61%, and these values were lower by about 10%, probably due to the official standard method used (glucoamylase method) or use of aged enzymes. It was demonstrated that during the first four months with a high level of infestation it was possible to obtain a profitable level of ethanol production, and it begins to decline after the fifth month, and over the sixth month it was not profitable because the amounts of glucose declined dramatically. In conclusion, it was possible to obtain profitable levels of ethanol from infested corn even under a high density of infestation by maize weevil, and it will optimize the use of corn as raw material for worldwide ethanol production.

Maximizing Ethanol Production from Genetically Modified Cyanobacteria Grown Autotrophically

Delpino, C., Planta Piloto de Ingeniería Química (PLAPIQUI), Universidad Nacional del Sur-CONICET
Estrada, V., Planta Piloto de Ingenieria Quimica (PLAPIQUI), Universidad Nacional del Sur - CONICET
Diaz, M. S., Planta Piloto de Ingenieria Quimica (PLAPIQUI), CONICET - Universidad Nacional del Sur

In this work we address the optimization of ethanol production from a mutant strain of the cyanobacterium Synechocystis sp. PCC 6803 (Vidal Vidal, 2009), growing on carbon dioxide as carbon source. This modified strain harbors the genes pdc and adhB from Zymomonas mobilis under the control of the gene PetE promoter for ethanol production. We carry out dynamic flux balance analysis by integrating photobioreactor nonlinear dynamic models and metabolic network linear models. Basically, the model includes two major components: (a) a dynamic model with mass balances for biomass, ethanol, nitrate, phosphate, internal nitrogen and phosphorus, and (b) a steady state metabolic Linear Programming (LP) model. The biomass equation includes limiting functions for light, temperature and nutrients, kinetics of growth inhibition by ethanol toxicity and the decrease in the available light by biomass concentration increase. The control variables of the dynamic optimization problem are batch temperature, light intensity and phosphate concentration in the culture medium.

The resulting dynamic optimization problem for ethanol production maximization is a bilevel optimization problem, with an embedded LP. The problem is reformulated to a single level one, by replacing the LP by its optimality conditions. The dynamic optimization problem is fully discretized by orthogonal collocation on finite elements, rendering a large-scale nonlinear programming (NLP) problem. Complementarity constraints associated to first order optimality conditions in the inner LP are efficiently dealt with the Interior Point algorithm within the IPOPT solver (Waechter & Biegler, 2006) in GAMS (Brooke et al., 2013). The discretized model has 88337 constraints and 58063 variables (73 finite elements and two collocation points). The model has been previously calibrated with experimental data (Laiglecia et al., 2013). In these experiments, performed over 73 hours, the runs have been carried out using the genetic modified strain, and activating the Pet promoter from the very beginning of the runs (with copper); i.e., enabling ethanol production path throughout the entire time horizon.

Numerical results obtained in this work suggest modifications in the metabolic network during the fermentation, with the consequent increase in ethanol production. The optimal pdc pathway should be activated after 20 hours of fermentation, with a consequent 26% increase in ethanol production. Another important issue is that ethanol production increase does not affect biomass growth, as it has been previously shown in experiments carried out by Vidal (2009). Optimal profiles for light intensity, suggest keeping it constant at 80 µE/(m2.s) up to 40 h and increasing to double its value by the end of the fermentation. In this way, alternatives are suggested for the enhancement of ethanol production from a genetically modified cyanobacterium strain.

References

Brooke, Kendrick, Meeraus and Raman (2013) GAMS.A User's Guide.

Estrada, V., R. Vidal, J. Florencio, M. Garcia Guerrero, M.S. Diaz, Parameter estimation of bioethanol production model by a genetic engineered cyanobacterium, AIChE Annual Meeting, Oct 28- Nov.3, 2012, Pittsburgh, USA

Laiglecia, J., V. Estrada, R. Vidal Vidal, F.J. Florencio, M.G. Guerrero, M.S. Diaz (2013), Dynamic flux balance analysis of a genetic engineered cyanobacterium for ethanol production. Parameter estimation. Chemical Engineering Transactions, 32, 955-960.

Vidal Vidal, R., 2009, Producción fotosintética de etanol por la cianobacteria Synechocystis sp. PCC 6803. PhD. Thesis, Universidad de Sevilla, Sevilla, Spain.

Waechter, A., Biegler, L.T. (2006), On the Implementation of an Interior Point Filter Line Search Algorithm for Large-Scale Nonlinear Programming, Mathematical Programming, 106, 1, 25-57.(http://projects.coin-or.org/Ipopt).

Water Allocation and Resources Management Using a Hydro-Economic Optimization Model: The Case Study of the Sub-Middle São Francisco Watershed

Souza da Silva, G.,
Moraes, M. M. G. A., Universidade Federal de Pernambuco

The Northeastern semiarid region of Brazil is known of its irregular rainfalls and water scarcity. Reservoirs and perennialized rivers are the only water source in the dry season. Water disponibility is limited, already causing several conflicts between different water users and uses, like hydropower, irrigation, human consumption, animal consumption, industrial and aquaculture. Furthermore efforts are taken to exploit this region economically with some irrigation sites using mainly crops with a high water need and high drought sensitivity, like sugar cane. Conflicts can get even more challenging with the Integration Project of the San Francisco River (PISF), where two lines of water channels are being built to deliver water to the north and the northeast of the region. Water allocation rules and decision making based on land use change and climate change becomes more important. The aim of the research, linked with the INNOVATE project, is the development of a hydro economic model which uses optimization algorithms to determine the hydro-economical optimum for water allocation in the middle São Francisco River Basin. The hydro-economic optimization model is written in GAMS. Expected results are optimal water storage in reservoirs and allocation rules for irrigated production under different scenarios of water transfers, land use and climate changes.

Considering Water Quality and Energy Efficiency to Define Optimum Operation of Integrated Systems of Water Supply and Value of Use of Water Quality

Rodella, L., Federal University of Pernambuco
de Amorim Filho, C. A. G., Federal University of Pernambuco
Moraes, M. M. G. A., Universidade Federal de Pernambuco

According to Gunkel et al. (2007), all rivers in the coastal region of Pernambuco´s state are influenced by sugarcane cultivation. Effects are often magnified as these same rivers supply cities as well and are generally heavily dammed for both urban water supply and electrical power generation. These reservoirs might also contribute to further water quality deterioration through eutrophication processes. Seeking to contribute with a decision support tool in the water supply area with this kind of problem, this study presents an optimization model that identifies the volume of water to be distributed from a set of integrated water supply systems monthly to each municipality in the Recife's metropolitan region (Pernambuco´s state), in order to minimize expenses with electric power and water treatment. The optimization problem presented uses the linear programming techniques with constraints on the maximum supply capacity of each integrated water system, the installed pipeline network and the minimum volume of water to be distributed to each city. The model results showed that the optimization leads to a reduction of 4.82% of the costs, particularly from an increase of 20.73% in volume produced by Pirapama system, the newest and largest supply system of the state of Pernambuco, the standstill in some months of the smallest integrated systems in the region (Caixa D'Água and Marcos Freire), which also have the m^3 of treated water more expensive in the region. At the same time requires a reduction of 15.38% in the activities of Tapacurá system, which suffers from eutrophication problems in the dam of the same name, its main source. By the reduced cost is possible to know how much the cost of the systems that do not operate at the optimal solution must be reduced so that they start to be considered as an option for use in catchment. Indeed, it may identify the opportunity cost of each of these systems, which is an important concept of economic theory. One of the scenarios studied considered the opportunity cost of Tapacurá system and the results showed a reduction in the value of the objective function of 6.98% (R$ 3,138,175.76) in relation to the optimal solution in scenario 1 (current conditions) and 11.47% (R$ 5,414,987.10) compared with that achieved by COMPESA in 2013. These results show the economy that the supply company would have for a year if invested in Tapacurá system to achieve lower energy and treatment costs, by installing more efficient pumps or policies to improve the water quality of the Tapacurá basin, for example. From the analysis of scenarios that consider different conditions of water quality of the main reservoirs of water catchment for integrated systems in the region, Pirapama and Tapacurá, the study estimates the direct use value of water quality for the company responsible for supplying water in the region, presenting a proposal for value the quality of water based on the avoided cost method. The direct use value of water quality from the Tapacurá system to the company responsible for the supply is R$ 3,874,662.18/year, which represents 8.21% of current expenses of the company on electricity and chemical products. The water quality of the dam Pirapama has a higher value, R$ 5,468,122.15/year, corresponding to 11.58% of the expenses of the company in 2013 with electricity and chemical products. The difference between these values results from the differentiated proportional share of them in the water supply of the metropolitan region of Recife.

Utilization of Crab Shells As Heterogeneous Catalyst in the Synthesis of Methyl and Ethyl Biodiesel

Reis, G., Universidade Federal De Minas Gerais
Medeiros, F. D. S., Universidade Federal de Minas Gerais
S. Costa, J. B., UNIVERSIDADE FEDERAL RURAL DE PERNAMBUCO
S.Moura, A., UNIVERSIDADE FEDERAL RURAL DE PERNAMBUCO
Oliveira DA Silva, R., UNIVERSIDADE FEDERAL DE PERNAMBUCO
M.D. Pasa, V.,
C. C. Bejan, C.,

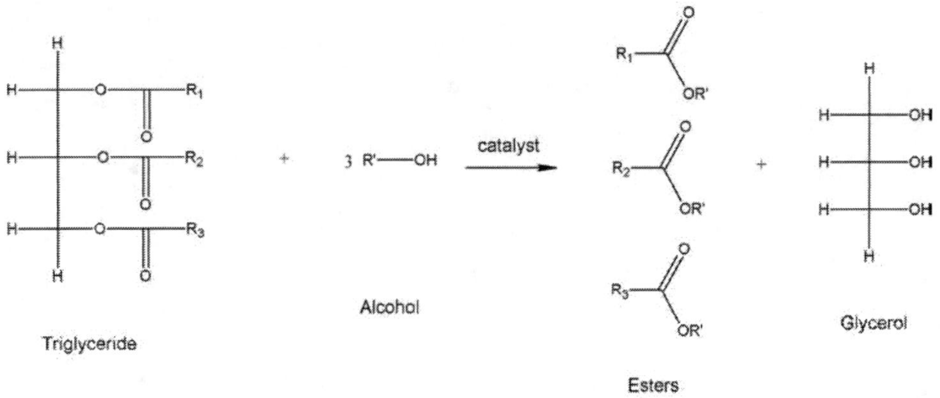

INTRODUCTION

Biodiesel is biodegradable and renewable fuel obtained through a chemical process called transesterification in which vegetable oil or animal fat (triglycerides) is reacted with an alcohol in the presence of a catalyst, and glycerin as a major co-product (Figure 1) is removed by decantation. Biodiesel has emerged as an alternative to replacing diesel oil by presenting similar combustion properties1 and minimize environmental pollution.2

Fig. 1. Transesterification reaction for biodiesel production. R1, R2 and R3 are a mixture of several fatty acid chains.

Among the catalysts, the mostly used in homogeneous phase is alkaline due to faster reaction and higher yields, requiring lower pressures, temperatures and molar ratio alcohol: oil.3 Nevertheless, the presence of free fatty acid (over 1.0 mg KOH/g) and water in the oil feedstock implies in soap formation, reducing the ester yield and inhibiting the biodiesel purification.3

As an alternative method of biodiesel production, the use of heterogeneous catalysts presents some advantages, such as easy separation from the reaction medium and reusability. The heterogeneous catalyst is separated by a filtration process instead of exhaustive washing, when homogeneous catalysis is used, reducing the volume of residual water and environmental impact.4 Nevertheless, heterogeneous catalysts have been proposed as a high operating cost.3, 5 Due to that, the biggest challenge in this area is to search for catalyst with a high performance and lower costs.

High interest has been shown regarding the use of CaO due to its economic advantages, high alkalinity, low solubility and easy handling.6 Despite recent and few numbers of papers published in this area, some of them presents as sources of CaO, the waste of natural products such as egg shells, bones, oyster shells of molluscs,7-9 resulted

from a calcination process since the bark of these species are composed of about 40-55% of CaCO3.10

Another challenge given to the productive sector of biodiesel takes place regarding the use of alcohol used in the transesterification reaction. On an industrial scale, methanol is the alcohol used for presenting a lower cost, and improved handling of the final product because of its low homogeneity biodiesel, thus allowing easy separation by settling.3 However it is a toxic alcohol, lethal, non-renewable, offers high risk of explosion and needs to be handled with great caution. Alternatively, ethanol has been studied as a promising possibility for not presenting these unwanted properties of methanol, furthermore ethanol is a renewable, providing greater environmental benefit to biodiesel.11

According to that, in this work we propose an environmentally friendly alternative for biodiesel production sector, studying the use of CaO obtained from food residue from crab shell as heterogeneous catalyst. A study of the kinetic of the reaction while using this catalyst in the synthesis of biodiesel, is done comparing both methyl and ethyl alcohol.

EXPERIMENTAL PROCEDURE

To obtain the catalyst, waste from crab shells obtained from local restaurants were used. The samples were previously washed, dried at 100 °C and ground in a blender resulting in a fine powder. Powder X-ray diffraction (PXRD) analyses of the samples were carried out on a Shimadzu XRD-6000, using a graphite crystal monochromator to select the Cu-Kα1 radiation source at λ = 1.5406 Å, with a step size of 0.02 s-1.

The transesterification reaction was carried out with 100 g of soybean oil, 5% (m/m) of the catalyst, a molar ratio alcohol/oil ratio of 12/1 (methanol or ethanol), magnetic stirrer and reflux temperature. Every hour during 8 hours, aliquots (6 mL) were picked and processed with a centrifugation up to remove the catalyst and formed glycerol, followed by evaporation of excess solvent emissions. The percentage of methyl and ethyl esters (% w/w) of the samples dissolved in CDCl3 was obtained by 1H NMR, using a 300 MHz Varian Unity Plus 300 spectrometer, as a relation between the converted esters and their respective triglycerides.

RESULTADOS E DISCUSSÕES: After calcination, the crab shells resulted in a white powder with a yield of around 41% (m/m). Analysis of diffraction patterns indicated that the crab shells before calcination has CaCO3 as the main constituent. while the crab shell after calcination at 900 ° C resulted in the CaO of high purity.

The samples collected during the reactions were analysed by 1H NMR and the percentage of fatty acid methyl esters (FAME) and fatty acid ethyl esters (FAEE) (% w/w) was determined as described at literature by Gelbard12 and Ghesti13, respectively.

From the analysis of the 1H NMR spectra of each sample was drawn a comparative curve of evolution of the levels of esters along the reaction when processed in methanol and ethanol (Figure 3). The content of the FAME reached 95% after 7 hours of reaction, while the level of FAEE even after 8 hours of reaction reached only 13%.

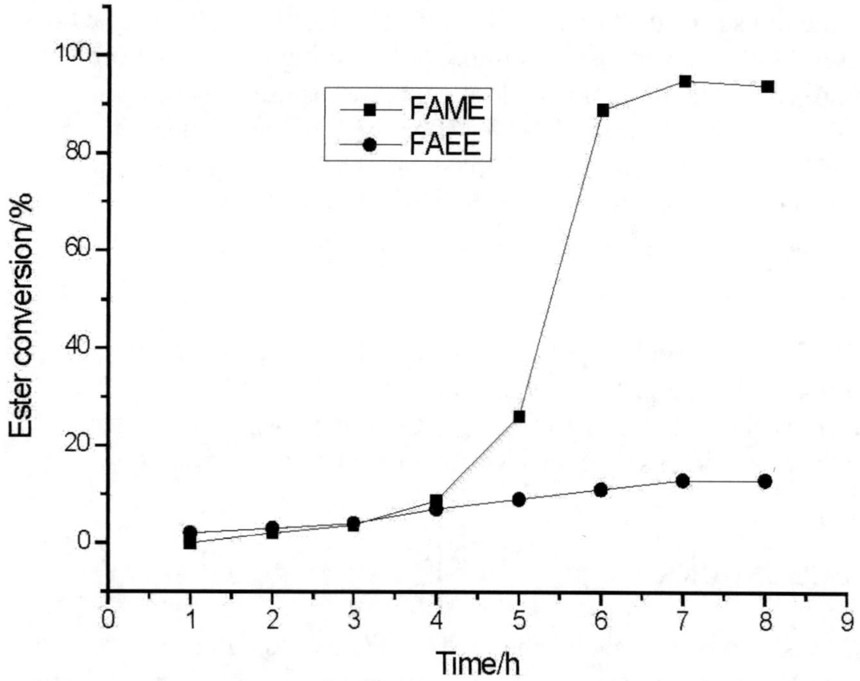

Fig. 3: Comparative kinetic study of the transesterification of soybean oil using CaO obtained from crab shell, producing the FAME and FAEE.

The minimum ester content proposed by EN 14214 is 96.5%. The results showed that the FAME obtained with the crab shell almost meet this standard specifications reaching 95% after 7 hours of reaction. Otherwise, the FAEE obtained with the same conditions, apart from the reflux temperature, presented an ester content of only 13% after 8 h of reaction. This result can be attributed to the lower reactivity of ethyl alcohol due to its lower acidity compared with methanol, and its greater steric hindrance.14, 15 However it can be seen that the kinetic reaction curve did not reach a threshold reaction time analysis, indicating that probably higher reaction time or in larger proportions of the catalyst, it is possible to obtain a better yield in the production of FAEE.

CONCLUSIONS

It was observed that CaO obtained through residue from crab shells has high purity after its calcination at 900 oC and revealed to be very efficient as heterogeneous catalyst at the transesterification reaction at the production of FAME. While used in a proportion of 5% led to a methyl ester content of 95% after 7 h of reflux. Nevertheless, the catalyst did not show the same performance in the production of FAEE, resulting in a conversion of only 13% after 8 h of reflux. These results indicate that the use of promising waste crab and probably several other crustaceans such as eco-friendly and sustainable alternative source for biodiesel production.

REFERENCES

1. Thaiyasuit P., Pianthong K., Worapun I., J. Oil Sci. 2012, 61, 81-88.

2. Hoekman S.K., Broch A., Robbins C., Ceniceros E., Natarajan M., Renew. Sustain. Energy Rev. 2012, 16, 143-169.

3. Lam M. K., Lee K. T., Mohamed A. R., Biotechnol. Adv. 2010, 28, 500-518.

4. Kawashima A., Matsubara, K., Honda K., Bioresour. Technol. 2008, 99, 3439-3443.

5. Refaat A. A., Int. J. Environ. Sci. Technol. 2009, 6, 677-694.

6. Kouzo M., Hidaka J., Fuel 2012, 93, 1-12.

7. Viriya-Empikul N., Krasae P., Puttasawat B., Yoosuk B., Chollacoop N., Faungnawakij K., Biores. Techonol. 2010, 101, 3765-3767.

8. Nakatani N., Takamori H., Takeda K., Sakugawa H., Bioresour. Technol. 2009, 100, 1510-1513.

9. Wei Z., Xu C., Li B., Bioresour. Technol. 2009, 100, 2883-2885.

10. Campana-Filho S. P., Britto D. D., Curti, E., Abreu F. R., Cardoso M. B., Battisti M. V., Sim P. C., Goy R. C., Signini R., Lavall R. L. Quím. Nova 2007, 30, 644-650. 11. Cavalcante K. S. B., Penha M. N. C., Mendonça K. K. M., Louzeiro H. C., Vasconcelos A. C. S., Maciel A. P., Souza A. G., Silva, F. C. Fuel 2010, 89, 1172.

12. Gelbard G., Brès O., Vargas R. M., Vielfaure F., Schuchardt U. E. J Am Oil Chem Soc. 1995, 72, 1239-1241.

13. Ghesti G. F., Macedo J. L., Resck I. S., Dias J. A., Dias S. C. L. Energy Fuels 2007, 21, 2475-2480.

14. Lang X., Dalai A. K., Bakhshi N. N., Reaney M. J., Hertz P. B. Bioresour. Technol. 2001, 80, 53-62.

15. Chemat F., Poux M., Galema S. A. J Chem Soc Perkin Trans 2, 1997, 2371-2374.

Enhanced Pyrolysis Oil Properties through Pretreatment of Aspen with Controlled Torrefaction

Klemetsrud, B., Michigan Technological University
Klinger, J., Michigan Technological University
Shonnard, D. R., Michigan Technological University
Bar Ziv, E., Michigan Technological University

Pyrolysis oil produced from thermochemical conversion of woody biomass has great potential as renewable fuels and feedstocks in chemical production. Perhaps one of the most exciting ways to utilize biomass resources is in the production of transportation fuels through fast pyrolysis and catalytic upgrading. Pyrolysis oil, however, has several disadvantages such as high oxygen content, acidity, corrosivity, low storage stability, and high water content. Our results indicate that these problems can be, at least partially, addressed through the use of torrefaction as a pretreatment method. During torrefaction, mainly the hemicelluloses fraction of biomass degrades resulting in a more uniform and desirable feedstock for pyrolysis. This poster presents preliminary data demonstrating that through careful control of torrefaction conditions, the oils formed during a subsequent fast pyrolysis step of aspen have significantly improved acidity (>40% reduction), corrosivity, water content(>25% removal), oxygen content(chromatographic shift), and thus improved storage stability and energy content. Pyrolysis (500°C for 20 seconds) and torrefaction (300°C for between 5 to 90 minutes) experiments were carried out in a high purity helium atmosphere on a 5200HP Pyroprobe (CDS Analytical), and the chemical analysis performed with a Trace GC coupled with a DSQII mass spectrometer (both Thermo-Fisher Scientific).

Energy-Water Nexus: An Optimal Control Model
Oliveira, G. A. D.,
Souza, F. M. C. D.,

Energy-Water Nexus: An Optimal Control Model

Geraldo Andrade de Oliveira and Fernando Menezes Campello de Souza

One may say that economics is the study of the allocation of scarce resources between competing uses. Scarcity is a characteristic of exhaustible resources.

Most of the world's energy sources known today are finite i.e., exhaustible, or limited.

In the global energy matrix, oil comes first, followed by coal and natural gas. These sources together account for approximately 80% of the world's energy supply. Coal is the source most used to generate electricity. It accounts for generating 41% of the world's electricity supply. The United States and China are examples of countries that are highly dependent on this resource. In Brazil, water is the resource most used to generate electricity and is followed by biomass. These resources, although renewable, are limited.

What oil, coal, natural gas, water and biomass have in common is that they are conventional. Conventional resources, represent stored energy; are found in specific and unchangeable locations; and offer a limited supply. Their scarcity and high demand create a capitalizable commodity and an avid and impatient market, i.e., these sources are goods with high marketability in the international market and thus subject to price variations.

The issue is that these natural resources are allocated both for producing energy and for producing non-energy goods. Therefore, they are used as an input for production and as an input to produce a different kind of input, namely power.

Energy and water are at the heart of any country's economy and way of life. National defense, food production, human health, manufacturing, recreation, tourism, and the daily functioning of households all rely on a clean and affordable supply of one or both of them.

It is known that the production and consumption of energy and water are closely intertwined. They are diversified. Energy includes electric energy, and fuels like gasoline, diesel, nafta, kerosene, alcohol, fuel oil, uranium, and the like. Water includes drinkable water (potable), water for irrigation, water for cooling, water for industrial processes, and so on. The end users of both, energy and water, are many. There are also several producers of both.

Energy and water are, for their turn, intrinsically related to the production and consumption of food and transport.

Keeping electric power plants cool requires lots of water. Keeping water safe takes lots of energy. Apparently this potentially forces a choice between the two.

Water is needed to generate energy. Energy is needed to deliver water. Both resources are limiting the other --- and both may be running short. Is there a way out? What would be the rational to deal with this problem?

In some countries, the two greatest users of freshwater are agriculture and power plants. Thermal power plants --- those that consume coal, oil, natural gas or uranium --- generate more than 90 percent of U.S. electricity, and they are water hogs. The sheer amount required to cool the plants impacts the available supply to everyone else. In other countries, like Brazil, for example, hydroelectric energy plays a major role, and it has been found that water to be used directly as water, as in agriculture, industrial processes, and human consumption, for example, worths twice the economic value it has after it has been transformed into electricity in a hydroelectric power plant.

At the same time, one uses a lot of energy to move and treat water, sometimes across vast distances. Health standards typically get stricter with time, too, so the degree of energy that needs to be spent per gallon will only increase.

A mathematical model is proposed here formulated in terms of an optimal control problem representing an evolving economy; an optimal economic growth model. It is written as a maximization of an intertemporal social welfare function, subject to constraints defined by income and investment identities, production technologies, the reserves consumption dynamics, the labor force growth dynamics, the energy balance and the labor force balance. The policy instruments are the investments in each sector, the consumption rate for the energy resources, the water usage rate, and the labor force growth rate. The model is treated via the Pontryagin maximum principle. The results obtained from the model are useful in the understanding of the sector as a whole, and as a support in establishing integrated policies in the context of the energy-water nexus.

Sorghum As an Advanced Biofuel: Price Effect on Wheat, Corn and Soybean Markets

Pokharel, K.,
Wibowo, R. P.,
Nti, F.,

BACKGROUND: The 2007 US Biofuel mandate requires total biofuel production to increase to 36 billion gallons by 2022. The Energy Independence and Security Act further specify that more than 55% of total biofuel production in 2022 must come from sources – grain, sorghum, wheat, etc – other than cornstarch. In 2012, the EPA announced that grain sorghum-based ethanol qualifies as a renewable fuel with more carbon credits than the corn-based ethanol. This may result in land use changing from the planting of wheat and other grains to planting grain sorghum. However, the problem of increasing sorghum production could cause a decrease in food supply leading to an upward pressure on grain prices. This study examines impact of the biofuel mandate on wheat, corn, soybean and sorghum prices considering the new situation where sorghum is considered an advanced biofuel feedstock.

METHOD: A stochastic partial equilibrium model is used to evaluate the short-run implications from demand and supply shocks from having grain sorghum approved as a advanced biofuel feedstock on corn, wheat, sorghum, and soybean prices in the USA. Different scenarios are simulated and compared to a counterfactual and the simulated results are compared with results obtained from the observed data.

EXPECTED OUTCOME: The use of sorghum as an input for biofuel production could alter grain prices in USA. With higher competitive prices from sorghum, the possibility of decrease production in wheat and other grains is inevitable. Since the USA is a leading wheat producer, any decline in its production will have effect on the global wheat market.

ACKNOWLEGDEMENT: This material is based upon work supported by National Science Foundation Grant: From Crops to Commuting: Integrating the Social, Technological, and Agricultural Aspects of Renewable and Sustainable Biorefining (I-STAR); NSF Award No.: DGE-0903701.

Assessment of an Power Generation Industrial Unit from Vinasse Treatment in Uasb Reactor

Santos, A. F. D. M. S., Cetrel
Sobrinho, M. A. D. M., UFPE
Handel, A. C. V., UFCG

ASSESSMENT OF AN POWER GENERATION INDUSTRIAL UNIT FROM VINASSE TREATMENT IN UASB REACTOR

André Felipe de Melo Sales Santos1,2, Maurício Alves da Mota Sobrinho2, Adrianus Cornelius van Handel3

1Cetrel S.A.

2Environmental Process and Technology Group – DEQ – UFPE

3Department of Civil and Environmental Engineering - UFCG

The UASB reactors (upflow anaerobic sludge bed) represent the type most widely disseminated and applied in the world of high rate anaerobic system. They have the advantage of treating wastewater diluted or concentrated, dissolved or particulate material, and simple or complex, with high efficiency under conditions of relatively simplified operation.Are applicable to sanitary sewers and industrial effluents that have organic nature, biodegradable, with low concentrations of oil and grease, and toxic compounds. Their cost/benefit ratio make it attractive for industrial uses with advantages of high biomass retention, low sludge production, high efficiency and the biogas exploitation possibility.The anaerobic digestion of vinasse, particularly in the case of Brazil, represents great opportunities for expanding the national energy matrix,taking views of the large volumes produced alcohol. Each liter of alcohol produced can generate from 10 to 18 L of vinasse with a BOD of around 15,000 to 50,000 mg O2/L. Cetrel, from a research project in its program of Technological Innovation in 2009 startedthe project Energy appreciation of vinasse for the generation of decentralized electricity obtained from biogas through anaerobic biodegradation. The project has evolved from the laboratory scale, pilot reaching an industrial unit currently operating in the state of Pernambuco, in Vitória de Santo Antão. The project consists of a modified anaerobic UASB reactor for the treatment of vinasse, with capacity of 1000m3, with organic loading volume (COV) projected in the range from 16 to 20 kg COD/m3.d, and maximum flow rate of 60m3 / h of raw vinasse. The production of electricity is approximately 0.87 MWh using a power generator biogas GUASCOR, SFG-LD 480, which connects directly with the local utility.The plant started operation at the end of the year 2012, with the data presented in this article concerning the 2012-2013 and 2013-2014 harvests. At the beginning of the operation was necessary to inoculate anaerobic flocculent sludge originating from UASBused in urban sewage treatment, but this was not appropriate to the high organic loads applied, especially when applied loads above 4 kg COD/m3. In this case it was necessary to acquire industrial granular anaerobic sludge of high activity (0.44 kg COD/kg TSS.d) to increase the treatment capacity.The unit began operating with low VOC of approximately 1 kg COD / m3.d in increments of 0.5kg COD / m3.d to promote the process of acclimatization of the anaerobic biomass. The performance of the unit in the second harvest has exceeded expectations of project reaching about 91% in COD removal (daily average), reaching levels of 71% methane in biogas. However, when using the vinasseoriginated from the dilution of molasses, there was great difficulty in their treatment compared to vinasse coming directly from sugar cane juice.This problem was due to the presence of sulfatewith concentrations in the range of 700-900 mg / L, which may interfere with the microbiological equilibrium of the system. Sulfate is used in control of indigenous bacteria in the fermentation of the must, and when in an anaerobic environment, the sulfidogenesis competes with methanogenesis interfering with the balance of alkalinity and volatile fatty acids (ratio AT / AGV) leading to operating instability. The vinasse broth does not present as great challenges.Another drawback of the sulfate is your anaerobic degradation leads to H2S, which is highly toxic and corrosive. Were detected sulphate content of the order of 2000-5000 mg/L in the biogas removal system is required to prevent deterioration and shortening the life of equipment. The technology proved to be technically feasible even requiring optimization step.Challenges of economic viability also point to the energy market taking into view that this type of sustainable energy is not encouraged by the government, and the amounts paid by the MWh currently prohibitive to enable the business. However in other applications such as generating isolated systems or other applications can be economically viable.

Key-words: Biogas, Vinasse, UASB, Electric Power

Comparison of Anaerobic Reactors for Electricity Production in PILOT Scale from Vinasse

Santos, A. F. D. M. S., Cetrel
Motta Sobrinho, M. A. D. Sr., Universidade federal de Pernambuco
van Handel, A. C.,

COMPARISON OF ANAEROBIC REACTORS FOR ELECTRICITY PRODUCTION IN PILOT SCALE FROM VINASSE

André Felipe de Melo Sales Santos1,2, Maurício Alves da Mota Sobrinho2, Adrianus Cornelius van Handel3

1Cetrel S.A.

2Environmental Process and Technology Group – DEQ – UFPE

3Department ofCiviland Environmental Engineering - UFCG

Among the liquid effluents of the sugarcane industry, the vinasse is which has higher pollution potential. In terms of BOD value ranges can range from 15,000 to 50,000 mg.L-1, and its production in conventional distilleries can reach 10-18 L of vinasse per liter of ethanol produced, according to the process and quality of sugarcane. The high amounts of biodegradable organic matter in the raw vinasse indicate its potential for use in biological processes using more rationally this energetic potential. One of the most promising technologies is the anaerobic digestion, which produces methane that can be applied in decentralized generation of electric power or burning processes in the unit. In this work three models of anaerobic reactors were evaluated in pilot scale, verifying the increase of the volumetric organic load on the stability of the reactor in its ability to withstand shock loading and its repercussion on the quality of the biogas generated. Were compared to the UASB reactor (R-11) two other models of anaerobic reactors: reactor IC (internal circulation) (R-21) and reactor (external settler) (R31) that present changes in concept and design in relation to reactor UASB.The reactors were built in fiberglass, in a cylindrical shape with a diameter of 1 m and 4m height (working volume 3.5 m3). Was utilized industrial granular anaerobic sludge of high activity (SMA = 0.7 kg.COD.CH4/kg.VTS.d) as inoculum to reduce the adaptation time of the anaerobic microbiotafor vinasse. As alkalizing, was made use of agronomic urea in aqueous solution, dosed at the entrance of the reactor by a metering pump. Recirculation of treated effluent (of the order of 1:3) was also used to reduce the consumption of alkalizing the use of alkalinity generated in the reactor.The criterion of stability and load increase was the maintenance of the relationship TA/VFA around 3.5 -. 4.0 with load increments of 0.5 to 1.0 kg.COD/m3.d until it reaches the design load of 20 kg COD.m3/d. The reactors operated continuously for 2 consecutive harvest seasons, with the first harvest (January 2011) the maximum load was reached only after 6 months for the three reactors, which presented similar behavior. In the first harvest the average removal efficiencies of COD were respectively 75%, 79% and 79% and the concentration of methane in the biogas on average 88%, 84%, 84% for R-11, R-21 and R-31 reactors, respectively. In the second harvest (August 2011 to April 2012) the maximum load was achieved in only 1 month.This rapid staggering load is due, probably,the adaptation of anaerobic sludge to the vinasse,reducing by one sixth the time required in the previous harvest. The average removal efficiencies of COD were respectively 85%, 87% and 88% and the concentration of methane in the biogas was on average 82%, 65%, 77% for R-11, R-21 and R-31 reactors, respectively in the second harvest. It was found that, in terms of applied load,that the second season the R-21 and R-31 reactors exceeded the design load reaching 30 kg COD.m3/d,however for the reactor R-11 reached only the load of 25 kg.COD.m3/d.Despite this load shedding, the reactor R-11 proved to be robust shown very satisfactory performance with respect to COD removal efficiency, stability and quality of the generated biogas (high methane content). Compared to the other two models tested, although these present technological improvements over contact sludge-substrate and sludge retention, the UASB technology was adequate for the treatment of vinasse, with lower cost to other models of anaerobic reactors studied.

Key-words: Biogas, Vinasse, UASB, internal circulation anaerobic reactor, external settler anaerobic reactor.

Switchgrass Genotype Study As a Bioenergy Crop in Kansas, USA

Min, D., Kansas State University
Prasad, V.,

Due to increasing fuel costs and uncertainty of fossil fuel supply on the planet, developing bioenergy crop such as switchgrass might play an important role in terms of diversifying energy sources and increasing energy sustainability. There are many potential bioenergy crops and switchgrass is one of them. Switchgrass is a native warm-season grass and once it's established, it can last several years with low inputs and maintenance. The objectives of this study were to: 1) assess the best performing switchgrass genotype suitable for Kansas soil and climatic condition in the USA, and 2) determine the correlation between plant height or tiller numbers per plant and dry biomass of various switchgrass genotypes. Seeds of twenty accessions of switchgrass were sown in trays under greenhouse conditions and transferred into cones after emergence. Twenty two different genotypes (i.e., Alamo, Cave-in-Rock, Kanlow, SL 93 C2-1, SL 93 C2-2, SL 93 C2-3, SL 93 C2-4, NL 94 C2-1, NL 94 C2-2, NL 94 C2-3, NL 94 C2-4, NSL 2009-1, NSL 2009-2, NSL 2009-3, NSL 2009-4, SWG 2007-1, SWG 2007-2, SWG 2007-3, SWG 2007-4, SNU 98 LMBP C1-1, SNU 98 LMBP C1-2, SNU 98 EMBP C1-1) of seedlings were allowed to grow in cones for 30 days under controlled environments. Thereafter, the seedlings were transplanted into the field at the Kansas State University North Research Farm in Manhattan, Kansas, USA. The growth and yield components of various switch grass genotypes were measured. Plant height and number of tillers per plant was measured in five randomly selected plants from each replication. Single plant was harvested and dried in oven at 50°C for a week and dry weight was recorded and expressed as g plant^{-1}. Plants in a meter square were hand harvested and dried in oven at 50°C for a week and dry weight was recorded and expressed as g m^{-2}. Significant difference in plant height was observed among the genotypes. The genotypes, SL 93 C2-2 was the tallest (193.9 cm) and there was no significant difference between SL 93 C2-2 and NL 94 C2-1, NL 94 C2-2, NL 94 C2-3, NL 94 C2-4, NSL 2009-1, NSL 2009-3, NSL 2009-4, SL 93 C2-1, SL 93 C2-3 and SWG 2007-2. The genotype Cave-in-Rock was the shortest (124.2 cm) among the genotypes. Significant difference in number of tillers per plant was observed among the genotypes. The genotypes Alamo recorded the highest numbers of tiller plant^{-1} (24.4) which was on par with NL 94 C2-1, NL 94 C2-4, NSL 2009-2, NSL 2009-3, SL 93 C2-1, SL 93 C2-2, SL 93 C2-3, SL 93 C2-4, SWG 2007-1 and SWG 2007-2. The genotype Cave-in-Rock had the lowest numbers of tiller plant^{-1} (14.3) compared with other genotypes. The genotypes Alamo, NL 94 C2-2, NL 94 C2-3, NSL 2009-1 and NSL 2009-1 had increased above ground biomass compared with other genotypes. The genotypes SWG 2007-3, SNU 98 LMBP C1-2, SNU 98 EMBP C1-1, Cave-in-Rock and SWG 2007-4 had lower above ground biomass than other genotypes of switchgrass. The correlation study indicates that there was a significant positive correlation between number of tillers per plant and per plant dry weight (R^2=0.93), number of tillers per plant and plant height (R^2=0.94), and plant height and per plant dry weight (R^2=0.82). This study might help bioenergy crop breeders develop certain genotypes that can have high biomass with both high number of tillers per plant and taller plant characteristics.

A Prototype Methodology for Incorporating Sustainability Indicators in Biorefinery Process Design

Sacramento Rivero, J. C.,
Vilchiz Bravo, L. E.,

Biorefineries are industrial plants that integrate equipment and processes to transform biomass into a range of biofuels, platform chemicals, end-use products, and energy. The biorefinery concept rises to address two concerns, mainly: assuring energy security (by displacing fossil fuels usage, and adequate stewardship of biomass resources) and reducing greenhouse gases emissions (to meet well specified targets). Traditional methodologies for chemical processing plants are commonly used to design biorefineries, but these fall short on considering these basic criteria systematically during the design process. In this work, a methodology for including various sustainability indicators into the traditional design of biorefinery plants. The two main drivers of biorefineries are included, along with other sustainability indicators, so that a quantitative 'sustainability footprint' can be calculated for each design as a standalone evaluation. This methodology consists of 9 indicators for the design stage, and 5 more for the ex-post evaluation of existing biorefineries. For its application it requires performing a Life Cycle Assessment during the conceptual design stage, and a techno-economical pre-feasibility study during the basic engineering stage. A normalized scale is proposed for all the indicators, requiring the identification of the 'ideal sustainability' and 'critical' states for each one. A case study illustrating the resulting sustainability footprint is presented and discussed.

Optimal Control Strategies for Different Stabilization Pond Systems

Ochoa, M. P., PLAPIQUI - CONICET - UNS
Estrada, V., Planta Piloto de Ingenieria Quimica (PLAPIQUI), Universidad Nacional del Sur - CONICET
Hoch, P. M., Plapiqui - UNS - Conicet

Wastewater generation is inevitable and its discharge into surface waters leads to environmental problems and health risks. To avoid this, standards for wastewater discharge have been enforced and are expected to become stricter, this resulting in a growing interest area of study [1]. Wastewater treatment in stabilization ponds mainly results from settling and complex symbiosis of bacteria and algae where the oxidation of organic matter is accomplished by bacteria in presence of dissolved oxygen supplied by algal photosynthesis and surface re-aeration [2].

In this work different configurations of stabilization ponds are considered. Dynamic modeling of each pond was implemented within a dynamic optimization environment and the whole system was simulated during a fixed time horizon. A detailed mechanistic model is constructed, based on first principles of mass conservation, of different types of systems of anaerobic, aerobic and facultative ponds in series.

We address the control problem of wastewater stabilization ponds of the differents systems by formulating an optimal control problem considering electrical motor power for mixers in the aereted pond and nutrient addition rate as control variables (degrees of freedom of the problem)[3]. Constraints are embedded in the DAE model and boundaries on the control variables. As the specification on chemical oxygen demand (COD) in the outlet stream is far from the target one, the objective is to minimize the offset between the desired value and the current one, along a time horizon of a year. As a result of the dynamic optimization problem the optimal time profiles of motor power and nutrient addition rates are obtained for the time horizon. The main objective of the work is to compare the performance of the different configuration pond system under different control strategies, by the amount of organic matter in the effluent of the treatment plant.

The model takes into account dynamic mass balances of biomass of algae, the main groups of bacteria: heterotrophic bacteria, autotrophic bacteria, fermenting bacteria, acetotrophicsulphate reducing bacteria and acetotrophicmethanogenic bacteria. Also, mass balances for organic load are formulated, such as slowly biodegradable particulate COD, inert particulate COD, fermentation products, inert soluble COD, and fermentable readily biodegradable soluble COD. For nutrients, ammonium and ammonia nitrogen, nitrate and nitrite nitrogen, sulphatesulphur and dissolved oxygen. Finally, molecular nitrogen and methane emissions are considered in the model.

The results provide useful information on the complex relationship among micro-organisms, nutrients and organic matter concentration, as well as information about the impact of modification in the pond system that can be used to improve the control of the effluent composition.

Keywords: Optimal Control - Wastewater Treatment- Stabilization Ponds System – DAE

Biogas in the Brewery: An Exploration of Practice and Sustainability

Jurado, J.,

RCN Conference Abstract - J. Jurado

In recent years, more breweries in the USA have installed anaerobic waste water treatment plants as an alternative to traditional aerobic plants. The biogas byproduct can be used as boiler or CHP fuel, and one brewery evaluated its use in fuel cell electricity generation. Admittedly, a 'micro-' scale biofuel solution which, at best, supplies 20% of the energy required by the brewery... yet which is like PV and solar heating applications in that individual adaptation may seem to be inconsequential, but industry-wide, it is additive and helps change energy benchmarks. A challenge has been to deploy anaerobic treatment systems for the burgeoning artisanal breweries. Once this technological approach was only viable for very large breweries, but then smaller regional breweries had successes with their implementations. The proliferation of new, very small breweries suggests that there exists a genuine opportunity for small-scale anaerobic digester systems. An exploration of successful existing installations lends a framework of what makes sense for smaller, unitized systems that must be more affordable for the small brewery. ROI is the final determinant which will be discussed from the lens of the brewery owner.

Polymeric Membranes and Gas Separation: Homogenous Blends of Matrimid®5218 and P84

Carson, J.,

Polymeric Membranes and Gas Separation

Homogenous Blends of Matrimid®5218 and P84

Jared C. Carson1 with **John P. Stanford2** and **Mary E. Rezac2**

1. REU in Sustainable Bioenergy at Kansas State University, Manhattan, KS. Summer 2013

2. Chemical Engineering Department, Kansas State University, Manhattan, KS.

Abstract: A study has been conducted to determine if a polymeric membrane consisting of a homogenous blend of the polyimides Matrmid® 5218 and P84 can be created, and if the H2/N2 selectivity of this membrane would compare to that of pure Matrimid® 5218. This research shows that by using NMP as a common solvent, a homogenous blend of both polyimides can be achieved. The experiment outlined in this report achieved a maximum H2/N2 selectivity of 50.4 using a 50/50wt% blended membrane, and a maximum H2/N2 selectivity of 51.6 for a pure Matrimid® membrane. Further investigation was conducted into the effects of annealing and cross-linking these membranes, though the results are inconclusive.

Incorporation of Sweet Sorghum Juice into Current Dry-Grind Ethanol Process
Saul, K., North Carolina State University

Incorporation of Sweet Sorghum Juice into Current Dry-Grind Ethanol Process

Kaelin Saul, Nana Baah Appiah-Nkansah, and Donghai Wang

Abstract

With an increase in demand for ethanol as a source of biofuel, there are many different biomass solutions being explored, such as utilizing grain sorghum and sweet sorghum juice as opposed to corn and water, for the dry-grind process. This paper conveys the optimum conditions for flour loading as well as for the hydrolysis time that will result in high ethanol yield and high fermentation efficiency. Two sets of experiments were conducted. The first consisted of varying the quantities of grain sorghum flour of 6 grams, 9 grams, 12 grams, and 15 grams whiles maintaining a constant amount of sweet sorghum juice concentrate. The results of the first study showed that the best flour loading was 6 grams of flour added to 100 milliliters (mL) of sweet sorghum juice. With optimal flour to juice ratio, the hydrolysis time can be significantly reduced from 60 min to 30 min. Future research will study the effectiveness of different enzymes (e.g. Stargen 002, alpha-amylase, etc.) on the ethanol yield and the fermentation efficiency.

Analysis of the Solid Waste Management of the Sugarcane Industry in Pernambuco

Souza Melo, M. G., IFPB
Sobral, M. D. C. M., UFPE
Ferreira, A. L. N., UFPE

Analysis of the solid waste management of the sugarcane industry in Pernambuco

Maiara Gabrielle de Souza Melo 1

1 Doctoral Graduate Program in Civil Engineering. Professor at Federal Institute of Education, Science and Technology of Paraiba. E-mail: @ maiara.melo ifpb.edu.br

2 Professor at Civil Enfineering of Federal University of Pernambuco 3 Doctoral Graduate Program in Civil Engineering of Federal University of Pernambuco

Maria do Carmo Sobral2

André Ferreira3

- Introduction

The sugarcane agribusiness stands out due to the large amount of solid waste generated in their production processes. One of the main problems caused by industrialization is precisely the disposal of waste resulting from the production process, which affect the natural environment and human health.

Most of the sugarcane waste activity needs an adequate destination; it cannot be accumulated indefinitely at the site where it was produced. The disposal of waste in the environment, by means of matter and energy emissions released into the atmosphere, waters, or soil must occur after being treated and must be framed on standards established in environmental legislation in order not to cause pollution (AQUARONE, 1990 PELIZER, 2007).

On the other hand, the processes of production and processing of sugarcane have special features, because their residues are not considered as waste, but as byproducts and thus are valued by the industrial sector (BNDES, 2008).

In this sense, the purpose of this work is to describe the legislation on solid waste management in Pernambuco and analyze how it is being applied to the sugarcane sector of the state. The methodology consisted of both bibliographic and documentary research.

- Legislation Related to Solid Waste Management in Pernambuco

In 2001 Pernambuco put state law No. 12.008 into effect, which regulates the State Solid Waste Policy. This law stipulated that requests for environmental permits are made to conform to the Solid Waste Management Plan (SWMP).

Subsequently, the CONAMA Resolution 313/2002 established the National Inventory of Industrial Solid Waste as part of the licensing process and therefore mandatory.

The CPRH Normative Instruction 03/2006 provides for the submission of the Annual Report of Solid Waste generated and provides:

Article 1 - all industrial enterprises subject to environmental licensing on The CPRH must submit the Annual Declaration of Industrial Solid Waste for the previous year,

I- Which produce an amount of annual waste exceeding 50 tonnes / year; and / or

II-Which produce any kind of waste classified as dangerous (Class I to ABNT NBR 10.004:2004).

Already the IN 04/2006 of The CPRH established the criteria for the submission of the Industrial Solid Waste Management Plan, making it obligatory for: industries with total waste generation above 1,000 tons per year; and/or industries that generate dangerous waste totaling over 250 tons per year. The Solid Waste Management Plan should be updated whenever there are changes in the production process that affect the generation or the management of waste.

In 2010 it established the New Solid Waste Policy of Pernambuco, Law n ° 14,236, which brought into their instruments:

I - State Program for Solid Waste Management

II - Plans for Integrated Management of Solid Waste

III - State Information System about Solid Waste

IV - inventories of solid waste;

V - environmental licensing;

VI - environmental monitoring and oversight;

VII - technical and financial cooperation between the public and private sectors for implementation;

VIII - scientific and technological research;

IX - reverse logistics;

X - environmental education;

XI - fiscal, financial and credit incentives.

- Management of agro-industrial wastes from Sugar Alcohol Sector in Pernambuco

In Pernambuco, until the publication of the first inventory of Industrial Solid Waste in 2001, there was no control over the waste generated by industries. The data from this inventory indicated that residues of the sugarcane agro-industry accounted for more than 90% of industrial waste generated in the state.

In relation to waste measured in tons, sugarcane bagasse with 42% of the total generated in t/year is highlighted; organic process wastes representing about 29%; and filter cake with 20%. Although it does not appear among the main waste, boiler ash corresponds to 2.8% of the total. Of this total, 99% are characterized as waste of Class II and 1% of Class I.

With respect to waste measured in tons, the organic waste process, with emphasis on the vinasse "liquid residue", representing 91% of the total, followed by waste oils which represent 6% and solid waste composed of non-toxic metals, with 2% of the total. Of this total, 93% of waste is Class II and 7% in class I.

The same inventory, the main waste from sugar and alcohol industries were listed (Table 01)

Table 01: Distribution of the total generation of the types of industrial solid waste by type.

Waste of sugar and alcohol industry	Amount (t / year)	%
Cane bagasse	568.021,08	45,913
Boiler ash	21.714,70	1,755
Empty containers of agricultural pesticides	5,04	0,00
used oils	722,31	0,058
Restaurant waste	1,00	000
No dangerous waste removal	1.171,00	0,095
Organic Waste of the processes (vinhoto, tallos, grease, etc) 2	377.125,00	30,483
Scrap iron	589,00	0,048
Filter Pie	267.818,40	21,648

2 In the assessment related to m^3 / year, organic process wastes (vinhoto, tallow and grease) corresponds to 99.99%.

Total 1.237.167,53 100,00

source: CPRH/GTZ, 2001.

About the final destination, were described as follows:

Table 02: Distribution of final disposal of industrial solid waste by type in 2001.

Final Destination	amount		%	
	t/ **year**	m^3/ **year**	t/ **year**	m^3/ **year**
Own Industrial Solid Waste Landfills	18.648,00		1,51	
fertirrigation	167.286,48	2.220.112,00	13,52	99,99
Incorporation into the soil	450.638,70	199,00	36,43	0,01
City dump	1.135,00		0,09	
other waste disposal	360,00		0,03	
Open burning	1,20		0,00	
Burning in boiler	321.557,00		25,99	
Reprocessing and external recycling	277.358,84	10,0	22,48	0,00
Reprocessing and recycling internal	182,31	6,0	0,01	0,00
Total	1.237.167,53	2.200.327,00	100,00	100,00

source: CPRH/GTZ, 2001.

Since 2006, the Annual Statement of Industrial Solid Waste and the management of industrial solid waste plan became obligatory for sugar and ethanol enterprises, thus normalizing access to information that previously was only possible through the collection for inventory. The annual declarations of Industrial and Solid Waste must be submitted each year as a condition for the granting of operating licenses, and a management plan for industrial solid waste once obtained, is valid for up to 5 years and needs only be changed if there is a significant modification in amount, type and destination of waste.

In these 41 annual declarations of Industrial and Solid Waste that were analyzed, the difficulty was on the part of companies to provide the information requested. The most frequent facts highlighted the error in measurement units of some wastes, the omission of many wastes generated, and lots of forms of storage and final disposal not specified being specified.

The diversity of waste generated, as well as the frequency with which they are declared by enterprises, indicates that there is very little uniformity in the annual declarations of

Solid and industrial Waste by the sugar and ethanol companies. It highlights the fact that many residues that are always generated in the industries were not declared .The Normative Instruction 03/2006 provides that in these cases it is necessary to correct the declarations, so this error can also be considered as an environmental administrative violation.

Wastewater composed of cane wash waters, floors and equipment, and the waters of barometric columns are usually mixed with vinhoto for fertirrigation of sugarcane fields, and so there is often no monitoring of the composition of these effluents. It is evident that these waters, mainly from barometric columns, are discarded at high temperatures and therefore can not be discharged directly into the water body. The cases in which these waters are discharged into the water body must attend CONAMA Resolution 357/05.

For the vinhoto, only five sugarcane mills declared their production, even with all these new developments being alcohol producers. The final destination is through fertirrigation .

Although all companies manage agrochemical packaging, only the five plants were cited. It is emphasized that only one company produced 6 tons of agrochemical packaging during the previous harvest, showing the large amount of chemicals used. The Federal Law n º 9.974/00 regulates the disposal of empty agrochemical packaging. These packages must be properly disposed.

The disposal of batteries, also cited by five companies, should follow CONAMA Resolution 257/1999. Several wastes such as plastic packaging, sweeping waste, waste paper and cardboard, and food residues were sent to municipal garbage dumps, although many of them could be sent for composting.

The straw from sugar cane, although generated by all projects analyzed, was announced only by one sucarcane mill.

In the annual declarations of Industrial and Solid Waste of previous years the straw had not been cited due to the practice of burning, which already was incorporated into in the habits of Pernambuco sugar cane industry.

The Sugarcane bagasse, the boiler ash, and filter pie continue to be the wastes generated in great quantities, as also is demonstrated by the data from the inventories of 2001 and 2003.The Boiler ash and filter pie are incorporated into agricultural soil, whereas all bagasse is used in caldeiras. Possibly the vinhoto is still one of wastes generated in larger quantities, but this fact cannot be affirmed from the annual declarations of Industrial and Solid waste in 2008 because only five projects declared their generation.

To Jendiroba (2006, p. 351-352), depending on the massive amount of wastes generated in the sector, the situation is complex, and requires planning for disposal during the entire harvest. Therefore, the practice of using the main waste producing areas cane sugar was developed.

AUTHOR INDEX

AUTHOR INDEX

AUTHOR INDEX

AIChE
120 Wall Street, FL 23
New York, NY 10005-4020

ISBN 978-1-5108-1574-2

CURRAN ASSOCIATES INC.
proceedings.com